FUNDAMENTALS
OF INTERFACIAL
ENGINEERING

ADVANCES IN INTERFACIAL ENGINEERING SERIES

Microstructures constitute the building blocks of the interfacial systems upon which many vital industries depend. These systems share a fundamental knowledge base—the molecular interactions that occur at the boundary between two materials.

Where microstructures dominate, the manufacturing process becomes the product. At the Center for Interfacial Engineering, a National Science Foundation Research Center, researchers are working together to develop the control over molecular behavior needed to manufacture reproducible and reliable interfacial products.

The books in this series represent an intellectual collaboration rooted in the disciplines of modern engineering, chemistry, and physics that incorporates the expertise of industrial managers as well as engineers and scientists. They are designed to make the most recent information available to the students and professionals in the field who will be responsible for future optimization of interfacial processing techniques.

Other Titles in the Series

FUNDAMENTALS OF INTERFACIAL ENGINEERING

Robert J. Stokes
D. Fennell Evans

WILEY-VCH

New York • Chichester • Weinheim • Brisbane • Singapore • Toronto

Robert J. Stokes
Center for Interfacial Engineering
University of Minnesota
179 Union Street S.E.
Minneapolis, MN 55455

D. Fennell Evans, Director
Center for Interfacial Engineering
University of Minnesota
179 Union Street S.E.
Minneapolis, MN 55455

This book is printed on acid-free paper. ♾

Library of Congress Cataloging-in-Publication Data
Stokes, Robert J.
    Fundamentals of interfacial engineering / Robert J. Stokes, D.
Fennell Evans.
        p.   cm. — (Advances in interfacial engineering series)
    Includes bibliographical references and index.
    ISBN 0-471-18647-3 (alk. paper)
    1. Surfaces (Technology) 2. Interfaces (Physical science)
3. Surface chemistry. I. Evans, D. Fennell. II. Title.
III. Series.
TA418.7.S69 1996
620′.44—dc20   96-10776
                                                       CIP

ISBN 0-471-18647-3 Wiley-VCH, Inc.

DEDICATION

We dedicate this book to our wives and children (Audrey Stokes, Neil Stokes, and Sandra Johnson; and Joyce and Christopher Evans). We appreciate their patience, understanding, and support during the period in which this book was developed.

PREFACE

Understanding and utilizing the special properties of molecules at interfaces constitutes one of the scientific and technological triumphs of the twentieth century. In the early 1900s, thermodynamic treatments of surface tension, surface chemistry, and adsorption by Poiseuille, Gibbs, Langmuir, and others established the fundamental base for describing surface phenomena. These conceptual advances enabled scientists and engineers to develop industrial processes employing surface active components such as catalysts, colloids and surfactants through the 1920s and 1930s. Demands for alternative materials in the 1940s and 1950s drove the development of tailor-made chemical materials and led to the manufacture of new polymers and surfactants with interfacial applications. The trend to microminiaturization of both "soft" (liquid and polymeric) and "hard" (solid state) material systems in the 1960s and 1970s meant that interface phenomena increasingly dominated the manufacture and function of these systems. At the same time new tools for surface and interface analysis met the burgeoning needs for interface characterization. Recently the unique properties of interfaces have been exploited in products as diverse as photographic films, adhesives, controlled release membranes, semiconductors, composites and ceramic sensors. Now the properties and stability of interfaces determine the operation and long term stability of many "high-tech" materials and devices. The engineering of interfacial systems has become a vital technology in its own right.

The objective of this book is to develop a unified approach to Interfacial Engineering, a coherent interdisciplinary activity that focuses on the design and manufacture of products and processes whose performance depends on interfacial forces and transport phenomena. While the textbook is designed for first year graduate students it should also provide a useful background for industrial engineers. It integrates existing knowledge from a number of disciplines, and presents the basic principles that guide and govern interfacial behavior. As outlined in Chapter 1 the book consists of two main parts, Chapters 3 through 8 define the behavior and properties of "soft" materials (liquids and polymers) and Chapters 9 through 11 the behavior and properties of "hard" materials (metals, ceramics and semiconductors). Emphasis is laid on the fact that the same fundamental principles govern the behavior of surfaces and interfaces in all materials. The text is deliberately designed to draw comparisons rather than distinctions in behavior.

Robert J. Stokes
D. Fennell Evans

ACKNOWLEDGMENTS

To develop a text as diverse as this it has been absolutely essential to call on many friends and colleagues for their help and expertise. We have woven much of their original material and many of their ideas into the final manuscript. Contributions have come in many forms, from written contributions, to reviews, to corrections, to editing.

First we must acknowledge the following authors for their original written contributions to chapters and sections of chapters as follows:

Chapters 2 through 6: Intermolecular Forces, Interfacial Properties of Fluid Systems, Colloids, Amphiphiles, and Polymers
Professor Håkan Wennerström, Chemical Center, University of Lund, Sweden

Chapter 7: Polymer Composites
Dr. Gibson Batch, 3M Company, St. Paul, Minnesota

Chapter 8: Liquid Coating Processes
Dr. Mario Errico, Center for Interfacial Engineering, University of Minnesota

Chapter 10: Thin Films
Professor Alfonso Franciosi, Department of Chemical Engineering and Materials Science, University of Minnesota

Professor William Gerberich, Department of Chemical Engineering and Materials Science, University of Minnesota

Professor Wayne Gladfelter, Department of Chemistry, University of Minnesota

Professor Vern Lindberg, Department of Physics, Rochester Institute of Technology, Rochester, New York

Chapter 11: Grain Boundary Surfaces and Interfaces
Professor William Gerberich, Department of Chemical Engineering and Materials Science, University of Minnesota

Chapter 4, Section 4.7.1: Paints
Dr. Jerry Seiner, PPG Industries, Inc., Pittsburgh, Pennsylvania

Chapter 4, Section 4.7.2: Protective Latex Coatings
Professor Dennis Prieve, Department of Chemical Engineering, Carnegie-Mellon University, Pittsburgh, Pennsylvania

Chapter 4, Section 4.7.4: Zeolite Synthesis
Professor Alon McCormick, Department of Chemical Engineering and Materials Science, University of Minnesota

Chapter 5, Section 5.5.1: Detergency
Dr. Thomas Oakes, Ecolab Inc., St. Paul, Minnesota

Chapter 5, Section 5.5.2: Pressure Sensitive Adhesives
Mr. Richard Bennett, 3M Co., St. Paul, Minnesota

Dr. Richard Greiner, Hercules Inc., Wilmington, Delaware

Chapter 5, Section 5.5.3: Low Density Foams
Dr. Blanca Haendler, Chemistry and Materials Science
Department, Lawrence Livermore National Laboratory,
Livermore, California

Chapter 6, Section 6.7.4: Disposable Diapers
Mr. John F. Reed, 3M Company, St. Paul, Minnesota

Chapter 8, Section 8.2.4: Real Time Imaging
Professor Fred Waltz, Center for Interfacial Engineering,
University of Minnesota

Chapter 8, Section 8.4.2: Photographic Films
Dr. David Locker, Eastman Kodak Company, Rochester,
New York

Chapter 8, Section 8.4.3: Magnetic Recording Tapes
Dr. Edgar Gutoff, Consulting Engineer, Brookline, Massa-
chusetts

Chapter 10, Section 10.8.1: Magneto-Optical Data Storage Disks
Dr. Charles Brucker, Eastman Kodak Company, Rochester,
New York

Professor Vern Lindberg, Department of Physics, Rochester
Institute of Technology, Rochester, New York

Second, we acknowledge the help of Professors Frank Bates,
Barry Carter, Timothy Lodge, Skip Scrivens, Matthew Tirrell, and
John Weaver, all from the University of Minnesota, and Emeritus
Professor Martin Allen, from the University of St. Thomas, for
their reviews, comments, and corrections. We want to thank the
graduate students and teaching assistants who helped us through
the early development of the course material on which this book
is based. We have also benefited from the presence and advice of
a number of Industrial Fellows at the Center for Interfacial Engineering, University of Minnesota.

Third, we are especially grateful to Hertha Schulze, who
brought discipline, consistency, and style to our diverse writing
efforts, to Mark Swanson, who prepared many of the original
drawings, and to Karen J. Hesla for secretarial support.

Finally we acknowledge the facilities provided by the Center for Interfacial Engineering, University of Minnesota, supported by the Engineering Research Centers Division of the
National Science Foundation.

CONTENTS

6 / Polymers 271

7 / Polymer Composites 347

10 / Thin Films—Solid–Solid Interfaces Processed from the Vapor Phase 527

11 / Grain Boundary Surfaces and Interfaces in Crystalline Solids 623

FUNDAMENTAL CONSTANTS

e   Electronic charge $= 1.602 \times 10^{-19}$ C — Coulombs

$\hbar$ Planck's constant $= 6.626 \times 10^{-34}$ J s — joule second

k   Boltzmann's constant $= 1.381 \times 10^{-23}$ J K^{-1} — joules per deg K

     $= 8.62 \times 10^{-5}$ eV K^{-1} — electron volts per deg K

N_{Av} Avogadro's number $= 6.022 \times 10^{23}$ — particles (atoms) per mole

ε_o Electrical permitivity of free space $= 8.854 \times 10^{-12}$ C^2 J^{-1} m^{-1}

R   Molar gas constant ($=kN_{Av}$) $= 8.31$ JK^{-1} — joules per deg K per mole

F   Faraday constant ($= eN_{Av}$) $= 9.648 \times 10^4$ C — Coulombs per mole

CONVERSIONS AND USEFUL RELATIONS

1 cal (calorie) = 4.184 joules
1 eV (electron volt) = 1.602×10^{-19} joules
1 J (joule) = 6.242×10^{18} eV = 0.239 calories
1 J (joule) = 10^7 ergs
1 J (joule) = 1 Pa m^3 (1 pascal meter cubed)

1 kcal mole^{-1} (kilocalorie per mole) $\equiv$ 0.0434 eV per particle
   (atom)
1 kilojoule mole^{-1} $\equiv$ 0.0104 eV per particle
1 eV particle^{-1} $\equiv$ 23.06 kilocalories per mole = 96.5 kilojoules
   per mole

kT at room temperature (300 K) = 4.14×10^{-21} joules =
   0.0258 eV ($\approx$ 1/40 eV)

1 mJ/m^2 (millijoule per square meter) = 1 erg/cm^2 = 1 dyn/cm

κ^{-1} (Debye length) = $0.304/\sqrt{M}$ nm for a 1:1 electrolyte at
   room temperature (300 K), where M = molar solution
   concentration
1M — one molar = one gram molecular weight per liter
   (1000 cm^3 = 10^{-3} m^3) of solution
1mM — one millimolar = one milligram molecular weight per
   liter of solution

1 N m^{-2} (newton per square meter) = 1 Pa (pascal)
1 GPa = 10^9 Pa = 10^{10} dyn cm^{-2} = 1.45×10^5 psi (pounds per
   square inch)
1Pa = 1 J m^{-3} (joule per cubic meter)

1 bar = 10^5 Pa; 1 mbar = 10^2 Pa
1 atm (atmospheric pressure at sea level) = 760 mm
   (millimeters) of Hg = 760 Torr
1 Torr = 133.32 Pa; 1 atm = 1.0132×10^5 Pa

Impingement flux = $2.63 \times 10^{20} \dfrac{P(\text{in pascals})}{\sqrt{MT}}$

$\qquad = 3.51 \times 10^{22} \dfrac{P(\text{in torr})}{\sqrt{MT}}$ molecules cm^{-2} s^{-1}

1

DEFINING INTERFACIAL ENGINEERING

1.1 What Is Interfacial Engineering?

Recent developments in many modern technologies involve new materials or processes in which interfaces or surfaces play a crucial role. These technologies impact basic industries, such as data processing, medicine, agriculture, and transportation. Interfaces dominate many modern products. For example, they are present at every stage of microelectronic fabrication, from device junctions to the final packaged assembly. Interfacial systems are important in coatings for optical and magnetic recording, controlled-release products (for example, medicines and agricultural herbicides), selective membranes, and adhesives. Interfacial engineering of new composite materials systems provides high strength-to-weight ratio for efficient transportation systems.

In the past chemists, chemical engineers, materials scientists, and mechanical and electrical engineers worked independently to develop these systems. Today the challenge is to optimize existing systems and processes by drawing all these disciplines together and using their respective insights to develop new products. We have named this cross-disciplinary approach *interfacial engineering.*

1.2 Trends in Interfacial Products and Processes

An inexorable trend for devices to become smaller, more compact, and more complex, and to perform more functions in a shorter period of time, is the driving force behind interfacial engineering. As the ratio of interfacial area to system volume increases, the integrity of the interfacial region is of paramount importance. The following examples illustrate these features.

1.2.1 Computer Chips and Packaging

In the silicon semiconductors of the 1960s, individual memory chips were limited to 16–64 bits of memory with 4 transistor devices per bit. By 1970, bit density per chip had increased to 1000. Currently, the most advanced commercially available chips contain 4 megabits, and the next generations are projected to contain 16 and 64 megabits. During the same time, semiconductor logic also grew from single gate to 40 to 10,000 logic gates per chip, and speed improved from milliseconds to nanoseconds per operation. As a rule of thumb, every 4 years produced an order of magnitude improvement in performance while the costs per chip have remained constant or decreased. A comparable change in the automobile industry would give us cars costing 5 cents and averaging 300,000 miles per gallon!

This remarkable progress has been achieved through the design of novel and more complex electronic circuitry and through processing developments. The dimensions of the circuit elements have shrunk from millimeters to micrometers. Present design rules allow line widths and spacings with a characteristic size of 0.25 μm and reach the limits of optical lithography. Thicknesses of the respective semiconductor, metal, and insulator layers have been reduced to nanometer dimensions. Indeed, the density and speed of circuitry on the chip have improved to such an extent that the primary factor limiting system performance has now become the length of the interconnects from chip to chip.

Increasing the density and speed on individual chips required innovations in the technology to package them. Packaging evolved from single chips mounted on a substrate and inserted into a simple printed circuit board, to many chips mounted on multilayer printed circuit boards, to the modern multichip module in which as many as 9–36 ultra-high-performance chips are mounted and interconnected on a single multilayer ceramic package. Design rules for modern packaging are equivalent to those set for chips about 20 years ago.

Because the interface-to-volume ratio has increased by orders of magnitude with each developmental step, the performance of the whole system is characterized by interfacial rather than bulk behavior. Constant scaling of the product to finer and finer dimensions means that processing of the chip must be considered on an atomic rather than a macroscopic scale. The metal, semiconductor, and insulator layers are thin films a few atoms thick, deposited using high-vacuum vapor deposition techniques. Lithographic techniques to generate submicrometer patterns require spin-coated high-quality photoresists of very uniform thicknesses to avoid light scattering and reflections. Etching the patterns has evolved beyond straightforward chemical etching to precision plasma etching techniques. In the future, as more and more layers of smaller and smaller dimensions are introduced, the challenge will move from a continuum to a molecular processing perspective. From the point of view of interfacial processing, the key issues will be to control film uniformity and adhesion with an efficient production process.

Likewise in packaging, multichip modules require advanced processing. The ceramic multichip module consists of about 40 alternate layers of alumina ceramic and patterned molybdenum metal interconnects cofired to produce the finished module with no loss of electrical continuity—an epic achievement in interfacial processing.

1.2.2 Photographic Films

Photography has come a long way from the glass plates coated with albumen with which Matthew Brady photographed scenes during the Civil War to modern Polaroid color film. All photography depends on the activation of a latent image in silver bromide. The latent image consists of clusters of silver atoms generated by the photodecomposition of silver halide particles responding to the incident radiation. During "development," further chemical conversion of the particle to silver brings out the image, while "fixing" stops the reaction. In modern photographic film the particles are held in gelatin, which prevents them from moving and improves the resolution of the image. For color photography, three photosensitive layers stacked on top of each other each respond to a selective range of wavelengths. They must be separated by filters.

In photographic films, both the product and the manufacturing process depend on interfaces. Progress in film "speed" (responsivity), resolution, and instant image development has been achieved through innovations in photochemistry and enhancement of photosensitivity through adsorption of sensitizing agents onto the surface of the silver halide particles. Advances in processing have led to purer, smaller, photosensitive silver halide particles and improved gelatin properties. Trends have been to decrease the size of the halide particles down to 0.5 μm and to increase the chemical complexity of the product. Thus photographic films have followed an evolution similar to that of the manufacture of computer chips: the ratio of interfacial area to volume has increased, and the scale of concern has become molecular.

A typical black and white negative film consists of 4 layers, a color negative film of at least 13 layers, and instant color print film of 18 layers. Each layer contains different materials, each with a specific function. The film resembles a miniature chemical factory organized so that a series of chemical processes occur in a controlled sequence. A key step in the evolution of modern photographic film has been the development of the multilayer slide coating process in which layers are deposited sequentially in a continuous high-speed process. Obviously products must be free from flaws that could create a false image. This is particularly critical for x-ray film used in medical diagnosis. The potential sources of defects in a high-speed multilayer coating are many, and disposal of deficient film constitutes a major environmental problem for the entire industry.

1.2.3 Polymer Composites

The first widely used polymer composites were fiberglass-reinforced epoxy laminates fabricated by laying down one layer of composite material, allowing it to cure, and repeating the process until the desired thickness was obtained. Small boat hulls are still constructed from such fiber-reinforced composites. The fabrication process is tedious, and it is difficult to mold intricate shapes. Although the resulting composite possesses an acceptable degree of toughness, it is still relatively brittle and tends to crack upon impact.

High-impact polyurethane (HIP) represents a modern polymer composite material that can be molded in a single-step process to produce tough, flexible automobile bumpers. Two reactive oligomer liquids, an isocyanate (I) and a diol (D), are mixed and injected into a mold and polymerized to form a multiblock polymer with a repeating structure—$[I\text{-}D]_N$. The individual polymer chains line up in register to form two types of microdomains—one containing I molecules, which crystallize, and the other containing D molecules, which remain in a rubbery state. The microdomains are typically 10 nm in size. By controlling the relative amounts of I and D molecules, we can fabricate tough, solid, yet flexible materials. Bumpers made from polyurethanes are 8 times lighter than steel, can be more readily formed into complex shapes in a single-step process, and do not corrode.

Injection molding of HIP exemplifies an interfacial process in which a two-component composite material is generated in situ by the "phase separation" of two incompatible microdomains. Simply changing the composition of the reactants can change the properties of the resulting material. The trend to smaller microdomain size and thus greater area-to-volume ratio calls for better understanding and control of the interface region.

1.2.4 Advanced Ceramics

Modern ceramics are finding new applications in situations in which reliability is absolutely essential, such as in gas turbine engines. The familiar tendency for ceramics to rupture in a brittle manner must be overcome by eliminating flaws. Conventional processes for fabricating ceramics by blending and firing of compacted powders no longer suffice. Although these techniques are good enough for the manufacture of pottery, porcelain, and whitewares, they are unacceptable for silicon nitride gas turbine rotors, alumina/zirconia cutting tools, or lead zirconate/titanate piezoelectric sonars.

Among the new approaches to ceramic powder processing are those based on sol-gel techniques. These techniques rely on advances in chemical synthesis to produce high-purity ceramic oxide or hydrated oxides in the form of small colloidal particles called *sols*. Depending on the application, the sol can be allowed to polymerize to form a gel, or the oxide particles can be precipitated to form a monodisperse colloid. Stabilization and flow of the ceramic colloid while it is being formed into the desired shape depends on the characteristic behavior of the solid/liquid interface. Drying the powder before firing to form the finished ceramic piece and controlling the microstructure of the ceramic material also require control over interfacial reactions.

Trends in these approaches to ceramic processing are to reduce the powder size down to submicrometer dimensions and to manipulate the chemistry of ceramic/liquid or ceramic/organic molecule interfaces to achieve precise shapes of uniform density free from flaws. These steps emphasize the critical role of interfacial phenomena and interfacial processing in the manufacture of modern devices.

TABLE 1.1 Interfacial Engineering Applications

Interface	Product	Industry
Liquid–vapor	Foams	Cosmetics, food
Liquid–liquid	Emulsions	Food, cosmetics
Liquid–liquid/solid	Films, tapes	Photographic, data recording
Liquid–solid	Pastes	Pharmaceuticals, food
Liquid/solid–liquid/solid	Polymers	Adhesives
Liquid/solid–solid	Composites	Transportation
Solid–solid	Thin films	Electronics
	Coatings	Machine tools
Solid–liquid	Surfactants	Oil recovery,
	Membranes	drugs, agriculture
Solid–vapor	Membranes	Air quality

1.2.5 Breadth of Field

With these specific examples in mind, we can generalize the field of interfacial engineering to encompass all possible combinations of vapor—liquid—solid interfaces and a wide range of processes, systems, and devices. Table 1.1 indicates some of this diversity.

1.3 Reproducibility and Reliability

Trends towards compression of size and increase of complexity make control of interface stability an even more crucial factor in our ability to manufacture interfacial products. The process must deliver the desired product with good reproducibility; in a batch process each run must be consistent; in a continuous process fluctuations must not yield unusable product even though efficiency requires the process to run as fast as possible. Manufacture also includes the development of a process that delivers the desired product with good reliability; it must contain no inherent flaws, and its interfacial structure must not collapse or "smear" to deteriorate with time.

Failure of just one interface may be fatal, and if that interface lies many levels deep in the system, fault finding and replacement may be impossible. Because a manufacturer's competitive edge depends on cost as well as performance, reproducible processes are imperative, and the product must be totally reliable. Reproducibility and reliability, therefore, constitute a major challenge in interfacial engineering. The ever-increasing importance of interfaces in a wide range of processes, systems, and devices demands that interfacial engineers be thoroughly knowledgeable about interfacial phenomena and skilled in controlling interfacial systems. This book is directed towards achieving that end.

The text emphasizes the fundamentals underlying the processing of interfacial systems. Each topic deals with the fundamental requirements of the interfacial system, its scale and size, the factors that determine stability, and the processing methods in use or under development for its manufacture.

1.4 Organization and Goals of This Book

Interfacial engineering is clearly a broad-ranging subject. Applications are found within electrical, optical, and mechanical disciplines; the systems involved may combine vapor, liquid, and solid materials, and they must be dealt with at both the atomic and the molecular level. Obviously, one book cannot treat all this subject matter in great depth. Rather our objective is to introduce the fundamental concepts that relate to interfacial phenomena and then apply them to different process and product situations.

Figure 1.1 illustrates the structure followed in organizing this material. Intermolecular forces embrace the whole subject. Within a bulk phase they are the source of cohesion; across an interface they hold materials together. Chapter 2 describes the nature of intermolecular forces that occur between molecules in the various states of matter. It outlines the fundamental intermolecular interactions that prevail in all interfacial systems. The analysis proceeds from individual molecules to molecular assemblies and is extended to introduce the nature of forces across the interface between particles.

1.4.1 Interfacial Phenomena in Fluids and Solids Are Treated Separately

Although we contend that interfacial behavior has the same fundamental origins in all systems and that the subject is truly interdisciplinary, we find it convenient to divide the subject into two parts, interfaces in fluids and interfaces in solids. Chapter 3 is concerned with the general properties of fluid surfaces and interfaces, that is, interfaces involving gas, vapor, or liquid phases.

FIGURE 1.1
Organization of the text.

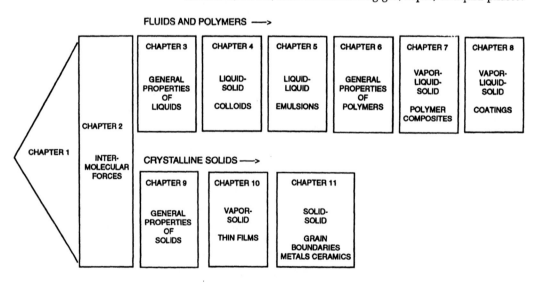

It deals with the general thermodynamic principles, kinetics and transport phenomena needed to describe the behavior of fluid systems. The advantage of dealing with fluid systems first is that ambient temperature is well above their melting point. They readily achieve thermodynamic equilibrium and fundamental thermodynamic principles can be demonstrated in real systems. Chapters 4 and 5 deal with the specific properties, processing, and behavior of fluid interfacial systems involving liquid/solid, liquid/liquid, and vapor/liquid/solid interfaces. Chapter 6 describes the general structure and properties of polymer fluids both in solution and melt form, and Chapters 6, 7, and 8 give examples of processing interfacial systems involving polymer fluids for a wide variety of applications.

Starting with Chapter 9, our attention switches to the solid state. Chapter 9 is concerned with the general properties of solid surfaces and interfaces, particularly crystalline solid interfaces. The text parallels the outline used for fluids in Chapter 3. While the properties of solid surfaces and interfaces have the same fundamental origins as those discussed for fluids, the kinetic approach to equilibrium is limited by solid-state diffusion and the complications due to elastic strain. Chapter 10 deals with the properties, processing, and behavior of solid thin films deposited on solid substrates. Chapter 11 is concerned with grain boundaries; these internal solid surfaces lead to unique interfacial phenomena in crystalline solid systems. The chapter also deals with the interfacial equilibrium between grain boundaries and different phases in metallurgical and ceramic materials. Manipulation of interphase boundaries to produce desired microstructures is critical to the application of these "interfacial" products.

Each chapter refers to appendices at the end of the chapter that provide further background to the theoretical equations or concepts used in the main body of the chapter.

1.4.2 Interfacial Engineering Embodies Both Macroscopic and Microscopic Levels of Understanding

In many systems, interfacial engineering requires a simultaneous focus on phenomena at both the macroscopic and the microscopic level. For the purposes of this text, we assume the reader is familiar with fundamental macroscopic phenomena, such as mass flow, viscosity, heat transfer, and conditions described by the Reynolds number, Stokes-Navier equations, capillarity, Peclet number, and so forth, although these phenomena are briefly defined in Chapter 3.

At the microscopic, or atomic, level, we include topics such as intermolecular forces and kinetics, crystalline imperfections and electron microscopic views of atomic arrangements. Many interfacial applications involve atomic or molecular monolayers or other nanometer-sized microdomains. Understanding intermolecular forces is particularly important because they provide the means by which we manipulate large interfacial areas in many interfacial processes.

Similarly the tools for characterizing interfacial phenomena range from those measuring macroscopic properties (such as rheometers) to those providing the highest level of atomic resolution (such as the scanning tunneling microscopes). The need to characterize the process and the product at all stages of manufacture is an essential component of interfacial engineering. The interfacial engineer must be thoroughly acquainted with the range of characterization tools available, their accuracy, sensitivity, and resolution.

1.4.3 Subject Headings and Concept Maps Provide Different Levels of Overview of the Subject Matter

In writing this book we have used sentences rather than words as subject headings and subheadings. The idea is that these sentences, when gathered together in the Contents, give a brief overview of the individual chapters.

In addition, each chapter beings with a concept map. These maps resemble the tree diagrams computer programmers or systems analysts use to analyze complex tasks. The linear nature of these maps has limitations with respect to establishing the links between subtopics; a three-dimensional globe would be needed to do that. Nevertheless they do present the logical sequencing of the salient points made in each chapter and contain much more information than the table of contents. The success of a concept map can be measured by whether or not we can use it to reconstruct the original text. If we can, then it may provide a useful tool for review.

A list of pertinent symbols is included at the beginning of each chapter. This is to avoid ambiguity when similar symbols are used for different properties in other chapters. We have tried to use accepted symbols where possible.

Finally this text is written in the first person plural to emphasize the fact that learning, particularly for an interdisciplinary field such as interfacial engineering, is a cooperative experience. Different backgrounds and experiences have been brought together in writing this book. Names of colleagues who have made major contributions to individual chapters and applications sections are included under the Acknowledgments to recognize the different and valuable insights they have provided into the subject. The course on which this book is based required active participation by students in the preparation and presentation of additional material to illustrate the fundamental principles of interfacial engineering, and indeed some of the applications sections are based on student papers. Together *we* can explore and enlarge on the topics presented here.

2

INTERACTION FORCES IN INTERFACIAL SYSTEMS

with H. Wennerström

LIST OF SYMBOLS

e	electronic charge
E	electrical field
F	force
G	Gibbs free energy; Δ**G**, change in free energy
h	separation distance between two colloidal particles
$\hbar$	Planck's constant
H	enthalpy; Δ**H**, change in enthalpy
H_{ij}	Hamaker constant (eq. 2.6.6)
l	spacing between electrical charges in a dipole
m	dipole moment
M	molecular weight
q	charge
R	radius (atom radius, spherical particle radius)
R_{12}	distance between atoms, ions, or molecules 1 and 2 in a gas
V_{ij}	potential energy of interaction between ions and dipoles in gas phase; V_{vdW} van der Waals attractive component, $V_{overlap}$ repulsive component
z	valence; z_i, valence including sign ($z = |z_i|$)
α	polarizability
ε_o	electrical permittivity of vacuum
ε_r	relative dielectric permittivity—the dielectric constant
λ	interatomic spacing in a condensed crystalline solid
ν	characteristic ionization frequency; $\hbar\nu$, ionization energy
ρ	density
ϕ, Ω	solid angles

CONCEPT MAP

- In principle, we can obtain the interaction energies V_{ij} between two molecules by solving the Schrödinger equation.

- However, it is more useful to write V_{ij} as the sum of five interactions: (1) charge transfer (ionic and covalent bonding described in Chapter 9); (2) charge-charge; (3) charge-induced polarization; (4) induced polarization-induced polarization attractions; and (5) overlap repulsion.

Intermolecular Interactions in the Vapor Phase

Interactions between isolated molecules in the vapor phase provide the simplest way to understand the basic concepts. Table 2.1 summarizes the essential equations.

(2) Charge-Charge (Coulombic) Interactions:

- Arise when the electrical field emanating from one charge does not perturb the charge distribution of other molecules.

- Can be attractive or repulsive depending on charge sign and orientation of the molecules.

- $V_{\text{ion-ion}}(R) \propto 1/R$ and are long-ranged interactions.

- $V_{\text{ion-dipole}}(R,\theta) \propto 1/R^2$ and depends on the orientation and magnitude of the dipole moment m (Fig. 2.1 and Table 2.2).

- $V_{\text{dipole-dipole}}(R,\theta,\phi) \propto 1/R^3$ and depends on θ and ϕ as depicted in Fig. 2.2.

- Dipole interactions are often less than kT, and we must average over all angles using Boltzmann statistics.

- Angle-averaged dipole-dipole interactions vary as $1/R^6$. Called the *Keesom* interaction, one of three van der Waals attractions.

(3) Charge-Induced Polarization Interactions:

- Arise when the electrical field emanating from one molecule induces charge polarization (dipoles) in other molecules.

- Are always attractive and more short-ranged than the direct Coulombic interactions.

- $V_{\text{ion-ind dipole}}(R) = -(1/2)\,\alpha\,E^2 \propto 1/R^4$, where α is the polarizability of the molecule and E is the electrical field of the ion (Table 2.3).

- $V_{\text{dipole-ind dipole}}(R) \propto 1/R^6$ and depends on the product $m_1^2 \alpha_2$, where m is the dipole moment. This angle-averaged dipole–induced dipole interaction is called the *Debye* interaction. One of three van der Waals interactions.

- If $V_{coulombic} \propto 1/R^n$, then $V_{induced} \propto 1/R^{2n}$ because one $1/R^n$ term induces the response while the second produces the interaction.

(4) Induced Polarization–Induced Polarization Interactions:

- Arise when instantaneous polarization (charge fluctuations) in one molecule induce and couple to instantaneous dipoles in other molecules.

- Are always attractive.

- $V_{ind\ dipole\text{-}ind\ dipole} \propto 1/R^6$ also depends on α^2 and $\hbar v$, where $\hbar v$ is equated to the ionization energy of the molecule.

- These interactions require a quantum-mechanical description and are called *London* dispersion interactions. One of three van der Waals interactions.

- Like gravitational forces, dispersion forces operate between all molecules.

Van der Waals Attractions

Keesom Interactions ($V_{dipole\text{-}dipole}$)

- $V_{att} \propto -\dfrac{m_1^2 m_2^2}{kTR^6}$

Debye Interactions ($V_{dipole\text{-}induced\ dipole}$)

- $V_{att} \propto -\dfrac{m_1^2 \alpha_2 + m_2^2 \alpha_1}{R^6}$

London Dispersion ($V_{induced\ dipole\text{-}induced\ dipole}$)

- $V_{att} \propto -\dfrac{\alpha^2 v}{R^6}$

Interaction Forces in Gases and van der Waals Solids

- van der Waals attractive interactions, V_{vdW}, are obtained by combining the Keesom, Debye, and London expressions (Table 2.4).

- In the vapor phase, Lennard–Jones (6–12) potential energy curves, $V(R)$ versus R (Fig. 2.5), are generated from $V(R) = V_{vdW} + V_{overlap}$ with $n = 12$ in eq. 2.1.2.

- We can calculate the lattice energy of van der Waals solids (inert spherical gas atoms) using V_{att}(dispersion). More complex solids will be considered in Chapter 9.

Attractive Forces between Particles and Macroscopic Bodies

- London dispersion energies between macroscopic bodies can be calculated by summing over all pairs of molecules, one in each body.

- For two planar blocks separated by a distance h, $V_{att} = -H_{ij}/12\pi h^2$, where the Hamaker constant $H_{ij} \propto$ (number of atoms per unit volume)$^2[\alpha^2(\hbar v)/(\varepsilon_o)^2]$. Force of attraction $F_{att} = H_{ij}/6\pi h^3$.

- Thus V_{att}(dispersion) between particles or macroscopic bodies is a long-range attractive interaction.

Interaction between Curved Surfaces in Vacuum

- According to the Derjaguin approximation, interaction forces F between two uniformly curved surfaces are directly related to interaction energies V between two plates. For spheres, F_{att}(spheres) = $-\pi R V_{att}$(plates).

- For two identical spheres, radius = R, separated by a distance h, V_{att}(spheres) = $-H_{ij} R/12h$ and F_{att}(spheres) = $-H_{ij} R/12h^2$ provided that R >> h.

- Table 2.5 gives V_{att} for other particle geometries.

Interactions between Particles in Condensed Phases

- The effective Hamaker constant for the interaction of two particles (phase 1) in a solvent (phase 2) is $H_{121} = [H_{11}^{1/2} - H_{22}^{1/2}]^2$, where H_{11} and H_{22} are the Hamaker constants for the individual materials.

- H_{121} is always attractive, but greatly diminished compared to homogeneous interactions in vacuum.

- H_{123} for two different particles (1 and 3) in medium (2) can be attractive or repulsive.

Interface Structure and the Boltzmann Equation

- The Boltzmann equation provides the guide for deciding whether an interaction energy will be effective in organizing molecules compared to the randomizing influence of thermal motion.

- At high temperatures, thermal motion randomizes molecules; as a consequence, concentration, pressure, etc., profiles are uniform.

- At low temperatures, interaction energies dominate and make it possible to obtain highly ordered systems.

2

Apart from the minor contribution due to gravity, *all forces acting between atoms, ions, and molecules arise as a consequence of charge interactions.* These range from simple Coulombic interactions between ions to the more subtle mutual polarization of atoms resulting from instantaneous displacements of electrons and nuclei. Operating at surfaces, across interfaces, and between microdomains, these intermolecular forces are the origin of interfacial energies that guide interfacial processing. Our ability to understand and control them often determines our success or failure in interfacial engineering. Therefore, it is essential at the outset to review the origin, magnitude, and directionality of intermolecular forces.

As portrayed in the concept map, the first part of this chapter focuses on pairwise interactions between molecules in the vapor phase. Vapor-phase interactions are important for two reasons. First, a number of interfacial processes actually involve the vapor phase, and pairwise interactions are the primary sources of attraction defining the molecular state and bonding energy of the vapor species present. Second, analyzing these relatively simple unfettered interactions establishes fundamental concepts that prevail with the more complex situations we will encounter later in condensed liquid and solid phases. The second part of the chapter focuses on the van der Waals attraction and the short-ranged repulsion that occurs between molecules in all phases of matter. In the third part of the chapter we develop an understanding of the attraction between particles containing many molecules in vacuum and in liquid media. Later chapters will develop our understanding of the more specific interparticle interactions that are important in different interfacial systems.

2.1 Interaction Forces between Molecules Can Be Most Simply Described as a Sum of Five Contributions

Our notion of intermolecular interactions rests on the idea that we can identify molecular units that remain largely unaffected by their immediate environment. Atoms form molecular units by sharing electrons (as in *covalently bonded molecules*) or by transferring electrons (as in *ionically bonded molecules*).

Covalent bonds are short-ranged, highly directional, and strong. Typical energies of covalent bonds range from 5 to 10 eV/molecule (200 to 400 kT/molecule at room temperature). The covalent bond is so strong that extreme temperatures or strong electrostatic energy sources are required to disassociate bound molecules into their constitutive atoms. For example, converting nitrogen gas molecules back to nitrogen atoms requires a high-energy microwave plasma because the enthalpy of formation, $\Delta H_{form} \approx 10$ eV, is so high. By comparison, ionic bonds are long-ranged, nondirectional, and not so strong. For example, the transfer of electrons from sodium metal atoms to chlorine gas atoms generates ionically bonded sodium chloride molecules for which $\Delta H_{form} \approx 4.8$ eV/molecule or 185kT/molecule at room temperature.

For simple molecules, such as N_2 and NaCl, we can solve the Schrödinger equation to obtain electron density maps, and we can calculate *intra*molecular energies, bond lengths, and angles from first principles. Obtaining similar information on more complex molecules requires a combination of theoretical and experimental information.

In principle, we can extend the approach and solve the Schrödinger equation to describe electron density between any two molecules and determine their interaction, the *inter*molecular potential V_{ij}. However, in practice it is more convenient and conceptually simpler to represent the total intermolecular potential energy as a sum of five parts:

V_{ij} = charge transfer interaction (1)
+ electrical multipole/electrical multipole interaction (2)
+ electrical multipole/induced multipole interaction (3)   (2.1.1)
+ dispersion interaction (4)
+ overlap interaction (5)

Different interactions dominate the total potential for different molecules. Here the general term *multipole* describes the charge distribution of individual molecules. It includes monopoles, where the charge distribution associated with the electrons and nuclei in the molecule can be represented by a single positive or negative point charge as in an ion; dipoles, where the charge distribution is asymmetrical and can be represented by two point charges, one positive the other negative, as in a polar molecule; and quadrupoles, an asymmetrical distribution where the charge distribution can be represented by four point charges, two positive and two negative.

Although each of the five terms depends on molecular properties that can be derived from the Schrödinger equation, organizing V_{ij} in this way simplifies the task of identifying their origin. Some of the terms can be more easily understood in their classical form rather than in their quantum-mechanical form. Briefly the five terms are:

1. **Charge transfer interaction**—occurs when one atom or molecule donates excess electrons to, or shares electrons with (covalency), an acceptor with an electron deficiency. Interactions of this type occur when molecules or atoms are in close proximity. They are important in the organization of crystalline solids and in interfaces between them. They do not typically act across interfaces in fluid systems, and for this reason we defer further discussion of charge transfer interaction until Chapter 9.

2. Electrical **multipole/electrical multipole interaction**—occurs when ions and molecules possess either net ionic charge or a permanent asymmetrical charge distribution of electrons and nuclei (as with polar molecules such as water). As shown in Section 2.2, we can describe this interaction by classical electrostatics.

3. Electrical **multipole/induced multipole interaction**—occurs when one polar molecule with a permanent electrical multipole induces a dipole in another molecule that is polarizable. Again, as shown in Section 2.3, we can describe this term by classical electrostatics.

4. **Dispersion interaction**—occurs when a multipole, induced by an electronic fluctuation in one molecule, in turn induces a multipole in another molecule to give an induced multipole/induced multipole interaction. These dispersion interactions can be described only in quantum-mechanical terms, as presented in Section 2.4. They are ubiquitous and occur between molecules of all forms.

5. **Overlap interaction**—occurs when the electron clouds of two closed-shell molecules overlap. Due to the Pauli exclusion principle, guest electrons forced into excited-state orbitals of the host molecule produce a strong energy increase. The repulsion that results is approximately proportional to the square of the overlap, and the electron density increases at least exponentially as the separation between molecules decreases. This effect determines a molecule's size; for simple monatomic species such as ions, the size is represented by a sphere of radius R. Like dispersion interactions, overlap repulsions are ubiquitous and occur between molecules of all forms.

We can conveniently express the repulsive interaction potential due to overlap for two spherical atoms or molecules as a function of their separation, r, by the power law

$$V(r) = \left(\frac{\sigma}{r}\right)^n \tag{2.1.2}$$

Here n is an integer and σ is a constant related to molecular size. When $n = \infty$, we obtain an expression for infinitely hard spheres,

TABLE 2.1 Potential Energy of Interactions between Molecules in the Vapor Phase

Description		V	Definitions and restrictions	Value of n in $V \propto R^{-n}$
Coulomb $V_{\text{ion-ion}}$	$(2)^a$	$\dfrac{z_1 z_2 e^2}{4\pi\varepsilon_o R}$	z = valence. Attractive or repulsive depending on z value. Pairwise additive.	1
Coulomb $V_{\text{ion-dipole}}$	(2)	$\dfrac{(z_1 e)\, m_2 \cos\theta}{4\pi\varepsilon_o R^2}$	m = dipole moment, θ defined in Fig. 2.1. Attractive or repulsive depending on z and θ. Pairwise additive.	2
Coulomb $V_{\text{dipole-dipole}}$	(2)	$-\dfrac{\text{const.}\, m_1 m_2}{4\pi\varepsilon_o R^3}$	Const. = $(2\cos\theta_1 \cos\theta_2 - \sin\theta_1 \sin\theta_2 \cos\phi)$ as defined in Fig. 2.2. Attractive or repulsive depending on values of θ_1, θ_2, ϕ. Pairwise additive.	3
Keesom $V_{\text{dipole-dipole}}$	(2)	$\dfrac{-2m_1^2 m_2^2}{3(4\pi\varepsilon_o)^2 kT\, R^6}$	Obtained by averaging Coulomb interactions over all angles, strong dependence on T. Always attractive. Pairwise additive.	6
Debye $V_{\text{dipole-induced dipole}}$	(3)	$-\dfrac{(m_1^2 \alpha_2 + m_2^2 \alpha_1)}{(4\pi\varepsilon_o)^2 R^6}$	α = polarizability. Always attractive. Not pairwise additive.	6
London $V_{\text{induced dipole-induced dipole}}$	(4)	$-\dfrac{3}{2}\dfrac{\alpha_1 \alpha_2}{(4\pi\varepsilon_0)^2 R^6}\left(\dfrac{\hbar v_1 v_2}{v_1 + v_2}\right)$	v = characteristic vibrational frequency of electrons. Always attractive. Assume pairwise additive.	6
Pauli principle $V_{\text{repulsion}}$	(5)	$\left(\dfrac{\sigma}{R}\right)^n$	σ = determined by molecule's size, n ranges from 9 to 15. Always repulsive.	12

From P. C. Hiemenz, *Principles of Colloid and Surface Chemistry*, 2nd ed., Marcel Dekker, New York, 1986, pp. 618–19.

^{a}Refers to different interactions in eq. 2.1.1

because when $r > \sigma$, the value of $V(r)$ is effectively zero, while when $r < \sigma$, it is infinite. In practice n values between 10 and 15 provide a more realistic description (see Section 2.5.1).

Because dispersion and overlap interactions are common to all molecules (both polar and nonpolar), they are not so useful for the manipulation of interactions across interfaces. Multipole and induced-multipole interactions occur more selectively and are potentially more effective in promoting selective interaction in interfacial systems like colloids and emulsions.

Many authors treat the hydrogen bond as a special type of interaction and include an additional term to represent it in the expression for V_{ij}. We will consider it as a special case of electrostatic interaction for the following reason. The low electron density on a hydrogen atom bonded to an electronegative atom, such as oxygen or nitrogen, weakens the Pauli exclusion repulsive force and enables the hydrogen ion in the molecule to come into close proximity to the polar center of another molecule. This causes a particularly strong electrostatic bond between the two molecules.

In discussing molecular interactions in the gas phase, it is useful to divide molecules into three categories: (a) permanently charged; (b) polar—that is, uncharged but possessing a permanent dipole moment; and (c) nonpolar—that is, uncharged and possessing no permanent dipole moment but capable of being polarized. Interactions between various pair combinations of these types of molecules lead to components of V_{ij} of the kind listed in (2), (3), and (4). We will analyze them in turn in the next three sections of the chapter. They are summarized in Table 2.1.

2.2 Interactions between Permanently Charged Molecules and Polar Molecules—Coulomb's Law Describes Attractive and Repulsive Interactions

We can use Coulomb's law to determine interactions between permanently charged or permanently polarized molecules whenever the electrical field emanating from one charged or polar molecule does not perturb the charge distribution of adjacent molecules. Under these conditions, we can add up the individual contributions from each charge center to obtain the total interaction, following the principle of linear superposition of potentials; in other words, we can sum all the pair potentials. This *pairwise additive* approach permits us to describe ion-ion, ion-dipole, and dipole-dipole interactions.

2.2.1 Depending on the Sign of the Charges, Ion-Ion Interactions Can Be Attractive or Repulsive

According to Coulomb's law, the magnitude of the electric field E_1 in the radial direction at a distance r from a charge q_1 is

$$E_1 = \frac{q_1}{4\pi\varepsilon_o\varepsilon_r r^2} \qquad (2.2.1)$$

where ε_o is the electrical permittivity of vacuum ($\varepsilon_o = 8.854 \times 10^{-12}$ $C^2 \, J^{-1} \, m^{-1}$, coulombs^2 per joule per meter) and ε_r is the relative dielectric permittivity, or the dielectric constant, of the medium surrounding the charge.[1] In free space or vacuum $\varepsilon_r = 1$. The ionic charge is given by $q = z_i e$, where e is the elementary charge (e = 1.602×10^{-19} C) and z_i is equal to the valence (z) multiplied by ± 1 according to the *sign* of the ion; for cations the charge is positive, and for anions the charge is negative. (For example, Na^+, $z_i = +1$; Mg^{2+}, $z_i = +2$; and Cl^-, $z_i = -1$.)

---

[1] *Note that in this notation ε_o is the dielectric permittivity of vacuum, ε_r is the relative permittivity or the dielectric constant of the medium between the charged particles. $\varepsilon_r\varepsilon_o$ is the dielectric permittivity of the medium. When we are dealing with the interaction between charged particles in vacuum or free space, as is the case in this chapter, $\varepsilon_r = 1$.*

The Coulombic force exerted on a second charge q_2 at distance R_{12} is

$$F = q_2\,E_1 = \frac{q_2\,q_1}{4\pi\varepsilon_o\varepsilon_r R_{12}^2} \qquad (2.2.2)$$

We can obtain the corresponding expression for the potential energy by integrating the force as a function of distance

$$V = -\int_{R_{12}}^{\infty} F\,dr$$

to yield

$$V_{ion-ion}(R_{12}) = \frac{q_1 q_2}{4\pi\varepsilon_o\varepsilon_r R_{12}} = \frac{z_1 z_2 e^2}{4\pi\varepsilon_o\varepsilon_r R_{12}} \qquad (2.2.3)$$

where R_{12} is the separation of the two charge centers. For like charges, both F and $V_{ion-ion}$ are positive, which corresponds to a repulsive interaction, while for unlike charges they are negative, corresponding to an attractive interaction.[2]

The maximum value of $V_{ion-ion}$ for an isolated sodium ion in contact with a chlorine ion occurs when R_{12} is the sum of the ionic radii, $R_{Na} + R_{Cl}$

$$V_{ion-ion} = -\frac{(1.602\times10^{-19}\mathrm{C})^2}{4\pi(8.85\times10^{-12}\ \mathrm{C^2 J^{-1}\ m^{-1}})(0.276\times10^{-9}\ \mathrm{m})} = -8.4\times10^{-19}\ \mathrm{J}$$

We can compare this value to the thermal energy as measured by the product kT, where k is Botzmann's constant $(1.381\times10^{-23}\ \mathrm{J\ K^{-1}})$. At room temperature (300 K) $kT = 4.14\times10^{-21}$ J, so that for room temperature sodium chloride $V_{ion-ion}/kT \approx 200$. Only at separations with R_{12} greater than 56 nm will $V_{ion-ion}$ be less than kT at room temperature. Thus ionic Coulomb interactions are very strong compared to ambient thermal energy.

2.2.2 Ion-Dipole Interactions Depend on the Orientation of the Ion and the Dipole

Many neutral molecules, such as water, ammonia, and HCl, possess a permanent separation of electronic and nuclear charge that gives rise to an electric dipole. They are said to be polar molecules. Because permanent dipoles arise from the asymmetric displacements of electrons along covalent bonds between different atoms, they are not observed in single atoms.

---

[2]*In describing intermolecular interactions, we must specify the reference state we are using. Values of energies and free energies always refer to the difference between the actual state and a reference state. In a gas the reference state is usually taken to be at separation distance $R = \infty$. In eq. 2.2.3. this corresponds to $V = 0$, and so a negative value for V corresponds to an attractive interaction energy and a positive value for V corresponds to a repulsive interaction energy. Similarly, by our definition, negative values of F correspond to an attractive force and positive values of F to a repulsive force.*

Figure 2.1
Coordinates for defining the interaction between an ion of charge ze, and a dipole containing charges of $+q$ and $-q$ separated by a distance l. The distance between the ion and the center of the dipole is represented by R_{12}, and θ stands for the angle between R_{12} and l.

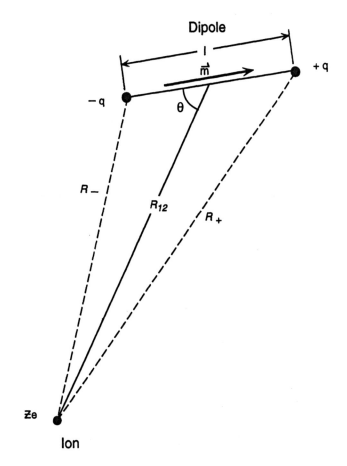

The magnitude of the dipole moment of a polar molecule is defined by

$$m = ql \qquad (2.2.4)$$

where l represents the distance between two charges q^+ and q^- as shown in Figure 2.1. For two electronic charges of opposite sign, $q = \pm e$, separated by $l = 0.1$ nm, the dipole moment is $m = (1.602 \times 10^{-19}\,\text{C})(10^{-10}\,\text{m}) = 1.6 \times 10^{-29}\,\text{C m} = 4.8\,\text{D}$. Here D (Debye) represents the basic unit of dipole moment, defined by $D = 3.336 \times 10^{-30}\,\text{C m}$. Table 2.2 gives the dipole moments of small polar molecules.

TABLE 2.2 Dipole Moments (m) of Molecules in Debye Units (1 Debye = 3.336×10^{-30} C m)

Molecule	m	Molecule	m
Alkanes	0	NH_3	1.47
C_6H_6 (benzene)	0	CH_3OH, C_2H_5OH	1.69
CCl_4	0	CH_3COOH	1.7
CO_2	0	H_2O	1.85
$CHCl_3$ (chloroform)	1.06	C_2H_4O (ethylene oxide)	1.9
HCl	1.08	CH_3COCH_3 (acetone)	2.85

J. N. Israelachvili, *Intermolecular and Surface Forces*, Academic Press, London, 1985, p. 37.

We can use Coulomb's Law to calculate the potential energy between an ion and a dipole using the geometry illustrated in Figure 2.1. Here an ion, charge $q = z_i e$, is separated by a distance R_{12} from the center of a dipole of length l with charges $\pm q$. The orientation of the dipole with respect to a line joining the centers of the two molecules is given by θ. The total energy is the sum of the Coulombic energies of ze interacting with $-q$ and with $+q$.

$$V(R_{12}) = \frac{z_i e\, q}{4\pi\varepsilon_o\varepsilon_r}\left(\frac{1}{R_+} - \frac{1}{R_-}\right) \tag{2.2.5}$$

This expression can be simplified using the cosine theorem

$$R_\pm = R_{12}[(1 \pm l\, \cos\, \theta/2R_{12})^2 + (l\, \sin\, \theta/2R_{12})^2]^{1/2} \tag{2.2.6}$$

Neglecting $(l/R_{12})^2$ terms, because $R_{12} \gg l$, the ion-dipole interaction becomes

$$V_{\text{ion-dipole}}(R_{12},\theta) = - \frac{z_i e\, m\, \cos\, \theta}{4\pi\varepsilon_o\varepsilon_r R_{12}^2} \tag{2.2.7}$$

Using eq. 2.2.1, $E(R_{12}) = z_i e/4\pi\varepsilon_o\varepsilon_r R_{12}^2$ for the field due to the ion, we can express eq. 2.2.7 as

$$V_{\text{ion-dipole}}(R_{12},\theta) = - mE(R_{12})\, \cos\, \theta \tag{2.2.8}$$

With a cation, z_i is positive and the maximum attraction occurs when the dipole points away from the ion, $\theta = 0°$, while maximum repulsion occurs when $\theta = 180°$. With an anion, z_i is negative and the opposite is true: maximum attraction occurs when $\theta = 180°$ and the dipole points toward the ion.

The maximum value of $V_{\text{ion-dipole}}$ for the interaction of a sodium ion, radius 0.095 nm, and a water molecule, radius 0.14 nm and dipole moment 1.85 D, in vacuum ($\varepsilon_r = 1$) is

$$V_{\text{ion-dipole,max}} = - \frac{(1.602 \times 10^{-19}\ \text{C})(1.85 \times 3.336 \times 10^{-30}\ \text{C m})}{4\pi(8.85 \times 10^{-12}\ \text{C}^2\ \text{J}^{-1}\ \text{m}^{-1})(0.235 \times 10^{-9}\ \text{m})^2}$$

$$= - 1.6 \times 10^{-19}\ \text{J}$$

Compared to kT at room temperature, this interaction is strong, $V_{\text{ion-dipole}}/kT = 39$. For a divalent cation, such as magnesium ($z = +2$, radius 0.065 nm), the interaction is even stronger and $V_{\text{ion-dipole}} = 150kT$. Thus ion-dipole interactions typically exceed thermal energy, and the ion's electrostatic field exerts a pronounced effect on the alignment of the dipole.

2.2.3 Dipole-Dipole Interaction Energies Are Often Less Than Thermal Energy

By a procedure similar to that used for evaluating ion-dipole interactions, we can derive the interaction energy for two dipoles of moments m_1 and m_2 at a distance R_{12} and oriented relative to each other as given in Figure 2.2. We find

$$V(R_{12}, \theta_1, \theta_2, \phi) = -\frac{m_1 m_2}{4\pi\varepsilon_o\varepsilon_r R_{12}^3} (2\cos\theta_1\cos\theta_2 - \sin\theta_1\sin\theta_2\cos\phi) \qquad (2.2.9)$$

Thus dipole-dipole interactions are strongly angle dependent, and the maximum attraction occurs when the two dipoles are in line and pointing in the same direction

$$V(R_{12}, 0, 0, \phi) = -\frac{2m_1 m_2}{4\pi\varepsilon_o\varepsilon_r R_{12}^3} \qquad (2.2.10)$$

Again we are interested in the strength of the dipole-dipole interaction. We ask, at what separation of two aligned dipoles with m = 1 D does the interaction energy equal kT at ambient temperature? According to eq. 2.2.10 (omitting the negative sign)

$$R_{12} = \left(\frac{2(1 \times 3.336 \times 10^{-30}\ \text{C m})^2}{4\pi(8.85 \times 10^{-12}\ \text{C}^2\text{J}^{-1}\ \text{m}^{-1})(4.12 \times 10^{-21}\ \text{J})} \right)^{1/3}$$

$$= 0.36\ \text{nm}$$

Because this distance equals the size of a typical molecule, the result means that dipole-dipole interactions influence alignment only between very polar molecules with high dipole moments in the gas phase at room temperature. Lesser interactions fail to force dipole alignment, and individual dipoles are able to rotate.

In liquid solutions the permittivity of the condensed phase is $\varepsilon_r\varepsilon_o$, and the strength of Coulombic interactions is reduced by the dielectric constant ε_r. Dipole-dipole interactions are then diminished to the point where they are never strong enough to overcome thermal energy to lead to substantial mutual alignment of polar molecules. In these situations the dipoles are relatively free to rotate with respect to each other, and we must determine the electrostatic contribution to the potential energy by an *angle-averaged potential*, as described in the next section.

The next higher term in the general multipole-multipole interaction, term (2) of eq. 2.1.1, involves dipole-quadrupole and

Figure 2.2
Coordinates for defining the interaction between two dipoles, m_1 and m_2, separated by a distance R_{12}. The angles between R_{12} and the axes of the dipoles are defined by θ_1 and θ_2. ϕ gives the relative dipole orientations with respect to rotation around the R_{12} axis.

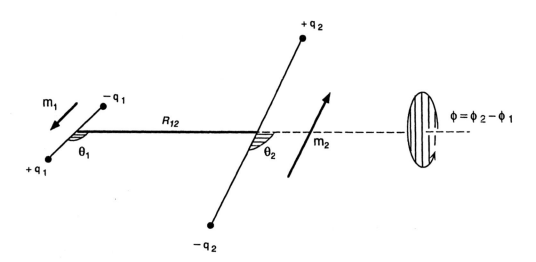

quadrupole-quadrupole interactions. Quadrupolar interactions are shorter-ranged and more orientation-dependent than dipole-dipole interactions. They too are weak relative to thermal energy and must be angle averaged. Quadrupolar interactions are of utmost importance in understanding the properties of phases containing quadrupolar molecules like CO_2, N_2, or C_6H_6, which do not possess permanent dipole moments. For example, one major reason why benzene is more soluble in water than saturated hydrocarbons lies in the high quadrupolar moment of the benzene molecule that interacts favorably with water molecule dipoles.

2.2.4 When Electrostatic Interactions Are Comparable to the Thermal Energy, We Use Angle-Averaged Interaction Potentials to Evaluate the Interaction Energy

As we have just seen, interactions involving dipoles and higher multipoles are often less than kT at room temperature and involve no strong alignment. When this is true, molecules will rotate more or less freely. The values of sin θ, cos θ, and cos φ in Figure 2.2 averaged over all angles are zero. But the values of the dipole-dipole interaction potential of eq. 2.2.9 averaged over all angles are *not* zero. Because the alignments are not totally random, they are subject to Boltzmann statistics. The Boltzmann weighting factor must be included, and it gives more value to orientations that have lower energy.

We can evaluate angle-averaged values of the interaction energy $V(R_{12})$ from the classical Boltzmann distribution

$$V(R_{12}) = \frac{1}{4\pi} \int V(R_{12},\Omega) \exp\left(\frac{-V(R_{12},\Omega)}{kT}\right) d\Omega \qquad (2.2.11)$$

where the integration is performed over the solid angle $d\Omega = \sin\theta \, d\theta \, d\phi$ (see Figure 2.3). When integrated over all angles $\int d\Omega = 4\pi$, so the integral is normalized by the factor $1/4\pi$.

When $V(R_{12},\Omega)$ is less than kT, we can expand the exponential term in eq. 2.2.11

$$\exp\left(\frac{-V(R_{12},\Omega)}{kT}\right) = 1 - V(R_{12},\Omega)/kT + \frac{1}{2}[V(R_{12},\Omega)/kT]^2 - \cdots \qquad (2.2.12)$$

and substitute it into eq. 2.2.11 to give

$$V(R_{12}) = \frac{1}{4\pi} \int [V(R_{12},\Omega) - V(R_{12},\Omega)^2/kT + \cdots] \, d\Omega \qquad (2.2.13)$$

By substituting eq. 2.2.7 for $V_{ion\text{-}dipole}(R_{12},\theta)$ in eq. 2.2.13, we can obtain the angle-averaged interaction energy for the ion-dipole interaction

$$V(R_{12}) = \frac{1}{4\pi} \int \left[\frac{z_i e \, m \cos\theta}{4\pi\varepsilon_o\varepsilon_r R_{12}^2} - \left(\frac{z_i e \, m}{4\pi\varepsilon_o\varepsilon_4 R_{12}^2} \right)^2 \frac{\cos^2\theta}{kT} + \cdots \right] d\Omega \qquad (2.2.14)$$

Substituting the spatially averaged value of $<\cos\theta> = 0$ and $<\cos^2\theta> = 1/3$ gives

Figure 2.3
Schematic for defining the
coordinates for calculating
angle-averaged interactions
using eq. 2.2.11.

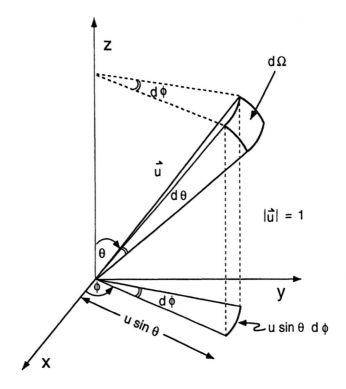

$$V_{ion-dipole}(R_{12}) = -\frac{(z_i e)^2 m^2}{3(4\pi\varepsilon_o\varepsilon_r)^2 kT\ R_{12}^4}, \quad \text{for } kT > \frac{z e\ m}{4\pi\varepsilon_o\varepsilon_r R_{12}^2} \quad (2.2.15)$$

Because of the squared terms, this interaction energy is always attractive irrespective of the ion sign. The magnitude is inversely dependent on temperature. Note also that the interaction energy varies as R_{12}^{-4} compared to R_{12}^{-2} (eq. 2.2.7) for the direct electrostatic interaction potential. This doubling of the exponent is general for angle-averaged interaction energy versus direct electrostatic interactions.

2.2.5 When Angle-Averaged Interaction Potentials Are Computed for Dipole-Dipole Interactions, We Obtain the Keesom Attractive Interaction Energy

We can obtain the angle-averaged potential for the dipole-dipole interaction by substituting eq. 2.2.9 into 2.2.13 to give

$$V_{dipole-dipole}(R_{12}) = -\frac{2m_1^2 m_2^2}{3(4\pi\varepsilon_o\varepsilon_r)^2 kT\ R_{12}^6}, \quad \text{for } kT > \frac{m_1 m_2}{4\pi\varepsilon_o\varepsilon_r\ R_{12}^3} \quad (2.2.16)$$

which is also always attractive and also dependent on temperature. The Boltzmann-averaged interaction between two permanent dipoles is known as the *Keesom interaction*. It shows a R_{12}^{-6} dependence and constitutes one of the three important contributions to van der Waals interactions.

For two water molecules with a dipole moment of 1.85 D and 0.5 nm apart, the Keesom interaction is given by

$$V(R_{12}) = -\frac{2(1.85 \times 3.336 \times 10^{-30}\ \text{C m})^4}{3(4\pi \times 8.85 \times 10^{-12}\ \text{C}^2\ \text{J}^{-1}\ \text{m}^{-1})^2(4.12 \times 10^{-21}\ \text{J})(0.5 \times 10^{-9}\ \text{m})^6}$$

$$= -1.2 \times 10^{-21}\ \text{J} \quad \text{or approx. 0.3 kT}$$

2.2.6 The Interaction Energy for Angle-Averaged Potentials Is Twice the Interaction Free Energy

A point of confusion that often arises in discussing electrostatic interactions is whether or not the expressions for the interaction potentials are energies or free energies. Equations 2.2.15 and 2.2.16 were derived using eq. 2.2.11 that took the Boltzmann average of the interaction potential; they both represent the interaction energy. But there is an entropic penalty to be paid for obtaining this nonrandom value, and it can be shown to equal one half the interaction energy. So the *interaction free energy* (the energy available for doing work and therefore available to exert the attractive force) is one-half the value of eqs. 2.2.15 and 2.2.16. Thus the values quoted in eq. 2.5.1 (later) and in Table 2.4 for the $V_{\text{dipole-dipole}}$ contribution to the van der Waals interaction (the *free energy of interaction*) is one-half the result in eq. 2.2.16.

2.3 Interactions between Permanently Charged Molecules or Polar Molecules and Induced Dipoles—Attractive Interactions Occur When Electrical Fields Emanating from a Polar Molecule Induce Polarization in Nearby Molecules

2.3.1 Induced Dipole Interactions Exhibit Only Half the Energy of Direct Interactions

In our discussion so far, we have assumed that molecules are nonpolarizable, that is, that they interact without perturbing each other's charge distributions. However, in real systems, *induced polarization* results when one molecule responds internally to electrical fields emanating from nearby molecules. The two most important induced electrostatic interactions are ion-induced dipole and dipole-induced dipole interactions.

We define a molecule's polarizability α by the strength of the induced dipole moment m_{ind} acquired in an electrical field

$$m_{\text{ind}} = \alpha E \qquad (2.3.1)$$

Table 2.3 gives values of α for a number of molecules.

TABLE 2.3 Polarizabilities (α) of Atoms and Molecules, Expressed in Units of $(4\pi\varepsilon_0)\ 10^{-30}\ m^3 = 1.11 \times 10^{-40}\ C^2\ m^2 J^{-1}$

Molecule	α	Molecule	α	Molecule	α
He	0.20	NH_3	2.3	$CH_2=CH_2$	4.3
H_2	0.81	CH_4	2.6	C_2H_6	4.5
H_2O	1.48	HCl	2.6	Cl_2	4.6
O_2	1.60	CO_2	2.6	$CHCl_3$	8.2
Ar	1.63	CH_3OH	3.2	C_6H_6	10.3
CO	1.95	Xe	4.0	CCl_4	10.5

J. N. Israelachvili, *Intermolecular and Surface Forces*, 2nd ed., Academic Press, London, 1992, p. 69.

To understand the nature of polarizability in more detail, we analyze the interaction between an electrical field and a nonpolar molecule. As a specific example, Figure 2.4a shows a Bohr atom in which an electron ($q = -e$) moves around a nucleus ($q = +e$) with a circularly symmetrical orbit, radius r, in the absence of an external field. We picture the Bohr atom as two opposite charges, $\pm e$, interacting through a spring with a spring constant K. Normally the centers of charge coincide, but in the presence of an external electrical field, they separate by a distance l, as shown in Figure 2.4b. This leads to an induced dipole moment $m_{ind} = ql$. By analogy with Hooke's law, we can write the internal energy for a small deformation as

$$V_{int} = \frac{1}{2} K\, l^2 \qquad (2.3.2)$$

A second contribution to the energy exists in the presence of an electric field E, which from eq. 2.2.8 amounts to

$$V_{field} = -\,Eql \qquad (2.3.3)$$

The total energy $V_{total} = V_{int} + V_{field}$.

At equilibrium, a balance exists between the externally imposed force and the restoring force, and by differentiating and setting $(dV_{total}/dl) = 0$, we obtain

$$l_0 = Eq/K \qquad (2.3.4)$$

Using eq. 2.2.4, the induced moment becomes $m_{ind} = ql_0 = Eq^2/K$. Comparison with eq. 2.3.1 shows that the polarizability equals

$$\alpha = q^2/K \qquad (2.3.5)$$

and the total energy

$$V_{total} = \frac{K}{2}\frac{E^2 q^2}{K^2} - Eq\frac{Eq}{K} = -\frac{E^2 q^2}{2K} = -\alpha\frac{E^2}{2} \qquad (2.3.6)$$

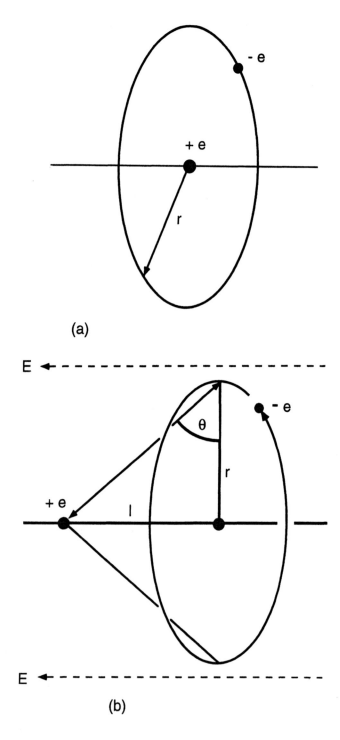

Figure 2.4
Schematic illustrating how polarization of a Bohr atom (a) by an electrical field E forms an induced dipole (b).

(a)

(b)

which is *one-half* the interaction energy between an ion and a *permanent* dipole of moment αE, as can be seen by substituting $m = \alpha E$ in eq. 2.2.8. This factor of one-half appears because it takes energy to generate the displacement, and the amount expended equals one-half the energy of direct interaction.

Polarizability has the dimensionality of ε_o times volume, and the magnitude of the volume typically corresponds to half of the molecular volume. For example, in the case of water, $\alpha/\varepsilon_o = 1.86 \times 10^{-29}$ m^3, while the value derived from liquid density is 3.0×10^{-29} m^3.

2.3.2 Comparing Ion-Induced Dipole Interactions with Ion-Dipole Interactions Shows That Induced Interactions Are More Short-Ranged

Now we can calculate $V_{\text{ion–induced dipole}}$ by combining eq. 2.3.6 and the equation for the electric field associated with an ion of charge $z_i e$

$$E = \frac{z_i e}{4\pi\varepsilon_o \varepsilon_r R_{12}^2} \qquad (2.3.7)$$

to give

$$V_{\text{ion–induced dipole}}(R_{12}) = -\frac{\alpha(z_i e)^2}{2(4\pi\varepsilon_o \varepsilon_r)^2 R_{12}^4} \qquad (2.3.8)$$

The general expression for the ion-induced dipole interaction between two differently charged molecules, one charge $z_1 e$ and polarizability α_1 and the second charge $z_2 e$ and polarizability α_2 is

$$V_{\text{ion–induced dipole}}(R_{12}) = -\frac{1}{2(4\pi\varepsilon_o \varepsilon_r)^2 R_{12}^4}(z_1^2 e^2 \alpha_2 + z_2^2 e^2 \alpha_1) \qquad (2.3.9)$$

which reduces to eq. 2.3.8 when one of the charges ($z_i e$) is zero. Note that the dependence on R_{12} is more short-ranged than the ion-permanent dipole interaction in eq. 2.2.7. The interaction potential depends on distance as R_{12}^{-4} rather than R_{12}^{-2}, because one R_{12}^{-2} term induces the response while a second R_{12}^{-2} term produces the interaction. The exponent for induced interaction is generally double that for direct interaction with a permanent dipole.

2.3.3 When Angle-Averaged Potentials Are Computed for Dipole-Induced Dipole Interactions We Obtain the Debye Attractive Interaction

We can obtain dipole-induced dipole interactions by combining eq. 2.3.6, which gives the potential energy of an induced interaction, with the electrical field associated with a permanent dipole. Both the magnitude and the orientation of the field at the polarizable molecule depend on the angle θ between the dipole moment and the vector joining the two species. The magnitude of the field is

$$E(R_{12}, \theta) = \frac{m(1 + 3\cos^2\theta)^{1/2}}{4\pi\varepsilon_o \varepsilon_r R_{12}^3} \qquad (2.3.10)$$

The resulting interaction is

$$V(R_{12}, \theta) = -\alpha\frac{E^2}{2} = -\frac{m^2\alpha(1 + 3\cos^2\theta)}{2(4\pi\varepsilon_o \varepsilon_r)^2 R_{12}^6} \qquad (2.3.11)$$

In the general case of two different molecules, each of which possesses a permanent dipole moment m_1 and m_2 and polarizability α_1 and α_2, the total interaction is given by

$$V(R_{12},\theta_1,\theta_2) = -\frac{1}{2(4\pi\varepsilon_0\varepsilon_r)^2}\frac{1}{R_{12}^6}[m_1^2\,\alpha_2(3\cos^2\theta_1 + 1)$$

$$+ m_2^2\,\alpha_1(3\cos^2\theta_2 + 1] \tag{2.3.12}$$

Since the strength of this interaction is not sufficient to mutually align the dipoles, we must obtain an angle-averaged value. The angle-averaged energy follows directly from eq. 2.3.12 by using $<\cos^2\theta> = 1/3$ to give

$$V_{\text{dipole–induced dipole}}(R_{12}) = -\frac{(m_1^2\,\alpha_2 + m_2^2\,\alpha_1)}{(4\pi\varepsilon_0\varepsilon_r)^2 R_{12}^6} \tag{2.3.13}$$

The Boltzmann-averaged interaction between a permanent dipole and an induced dipole is known as the *Debye interaction*. It is always attractive and shows an R_{12}^{-6} dependence and constitutes the second of the three inverse-sixth-power contributions to the total van der Waals interaction energy between molecules.

Induced interactions are *not pairwise additive,* and this fact adds an important complication to our calculations. For permanent moments, the interaction V_{123} between three molecules at fixed orientations is simply

$$V_{123} = V_{12} + V_{13} + V_{23} \tag{2.3.14}$$

or the sum over the individual pairs. But this simplicity does not hold true for induced interactions. For example, a molecule midway between two similar ions does not experience any electrical field, and the interaction potential from the ion-induced dipole term is zero. If we use eq. (2.3.9) and assume pairwise additivity in such a case, we will get a large and quite erroneous result for the interaction energy—one that implies an artificial stabilization of the particular molecular configurations.

2.4 Interactions between Induced Dipoles and Induced Dipoles— Dispersion Forces—Lead to Attractive Interactions between All Molecules

We can understand the origin of dispersion interactions from the following argument. For nonpolar atoms, such as the rare gases, the time-averaged dipole moment is zero. However, at any instant there exists an instantaneous dipole moment determined by the location of the electrons around the nucleus. This dipole generates an electric field, which in turn induces a dipole in nearby neutral atoms. The resulting interaction gives rise to an attractive force between the two atoms whose time average is finite. The same argument applies for the attraction of two nonpolar molecules.

Dispersion (or *London*) forces require a quantum-mechanical description. In 1933 London derived an expression for the attraction between a pair of atoms by solving the Schrödinger equation. He modeled each of the atoms as a charged harmonic oscillator with a characteristic frequency ν and obtained

$$V(R_{12}) = -\frac{3}{4} \frac{\alpha^2 h\nu}{(4\pi\varepsilon_0\varepsilon_r)^2 R_{12}^6} \qquad (2.4.1)$$

for two similar atoms and

$$V(R_{12}) = -\frac{3}{2} \frac{\alpha_1\alpha_2}{(4\pi\varepsilon_0\varepsilon_r)^2 R_{12}^6} \left(\frac{h\nu_1\nu_2}{(\nu_1 + \nu_2)} \right) \qquad (2.4.2)$$

for dissimilar atoms. In these equations, $\hbar$ is Planck's constant and $\hbar\nu$ generally equals the ionization energy of the atoms.

London's expression provides a useful basis for calculating dispersion forces, particularly in the gas phase. In condensed phases, the Lifshitz theory gives a more general description. It invokes a continuum electrodynamics point of view in which each material is characterized by a frequency-dependent dielectric permittivity. To pursue this distinction in detail would carry us far beyond the scope of this text. Instead we note that the London and Lifshitz approaches yield similar results for insulators, but we must use the more general Lifshitz approach for metals.

Dispersion forces, like gravitational forces, operate between *all* atoms or molecules. They too are always attractive and constitute the third and often the major contribution to van der Waals forces. The next section summarizes all the van der Waals interactions. They play an important role in many interfacial phenomena —adhesion, surface tension, physical adsorption, coagulation of colloidal particles, and in structures of macromolecules, such as proteins and polymers.

2.5 We Can Obtain the van der Waals Attractive Force by Combining Expressions for the Keesom, Debye, and London Forces

2.5.1 Potential Energy Curves Describing Intermolecular Interactions in the Vapor Phase Combine Expressions for van der Waals Attractions and Hard Sphere Repulsions

Three forces contribute to the attractive interactions for molecules in the vapor phase: Keesom orientation forces, Debye inductive forces, and London dispersion forces. Collectively they are known as the van der Waals attraction, V_{vdW}, and the interaction energy of each one varies with the inverse sixth power of the distance. For two dissimilar molecules, therefore, we can write for the *attractive* interaction energy

$$V_{vdW} = (V_{orient} + V_{ind} + V_{disp})$$

$$V_{vdW} = -\left(\frac{m_1^2 m_2^2}{3kT} + (m_1^2\alpha_2 + m_2^2\alpha_1) + \frac{3\alpha_1\alpha_2\hbar v_1 v_2}{2(v_1 + v_2)}\right)\left(\frac{1}{(4\pi\varepsilon_o\varepsilon_r)^2 R_{12}^6}\right) \qquad (2.5.1)$$

Table 2.4. shows the relative contribution of each component to the total van der Waals energies of various molecules. Dispersion attractions are more important than either orientation or induced interactions for all molecules except highly polar molecules like water, as can be seen from the ratio expressed as a percentage in the final column. In interactions between dissimilar molecules, dispersion forces dominate almost completely if one of the molecules is nonpolar.

Inspection of Table 2.4 shows that V_{vdW} for two dissimilar molecules A and B is usually intermediate between the values for A-A and B-B interactions. Often we can estimate the value of V_{vdW} for A-B by determining the geometric mean for A-A and B-B. For example, the geometric mean for Ne-CH$_4$ is $(4 \times 102)^{1/2} = 20$, while the directly computed value is 19. For HCl-HI, we obtain $(123 \times 372)^{1/2} = 214$ compared to 205. Interactions involving water provide an exception to this generalization. For example, the computed value for the CH_4–H_2O interaction is actually less than the values of either H_2O–H_2O or CH_4–CH_4.

As noted at the beginning of this chapter, we can obtain expressions for the total intermolecular pair potential by adding together the attraction and repulsion potentials. Figures 2.5a and b show repulsion curves for the hard sphere repulsion potential ($n = \infty$ in eq. 2.1.2) and the soft sphere repulsion potential ($n = 12$ in eq. 2.1.2), respectively. Figures 2.5c and d show the result of adding the van der Waals attraction term to the repulsion term for both cases.

One of the simplest and most widely used expressions for combining these values is the Lennard–Jones 6–12 potential, which assumes the attraction to be van der Waals, and therefore to vary as $1/R_{12}^6$, and the repulsion to vary as $1/R_{12}^{12}$ as in Figure 2.5d. This expression may be written as

$$V(R_{12}) = -\frac{C_1}{R_{12}^6} + \frac{C_2}{R_{12}^{12}} = 4\delta\left[-\left(\frac{\sigma}{R_{12}}\right)^6 + \left(\frac{\sigma}{R_{12}}\right)^{12}\right] \qquad (2.5.2)$$

where the parameters σ and δ are adjustable. It is important to note that the parameter σ is related to but does not equal the diameter of a molecule because when $R_{12} = \sigma$, $V(R_{12}) = 0$. The energy minimum, where $dV(R_{12})/dR_{12} = 0$, occurs at $R_{12} = (2)^{1/6}\sigma = 1.12\sigma$ and at this value of R_{12}, $V(R_{12}) = -\delta$, (Figure 2.5d), which results from a contribution of -2δ from the attraction term and $+\delta$ from repulsion. Thus the soft repulsive $1/R_{12}^{12}$ term decreases the strength of binding by 50 percent. By comparison for the hard sphere repulsive potential, the binding energy at contact ($R_{12} = \sigma$) equals the van der Waals attractive interaction energy (Figure 2.5c).

TABLE 2.4 Keesom, Debye, and London Dispersion Free Energy Contributions to van der Waals Energy for Various Pairs of Molecules in Vacuum at 293 K

Interacting molecules	Polarizability $\frac{\alpha}{4\pi\varepsilon_o}$ $(10^{-30}$ m$^3)$	Dipole moment m (D)[a]	Ionization potential $h\nu$ (eV)[b]	van der Waals energy coefficients $(10^{-79}$ J m$^6)$			Total V_{vdW} energy eq. 2.5.1 (theoretical)	$V_{ind.\,dipole-ind.\,dipole}/V_{vdW}$ (%) (theoretical)
				$V_{dipole-dipole}$ $\frac{m^4/3kT}{(4\pi\varepsilon_o)^2}$	$V_{dipole-induced\,dipole}$ $\frac{2m^2\alpha}{(4\pi\varepsilon_o)^2}$	$V_{ind.\,dipole-induced\,dipole}$ $\frac{3\alpha^2 h\nu}{4(4\pi\varepsilon_o)^2}$		
Ne – Ne	0.39	0	21.6	0	0	4	4	100
CH$_4$ – CH$_4$	2.60	0	12.6	0	0	102	102	100
HCl – HCl	2.63	1.08	12.7	11	6	106	123	86
HI – HI	5.44	0.38	10.4	0.2	2	370	372	99
CH$_3$Cl – CH$_3$Cl	4.56	1.87	11.3	101	32	282	415	68
H$_2$O – H$_2$O	1.48	1.85	12.6	96	10	33	139	24
Dissimilar molecules				$\frac{m_1^2 m_2^2/3kT}{(4\pi\varepsilon_o)^2}$	$\frac{m_1^2\alpha_2 + m_2^2\alpha_1}{(4\pi\varepsilon_o)^2}$	$\frac{3\alpha_1\alpha_2 h\nu_1\nu_2}{2(4\pi\varepsilon_o)^2(\nu_1+\nu_2)}$		
Ne – CH$_4$				0	0	19	19	100
HCl – HI				1	7	197	205	96
H$_2$O – Ne				0	1	11	12	92
H$_2$O – CH$_4$				0	9	58	67	87

J. N. Israelachvili, *Intermolecular and Surface Forces*, 2nd ed., Academic Press, London, 1992, p. 95.

[a] 1 D = 3.336×10^{-30} C m.

[b] 1 eV = 1.602×10^{-19} J

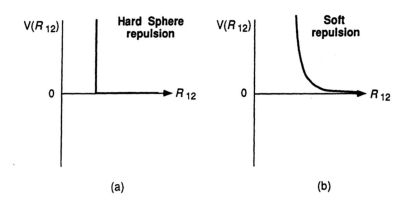

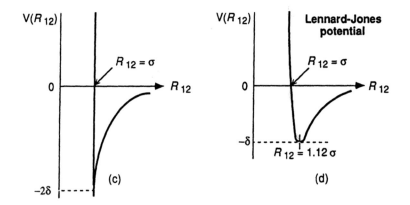

Figure 2.5
Potential energy curves for: (a) hard sphere repulsion (n = ∞ in eq. 2.1.2); (b) soft repulsion (n = 12 in eq. 2.1.2); (c) hard sphere repulsion combined with van der Waals attraction; and (d) Lennard–Jones 6–12 potential as given by eq. 2.5.2. (J. N. Israelachvili, *Intermolecular and Surface Forces*, 2nd ed., Academic Press, London, 1992, p. 114.)

2.5.2 Dispersion Forces Determine the Properties of van der Waals Solids Composed of Inert, Spherical Molecules

Inert spherical atoms and molecules, such as neon, argon, and methane, provide examples of van der Waals solids. Held together solely by dispersion forces, they have low melting points and low latent heats of melting. Van der Waals bonds in this case are nondirectional; that is, they are exerted equally in all directions. As a result, in the crystalline state, van der Waals molecules surround themselves with the highest possible number of nearest neighbors. Typically they form face-centered cubic or close-packed hexagonal structures (see Figure 9.2) with a coordination number of 12 nearest neighbors, which gives 12 shared interactions or 6.0 full interactions. When the contribution of next-nearest and more distant neighbors is included, the equivalent number of full interactions rises from 6.0 to 7.22.

We can estimate the lattice energy U per mole of van der Waals solids by summing the pair potentials of all N_{Av} atoms in a mole, where N_{Av} is Avogadro's number. This process modifies eq. 2.4.1 to give

$$U = 7.22N_{Av}\left(\frac{3}{4}\frac{\alpha^2\hbar v}{(4\pi\varepsilon_0\varepsilon_r)^2\lambda^6}\right) \qquad (2.5.3)$$

where λ replaces R_{12} and represents the equilibrium interatomic spacing in the crystalline solid. Because the atoms interact through free space, $\varepsilon_r = 1$. For argon, using $v = 2.52 \times 10^{-8}$ J, $\lambda = 0.376$ nm, and $\alpha/(4\pi\varepsilon_0) = 1.63 \times 10^{-30}$ m^3, we calculate $U = 7.7$ kJ per mole. The sum of the latent heats of melting and vaporization is 7.7 kJ. Agreement between calculated and measured values of U is quite satisfactory for van der Waals solids, but predictions based on eq. 2.5.3 for liquids and solids containing nonspherical apolar, polar, or hydrogen-bonded molecules show discrepancies from the experimental data that increase as the molecular interactions become more complex. Some reasons for the larger discrepancies between theoretical and experimental values arise as follows:

1. In defining the polarizability of molecules in eq. 2.3.1, we treated it as a molecular constant. However, except for spherical molecules, all polarizabilities are anisotropic and exhibit different values along different molecular directions. As a consequence, the dispersion forces between molecules depend on their mutual orientation. In the gas phase, or in many liquids in which molecules rotate rapidly, we can base our calculations on a mean polarizability (Table 2.3). However, in solids and some liquids, anisotropic forces can play an important role in determining specific configurational orientation. Polarization anisotropy influences the aggregated configuration of many macromolecules, such as the polymers, proteins, and liquid crystals described in subsequent chapters.

2. The interaction between two molecules that leads to London dispersion forces occurs as the result of the coupling of instantaneous dipoles. When the time required for the electric field of one atom to reach another becomes comparable to the period of the fluctuating dipole, synchronization between dipoles is diminished, which leads in turn to a decrease in the attractive force. Called the retardation effect, this decrease changes the distance dependence of the attractive energy from $1/R_{12}^6$ to $1/R_{12}^7$. Retardation effects begin at separations of approximately 5 nm and thus are not very important for two molecules in vacuum or the gas phase. However, in condensed systems or larger systems containing macroscopic surfaces or large particles, retardation can become an important factor.

Electrostatic bonds, metallic bonds, and covalent bonds important in the cohesion of other crystalline solids will be discussed in Chapter 9.

2.6 We Can Express Interaction Forces between Particles by a Sum of Terms Similar to That Used to Express the Interaction between Molecules

We now move on from our analysis of the interaction between two molecules in a gas to consider interaction potentials and forces between two *particles* that contain many atoms. By analogy with eq. 2.1.1, we can obtain the total interaction by summing all the possible contributions. Some of the terms are well understood theoretically and well verified by experiments; others have a more empirical basis. During the past decade, our ability to manipulate interfacial systems has made major advances through the growing realization that a large number of interaction forces exist, each with a characteristic magnitude and dependence on distance. Furthermore, we can modulate some of these forces between particles by small changes in temperature, pH, salt concentration, or solvent composition, features that are extremely significant in process control. We will pursue details of these other forces as they pertain to the behavior of colloids and amphiphiles in Chapters 4 and 5, respectively.

To conclude this chapter, we focus on the two major interaction forces that operate in all particle systems. These are the dispersion attractive forces, which take on quite a surprising distance relationship when integrated for macroscopic bodies, and the overlap repulsive force, which takes the same $1/r^{12}$ form for macroscopic bodies as for individual molecules.

2.6.1 We Can Obtain the Dispersion Forces Acting between Two Particles or Macroscopic Bodies in Vacuum by Summing the London Dispersion Forces over All Molecules Contained in the Two Bodies on a Pairwise Basis

If we assume pairwise additivity between molecules, we can sum over all possible pairs, one molecule in each body, to obtain the dispersion force between two macroscopic bodies or particles. The discussion that follows represents the two particles as two blocks with planar surfaces of infinite extension separated by a distance h in vacuum ($\varepsilon_r=1$). In calculating the attractive interaction between pairs of molecules, we use eq. 2.4.1

$$V(R) = \frac{C_{disp}}{R_{12}^6}$$

where $C_{disp} = (3/4)\alpha^2 \hbar \nu / (4\pi\varepsilon_o)^2$.

We consider first the interaction between a single molecule (or atom) and a block (block 1) where the normal distance from the molecule to the surface of the block is z, as shown in Figure 2.6a.

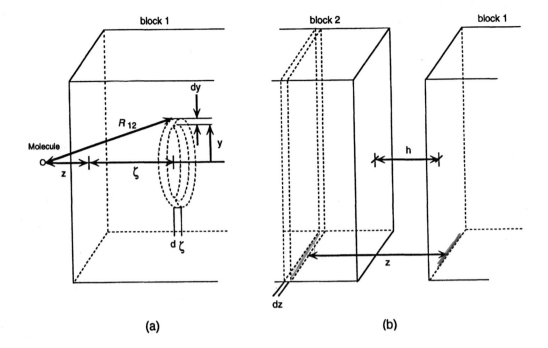

block 1   block 2   block 1

(a)   (b)

Figure 2.6
Schematic for coordinates used in calculating London dispersion forces (a) between a molecule located at a distance z away from a block and (b) between two blocks separated by a distance h. (P. C. Hiemenz, *J. of Chemical Education* **49**, 164, 1972.)

Integration will replace the summation between this molecule and all the molecules in the block. For the ring-shaped volume element shown in Figure 2.6a, the volume equals $2\pi y \, dy \, d\zeta$, and the interaction potential between the single molecule and all the molecules in this small volume is given by

$$dV(R_{12}) = -\frac{C_{disp}}{R_{12}^6} \frac{\rho N_{Av}}{M} 2\pi \, y \, dy \, d\zeta \qquad (2.6.1)$$

where C_{disp}/R_{12}^6 gives the pairwise interaction potential, $\rho N_{Av}/M$ (ρ, density; M, molecular weight) is the number of molecules per unit volume in the block, and $2\pi y \, dy \, d\xi$ the differential volume. When the ring is located a distance ζ inside the block, $R_{12}^2 = (z + \zeta)^2 + y^2$ and eq. 2.6.1 becomes

$$dV(z) = -C_{disp}\frac{\rho N_{Av}}{M} \frac{2\pi \, y \, dy \, d\zeta}{[(z + \zeta)^2 + y^2 \,]^3}$$

The double integral of dV(z) for all values of y and ζ from zero to infinity gives the interaction between the single molecule and the particle at a normal distance z

$$V(z) = -\int_0^\infty \int_0^\infty \frac{C_{disp} \, \rho \, N_{Av}}{M} \frac{2\pi \, y}{[(z + \zeta)^2 + y^2]^3} \, dy \, d\zeta \qquad (2.6.2)$$

Carrying out this integration gives

$$V(z) = -\frac{\rho \, N_{Av}}{M} \frac{C_{disp} \, \pi}{6 \, z^3} \qquad (2.6.3)$$

Note that the potential energy of interaction between a molecule (or atom) and a particle varies with the distance from the surface as $1/z^3$.

Now we can determine the interaction energy between two blocks by considering the molecule (or atom) to be one of many located inside a second block (block 2) as in Figure 2.6b. All the atoms in a slice of block 2 located at a distance z from the first block have an energy given by eq. 2.6.3. In a volume element dz we will have $(\rho N_{Av}/M)dz$ molecules per unit area, and the interaction potential due to that element per unit area will be

$$dV(z) = -\left(\frac{\rho N_{Av}}{M}\right)^2 \frac{C_{disp}\pi}{6} \frac{dz}{z^3} \tag{2.6.4}$$

Integrating eq. 2.6.4 from $z = h$ to ∞, where h represents the normal separation between the planar surfaces of the two particles, gives the attractive interaction energy *per unit surface area* of the particles $V_{att,p}(h)$,

$$V_{att,p}(h) = -\left(\frac{\rho N_{Av}}{M}\right)^2 \frac{C_{disp}\pi}{12} \frac{1}{h^2} = -\frac{H_{11}}{12\pi} \frac{1}{h^2} \tag{2.6.5}$$

In this equation, H_{11} equals the *Hamaker constant,* a material constant that measures the attraction between two particles of that material in vacuum. Using the London model for dispersion attraction (eq. 2.4.1) to substitute for C_{disp}, we can estimate the value of H_{11} to be

$$H_{11} = \left(\frac{\rho N_{Av}\pi}{M}\right)^2 \frac{3}{4} \frac{\alpha^2 \hbar v}{(4\pi\varepsilon_0)^2} \tag{2.6.6}$$

Note that the potential energy of interaction between the two flat particles varies with the separation between them as $1/h^2$, a surprisingly long-range interaction.

Differentiating eq. 2.6.5 with respect to separation h gives the attractive force per unit area between two flat coplanar surfaces

$$F_{att,p} = -\frac{H_{11}}{6\pi} \frac{1}{h^3} \tag{2.6.7}$$

2.6.2 Interaction Forces between Spheres Can Be Obtained from the Interaction Energy between Parallel Plates

The interaction between two infinite blocks separated by a normal distance h provides the simplest geometry with which to evaluate attractive forces. However, we are often interested in interactions between curved surfaces, particularly spherical particles. Derjaguin realized that when surfaces of particles are uniformly curved, the interaction can be derived simply by breaking the particles down into a series of planar forms and using the interaction energy between two planar surfaces, $V_{att,p}$.

To illustrate this idea, Figure 2.7 shows two equal-sized spheres of radius R separated by a distance of closest approach h. We can approximate the sphere by a series of circular (anular) rings possessing planar faces. The planar faces are separated by a distance z, assuming that the main line of interaction occurs along the line joining the centers of the two spheres.

The interaction energy between the pair of planar rings at a distance y from the center of the sphere is

$$V_{att} \, dA = V_{att} \, (2\pi y \, dy) \qquad (2.6.8)$$

where V_{att} is the attractive interaction energy per unit area (given by eq. 2.6.5) and dA is the area of the ring, which we can write as $2\pi y \, dy$. We can relate the separation of the ring to the distance of closest approach by

$$y^2 + \left(R - \frac{(z-h)}{2} \right)^2 = R^2 \qquad (2.6.9)$$

Differentiating eq. 2.6.9 (with R constant) and rearranging gives

$$2y \, dy = R(1 - y^2/R^2)^{1/2} \, dz \qquad (2.6.10)$$

Substituting eq. 2.6.10 into eq. 2.6.8, we obtain

$$V_{att} \, dA = \pi R(1 - y^2/R^2)^{1/2} \, V_{att} \, dz \qquad (2.6.11)$$

If the spheres are large compared to their separation, then the interactions between them occur primarily in the regions for which $y \ll R$. So we can assume $y^2/R^2 \ll 1$, and eq. 2.6.11 simplifies to

$$V_{att} \, dA = \pi R \, V_{att} \, dz \qquad (2.6.12)$$

We obtain the attractive potential energy of interaction for two identical spheres $V_{att,s}$ by integrating eq. 2.6.12 over all values of z from h to infinity (a limit that is justified when the spheres are large compared to their separation). Substituting the value for V_{att} from eq. 2.6.5 gives

$$V_{att,s} = \int_{h}^{\infty} \frac{H_{11}R}{12} \frac{dz}{z^2} = -\frac{H_{11}R}{12h}, \quad h \ll R \qquad (2.6.13)$$

Figure 2.7
The potential energy between two equal-sized spheres of radius R separated by a distance of closest approach h is obtained by approximating the spheres as stacks of circular rings and summing over them. (P. C. Hiemenz, *Principles of Colloid and Surface Chemistry,* 2nd ed., Marcel Dekker, New York, 1986, p. 715.)

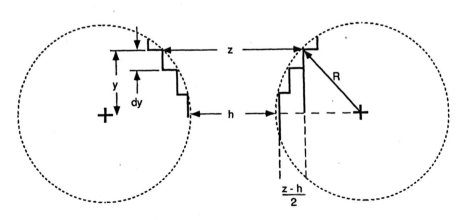

Note that the attractive potential energy of interaction between two identical spheres varies with the separation between them as 1/h.

When the approximation h << R cannot be made, eq. 2.6.11 must be integrated directly. Table 2.5 gives this more complex result. It also includes results for two spheres of different sizes.

Differentiating eq. 2.6.13 with respect to separation h gives the attractive force between two identical spheres

$$F_{att,s} = -\frac{H_{11}R}{12}\frac{1}{h^2} \tag{2.6.14}$$

a function that varies as the inverse of the distance squared. This attractive force is long-ranged, and a powder consisting of many spherical particles will be tightly held together. The particles attempt to pack together with maximum coordination in the same manner as the inert spherical molecules in van der Waals solids described in Section 2.5.2. Figure 4.33 illustrates close-packed coordination of spherical powder particles. Agglomeration of tiny particles due to attractive dispersion forces is a major issue in the behavior of colloidal systems to be discussed in Chapter 4.

We can use this same approach to derive the interaction energy between a sphere and a flat surface or between two crossed

TABLE 2.5 Potential Energy of Attraction between Two Particles with Various Geometries

Particles	V_{att}	Definitions/limitations
Two spheres	$-\dfrac{H_{12}}{6}\left[\dfrac{2R_1R_2}{h^2+2R_1h+2R_2h}+\dfrac{2R_1R_2}{h^2+2R_1h+2R_2h+4R_1R_2}\right.$ $\left.+\ln\left(\dfrac{h^2+2R_1h+2R_2h}{h^2+2R_1h+2R_2h+4R_1R_2}\right)\right]$	R_1, R_2 = radii; h = separation of surfaces along line of centers
Two spheres of equal radius	$-\dfrac{H_{12}}{6}\left[\dfrac{2R^2}{h^2+4Rh}+\dfrac{2R^2}{h^2+4Rh+4R^2}+\ln\left(\dfrac{h^2+4Rh}{h^2+4Rh+4R^2}\right)\right]$	$R_1=R_2=R$
Two identical spheres of equal radius	$-\dfrac{H_{11}R}{12h}$	$R \gg h$
Two spheres of unequal radius	$-\dfrac{H_{12}R_1R_2}{6h(R_1+R_2)}$	R_1 and $R_2 \gg h$
Two plates of equal thickness	$-\dfrac{H_{12}}{12\pi}\left\{\dfrac{1}{h^2}+\dfrac{1}{(h+2\delta)^2}-\dfrac{2}{(h+\delta)^2}\right\}$	δ = thickness of plates
Two identical blocks	$-\dfrac{H_{11}}{12\pi h^2}$	$\delta \to \infty$

P. C. Hiemenz, *Principles of Colloid and Surface Chemistry*, 2nd ed., Marcel Dekker, New York, 1986, p. 648.

(a)

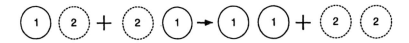

(b)

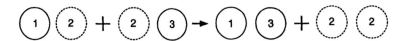

Figure 2.8
Schematic for calculating the Hamaker constant (a) H_{121} between two identical particles (numbered 1) separated by solvent (numbered 2) and (b) H_{123} between two different particles (1 and 3) separated by solvent (2).

cylinders of radius R. Table 2.5 summarizes values of the attractive interaction V_{att} for various geometries.[3]

2.6.3 Attractive Energies between Particles Are Much Smaller in a Condensed Phase Than in Vacuum

So far our discussion applies to the interaction of two particles or macroscopic bodies in a vacuum—a reasonable approximation for gas-phase interactions. However, many interfacial engineering applications involve interactions in a liquid medium. The attraction between particles follows relationships similar to the ones we have just derived, and the Hamaker constant applies in such cases. We can calculate its effective value in terms of the interactions shown in Figure 2.8.

Consider the interaction of two particles (numbered 1) in a solvent (numbered 2). In the initial condition, each particle and its satellite solvent comprise an independent unit. The process depicted in Figure 2.8a involves the two particles coming together and squeezing out the satellite solvent, which then interacts with itself.

We can write the change in potential energy ΔV for this process as

$$\Delta V = V_{11} + V_{22} - 2V_{12} \qquad (2.6.15)$$

Each term in eq. 2.6.15 exhibits an identical dependence on size and distance parameters. They differ from one another only in the molecular parameters contained in the Hamaker constants. Therefore, we can write

$$H_{121} = H_{11} + H_{22} - 2H_{12} \qquad (2.6.16)$$

where the subscript 121 indicates two particles of type 1 separated by a medium of type 2.

---

[3]Note that the different expressions for V_{att} and F_{att} have different dimensions. In eq. 2.6.5, V_{att} stands for the potential energy per unit surface area, while in eq. 2.6.13, it represents the potential energy per pair of spheres. In eq. 2.6.7, F_{att} stands for the force per unit surface area, while in eq. 2.6.14, it represents the force per pair of spheres.

We approximate

$$H_{12} \approx (H_{11}H_{22})^{1/2} \qquad (2.6.17)$$

using the geometric mixing rule, which says that the interaction between two dissimilar bodies is given by the geometric mean of the homogeneous interactions for two species separately (already used to estimate the attraction of two dissimilar molecules in Section 2.5.1). Combining eqs. 2.6.16 and 2.6.17 gives

$$H_{121} = (H_{11}^{1/2} - H_{22}^{1/2})^2 \qquad (2.6.18)$$

If $H_{11} \neq H_{22}$, then H_{121} is always positive, and identical particles in the medium exert a net attraction on each other. Note that the squared power of eq. 2.6.18 makes $H_{121} = H_{212}$. For example, water droplets suspended in oil exert the same attractive interaction between droplets as oil droplets in water. Also, if $H_{11} = H_{22}$, then H_{121} is zero and no net attraction exists between the particles.

When particles are immersed in a medium, the attractive interaction is greatly diminished. As Table 2.6 shows, the values of H_{11} for homogeneous interactions in a vacuum (or air) are generally on the same order of magnitude for various substances. Because the magnitude of the interaction in a medium (represented in eq. 2.6.18) depends on the difference in values of H_{11} and H_{22}, then the value for H_{121} will be much smaller.

TABLE 2.6 Values of Hamaker Constants.
Hamaker Constants for Some Common Materials, in Units of 10^{-20} J

Material (M)	M l air l M	M l water l M	M l water l air	M l air l water
Water	3.70	0	0	3.70
Alkanes				
n = 5	3.75	0.336	0.153	3.63
6	4.07	0.360	−0.00368	3.78
10	4.82	0.462	−0.344	4.11
14	5.10	0.514	−0.464	4.23
16	5.23	0.540	−0.518	4.28
Fused quartz	6.50	0.833	−1.01	4.81
Fused silica	6.55	0.849	−1.03	4.83
Sapphire	15.6	5.32	−3.78	7.40
Poly(methylmethacrylate)	7.11	1.05	−1.25	5.03
Polystyrene	6.58	0.950	−1.06	4.81
Poly(isoprene)	5.99	0.743	−0.836	4.59
Poly(tetrafluoroethylene)	3.80	0.333	0.128	3.67
Mica (green)	−	2.14	−	−

D. F. Evans and H. Wennerström, *The Colloidal Domain: Where Physics, Chemistry, Biology, and Technology Meet*, VCH Publishers, New York, 1994, p. 221. Reprinted with permission by VCH Publishers © 1994.

In Figure 2.8b, we also consider the interaction between two different particles (1,3) separated by a medium (2). The change in potential energy is then

$$\Delta V = V_{22} + V_{13} - V_{12} - V_{32} \qquad (2.6.19)$$

or

$$H_{123} = H_{22} + H_{13} - H_{12} - H_{32} \qquad (2.6.20)$$

Using eq. 2.6.17 in the general form $H_{ij} \approx (H_{ii}H_{jj})^{1/2}$, we can rewrite eq. 2.6.20 as

$$H_{123} = (H_{11}^{1/2} - H_{22}^{1/2})(H_{33}^{1/2} - H_{22}^{1/2}) \qquad (2.6.21)$$

If the Hamaker constant for the solvent H_{22} is intermediate between those of the particles, that is, $H_{11} > H_{22} > H_{33}$ or $H_{33} > H_{22} > H_{11}$, then the effective Hamaker constant is negative. This result corresponds to a repulsive interaction between the two particles 1 and 3.

2.7 The Boltzmann Equation Provides the Basis for Deciding When Interfacial Energies Are Important

This chapter has focused on understanding the physics underlying attractive and repulsive interactions between molecules in the gas phase as well as on dispersion force interactions between particles and macroscopic bodies. However, our goal in this book is much more ambitious. We want to learn how to manipulate and control interaction forces during interfacial processing.

What criteria can we use to decide when an interaction force will be important? We started down the right path by comparing the interaction energies V_{ij}, which organize molecules at a molecular level, to the thermal motion, kT, which randomizes molecules. The net ordering described by the Boltzmann equation represents a compromise between these opposing interactions.

A simple example is the barometric equation, which gives the change in pressure (concentration) of an ideal gas due to the influence of gravity at various distances from the earth's surface

$$P_h = P_o \exp\left(\frac{-mgh}{kT}\right) \qquad (2.7.1)$$

Here mgh equals the potential energy of molecules at the height h and P_o is the pressure at a reference plane taken as h = 0. In the limit of high temperature, where thermal motions dominate, exp $-(mgh/kT) \to 1$ and the gas molecules are uniformly distributed. At low temperatures, where the ordering effect of gravity is most pronounced, the pressure decreases exponentially with distance. For values of h > mg/kT, it rapidly approaches zero.

The general form of the Boltzmann equation is exp $-(V/kT)$. We will encounter various forms of it throughout this book because whenever we seek to determine the importance of an interaction between molecules, we must also consider the effect of thermal motion. Therefore, every time we use the ratio between interaction energy and **k**T, we are implicitly thinking in terms of the Boltzmann equation.

Bibliography

J. Israelachvili, *Intermolecular and Surface Forces,* London: Academic Press Ltd., 1985.

M. Rigby, E. B. Smith, W. A. Wakeham, and G. C. Maitland, *The Forces between Molecules,* Oxford: Oxford University Press, 1986.

Exercises

2.1 Is the force of attraction between Na^+ and Cl^- in the vapor
equal to _ ; less than _ ; or greater than _
the force of attraction between these two ions in water when they are separated in both instances by 10 nm? Why?

2.2 Is the solubility of HCl in benzene
equal to _ ; less than _ ; or greater than _
that of H_2O? Why?

2.3 The polarizability of a free Cl^- ion can be estimated to be 5.0×10^{-39} C m^2/V. What is the polarization interaction in a Na^+Cl^- pair when the distance between the charges is 0.28 nm? (Neglect the polarization of Na^+).
What is the polarization interaction in the linear complex $Na^+ Cl^- Na^+$?

2.4 Given the following data

Compound	Dipole moment m (Debye)	Polarizability $\alpha/4\pi\varepsilon_o$ $(10^{-30}\ m^3)$	Ionization potential v (eV)
Ethanol	1.69	5.49	10.49
Benzene	0	10.3	9.24

Calculate the different van der Waals energy coefficients (Table 2.4 and eq. 2.5.1) for interaction (a) between two ethanol molecules and (b) between ethanol and benzene molecules.

2.5 Gold is thermally evaporated at room temperature with a contaminating background pressure of 10^{-6} Torr. The contamination is due primarily to H_2O.
Do Au and H_2O interact? What are the interaction forces? At what distance does this interaction compare with **k**T? Comment on the likelihood of the H_2O affecting the deposition.

2.6 Calculate the cohesive energy of a (1D) line of ions of alternating charge. Use only two terms, the Coulombic and the hard sphere interaction.

2.7 Speculate and discuss some of the interfacial concerns that might be encountered between a medical (implanted) device and the body. Focus on the intermolecular forces that would be present.

2.8 What are the three material properties that determine the theoretical Hamaker constant (eq. 2.6.6)?
Would you expect the experimental value for the Hamaker constant to be
equal to _ ; less than _ ; or greater than _
the theoretical value for water? For benzene? Why?

2.9 Use eq. 2.6.6 to estimate the Hamaker constant for water assuming v to be about 10 eV.
Calculate the approximate force of attraction

(a) between two thick slabs of ice 100 Å apart (per unit area);
(b) between two spheres of ice of radius 1 μm and 100 Å apart.

2.10 Is the force of attraction between two identically shaped tiny particles of quartz placed in vacuum
equal to _ ; less than _ ; or greater than _
the force of attraction when they are placed in water? Why?

2.11 Assuming they are uncharged, do dust particles and mist droplets
attract _ ; or repel _
each other in air?

2.12 Would you expect tiny spheres of polystyrene and PTFE (polytetrafluoroethylene) to
attract _ ; or repel _
(a) themselves, or (b) each other in octane?

3

GENERAL PROPERTIES OF SYSTEMS CONTAINING FLUID INTERFACES

with H. Wennerström

LIST OF SYMBOLS

a_o	area per molecule at surface (eq. 3.6.2)
a_i	chemical activity of component i (eq. 3.5.11)
c_i	molar concentration of solute i in solution
c_i^σ	molar concentration at surface
D	diffusion coefficient (eq. 3.8.26)
f	friction factor (eq. 3.8.22)
G	Gibbs free energy; G^σ surface free energy; ΔG change in free energy
ΔG_{mix}	free energy of mixing in a real solution
H	enthalpy; ΔH change in enthalpy
ΔH_{vap}	heat of vaporization
I	nuclei formation rate (eq. 3.4.8)
J	flux of material through unit cross-sectional area per unit time (eq. 3.8.26)
K_{ads}	equilibrium constant for adsorption reaction (eq. 3.7.11)
M	molar solution concentration ($1M$ – one molar = one gram molecular weight per liter of solution)
n_i	number of moles of component i ($n_i = N_i/N_{Av}$)
N_i	number of molecules of component i
N	number of molecular sites
N_{Re}	Reynolds number (eq. 3.8.11)
P_o	equilibrium vapor pressure
S	spreading coefficient (eq. 3.2.5)
S	entropy; S^σ surface entropy; ΔS change in entropy
ΔS_{mix}	entropy of mixing in a real solution (eq. 3.7.6)

U	internal energy; $\mathbf{u}$ internal energy per unit volume
V_L	molar volume of phase L; v_m molecular volume of molecule m
V	volume; dV/dt volume flow rate (eq. 3.8.9)
v	velocity
w	work; $w_{cohesion}$ work of cohesion (eq. 3.2.1)
w	effective molar interaction parameter (eq. 3.1.8)
W	effective molecular interaction parameter (eq. 3.1.8)
W	molecular pair interaction energy in condensed, liquid or solid, phase
X_i	mole fraction of i
z	coordination number; z_b bulk (b), z_s surface (s) cooordination number
$\mathbf{z}$	valence; $\mathbf{z}_i$, valence including sign ($\mathbf{z} = |\mathbf{z}_i|$)
Z	compressibility factor for surface monolayer (eq. 3.6.2)
γ_{yx}	shear strain; displacement in x direction per length in y direction
γ_{12}	surface (interface) tension between phases 1 and 2 (eq. 3.1.7)
Γ_i	surface excess concentration of i (eq. 3.5.13)
η	viscosity (eq. 3.8.1); η_s viscosity of surface monolayer (eq. 3.8.13)
$[\eta]$	intrinsic viscosity (eq. 3.8.19)
θ	contact angle (eq. 3.2.4)
Θ	fractional surface coverage (eq. 3.7.11)
μ_i	chemical potential of component i; μ_i^σ chemical potential at surface
Π_s	spreading pressure for surface monolayers (eq. 3.6.1)
σ_{xx}	tensile stress; force in x direction divided by area normal to x axis
τ_{yx}	shear stress; force in x direction divided by area normal to y axis
ϕ	volume fraction of spheres
Ω	number of possible states (eq. 3.7.3)

CONCEPT MAP

- Distinctive features of fluid interfaces are:
  1. They flow in response to shear stresses;
  2. Rapid exchange of molecules leads to rapid attainment of equilibrium;
  3. Free volume dominates their kinetic transport properties.
- By analogy with the thermodynamic equation for bulk material, the excess surface free energy can be written

$$dG^\sigma = \gamma\, dA + \Sigma\, \mu_i^\sigma dn_i^\sigma$$

$dG^\sigma = \gamma\, dA$ (Physical Effects)

- Surface tension gives the change in free energy per unit area (eq. 3.1.4) or the force per unit length (eq. 3.1.1).

- Surface tension reflects the unfavorable change in interaction energy that occurs upon moving a molecule from the bulk to the surface.

- The difference between the work of cohesion (creating surface in a single liquid) and adhesion (creating interfacial area between two phases) predicts when one liquid will spontaneously spread onto another.

- The Young-Laplace equation (eq. 3.3.8) relates pressure differences across a curved interface to radii of curvature and surface tensions and explains why liquids rise in capillary tubes.

- The Kelvin equation gives the degree of supersaturation (P/P_o, eq. 3.4.4) required to initiate homogeneous nucleation and also describes condensation in porous material.

- Combining the Kelvin equation with a rate equation gives an expression for the rate of formation of homogeneous nuclei.

- In systems containing nucleating surfaces, such as dust particles, heterogeneous nucleation leads to phase transformations at their thermodynamic equilibrium values.

$dG^\sigma = \Sigma\,\mu_i^\sigma dn_i^\sigma$ (Chemical Effects)

- It is difficult to assign a volume to an interfacial region because physical properties change continuously across the interface.

- In the Gibbs model, we treat the interface as a mathematical plane of area A and define a surface concentration by $\Gamma_i = n_i^\sigma/A$ (eq. 3.5.4).

- In a binary solution, if we choose $\Gamma_1 = 0$, we focus on the behavior of the solute.

- We can relate the change in surface tension with ln c of an added solute to a surface excess concentration Γ_i (eq. 3.5.13).

- Γ_i is an algebraic quantity reflecting solute concentration or depletion at the interface.

- Insoluble monolayers are two-dimensional analogs of bulk systems displaying vapor, liquid, and solid transformations.

In multicomponent interfacial systems both curvature and composition can change at a fluid interface to minimize ΔG.

Entropy of Mixing and Adsorption at Solid/Fluid Interfaces

- Entropy of mixing, $\Delta S_{mix} = -k\,\Sigma N_i \ln X_i$, which we can determine from elementary statistical mechanics, provides a valuable guide in analyzing many interfacial processes.

- In an ideal solution, $\Delta H_{mix} = 0$ and $\Delta V_{mix} = 0$ and $\Delta G_{mix} = -T\,\Delta S_{mix}$.

- The Langmuir adsorption isotherm (eq. 3.7.11) often is used to describe adsorption of solutes onto solid surfaces. It adopts ideal solution assumptions to model the equilibrium concentration of molecules on the surface.

Fluid Transport Processes

Fluid flow and mass transfer play an important role in many interfacial processes.

Fluid Flow

- Newtonian fluids obey Newton's law of viscosity (eq. 3.8.1) and the coefficient of viscosity, η, is independent of shear rate (Fig. 3.19). Non-Newtonian fluids do not follow this simple relationship.

- Solving equations of continuity and motion provides velocity profiles and average velocities of fluids in ducts and pipes (Figs. 3.20, 3.21) in terms of geometry, applied pressure, and viscosity.

- We can measure surface viscosities in a modified Langmuir balance.

- The Einstein equation (eq. 3.8.17) relates a solution's viscosity to the volume fraction of spherical particles.

- The terminal velocity of particles in fluids equals the applied force divided by the friction factor f (eq. 3.8.22). For spheres of radius R, $f = 6\pi\eta R$.

- Many fluids containing colloids and polymers display non-Newtonian behavior (Fig. 3.19).

Diffusion

- Self-diffusion coefficients describe the motion of particles arising from thermal energy. Transport by self-diffusion is measured by the Einstein equation (eq. 3.8.25).

- Mutual diffusion between two solutions due to a difference in solute concentration leads to a flux of solute determined by Fick's laws (eqs. 3.8.26 and 3.8.27).

- The diffusion coefficient D for the diffusion of particles through a solution is related to the friction factor by $D = kT/f$ (eq. 3.8.35). For spheres Stokes law gives $D = kT/6\pi\eta R$ (eq. 3.8.36).

- Solvation and asymmetry both increase the friction factor, resulting in smaller particle diffusion coefficients.

3

Although the same fundamental principles apply to both fluid and solid interfaces, we discuss them separately because differences in response time lead to a rather different set of phenomena. The separation is somewhat arbitrary, but we use it to assist in the organization of the material. This chapter presents the general properties of systems containing fluid interfaces, while Chapter 9 does the same for solid interfaces.

Fluids include liquid and gas phases. These phases exhibit three characteristics that differ from solids: (1) fluids cannot support a shear stress, whereas solids can; (2) fluid kinetic transport phenomena—such as diffusivity, viscosity, and thermal conductivity—show a temperature dependence different from that found in solids; (3) fluids possess free volume that facilitates rapid exchange of molecules and speeds the approach to thermodynamic equilibrium. In solid systems where free volume is almost nonexistent, the approach to equilibrium is much slower at ambient temperatures.

This chapter is organized into three parts: physical effects at fluid surfaces, chemical effects at fluid surfaces, and transport phenomena in fluids. By analogy with the thermodynamic equation for a bulk system at constant temperature and pressure, $dG = -P\,dV + \Sigma\mu_i\,dn_i$, we can write an equation for the Gibbs free energy of an interface as

$$dG^\sigma = \gamma\,dA + \sum_i \mu_i^\sigma\,dn_i^\sigma$$

where the superscript σ denotes a surface thermodynamic quantity, γ the surface tension, A the surface area, and μ_i^σ and n_i^σ stand for surface chemical potentials and number of moles of component i at the surface. This equation tells us that an interface can minimize its free energy by changes in area and composition. The first part of the chapter, Sections 3.1 through 3.4, deals with structural or physical phenomena, such as capillarity, associated with the first term in the equation; the second part of the chapter, Sections 3.5 through 3.7, deals with chemical phenomena, such as adsorption, associated with the second term in the equation. The third part of the chapter, Section 3.8, covers transport processes in fluids.

3.1 Surface Tension Is the Key Concept in Characterizing Fluid Interfaces

A useful way to conceptualize surface tension is in terms of the simple mechanical device consisting of a wire loop with a movable slide shown in Figure 3.1. We assume that the device behaves as an idealized frictionless apparatus.

We can form a liquid film by dipping the wire loop into a liquid and withdrawing it. Unless we apply an opposing force to the movable slide, it retracts so as to decrease the surface area of the liquid film. The force F required to prevent this retraction varies linearly with the length l. We define surface tension γ in terms of our apparatus by

Figure 3.1
The wire loop with a movable slide of length l is a simple mechanical device illustrating many of the features of surface tension γ. When a liquid film is contained between the loop and the slide, a force $F = 2\gamma l$ must be exerted to prevent the film from contracting.

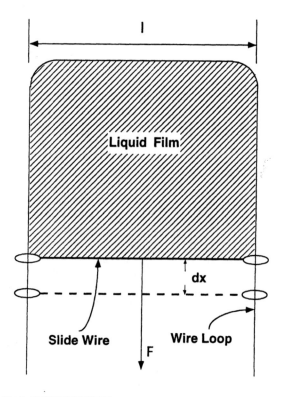

$$\gamma = \frac{F}{2l} \qquad (3.1.1)$$

Here the total length $l = 2l$ because the film has two sides or two surfaces, front and back, and therefore two lines of length l that intersect the wire. As an intrinsic property of a liquid surface, surface tension has a characteristic value for each liquid.

This loop and wire device is a two-dimensional analog of the ideal cylinder-piston apparatus used to evaluate PV products in gas cycles in thermodynamics. The analogy suggests that γ is a two-dimensional pressure, a point to which we return in Section 3.6. One major difference between the two systems occurs when we remove the restraining force; in the three-dimensional system the piston expands; in the two-dimensional system the slide wire retracts.

Figure 3.1 also indicates the work done to expand the interfacial area. We can write the work done as

$$dw = F \, dx = \gamma \, 2l \, dx = \gamma \, dA \qquad (3.1.2)$$

where dA is the total increase in surface area, both front and back. If we apply a force to the slide wire that is only infinitesimally larger than the equilibrium force, we can evaluate the work done in terms of a reversible path and estimate the change in the Gibbs free energy (dG). At constant temperature and pressure, the change equals the reversible nonpressure-volume work. Therefore, we can relate eq. 3.1.2 to thermodynamic properties and write

$$(dG)_{T,P} = \gamma \, dA \qquad (3.1.3)$$

or

$$\gamma = \left(\frac{\partial G}{\partial A} \right)_{T,P} \qquad (3.1.4)$$

This expression identifies surface tension as the increase in Gibbs free energy per unit increment in surface area.

We have defined surface tension in two ways: as the free energy per unit area (force $\times$ length)/(length)^2 as in eq. 3.1.3; or as the force per unit length as in eq. 3.1.1. Historically these two different perspectives have posed a dilemma, but they really represent two ways of looking at the same thing. Both have their uses. Understanding them allows us to determine when one point of view works better and when the other provides a more effective cognitive interpretation of the data.

3.1.1 Molecular Origins of Surface and Interface Tension Can Be Understood in Terms of Differences in Interaction between Molecules in the Bulk and at the Interface

Our mechanical analysis leads to useful relations between surface tension, reversible work, and free energy. However, it provides no insights into the molecular origins of surface tension. We can approach that question by analyzing the differences between the energies of molecules located at the surface and the energies of molecules located in the bulk phase.

In the case of a pure liquid in equilibrium with its vapor, we assume that the interaction energies between molecules in the bulk phase are pairwise additive. We can write the interaction energy $E_{A,bulk}$ per A molecule in the bulk (subscript b) phase as

$$E_{A,b} = z_b \frac{W_{AA}}{2} \tag{3.1.5}$$

where W_{AA}[1] is the pair interaction energy and z_b represents the number of nearest neighbors in the bulk phase—the coordination number. The factor 2 accounts for the fact that two atoms interact and the potential due to each is $W_{AA}/2$.

The corresponding interaction energy per A molecule at a free surface, $E_{A,s}$, is

$$E_{A,s} \approx z_s \frac{W_{AA}}{2} \tag{3.1.6}$$

where z_s is the number of nearest neighbors for a surface molecule.

The difference in energy for a molecule located at the surface rather than in the bulk per unit area is

$$\frac{1}{a_o} (E_{A,s} - E_{A,b}) = \frac{W_{AA}}{2} \frac{z_s - z_b}{a_o} = \gamma_{\alpha v} \tag{3.1.7}$$

where a_o is the area per molecule at the surface. This expression is the surface energy or surface tension (the same as γ in eq. 3.1.4) for the condensed phase α. It is represented here by $\gamma_{\alpha v}$, the extra subscripts defining the liquid vapor surface precisely.[2] Because W_{AA} is *negative* and $z_b > z_s$, the expression for $\gamma_{\alpha v}$ is *always positive*. Moving a molecule from the bulk to the surface increases the internal energy; work must be done to create a new surface, and that is the surface energy.

We can illustrate these ideas with the following estimate of $\gamma_{\alpha v}$. To obtain W_{AA} we need a way to estimate $E_{A,b}$. $E_{A,b}$ corresponds to the energy required to move molecules far enough apart enough so that their interactions are effectively zero. To a good approximation, we can use the heat of vaporization, ΔH_{vap}, for this purpose because vaporization involves the transformation from a liquid where molecules are in close contact to the vapor state where they are separated by large distances.

If we use carbon tetrachloride as an example, ΔH_{vap} equals 29.7 kJ/mole. Dividing by Avogadro's number, N_{Av}, this gives

$$E_{A,b} = -29.7 \times 10^3/6.022 \times 10^{23} = -4.932 \times 10^{-20} \text{ J per molecule}$$

---

[1]*Following convention, we use W to represent the pair potential in condensed phases; it is identical in origin to V used for the interaction potential in the gas phase in Chapter 2.*

[2]*We adopt the convention that liquid phase α contains mostly A molecules, phase β mostly B molecules, etc. αv denotes the interface between the liquid phase α and vapor v.*

which corresponds to -12 kT at room temperature ($kT = 4.14 \times 10^{-21}$ J at 300 K). Assuming $z_b = 6$, eq. 3.1.5 leads to a value of $W_{AA} \approx -4kT$ per molecule pair interaction.

To estimate the surface tension, we view the molecules as simple cubes. The density of CCl_4 is 1.6×10^3 kg/m³. Using the cube model, this density gives an area of 3×10^{-19} m² per molecule for each side of the cube, which equals the exposed area per molecule at the surface. Assuming $z_b = 6$ and $z_s = 5$ (because there is one free cube side at the surface) and taking $W_{AA} = -1.64 \times 10^{-20}$ J from the previous estimate, we obtain a value for the energy required per unit surface area from eq. 3.1.7 of

$$\frac{-1.64 \times 10^{-20}}{2} \frac{(-1)}{3 \times 10^{-19}} = 24.7 \text{ mJ/m}^2$$

which is fortuitously close to the experimental value of 26.4 mJ/m² (or 26.4 mN/m) for the surface tension.[3]

If we consider an interface between two condensed phases, α and β, containing molecules of type A and type B, we can see that when an A molecule moves to the interface, it loses interactions with A but gains about an equal number of interactions with B. Similarly, B molecules lose interactions with B but gain interactions with A. The net energy change for N_{int} interface molecules of each kind is

$$E_{int} = N_{int} (z_b - z_{int}) (W_{AB} - \frac{1}{2}W_{AA} - \frac{1}{2}W_{BB})$$

$$= N_{int} \frac{z_b - z_{int}}{z_b} W = N_{int} \frac{z_b - z_{int}}{z_b} \frac{w}{N_{Av}} \qquad (3.1.8)$$

where $W = z_b [W_{AB} - (W_{AA} + W_{BB})/2]$ is the *molecular effective interaction parameter* and $w = z_b N_{Av} [W_{AB} - (W_{AA} + W_{BB})/2]$ is the *molar effective interaction parameter*. We have assumed that the coordination number is approximately the same for both phases in the bulk as well as at the interface—a reasonable assumption for molecules of similar size. The energy per unit area (A) of interface is then

$$\frac{E_{int}}{A} = \frac{1}{a}\left(\frac{z_b - z_{int}}{z_b}\right) W = \gamma_{\alpha\beta} \qquad (3.1.9)$$

This equals the interfacial energy or interfacial tension between phases α and β, $\gamma_{\alpha\beta}$. Note that the average of W_{AA} plus W_{BB} is always more *negative* than W_{AB}, so W is always *positive*. W and thus $\gamma_{\alpha\beta}$ approach zero as the affinity of the dissimilar molecules increases.

For strongly immiscible liquids, the average of W_{AA} plus W_{BB} is considerably more negative than W_{AB}. If we assume the

---

[3] *Note the various surface tension and surface energy units that appear in the literature, dyne cm^{-1}, ergs cm^{-2}, mN m^{-1}, mJ m^{-2}, are all numerically equivalent.*

difference to be kT, then $W = 6kT$. If we also assume the interfacial coordination number z_{int} is 5, and the area per molecule is 3×10^{-19} m^2, then the interface energy is ~14 mJ/m^2. Comparing this estimate with the previous estimate of the surface tension of CCl$_4$ illustrates the general rule that interfacial energies are smaller than surface energies. These points are illustrated for the surface and interface tension between water and some organic liquids in Table 3.1.

In deriving eqs. 3.1.8 and 3.1.9, we assumed that α and β were pure liquids. $\mathbf{E}_{int}$ represents the energy at the instant the interface is formed. However, at equilibrium some B molecules will mix into the α liquid and vice versa. This process decreases the interface energy. For miscible liquids ($W < 2kT$), the interface eventually disappears altogether and $\mathbf{E}_{int}$ drops to zero.

In eqs. 3.1.7 and 3.1.9, we estimated only the contribution of interaction energy ($\mathbf{E}^\sigma$) to the formation of a surface or an interface. To calculate the surface free energy or surface energy ($\mathbf{G}^\sigma$), we also must consider the entropy contribution ($\mathbf{S}^\sigma$). For example, entropy contributions could arise from the ordered alignment of molecules at the surface or from capillary waves at the surface. Normally the entropy contribution is not large and can be neglected, leaving interaction energy as the major contributor to surface free energy.

Although we have restricted our analysis to surface and interface energy of liquids, a similar approach is used for other types of interfaces. The same concepts reappear for solids in Chapter 9.

TABLE 3.1 Surface Tension of
Selected Liquids at Room Temperature
(mJ/m^2 = ergs/cm^2 = dyn/cm)

Liquid–vapor surface	γ_{lv}
Water	72.5
Benzene	28.9
Carbon tetrachloride	26.4
Methanol	22.5
Ethanol	22.4
Octane	21.6
Heptane	20.1

Liquid–liquid interface	$\gamma_{\alpha\beta}$
Water–benezene	35
Water–carbon tetrachloride	45
Water–heptane	50

3.2 The Relation between Surface Tension and Work of Cohesion and Adhesion Affects the Spontaneous Spreading of One Liquid upon Another

3.2.1 Work of Cohesion and of Adhesion Are Key Concepts in Understanding Many Interfacial Processes

In the case of a single liquid, the work of cohesion corresponds to the work required to pull apart a volume of unit cross-sectional area, as shown in Figure 3.2a. Thus the work of cohesion, w_{cohesion}, is given by the relation

$$w_{\text{cohesion}} = 2\,\gamma_{\alpha v} \qquad (3.2.1)$$

Interpreting $\gamma_{\alpha v}$ as half the work of cohesion per unit area is consistent with the fact that surface tension measures the change in free energy that occurs when molecules are moved from the bulk of a sample to its surface. Decohesion results in the formation of two free surfaces.

The work of adhesion between two immiscible liquids equals the work required to separate the unit area of the interface

Figure 3.2
A conceptual experiment illustrating (a) the work of cohesion, $w_{\text{cohesion}} = 2\gamma_{\alpha v}$, which corresponds to the work required to create two vapor–liquid interfaces of unit area in a homogeneous phase, and (b) the work of adhesion,

$w_{\text{adhesion}} = \gamma_{\alpha v} + \gamma_{\beta v} - \gamma_{\alpha\beta}$, which is given by the work required to separate a unit area of interface between two phases α and β to form two vapor–liquid interfaces.

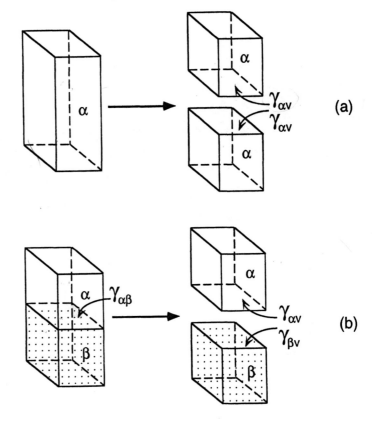

into two liquid–vapor interfaces, as shown in Figure 3.2b and expressed by the Dupré equation

$$W_{\text{adhesion } \alpha\beta} = \gamma_{\alpha v} + \gamma_{\beta v} - \gamma_{\alpha\beta} \qquad (3.2.2)$$

3.2.2 The Spreading Coefficient Predicts Wetting of One Liquid by Another

To understand what happens when two liquids come into contact with one another, we apply the concepts of cohesion and adhesion to different configurations. When we place a drop of an insoluble liquid, such as oil, on a clean liquid surface, such as water, it may behave in one of three ways:

1. Remain as a nonspreading lens as in Figure 3.3a

Figure 3.3
A drop of an insoluble liquid placed on a clean liquid surface may: (a) remain as a nonspreading lens; (b) spread uniformly over the surface; or (c) spread as a monolayer leaving excess lenses floating on the surface. (D. J. Shaw, *Introduction to Colloid and Surface Chemistry,* 3rd ed., Butterworth–Heinemann, London, 1980, pp. 88, 90.)

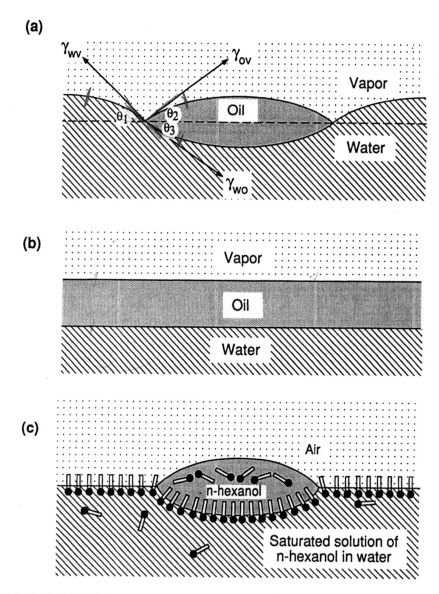

(a)

(b)

(c)

2. Spread uniformly over the surface as a duplex film thick enough for the liquid–liquid interface and the liquid–air interface to be independent of one another and to retain their characteristic surface (interface) tensions as in Figure 3.3b;

3. Spread as a monolayer, leaving excess liquid to form lenses as in Figure 3.3c.

The Young equation describes the overall shape of an oil (o) droplet on water (w) in air (v), such as the one shown in Figure 3.3a. It is derived by applying the boundary condition of tension balance at the three-phase contact line[4]

$$\gamma_{wv} \cos \theta_1 = \gamma_{ov} \cos \theta_2 + \gamma_{ow} \cos \theta_3 \qquad (3.2.3)$$

If the lower phase is a solid and remains unperturbed by the liquid, as shown in Figure 3.4, eq. 3.2.3 becomes the familiar version of the Young equation

$$\gamma_{sv} = \gamma_{sl} + \gamma_{lv} \cos \theta \qquad (3.2.4)$$

where θ is referred to as the *contact angle*.

What determines whether the lens shown in Figure 3.3a spreads to cover the surface of the lower phase completely or remains as an isolated droplet? When the oil spreads, the oil-water interface increases at the expense of the water-vapor surface. If the interfacial area increases by dA, the change in free energy of the system is given approximately by $(\gamma_{ov} + \gamma_{ow} - \gamma_{wv})$ dA. Spontaneous spreading occurs when this change in free energy is negative or when the quantity $\gamma_{wv} - (\gamma_{ov} + \gamma_{ow})$ is positive.

We define this quantity as the *initial spreading coefficient*

$$S = \gamma_{wv} - (\gamma_{ov} + \gamma_{ow}) \qquad (3.2.5)$$

where the interfacial tension of the two liquids, γ_{ow}, is measured at the instant the interface is formed, before any mutual solubilization of the two liquids takes place.

The spreading coefficient is related to the work of adhesion and cohesion, eqs. 3.2.1 and 3.2.2, through

$$S = W_{(adhesion\ oil-water)} - W_{(cohesion\ oil)} \qquad (3.2.6)$$

Thus, for S to be positive or 0 and for spontaneous spreading to occur, the oil must adhere to the water more strongly than it coheres to itself.

Spreading behavior can be very different before and after mutual solubilization. The difference between the initial and final behavior of benzene or hexanol on water illustrates the importance of drawing such distinctions. Initially benzene spreads over water because S is positive:

$$S(init) = 72.8 - (28.9 + 35.0) = 8.9 \text{ mJ/m}^2$$

---

[4]*Interfacial tension between two phases is represented by a vector tangential to the interface at the line of contact.*

Figure 3.4
When a drop of liquid is
placed on a surface, a force
balance at the three-phase
contact line leads to the
Young equation (3.2.4), in
which θ represents the
contact angle.

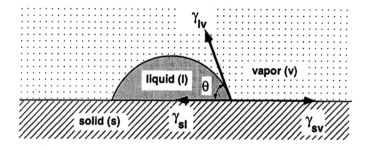

but after equilibration, some benzene dissolves in the water, and reduces the surface tension of water from 72.8 to 62.4 mN/m^2 so that S becomes negative

$$S(\text{final}) = 62.4 - (28.9 + 35.0) = -1.4 \text{ mJ/m}^2$$

For hexanol on water

$$S(\text{init}) = 72.8 - (24.8 + 6.8) = 41.2 \text{ mJ/m}^2$$

and

$$S(\text{final}) = 28.5 - (24.8 + 6.8) = -3.0 \text{ mJ/m}^2$$

In both cases the final state consists of a monolayer of benzene or hexanol that covers the water, and the excess benzene or hexanol forms flat lenses, as illustrated in Figure 3.3c. Such layers give rise to the familiar interference colors of oily water.

Trace amounts of impurity also can dramatically change spreading behavior. Impurities in the oil phase can reduce γ_{ow} enough to make S positive and permit spreading. On the other hand, impurities in the aqueous phase generally reduce S because the impurity lowers γ_{wv} more than it lowers γ_{ow}, especially if γ_{ow} is already low. Thus n-octane will spread on a clean water surface, but not on a contaminated one.

(See Appendix 8A for further discussion of liquid–solid contact under conditions where the liquid is moving over the solid surface—the dynamic contact angle.)

3.3 The Young-Laplace Equation Relates Pressure Differences to Curvature across a Surface

Surface tension operates in a liquid film and must be balanced by some equal and opposite force to obtain mechanical equilibrium. For example, to blow a soap bubble, we must apply an excess pressure on the inside.

How does excess pressure relate to surface tension and curvature? Figure 3.5 illustrates a circular cross section of a bubble, radius R, confined within a liquid. When the bubble is expanded infinitesimally to a radius R + dR, the concomitant surface area change is

$$dA = 4\pi \left[(R + dR)^2 - R^2\right] = 8\pi R \, dR \qquad (3.3.1)$$

Figure 3.5
The solid circular line in this figure represents a cross section of a gas bubble contained in a liquid. Balancing the work of increasing the surface area as the radius increases from R to R + dR with the pressure–volume work leads to the Young–Laplace equation (3.3.4) for the pressure difference across a spherical interface.

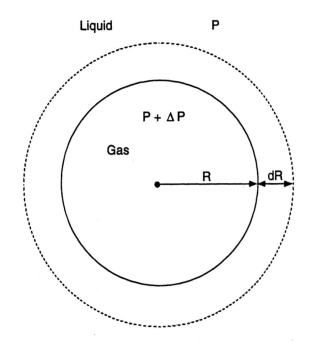

and the corresponding change in free energy is

$$dG = \gamma \, dA = \gamma \, 8\pi \, R \, dR \tag{3.3.2}$$

At equilibrium this free energy is balanced by a pressure-volume work, $dw = \Delta P \, dV$, due to the pressure difference, ΔP, across the film. The infinitesimal volume change is

$$dV = \frac{4\pi}{3}[(R + dR)^3 - R^3] = 4\pi \, R^2 \, dR$$

and so

$$dw = \Delta P \, dV = \Delta P \, 4\pi \, R^2 \, dR \tag{3.3.3}$$

Balancing the work of increasing the surface area (dG) with the pressure-volume work (dw) leads to an expression for the pressure difference per interface

$$\Delta P = \gamma \frac{2}{R} \tag{3.3.4}$$

For a soap bubble floating freely in air, the pressure difference is twice this value, $\Delta P = 4\gamma/R$, because liquid/air interfaces exist on the inside as well as on the outside of the bubble.

A liquid droplet suspended in air also experiences an internal increase in pressure given by eq. 3.3.4. The smaller the droplet, the greater the hydrostatic pressure within it.

If we consider a uniform cylinder of length L, as shown in Figure 3.6, and neglect end effects, the area change becomes

$$dA = 2\pi \, L \, [(R + dR) - R] = 2\pi \, L \, dR \tag{3.3.5}$$

and the volume change is

Figure 3.6
Balancing the work of increasing the surface area as the radius increases from R to R + dR with the corresponding pressure–volume work leads to the Young–Laplace equation (3.3.7) for the pressure difference across a cylindrical interface.

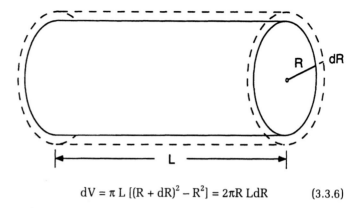

$$dV = \pi L [(R + dR)^2 - R^2] = 2\pi R\, L dR \qquad (3.3.6)$$

In the cylindrical geometry, a balance between the surface and the pressure–volume work results in

$$\Delta P = \frac{\gamma}{R} \qquad (3.3.7)$$

per interface.

If we generalize these equations to interfaces of arbitrary shape, the result is the *Young–Laplace equation*

$$\Delta P = \gamma \left(\frac{1}{R_1} + \frac{1}{R_2} \right) \qquad (3.3.8)$$

where R_1 and R_2 equal the principal radii of curvature of the interface. For a sphere, $R_1 = R_2 = R$, while for a cylinder, $R_1 = R$

Figure 3.7
When a soap film is formed between two open wire loops, no pressure difference exists between the inside and the outside of the film. The curvature of the surface must everywhere be zero; that is, $R_1 = -R_2$ everywhere, a condition satisfied by the catanoid surface shown here.

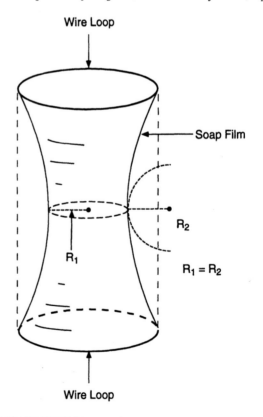

and $R_2 = \infty$, and eq. 3.3.8 reduces to eqs. 3.3.4 and 3.3.7 for spheres and cylinders, respectively.

If we take two wires and form a tubelike soap film between them, the pressure difference across the film is zero because the ends of the tube are open. Then according to eq. 3.3.8, the mean curvature of the film is zero everywhere, and we obtain the catanoid surface illustrated in Figure 3.7.

We have assumed that the weight of the liquid does not affect these equilibrium shapes. In reality, we cannot ignore the effect of gravity in most practical applications; so in the Young–Laplace equation, ΔP, R_1, and R_2 become functions of the location in space for a given surface. Although this gravitational contribution leads to rather tedious differential equations, solutions are available, particularly for surfaces that possess axial symmetry.

3.3.1 The Young–Laplace Equation Accounts for Capillary Rise

The pressure difference across a curved surface has a number of significant consequences. One is the existence of a capillary pressure gradient. We can use the Young–Laplace equation to relate the phenomenon of *capillary rise* to the surface tension of a liquid, as illustrated in Figure 3.8. A simple, albeit somewhat approximate, relation exists between the capillary rise h, the capillary radius r, the surface tension γ, and contact angle θ.

Figure 3.8
Schematic illustration of the relationship between the wetting of a capillary tube by a liquid and capillary rise h (eq. 3.3.12). The radius of the meniscus R and the radius of the capillary tube r are related by $R \cos \theta = r$; $R = r$ only when the contact angle θ equals zero.

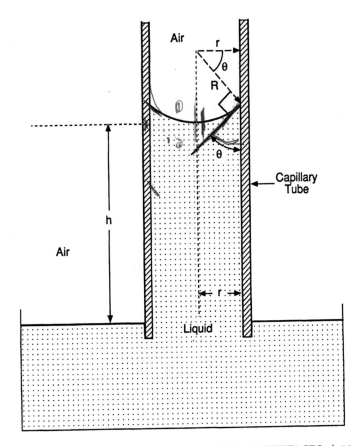

The liquid surface in the capillary tube has a spherical cap. As shown in Figure 3.8, the radius of curvature R of the liquid surface relates to the radius of the tube r and the contact angle θ by

$$R \cos \theta = r \qquad (3.3.9)$$

A Young–Laplace pressure difference exists across the interface. Because the surface curvature for the liquid is negative, the pressure in the liquid immediately below the interface is *less than* the air pressure outside by the amount

$$\Delta P = -2\gamma/R = -2\gamma \cos \theta/r \qquad (3.3.10)$$

This capillary pressure difference causes the liquid to rise up the capillary tube until the pressure at the base of the liquid column equals that of the air outside. The hydrostatic pressure due to the weight of the column, height h, is

$$\Delta P = \Delta \rho \, gh \qquad (3.3.11)$$

where $\Delta \rho$ represents the density difference between the liquid and the vapor, $\rho_{liq} - \rho_{vap}$ ($\approx \rho_{liq}$), and g is the acceleration due to gravity. Balancing the two pressures, we find

$$\gamma = \frac{\Delta \rho \, ghr}{2 \cos \theta} \qquad (3.3.12)$$

This equation is particularly simple to apply when the liquid wets the solid surface, that is, when θ is zero and cos θ = 1. In that case, if we know ρ_{liq} and r and can measure h, we can calculate the liquid–vapor surface tension.

Effects of gravity on the shape of the interface and the contribution of the small amount of liquid contained above the meniscus in Figure 3.8 lead to complications. For this reason the analysis must be refined to achieve a precise determination of γ.

3.4 The Kelvin Equation Describes Vapor Pressure of Small Droplets and Explains Homogeneous Nucleation and Capillary Condensation

3.4.1 The Kelvin Equation Describes the Critical Condition for Homogeneous Nucleation of Liquid Droplets from a Vapor

The pressure change across a curved surface also is significant in the nucleation of liquid droplets from a supersaturated vapor. Let us consider the nucleation of a liquid droplet from a vapor at pressure P, where P exceeds the equilibrium vapor pressure P_o, and P/P_o represents the supersaturation. Assuming an ideal vapor, the free energy change per mole needed to raise the pressure from P_o to P is $\mathbf{R}T \ln(P/P_o)$. Because a liquid droplet of radius R contains

$(4/3)\pi R^3/V_L$ moles (where V_L represents the liquid molar volume), the free energy of condensation of the droplet is given by

$$-\frac{4}{3}\frac{\pi R^3}{V_L}\,\mathbf{RT}\,\ln(P/P_0).$$

The surface energy of the droplet is $4\pi R^2\gamma$. So we can write the total free energy change required to form a liquid drop of radius R as

$$\Delta G = -\frac{4}{3}\frac{\pi R^3}{V_L}\mathbf{RT}\,\ln(P/P_0) + 4\pi R^2\gamma \qquad (3.4.1)$$

The first term on the right-hand side of this equation varies as R^3, and the second term varies as R^2. Thus, as R increases, the value of ΔG initially increases, but then it decreases as the R^3 term dominates. As a result, a maximum in ΔG occurs at some intermediate value that we designate as R_c. Figure 3.9 illustrates the variation of ΔG with R. At the maximum, $d(\Delta G)/dR = 0$. Differentiating eq. 3.4.1 and equating to zero gives, for molar quantities,

$$\mathbf{RT}\ln P/P_0 = \frac{2\gamma V_L}{R} \qquad (3.4.2)$$

and, for molecular quantities,

Figure 3.9
Change in free energy ΔG, with radius R of a spherical nucleus. The critical nucleus size, R_c, is given by eq. 3.4.5, and the corresponding ΔG_{max} by eq. 3.4.6. Data for water droplet at 100°C and $P/P_0 = 3$. (D. F. Evans and H. Wennerström, *The Colloidal Domain: Where Physics, Chemistry, Biology, and Technology Meet*, VCH Publishers, New York, 1994, p. 59. Reprinted with permission of VCH Publishers © 1994.)

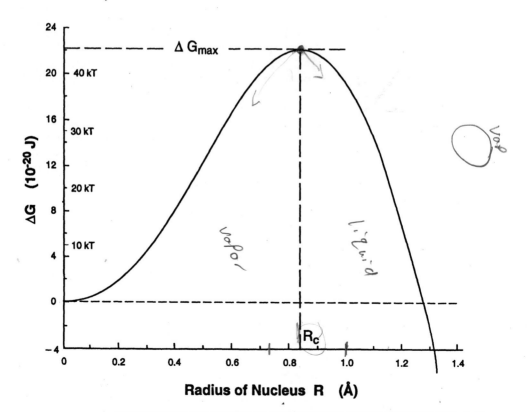

$$kT \ln P/P_o = \frac{2\gamma v_m}{R} \qquad (3.4.3)$$

where v_m is now the molecular volume. Rearranging eq. 3.4.3, we obtain

$$\frac{P}{P_o} = \exp\left(\frac{2\gamma v_m}{kTR}\right) \qquad (3.4.4)$$

which is the *Kelvin equation*. The value of P given by the Kelvin equation may be interpreted as the equilibrium vapor pressure above a surface of curvature 1/R and surface tension γ.

For a droplet of radius R_c

$$\frac{P_c}{P_o} = \exp\left(\frac{2\gamma v_m}{kTR_c}\right) \qquad (3.4.5)$$

the pressure P_c represents a threshold. For vapor pressures below P_c drops of radius R_c will re-evaporate, while for vapor pressures above P_c drops of radius R_c will grow by condensation to form a liquid phase. What will be the fate of a drop of radius R_1 if we introduce it into a vapor of a given vapor pressure P? Figure 3.9 shows that when R_1 is smaller than the critical radius R_c given by the Kelvin equation and lies to the left-hand side of the maximum, free energy is gained by vaporizing the liquid droplet. When R_1 is greater than the radius R_c, free energy is gained by condensing more vapor onto the drop to make it larger, which leads to the formation of a bulk liquid phase.

Droplets with smaller R require a greater degree of supersaturation to exist in metastable equilibrium. Using eq. 3.4.4, we can calculate the maximum vapor pressure as a function of the sphere's radius. For example, in the case of water at 20°C, drops with radii of 10^{-6}, 10^{-7}, 10^{-8}, and 10^{-9} m require supersaturation values of P/P_o equal to 1.0011, 1.0184, 1.1139, and 2.94, respectively.

We can also apply the Kelvin equation to determine the vapor pressure inside a gas bubble in a liquid. Because the bubble radius is measured on the inside (concave side) of the surface, R_c is negative and a minus sign enters eq. 3.4.5. For bubbles in water with radii of 10^{-6}, 10^{-7}, 10^{-8}, and 10^{-9} m, calculated values of P/P_o are 0.9989, 0.9803, 0.8976, and 0.339, respectively.

In some instances in this section (3.4) and the previous section (3.3), we have used conventional equations to analyze systems with extremely small radii of curvature, whose structures approach the dimensions of individual molecules. What is the limit of application of these equations, given that they employ macroscopic concepts such as density, surface tension, and radius of curvature?

To answer this question, we must determine the point at which microscopic quantities become important. Recent experiments using the Surface Forces Apparatus (SFA), which measures interparticle surface forces, show that for simple liquids, such as cyclohexane, the Kelvin equation is valid down to a radius of curvature that corresponds to about seven times the molecular diameter. Experimental tests confirm this finding for other simple liquids. An important exception is water, where significant deviations from Kelvin predictions occur for larger radii of curvature. This exception may reflect the fact that surface tension in water involves more water molecules than those in the closest coordination. As demonstrated in eq. 2.3.10, interactions between water molecule dipoles are long-range and vary as $1/R_{12}^3$.

3.4.2 Combining the Kelvin Equation and an Arrhenius Rate Equation Provides an Expression for the Rate of Homogeneous Nucleation

The Kelvin equation establishes the basis for a general understanding of subcooling, superheating, and supersaturation phenomena that are sometimes observed in phase transitions. Such behavior characterizes *homogeneous nucleation* in pure systems. Homogeneous nucleation occurs only when the system is free from dust or any other extraneous material and the container walls do not serve as nucleation sites. For example, a dust-free sample of water can be cooled to $-48°C$ before it freezes or heated to considerably above $100°C$ before it vaporizes.

By combining eq. 3.4.1 and 3.4.2, we can obtain the value for ΔG_{max} in terms of R_c as

$$\Delta G_{max} = \frac{4\pi R_c^2 \gamma}{3} \tag{3.4.6}$$

which by substituting for R_c gives

$$\Delta G_{max} = \frac{16\pi \gamma^3 V_L^2}{3[RT \ln(P/P_o)]^2} \tag{3.4.7}$$

So far, our discussion of homogeneous nucleation has been concerned only with the energetics of nuclei formation. Two additional factors dictate whether they do or do not grow from the embryonic stage to form stable nuclei. First, the concentration of embryonic nuclei is determined by thermal fluctuations (described by Boltzmann statistics); second, their growth is determined by the rate at which molecules arrive from the vapor to add to the nucleus (described by the gas-phase collision frequency f, the number of vapor molecules colliding with unit surface area per second; see Section 10.1.4). The formation rate I for nuclei of size R_c is then given by the Arrhenius-type rate expression

$$I = f \exp\left(-\frac{\Delta G_{max}}{RT}\right) \tag{3.4.8}$$

where $f = 10^{23} P^2$ when P is measured in Torr (see Section 10.1.4). More exact expressions for eq. 3.4.8 involve more complex preexponential terms, but the numerical value does not change significantly; so the approximate term suffices for our purposes. By employing the expression for ΔG_{max} and using the ideal gas relationship, we can rewrite eq. 3.4.8 as

$$I = 10^{23} P^2 \exp\left(-\frac{17.5 \, V_L^2 \gamma^3}{T^3 [\ln(P/P_o)]^2}\right) \qquad (3.4.9)$$

which gives I in nuclei per cubic centimeters per second.

For water vapor at 0°C, $V_L = 20$ cm^3/mole, $\gamma = 72$ mJ/m^2, and $P_o = 4.6$ Torr. Equation 3.4.9 becomes

$$I = 2 \times 10^{24} \left(\frac{P}{P_o}\right)^2 \exp\left(-\frac{118}{[\ln(P/P_o)]^2}\right) \qquad (3.4.10)$$

Figure 3.10 plots values of I from this equation as a function of P/P_o and shows how rapidly the rate of nuclei formation increases with increasing vapor pressure. The critical supersaturation pressure is arbitrarily selected as the value for P/P_o that gives a value for ln I equal to unity. For water vapor $P_c/P_o = 4.18$.

One important point to note in this brief introduction to homogeneous nucleation is the critical dependence of the phase transformation on the interfacial tension γ between the new and the parent phase.

We can develop equations parallel to eqs. 3.4.7 and 3.4.8 for vapor–solid (Chapter 10), liquid–solid and solid–solid (Chapter 11) phase transformations and for crystallization of supersaturated solutes, although modifications must be made to account for diffusion-controlled kinetics of cluster formation in liquids and solids, and anisotropic growth and lattice strain in crystalline materials.

Figure 3.10
Variation of the rate of nucleation I as a function of the degree of supersaturation P/P_o, where P_o represents the equilibrium vapor pressure above a liquid or solid flat surface. For water at 0°C homogeneous nucleation becomes an effective mechanism for initiating a phase transformation when $P/P_o \approx 4$.

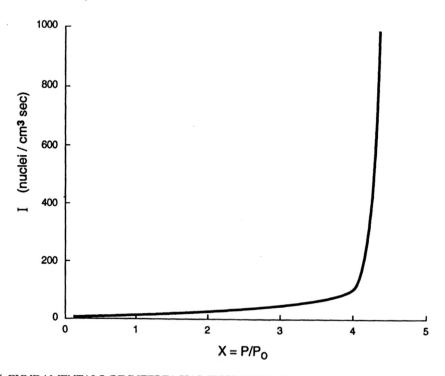

3.4.3 Heterogeneous Nucleation Involves the Interfacial Free Energy between the New Phase and the Substrate

Equations 3.4.6–3.4.10 deal with homogeneous nucleation. They explain why phase transition temperatures or pressures in clean systems may differ significantly from the thermodynamic equilibrium values. But most systems are not perfectly clean; surfaces, dust, or seed particles provide a template for the initial growth of a new phase. *Heterogeneous nucleation* presents an alternative and easier path for phase transformation. Lower interfacial energy between the new condensing phase and the template considerably reduces the energetics of nuclei formation, and then phase transformation occurs at or close to the equilibrium temperature.

For heterogeneous nucleation, the value for ΔG_{max} is reduced from the value for homogeneous nucleation (given in eq. 3.4.6) by a factor that is a function of the contact angle θ between the condensing liquid and the solid template

$$\Delta G_{max}(\text{hetero}) = \Delta G_{max}(\text{homo}) \frac{(2 + \cos\theta)(1 - \cos\theta)^2}{4} \quad (3.4.11)$$

This factor takes care of the volume of liquid in the spherical caplet (in contrast to the volume of the spherical droplet in homogeneous nucleation) and, through the contact angle, of both the liquid–solid and liquid–vapor interfacial energy (in contrast to just the liquid–vapor interfacial energy in homogeneous nucleation). If the liquid wets the solid and θ approaches zero, the value for $\Delta G_{max}(\text{hetero})$ also approaches zero, and nucleation occurs spontaneously at the equilibrium temperature with no supersaturation. Once again we note the critical dependence of phase transformations on interfacial energy.

Control over *nucleation and growth* plays an important role in many aspects of interfacial processing. Smoke particles nucleate smog; silver halide particles are used to seed clouds. To produce the monodisperse sols described in Chapter 4, we must be able to control the initial nucleation event and subsequent growth so that all the colloidal particles will be the same size. Chapter 5 demonstrates that fundamental understanding of the nucleation and growth of self-assembly systems requires deep insight into the formation of the amphiphilic clusters that produce micelles, vesicles, bilayers, and other liquid–liquid colloidal structures. Successful growth of thin films from the vapor, described in Chapter 10, also relies on control over nucleation and growth.

3.4.4 Due to Capillary Effects, Surface Energy Can Cause a Liquid to Condense on a Rough Surface Prior to Saturation in the Bulk Phase

Figure 3.11 illustrates the capillary condensation process through the example of a conical capillary tube placed in a vapor with pressure $P < P_o$. In eq. 2.7.3 we saw that a molecule is attracted to a bulk flat surface with a potential that varies as $1/z^3$,

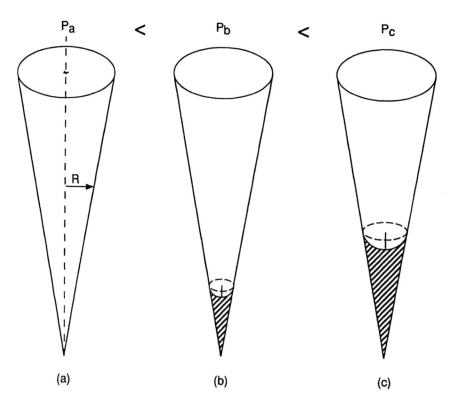

P_a < P_b < P_c

(a)   (b)   (c)

Figure 3.11
Vapor condensing into a conical tube illustrates capillary condensation for vapor pressures less then the equilibrium value, P_o. In the figure $P_a < P_b < P_c < P_o$. The extent to which the liquid fills the tube to a level of radius R is given by the Kelvin equation (3.4.4).

where z is the distance from the molecule to the surface. This effect is particularly strong at the apex of the cone, where z reduces to zero, and vapor molecules will be attracted to this region. If the outside vapor pressure is sufficiently close to saturation, increased local concentration will cause condensation at the apex.

To what extent will this condensate grow in the conical tube? As the cone fills up, equilibrium is established at the height where the gain in surface and interface energy is balanced by the free energy decrease in condensing the vapor. Thus the interfacial energies between liquid, vapor, and substrate have an important impact on capillary condensation. The equilibrium height of the liquid in the cone depends on the contact angle θ contained in the Young equation 3.3.9. In particular, when $\theta = 0$ and the liquid wets the solid, some capillary condensation takes place for all vapor pressures. Condensation fills the tube until the radius of curvature of the liquid meniscus R (negative) equals the radius given by the Kelvin equation 3.4.4 for the existing vapor pressure P. When $\theta > 0$, capillary condensation does not start until the vapor pressure exceeds a certain value. When $\theta > 90°$, the liquid does not wet the solid, and then the capillary is the last place for condensation to occur.

One consequence of capillary condensation is that heterogeneous nucleation and growth of a condensing phase, which we discussed in the previous section, is sensitive to the roughness of the template. Heterogeneous nucleation is more likely to occur preferentially along scratches in a surface, as demonstrated by the formation of frost on the windshield of a car. Capillary condensation also plays a determining role in adsorption of vapors in porous materials, such as catalysts, and in adhesion of dust and powders to surfaces.

The same principles described for capillary condensation also apply during drying to the evaporation of liquid from porous materials. Drying processes also depend critically on interfacial energies. If the liquid wets the solid, the last liquid residue will be drawn from the smallest pores. To overcome this difficulty in drying porous compacts, we often place them in an autoclave and raise the pressure to the critical point where the density of the vapor and liquid are equal. The surface tension between them drops to zero, and the compact can be heated and dried equally from all porous regions irrespective of the pore size. This procedure is called *critical point drying*. If the liquid does not wet the solid, pore size has no major impact on the drying process.

3.5 Thermodynamic Equations That Include Surface Chemistry Contributions Provide a Fundamental Basis for Characterizing Fluid–Fluid Interface Behavior

In the previous sections, we developed the equations that describe the physical behavior of liquid interfacial systems. For the most part, these equations describe the behavior of pure, single-component systems. Now we extend the thermodynamics to the more general condition where a second component (an impurity) is present and examine the consequences with respect to the chemical composition of the interface. The Gibbs approach gives us a workable model with which to conceptualize small-scale fluid–fluid interface systems in thermodynamic terms.

3.5.1 The Gibbs Model Provides a Powerful Basis for Analyzing Interface Phenomena by Dividing a System into Two Bulk Phases and an Infinitesimally Thin Dividing Interface

In developing thermodynamic equations of state, we usually can ignore the contributions of surfaces and interfaces, which are often small compared to the contributions of the bulk phases. However, when interfacial areas are large, or when we are particularly interested in analyzing the effect of an interface, this approximation is no longer satisfactory. A straightforward approach to handle interface thermodynamics assigns a volume to the bulk phases and a separate volume to the interface. But in fluids this approach is not so tractable because the properties (P in Figure 3.12a) change continuously over a small distance of the order 1 to 3 nm (ΔX in Figure 3.12a). For example, at an oil—water interface, the dielectric constant changes from 2 to 78.5. Consequently, it is not so clear where to establish the interface, and we cannot unambiguously assign a volume to it. Gibbs avoided this dilemma by treating the interface as a mathematical plane pos-

Figure 3.12
(a) Variation of a property *P* across an interface between two phases α and β. The actual thickness of the interface, ΔX, is on the order of angstroms, but in the Gibbs model we treat this dividing surface as a mathematical plane at X_o. (b) We can locate the Gibbs dividing surface *S–S* so that the excess adsorption of either the solvent (as illustrated in b) or the solute is zero.
(P. C. Hiemenz, *Principles of Colloid and Surface Chemistry*, 2nd ed., Marcel Dekker, New York, 1986, p. 388.)

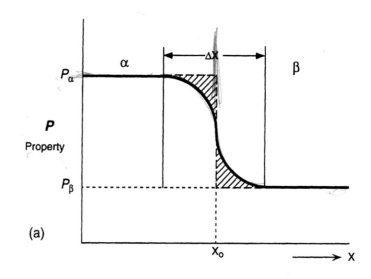

(a)

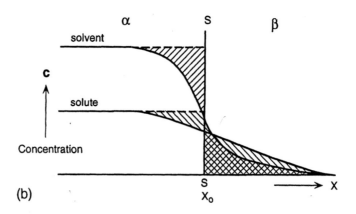

(b)

sessing area without thickness. Upon first encounter, this abstraction seems baffling, but as we shall see, it provides a powerful way to analyze interfaces.

We divide the system into three parts: the volumes of the bulk phases, V_α and V_β, plus the interface that separates them. Any extensive thermodynamic property of a system can be apportioned between these three parts, which enables us to write the energy ascribed to the surface or interface, $\mathbf{U}^\sigma$, as

$$\mathbf{U}^\sigma = \mathbf{U} - \mathbf{u}_\alpha V_\alpha - \mathbf{u}_b V_\beta \qquad (3.5.1)$$

where $\mathbf{U}$ equals the total energy of the system and $\mathbf{u}_\alpha$ and $\mathbf{u}_\beta$ are the energies per unit volume in the two phases, α and β. Similar surface quantities can be written for the other thermodynamic quantities.

By dividing the total volume of the bulk phases into the precise quantities V_α and V_β, while separating the two phases by an infinitesimally thin dividing interface, we can construct an imaginary system that displays the thermodynamic properties of a real system. In this imaginary system, all the physical and thermodynamic properties remain constant up to the interface, whereas in a real system they change rapidly (but not discontinuously) at the interface. Replacing the real system with the model makes it easier to think of extensive properties, such as energy and volume, as discrete segments rather than as continuously varying quantities.

Molecular composition also changes across the interface. In a two-phase multicomponent system with volumes V_α and V_β, we can write $c_{i\alpha}$ and $c_{i\beta}$ as the molar concentration (per unit volume) of the second component i in each phase α and β. Then the number of moles of component i in each phase becomes

$$n_{i\alpha} = c_{i\alpha} V_\alpha \text{ and } n_{i\beta} = c_{i\beta} V_\beta \qquad (3.5.2)$$

The amount of component i in the surface or interface is

$$n_i^\sigma = n_i - c_{i\alpha} V_\alpha - c_{i\beta} V_\beta \qquad (3.5.3)$$

where n_i equals the total number of moles of i in the whole system. We can define a surface or interface concentration (per unit area) for component i as

$$\Gamma_i = \frac{n_i^\sigma}{A} \qquad (3.5.4)$$

As we shall see later, the magnitude and sign of Γ_i is sensitive to the chosen location of the interface, and we need to be arbitrary in setting its position.

3.5.2 The Gibbs Adsorption Equation Relates Surface Excess to Surface Tension and the Chemical Potential of the Solute

We can develop thermodynamic relations for the surface or interface variables in a way that is fully analogous to those used for bulk phases. For example, the differential of the "internal" energy at the interface is

$$dU^\sigma = T \, dS^\sigma + \gamma \, dA + \sum_i \mu_i^\sigma(T, n_i^\sigma) \, dn_i^\sigma \qquad (3.5.5)$$

where the bulk work term, $P \, dV$, has been replaced with the surface work term, $\gamma \, dA$. The surface chemical potential of the component i is μ_i^σ. Using the same manipulations that lead to the

relation $U = TS + PV + \sum_i \mu_i n_i$ for the bulk yields

$$U^\sigma = T \, S^\sigma + \gamma A + \sum_i \mu_i^\sigma n_i^\sigma \qquad (3.5.6)$$

for the surface quantities. If we then differentiate eq. 3.5.6 and compare it with eq. 3.5.5, we obtain the relation

$$S^\sigma dT + Ad\gamma + \sum_i n_i^\sigma d\mu_i = 0 \tag{3.5.7}$$

At a constant temperature $dT = 0$ and eq. 3.5.7 becomes the *Gibbs adsorption isotherm*

$$-d\gamma = \sum_i \frac{n_i^\sigma}{A} d\mu_i = \sum_i \Gamma_i \, d\mu_i \tag{3.5.8}$$

Implicit in this equation is the *concept of surface concentration*, that is, an above-average or below-average concentration of solute in the bulk located at the interface. We want to estimate the magnitude of this excess or deficiency.

Figure 3.12a indicates with a solid line how a general property P actually changes between phase α and phase β in a real system. For a solid/vapor interface the dividing line between phases and physical properties is easy to establish—it is at the surface of the solid, but for liquid/vapor interfaces the dividing line is not so clearcut. The interface itself is a transition zone with thickness ΔX measured in the direction X perpendicular to the interface. The zone thickness is on the angstrom scale. Across it the properties of the system vary continuously from those characteristic of phase α to those characteristic of phase β. Figure 3.12a also indicates with a dashed line how the property varies in the model or imaginary system. We do not assign any thickness or volume to the interface; instead we treat it as a dividing plane situated at some specific location, X_o. The properties P of the individual phases, α and β, extend right up to this imaginary interface.

How do we choose the location of X_o? Suppose we locate it so that the shaded areas are equal. The shaded area to the left of X_o represents the underestimation of P in phase α, while the corresponding shaded area to the right of X_o represents the overestimation of P for phase β. The problem with this choice is that the profile sketched in Figure 3.12a will differ for each property, so that a line that divides the profile of one property evenly will not divide the profile of other properties in the same way. From the standpoint of thermodynamics, molecular details have no significance; so we can position it wherever is most convenient.

In terms of the Gibbs adsorption isotherm, setting X_o so that the solvent (component 1) composition profile is equally compensated is a particularly convenient choice. Figure 3.12b is drawn for this condition. This division of surface means

$$\Gamma_1 = 0 \tag{3.5.9}$$

and so eq. 3.5.8 becomes

$$d\gamma = -\Gamma_2 \, d\mu_2 \tag{3.5.10}$$

Thus Γ_2 equals the algebraic difference between the overestimated and underestimated portions of the curve; it describes the moles of solute (component 2) in excess (or deficiency) at the surface when a location X_o is chosen so as to make the surface excess of the solvent (component 1) equal to zero. Figure 3.12b illustrates this effect.

Now we can relate eq. 3.5.8 to accessible experimental quantities. At constant temperature, differentiating the equation for the chemical potential

$$\mu_2 = \mu_2^o + RT \ln a_2 \qquad (3.5.11)$$

where the superscript o signifies the standard state defined by an infinitely dilute solution and a_2 represents the solute chemical activity, yields

$$d\mu_2 = RT \, d(\ln a_2) \qquad (3.5.12)$$

By combining this eq. with equation 3.5.10, we obtain

$$\Gamma = -\frac{1}{RT}\left(\frac{d\gamma}{d(\ln a)}\right)_T \approx -\frac{1}{RT}\left(\frac{d\gamma}{d(\ln c)}\right)_T \qquad (3.5.13)$$

sometimes written in the form

$$\Gamma = -\frac{a}{RT}\left(\frac{d\gamma}{da}\right)_T \approx -\frac{c}{RT}\left(\frac{d\gamma}{dc}\right)_T$$

where substituting molar concentration (c) for activity (a) assumes an ideal solution. We have also dropped the subscript so that c is now the solute concentration. This is a more useful form for the Gibbs adsorption isotherm. When the second phase β is a gas or air, eq. 3.5.13 expresses the concentration of solute at the liquid surface in terms of the variation in surface tension with composition. Experimental measurement of the surface tension of a liquid γ as a function of concentration of solute or impurity c at a given temperature leads to an estimate of the surface excess Γ present at the liquid surface at that temperature.

The surface excess concentration Γ represents an algebraic quantity that can be either positive or negative. For example, Γ is *negative* for common electrolytes, such as NaCl, at the air-water interface, so the surface concentration of sodium or chlorine ions is *less* than the bulk value. For other solutes, such as surfactants, Γ is *positive* and their surface concentration is *greater* than the bulk value. In amphiphilic sytems, to be discussed later and which are the subject of Chapter 5, surfactants concentrate extensively at oil-water interfaces.

As we have noted, the location of the interface plane is arbitrary. In applying eq. 3.5.8, we could de-emphasize the solute by locating X_o so that Γ_2 equals zero; this emphasizes Γ_1, which then would not be zero. Alternately, we could choose to show a zero surface excess determined by the total number of moles or the total volume. In this case, both Γ_1 and Γ_2 would have nonzero values. Because all these conventions emphasize different experimental or mathematical features of interfacial systems, it is important to check the conventions used to calculate and apply particular values of Γ.

3.6 Monolayers Formed by Insoluble Amphiphiles Behave As a Separate Phase and Are Most Readily Characterized Using a Langmuir Balance

Many interfacial systems involve monolayers of molecules preferentially adsorbed at fluid–fluid interfaces. These molecules typically possess well-defined polar and nonpolar regions and are called amphiphilic molecules (see Chapter 5 for a detailed discussion). Amphiphilic molecules form monomolecular interfacial layers that separate polar (H_2O) and nonpolar (oil or vapor) phases. At air–water interfaces, such monolayers can dramatically lower evaporation rates or reduce the formation of surface waves—pouring oil on troubled waters! Amphiphilic molecules located at oil–water interfaces play a key role in the stability of emulsions, as we shall discuss in Chapter 5. Much of our knowledge of such monolayers comes from using a Langmuir balance to study them.

The usual way to prepare a monolayer for study in a Langmuir balance is to dissolve the amphiphilic molecules in a volatile organic solvent and disperse drops of the solution onto the air–water interface. Initially the spreading coefficient is positive and the solution spreads all over the water surface. The amphiphilic molecules concentrate at the interface in accordance with the behavior we would expect from the Gibbs adsorption isotherm, but when the solvent evaporates, an amphiphilic layer forms. If the amphiphiles possess sufficiently large nonpolar groups, they are insoluble in water and remain confined to the surface. Figure 3.3c illustrates the configuration. Under these conditions, the amphiphiles can be treated as a separate phase.

We represent the lowering of the surface tension of the interface by the adsorbed molecules through the equation

$$\Pi_s = \gamma_o - \gamma_a \qquad (3.6.1)$$

which represents the difference between γ_o, the surface tension with no amphiphile present, and γ_a, the value with adsorbed amphiphile. Π_s represents the force per unit length needed to prevent the film from spreading (or from the other point of view, the force per unit length that drives the film to spread). It is the two-dimensional equivalent of pressure, and in fact, Π_s is called the *spreading pressure.*

Π_s varies with the surface concentration of the amphiphilic molecules. We can manipulate the amphiphilic concentration by adding or subtracting amphiphilic molecules to and from the system, or by taking those already present and compacting or expanding them. The latter approach lies behind the Langmuir trough. In this apparatus the surface film of amphiphilic molecules is swept up and contained with a fine wire or barrier (like an oil spill boom).

We can measure Π_s using a Langmuir balance (Figure 3.13). The monolayer is contained between a movable barrier (2) and a

Figure 3.13
In the Langmuir balance, the relation between surface pressure and surface areas of an insoluble solute adsorbed at the liquid–vapor interface is determined by moving the sweep barrier (2) toward the float (3). A torsion wire (4) attached to the float measures the deflection. (A. W. Adamson, *Physical Chemistry of Surfaces,* 4th ed., © 1982, John Wiley & Sons, p. 113. Reprinted by permission of John Wiley & Sons, Inc.)

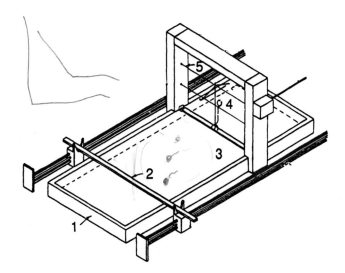

float (3) attached to a torsion wire arrangement (4). The surface pressure of the monolayer is directly measured by the horizontal force exerted on the float. Moving the barrier varies the area covered by the film (A). In an experiment, the spreading pressure, Π_s, is measured as a function of the area A at constant temperature (the two-dimensional equivalent of a P–V curve). To make reliable measurements the water must be extremely pure, and precautions must be taken to avoid contaminating the monomolecular film by ubiquitous surface-active material. A single human fingerprint contains enough surface-active material to form numerous monolayers. The Langmuir balance is normally operated in a thermostatically controlled environment.

Figure 3.14 displays typical features of pressure–area isotherms observed for insoluble monolayers. For large surface area the amphiphilic molecule concentration is low and the spreading pressure is very low. The monolayers exhibit 'gas-like' (G) behavior. That is to say, as Π_s approaches zero, the pressure-area isotherms follow a hyperbolic curve consistent with a two-

Figure 3.14
Schematic of a two-dimensional spreading pressure, Π_s, versus area per molecule, a_0, isotherm. The transformation from gas phase to solid-phase behavior involves two liquid phases, L_1 and L_2. Dotted lines represent extrapolations that give the area per molecule at the phase transformations. At sufficiently high pressures Π_c, the solid monolayer film collapses. (P. C. Hiemenz, *Principles of Colloid and Surface Chemistry,* 2nd ed., Marcel Dekker, New York, 1986, p. 364.)

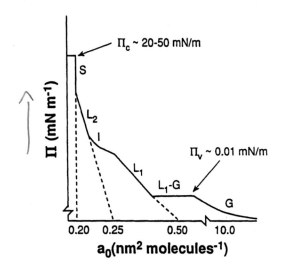

dimensional equivalent of the ideal gas law

$$\Pi_s A/N_s = \Pi_s a_o = kT \qquad (3.6.2)$$

where N_s is the number of amphiphilic molecules on the surface and a_o the effective area per molecule. In this region the hydrocarbon chains lie flat on the surface, and the effective size of the molecule is πl_{hc}^2, where l_{hc} represents the length of the hydrocarbon chain.

Plots of the compressibility factor Z (defined as $\Pi_s a_o/kT$) reveal deviations from "ideal" conditions, for which Z = 1. Figure 3.15 shows Z plots for a series of carboxylic acids at 25°C. The ideal condition is represented by the horizontal dashed line at Z = 1. "Negative" deviations (Z < 1) occur at low spreading pressures, just as they do for analogous three-dimensional pressure–volume plots for a gas, due to the attraction between the amphiphilic molecules. Because the strength of the attraction increases with molecular size, the deviations become more pronounced as the alkyl chain lengthens (the length increases from curves 1 to 6). "Positive" deviations (Z > 1) occur at high spreading pressures. This deviation reflects the excluded area, the area taken up by molecules, analogous to the excluded volume in three-dimensional gases.

At the other end of the isotherm in Figure 3.14, the pressure Π_s is high and a_o is low, and we observe "solid-like" (S) behavior. In this case, the hydrocarbon chains are oriented vertically and closely packed together. The film is now relatively incompressible, and attempts to increase Π_s still further cause the monolayer film to buckle or collapse above Π_c.

Figure 3.15
Plots of the compressibility, $Z = \Pi_s a_o/kT$, versus Π_s for n alkyl carboxylic acids ($C_n H_{2n+1} COOH$). In curve (1) n = 4, (2) n = 5, (3) n = 6, (4) n = 8, (5) n = 10, and (6) n = 12. Ideally the compressibility Z equals 1, given by the horizontal dashed line. Deviations from ideality increase as the alkyl chain lengthens. (P. C. Hiemenz, *Principles of Colloid and Surface Chemistry*, 2nd ed., Marcel Dekker, New York, 1986, p. 367.)

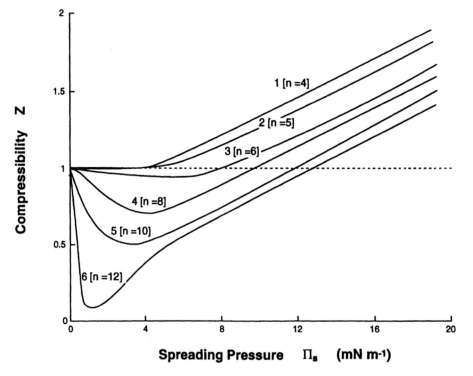

A monolayer's transition from "gas" to "solid" is more complex than the behavior in bulk phases. The transition from G to L_1 occurs with significant compression of area at a constant pressure, Π_v, as it does in a bulk gas–liquid transition. The L_1–G region also shows the two-dimensional equivalent of the critical temperature that we find in three-dimensional fluids. As the temperature is raised, the length of the L_1–G region decreases until at the critical temperature, it vanishes. Unlike bulk phases there are two liquid phases, L_1 (the liquid-expanded state) and L_2 (the liquid-condensed state), separated by a transition region I. We do not yet understand the transition from L_1 to L_2 in any detail.

3.7 Thermodynamics of Ideal Solutions Provide a Fundamental Basis for Characterizing Adsorption in Fluid–Solid Systems

In the Gibbs model for a fluid interface discussed in Section 3.5, we developed the appropriate thermodynamic expressions on the implicit assumption that phases α and β exist in equilibrium. By their very nature, such thermodynamic expressions provide no physical insight into the molecular interactions that determine whether components form miscible or immiscible solutions. Such insights are clearly important in designing and controlling interfacial processes, and they underlie many of the concepts to be developed in the next three chapters. So in this section we review the features associated with the entropy of mixing A and B molecules and show how this leads to the definition of an ideal solution and thence to the Langmuir adsorption isotherm that describes adsorption at fluid–solid interfaces.

3.7.1 Entropy of Mixing Provides the Basis for the Ideal Solution Model and Very Dilute Real Solutions

We can develop an explicit expression for the entropy of mixing, ΔS_{mix}, in terms of elementary statistical mechanics. The entropy of a system at constant energy and volume is

$$\mathbf{S} = \mathbf{k} \ln \Omega \qquad (3.7.1)$$

where $\mathbf{k}$ represents Boltzmann's constant and Ω stands for the number of states accessible to the system under the given constraints.

The simplest approach to the task of calculating the entropy change in mixing N_A molecules of A with N_B molecules of B in the liquid state is to adopt a so-called lattice model. Figure 3.16 shows a two-dimensional analog of this model. We divide the total volume V into $N = N_A + N_B$ cells and count the number of accessible states by determining how many ways we can distribute the molecules in the different cells. Each cell can contain either an A molecule or a B molecule, with a probability depen-

Figure 3.16
Two-dimensional lattice model used to calculate the entropy of mixing of two components possessing mole fractions $X_A = n_A/n \; (=N_A/N)$ and $X_B = n_B/n \; (=N_B/N)$. [The figure can also be used to represent the adsorption of solute atoms (the X's) onto a two-dimensional solid surface as described by the Langmuir isotherm, eq. 3.7.11.]

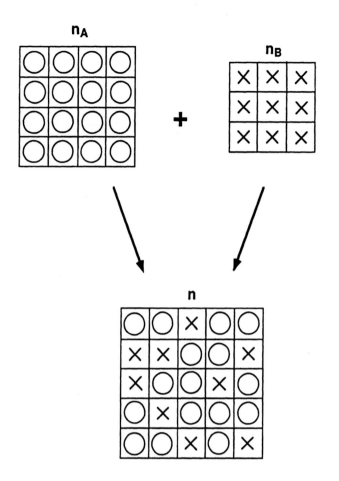

dent only on the mole fractions, $X_A \; (= n_A/n = N_A/N)$ or $X_B \; (= n_B/n = N_B/N)$. We assume A molecules have no preference for either their A or B neighbors, and vice versa.

Given these assumptions, calculating Ω becomes an exercise in statistics. We start by distributing the A molecules. There are

$$N! \, / \, (N_A)! \, (N - N_A)! \qquad (3.7.2)$$

ways to distribute N_A identical objects on N identical positions.

Then we consider the options for placing the B molecules. Because they are equivalent and only N_B positions are left, we have no choice except to put one B molecule on each of the empty positions, and there is only one way to do this. Thus we obtain the total number of ways of distributing A and B molecules by multiplying eq. 3.7.2 by 1 to give

$$\Omega = N! \, / \, (N_A)! \, (N - N_A)! = N! \, / \, (N_A)! \, (N_B)! \qquad (3.7.3)$$

Using eq. 3.7.1, the entropy in the mixed system becomes

$$\Delta S_{mix} = k \, (\ln N! - \ln N_A! - \ln N_B!) \qquad (3.7.4)$$

We can simplify this expression further by using the Stirling approximation, which states that for large N

$$\ln N! = N \, (\ln N - 1) \qquad (3.7.5)$$

Thus eq. 3.7.4 becomes

$$\Delta S_{mix} = - k \left(N_A \ln \frac{N_A}{N} + N_B \ln \frac{N_B}{N} \right)$$

and

$$\Delta S_{mix} = - k \, (N_A \ln X_A + N_B \ln X_B)$$

or

$$\Delta S_{mix} = - R \, (n_A \ln X_A + n_B \ln X_B) \qquad (3.7.6)$$

Equation 3.7.6 represents the desired expression for the ideal entropy of mixing of species A and B.

Equation 3.7.6 correctly describes the entropy of mixing of two ideal gases by virtue of the following argument. In the derivation, we used the constraint that only one molecule could be placed into each cell, with no correlations, and the constraint applied to both A and B molecules. For an ideal gas mixture, we can place any molecule, either A or B, anywhere within the volume V, irrespective of the location of other molecules. For this reason eq. 3.7.6 applies to an ideal gas mixture.

Equation 3.7.6 likewise correctly describes the entropy of mixing of an *ideal solution*. The ideal solution model assumes

1. $\Delta H_{mix} = 0$, which implies that all pairwise interaction energies are equal; that is, $W_{AA} \approx W_{BB} \approx W_{AB}$;

2. $\Delta V_{mix} = 0$, which implies that molecules A and B are the same size;

3. Given (1) and (2), then $\Delta G_{mix} = - T\Delta S_{mix}$, which states that the free energy associated with mixing derives entirely from the entropy of mixing.

Equation 3.7.6 does not rigorously describe the entropy of mixing of a real (nonideal) solution, because we made a number of assumptions in deriving it that do not apply to real solutions. However, eq. 3.7.6 often provides a useful approximation for the entropy of mixing of *very dilute real solutions*.

3.7.2 Based on an Ideal Mixing Model, the Langmuir Equation Describes Adsorption of Fluids at Solid Surfaces

Solubilization, detergency, and flotation (see Chapters 4 and 5) are among the many applications of interfacial products that involve adsorption of surfactant molecules from a fluid onto a solid surface. Because we cannot directly measure changes in the surface tension of solids that occur as a result of solute or impurity adsorption, as we can for fluids, we cannot use the Gibbs adsorption isotherm (eq. 3.5.13) to describe adsorption from a fluid onto a solid. Instead we use the Langmuir adsorption isotherm. This isotherm relates the concentration of adsorbed molecules (the *adsorbate*) on a solid surface (the *adsorbent*) as a function of their concentration in the surrounding fluid.

The Langmuir isotherm can be derived using a two-dimensional analog of the ideal mixing lattice model (Figure 3.16). The model consists of solvent and solute molecules adsorbed onto the solid surface, and its basic assumptions are that

1. Adsorbed solvent and solute molecules obey the ideal mixing model; that is, pairwise interactions are equal.

2. Adsorbed solvent molecules (1) and solute molecules (2) occupy equal surface areas.

Following arguments similar to those used to derive eq. 3.7.6, we derive an expression for the entropy at the surface

$$\Delta S^\sigma = -k \ [N_2 \ln X_2 + (N - N_2) \ln(1 - X_2)] \qquad (3.7.7)$$

where N_2 represents the number of adsorbed solute molecules, N is the total number of sites, and $X_2 = N_2/N = n_2/n$ is the concentration of the adsorbate at the surface. The free energy (G^σ) of the surface expressed in molar quantities is then

$$G^\sigma = G^{\sigma o} + n_2 \mu_2^{\sigma o} + RT \ [n_2 \ln X_2 + (n - n_2) \ln(1 - X_2)] \qquad (3.7.8)$$

where the superscript o signifies the properties of the standard state defined by an infinitely dilute solution. The chemical potential of the solute at the surface, μ_2^σ, is

$$\mu_2^\sigma = \left(\frac{dG^\sigma}{dn_2} \right)$$

and by differentiating eq. 3.7.8 we get

$$\mu_2^\sigma = \mu_2^{\sigma o} + RT \ln \left(\frac{X_2}{(1 - X_2)} \right) \qquad (3.7.9)$$

The chemical potential of the solute in the bulk solution

$$\mu_2 = \mu_2^o + RT \ln c_2 \qquad (3.7.10)$$

where c_2 stands for the molar concentration of solute in the bulk solution.

When the solution and the surface are in equilibrium, the two chemical potentials μ_2^σ and μ_2 must be equal. Equating the right-hand side of eqs. 3.7.9 and 3.7.10 and rearranging yields

$$X_2 = \Theta = \frac{K_{ads} \ c}{K_{ads} \ c + 1} \qquad (3.7.11)$$

where $K_{ads} = \exp[(\mu_2^{\sigma o} - \mu_2^o/RT]$. K_{ads} is the *equilibrium constant for the adsorption reaction*, and the term $(\mu_2^{\sigma o} - \mu_2^o)$ measures the relative affinity of the solute for the surface as compared to the bulk. Again we have dropped the subscript so that c stands for the solute concentration.

Equation 3.7.11 is the *Langmuir adsorption isotherm*. It contains two parameters, Θ and c, that can be measured experimentally, and the third parameter, K_{ads}, can be deduced from them. $\Theta = X_2 = N_2/N$ is the fraction of the surface sites occupied by the adsorbed solute molecules (also known as the *fractional surface coverage*) when the concentration of solute in the bulk solution is c. (It is physically similar to the fraction of X sites in the lower diagram of Figure 3.16.) Experimentally Θ is measured as a function of c to determine K_{ads}. At high solute concentrations (c large) or high surface affinities (K_{ads} large), $K_{ads}c \gg 1$ and $\Theta \approx 1$, and the surface is saturated. At the other extreme, $K_{ads}c \ll 1$, and Θ becomes proportional to c, and the constant of proportionality is K_{ads}. Thus the characteristic features of the Langmuir adsorption isotherm, shown in Figure 3.17, are a surface concentration that initially increases linearly with solute concentration and gradually switches to an asymptotic approach to monolayer saturation ($\Theta = 1$).

(The Langmuir adsorption isotherm also describes the adsorption of the first monolayer of molecules from a vapor onto a solid surface. As we shall find in Section 10.6.2.2, the solute concentration c is then replaced by the vapor pressure P of the adsorbate.)

Rather stringent assumptions are used to derive the Langmuir equation. For example, ideal behavior on the surface implies a homogeneous surface that lacks surface steps, dislocations, or any other structural nonidealities that might induce preferred adsorption. Interactions between adsorbed molecules are ignored, and *the equation is only applicable to the formation of the first monolayer*. Although these ideal conditions are almost never realized in practice, the Langmuir relationship nevertheless provides a conceptual basis for thinking about adsorption on solid surfaces as well as a basis for modeling the adsorption process. It provides a convenient form for plotting data and to identify deviations from ideality.

Figure 3.17
Fractional coverage of a surface Θ by adsorbate as a function of its molar concentration, c, in solution predicted by the Langmuir adsorption equation (3.7.11). The initial slope gives the equilibrium constant for adsorption, K_{ads}.

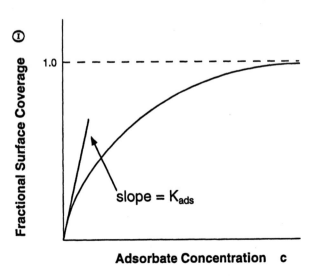

3.8 Transport Processes Play a Fundamental Role in Interfacial Process Engineering

Many interfacial processes involve manipulation of fluids. This section reviews the fundamental relationships describing transport phenomena in fluids. These relationships are based on equations of continuity of mass and motion included in Appendix 3.A. Here we focus primarily on fluid flow, sedimentation, and diffusion. Other aspects of transport properties will be developed as we need them throughout the rest of the book. References cited at the end of this chapter provide more extended discussions.

As noted at the outset, in principle we can distinguish a fluid from a solid by its response to an applied force or stress (force per unit area). A purely elastic solid deforms by an amount proportional to the applied stress, and the deformation is reversible. In contrast, a pure fluid deforms continuously under the influence of applied stress, and the velocity at which it flows increases with increasing stress. Although this distinction between fluids and solids is generally useful, many materials encountered in interfacial processes display complex responses to stress, behavior that is intermediate between that of pure elasticity and pure fluidity, and we will account for these complications both in this section and more specifically for polymers in Section 6.6.4.

3.8.1 Viscosity and Fluid Flow Provide Understanding Necessary for Analyzing Many Interfacial Processes

3.8.1.1 In Newtonian Fluids Shear Rate Is Proportional to Shear Stress—the Proportionality Constant Is the Coefficient of Viscosity

Fluid viscosity is defined most conveniently in terms of the apparatus shown in Figure 3.18, in which the fluid is sandwiched between two parallel plates of area A separated by a distance y. If we exert a shear stress (τ_{yx} = F/A) on the top plate, it moves with a velocity v relative to the velocity of the bottom plate (which for convenience we can set equal to zero). We assume the "stick" or "no slip" boundary condition, in which the fluid immediately adjacent to each plate moves with the same velocity as the plate. Newton's law of viscosity states that the stress, F/A, relates to the velocity gradient, dv/dy, by

$$\frac{F}{A} = \eta \frac{dv}{dy} \qquad (3.8.1)$$

where the proportionality constant η is defined as the *coefficient of viscosity*. Viscosity dimensions are mass $\times$ length^{-1} $\times$ time^{-1}, and the practical units are poise (p) measured in grams per centimeter per second, or centipoise, cp = 10^{-2}p. For liquids at room temperature, viscosity values range from 10^{-2} poise for water, to 10^{3} for molasses, to 10^{10} for pitch.

Figure 3.18
Schematic used to define Newton's law of viscosity (eq. 3.8.1) in which a liquid is contained between two plates of area A separated by a distance y. A shear stress, F/A, exerted on the top plate generates a velocity gradient or shear rate dv/dy, and the proportionality constant between these two quantities is the viscosity η.

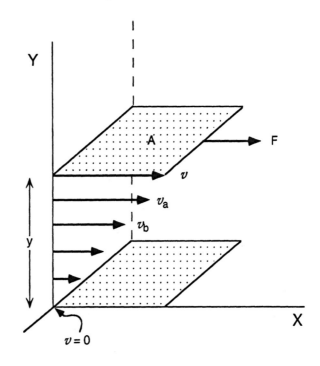

Shear strain is defined by $\gamma_{yx} = dx/dy$; so we can write the velocity gradient, dv/dy, as $d\gamma_{yx}/dt$, that is, the shear rate. Equation 3.8.1 can be rewritten as $F/A = \eta \, d\gamma/dt$, a linear relationship between shear stress and shear rate with a slope equal to η and a zero intercept. As shown in Figure 3.19, fluids that obey this linear relationship are called *Newtonian fluids*. Non-Newtonian fluids yield nonlinear plots or give nonzero intercepts, and Figure 3.19 displays several examples. For the time being we post-

Figure 3.19
Comparison between Newtonian viscous flow and several forms of non-Newtonian behavior.

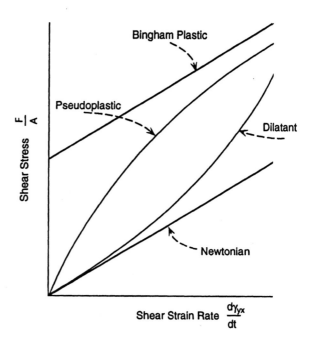

pone a discussion of this more complex non-Newtonian behavior to Section 3.8.3.2 and focus on Newtonian fluids.

By rearranging the units for the shear stress,

$$\tau_{yx} = \frac{F}{A} = \frac{momentum}{area \times time} \tag{3.8.2}$$

we see that shear stress can be related to the transfer of momentum per unit area per unit time, that is, to momentum flux. We can interpret this result by considering the interaction between two adjacent layers of the fluid (Figure 3.18) that have different average velocities $v_a > v_b$ and hence possess different momentum in the x direction. Brownian motion creates an exchange of molecules between these layers. When molecules move from faster-moving layers to slower ones, they collide with the slower-moving molecules and tend to speed them up so that their average momentum increases in the x direction. Molecules transported from b to a have the opposite effect. This exchange of molecules between layers produces a transfer of momentum from high-velocity to low-velocity layers, that is, a flux of x-directed momentum in the y direction. The coefficient of viscosity expresses the momentum flux per unit velocity gradient.

A further extension of these ideas links the coefficient of viscosity to energy dissipated per unit volume per velocity gradient squared. This relationship is important because it indicates that the viscosity of a fluid system determines the rate at which energy is dissipated. Indeed the rate at which oscillations are damped by a fluid contained between two concentric cylinders is one method used to characterize the coefficient of viscosity of a fluid.

3.8.1.2 The Hagen-Poiseuille Equation Describes Laminar Flow through an Open Circular Pipe

A fluid system's flow behavior under the influence of an external pressure depends on the fluid velocity. At low velocities, adjacent layers slide past one another, and fluids flow without macroscopic lateral mixing, as depicted in Figure 3.18. This flow regime is called *laminar flow*. At higher velocities, eddies spontaneously form, leading to lateral mixing and *turbulent flow*.

For the laminar flow regime, we can solve conservation equations for matter and momentum to obtain expressions for the fluid velocity profiles, average velocities, and volumetric flow rates. This is done in Appendix 3.A.

Figure 3.20 illustrates the velocity profile for flow between the two parallel plates. The velocity of the different layers as a function of their distance from the stationary plates is given by

$$v_x = \frac{1}{2\eta} \frac{\Delta P}{\Delta x} (y^2 - y_o^2) \tag{3.8.3}$$

where $\Delta P/\Delta x$ is the pressure gradient and $2y_o$ is the separation between the plates. The maximum velocity, $v_{x\ max}$, occurs at the center where $y = 0$

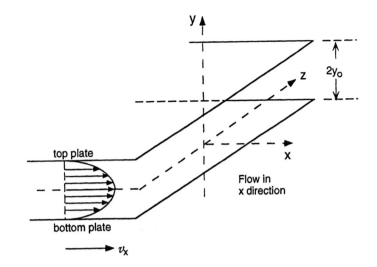

Figure 3.20
Coordinate systems and velocity profile for laminar flow of a liquid between two stationary parallel plates (eq. 3.8.3). (C. J. Geankoplis, *Transport Processes and Unit Operations*, © 1978, p. 118. Reprinted by permission of Prentice Hall, Englewood Cliffs, New Jersey.)

$$v_{x\,max} = \frac{1}{2\eta} \frac{\Delta P}{\Delta x} y_o^2 \qquad (3.8.4)$$

and thus

$$v_x = v_{x\,max}\left[1 - \left(\frac{y}{y_o}\right)^2\right] \qquad (3.8.5)$$

The velocity profile is parabolic. By connecting all points that have the same velocity, we can construct stream lines to provide a visual map of the fluid's flow. For laminar flow between parallel plates, the stream lines lie parallel to the surfaces of the plates.

Laminar flow of a fluid through a pipe (radius R, length L) due to a pressure drop ΔP displays characteristics similar to those of a parallel plate. Solving the equations of continuity and motion (Appendix 3.A) gives the following expression for the velocity profile of the flow,

$$v_x = \frac{1}{4\eta} \frac{\Delta P}{L} R^2\left[1 - \left(\frac{r}{R}\right)^2\right] \qquad (3.8.6)$$

which again is parabolic, as shown in Figure 3.21. We find the maximum value of the velocity at the center of the pipe where $r = 0$,

$$v_{x\,max} = \frac{\Delta P}{4\eta L} R^2 \qquad (3.8.7)$$

By summing the velocities over the entire cross section of the conduit and dividing by the cross-sectional area, we can obtain a value for the average velocity $v_{x\,av}$

$$v_{x\,av} = \frac{\Delta P}{8\eta L} R^2 \qquad (3.8.8)$$

which is the *Hagen-Poiseuille equation*. This equation relates pressure drop to the average velocity for laminar flow in a

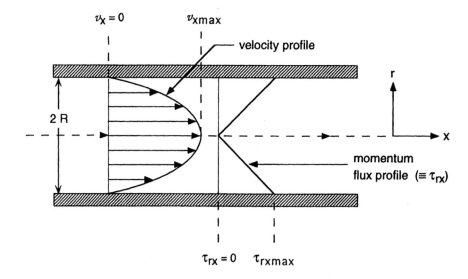

Figure 3.21
Velocity (v_x) and momentum flux ($\tau_{rx} \sim dv_x/dr$) profiles for laminar flow of a liquid in a conduit of radius R (eq. 3.8.6). (C. J. Geankoplis, *Transport Processes and Unit Operations,* © 1978, p. 71. Reprinted by permission of Prentice Hall, Englewood Cliffs, New Jersey.)

horizontal pipe. Comparing eqs. 3.8.7 and 3.8.8 shows that $v_{x\,av} = \frac{1}{2}v_{x\,max}$.

The volumetric flow rate, dV/dt, is given by

$$\frac{dV}{dt} = \pi R^2 v_{x\,av} = \frac{\pi}{8\eta L}\Delta P \, R^4 \qquad (3.8.9)$$

If the pipe is held vertically rather than horizontally, the volumetric flow rate is given by

$$\frac{dV}{dt} = \frac{\pi}{8}\frac{(\rho gL + \Delta P)}{\eta L}R^4 \qquad (3.8.10)$$

where the ρgL term accounts for the gravitational force. This relationship is important because it provides another basis for measuring viscosity, namely, the capillary viscometer illustrated in Figure 3.22a. Pressure is applied by the liquid head; so $\Delta P = \rho gh$ in eq. 3.8.10, and the time taken for a precalibrated volume of fluid to pass through the capillary of known radius and length L is used to determine the viscosity.

Figure 3.21 also shows the momentum flux ($\tau_{rx} \sim dv_x/dr$) profile for a circular pipe. The momentum flux varies linearly with the radius and attains a maximum at the wall.

3.8.1.3 The Reynolds Number Determines the Transition from Laminar Flow to Turbulent Flow

The transition from laminar flow to turbulent flow in a pipe is governed not only by the average velocity (v) but also by the density (ρ) and viscosity (η) of the fluid and the diameter D of the conduit. The *Reynolds number, N_{Re},* provides a guide to the transition between these two flow regimes. It is a dimensionless number defined by

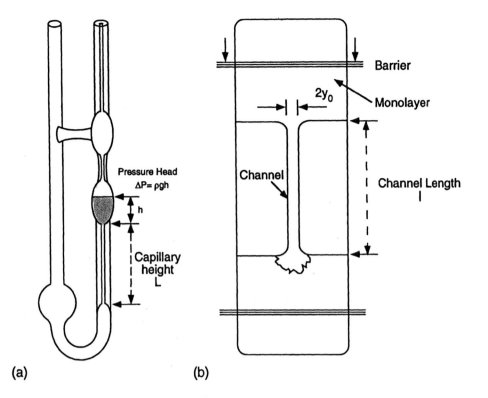

Figure 3.22
Schematics of (a) capillary viscometer and (b) a Langmuir balance trough containing two movable barriers and a channel for measuring surface viscosity. (P. C. Hiemenz, *Principles of Colloid and Surface Chemistry*, 2nd ed., Marcel Dekker, New York, 1986, pp. 182, 379.)

$$N_{Re} = \frac{Dv\rho}{\eta} \qquad (3.8.11)$$

and represents a ratio of kinetic or inertial forces, proportional to ρv^2, and viscous forces, proportional to $\eta v/D$. Laminar flow occurs when viscous forces > inertial forces, while turbulent flow occurs when inertial forces >> viscous forces. For a straight circular pipe we can observe laminar flow when $N_{Re} < 2100$. When $N_{Re} > 4000$, the flow will be turbulent. In the transition region that lies between these two values, the flow may be laminar or turbulent, depending on details of the pipe and apparatus such as surface roughness that cannot be predicted. The Reynolds number for the transition from laminar to turbulent flow also depends on the cross-sectional shape of the conduit.

3.8.1.4 The Bernoulli Equation Describes Flow Through a Conduit Due to an External Force

In many processes, we are interested in the flow of a fluid through a conduit caused by the application of an external force. Mechanical energy balance provides a way to analyze this situation. For an incompressible fluid of unit mass, we can write

$$\frac{1}{2\alpha}(v_{2av}^2 - v_{1av}^2) + g(h_2 - h_1) + \frac{P_2 - P_1}{\rho} + W_s + \sum E = 0 \qquad (3.8.12)$$

The first term in this equation gives the change in kinetic energy associated with changes of the fluid velocity from v_{1av} to v_{2av}. α represents the velocity correction factor and has a value of 0.5 for laminar flow and 1.0 for turbulent flow. These changes in fluid velocity arise from variations in the next three terms: the potential energy, $g \Delta h$, where Δh is the vertical drop; $\Delta P/\rho$, where ΔP is the pressure drop across the system and W_s, which represents shaft work, the mechanical energy added to the system (generally by a pump) or extracted from it by a turbine. The last term, ΣE, corresponds to the sum of all sources of energy dissipation due to viscosity and to losses in mechanical energy caused by frictional resistance to flow. It is converted to heat.

If we write eq. 3.8.12 without the W_s and ΣE terms, we obtain *Bernoulli's equation*. Bernoulli's equation describes the flow of an ideal fluid in which no shear forces operate.

3.8.2 Viscosities of Surface Films Can Be Described by a Two-Dimensional Form of Poisseuille's Equation and Measured Using a Modified Langmuir Balance

Concepts of viscosity in bulk fluids can be extended to the surface domain. We can define a surface viscosity η_s using a two-dimensional form of eq. 3.8.1

$$\frac{F}{l} = \eta_s \frac{dv}{dy} \tag{3.8.13}$$

where F/l represents the force per unit length along the edges of an element of surface area and dv/dy is the velocity gradient between the edges. The units of η_s are mass $\times$ time^{-1}.

Using the surface "capillary" viscometer illustrated in Figure 3.22b, we can measure the surface viscosity for monolayers. The capillary tube is replaced by a narrow channel of width $2y_0$ and length l mounted in a Langmuir trough. Moving the barrier pushes a monolayer through the channel. We can measure the "area rate" at which the monolayer emerges from the channel under a constant applied pressure Π controlled by keeping the torsion force on the barrier in Figure 3.13 constant.

By solving the conservation equations, we obtain a two-dimensional form of the Hagen-Poiseuille equation. The velocity of the monolayer varies across the channel from zero at the edges to a maximum at the center

$$v_{max} = \frac{\Pi}{2\eta_s l} y_0^2 \left[1 - \left(\frac{y^2}{y_0^2} \right) \right] \tag{3.8.14}$$

Integrating to find the average velocity gives

$$v_{av} = \frac{\Pi y_0^2}{3\eta_s l}$$

and the corresponding area flow rate is

$$\frac{dA}{dt} = \frac{2}{3} \frac{\Pi}{\eta_s l} y_0^3 \tag{3.8.15}$$

which should be compared with eqs. 3.8.8 and 3.8.9, respectively. Because surface viscosities can be studied as a function of monolayer concentration, we can measure surface viscosities for the "gaseous" state ($\eta_s \approx 10^{-5}$ g s^{-1}) and for the "condensed" state ($\eta_s \approx 10 - 10^{-2}$ g s^{-1}).

It is difficult to get a feeling for what these values of η_s mean, but by assigning a thickness d to the surface, we can compare η_s (mass time^{-1}) and η (mass time^{-1} length^{-1})

$$\eta = \eta_s/d \qquad (3.8.16)$$

where d represents the thickness of the interfacial region. If we assume that d $\approx$ 1.0 nm, a surface viscosity of 10^{-4} g s^{-1} equals a bulk viscosity of

$$\eta = \frac{10^{-4} \text{ g s}^{-1}}{10^{-7} \text{ cm}} = 10^3 \text{ g cm}^{-1}\text{sec}^{-1} = 10^3 \text{ poise}$$

This value is comparable to the viscosity of molasses.

3.8.3 Viscosity and Flow of Fluids Containing Particles Are Important for Many Interfacial Processes

3.8.3.1 The Einstein Equation Describes the Viscosity of a Solution Containing a Dilute Dispersion of Spheres

We now move to the transport of fluids containing particles or solutions containing macromolecules. Hydrodynamic interactions between particles and the fluid can dramatically increase viscosity. To gain insight into these interactions, we begin by modeling a fluid containing a dilute suspension of spheres.

When a particle of any shape is placed in the velocity gradient of a flowing fluid, two effects contribute to the increase in viscosity. First, if the particle is not rotating, as indicated in Figure 3.23b, the velocity profile must change because fluid layers on the top and bottom sides of the particle must have the same velocity as the particle itself. This change reduces the overall velocity gradient and results in an increase in the fluid's viscosity.

Second, if the particle is free to rotate, as indicated in Figure 3.23c, then rotation consumes energy intended to produce translational motion in the fluid, and constitutes a dissipative loss that also increases the viscosity. We can determine the sense of rotation by placing a coordinate system at the center of a particle located in the layer normally moving with velocity v_p and then subtracting v_p from each velocity vector intercepting the particle. This process measures the particle's motion relative to the fluid. An induced rotation, counterclockwise in this instance, due to the velocity gradient develops as shown in Figure 3.23c.

Estimating how these two effects combine to increase the viscosity requires the solution of complex hydrodynamic differential equations. However, by assuming that the particles are spheres, that the concentration of spheres is so dilute that they do not interact with one another, and that the small size of the spheres makes it possible to treat the fluid as a continuum,

Figure 3.23
The velocity gradient in a flowing stream (a) is perturbed by the presence of a particle (b). If we place coordinate systems at the center of the particle (c), we can see how the fluid's velocity gradient induces rotation in the particle. Rotation leads to additional dissipative loss and therefore to an increase in the viscosity of the fluid. (P. C. Hiemenz, *Principles of Colloid and Surface Chemistry*, 2nd ed., Marcel Dekker, New York, 1986, p. 188.)

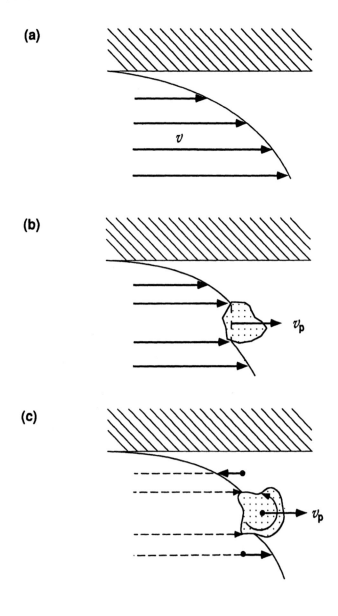

Einstein obtained a remarkably simple result. The viscosity increases from η_0, the viscosity of the pure liquid (the solvent), to η according to the relation

$$\frac{\eta}{\eta_0} = 1 + 2.5\phi \qquad (3.8.17)$$

where ϕ is the *volume fraction* of spheres. Note that eq. 3.8.17 depends only on the volume fraction of spheres, not on their size. Figure 3.24 shows a plot of the *relative viscosity*, η/η_0, for solutions containing spherical particles of different sizes that provides experimental verification of the Einstein viscosity equation at low volume fractions.

With increasing volume fraction of spheres we expect to see departures from eq. 3.8.17 when interactions between the spheres

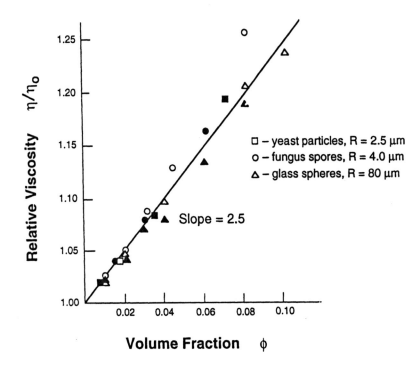

Figure 3.24
Plot of the relative viscosity η/η_o versus volume fraction ϕ for spherical particles of different sizes. The slope and intercept are independent of particle size, providing an experimental verification of the Einstein viscosity equation (3.8.17). (P. C. Hiemenz, *Principles of Colloid and Surface Chemistry,* 2nd ed., Marcel Dekker, New York, 1986, p. 192.)

become important. We can extend eq. 3.8.17 and introduce non-linear terms to represent the interactions, so that

$$\frac{\eta}{\eta_o} = 1 + 2.5\phi + C_1\phi^2 + C_2\phi^3 + \cdots$$

For convenience in analyzing data, we can rewrite this equation as

$$\left(\frac{\eta}{\eta_o} - 1\right)\bigg/\phi = 2.5 + C_1\phi + C_2\phi^2 + \cdots \tag{3.8.18}$$

where $(\eta/\eta_o - 1)/\phi$ is called the *reduced viscosity*. Figure 3.25 shows a plot of the reduced viscosity versus ϕ for glass spheres. The intercept agrees with the predicted value of 2.5, and the initial slope suggests a value of 10.0 for C_1. At high volume fractions, the reduced viscosity increases rapidly as higher-order interactions become important.

So far we have limited our attention to unhydrated spherical particles. If solvent molecules are attached to spherical particles such as spherical polymer molecules to form a solvation shell, then their effective volume fraction, $\phi_{eff} = \phi_{sphere} + \phi_{solvation}$, increases and so does η.

We also consider how particle shape affects η. Asymmetric particles are most readily characterized by ellipsoids defined by the two major axes of revolution, a and b. Large asymmetric particles will orient themselves in a flowing liquid so that their longest axis lies parallel to the flow direction. But if the particles are so small that they are swept through all orientations by Brownian motion, then η will be larger than it is for a spherical

Figure 3.25
Plot of the reduced viscosity, $(\eta/\eta_0 - 1)/\phi$, versus the volume fraction ϕ of glass spheres, in water, gives an intercept of 2.5, in agreement with the modified Einstein equation (3.8.19). (P. C. Hiemenz, *Principles of Colloid and Surface Chemistry,* 2nd ed., Marcel Dekker, New York, 1986, p. 195.)

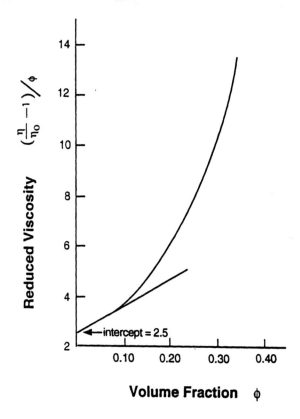

particle of the same mass. Figure 3.26 shows how the *intrinsic viscosity,* [η], defined as

$$[\eta] = \frac{(\eta/\eta_0 - 1)}{c} \tag{3.8.19}$$

where c is the *molar* concentration of solute, depends on both solvation and particle asymmetry for an ideal protein–water solution.

3.8.3.2 Many Fluids Containing Colloidal Particles or Polymers Display Non-Newtonian Behavior

As we noted in connection with Figure 3.19, many fluids do not show a linear relationship between shear rate and shear stress. They are known collectively as non-Newtonian fluids. Non-Newtonian fluids fall into two broad categories based on their shear stress/shear rate behavior: those whose shear stress is independent of the duration of shear (time independent) and those whose shear stress depends on the duration of shear (time dependent).

Most non-Newtonian fluids exhibit time-independent flow properties. They are further divided into three classes, as indicated in Figure 3.19.

1. *Bingham fluids* differ from Newtonian fluids only in that they require a finite shear stress to initiate flow. Once flow begins, they exhibit Newtonian behavior. Examples of Bingham plastic fluids include drilling muds, greases, soaps, toothpaste, paper pulp, and sewage sludge.

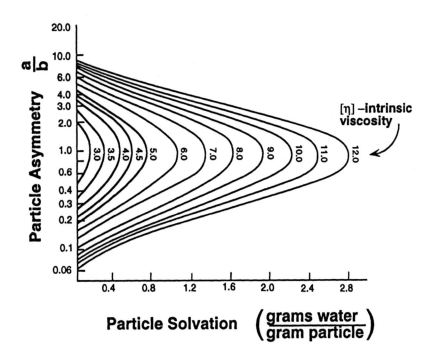

[η] –intrinsic viscosity

Particle Asymmetry $\frac{a}{b}$

Particle Solvation $\left(\dfrac{\text{grams water}}{\text{gram particle}}\right)$

Figure 3.26
Variation of the intrinsic viscosity, $[\eta] = (\eta/\eta_o - 1)/c$, with the extent of particle solvation (grams of water/grams of particle) and particle asymmetry (axial ratio a/b for an ellipsoid) for an ideal protein–water solution. [L. Oncley, *Annals of the New York Academy of Sciences* **41**, 121 (1941).]

2. *Shear-thinning fluids* (or pseudoplastic fluids) exhibit a viscosity that decreases with increasing shear rate. The shape of the flow curve can be represented by a power law equation

$$\frac{F}{A} = \kappa\left(-\frac{dv}{dy}\right)^n, \quad n < 1 \tag{3.8.20}$$

where κ stands for the *consistency index* and n the *flow behavior index*. Most non-Newtonian fluids fall into this category, and they include polymer solutions or melts, paints, starch suspensions, and detergent slurries.

3. *Shear-thickening fluids* (or dilatant fluids) exhibit an apparent viscosity that increases with shear rate. The same power law equation, 3.8.20, often applies to them, but with $n > 1$. Shear-thickening fluids are far less common than shear-thinning fluids. Examples include cornflour—sugar solutions, starch in water, wet beach sand, and some solutions containing a high concentration of powder in water.

Time-dependent fluids are further divided into two classes:

4. *Thixotropic fluids* exhibit a decrease in viscosity under shear. The shear stress decreases asymptotically with time until it reaches a limiting value that depends on the shear rate. The higher the shear rate, the lower the ultimate shear stress (within limits). Even more important, this property is reversible. When shearing stops, the viscosity recovers and returns to its original value. As described in Section 4.7, paints are designed to display this behavior because they must flow when brushed on, yet recover a high enough

viscosity so that they do not flow down vertical surfaces when brushing stops.

5. *Rheopectic fluids* exhibit an increase in viscosity under shear. The shear stress increases with time at a constant rate of shear. Again, the phenomenon is reversible. Such fluids are rare, but colloidal suspensions containing bentonite clay or gypsum are examples.

We have developed an empirical basis for classifying non-Newtonian fluids and provided examples of such systems based on their response to an applied shear stresss. What molecular interactions differentiate Newtonian from non-Newtonian fluids? To answer this question we can refer to the statement in Section 3.8.1.1 that viscosity involves the transfer of momentum from one plane of molecules to an adjacent plane through Brownian motion. This momentum exchange is a relaxation process, and we can associate a characteristic *relaxation time* τ, with a molecule's movement from one configuration to another. Fluids containing simple molecules display Newtonian behavior across a wide range of shear rates because their relaxation times are rapid. Solutions containing larger molecules or particles have longer relaxation times and exhibit non-Newtonian behavior. Amorphous solids like glass have extremely long relaxation times, and they generally show an elastic response to an applied stress rather than viscous flow. Glass windows in medieval cathedrals are thicker at the bottom than at the top, an effect that has taken centuries to produce because the relaxation times of glass are so very long.

3.8.4 Sedimentation Rate of Particles Depends on Particle Size and Solution Viscosity

An important phenomenon affecting our ability to manipulate and transport fluids containing particles or polymer molecules is their tendency to separate with time via the process of *sedimentation*. Sedimentation rate is determined by the velocity of particles moving through a liquid solution under an external force, such as gravity. Our goal is to ascertain the dependence of particle velocity on size and shape, on external forces, and on the solution viscosity.

As a specific example, we consider sedimentation resulting from gravity acting on a spherical particle of volume V and density ρ_2, immersed in a fluid of density ρ_1, as shown in Figure 3.27. The particle experiences a force F_g due to gravity in the downward direction (taken as positive) and a buoyancy force F_b acting in the opposite direction. The net force is

$$F_{net} = F_g - F_b = V(\rho_2 - \rho_1)g \qquad (3.8.21)$$

We note that when $\rho_2 > \rho_1$, then F_{net} and F_g have the same sign and the particle settles or "sediments" to the bottom (metal balls settle to the bottom of a water bath or paint pigments settle to the bottom of a paint can). When $\rho_2 < \rho_1$, the converse is true and the particle rises, a process called "creaming" (cream rises to the top of a bottle of milk).

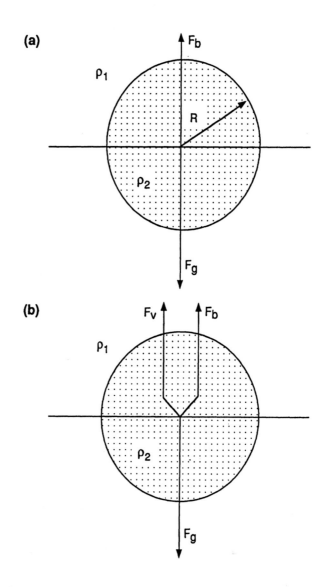

Figure 3.27
Forces acting on a falling spherical particle density ρ_2 submerged in a fluid of density ρ_1 where $\rho_2 > \rho_1$. (a) Forces due solely to gravity, F_g, and buoyancy, F_b; (b) with the addition of viscous forces, F_v. (P. C. Hiemenz, *Principles of Colloid and Surface Chemistry*, 2nd ed., Marcel Dekker, New York, 1986, p. 61.)

The net force causes the particle to accelerate. As the particle's velocity increases, the viscous force F_v opposing its motion also increases—rapidly reaching a steady state for which $F_g - F_b = F_v$, the particle experiences no net force and moves with a steady state or terminal velocity v_s. For a low terminal velocity we can write

$$F_v = f\, v_s \tag{3.8.22}$$

That is, the viscous force of resistance and the terminal velocity are related by a proportionality constant f, called the *friction factor*.

In 1850, G. G. Stokes solved the continuity equations and obtained expressions for the fluid velocity profile, or streamlines, around a sphere and the friction factor for laminar flow conditions (see Appendix 3.A.4). Laminar flow occurs for a sphere when the Reynolds number $2R\rho v/\eta < 1$, where R equals the particle's radius.

Figure 3.28 shows a schematic of the streamlines resulting from Stokes' solution. It illustrates how fluid velocity varies as a function of distance r from the sphere surface located at R. The important quantity is the relative motion between the sphere and the fluid; so we can interpret Figure 3.28 either in terms of a sphere moving downward in a quiescent fluid or in terms of a fluid flowing upward around a stationary sphere. The latter situation is somewhat easier to analyze.

When r = ∞, the disturbing influence of the particle has been damped away and the streamlines behave as if the particle were not present. At the particle's surface, the velocity is zero. Clearly, a velocity gradient exists in the vicinity of the particle that is a complex function of r and θ and results in a flux of momentum and a corresponding viscous drag. Stokes showed that in this situation, the friction factor defined in eq. 3.8.22 is given by

$$f = 6\pi \eta R \tag{3.8.23}$$

where R is the sphere's radius.

Substituting this result for f in eq. 3.8.22 and equating F_v to F_{net} gives the terminal sedimentation velocity

$$v_{sedimentation} = \frac{2}{9} \frac{R^2(\rho_2 - \rho_1) g}{\eta} \tag{3.8.24}$$

3.8.5 Material Transport in Many Interfacial Processes Is Diffusion Controlled

3.8.5.1 Self-Diffusion of Particles in Fluids or Molecules in Solution Occurs by Brownian Motion

In 1827, Robert Brown studied pollen grains under a light microscope and observed that the particles displayed continuous random motion (Brownian motion). In 1888, G. Gouy suggested that such particles were propelled by collisions with the rapidly moving molecules of the suspension liquid. In terms of modern concepts, we now understand that although a liquid is totally homogeneous on a macroscopic scale, it undergoes continuous fluctuations at a molecular level. While the mean density of the

Figure 3.28
Distortion of flow or streamlines around a spherical particle. Streamlines are constructed by connecting points with identical velocities.
(P. C. Hiemenz, *Polymer Chemistry: The Basic Concepts,* Marcel Dekker, New York, 1984, p. 586.)

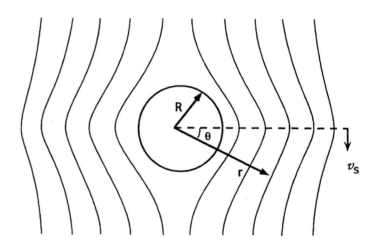

sample retains its bulk value, the density of molecules at any location varies with time. Fluctuating molecular densities result in randomly varying pressures, which in turn can cause unbalanced forces on a particle's surface. As a consequence, particles move in a random manner through the solution. Such motions result in self-diffusion of the particles in the fluid.

About 1905 Einstein derived a relationship between the distance a particle moves as a result of Brownian motion and its diffusion coefficient D. He employed a model that treated Brownian motion as a random walk process and obtained

$$x^2 = 2 D t \qquad (3.8.25)$$

where x equals the average displacement of the particle during a period t.

3.8.5.2 Interdiffusion or Mutual Diffusion between Two Fluids of Different Concentration Is Described By Fick's Laws

Our discussion of Brownian motion provides a way to understand how differences in concentration lead to a flux of matter across interfaces at a molecular level. Consider the situation at t = 0, as depicted in Figure 3.29a where two fluids of solute concentrations c_1 and c_2 are brought together at the plane x = 0. (Note that subscripts 1 and 2 now distinguish between two different concentrations of the same solute.) Because Brownian motion has no preferred direction, the number of molecules crossing the plane will be proportional to the concentration. Assuming $c_1 > c_2$, the net result will be that more solute will move from c_1 to c_2 in a manner proportional to the concentration difference.

For many interfacial processes, we need to know how rapidly molecules move across the interface under the influence of a concentration gradient. Fick's laws are phenomenological equations that provide the natural starting point for this discussion.

Fick's first law

$$J = -D\frac{dc}{dx} \qquad (3.8.26)$$

relates the flux J of a material to the change of concentration with distance.

This law applies most readily to a steady-state diffusion process in which the concentration gradient is constant and the composition also remains constant at all positions in the system. Although rarely true in a rigorous sense, the approximation is useful. The rate of coagulation of colloids discussed in Section 4.5.2 is calculated using Fick's first law.

Fick's second law

$$\frac{dc}{dt} = \frac{dJ}{dx} = \frac{d}{dx}\left(-D\frac{dc}{dx}\right) \approx -D\frac{d^2c}{dx^2} \qquad (3.8.27)$$

relates the change in concentration with time and position.

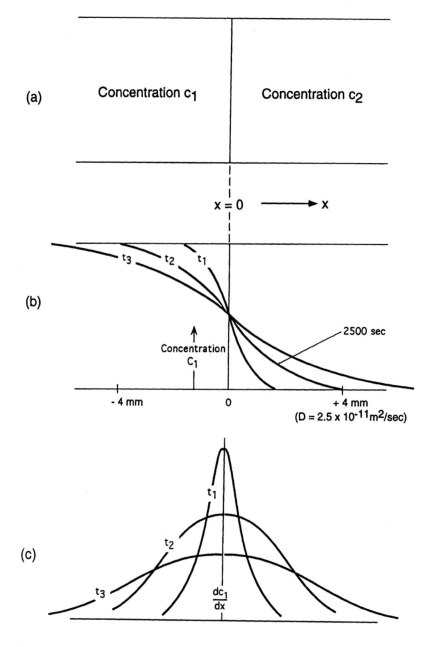

Figure 3.29
(a) Schematic of a diffusion cell with a sharp interface initially formed between two solutions with solute concentrations c_1 and c_2. (b) The change in concentration with time and location away from an interface. (c) The change in concentration gradient with time and location away from an interface. The concentration gradient forms a Gaussian curve that broadens with time and finally decays to a horizontal line as $t \to \infty$.

The diffusion cell shown in Figure 3.29 provides a helpful way to visualize this situation. We form a sharp boundary between two semi-infinite columns of liquid with initial concentrations c_1 and c_2 in Figure 3.29a. At time $t = 0$, we have $c = c_1$ for $0 > x > -\infty$ and $c = c_2$ for $0 < x < +\infty$; when $t > 0$, we have $c = c_1$ at $x = -\infty$, and $c = c_2$ at $x = +\infty$. Solving equation 3.8.27 using these boundary conditions gives

$$\frac{dc}{dx} = -\frac{c_1 - c_2}{2\sqrt{\pi Dt}} \exp\left(-\frac{x^2}{4Dt}\right) \qquad (3.8.28)$$

for the change in the concentration gradient dc/dx with time and location and

$$c = \frac{c_1 + c_2}{2} - \frac{c_1 - c_2}{2} \operatorname{erf}\left(\frac{x}{2\sqrt{Dt}}\right) \qquad (3.8.29)$$

for the change in composition with time and location. Note that the concentration at the original plane of demarcation, $x = 0$, remains fixed at $(c_1 + c_2)/2$. Figure 3.29b shows the concentration distribution after different times t_1, t_2, and t_3, where $t_3 > t_2 > t_1$. If the diffusion coefficient is 2.5×10^{-11} m^2/s, a value characteristic of a slowly diffusing fluid, then the t_2 curve corresponds to the situation after about 2500s or 0.7h. Figure 3.29c shows the corresponding concentration gradients at times t_1, t_2, and t_3.

3.8.5.3 Diffusion of Particles through Viscous Fluids or Molecules through Solutions Depends on Their Size and Shape

The diffusion of particles (or polymer molecules) through a solution depends on their geometry. We want to develop an explicit relationship for the diffusion coefficient in terms of particle size and shape. The analysis is similar to that used to derive the sedimentation velocity in Section 3.8.4 except that the force acting on the particle is now the negative gradient of the chemical potential rather than the gravitational field.

The force acting on the particles (or polymer molecules) due to the negative gradient of the chemical potential F_{diff} is

$$F_{\text{diff}} = -\frac{1}{N_{Av}} \frac{d\mu_i}{dx} \qquad (3.8.30)$$

We divide by Avogadro's number, N_{Av}, because the chemical potential is a molar quantity. The chemical potential of the particles i is related to their concentration by

$$\mu_i = \mu_i^o + RT \ln a_i \approx \mu_i^o + RT \ln c_i \qquad (3.8.31)$$

Substituting eq. 3.8.31 into 3.8.30 gives

$$F_{\text{diff}} = -kT \frac{d \ln c_i}{dx} = -\frac{kT}{c_i} \frac{dc_i}{dx} \qquad (3.8.32)$$

since $\mathbf{k} = \mathbf{R}/\mathbf{N}_{Av}$. Under steady state conditions, F_{diff} equals the viscous force of resistance $F_v = fv$ (eq. 3.8.22), and the diffusion velocity is

$$v_{diff} = -\frac{kT}{fc}\frac{dc}{dx} \tag{3.8.33}$$

where we drop the subscript because all quantities in the relationship refer to the particles.

The flux of material through a given cross section equals the product of its concentration and diffusion velocity

$$J = cv_{diff} \tag{3.8.34}$$

Substituting eq. 3.8.33 into eq. 3.8.34 and comparing with eq. 3.8.26 leads to

$$D = kT/f \tag{3.8.35}$$

which provides a simple relationship between the diffusion coefficient and the friction factor.[5]

In this derivation effects due to the geometry of the particle are contained in the friction factor. If we insert the Stokes expression for the friction factor for a spherical particle (eq. 3.8.23), we obtain

$$D = kT/6\pi\eta R \tag{3.8.36}$$

which provides a relation between the diffusion coefficient D, the radius of the spherical particle R, and the viscosity η of the solution.

A particle's friction factor, f_o, is the smallest when the particle is in the form of an unsolvated sphere. Solvation, asymmetry, or a combination of the two, results in values of $f > f_o$. A solvated spherical polymer molecule carries solvent as it diffuses through the solution, making the hydrodynamic radius, R_h, larger than the particle radius, R.

The friction coefficient of an asymmetric particle (or polymer molecule) depends on its orientation. At low velocity, such particles have random orientations. The value of f is smallest when an asymmetrical particle is end-on to the direction of flow, but it increases for side-on orientations. If we obtain a value of f by averaging over all orientations, we find that f is always larger than f_o. For particles of equal volume, f increases with increasing asymmetry. If we construct a contour diagram in which we plot f/f_o versus extent of hydration, we obtain a diagram very similar to Figure 3.26, in which f/f_o replaces [η].

---

[5]Note that the absolute mobility B of a molecule or particle is defined as its velocity per unit force—the inverse of the friction factor. When we substitute B = 1/f in eq. 3.8.35, we obtain B = D/kT. This is the well-known Nernst–Einstein relationship between the absolute mobility and the diffusion coefficient.

Bibliography

A. W. Adamson, *Physical Chemistry of Surfaces* (5th edition), New York: John Wiley & Sons, Inc., 1990.

R. B. Bird, W. E. Stewart and E. N. Lightfoot, *Transport Phenomena*, New York: John Wiley & Sons, Inc., 1960.

H. T. Davis, *Statistical Mechanics of Phases, Interfaces and Thin FIlms*, New York: VCH Publishers Inc., 1996.

C. J. Geankoplis, *Transport Processes and Unit Operations*, Boston: Allyn and Bacon Inc., 1978.

P. C. Heimenz, *Principles of Colloid and Surface Chemistry* (2nd edition), New York: Marcel Dekker, Inc., 1986.

Exercises

3.1 Repeat the procedure used in Section 3.1.1 to estimate the surface tension of water (ΔH_{vap}= 40.67 kJ/mole). Why does this estimate differ so greatly from the experimental value (72.8 mJ m^{-2}), while the value derived for CCl_4 in the text is so close to the experimental value?

3.2 At 20°C the surface tensions of water and n-octane are 72.8 and 21.8 mJ m^{-2}, respectively, and the interfacial tension of the n-octane–water interface is 50.8 mJ m^{-2}. Calculate (a) the work of cohesion for (i) n-octane and (ii) water; (b) the work of adhesion between n-octane and water; and (c) the initial spreading coefficient of n-octane on water.

3.3 For n-hexadecane and n-octanol the surface tensions are 30.0 and 27.5 mJ m^{-2}, and the interfacial tensions on water are 52.1 and 8.5 mJ m^{-2}, respectively. Comment on the long-term spreading behavior of the three hydrocarbons, n-octane (see Exercise 3.2), n-hexadecane, and n-octanol on water.

3.4 A highly viscous silicone oil has a surface tension of about 20 mJ m^{-2}. A small drop of water is placed on a film of the oil. The contact angle measured immediately after the water is placed on the film is 110 degrees. What is the interfacial tension between the water and the oil?

3.5 The pressure inside an air bubble in water is
equal to _ ; less than _ ; or greater than _ ;
the hydrostatic pressure inside a droplet of water in air when both have the same diameter. Why?

3.6 Two soap bubbles of radius R_1 and R_2 are joined by a common boundary. Derive the radius of curvature of this boundary.

3.7 The capillary rise between two parallel plates of separation d is
equal to _ ; less than _ ; or greater than _ ;

the capillary rise in a tube, diameter d, of the same material. Why?

3.8 A glass sphere, radius R, is placed on a flat glass plate and a drop of liquid, surface tension γ, spreads between them. Assuming that the liquid completely wets the glass and that the radius of the sphere is much larger than the thickness of the liquid, show that the capillary force of adhesion is $4\pi R\gamma$.

3.9 It is desired to use the capillary rise method to measure the surface tension of a large number of organic and aqueous liquids whose surface tension is anticipated to range from 15 to 70 mJ m^{-2}. The density of the liquids will range from 0.6 to 1.23 g cm^{-3}. What diameter should the capillary be so that the error in surface tension due to the spherical cap at the liquid–air interface within the capillary is no greater than 1%? Assume a 0° contact angle.

3.10 A conical glass tube is held vertically with its lower end, of diameter 0.5 cm, just touching the surface of water, surface tension 72.8 mJ m^{-2}. The tube is 20 cm long and the diameter at the upper end is 0.1 cm. If the water completely wets the glass, how high will it rise up the tube?

3.11 The pressure required to prevent liquid from entering a plug of finely divided solid is twice as great for a liquid of surface tension 50 mJ m^{-2}, which completely wets the solid, as it is for a liquid of surface tension 72.8 mJ m^{-2}, which has a finite contact angle with the solid. Calculate this contact angle.

3.12 A small spherical crystal of radius R in equilibrium with its liquid melts at a temperature
equal to _ ; less than _ ; or greater than _ ;
the melting temperature of a massive crystal.

3.13 Rewrite eq. 3.4.1 for the case of bubble nucleation in a liquid. What qualitative changes are there to the curve in Figure 3.9?

3.14 The windward sides of the Hawaiian Islands are characterized by frequent rains that are heavy but brief in duration. The skies of Southern California sometimes experience a heavy overcast for weeks at a time. Give a reason for this difference in climate.

3.15 The plot of surface tension versus the natural logarithm molar concentration for a surfactant in water at 25°C is linear with a slope of -7.25 mJ m^{-2}. Calculate the area per surfactant molecule at the surface in inverse square nanometers.

3.16 The surface tensions of solutions of sodium dodecyl sulfate (SDS) in 0.1 NaCl at 20°C are as follows:

SDS (mM)	2.0	1.0	0.5	0.2	0.1	0.04	0.02	0.01
γ (mJ m^{-2})	37.90	38.26	45.28	53.87	60.04	66.13	69.31	71.44

The surface tension of a 0.1 NaCl solution is 72.93 mJ m^{-2}
(a) Plot the data and use the Gibbs adsorption isotherm

equation to calculate the surface excess concentration of SDS, Γ_{SDS}, in molecules per 1000 Å^2 for each concentration. (b) Plot the spreading pressure, Π_s versus effective area per molecule, a_0, and plot the compressibility factor ($\Pi_s\, a_0/kT$) versus spreading pressure for this system. Comment on the deviation from ideality.

3.17 When 2.0 g of finely divided bone charcoal is immersed in 100 cm^3 of a $10^{-4}M$ dye solution and brought to equilibrium, the molar concentration of the dye solution drops to $0.4 \times 10^{-4}M$. When an additional 2.0 g is added, the concentration drops to $0.2 \times 10^{-4}M$. Calculate the specific surface area of the powder in square meters per gram, assuming the adsorption obeys the Langmuir isotherm. The effective surface area of the dye molecule is 65 Å^2.

3.18 The coefficient of viscosity of a solution containing spherical particles, average radius R_1, is
equal to _ ; less than _ ; or greater than _ ;
the coefficient of viscosity of a solution containing an equal number of spherical particles per unit volume of average radius R_2, where $R_2 > R_1$.

3.19 The sedimentation rate of a solution containing spherical particles, average radius R_1, is
equal to _ ; less than _ ; or greater than _ ;
the sedimentation rate of a solution containing an equal number of spherical particles per unit volume of average radius R_2, where $R_2 > R_1$.

3.20 Calculate the average displacement in 1 min along a given axis produced by Brownian motion of a spherical particle of radius 0.1 μm suspended in water at 25°C ($k = 1.3805 \times 10^{-23}$ J K^{-1}, coefficient of viscosity of water at 25°C $= 8.9 \times 10^{-4}$ Pa s). (Note 1 J = 1 Pa m^3; 1 Pa s = 10 poise.)

APPENDIX 3A
EQUATIONS OF
CONTINUITY AND MOTION

Our goal is to use the equations of continuity and motion to calculate velocity profiles and average velocities for fluid flow both between parallel plates and through a conduit and to analyze the equations leading to the Stokes friction factor. To attain that goal we will: (1) describe the physical basis for the equations of continuity and motion; (2) give the general results for steady-state incompressible flow; (3) cite the specific equations for rectangular and cylindrical coordinates; and (4) illustrate their use with specific examples.

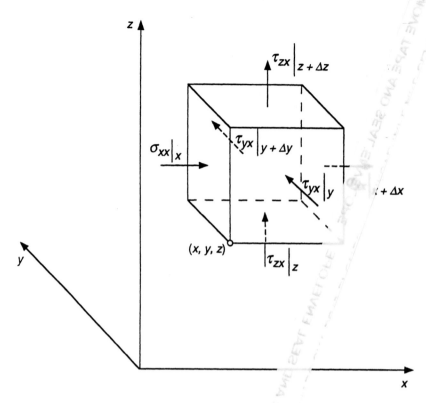

Figure 3A.1
Volume element $\Delta x\,\Delta y\,\Delta z$ with arrows indicating direction in which the x component of momentum is transported through the surfaces.

3A.1 Equation of Continuity

The chosen control volume is taken as a stationary cube of dimensions $\Delta x\,\Delta y\,\Delta z$ (Figure 3.A1). We write a mass balance for the control volume through which fluid is flowing as

$$\begin{bmatrix} \text{rate of mass} \\ \text{accumulation} \end{bmatrix} = \begin{bmatrix} \text{rate of mass} \\ \text{in} \end{bmatrix} - \begin{bmatrix} \text{rate of mass} \\ \text{out} \end{bmatrix} \qquad (3A.1)$$

The rate of mass accumulation within the control volume is $(\Delta x\,\Delta y\,\Delta z)(\partial\rho/\partial t)$, the rate of mass transport through the face at x is $(\rho v_x)|_x\,\Delta y\,\Delta z$, and the rate of mass out through the face at $x + \Delta x$ is $(\rho v_x)|_{x+\Delta x}\,\Delta y\,\Delta z$. Similar expressions can be written for the other two faces. The mass balance then becomes

$$\Delta x\,\Delta y\,\Delta z\,\frac{\partial\rho}{\partial t} = \Delta y\,\Delta z\,[(\rho v_x)|_x - (\rho v_x)|_{x+\Delta x}] + \Delta x\,\Delta z\,[(\rho v_y)|_y - (\rho v_y)|_{y+\Delta y}]$$

$$+ \Delta x\,\Delta y\,[(\rho v_z)|_z - (\rho v_z)|_{z+\Delta z}] \qquad (3A.2)$$

Dividing this equation by $(\Delta x\,\Delta y\,\Delta z)$ and taking the limit as the dimensions approach zero gives

$$\frac{\partial \rho}{\partial t} = -\left(\frac{\partial}{\partial x}\rho v_x + \frac{\partial}{\partial y}\rho v_y + \frac{\partial}{\partial z}\rho v_z\right) \qquad (3A.3)$$

which is the equation of continuity. Under steady-state conditions, $\partial\rho/\partial t = 0$, and with incompressible flow, ρ is constant; so we obtain

$$\left(\frac{\partial v_x}{\partial x} + \frac{\partial v_y}{\partial y} + \frac{\partial v_z}{\partial z}\right) = 0 \qquad (3A.4)$$

which we can write succinctly in vector notation as

$$\left(\mathbf{i}\frac{\partial}{\partial x} + \mathbf{j}\frac{\partial}{\partial y} + \mathbf{k}\frac{\partial}{\partial z}\right)(\mathbf{i}\,v_x + \mathbf{j}\,v_y + \mathbf{k}\,v_z) = (\nabla \cdot \mathbf{v}) = 0 \qquad (3A.5)$$

where $\mathbf{i}, \mathbf{j},$ and $\mathbf{k}$ are the unit vectors in the x, y, and z directions, $\mathbf{i} \cdot \mathbf{i} = \mathbf{j} \cdot \mathbf{j} = \mathbf{k} \cdot \mathbf{k} = 1$, and all the cross-dot products, such as $\mathbf{i} \cdot \mathbf{j}, = 0$.

3A.2 Equation of Motion for a Fixed Control Volume

For a stationary volume element we can also write an equation for momentum balance

$$\begin{bmatrix} \text{rate of} \\ \text{momentum} \\ \text{accumulation} \end{bmatrix} = \begin{bmatrix} \text{rate of} \\ \text{momentum} \\ \text{in} \end{bmatrix} - \begin{bmatrix} \text{rate of} \\ \text{momentum} \\ \text{out} \end{bmatrix} + \begin{bmatrix} \text{sum of} \\ \text{forces acting} \\ \text{on system} \end{bmatrix} \qquad (3A.6)$$

Momentum is transferred in and out of the control volume by two mechanisms: by convection, that is, by bulk fluid flow, and by molecular transport arising from velocity gradients, as described in Section 3.8.1.1.

To understand the form of eq. 3A.6 in more detail, we evaluate the individual terms. First, the term on the left-hand side of eq. 3A.6, the rate of momentum accumulation in the x-direction in the control volume is given by

$$\Delta x\, \Delta y\, \Delta z\, \frac{\partial \rho v_x}{\partial t} \qquad (3.A.7)$$

The first two terms on the right-hand side represent the momentum transfer by *convection* and are evaluated as follows. Consider the face normal to the y direction of the control volume shown in Figure 3A.1. Mass enters this face at a rate $\rho v_y\, \Delta x\, \Delta z$ and with an x direction velocity v_x. So the rate at which the x component of momentum enters the face at y by *convection* is $\Delta x\, \Delta z\, \rho v_y v_x|_y$, while the rate at which it leaves at y + Δy is $\Delta x\, \Delta z\, \rho v_y v_x|_{y+\Delta y}$. Similar terms apply at other faces of the control volume. By summing over all six faces, we obtain the rate of momentum gain due to convection in the x direction

$$\Delta y\, \Delta z\, [\rho v_x v_x|_x - \rho v_x v_x|_{x+\Delta x}] + \Delta x\, \Delta z\, [v_y v_x|_y - \rho v_y v_x|_{y+\Delta_y}]$$

$$+ \Delta x\, \Delta y\, [\rho v_z v_x|_z - \rho v_z v_x|_{z+\Delta z}] \qquad (3A.8)$$

The rate at which the x component of momentum enters the face at x by *molecular transport* is $\sigma_{xx}|_x \Delta y\, \Delta z$ and the rate that it leaves the face at $x + \Delta x$ is $\sigma_{xx}|_{x + \Delta x} \Delta y\, \Delta z$. σ_{xx} is the normal stress on the x face. Fluxes of x momentum through the faces perpendicular to the y and z axes are represented by the shear stress τ_{yx} and τ_{zx}, as discussed in Section 3.8.1.1.[1] The rate at which the x-directed momentum enters the face at y is $\tau_{yx}|_y \Delta x\, \Delta z$, and at the z face is $\tau_{zx}|_z \Delta x\, \Delta y$. By summing over all six faces, we obtain the rate of momentum gain due to viscous material transport in the x direction

$$\Delta y\, \Delta z\, [\sigma_{xx}|_x - \sigma_{xx}|_{x + \Delta x}] + \Delta x\, \Delta z\, [\tau_{yx}|_y - \tau_{yx}|_{y + \Delta y}] +$$

$$\Delta x\, \Delta y\, [\tau_{zx}|_z - \tau_{zx}|_{z + \Delta z}] \tag{3A.9}$$

With respect to the third term on the right-hand side of eq. 3A.6, pressure and gravity are the chief forces acting on the control volume, and their contributions to the rate of increase of momentum per unit volume in the x direction are

$$\Delta y\, \Delta z\, (P|_x - P|_{x + \Delta x}) + \Delta x\, \Delta y\, \Delta z\, \rho g_x \tag{3A.10}$$

Substituting eqs. 3A.7–3A.10 into eq. 3A.6, dividing the resulting equation by $\Delta x\, \Delta y\, \Delta z$, and taking the limit as Δx, Δy, and Δz approach zero gives

$$\frac{\partial}{\partial t} \rho v_x = -\left(\frac{\partial}{\partial x} \rho v_x v_x + \frac{\partial}{\partial y} \rho v_y v_x + \frac{\partial}{\partial z} \rho v_z v_x \right) - \left(\frac{\partial}{\partial x} \sigma_{xx} + \frac{\partial}{\partial y} \tau_{yx} + \frac{\partial}{\partial z} \tau_{zx} \right)$$

$$- \left(\frac{\partial P}{\partial x} \right) + \rho g_x \tag{3A.11}$$

which is the x component of the equation of motion. In this equation, $\sigma_{xx} = -2\eta(\partial v_x/\partial x) + 2/3\, \eta(\nabla \bullet v)$, $\tau_{yx} = -\eta(\partial v_x/\partial y)$, and $\tau_{zx} = -\eta(\partial v_x/\partial z)$. A similar analysis for the y and z coordinates yields equations of motion in the y and z directions.

Summing the x, y, and z components of the equation of motion for a stationary volume element and using vector notation, we can write

$$\frac{\partial}{\partial t} \rho v = -(\nabla \bullet \rho v\, v) - (\nabla \bullet \tau) - (\nabla P) + \rho g \tag{3A.12}$$

We illustrate the use of these equations by calculating the velocity profile for laminar fluid flow between two parallel plates as in Figure 3.20. We proceed by a three-step process:

1. Postulate by symmetry or intuition the dependence of the velocity upon the variables. Figure 3.20 suggests that $v_x = f(y)$, $v_y = 0$, and $v_z = 0$.

2. Substitute the postulates into the equation of continuity, eq. 3A.3. In steady state, $\partial \rho/\partial t = 0$, and, by our postulates, the other three terms equal zero. In this case, we obtain no useful information from the equation of continuity.

---

[1] *In these equations τ_{yx} represents the shear stress due to a force in the x direction acting over an area normal to the y axis.*

3. Substitute the postulates into the equations of motion, eq. 3A.11, and integrate the resulting equations. All terms involving v_y and v_z and $\partial v_x/\partial x$ and $\partial v_x/\partial z$ are zero; and because the flow is in the horizontal direction, no gravity forces affect the flow. As a result

for the y component, $\qquad -\dfrac{\partial P}{\partial y} = 0, \qquad$ P is independent of y

for the z component, $\qquad -\dfrac{\partial P}{\partial z} = 0, \qquad$ P is independent of z

and for the x component, $\eta\dfrac{\partial^2 v_x}{\partial y^2} = -\dfrac{\partial P}{\partial x}$ $\qquad$ (3A.13)

Because $P = f(x)$ and v_x is not a function of x, $\partial P/\partial x$ must be equal to a constant. As a consequence, we can write eq. 3A.13 using derivatives rather than partial derivatives and

$$\eta\frac{d^2 v_x}{dy^2} = \frac{dP}{dx} = A_1 \qquad (3A.14)$$

where the pressure drop along the plate, length Δx, is ΔP, and $\Delta P/\Delta x = A_1$. Integrating eq. 3A.14 once and using the condition $dv_x/dy = 0$ at $y = 0$ by virtue of symmetry gives

$$\frac{dv_x}{dy} = \frac{A_1 y}{\eta} \qquad (3A.15)$$

Integrating again and using the condition $v_x = 0$ at $y = y_0$ gives

$$v_x = \frac{\Delta P}{2\eta\,\Delta x}\,(y^2 - y_0^2) \qquad (3A.16)$$

which is eq. 3.8.3.

To determine flow through a conduit of circular cross section, we need to express the equations of continuity and motion in cylindrical coordinates. We substitute the relations $x = r\cos\theta$, $y = r\sin\theta$ and $z = z$ and obtain the results included in Table 3A.1. We illustrate the use of these equations by calculating the velocity profiles for a fluid flowing in a horizontal conduit of radius R and length L under steady-state conditions as in Figure 3.21. Following the three-step procedure outlined above, and neglecting gravity, we

1. Postulate $v_r = 0$ and $v_\theta = 0$ and that $v_z = f(r,z)$.

2. Substitute the postulates into the equation of continuity. In steady state, $\partial\rho/\partial t = 0$, and, by our postulates, $(\partial\rho v_r/\partial r) = 0$, and $(\partial\rho v_\theta/\partial\theta) = 0$; so from the equation of continuity we obtain $\partial\rho v_z/\partial z = 0$. This relation states that v_z is independent of z; so v_z must be a function solely of the radius r.

3. Substitute the postulates into the equation of motion. As a result

for the r component, $\qquad -\dfrac{\partial P}{\partial r} = 0, \qquad$ P is independent of r

for the θ component, $\qquad -\dfrac{1}{r}\dfrac{\partial P}{\partial\theta} = 0, \qquad$ P is independent of θ

TABLE 3A.1 Equations of Continuity and Motion (Cylindrical Coordinates)

Equation of Continuity

$$\frac{\partial \rho}{\partial t} + \frac{1}{r}\frac{\partial}{\partial r}(\rho r v_r) + \frac{1}{r}\frac{\partial}{\partial \theta}(\rho v_\theta) + \frac{\partial}{\partial z}(\rho v_z) = 0$$

Equations of Motion

r component

$$\rho\left(\frac{\partial v_r}{\partial t} + v_r\frac{\partial v_r}{\partial r} + \frac{v_\theta}{r}\frac{\partial v_r}{\partial \theta} - \frac{v_\theta^2}{r} + v_z\frac{\partial v_r}{\partial z}\right) = -\frac{\partial P}{\partial r}$$

$$+ \eta\left[\frac{\partial}{\partial r}\left(\frac{1}{r}\frac{\partial}{\partial r}(rv_r)\right) + \frac{1}{r^2}\frac{\partial^2 v_r}{\partial \theta^2} - \frac{2}{r^2}\frac{\partial v_\theta}{\partial \theta} + \frac{\partial^2 v_r}{\partial z^2}\right] + \rho g_r$$

θ component

$$\rho\left(\frac{\partial v_\theta}{\partial t} + v_r\frac{\partial v_\theta}{\partial r} + \frac{v_\theta}{r}\frac{\partial v_\theta}{\partial \theta} + \frac{v_r v_\theta}{r} + v_z\frac{\partial v_\theta}{\partial z}\right) = -\frac{1}{r}\frac{\partial P}{\partial \theta}$$

$$+ \eta\left[\frac{\partial}{\partial r}\left(\frac{1}{r}\frac{\partial}{\partial r}(rv_\theta)\right) + \frac{1}{r^2}\frac{\partial^2 v_\theta}{\partial \theta^2} + \frac{2}{r^2}\frac{\partial v_r}{\partial \theta} + \frac{\partial^2 v_\theta}{\partial z^2}\right] + \rho g_\theta$$

z component

$$\rho\left(\frac{\partial v_z}{\partial t} + v_r\frac{\partial v_z}{\partial r} + \frac{v_\theta}{r}\frac{\partial v_z}{\partial \theta} + v_z\frac{\partial v_z}{\partial z}\right) = -\frac{\partial P}{\partial z}$$

$$+ \eta\left[\frac{1}{r}\frac{\partial}{\partial r}\left(r\frac{\partial v_z}{\partial r}\right) + \frac{1}{r^2}\frac{\partial^2 v_z}{\partial \theta^2} + \frac{\partial^2 v_z}{\partial z^2}\right] + \rho g_z$$

R. B. Bird, W. E. Stewart, and E. N. Lightfoot, *Transport Phenomena*, © 1960, John Wiley & Sons, pp. 83, 85. Reprinted by permission of John Wiley & Sons, Inc.

and for the z component, $\dfrac{\partial P}{\partial z} = \eta\left[\dfrac{1}{r}\dfrac{\partial}{\partial r}\left(r\dfrac{\partial v_z}{\partial r}\right)\right]$ (3A.17)

Using the same reasoning that led to eq. 3A.14, we can write eq. 3A.17 as

$$\frac{\eta}{r}\frac{d}{dr}\left(r\frac{dv_z}{dr}\right) = \frac{dP}{dz} = A_1$$ (3A.18)

where the pressure drop along the pipe, length L, is ΔP and $A_1 = \Delta P/L$. Integration of eq. 3A.18 gives

$$r\frac{dv_z}{dr} = \frac{A_1}{2\eta}r^2 + A_2$$ (3A.19)

Using the boundary conditions $v_z = 0$ when $r = R$ and integrating eq. 3A.19 gives

$$v_z = \frac{A_1}{4\eta}(r^2 - R^2) + A_2\ln\frac{r}{R}$$ (3A.20)

where we can set $A_2 = 0$ because when $r = 0$, v must be finite. Thus we obtain eq. 3.8.6.

$$v_z = \frac{\Delta P}{4\eta L} R^2 \left[1 - \left(\frac{r}{R} \right)^2 \right]$$ (3A.21)

Maximum velocity occurs when $r = 0$ and $v_z = (\Delta P/4\eta L)R^2$ (eq. 3.8.7). Average velocity is obtained by summing all velocities over a cross section and dividing by the cross-sectional area

$$\langle v_z \rangle = \frac{\int_0^{2\pi} \int_0^R v_z\, r\, d\theta\, dr}{\int_0^{2\pi} \int_0^R r\, d\theta\, dr} = \frac{\Delta P\, R^2}{8\eta L}$$ (3A.22)

which is the Hagen–Poiseuille eq. 3.8.8.

The volumetric flow rate through the conduit is the product of cross-sectional area and average velocity

$$\frac{dV}{dt} = \frac{\pi \Delta P\, R^4}{8\eta l}$$ (3A.23)

which is eq. 3.8.9.

3A.3 Equation of Motion for a Moving Control Volume

So far we have confined our attention to systems in which the control volume remains fixed in space and we analyze fluid flowing through it. Sometimes it is more useful to have the control volume move along with the stream. If we wish to evaluate how some property of the system $c = f(x,y,z,t)$ changes as we move along with the fluid flow, we can write

$$\frac{Dc}{Dt} = \frac{\partial c}{\partial t} + v_x \frac{\partial c}{\partial x} + v_y \frac{\partial c}{\partial y} + v_z \frac{\partial c}{\partial z}$$ (3A.24)

where Dc/Dt is the substantial time derivative of c; v_x, v_y, and v_z are the components of the local fluid velocity v; and $\partial c/\partial t$ is the partial derivative of c with respect to time t, holding x, y, z constant. Equation 3A.11 becomes simply a statement of Newton's second law, mass times acceleration equals a sum of forces.

For an incompressible Newtonian fluid ρ and η are constant and the equation of motion, eq. 3A.12, becomes

$$\rho \frac{Dv}{Dt} \quad = \quad -\nabla P \quad + \quad \eta \nabla^2 v \quad + \quad \rho g$$ (3A.25)

$$\begin{bmatrix} \text{mass per unit} \\ \text{volume times} \\ \text{acceleration} \end{bmatrix} = \begin{bmatrix} \text{pressure} \\ \text{force on} \\ \text{element} \\ \text{per unit} \\ \text{volume} \end{bmatrix} + \begin{bmatrix} \text{viscous force} \\ \text{on element} \\ \text{per unit} \\ \text{volume} \end{bmatrix} + \begin{bmatrix} \text{gravitational} \\ \text{force on element} \\ \text{per unit volume} \end{bmatrix}$$

where ∇^2 is the operator $\nabla^2 = \partial^2/\partial x^2 + \partial^2/\partial y^2 + \partial^2/\partial z^2$. Equation 3A.25 is the *Navier–Stokes* equation.

When we can ignore the contribution of shear force compared to other forces, the substantial time derivative becomes

$$\rho \frac{Dv}{Dt} = -\nabla P + \rho g \qquad (3A.26)$$

which is the *Euler equation*. It is used for calculating streamlines in flow systems when viscous effects are relatively unimportant.

3A.4 Flow around a Sphere As an Example of Two-Dimensional Flow

The examples we have considered so far were chosen so that only one component of the velocity remained nonvanishing in the equation of motion. However, in many situations velocity calculations depend on two or more variables and involve solving partial differential equations. An example is flow around a sphere where the velocity changes with both distance and angle. In the discussion that follows, we begin with the solutions to the Navier–Stokes equation and show how we can use this information to obtain the Stokes friction factor eq. 3.8.23.

Figure 3A.2 shows a fluid with a uniform velocity, v_∞, approaching a sphere of radius R. The velocity components in spherical coordinates are

Figure 3A.2
Coordinate system for describing fluid flow around a sphere.

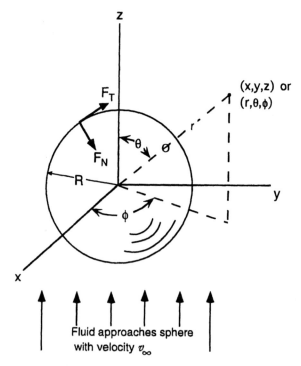

Fluid approaches sphere with velocity v_∞

$$v_r = v_\infty \left[1 - \frac{3}{2}\left(\frac{R}{r}\right) + \frac{1}{2}\left(\frac{R}{r}\right)^3 \right] \cos\theta \qquad (3A.27)$$

$$v_\theta = -v_\infty \left[1 - \frac{3}{4}\left(\frac{R}{r}\right) - \frac{1}{4}\left(\frac{R}{r}\right)^3 \right] \sin\theta \qquad (3A.28)$$

$$P = P_0 - \rho gz - \frac{3}{2}\frac{\eta v_\infty}{R}\left(\frac{R}{r}\right)^2 \cos\theta \qquad (3A.29)$$

$$\tau_{r\theta} = \frac{3}{2}\frac{\eta v_\infty}{R}\left(\frac{R}{r}\right)^4 \sin\theta \qquad (3A.30)$$

where P_0 stands for the pressure in a plane far removed from the sphere and $-\rho gz$ represents the contribution of the fluid weight.

These equations are valid only in the limit of what is called creeping flow, which corresponds to a Reynolds number of

$$N_{Re} = \frac{2R\rho v_\infty}{\eta} < 0.1 \qquad (3A.31)$$

where R equals the radius of the sphere and v_∞ is the unperturbed velocity. With increasing N_{Re}, eddies form downstream from the sphere and constitute another source of dissipative energy loss.

We can gain some insight into eqs. 3A.27–3A.30 by examining their behavior at various limits. At the surface of the sphere r = R and v_r and v_θ = 0 in accord with the "stick" condition. Far from the sphere's surface r >> R, v_z approaches v_∞ and the pressure approaches $P_0 - \rho gz$. We are assuming that gravity operates in the −z direction, and $P_0 - \rho gz$ gives only the change in hydrostatic pressure with height.

Now we can calculate the net force exerted by the fluid on the sphere by integrating the normal and tangential forces over the sphere's surface.

3A.4.1 Integration of the Normal Force

We can understand the origin of the normal force by considering the flow of a fluid in a vertical direction past an immersed object, as shown in Figure 3.28. Bernoulli's equation, eq. 3.8.12 without the W_s and ΣE terms, gives

$$\frac{1}{2\alpha}[(v_{2av})^2 - (v_{1av})^2] + \frac{P_2 - P_1}{\rho} = 0 \qquad (3A.32)$$

At the surface of the immersed object, $v_2 = 0$ and

$$\frac{P_2 - P_1}{\rho} = \frac{1}{2\alpha}(v_{1av})^2 \qquad (3A.33)$$

In other words, when the fluid velocity increases, kinetic energy is converted to an increase in pressure that exerts a force on the immersed object. This type of force is called the "form" drag.

At each point of the sphere's surface (Figure 3A.2), a force per unit area P is exerted perpendicular to the surface. The z component of this force is $-P\cos\theta$. By multiplying the local force per unit area by the surface area on which it acts, $R^2 \sin\theta\, d\theta\, d\phi$ and integrating over the surface of the sphere, we can obtain the net force on the sphere

$$F_N = \int_0^{2\pi} \int_0^{\pi} (-P|_{r=R} \cos \theta) R^2 \sin \theta \, d\theta \, d\phi \qquad (3A.34)$$

The pressure at the surface of the sphere is

$$P|_{r=R} = P_0 - \pi g R \cos \theta - \frac{3}{2} \frac{\eta v_\infty}{R} \cos \theta \qquad (3A.35)$$

By substituting eq. 3A.35 into eq. 3A.34 and integrating, we obtain

$$F_N = \frac{4}{3} \pi R^3 \rho g + 2\pi \eta R v_\infty \qquad (3A.36)$$

The integral involving P_0 vanishes identically, the integral involving $-\pi g R \cos \theta$ gives the buoyant force of the fluid on the solid, and the integral involving the velocity gives the "form" drag.

3A.4.2 Integration of the Tangential Force

Each point on the surface also experiences a tangential shear stress, $-\tau_{r\theta}$, which operates in the θ direction on the unit area of the sphere's surface. The z component of this force is $(-\tau_{r\theta})(-\sin \theta)$ and the surface area on which it acts is $R^2 \sin \theta \, d\theta \, d\phi$. We can write the resultant force in the z direction by

$$F_T = \int_0^{2\pi} \int_0^{\pi} (\tau_{r\theta}|_{r=R})(R^2 \sin \theta \, d\theta \, d\phi) \qquad (3A.37)$$

where

$$\tau_{r\theta}|_{r=R} = \frac{3}{2} \frac{\eta v_\infty}{R} \sin \theta \qquad (3A.38)$$

Substituting eq. 3A.38 into eq. 3A.37 and integrating gives

$$F_T = 4\pi \eta R v_\infty \qquad (3A.39)$$

which is called the "friction" drag.

The total force of the fluid on the sphere is the sum of the normal and the tangential force

$$F = \frac{4}{3} \pi R^3 \rho g + 2\pi \eta R v_\infty + 4\pi \eta R v_\infty \qquad (3A.40)$$

$$\begin{array}{ccc} \text{(bouyant} & \text{(form} & \text{(friction} \\ \text{force)} & \text{drag)} & \text{drag)} \end{array}$$

or

$$F = \frac{4}{3} \pi R^3 \rho g + 6\pi \eta R v_\infty \qquad (3A.41)$$

Inspection of eq. 3A.41 shows that the first term on its right-hand side represents the buoyant force exerted even when the fluid is stationary. The second term, $f = 6\pi \eta R v$, is the Stokes friction factor, eq. 3.8.23 for flow of fluid around a sphere.

4

COLLOIDS

with H. Wennerström

LIST OF SYMBOLS

c_i	local molar concentration of solute i in electrolyte solution
c_{io}	bulk molar concentration of solute i in electrolyte solution
c_i^*	local concentration of solute ion i in electrolyte solution ($c_i^* = 1000\, N_{Av}\, c_i$ ions)
CCC	critical coagulation concentration (eqs. 4.5.8 and 4.5.9)
D	colloidal particles diffusion coefficient (eq. 4.5.13)
F_{el}; F_{os}	electrostatic force (el), osmotic force (os) in liquid electrolyte (eq. 4.4.2)
F	Faraday constant
h	distance between two colloidal particles
H_{ij}	Hamaker constant (eq. 2.6.6)
J	flux of colloidal particles during coagulation (eq. 4.5.11)
k_r	rapid (r) coagulation rate constant (eq. 4.5.14)
k_s	slow (s) coagulation rate constant (Appendix 4.E, eq. 4E.7)
M	molar solution concentration
$[P_m]$	concentration of coagulated colloidal particles, m degree of association (eq. 4.5.15)
R	spherical particle radius
$V_{att,p}$	attractive interaction potential between two flat particles (eq. 4.1.1)
$V_{att,s}$	attractive interaction potential between two spherical particles (eq. 4.1.2)
$V_{rep,p}$	repulsive interaction potential between two flat particles (eq. 4.4.9)
$V_{rep,s}$	repulsive interaction potential between two spherical particles (eq. 4.4.10)
W	stability ratio (=k_r/k_s) (eq. 4.5.19)
X_i	mole fraction of i
z	valence; z_i, valence including sign ($z = \mid z_i \mid$)
Γ_o	Gouy–Chapman coefficient (eq. 4.3.7)
δ	thickness of the Stern layer (eq. 4.6.4)
ε_o	electrical permittivity of vacuum; ε_r relative dielectric permittivity
ζ	zeta potential; measured value of Φ_δ
Θ	fractional surface coverage (eqs. 3.7.11 and 4.6.2)
κ^{-1}	Debye screening length (eq. 4.3.13)

Π_{osm} osmotic pressure (eq. 4.4.1)

σ_0 surface charge density (eqs. 4.3.19 and 4.3.20)

τ "half time" of rapid coagulation (eq. 4.5.17)

$\Phi(x)$ potential in electrolyte distance x from charged surface (eqs. 4.3.6 and 4.3.12)

Φ_0 surface potential; Φ_δ, potential at Stern layer

CONCEPT MAP

General Properties of Colloids

Colloidal systems contain insoluble particles (dispersed phase) ranging in size from 10 to 1000 nm dispersed in a continuous phase (dispersion medium).

- Large interfacial areas associated with the solid–liquid interface make surface chemistry important in colloidal systems.

- Colloidal systems are thermodynamically unstable, but they can be kinetically stabilized by steric (polymeric) or electrostatic forces.

- Coagulation leads to the irreversible formation of large aggregates of colloidal particles, which separate out of solution under the influence of gravity.

Formation of Colloidal Particles

- Because colloids are thermodynamically unstable, they must be prepared by condensation (exceeding an equilibrium solubility limit) or comminution (mechanical grinding) methods.

- To produce monodisperse particles, nucleation and growth steps must be separately controlled (Figure 4.2).

Charged Interfaces

- Virtually all vapor–liquid–solid interfaces acquire charge by dissociation or adsorption of ionic constituents.

- Because electrostatic forces are long-ranged, charged interfaces play an important role in many interfacial processes.

- The properties of a charged interface are described by the Gouy–Chapman theory, which relates surface charge density, σ_o, and surface potential, Φ_o (eq. 4.3.19). The theory also shows that the potential in an electrolyte solution decays exponentially with distance from the surface (eq. 4.3.12, Figure 4.7) with a decay constant given by the Debye length, $1/\kappa$.

- The Debye length (eq. 4.3.13) contains the valence and concentration of ions, the dielectric constant and temperature of the electrolyte solution. It constitutes a key parameter defining the interaction between charged particles and ions in solutions.

- Near a charged surface, counterions (of opposite charge to the surface) concentrate while coions (of the same charge) are repelled. The Gouy–Chapman theory permits us to calculate how ionic concentrations vary with distance in solution (Figure 4.8).

- When two charged surfaces approach, the electrostatic potential energy, V_R, increases exponentially with distance. The magnitude of the energy is determined by Φ_o and κ.

The DLVO Theory

- DLVO theory gives the total potential energy between two charged surfaces, $V_T(h)$ (eq. 4.5.2), by adding together expressions for their attraction, V_A (eq. 4.1.1), and electrostatic repulsion, V_R (eq. 4.4.9) (Figure 4.12).

- Key features of $V(h)$ are the values of the maxima and minima, which give the changes in potential energy associated with coagulation, $V_{pr. min}$, and the barrier to coagulation, V_{max}, in Figure 4.1.

Colloidal Stability

- When $V_{max} = 0$, coagulation becomes rapid.

- We obtain expressions relating the critical electrolyte concentration for coagulation (CCC) and Φ_o (eqs. 4.5.8 and 4.5.9).

- When $\Phi_o \gg kT$, the CCC is proportional to $1/z^6$ (the valence of the counterion) and independent of Φ_o (eq. 4.5.8).

- When $\Phi_o < kT$, the CCC becomes proportional to Φ_o^4/z^2, and the stability of the system becomes extremely sensitive to Φ_o (eq. 4.5.9).

Kinetics of Coagulation

- When $V_{max} = 0$, coagulation becomes a diffusion-controlled process.

- Dimer formation is a second-order rate process with $k_r = 4kT/3\eta$. It is independent of (spherical) particle size (eq. 4.5.14).

- The time of coagulation, τ (eq. 4.5.17), is the time required for the concentration of dispersed particles, $[P_0]$, to decrease by half. It takes seconds to minutes.

- With a coagulation barrier, $V_{max} > 0$, the coagulation rate constant, k_s, is slower. $k_s = k_r/W$, where W, the stability ratio, varies as $\exp(V_{max}/kT)$ (eq. 4.5.19).

Surface Chemistry in Colloidal Systems

- The Stern model adapts the Gouy–Chapman model to account for ionic size and specific adsorption at charged surfaces, but introduces a number of parameters that are difficult to evaluate.

- Colloidal stability correlates with the zeta potential, ζ, the surface potential measured in electrophoretic measurements. In most practical colloidal systems, ζ is the only experimentally accessible parameter characterizing the potential in the double layer.

- Surface chemistry identifies three specific ionic effects that modify particle–electrolyte interaction:

  1. Potential-determining ions that directly affect Φ_o; minute changes in their concentration can have a pronounced effect on coagulation processes.
  2. Indifferent electrolytes that change the electrostatic screening through their effect on the Debye length, $1/\kappa$.
  3. Charge-reversing ions that adsorb so strongly they reverse the potential on the colloidal particle.

- Heterocoagulation of dissimilar particles is more complex than that of identical particles because H_{121} can be attractive or repulsive and the usual assumptions involved in calculating V_R do not apply.

4

Colloidal systems contain one phase A dispersed in a second phase B. Substance A is called the dispersed phase, and substance B the dispersion medium. Table 4.1 illustrates the generality of this definition and shows that all possible combinations of insoluble gas, liquid, and solid phases form colloidal dispersions. This chapter concentrates on interfacial systems involving solids dispersed in liquids, liquid–solid colloids.

Liquid–solid colloidal systems play an important role in many interfacial industrial processes and products. For example, processing of clay–water dispersions produces ceramics ranging from delicate china figurines to toilet bowls and masonry bricks. Applying modern colloidal technology to silica and other oxide sols yields tough, fracture-resistant ceramics that find application in high-temperature automobile engines, rocket nose cones, and as longer-lasting medical prostheses. Drilling muds used in oil exploration are complex colloidal materials that are indispensable as lubricants and rheology control agents. Virtually all

TABLE 4.1 Types of Colloidal Systems with Some Common Examples

Dispersion medium	Dispersed particle	Technical name	Examples in nature	Examples in technology
Gas	Liquid	Aerosol	Mist; fog	Hairspray; smog
Gas	Solid	Aerosol	Volcanic smoke; dust	Pharmaceutical inhalants
Liquid	Gas	Foam	Foam on polluted rivers	Fire-extinguisher foam; porous plastics
Liquid	Liquid	Emulsion	Milk; biological membranes	Drug delivery emulsions; mayonnaise; adhesives
Liquid	Solid	Colloidal sol or dispersion	River water; muddy silt; clay	Printing-ink; paint; toothpaste
Solid	Gas	Solid foam	Pumice; loofah	Styrofoam; zeolites
Solid	Liquid	Solid emulsion	Opal; pearl; oil-bearing rocks	High impact plastics; bituminous road paving
Solid	Solid	Solid dispersion	Wood; bone	Composites; pigmented plastics

coating systems employ colloidal materials, and most industrial products are coated either for protection or decoration. Paints are colloidal suspensions containing titanium dioxide and latex particles. Paper making involves producing a meshwork from colloidal fibers in which clay particles are used as filler to improve print quality and produce a pleasing surface texture. Inks used in ordinary ballpoint pens, xerography, and high-speed printing presses are colloidal liquids or pastes. Scouring powders, toothpaste, and other heavy cleaning agents contain colloidal material such as pumice.

Control of colloidal systems is also a central issue in dealing with a host of environmental problems associated with our heavy use of technology. For example, smog consists of colloidal size particles generated by atmospheric photochemical reactions involving petroleum as well as natural products. Controlling the fines and other colloidal debris associated with processing at wood pulping plants, mineral flotation sites, coal grinding processing units, and asbestos plants requires application of colloidal chemical techniques. Some of the key steps in the purification of water and the treatment of sewage also depend on the adsorptive capacity of colloidal materials.

What size particles exhibit colloidal behavior? Dispersed particles must be larger than 1 nm in at least one dimension. Colloidal systems containing particles smaller than this become indistinguishable from true solutions. The upper limit for the solid particles is generally set at a radius of 1000 nm (1 μm), where Brownian motion keeps the solid particles in solution, known as a *sol*. Particles larger than this settle out under the influence of gravity although entities of larger size are encountered in some emulsions, mineral separations, and ceramic powders.

A simple geometric calculation illustrates an important aspect of colloidal systems. If we take a sphere with a radius of 1 cm and break it up into 10^{21} spheres, each with a radius of 1 nm, the total surface area equals 1.26×10^8 cm^2! As we have seen, surfaces are generally areas of high free energy, and since colloidal systems possess intrinsically large interfacial areas, interfacial forces and surface chemistry must play an important role in their behavior. This is indeed the case, and our major concern in this chapter will be to develop an understanding of the methods used to control interfacial interactions between solid particles held in electrolyte solutions.

Section 4.1 sets the stage for the strategies to be pursued for the control of interparticle interaction and colloid stabilization. Then we discuss the basic approaches to colloid particle preparation in Section 4.2. The next three sections of the chapter deal with the theory of electrostatic interaction between charged particles in an electrolyte, electrostatic stabilization, and the kinetics of coagulation. Section 4.6 discusses the surface chemistry of particles held in different kinds of solution. Finally, Section 4.7 includes some examples of colloidal systems used in various applications. The organization is presented in the concept map at the beginning of the chapter.

4.1 Colloidal Systems Are Thermodynamically Unstable, but Can Be Kinetically Stabilized by Steric or Electrostatic Repulsive Forces

First, it is important to remember that the attractive forces between particles in colloidal systems operate as long-range forces. As we demonstrated in Chapter 2.6, the potential energy of attraction, V_{att}, for two flat parallel particles in a liquid medium varies with separation distance h as $1/h^2$

$$V_{att(flat\ particles)} = -\frac{H_{121}}{12\pi}\frac{1}{h^2} \qquad (4.1.1)$$

and as $1/h$ for two spherical particles of radius R

$$V_{att(spheres)} = -\frac{H_{121}}{12}\frac{R}{h} \qquad h << R \qquad (4.1.2)$$

where H_{121} is the Hamaker constant for two particles (phase 1) in a liquid medium (phase 2).

Colloidal particles continually move around in solution as a result of Brownian motion. When two particles approach one another, attractive interactions draw them together until they come into contact, a process known as *coagulation*. As a result, the particles settle into the deep potential energy well, known as the primary minimum, $V_{pr\ min}$, in Figure 4.1a, defined by the combination of V_{att} and V_{CR}, where V_{CR} is the hard sphere core repulsion (CR) between molecules located at the surface of the particles, as defined in eq. 2.5.2. If the magnitude of $V_{pr\ min}$ is >> kT, coagulation is irreversible.

Figure 4.1a also plots the interaction forces F between particles. By defining F = –dV/dh, a negative force represents an attractive interaction and a positive force a repulsive interaction.

The initial coagulation event between two particles leads to the formation of a doublet; the doublet in turn combines with other particles to form large *agglomerates*. When agglomerates become large enough, they settle out of solution under the influence of gravity. In this way, the process of coagulation converts the sol, which was a homogeneous phase containing dispersed particles, into a two-phase system consisting of a solid mass at the bottom of a container and a liquid phase above it.

Later in this chapter, we will show that a typical colloidal system coagulates in seconds to minutes. For example, titanium dioxide colloidal particles are used in paints to give them high hiding power, as we will discuss in more detail in Section 4.7.1. If these TiO_2 particles were allowed to interact only via the normal attractive and repulsive forces, the shelf life of the paint would be so short that it would never make it out of the paint factory in a usable form. Clearly steps must be taken to prevent coagulation, that is, to *stabilize* the sol, to make it useful in a paint.

For this reason, optimization of colloidal systems focuses in large part on the introduction of long-range repulsive forces to

Figure 4.1
The total interaction energy V_T is plotted versus distance of separation h for two spheres of equal size. (a) V_T is obtained by adding together contributions from the van der Waals attractive interactions, V_{att}, and the core repulsive interactions, V_{CR}, arising from electron overlap of molecules on the surface of the spherical particles. Particles are tightly bound in the primary minimum $V_{pr\,min}$. (b) V_T is obtained by adding a second repulsive interaction, V_{rep}, associated with electrostatic or steric effects in solution. When V_{max} is higher than $2kT$, particles may be held apart at the secondary minimum $V_{sec\,min}$. (R. J. Hunter, *Foundations of Colloid Science*, Oxford University Press, Oxford, 1989, p. 419. By permission of Oxford University Press.)

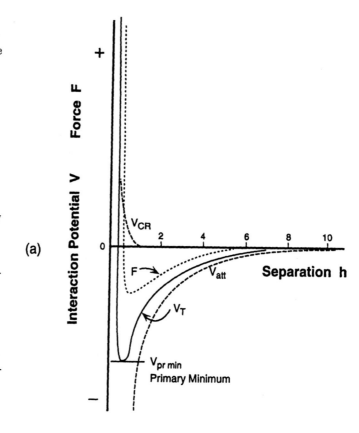

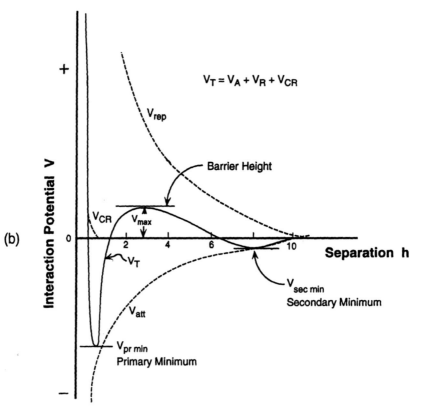

prevent coagulation. These additional repulsive forces have two major origins. The first leads to *steric stabilization*. It occurs when particle surfaces are covered with polymer molecules. If the polymer molecules extend sufficiently far out into solution, the distance of closest approach for two particles exceeds the distance where attractive interactions dominate. Because the configuration of polymer chains plays a decisive role in steric stabilization, we will postpone discussion of this topic until we review the interfacial properties of polymers in more detail in Section 6.5.3.

The second repulsive force leads to *electrostatic stabilization*. It occurs when the colloidal particles acquire a surface charge and are stabilized in an electrolyte solution. We shall show that this potential energy of repulsion between parallel plates or spheres, V_{rep}, has the functional form

$$V_{rep} = f[\sigma_o, \exp(-c^{1/2}h)] \qquad (4.1.3)$$

where σ_o represents the surface charge density (charge per unit area) on the particle, c is the electrolyte (salt) concentration in the solution, and h the distance of separation.

We can write the total interaction potential between two particles as the sum of the attractive (att), repulsive (rep), and core repulsive (CR) potentials

$$V_T = V_{att} + V_{rep} + V_{CR} \qquad (4.1.4)$$

Figure 4.1b shows how V_{att}, which has a power-law dependence on h, and V_{rep}, which varies exponentially with distance, add to give V_T. V_{CR} has such short range that it has little impact on the shape of the curve except in the immediate vicinity of the primary minimum. In later discussion, the contribution of V_{CR} is often neglected altogether. On the other hand, the effect of electrostatic repulsion, V_{rep}, is most important. It produces a repulsive maximum, V_{max}, in the total potential energy curve. Manipulation of this repulsive barrier is our primary concern. If $V_{max} \gg kT$, then it serves as a barrier to prevent the particles from moving together into the primary minimum.

We can summarize these concepts in three principles:

- Colloidal particles in a sol are thermodynamically unstable.

- Coagulation is usually irreversible because $V_{pr\,min} \gg kT$.

- Colloidal particles are kinetically stabilized when $V_{max} \gg kT$.

In the next sections, we describe how colloidal systems are prepared, develop an understanding of charged interfaces that permits us to derive eq. 4.1.3 and understand how colloidal systems can be stabilized, and obtain equations describing the coagulation process and show under what conditions coagulation becomes rapid.

4.2 Colloids Can Be Prepared in Two Ways

4.2.1 Preparation of Colloids by Precipitation–Nucleation and Growth Determine the Size, Shape, and Polydispersity of Colloids

Because colloidal dispersions are thermodynamically unstable, they must be prepared under nonequilibrium conditions. Two basic strategies for preparing them are *condensation* (precipitation)—the formation and growth of a new phase by exceeding an equilibrium condition such as the solubility limit—or *comminution*—the breaking of large particles into successively smaller ones.

Methods for forming colloidal particles by condensation include chemical reaction, condensation from the vapor, and dissolution and reprecipitation. We now consider several examples of these techniques.

Condensation processes can form new colloidal phases of a variety of materials. For example, reduction of the gold chloride complex by hydrogen peroxide

$$Au(Cl_4)^- + H_2O_2 \rightarrow Au_{(sol)}$$

forms colloidal gold. Adding hydroxide to aluminum ions

$$Al^{3+} + OH^- \rightarrow Al(OH)_{3(sol)}$$

leads to the formation of the aluminum hydroxide colloid. The chemical reaction

$$KI + AgNO_3 \rightarrow Ag_{(sol)} + K^+, I^-, NO_3^-, Ag^+ \rightarrow AgI_{(sol)}$$

can be driven to form a silver iodide sol by exceeding the solubility limit of AgI (see Sections 4.6.3.1 and 8.4.2.3). To prepare a stable colloid, we must remove the excess ions by dialysis.

All these reaction processes involve simultaneous nucleation and growth; so they produce colloidal particles that have a wide range of particle sizes. For example, when we simply mix solutions of $AgNO_3$ and KI, we create small regions in which the concentration of Ag^+ and I^- ions exceeds the solubility product. This condition initiates the nucleation process, and growth follows. As mixing continues, nucleation followed by growth begins in other regions of the solution. The sequence of nucleation events accompanied by continuous growth results in a wide distribution of particle sizes (*polydisperse*). While such colloids have many uses, colloids in which all the particles have the same size (*monodisperse*) are desirable in a number of applications.

The key to forming monodisperse colloids is to separate and control the nucleation and growth processes. As we saw in Section 3.4, nucleation occurs at concentrations or vapor pressures that exceed the equilibrium value by a considerable amount. For example, in dust-free conditions water vapor condenses at a pressure that is four times the equilibrium value. By analogy, we can also force precipitation of colloidal material to begin at a concentration of reactants that exceeds the solubility product by a considerable amount.

Figure 4.2 illustrates the strategy used to produce monodisperse colloids. Two important concentrations are involved; the nucleation concentration, above which nucleation and growth begin, and the saturation concentration, or solubility limit, below which growth stops. The idea is to adjust the temperature or composition so that the initial concentration exceeds the nucleation concentration. Nucleation occurs in a single short burst that causes the concentration of reactants to change and fall below the nucleation threshold. Control over subsequent growth can be achieved by maintaining the reaction concentration at a level below the nucleation threshold but above the solubility limit. The number of nuclei formed in the initial stage determines the number of particles; the length of the growth period determines their size.

Formation of monodisperse gold sols illustrates how sometimes we can completely separate the nucleation and growth steps. In the nucleation step, small gold particles are formed by reacting the gold chloride complex with red phosphorus

$$Au(Cl)_4^- + P_{(red)} \rightarrow Au_{(nuclei)}$$

The growth step involves adding a mild reducing agent, such as formamide, along with more $Au(Cl)_4^-$, under conditions that do not permit new nuclei to form:

$$Au_{(nuclei)} + Au(Cl)_4^- + H_2CO \rightarrow Au_{(sol)}$$

Matijevic and his co-workers have prepared a number of monodisperse metal oxide or metal hydroxide colloids using a controlled hydrolysis technique. Their approach consists of heating the transition metal complex with anions, such as chloride, sulfate, or phosphate ions, to accelerate the rate of deprotonation of coordinated water. Manipulating parameters, such as the rate of heating, concentration and purity of reactants, growth temperature, and time, controls the nucleation burst and the extent of

Figure 4.2
Producing monodisperse colloidal particles requires control and separation of nucleation and growth processes. The number of particles is set by increasing the concentration of reactants above the nucleation concentration for a brief period. Particle growth is controlled by maintaining the reactant concentration below the nucleation concentration but above the equilibrium concentration for a period of time sufficient to obtain the desired particle size. (D. F. Evans and H. Wennerström, *The Colloidal Domain: Where Physics, Chemistry, Biology, and Technology Meet*, VCH Publishers, New York, 1994, p. 370. Reprinted with permission by VCH Publishers © 1994.)

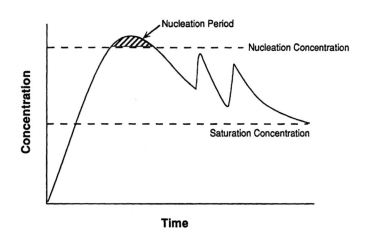

growth. It is also possible to alter the shape of the colloidal particles from their common spherical shape to cubic or "needle-like" acicular shapes. Figure 4.3 shows electron micrographs of several of Matijevic's colloids.

Organic polymer colloidal particles can be prepared by *emulsion polymerization*. For example, the process of preparing monodisperse latex spheres starts with an emulsion of a monomer, such as styrene, that is stabilized by a surfactant, such as sodium dodecyl sulfate (SDS). The concentration of SDS is adjusted so that the emulsion contains micelles that enclose some of the "solubilized" monomer (see Section 5.4). Then a water-soluble polymerization initiator, such a potassium persulfate, is added. Statistically, polymerization begins in the micelles rather than in the emulsion droplets. This step provides controlled nucleation. As the polymerization process depletes the number of monomers contained in the micelles, more monomers diffuse into them from the emulsion. This step provides controlled

Figure 4.3
Examples of monodisperse colloids prepared by Matijevic using the controlled hydrolysis technique: (a) zinc sulfide spherulites (diam. ≈ 0.5 μm) and (b) cadmium carbonate cubelets (edge ≈ 1.0 μm). Small variations in experimental nucleation and growth conditions lead to very different particle morphologies. (Photographs courtesy of Prof. Egon Matijevic, Clarkson University, Potsdam, New York.)

(a)

(b)

growth. Thus the number of activated micelles determines the number of particles, while the amount of monomer present in the system determines their size. This process is widely used in the manufacture of monodisperse latex spheres for various applications.

Latexes differ from inorganic colloidal particles in one important way. At a characteristic temperature for each type of latex, the polymer transforms from a solid to a glasslike material. Above this glass transition temperature T_g, coagulated latex particles can coalesce and fuse together upon drying. As we shall see in Section 4.7.1, this property plays an essential role in paints, where coalesence leads to the formation of a continuous protective latex film.

4.2.2 Colloids Also Can Be Prepared by Communition

The comminution method for preparing colloidal particles requires a mechanical process for breaking down bulk material into colloidal dimensions. For example, materials like the titanium dioxide used in paints are broken down by being tumbled together with ceramic or hardened metal balls in a comminution mill.

Unfortunately comminution can reduce size only to a limited degree because small particles tend to agglomerate during the grinding process and resist redispersion in subsequent processing steps. For this reason, condensation methods, which are more easily controlled and versatile, are much more widely used as the starting point for colloid preparation.

4.2.3 The Surfaces of Colloidal Particles Are Charged through Mechanisms Involving Surface Disassociation or Adsorption of Ionic Species

A major aspect of the electrostatic stabilization of colloids is concerned with manipulating the charge on the surface of colloidal particles. First we need to understand the origins of surface charge. A surface or an interface placed in a solution can become charged by two mechanisms: (1) surface ionization, in which ions dissociate from the particle surface and diffuse into the adjoining phase; or (2) preferential adsorption, in which one ionic species adsorbs onto the particle surface.

Surface Ionization. An example of surface ionization occurs when a clay is mixed with water. Clays are a major constituent of soil. Originally they solidified as crystalline silicates and were subsequently broken down to colloidal size by geological forces. In their crystalline form they are made up of stacked unit layers (Figure 4.4) containing silicon tetrahedrally coordinated with oxygen, and aluminum octahedrally coordinated with oxygen and hydroxide. In most clays, the tetrahedral and octrahedral sheets join together to form two- or three-layered sheets. Because the forces between sheets are much weaker than within them, clay minerals easily break up into platelets.

In many clays, atom substitution occurs in the tetrahedral layers, where trivalent Al replaces tetravalent Si, or in the octahedral layers, where divalent Mg replaces trivalent Al. If an atom

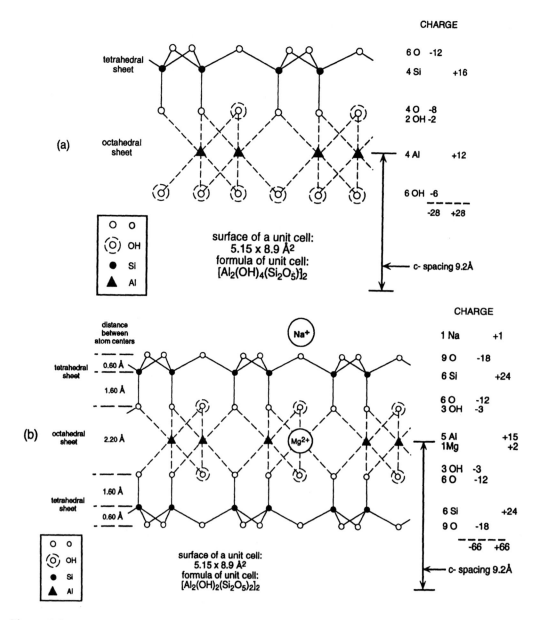

Figure 4.4
Clay particles consist of stacks of identical units held together by van der Waals forces. The units contain (a) two or (b) three layers composed of silicon and aluminum bonded to oxygen or hydroxide. Charged clay particles result from impurity substitution. The montmorillonite clay particle shown in (b) has the composition $(Si_8)(Al_{3.33}Mg_{0.67})(O_{20})(OH)_4Na_{0.67}$. One aluminum ion in every six has been replaced by a magnesium ion, and for charge compensation one sodium ion has been added at the surface of the stack. (b) Reproduces one and one-half unit cells of the basic crystal structure to demonstrate charge balance with substitution. (H. van Olphen, *An Introduction to Clay Colloid Chemistry: For Clay Technologists, Geologists, and Soil Scientists,* © 1963, John Wiley & Sons, pp. 64–65. Reprinted by permission of John Wiley & Sons, Inc.)

of lower valence (Mg^{2+}) replaces one of higher valence (Al^{3+}) without any other structural change, the layer acquires a net negative charge that must be neutralized by the incorporation of positive cations (such as Na^+ or K^+) into the structure. These cations cannot be accommodated within the layer and instead are located on a plane between the sheets as illustrated in Figure 4.4b. There they are ionically bonded to the layer. When the sheets are split apart, the positive charges are exposed, and when placed in water, the sodium (or potassium) ions readily dissolve, leaving the surface of the particle *negatively charged*.

The surface charge density, σ_0, can be estimated as follows. Figure 4.4b shows a montmorillonite clay structure that has a unit cell formula of $(Si_8)(Al_{3.33}Mg_{0.67})(O_{20})(OH)_4Na_{0.67}$. From x-ray measurements, we know that the surface area per unit cell is 5.15×8.9 Å^2. Since two surfaces share the $Na_{0.67}$, the surface cation density is $Na_{0.33}$ per unit cell, and the surface charge density is $0.33e/5.15 \times 8.9$ Å^2, or $\sigma_0 = 0.12$ C/m^2 (coulomb per meter squared).

Preferential Adsorption. We can illustrate preferential adsorption, the second mechanism whereby a particle surface becomes charged, by the following experiment. If we form air bubbles (small enough so that we can ignore gravitational forces) in an aqueous NaCl solution, place two electrodes into the solution, and apply a potential, we find that the bubbles move toward the positive electrode. This observation establishes the fact that chloride ions preferentially adsorb over sodium ions at the air–water interface. If we replace NaCl by a long-chain cationic surfactant, such as dodecylammonium chloride ($C_{12}H_{25}NH_3Cl$), and repeat the experiment, we find that the air bubbles are now positively charged and move toward the negative electrode. Thus, by selecting our electrolyte and varying its concentration, we can control the sign and magnitude of charges adsorbed at the air–water interface. Similar preferential adsorption of charged species from solution leads to charged surfaces in solid–liquid systems, as will be discussed in more detail in Section 4.6.3.

In many manufacturing processes, we purposely add materials that selectively adsorb onto interfaces to charge them, and then use these charged interfaces to control the process. Unfortunately, the accumulation of charged surface-active impurities sometimes creates unwanted charged interfaces resulting in the formation of new, sometimes troublesome, phases or microstructures. Many environmental problems arise from this adsorbed charge accumulation.

4.3 Charged Interfaces Play a Decisive Role in Many Interfacial Processes

Charged interfaces are ubiquitous in interfacial systems and often play critical roles in industrial and biological processes. In colloidal systems, we are concerned with the interactions between two charged particle surfaces that stabilize the colloid. In other interfacial systems, single charged interfaces are important. For example, living cells control the flow of material and information between their external environment and their interior by manipulating charge flow across their membranes. Many of the solid-state devices described in Chapters 10 and 11 operate by controlling charge at solid–solid interfaces. Thus the equations to be developed in the next section have a broader range of application than the stabilization of colloids. We will redevelop similar equations again for solids in Chapter 9.

4.3.1 The Gouy–Chapman Theory Describes How a Charged Surface and an Adjacent Electrolyte Solution Interact

In this section we focus on electrostatic interactions at solid–liquid interfaces. We start with the interaction between the charged particle surface and the ions in the solution surrounding it. If the initial solvent is pure water, then the ions in solution are hydroxyl ions plus those ions that have been dissolved from the particle surface. For example, when the clay particles in Figure 4.4b are immersed in water, they become negatively charged and the water contains positive sodium ions removed from the clay particle surface; in this instance the negatively charged clay particles repel each other. More often we are interested in the behavior of systems where a salt has been added to the water deliberately to form an *electrolyte* solution. The behavior of charged particles then depends critically on how much salt is added to the electrolyte solution. For low salt concentrations the clay remains dispersed and workable, but above a certain critical concentration it suddenly becomes an agglomerated mass. We are interested in the origins of this sudden transition. To achieve this we must analyze the charge distribution surrounding a charged surface immersed in an electrolyte, the approach that constitutes the *Gouy–Chapman theory*.

We derive the Gouy–Chapman equations using the model depicted in Figure 4.5. It involves a planar charged surface char-

Figure 4.5
Model used in deriving the Gouy–Chapman equations that describe the interaction between a planar charged surface and an adjacent electrolyte solution. The surface extends infinitely in the x and y directions and is characterized by a charge density σ_o and surface potential Φ_o. The solution contains positive and negative ions of valency z_i and concentration c_{oi}, which are treated as point charges. (D. F. Evans and H. Wennerström, *The Colloidal Domain: Where Physics, Chemistry, Biology, and Technology Meet*, VCH Publishers, New York, 1994, p. 112. Reprinted with permission by VCH Publishers © 1994.)

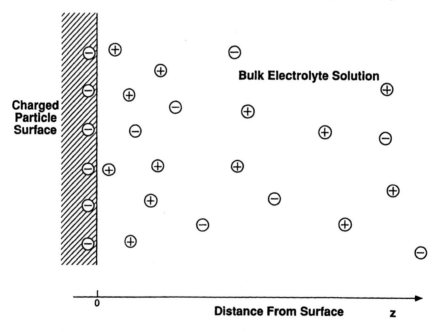

Charged Particle Surface

Bulk Electrolyte Solution

0

Distance From Surface   z

acterized by the surface charge density σ_o and surface potential Φ_o. The adjacent solution is an electrolyte characterized by the bulk concentration c_{io}, of ions, charge ze, where z is equal to the *valence multiplied by the sign* of the ion, and dielectric permittivity $\varepsilon_r\varepsilon_o$.[1] We assume that the ions can be approximated as point charges. Ions in the electrolyte solution bearing a charge *opposite* to that on the particle surface are known as *counterions*, while those bearing the *same* charge are known as *co-ions*.

Our goals are to determine: (1) how the electrical potential and distribution of ions in the electrolyte solution varies with distance z from the charged interface; and (2) the relationship between σ_o and Φ_o. Armed with these relationships we will then be able to analyze how interparticle repulsion or attraction depends on electrolyte type, concentration, and temperature.

4.3.1.1 The Poisson–Boltzmann Equation Is Used to Derive an Expression for the Distribution of Charged Ions in the Electrolyte and the Associated Electrical Potential

In Section 2.3.1, our discussion of ion–ion charge interactions was restricted to only a few fixed charges, and under those conditions Coulomb's law was applicable. However, in the present situation, in which many charges are free to move throughout the volume of the electrolyte in response to electrical fields and also are under the influence of thermal motion, we need to use a more general expression obtained by combining two fundamental equations, the Poisson equation and the Boltzmann equation, to describe the interaction.

The Poisson equation provides a relation between the electrical potential Φ and charge density in vacuum

$$\nabla^2\Phi = -\frac{\rho}{\varepsilon_o} \tag{4.3.1}$$

where ∇^2 stands for the operator $\partial^2/\partial x^2 + \partial^2/\partial y^2 + \partial^2/\partial z^2$, and ρ is the charge density obtained by summing all charges. (Equation 4.3.1 is written in SI units, and its left-hand side must be multiplied by $1/4\pi$ to convert it to cgs units. In other texts it is important to ascertain which units are being used in the electrostatic equations.)

Attempts to use eq. 4.3.1 to describe an electrolyte solution require simultaneous evaluation of all charge interactions (ion–ion, ion–dipole, dipole–dipole, etc.) and result in an intractable problem. We can circumvent this difficulty and write the Poisson equation in a more convenient form by applying the following arguments. Ion–ion interactions are stronger and longer-ranged than all other types of charged interactions. As a result, ion–ion interactions typically play a dominant role in electrolyte solu-

---

[1] *Note that in this notation ε_o is the dielectric permittivity of vacuum, ε_r is the relative permittivity of the solution between the charged particles, and so $\varepsilon_r\varepsilon_o$ is the dielectric permittivity of the electrolyte solution. ε_r is also known as the dielectric constant. In most electrolytes the solvent is water, for which $\varepsilon_r \approx 78$.*

tions. This fact suggests that we write the interaction between free charges explicitly while averaging over the solvent degree of freedom, thus eliminating the explicit consideration of ion–dipole and dipole–dipole interactions. We will not give the detail of this averaging process, but simply note that it transforms the Poisson equation in a deceptively simple way to

$$\nabla^2 \Phi = - \frac{\rho(\text{free ions})}{\varepsilon_r \varepsilon_o} \qquad (4.3.2)$$

where we account for the effect of the solvent through its dielectric constant ε_r.

The charge density per unit volume ρ at any location in the solution (Figure 4.5) is expressed as

$$\rho(\text{free ions}) = \sum_i z_i e\, c_i^* \qquad (4.3.3)$$

where z_i is the valence of the ion multiplied by ± 1 according to its sign, and c_i^* represents the local concentration of ions of type i, measured as the *number* of i ions per unit volume. (We use an asterisk to differentiate between number concentration c* *measured in ions per cubic meter* and molar concentration c *measured in moles per liter. Thus c* = 1000 N_{Av} c ions/m³.)

The solution's charge density ρ cannot be associated with a set of fixed charges because ions in solution are free to move in response to electrical fields. In addition, we must consider the interplay between electrostatic interactions that favor an ordered and localized ion arrangement, and entropic factors that strive to generate a random distribution of ions.

As we noted in Section 2.3.4 and 2.8, the Boltzmann distribution expresses the compromise between molecular order and disorder. For ions in solution the electrostatic energy of an ion of valence z_i at a point where the potential is Φ is represented by $z_i e\Phi$. So the Boltzmann equation is

$$c_i^* = c_{io}^* \exp\left(- \frac{z_i e \Phi}{kT} \right) \qquad (4.3.4)$$

In this equation, c_{io}^* equals the concentration of ion species when $\Phi = 0$, which we usually take as equal to the bulk ion concentration. Near positively charged surfaces Φ is positive, and near negatively charged surfaces Φ is negative.

Combining eqs. 4.3.2, 4.3.3, and 4.3.4 gives the Poisson–Boltzmann equation. Since we are interested in the potential variation $\Phi(z)$ in the direction z away from a charged flat surface, we write

$$\varepsilon_r \varepsilon_o \frac{d^2\Phi}{dz^2} = - \sum_i z_i e c_{io}^* \exp\left(\frac{- z_i e \Phi}{kT} \right) \qquad (4.3.5)$$

which expresses $\nabla^2 \Phi$ from eq. 4.3.2 in terms of the direction z normal to the surface in Figure 4.5.

At this juncture we can either solve eq. 4.3.5 completely or make simplifying assumptions that lead to solutions that are straightforward but approximate and therefore of more limited utility. In this text we do both. We derive the complete solution in Appendix 4A and summarize the results in eqs. 4.3.6 and 4.3.7. We derive the simpler solution starting at eqs. 4.3.8.

From the complete solution the change in potential with distance is given by

$$\Phi(z) = \frac{2kT}{ze} \ln\left(\frac{1 + \Gamma_o \exp(-\kappa z)}{1 - \Gamma_o \exp(-kz)}\right) \tag{4.3.6}$$

where the quantity $1/\kappa$ is the Debye screening length, to be defined later, and Γ_o contains the surface potential Φ_o in the form

$$\Gamma_o = \frac{\exp(ze\Phi_o/2kT) - 1}{\exp(ze\Phi_o/2kT) + 1} \tag{4.3.7}$$

When $\Phi_o = 0$, then $\Gamma_o = 0$; and when Φ_o becomes large, $\Gamma_o \to 1$. To obtain this solution to the Poisson–Boltzmann equation requires that the valencies of the counterions (cations) and co-ions (anions) be equal, that is, the electrolyte be symmetrical, such as Na^+Cl^-. Consequently, we write these equations in terms of $z = |z_i|$.

For the approximate solution we limit our interest to the situations in which $ze\Phi << kT$. We can then expand the exponential in eq. 4.3.5 and neglect high-order terms to give

$$\varepsilon_r\varepsilon_o \frac{d^2\Phi}{dz^2} = -\sum_i z_i e c_{io}^*(1 - z_i e\Phi/kT) \tag{4.3.8}$$

The electroneutrality condition means that the sum of positive and negative ion charges is zero

$$\sum_i z_i e c_{io}^* = 0 \tag{4.3.9}$$

leading to the cancellation of the first term displayed in eq. 4.3.8 and leaving

$$\varepsilon_r\varepsilon_o \frac{d^2\Phi}{dz^2} = \frac{\Phi}{kT} \sum_i (z_i e)^2 c_{io}^* \tag{4.3.10}$$

It is convenient to identify the cluster of constants in eq. 4.3.10 by the symbol $\kappa^2 = \sum_i (z_i e)^2 c_{io}^*/\varepsilon_r\varepsilon_o kT$. Then eq. 4.3.10 becomes

$$\frac{d^2\Phi}{dz^2} = \kappa^2\Phi \tag{4.3.11}$$

Using the boundary conditions $\Phi \to \Phi_o$ as $z \to 0$ and $\Phi \to 0$ as $z \to \infty$, we can solve eq. 4.3.11 to give

$$\Phi(z) = \Phi_o \exp(-\kappa z) \tag{4.3.12}$$

This result should be compared with the more complex complete solution of eq. 4.3.6. Equation 4.3.12 states that the *electrostatic potential drops away exponentially with distance from a charged surface in an electrolyte* at a rate determined by κ.

The quantity $1/\kappa$ has the dimension of length and is defined as the *Debye screening length*

$$\frac{1}{\kappa} = \left(\frac{\varepsilon_r \varepsilon_0 kT}{\sum_i (z_i e)^2 c_{io}^*} \right)^{1/2} = \left(\frac{\varepsilon_r \varepsilon_0 kT}{1000 e^2 N_{Av} \sum_i z_i^2 c_{io}} \right)^{1/2} \qquad (4.3.13)$$

when c_{io} is the concentration of counterions in the electrolyte measured in moles/liter denoted by the unit M.

For water at 25° C ($\varepsilon_r = 78.54$) containing a symmetrical monovalent salt such as Na^+Cl^-, $z_i = \pm 1$,

$$\frac{1}{\kappa} = \frac{3.043 \times 10^{-10}}{(c_{io})^{1/2}} \text{ m} \qquad (4.3.14)$$

With $c_{io} = 0.01M$, $1/\kappa = 3.043$ nm, a dimension comparable to the size of a colloidal particle. In an aqueous solution, $1/\kappa$ varies only

Figure 4.6
Decay in the potential in the double layer as a function of distance from a charged surface according to the limiting form of the Gouy–Chapman equation (4.3.12). (a) Curves are drawn for a 1:1 electrolyte of different concentration. (b) Curves are drawn for different 0.001 M symmetrical electrolytes. (P. C. Hiemenz, *Principles of Colloid and Surface Chemistry*, 2nd ed., Marcel Dekker, New York, 1986, p. 695.)

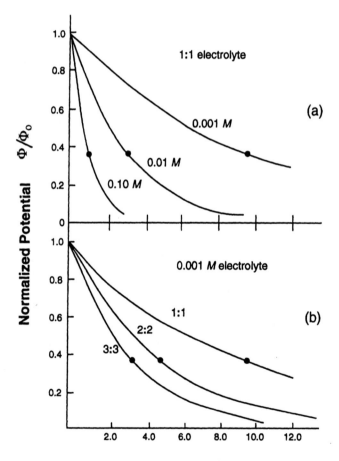

slowly with temperature because $\varepsilon_r\varepsilon_0 kT$ is almost constant over a broad temperature range.

Equation 4.3.12 demonstrates that the potential in the solution decays exponentially with distance from the particle, and the decay rate is set by the Debye length. In fact, when $z = 1/\kappa$, Φ has dropped to $\Phi_0/\exp(1)$. Figures 4.6a, 4.6b, 4.7a, and 4.7b illustrate the effect of concentration (c_{io}) and valence (z) of the ions in the electrolyte on $\Phi(z)$ as a function of z. As eq. 4.3.13 shows, *the higher the salt concentration and the higher the valence of the salt ions, the more rapidly the electrical potential decays away from the surface of the particle.*

We can gain further insight into the properties of the electrolyte in the vicinity of a charged surface by calculating how the concentration of both the counterions and the co-ions varies as a function of distance z from the surface. Assuming Φ_0 is constant, we first calculate Φ for different values of z and then use the Boltzmann equation 4.3.4 to calculate the concentration of positive and negative ions at those $\Phi(z)$ values. Figure 4.8 plots c_i versus z for a negatively charged surface. Figures 4.7 and 4.8 have been constructed using eqs. 4.3.6 and 4.3.7, although we can readily interpret the figures using the simple equations. In the plot the concentration of positively charged counterions increases from the bulk value c_{io} as we move toward the negatively charged surface. At the same time, the concentration of co-ions decreases below the bulk value. These results accord with our intuition that counterions concentrate at a charged surface, while co-ions are repelled. As the electrolyte concentration increases, the departures from c_{io} move closer to the surface, in accordance with eq.

Figure 4.7
Change in the potential as a function of distance (eq. 4.3.6) for two different electrolyte concentrations. (a) At constant surface potential, Φ_0, addition of electrolyte increases σ_0 and thus the slope β is greater than α. (b) At constant surface charge density, σ_0, the slopes α and β are identical (eq. 4.3.17); the addition of electrolyte decreases the surface potential Φ_0 (since κ increases, Φ_0 decreases from eq. 4.3.19). (H. van Olphen, *An Introduction to Clay Colloid Chemistry: For Clay Technologists, Geologists, and Soil Scientists,* © 1963, John Wiley & Sons, p. 34. Reprinted by permission of John Wiley & Sons, Inc.)

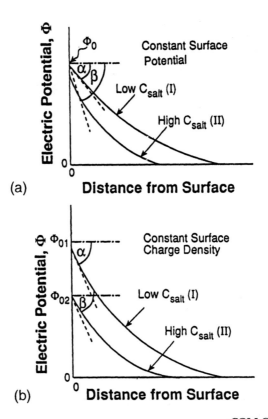

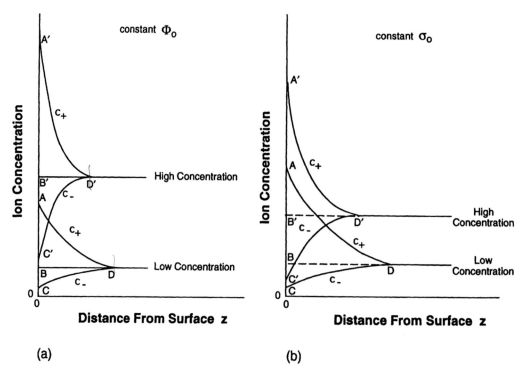

Figure 4.8

Charge distribution in the Gouy–Chapman double layer at (a) constant potential and (b) constant charge density for two different concentrations of added electrolyte. D and D′ correspond to the distances where the local concentrations of cations (c_+) as given by line AD and A′D′, and anions (c_-) as given by CD and C′D′, begin to depart from the bulk concentrations. The algebraic sum of the curves ACD and A′D′C′ are proportional to the net charge in the solution and thus equal the charge density on the surface. For constant σ_o, ACD = A′C′D′, while for constant Φ_o, A′C′D′ > ACD. (H. van Olphen, *An Introduction to Clay Colloid Chemistry: For Clay Technologists, Geologists, and Soil Scientists*, © 1963, John Wiley & Sons, pp. 32–33. Reprinted by permission of John Wiley & Sons, Inc.)

4.3.12. Figures 4.6, 4.7, and 4.8 illustrate how the potential and concentration of charged ions vary with distance into the electrolyte. These results achieve the *first goal* we set for ourselves in Section 4.3.1: to determine how the electrical potential and distribution of ions in the electrolyte solution varies with distance z from the charged interface.

4.3.1.2 The Poisson–Boltzmann Equation Also Leads to the Relationship between Surface Charge Density and Potential at the Charged Surface

We can obtain a relationship between surface charge density σ_o and the surface potential by realizing that in order to achieve electroneutrality, the charge per unit area on the surface must be equal and opposite to the charge contained in a volume element of solution of unit cross-sectional area extending from the surface to infinity. Stated as an equation, this equivalence becomes

$$\sigma_o = -\int_0^\infty \rho\, dz \tag{4.3.15}$$

By combining eq. 4.3.15 with the Poisson equation 4.3.2, we obtain

$$\sigma_o = \varepsilon_r \varepsilon_o \int_0^\infty \frac{d^2\Phi}{dz^2}\, dz \tag{4.3.16}$$

which is readily integrated to yield

$$\sigma_o = \varepsilon_r \varepsilon_o \frac{d\Phi}{dz}\bigg|_0^\infty = -\varepsilon_r \varepsilon_o \left(\frac{d\Phi}{dz}\right)_o \tag{4.3.17}$$

because $d\Phi/dz$ equals zero at infinity. Equation 4.3.17 tells us that the *surface charge density is proportional to the potential gradient in the vicinity of the surface;* that is, $-d\Phi/dz$ as $z \to 0$. This important general result is one we use repeatedly.

Using the approximate solution for $\Phi(z)$, eq. 4.3.12, we can evaluate $(d\Phi/dz)_0$ in the limit as $z \to 0$ and find

$$\left(\frac{d\Phi}{dz}\right)_0 = -\kappa\Phi_0 \exp(-\kappa z) = -\kappa\Phi_0 \quad \text{when } z \to 0 \quad (4.3.18)$$

Substituting eq. 4.3.18 into eq. 4.3.17 gives

$$\sigma_0 = \varepsilon_r\varepsilon_0\kappa\,\Phi_0 \quad (4.3.19)$$

which shows that the simple solution to the Poisson–Boltzmann equation predicts a linear relationship between surface charge density and surface potential.

The complete solution for the relationship between the surface charge density and the surface potential obtained in Appendix 4A gives

$$\sigma_0 = (2\varepsilon_r\varepsilon_0 kTc_{io}^*)^{1/2}\left[\exp\left(\frac{z_i e\Phi_0}{2kT}\right) - \exp\left(-\frac{z_i e\Phi_0}{2kT}\right)\right]$$

or

$$\sigma_0 = (8\varepsilon_r\varepsilon_0 kTc_{io}^*)^{1/2} \sinh\left(\frac{z e\Phi_0}{2kT}\right) \quad (4.3.20)$$

The results in eqs. 4.3.19 and 4.3.20 achieve the *second goal* set in Section 4.3.1: to determine the relationship between σ_0 and Φ_0. We now have the Gouy–Chapman expressions for the dependence of electrical potential (eqs. 4.3.6 and 4.3.12) and the distribution of ions away from the charged surface as well as for the relationship between surface charge and surface potential (eqs. 4.3.19 and 4.3.20). Appendix 4B gives some examples of calculations involving these formulae.

Now we can consider two limiting cases of these general relationships, either $\Phi_0 = $ constant or $\sigma_0 = $ constant. Figures 4.6 and 4.7a show how Φ changes with distance from the charged surface at three different electrolyte concentrations calculated on the assumption that Φ_0 remains constant. Since the surface charge density σ_0 is proportional to the limiting slope, $-d\Phi_0/dz$, from eq. 4.3.17, then, from Figure 4.7a, surface charge density must increase with added salt at constant surface potential. At constant surface charge density, Figure 4.7b, the surface potential decreases as the concentration of salt increases.

Figure 4.8 shows how the concentration of ions varies as a function of distance at either constant surface potential or constant surface charge density. While the plots for constant Φ_0 (Figure 4.8a) and constant σ_0 (Figure 4.8b) look similar, careful inspection proves they contain important differences. For electroneutrality, the net space charge concentration, depicted by the difference between the areas DAB (the cation excess) and DCB (the anion depletion), must be equal and opposite to the charge on the flat surface. With $\sigma_0 = $ constant, the difference between the areas

DAB and DCB must remain constant, irrespective of the concentration of ions in the electrolyte. With Φ_0 = constant, the difference in the areas, and consequently in σ_0, must increase as the concentration increases in accordance with eq. 4.3.19 with substitution for κ from eq. 4.3.13.

4.3.2 The Electrical Double Layer Is Equivalent to a Capacitor—with One Electrode at the Particle Surface and the Other in the Electrolyte at a Distance Equal to the Debye Length

Now we are in a position to gain a feeling for the significance of the Debye length, $1/\kappa$. We start by examining the expression for the capacitance C per unit area A of a parallel plate capacitor. We assume the capacitor to have a separation d between the plates and to be filled with a medium of dielectric constant ε_r, as illustrated in Figure 4.9a. The capacitance per unit area then equals $\varepsilon_r\varepsilon_0/d$. The capacitance per unit area is also the charge stored per unit area of the plates, σ_0, divided by the potential difference between them, Φ_0, so that

$$\frac{C}{A} = \frac{\sigma_0}{\Phi_0} = \frac{\varepsilon_r\varepsilon_0}{d} \qquad (4.3.21)$$

Comparison with eq. 4.3.19 ($\sigma_0/\Phi_0 = \varepsilon_r\varepsilon_0/\kappa^{-1}$) reveals that

Figure 4.9
(a) Parallel plate capacitor showing the variation of potential with distance between two charged plates, separation d, bearing equal but opposite charges $\pm\sigma_0$. (b) Schematic of a double layer as a capacitor in which one plate is the charged particle surface and the second plate corresponds to an imaginary surface placed at a distance $1/\kappa$ that carries all of the double layer charge.

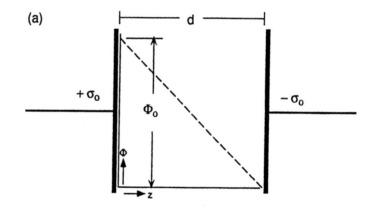

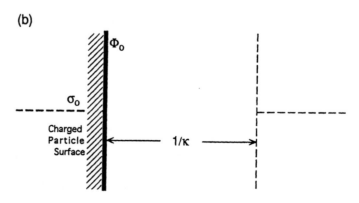

$$1/\kappa \equiv d \qquad (4.3.22)$$

Thus we can model the electrical interaction between the charged surface and the adjacent solution as if it were the capacitor shown in Figure 4.9b. One of the capacitor plates represents the surface of the charged particle, while the second plate represents an imaginary surface located at a distance $1/\kappa$ away from it. The net space charge resulting from the counterions and the co-ions behaves electrostatically as if all these ions were located on the imaginary surface. This picture gives rise to the notion of the *electrical "double layer."* But in no way should it be construed to mean that the ions physically lie on the imaginary plane of the double layer.

Often the more rapid decay of electrical potential in the solution with increased salt concentration and the corresponding decrease in Debye length is described as a more effective *screening* or shielding of the charged surface by the electrolyte. With the addition of more salt, the concentration of charge on the surface increases, κ increases, and the double layer narrows, so that the imaginary plate moves closer to the charged surface.

According to eq. 4.3.12, when the distance from the surface equals $1/\kappa$, the potential decreases to $\Phi = \Phi_o/\exp(1)$ or by a factor of 2.7 (37%). Viewed in this way the Debye length provides us with a convenient linear scale with which to assess the importance of electrostatic interactions in solutions. The Debye length correctly reflects the combined contribution of valence, concentration, and dielectric constant to the interaction of charges in solution. In the same way that we examine interaction energies by the ratio U/kT, we can assess the extent of electrostatic interactions by the ratio of distance to Debye length.

We conclude this section by considering the valence of the salt ions, a property that plays a decisive role in colloidal systems. If we have a solution containing equal bulk concentrations of monovalent and divalent counterions—for example, two solutions, one $c_{io}^*(Na^+)$, the other $c_{io}^*(Ca^{2+})$—what will be the relative concentration of those ions in the double layer region of a negatively charged surface? If we specify a potential, such as $\Phi = 154$ mV, for which $e\Phi/kT \doteq 6$, we can use the Boltzmann equation 4.3.4 to calculate the ratio of the concentration of the two ions in the double layer region

$$\frac{c_i^*(Ca^{2+})}{c_i^*(Na^+)} = \frac{\exp(2 \times 6)}{\exp(6)} \approx 400$$

With a trivalent ion, such as lanthium, $c_i^*(La^{3+})/c_i^*(Na^+) \approx 1.6 \times 10^5$. Thus we see that multivalent ions preferentially concentrate near charged surfaces and are very effective at screening the charged surface, a fact we can also ascertain simply by calculating the Debye length.

Several important industrial processes exploit this congregation of multivalent ions at charged interfaces. For example, the water softeners we use in our homes contain negatively charged polymer resin beads. Softening water involves exchanging the Na^+ initially loaded onto the resin with dissolved divalent ions like Ca^{2+}, which make water hard. The Ca^{2+} ions are preferentially concentrated in the vicinity of the polymer resin beads. When the resin becomes saturated with Ca^{2+} ions, then we have to recharge it by passing a concentrated solution of NaCl over the resin and forcing the equilibrium between Ca^{2+} and Na^+ in the opposite direction. The first commercial water softening processes used clay particles like those described in Section 4.2.3 as ion exchangers. Other processes that exploit this property of charged interfaces are discussed in Section 4.7.

4.4 The Repulsive Potential Energy of Interaction, V_{rep}, between Two Identical Charged Surfaces in an Electrolyte Increases Exponentially as the Surfaces Move Together

4.4.1 Repulsive Forces Originate Due to Electrostatic Interaction

In this section we move on to analyze the potential energy of interaction between *two* charged particles immersed in an electrolyte so that we can determine the value of V_{rep} in eq. 4.1.3. Figure 4.10 shows the configuration used to model the interaction. We assume that the particles are very large, parallel plates (so we can ignore edge effects) and that they are immersed in a bath containing solution with bulk concentration c_{io}. Associated with each plate is a potential that decays exponentially with distance. We also assume the plates have identical and fixed surface potentials Φ_o.

Figure 4.10
Overlap of two double layers between a pair of charged surfaces separated by a distance h. The total potential—obtained by adding the potentials from each of the double layers—displays a minimum at the midplane between the two surfaces. (P. C. Hiemenz, *Principles of Colloid and Surface Chemistry*, 2nd ed., Marcel Dekker, New York, 1986, p. 704.)

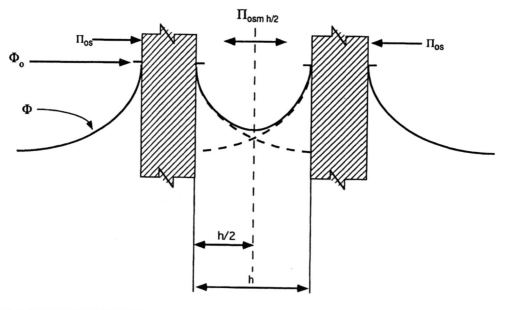

When the plates are separated by a large distance h, such that $h > 1/\kappa$ the electrostatic interaction between them is negligible. When the plates are brought together, electrostatic interactions between them become appreciable at separations of order $1/\kappa$. At this point, the electrical double layers overlap, and because both surfaces carry the same charge, they repel one another. We want to estimate the magnitude of this repulsive interaction as a function of the separation of the particles h.

To accomplish this goal, we consider the hydrodynamic stability of the electrolyte solution. For a liquid to be in equilibrium, the net force on any volume element of it must be zero, otherwise there will be flow from one volume element to another. That means the sum of the forces acting on a unit volume element in the equation of motion (the right-hand term in eq. 3A.6) must be zero.

If we focus our attention on the forces operating on volume elements in the region between the two plates, we find the electrical field emanating from the charged surfaces exerts an electrostatic force on the ions in solution. According to eq. 2.3.2, the electrical force exerted on an isolated charge by an electric field E is $F_{el} = (z_i e)E$. The corresponding expression for the force *per unit volume element* exerted on a volume element of the electrolyte in the z direction between the plates is $F_{el,z} = \rho(d\Phi/dz)$, where ρ is the net charge per unit volume.

4.4.2 Repulsive Forces Also Originate Due to Osmotic Pressure

A second force present in the electrolyte between the plates has an origin that may be less obvious. Due to the double layer, the concentration of ions in the vicinity of the plates is larger in that region than it is out in the bulk solution. Differences in concentration give rise to *osmotic pressure*. Because osmotic pressure plays such an important role in understanding repulsive forces here and in subsequent chapters, we will pause to review its origin and magnitude.

According to *Raoult's law*, when we add a nonvolatile solute to a solvent, we lower the solvent's vapor pressure by an amount $\Delta P_I = P_o X_I$, where P_o equals the vapor pressure of the pure solvent and X_I is the mole fraction of the solute. (We assume ideal behavior in this discussion.) If we place two beakers containing solutions with different amounts of solute in a desiccator, as indicated in Figure 4.11a, Raoult's law says solvent will evaporate from the more dilute solution (I) and condense in the more concentrated solution (II) until both solutions have identical composition.

We can carry out the same experiment with a rigid membrane that is permeable to solvent, but not to solute, using the apparatus shown in Figure 4.11b. If we place the two solutions in chambers on either side of the semipermeable membrane, solvent will flow from I to II. The solution in chamber II will rise up the capillary tube, generating a difference in hydrostatic pressure ΔP = (density) × gh between the two solutions. At a value of ΔP (= ΔP_I − ΔP_{II}) determined by the difference in concentration of solutes in I and II, the flow of solvent stops. Viewed in another way, we could

Figure 4.11
Two experiments illustrating osmotic pressure Π_{osm}. (a) At the start of the first experiment, two beakers containing solutions made up of solute (mole fraction $X_{II} > X_I$) are placed inside a thermostatted, evacuated chamber. Solvent evaporates from I and condenses in II until at equilibrium, $P_I = P_o X_I = P_o X_{II} = P_{II}$, where P_o represents the vapor pressure of the solvent and P_I and P_{II} are the partial pressures of solution I and II. (b) At the beginning of the second experiment, solvent is placed in compartment I and solution in compartment II. A rigid, semipermeable membrane, which admits only solvent, separates the two compartments. As solvent flows through the membrane, the solution rises in the capillary tube until the pressure head equals the osmotic pressure.

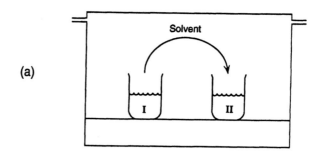

(a)

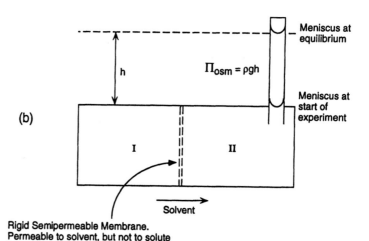

Meniscus at equilibrium

$\Pi_{osm} = \rho g h$

Meniscus at start of experiment

(b)

I   II

Solvent

Rigid Semipermeable Membrane.
Permeable to solvent, but not to solute

prevent solvent flow across the membrane at the beginning of the experiment by placing a small piston in the capillary and using it to exert a pressure difference of ΔP across the membrane. The difference in pressure is the *osmotic pressure* between the two solutions.

For an ideal dilute solution, we can define osmotic pressure by

$$\Pi_{osm} = kTc_{io}^* \qquad (4.4.1)$$

Osmotic pressure (like other colligative properties) depends on the number of solute particles per unit volume. When we add a salt, such as NaCl, we generate two particles of solute per molecule of salt, so eq. 4.4.1 becomes $\Pi_{osm} = 2kTc_{io}^*$.

Now we are in a position to explain how variations in osmotic pressure in the solution between the plates give rise to a repulsive force. The central point to bear in mind is that counterions are constrained to remain between the charged plates by their electrostatic interactions with the charged surfaces, and furthermore they are constrained to maintain a concentration gradient in the vicinity of the plates. The expression for the force *per unit volume element* exerted on a volume element of the electrolyte in the z direction due to the osmotic force in the z direction is $F_{osm,z} = d\Pi_{osm,z}/dz$.

4.4.3 The Total Repulsive Force between Two Charged Particles in an Electrolyte Is the Sum of the Electrostatic and the Osmotic Force

The total repulsive force on a volume element of the electrolyte is

$$F_{rep} = F_{el,z} + F_{osm,z} = \rho\frac{d\Phi}{dz} + \frac{d\Pi_{osm,z}}{dz} \qquad (4.4.2)$$

By examining Figure 4.10, we see that at the midpoint (h/2) between the two particles $d\Phi/dz = 0$; so the value of $F_{el,h/2} = 0$, and the only force acting on a volume element at that position is the osmotic force, $F_{osm,h/2}$. By arguments of continuity this same force must act on every volume element in the region between the plates. Thus the total hydrostatic force of repulsion F_{rep} *per unit surface area* of the plate (obtained by integrating $F_{osm,z}$ dz) equals the difference in osmotic pressure between the electrolyte at the midway point and the bulk solution.

$$F_{rep} = \Pi_{osm,h/2} - \Pi_{osm,bulk} = kT \,[(c^*_{ih/2,cations} + c^*_{ih/2,anions}) - 2c^*_{io}] \quad (4.4.3)$$

We can use the Boltzmann equation 4.3.4 to relate the local concentrations of ions, $c^*_{ih/2}$, to the potential at the midplane, $\Phi_{h/2}$, by

$$c^*_i = c^*_{io} \exp\left(-\frac{z_i e\Phi_{h/2}}{kT}\right)$$

Substituting this value into eq. 4.4.3 gives

$$F_{rep} = c^*_{io}\, kT \left[\exp\left(-\frac{ze\Phi_{h/2}}{kT}\right) + \exp\left(+\frac{ze\Phi_{h/2}}{kT}\right) - 2\right]$$

or
$$F_{rep} = 2c^*_{io}kT\left(\cosh\frac{ze\Phi_{h/2}}{kT} - 1\right) \qquad (4.4.4)$$

where we have used $z \equiv |z_1|$ and $\pm$ in the exponential to make a clear distinction between the cation and anion contribution. Equation 4.4.4 is valid only for symmetrical electrolytes, such as NaCl, $z_i = \pm1$, or $MgSO_4$, $z_i = \pm2$, not asymmetric ones like $MgCl_2$.

Before we proceed further, it is useful to remind ourselves of our goal. We want to obtain an expression for repulsive interaction energy between two particles as a function of their separation, $V_{rep}(h)$, where

$$V_{rep}(h) = -\int F_{rep}dh \qquad (4.4.5)$$

Equation 4.4.4 is not yet in a suitable form for integration because F_{rep} is written in terms of an unknown quantity, $\Phi_{h/2}$. We can relate $\Phi_{h/2}$ to Φ_o using the Gouy–Chapman theory with appropriate boundary conditions. Inserting the complete solution for Φ (eq. 4.3.6) leads to a differential equation so complex that it requires numerical integration. Instead we consider a simpler case where h/2 is large, that is, where $ze\Phi_{h/2} \ll kT$. We can then expand eq. 4.4.4 as a power series to obtain

$$F_{rep} = c_{io}^* kT \left(\frac{ze\Phi_{h/2}}{kT} \right)^2 \tag{4.4.6}$$

that still contains the unknown quantity $\Phi_{h/2}$. We now note that $\Phi_{h/2}$ between two particles is just twice the potential $\Phi(z)$ at $z = h/2$ from each of the individual surfaces. By expanding the terms involving Φ in the Gouy–Chapman equation 4.3.6, we obtain $\Phi(z)$ at h/2 for each of the individual surfaces and get

$$\Phi_{h/2} = 2\Phi(h/2) = 2 \, \frac{4kT\Gamma_o}{ze} \, \exp(-\kappa h/2) \tag{4.4.7}$$

Substituting eq. 4.4.7 into eq. 4.4.6 gives

$$F_{rep} = c_{io}^* kT [8\Gamma_o \exp(-\kappa h/2)]^2$$

or

$$F_{rep} = 64 c_{io}^* kT \Gamma_o^2 \exp(-\kappa h) \tag{4.4.8}$$

This result shows that the *repulsive force per unit area between two flat charged surfaces immersed in an electrolyte increases exponentially as the distance between them decreases.* The separation at which the repulsion starts to become significant equals the Debye length.

Now we can integrate eq. 4.4.8 to yield

$$V_{rep,p}(h) = \frac{64 c_{io}^* kT \Gamma_o^2}{\kappa} \, \exp(-\kappa h) \tag{4.4.9}$$

for the repulsive potential energy *per unit area* between two flat charged particles separated by a distance h in an electrolyte solution.

Using the Derjaguin approximation described in Section 2.6.2, we can obtain the value of $V_{rep,s}$ for two spherical particles of radius R separated by a distance h.

$$V_{rep,s}(h) = \frac{64\pi R c_{io}^* kT \Gamma_o^2}{\kappa^2} \, \exp(-\kappa h) \tag{4.4.10}$$

In this instance, V_{rep} is the repulsive potential energy *per pair* of identical spheres.

Equations 4.49 and 4.4.10 are the detailed expressions for the potential energy of repulsion, V_{rep}, between two particles as a function of their separation, a term introduced in eq. 4.1.3. Note in particular that the sensitivity of V_{rep} to electrolyte concentration is represented (through κ) by the exponential term; *the higher the concentration of counterions, the shorter the range of the repulsive interaction.* As the concentration increases, the charged particles come closer together. Thus, while the addition of salt is needed to stabilize a colloidal system, too much salt allows the particles to come so close together that they coagulate.

4.5 Electrostatic Stabilization of Colloidal Dispersions—Combining V_{att} and V_{rep} Leads to the DLVO Equation

In the 1940s Derjaguin and Landau in Russia and Verway and Overbeek in the Netherlands independently published a theory relating colloidal stability to the balance of long-range attractive and double-layer repulsive forces. The theory they proposed is known as the DLVO theory, from the initial letters of their names. They suggested that the total interaction energy V_T between two particles as a function of their separation h is the simple sum of the attractive and repulsive components

$$V_T(h) = V_{att}(h) + V_{rep}(h) \tag{4.5.1}$$

For parallel plates or flat particles, using eqs. 4.1.1 and 4.4.9

$$V_T(h) = \left(-\frac{H_{121}}{12\pi} \frac{1}{h^2} + \frac{64 c_{io}^* kT\Gamma_o^2 \exp(-\kappa h)}{\kappa} \right) \tag{4.5.2}$$

and for two spherical particles of radius R, using eqs. 4.1.2 and 4.4.10

$$V_T(h) = \pi R \left(-\frac{H_{121}}{12\pi} \frac{1}{h} + \frac{64 c_{io}^* kT\Gamma_o^2 \exp(-\kappa h)}{\kappa^2} \right) \tag{4.5.3}$$

These equations do not include the core repulsive terms for electron cloud overlap, V_{CR}, the $1/R^{12}$ term of eqs. 2.5.2 and 4.1.4, because it is so short ranged, and in general are not reliable for $h \ll \kappa^{-1}$.

Figure 4.12 shows $V_T(h)$ curves for parallel plates at two values of the Debye length, κ^{-1}. V_{att} dominates over V_{rep} when h

Figure 4.12
Total interaction energy V_T obtained by summing the van der Waals attractive energy V_{att} with either of two repulsion energies, $V_{rep}(I)$ or $V_{rep}(II)$. Curve $V_T(I)$ corresponds to a situation where there is a repulsive (positive) potential, which stabilizes the colloid if $V_{max} \gg kT$. Curve $V_T(II)$ corresponds to a situation in which the potential is just zero at the maximum. The absence of a repulsive interaction permits rapid coagulation. (D. J. Shaw, *Introduction to Colloid and Surface Chemistry,* 3rd ed., Butterworth–Heinemann, London, 1980, p. 192.)

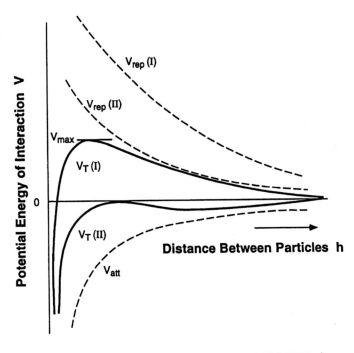

is either very large or very small. For intermediate separations, the double layer gives rise to a potential energy barrier if the surfaces are sufficiently charged (high Φ_o) or if the electrolyte concentration is so low (large Debye length) that it does not screen too much.

Three characteristic features of the total potential energy $V_T(h)$ curves shown in Figure 4.1 are extremely important in determining the behavior of a colloidal system. They are the primary minimum, $V_{pr\ min}$, the potential energy barrier, V_{max}, and the secondary minimum, $V_{sec\ min}$.

1. The total change in potential energy when particles coagulate is $V_{pr\ min}$. It is often so large that *coagulation is an irreversible process.*

2. The rate at which particles coagulate is determined by V_{max} (a topic to be pursued in the next section):

   a. When $V_{max} \gg kT$, particles are kinetically stabilized;
   b. When $V_{max} = 0$, coagulation becomes a rapid diffusion-controlled process.

3. $V_{sec\ min}$ is important only when it has a depth $> 5kT$ and when V_{max} is so large that the particles do not pass over it into the primary minimum. These conditions are met only with relatively large spheres, but when they occur, the particles move together until their average separation equals $h_{sec\ min}$. This process is called *flocculation,* and it *is reversible* because stirring easily separates the particles again.

4.5.1 We Can Use the DLVO Theory to Determine the Conditions under Which Coagulation Becomes Rapid

When the surface potential Φ_o is sufficiently reduced or when the salt concentration, represented by κ, is sufficiently increased, we reach a special case in which the barrier to coagulation vanishes and

$$V_{T(max)} = V_{att} + V_{rep} = 0 \tag{4.5.4}$$

as illustrated in Figure 4.12, curve $V_T(II)$. In this situation, coagulation occurs rapidly, and a previously stable sol separates into liquid and coagulated solid particles.

We can understand the features leading to this situation by noting that the maximum in the potential energy curve corresponds to $dV_T/dh = 0$; that is

$$\frac{dV_T}{dh} = \frac{dV_{att}}{dh} + \frac{dV_{rep}}{dh} = 0 \tag{4.5.5}$$

For parallel plates, we differentiate eq. 4.5.2 to obtain

$$\frac{dV_T}{dh} = +2\frac{V_{att}}{h} - \kappa V_{rep} = 0 \tag{4.5.6}$$

and because $V_{att} = -V_{rep}$ at the maximum

$$h_{max} = \frac{2}{\kappa} \text{ for plates; } h_{max} = \frac{1}{\kappa} \text{ for spheres} \quad (4.5.7)$$

Inserting the values for h_{max} for plates and spheres from eq. 4.5.7 into eqs. 4.5.2 and 4.5.3, respectively, and writing out the explicit dependence of κ on c, eq. 4.3.13, we obtain an expression for the concentration of salt that lowers V_T to zero and thus leads to rapid coagulation. This is the *critical coagulation concentration* (CCC), sometimes also known as the critical flocculation concentration (CFC). It is given by

$$CCC \approx 1 \times 10^5 \frac{(\varepsilon_r\varepsilon_o)^3(kT)^5}{z^6 e^6 H_{121}^2} \Gamma_o^4 \quad (4.5.8)$$

Equation 4.5.8 contains three key variables: T, z, and Γ_o. The important conclusions with respect to the first two variables are that coagulation can be stimulated by lowering the temperature and by increasing the valence of the counterions in the electrolyte. Equation 4.5.8 predicts that the CCC $\propto 1/z^6$ (when $\Gamma_o \approx 1$), a result observed quantitatively around 1900 that became known as the Schulze–Hardy rule. One of the great achievements of the DLVO theory was to provide a simple theoretical derivation for this rule.

Table 4.2a compares the ratios of the critical coagulation concentrations for counterions with various valencies to the theoretical predictions of the Schulze–Hardy rule. (Note that because the counterions are concentrated in the double layer, their valence, rather than the valence of the co-ions, is important in applying eq. 4.5.8.) The agreement is satisfactory for the low-valence counterions, but we observe significant deviations for tri- and tetravalent counterions. This discrepancy arises because the higher-valence multivalent ions, such as trivalent La^{3+}, associate with anions like chlorine to form divalent complexes, such as $(LaCl)^{2+}$, thereby reducing the concentration of the high-valence species.

With respect to the third variable Γ_o, we now consider how eq. 4.5.8 depends on the surface potential Φ_o. At high surface potentials, Φ_o is high and Γ_o approaches a value of unity (see Section 4.3.1); so the CCC becomes independent of potential and depends on $1/z^6$. At low surface potentials, we can expand the exponentials in eq. 4.3.7 to obtain $\Gamma_o = ze\Phi_o/4kT$. Substituting this result in eq. 4.5.8 gives

$$CCC \approx 4 \times 10^2 \frac{(\varepsilon_r\varepsilon_o)^3 kT\Phi_o^4}{z^2 e^2 H_{121}^2} \quad (4.5.9)$$

We have reduced the dependency on the counterion valence to $1/z^2$, but at the same time introduced an extreme sensitivity to potential $\propto \Phi_o^4$. Figure 4.13 plots CCC as a function of Φ_o and shows a transition from a z^{-6} dependence (in the "vertical" portions of the curves) to a z^{-2} dependence (in the "horizontal" portions of the curves) as the potential decreases.

Table 4.2b also contains a specific example of an important phenomenon involving the effect of counterions. For the Fe_2O_3 colloid system, the CCC for the hydroxide ion is considerably

TABLE 4.2 Critical Coagulation Concentrations (CCCs) for Counterions of Different Valence (z) in Negatively and Positively Charged Sols

(a) Comparison of CCCs for Three Negatively Charged Sols (As_2S_3, Au, and AgI) Containing Counterions of Different Valence with Theoretical Predictions of Eq. 4.5.8

Counter-ion valency (z)	As_2S_3 CCC (milli-moles per liter)	Ratio $\dfrac{CCC}{CCC_{z=1}}$	Au CCC (milli-moles per liter)	Ratio $\dfrac{CCC}{CCC_{z=1}}$	AgI CCC (milli-moles per liter)	Ratio $\dfrac{CCC}{CCC_{z=1}}$	Theoretical value of ratio ($=z^{-6}$)
+1	55.0	1.0	24.0	1.0	142.0	1.0	1.0
+2	0.69	0.013	0.38	0.016	2.43	0.017	0.0156
+3	0.091	0.0017	0.006	0.0003	0.068	0.0005	0.00137
+4	0.090	0.0009	0.0009	0.00004	0.013	0.001	0.00024

P. C. Hiemenz, *Principles of Colloid and Surface Chemistry*, 2nd ed., Marcel Dekker, New York, 1986, p. 718.

(b) Comparison of CCCs for Negatively and Positively Charged Sols Containing Counterions of Different Valence with Theoretical Predictions of Eq. 4.5.8

Negatively Charged As_2S_3 Sol

Counterion valency (z)	Electrolyte	CCC (millimoles per liter)	Ratio $\dfrac{CCC}{CCC_{z=1}}$	Theoretical value of ratio ($=z^{-6}$)	Zeta potential at CCC (ζ mV)
+1	K-Cl	40.0	1.0	1.0	44
+2	Ba-Cl$_2$	1.0	0.025	0.0156	26
+3	Al-Cl$_3$	0.15	0.00375	0.00137	25
+4	Th-(NO$_3$)$_4$	0.20	0.005	0.00024	27
+4	Th-(NO$_3$)$_4$	0.28	0.007	0.00024	26

Positively Charged Fe_2O_3 Sol

Counterion valency (z)	Electrolyte	CCC (millimoles per liter)	Ratio $\dfrac{CCC}{CCC_{z=1}}$	Theoretical value of ratio ($=z^{-6}$)	Zeta potential at CCC (ζ mV)
−1	K-Cl	100.0	1.0	1.0	33.7
−1	Na-OH	7.5	—	—	31.5
−2	Ca-SO$_4$	6.6	0.066	0.0156	32.5
−2	K$_2$-CrO$_4$	6.5	0.065	0.0156	32.5
−3	K$_3$-Fe(CN)$_6$	0.65	0.0065	0.00137	30.2

D. F. Evans and H. Wennerström, *The Colloidal Domain: Where Physics, Chemistry, Biology, and Technology Meet*, VCH Publishers, New York, 1994, p. 349. Reprinted with permission VCH Publishers © 1994.

lower than that of other monovalent anions, for example Cl$^-$. This phenomenon results from a specific interaction of hydroxide ions with the colloids containing iron or aluminum. Hydroxide ions adsorb on the colloidal particles to change the surface potential Φ_0. They are potential-determining ions—as we will see in Section 4.6.3.1—and this interaction lowers the CCC.

Figure 4.13
Critical coagulation concentrations CCC as a function of particle surface potential Φ_o. Solid lines calculated from eq. 4.5.8 using $H_{121} = 10^{-19}$ J, $\varepsilon_r =$ 78.5, and T = 298 K with a counterion valency of $z =$ 1, 2, and 3. The region above and to the left of each of the curves corresponds to the presence of a repulsive barrier to coagulation and is predicted to be a stable sol; the region to the right is predicted to coagulate. (D. J. Shaw, *Introduction to Colloid and Surface Chemistry*, 3rd ed., Butterworth–Heinemann, London, 1980, p. 198.)

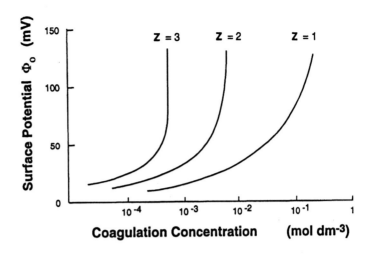

4.5.2 We Can Also Use the DLVO Theory to Determine the Rate at Which Colloidal Particles Coagulate

4.5.2.1 Rapid Coagulation of Colloidal Particles Is Limited Only by the Viscosity of the Solution

Having established in the previous section an understanding of the conditions for rapid coagulation, we must now ask how fast it actually occurs—the kinetics of coagulation. Contrary to the static image presented in our figures, it is important to emphasize that colloidal particles in a sol are in a continuous state of movement and agitation due to their thermal energy. They can approach each other with sufficient energy to overcome the energy barrier and bond together irreversibly. Coagulation involves a bimolecular association between two colloidal particles, P, to form a dimer, or $P + P \rightarrow P_2$. The decrease in the concentration of colloidal particles $[P]$ with time is

$$-\frac{d}{dt}[P] = k_r[P]^2 \qquad (4.5.10)$$

where k_r is the second-order rate constant for rapid coagulation (the subscript r stands for rapid).

We can evaluate k_r by considering a system of uniform spherical particles of radius R undergoing Brownian motion. We assume that the spheres interact only upon contact and adhere to form a doublet. This assumption is equivalent to replacing the potential energy curve $V_T(II)$ in Figure 4.12 by a square-well potential with an interaction distance equal to 2R, as shown in Figure 4.14. Although this assumption ignores interactions that occur at a distance greater than 2R, it provides a simple analytical model from which to develop more realistic models.

The relation between the rate of coagulation and particle diffusion assuming the square-well interaction was developed by Von Smoluchowski in 1917. He assumed that particles diffuse toward each other at a rate given by Fick's first law

$$J = -D \, \nabla[P] \qquad (4.5.11)$$

Figure 4.14
Square-well potential approximation. This approximation replaces the DLVO relation [corresponding to curve $V_T(II)$ in Figure 4.12] to permit calculation of coagulation rate under diffusion-controlled conditions. (D. F. Evans and H. Wennerström, *The Colloidal Domain: Where Physics, Chemistry, Biology, and Technology Meet*, VCH Publishers, New York, 1994, p. 339. Reprinted with permission by VCH Publishers, © 1994.)

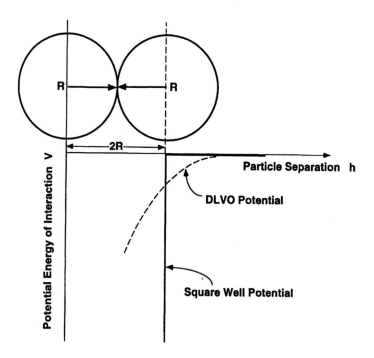

and obtained an expression for k_r

$$k_r = 8\pi R D \qquad (4.5.12)$$

(Appendix 4C gives the derivation.) The relationship between the diffusion coefficient D, the particle size, and the solvent viscosity η for uniform spheres with a radius R, is given by the Stokes expression (eq. 3.8.36)

$$D = \frac{kT}{6\pi\eta R} \qquad (4.5.13)$$

When substituted into eq. 4.5.12, this expression yields

$$k_r = \frac{4kT}{3\eta} \qquad (4.5.14)$$

Thus the *rate of rapid coagulation* of identical spherical particles at a given temperature *depends only on the viscosity* of the solution. It is independent of particle size because the increased probability that small particles will collide due to their increased mobility offsets the increased probability of large particles colliding due to their size. In water at 20°C, the diffusion-controlled binary association rate constant, k_r, equals 0.54×10^{-17} m^3 particle^{-1} s^{-1} or 3.25×10^9 M^{-1} s^{-1}.

During rapid coagulation, the association process does not stop with two particles coagulating, but continues to three, four, or more particles, until macroscopic particles precipitate. We derive an expression that describes the concentration of particles as a function of time in Appendix 4D. The relation shows that the concentration of particles P_m with different degrees of association m, where $m = 2$ for two coagulated particles, $m = 3$ for three, and so forth, is given by the general relation

Figure 4.15
Change in the total fractional number of particles, $\Sigma[P_m]/[P]_o$ (eq. 4.5.16) with time (upper curve). Curves plot the concentration of monomers, $[P_1]/[P]_o$, dimers, $[P_2]/[P]_o$, and trimers, $[P_3]/[P]_o$ as a function of the reduced time, t/τ (eq. 4.5.15); τ equals the half-time of coagulation (eq. 4.5.17). (J. Th. G. Overbeek, in *Colloid Science,* Vol. I, H. R. Kruyt, ed., Elsevier, Amsterdam, 1952, p. 282.)

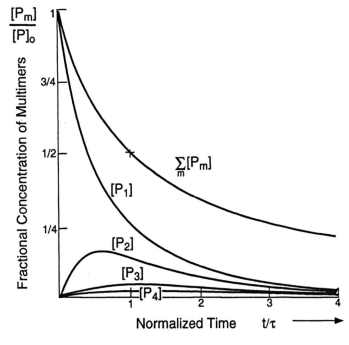

$$[P_m] = [P]_o \left(\frac{t}{\tau}\right)^{+m-1} \left(1 + \frac{t}{\tau}\right)^{-m-1} \qquad (4.5.15)$$

and the total concentration of all particles is

$$\sum_m [P_m] = \frac{[P]_o}{1 + t/\tau} \qquad (4.5.16)$$

When $t = \tau$, $\sum_m [P_m] = 1/2\,[P]_o$, half the original dispersed particles have coagulated, so τ equals the "half time" of rapid coagulation. The value of τ is given by

$$\tau = \frac{3\eta}{4kT[P]_o} \qquad (4.5.17)$$

For water and a starting concentration of 10^{10} particles per cubic centimeter, τ equals about 20 s.

Figure 4.15 shows the change in the total number of particles with time as well as the distribution of single particles, two particles, and so on during a rapid coagulation process.

4.5.2.2 Slow Coagulation Is Limited by the Height of the Energy Barrier

When the potential energy curve exhibits a barrier appreciably larger than kT, V_{max} on curve $V_T(I)$ in Figure 4.12, the rate of coagulation can slow down by orders of magnitude compared to the rate calculated by eq. 4.5.14. We can derive a rate expression for the initial stages of slow coagulation using a generalized form of Fick's first law that includes a term involving diffusion in the presence of an external barrier, given in Appendix 4E.

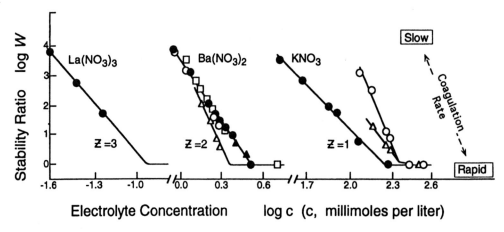

Figure 4.16
Measured values of the log of the stability ratio W versus log of the electrolyte concentration c for AgI sols of five different particle sizes (represented by different symbols on each curve). High values for log W correspond to very slow coagulation and low values to very rapid coagulation. Note in the transition regions represented by these curves that a change of electrolyte concentration by a factor of ten affects W by 10^8. The counterion valency z decreases from 3 to 2 to 1 for the three curves going from left to right. Note the pronounced dependence of W on counterion valency z. (H. Reerink and J. Th. G. Overbeek, *Discussions of the Faraday Society* **18**, 74 (1954).)

We can relate the slow rate constant k_s in the presence of a barrier to the rapid rate constant k_r in the absence of a barrier by

$$k_s = \frac{k_r}{W} \quad \text{or} \quad W = \frac{k_r}{k_s} \qquad (4.5.18)$$

where W is the *stability ratio*. When W is high, k_s is low, and the sol is extremely stable for a long period of time. W is expressed in terms of the maximum height of the potential barrier, V_{max}, by

$$W = \frac{\pi^{1/2} R}{h_{max}^2 p} \exp\frac{V_{max}}{kT} \qquad (4.5.19)$$

where $p = [(d^2V/dh^2)_{max}/2kT]^{1/2}$ and h_{max} represents the separation of the particles at which $V(h) = V_{max}$.

Figure 4.16 illustrates the important points concerning coagulation kinetics in electrostatically stabilized colloidal systems. *The switch from slow to rapid coagulation is sensitive to both the concentration and the valence of the counterion.* Theoretical analysis predicts a linear relationship between log W and log c in the slow coagulation region, as shown in Figure 4.16. In the region where the transition from slow to rapid coagulation is taking place, proceeding down the curves from left to right, the slope is such that increasing the electrolyte concentration by a factor of 10 decreases W by $\approx 10^8$ and therefore increases the coagulation rate by $\approx 10^8$. As can also be seen in the figure, the concentration at which the slow to rapid transition is complete increases by approximately four orders of magnitude as the valence of the counterion decreases from 3 to 2 to 1, so that a high-valence counterion is much more effective at initiating coagulation.

4.6 Surface Chemistry Plays an Important Role in Determining the Stability and Specific Properties of Colloidal Systems

One hallmark of a colloidal system is the large solvent–particle (liquid–solid) interfacial area. Interfacial chemistry must play a significant role in the behavior of colloidal systems. So far, we have characterized the electrical properties of this interface in great detail; we have analyzed the interactions between charged colloidal particles and their space-charge environment; we have seen how this interaction leads to an understanding of the stability of colloidal systems, the dependence of stability on the surface potential, and the dependence of stability on the concentration and valence of electrolyte. However, we have paid no particular attention to the details of the chemistry of the interface. We have treated ions as point charges with no chemical uniqueness other than their valence number. We have been concerned neither with the size nor the shape of the ions, nor with the nature of the chemical interaction between them and the surface of the charged particle. Yet, when ions are drawn toward charged surfaces, the finite size of the ions places an upper limit on the number that can be held at the surface. In addition, direct contact between ions and the surface promotes bonding or the formation of charge transfer complexes that must lead to effects dictated by the specific ion–surface interaction.

The next sections consider interface chemistry in more detail and how chemical aspects such as ion specificity and surface charge may affect colloid equilibrium conditions. We will find that attempts to incorporate surface chemical interactions into a theoretical model lead to intractable complications; we will show why the basic results of the electrostatic Gouy–Chapman and DLVO theories hold up remarkably well and how chemical effects may be understood in a qualitative rather than a quantitative way.

4.6.1 The Stern Model Provides a Way to Include Specific Ion Effects at Charged Interfaces

The Stern model provides one way to incorporate both finite ion size effects and ion–surface interactions into the electrostatic double layer model developed in the previous sections. It divides the double layer into two segments separated by a hypothetical boundary located at a distance δ from the surface. The segment immediately adjacent to the charged particle is known as the *Stern layer*. Usually the thickness of the Stern layer, δ, is taken to coincide with the centers of the first layer of counterions, as illustrated in Figure 4.17, and is therefore on the order of 1–3 Å thick. The segment beyond the Stern layer is treated as a Gouy–Chapman diffuse layer.

To evaluate the Stern layer we assume that the fraction of surface sites occupied by the counterion molecules, $\Theta = N_i/N$, can be expressed by the Langmuir isotherm, $\Theta = K_{ads} X_{io}/\{K_{ads} X_{io} + 1\}$ (eq. 3.7.11), where X_{io} is the bulk molar fraction or concentration of the counterion. In the Stern model, the equilibrium constant K_{ads}, is given by

$$K_{ads} = \exp -\left(\frac{z_i e\Phi_\delta + \Delta G_{ads\,i}}{kT}\right) \qquad (4.6.1)$$

where the energy term in the exponential is the sum of two contributions: the electrical energy of the ion in the Stern layer,

Figure 4.17
Schematic illustrating the Stern model that accounts for counterion size at the surface of a particle. The Stern layer extends a distance δ from the surface. The Gouy–Chapman diffuse layer starts at δ and extends beyond the Debye length (1/κ). The zeta potential, measured electrophoretically, is the potential at the shear surface between the particle and the solution. (D. J. Shaw, *Introduction to Colloid and Surface Chemistry*, 3rd ed., Butterworth–Heinemann, London, 1980, p. 156.)

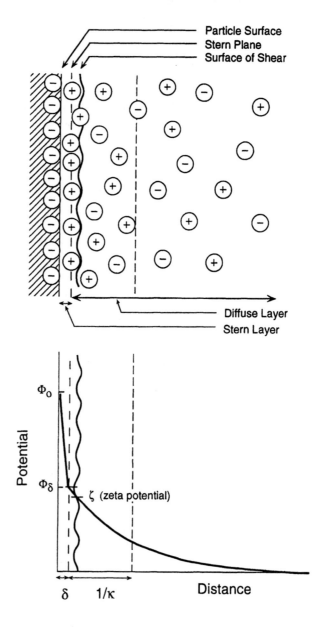

$z_i e\Phi_\delta$, and the specific chemical energy associated with the adsorption of ion i on the particle surface, $\Delta G_{ads, i}$.

We can also express Θ in terms of the charge densities associated with the adsorbed ions in the Stern layer, σ_δ, and the surface charge density on the particle, σ_o,

$$\Theta = \frac{\sigma_\delta}{\sigma_o} = \frac{\sigma_\delta}{z_i e N_{is}} \qquad (4.6.2)$$

N_{is} equals the total number of adsorption sites for ion i on the surface, and the quantity $z_i e N_{is}$ represents the charge density of the Stern layer if all the sites are occupied.

Combining eqs. 4.6.2 and 4.6.1 with eq. 3.7.11 and rearranging gives

$$\sigma_\delta = \frac{\text{zeN}_{cs}}{1 + X_{co}^{-1} \exp\left(\dfrac{\text{ze}\Phi_\delta + \Delta G_{ads\,c}}{kT}\right)} - \frac{\text{zeN}_{as}}{1 + X_{ao}^{-1} \exp-\left(\dfrac{\text{ze}\Phi_\delta - \Delta G_{ads\,a}}{kT}\right)} \tag{4.6.3}$$

where the first term on the left-hand side of eq. 4.6.3 gives the contribution of the counterions (c cations) and the second term that of the co-ions (a anions). When $\pm\,\text{ze}\Phi_\delta \gg kT$, one of these terms dominates over the other.

By analogy with the capacitor model for the electrical double layer in Section 4.3.2 and eq. 4.3.21, we can view the Stern layer as a molecular capacitor of thickness δ and dielectric constant ε_r', so that

$$\frac{\sigma_o}{\Phi_o - \Phi_\delta} = \frac{\varepsilon_r' \varepsilon_o}{\delta} \tag{4.6.4}$$

We can express electrical neutrality throughout the entire double layer by

$$\sigma_o + \sigma_\delta + \sigma_d = 0 \tag{4.6.5}$$

where σ_o and σ_δ have already been defined, and σ_d represents the charge density of the diffuse layer defined by the volume element of unit cross section extending from $x = \delta$ to infinity. Equation 4.3.20 gives σ_d with the sign reversed and with Φ_o replaced by Φ_δ. Substituting eqs. 4.6.4, 4.6.3, and the modified version of eq. 4.3.20 into eq. 4.6.5 gives the expression for the Stern model of the double layer

$$\frac{\varepsilon_r' \varepsilon_o}{\delta}(\Phi_o - \Phi_\delta) + \frac{\text{zeN}_{cs}}{1 + X_{co}^{-1} \exp\left(\dfrac{\text{ze}\Phi_\delta + \Delta G_{ads\,c}}{kT}\right)} - \frac{\text{zeN}_{as}}{1 + X_{ao}^{-1} \exp-\left(\dfrac{\text{ze}\Phi_\delta - \Delta G_{ads\,a}}{kT}\right)}$$

$$- (8kTc_{io}^* \varepsilon_r \varepsilon_o)^{1/2} \sinh(\text{ze}\Phi_\delta/2kT) = 0 \tag{4.6.6}$$

where

$$X_{io} \approx c_{io}^* V/N_{Av} \tag{4.6.7}$$

V is the solvent molar volume. This expression assumes that molar fraction is approximately (number of ions of solute/m^3)/(number of solvent particles/m^3), an approximation that is satisfactory for dilute solutions.

Equation 4.6.6 is the Stern extension of the Gouy–Chapman equations, specifically designed to take into account ion size and chemical interactions at the colloidal particle surface. The equation is conceptually useful, but it contains five adjustable parameters (ε_r', δ, N_{io}, Φ_δ, and ΔG_{ads}). Ways to estimate these parameters have been devised only in well-characterized systems. Although such ideal systems provide useful insights into the properties of the double layer (see the example in Appendix 4B), attempts to analyze most practical interfacial colloidal systems in terms of eq. 4.6.6 are not very useful simply because the systems are not sufficiently well characterized.

Actually things are not as hopeless as they might appear. After all, our main concern in colloid stability is with the overlap of the diffuse segment of the double layer. The Stern layer does not invalidate the expressions developed in Section 4.4.3 to describe the repulsive interaction energy between two particles. In fact, lowering the potential at the inner boundary of the diffuse layer to Φ_δ enhances the validity of the low-potential approximations that we used to derive expressions for the DLVO theory. All our conclusions remain intact if we simply replace Φ_0 in the equations by Φ_δ.

In previous discussion of the Gouy–Chapman and DLVO theory in Sections 4.3, 4.4, and 4.5, we tacitly assumed that Φ_0, the surface potential that sets the potential in the diffuse part of the double layer, is measurable. Actually only a few colloidal systems provide this information directly. In general Φ_0 is not available. On the other hand, Φ_δ is more accessible through electrophoresis measurements (see the next section). So replacing Φ_0 by Φ_δ makes the DLVO theory all the more tractable. DLVO continues to provide the most effective theoretical foundation for understanding electrostatic stabilization of colloidal systems.

We can measure the interaction forces between two mica surfaces in an electrolyte directly using the surface forces apparatus (SFA). Figure 4.18 shows the force curve in $1.2 \times 10^{-3} M\,HCl$. The solid line corresponds to the Gouy–Chapman equation 4.5.2, with a Hamaker constant $H_{121} = 2.2 \times 10^{-20}$ J, a Debye length $1/\kappa$ = 12.5 nm, and a potential $\Phi = -130$ mV. Good agreement between the theoretical curve and the actual measurements confirms that the interaction forces at normal separations are controlled by the

Figure 4.18
Force curve for the interaction of two mica surfaces in $1.2 \times 10^{-3} M\,HCl$ obtained by surface forces apparatus measurements. Normalized repulsive force, F/R, between two crossed cylindrical surfaces, radius R, as a function of separation h follows Gouy–Chapman theory (solid line) up to point where the two surfaces jump together. [R. M. Pashley, *J. of Colloid and Interface Science* **80**, 153 (1981).]

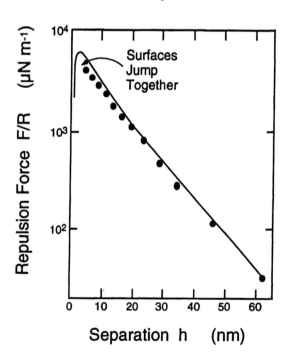

electrical potential in the diffuse layer and are not dominated by the surface. Unfortunately we cannot use the SFA to measure shorter-ranged interactions because the attractive forces become larger than the spring constant of the spring that holds the mica surfaces apart. As a result, the surfaces jump into contact. This occurs at approximately 3 nm (30Å) in Figure 4.18.

4.6.2 The Zeta Potential Provides a Useful Correlation between Double Layer Potentials and Colloid Stability

We can estimate the value of the Stern layer potential Φ_δ from electrophoresis measurements. In an electrophoresis experiment, we apply an electrical field to a colloidal system and measure the velocity of the colloidal particles directly by timing their motion between fixed points in an optical microscope. The velocity depends not on Φ_o, the potential on the particle surface, but on the potential at the interface of shear between the charged particle and the solution. We do not know the exact location of this surface, but usually it is assumed to correspond to a position equal to or slightly farther from the surface than the Stern layer δ. The potential at the shear interface obviously corresponds closely to the potential at the boundary between the Stern layer and the diffuse layer, represented by Φ_δ in the previous section. The actual measured value is called the *zeta potential* ζ, and it represents the effective surface potential that a colloidal particle carries as it moves through the sol.

As might be expected, the zeta potential is the most significant measurable property that can be used to categorize the stability of a colloid. In particular the zeta potential of a moving particle defines the propensity for coagulation. Table 4.2b shows that for a given sol the zeta potential at the critical coagulation concentration is approximately the same in different electrolytes, 30±5 mV. Significantly this result applies also for the Fe_2O_3 sol stabilized with monovalent OH^-. Thus zeta potential measurement is a useful probe in determining how close a sol is to coagulation.

4.6.3 In Colloidal Systems Many Specific Ion Effects Can Be Understood by the Action of Potential-Determining, Indifferent, and Charge-Reversing Ionic Constituents

In Section 4.2.3 we noted that surface charges in colloidal systems have a chemical source due to surface ionization and/or selective adsorption. As we might expect, the actual surface charge on a particle due to selective adsorption relates to the specific nature of the interaction between the colloidal particle and the electrolyte. We can distinguish three different kinds of chemical interaction, depending on whether the electrolyte contains potential-determining ions, indifferent ions, or charge-reversing adsorbed ions or complexes.

4.6.3.1 Potential-Determining Ions

Ag^+ or I^- ions are potential-determining ions for AgI particles. In describing the preparation of a AgI colloid by the precipitation technique in Section 4.2.1, we inferred that by varying the relative amounts of $AgNO_3$ or KI added initially to the solution, we could end up with an excess concentration of Ag^+ or I^- ions in solution. However, we cannot vary the concentrations of Ag^+ and I^- independently because their solubility product is constant at $c_{Ag^+}c_{I^-} = 7.5 \times 10^{-17}$ at 25°C. If the concentration of the two ions in solution is equal, then $c_{Ag^+} = c_{I^-} = 8.7 \times 10^{-9}$ M.

How does varying the concentration of Ag^+ and I^- in solution affect the charge on the surface of the AgI colloidal particles? We can determine their net charge by noting whether the particles move toward the positive or negative electrode during an electrophoresis experiment. When $c_{Ag^+} = c_{I^-}$, the particles move toward the positive electrode; that is, they carry a negative net surface charge.

If we vary the concentration of Ag^+ (or I^-) by adding $AgNO_3$ (or KI), we observe that when $c_{Ag^+} > 3.0 \times 10^{-6}$ M, the particles are positively charged; below this value, when $c_{Ag^+} < 3.0 \times 10^{-6}$ M, they bear a negative charge. Thus zero net surface charge occurs when $c_{Ag^+} = 3.0 \times 10^{-6}$ M and $c_{I^-} = 2.5 \times 10^{-11}$ M. This is the point where the curve in Figure 4.19 crosses the zero potential axis. Furthermore, in this concentration region, the potential on the particles, as determined by electrophoretic measurements, varies almost linearly as a function of the logarithm of added Ag^+ (or I^-) ions. Because the relative concentrations of Ag^+ or I^- ions in solution determine the charge and therefore the potential on the surface of the AgI particles, they are known as *potential-determining ions*.

We can understand this charge variation in terms of the Gibbs adsorption isotherm. At equilibrium, the chemical potentials of the Ag^+ and I^- ions in solution equal those on the colloidal particle. Each competes for adsorption sites on the surface of the particle, and if we add both Ag^+ and I^- ions to the solution, further

Figure 4.19
Comparison of the calculated surface potential Φ using the Nernst equation 4.6.9 (dashed line) with measured zeta potentials ζ (solid curve) as a function of silver ion concentration in a silver iodide colloidal system. (J. Th. G. Overbeek, in *Colloid Science,* Vol. I, H. R. Kruyt, ed., Elsevier, Amsterdam, 1952, p. 231.)

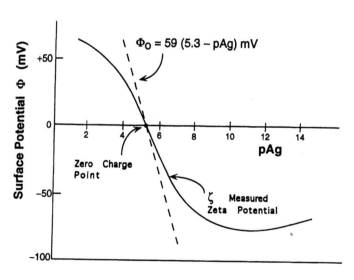

$\Phi_0 = 59 (5.3 - pAg)$ mV

Zero Charge Point

ζ Measured Zeta Potential

− Log Concentration (I⁻)

adsorption leads to growth of the colloidal particle. The concentration corresponding to zero surface charge, the *zero-charge point* c_{zp}, occurs when the two ions are adsorbed in equal numbers, and the fact that c_{zpI^-} (= 2.5×10^{-11} M) is much smaller than c_{zpAg^+} (= 3.0×10^{-6} M) simply reflects readier adsorption of I^- ions on the AgI surface.

Using the Nernst equation, we can develop a relationship between the chemical potential of the Ag^+ and I^- ions in solution and the surface potential Φ_o on the particles. In differential form, the Nernst equation is

$$d\Phi_o = -(2.303 \, RT/F) \, d(\log_{10} a) \approx -(2.303 \, RT/F) \, d(\log_{10} c) \quad (4.6.8)$$

where F represents the Faraday constant, R the gas constant, and a the activity. At 25°C, 2.303RT/F equals 59 mV. Integrating eq. 4.6.8, using the boundary condition of $\Phi_o = 0$ at $c_{Ag^+} = 3 \times 10^{-6}$ M or $c_{I^-} = 2.5 \times 10^{-11}$ M, gives

$$\Phi_o = 59 \log\left(\frac{c_{Ag^+}}{3.0 \times 10^{-6}}\right) = -59 \log\left(\frac{c_{I^-}}{2.50 \times 10^{-11}}\right) mV \quad (4.6.9)$$

Consequently, minute changes in the concentration of potential-determining ions dramatically affects the surface potential and thus the stability of colloidal systems. For example, when the concentration of Ag^+ ions varies from 3×10^{-7} to 3×10^{-5} M, the surface potential according to the Nernst equation changes from −59 to 59 mV.

Figure 4.19 compares values of Φ_o calculated from the Nernst equation (the dashed line in the figure) with measured zeta potentials (solid line) for a silver iodide sol. Equation 4.6.9 predicts a linear relationship between Φ_o and log c with a slope of 59 mV passing through zero at pAg = 5.3. Experimentally the slope of the solid line, $(d\zeta/d \log c_{Ag^+})_{\zeta \to 0}$, is −40 mV and not −59 mV. This difference reflects the effect of the Stern layer.

Potential-determining ions play a key role in the processing and stabilization behavior of many important colloidal systems (see Table 4.3). Hydroxide and hydrogen ions serve as potential-determining ions for colloidal molecules or particles containing weak basic or acidic groups. In these colloids, the pH of the solution is the critical variable determining the surface charge. *To maintain stability, the pH should never be allowed to vary through the zero-charge point because at this pH value electrostatic stabilization does not exist, attractive forces dominate, and irreversible agglomeration takes place.* Colloidal systems sensitive to pH

TABLE 4.3 Potential-Determining Ions

System	Potential-determining ion
Al (OH)$_3$	Al^{3+}, H^+, OH^-
Fe (OH)$_3$	Fe^{3+}, H^+, OH^-
AgI	Ag^+, I^-
Au	Cl^-
Proteins	H^+, OH^-

include ionic polymers, latex spheres, amphiphilic aggregates containing carboxyl and/or amino groups, and most proteins. Among the other colloids noted in Section 4.2, the chloride ion is a potential-determining ion for gold particles because it forms a particularly strong complex. Hydrogen, hydroxide, and metal ions, such as Al^{3+} or Fe^{3+} and their amphoteric complexes, are potential-determining ions for $Al(OH)_3$ and $Fe(OH)_3$, as well as for many ceramic colloidal systems.

4.6.3.2 Indifferent Ions

In the AgI system, KNO_3 functions as an indifferent electrolyte. Neither the K^+ nor the NO_3^- ions preferentially adsorb on the AgI particles to change the surface charge density. *Indifferent ions* in the electrolyte do not directly affect the surface charge or surface potential on the colloidal particles. However, they do continue to function as counterions and co-ions in the diffuse layer of the electrolyte solution, and their concentration does change the screening interactions between colloidal particles, as measured by the Debye length $1/\kappa$. They also change the activity of potential-determining ions, but that effect is generally a secondary one.

4.6.3.3 Charge-Reversing Adsorbed Species

The third kind of interaction between colloidal particles and the electrolyte involves the manipulation of surface charge through the selective adsorption of other molecules or complexes from solution. The net surface charge depends on the concentration of potential-determining ions, the concentration of adsorbed species, and the pH of the solution.

To illustrate these ideas, we examine a system involving silver iodide sols where we vary the concentration of Al^{3+} ions and the pH. Figure 4.20 shows the stabilization-coagulation behavior for silver iodide particles suspended in silver nitrate–potassium iodide solutions. Aluminum nitrate is added to these solutions, and the pH is changed to determine the critical coagulation concentrations (CCC). Electrophoresis measurements determine the charge on the particles.

To understand the results plotted in Figure 4.20, we start with the system containing an aluminum nitrate concentration given by $\log[Al(NO_3)_3] = -4.7$. Increasing the pH while holding the aluminum nitrate concentration fixed, that is, traversing the path XX', leads to the following sequence of events. At pH = 3, we obtain a negatively charged stable colloid, a negative sol. At pH 4.4, we encounter a critical coagulation concentration, and the particles precipitate out of the colloidal suspension. If we increase the pH to values ranging from 5.5 to 5.9 (which must be done rapidly to avoid coagulation), we regain a stable system, but electrophoretic measurements show that it has undergone charge reversal—it is now a positive sol. At pH 5.9, coagulation reoccurs, and above pH ≈ 7.6, the system returns to a stable negative sol.

Figure 4.20 plots the loci of the CCCs for different aluminum nitrate concentrations as a function of pH and establishes the "phase field" for the three different situations. Coagulation occurs

Figure 4.20
Change in the critical coagulation concentration CCC for a AgI sol as a function of the concentration of $Al(NO_3)_3$ and pH. Below the lower curve, the colloid is negatively charged and stable. Above the upper curve, the colloid is positively charged due to adsorption of aluminum complexes and stable. In the shaded "field" between the two curves, the concentration exceeds the CCC, and the AgI sol precipitates. [E. Matijevic, K. G. Mathai, R. H. Ottewill, and M. Kerker, *J. of Physical Chemistry* **65**, 826 (1961). Reprinted with permission from *J. of Physical Chemistry*. Copyright 1961 American Chemical Society.]

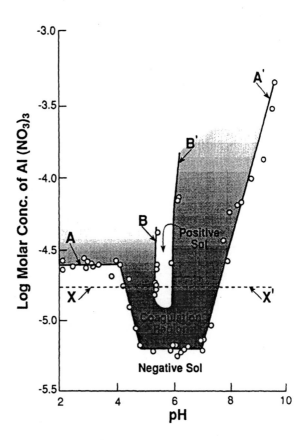

in the "field" between the two curves AA′ and BB′. The sol is stabilized in the region outside the two curves; for the large "field" below AA′ the particles are negatively charged as they normally should be, but in the narrow "field" above BB′ they are positively charged—a charge reversal is observed.

This multifaceted stability results from a complex interplay between the colloidal particles and a variety of hydrated aluminum species. We can summarize the features of this AgI colloid system in the following way:

1. When the pH is less than 4, the CCC is constant and equal to 2.5×10^{-5} *M*, a value characteristic for destabilization of the colloid by the hydrated trivalent Al^{3+} species. Zeta potential measurements show that the AgI sol remains negatively charged below the CCC in acidic solutions in the presence of $Al(NO_3)_3$ concentrations.

2. At pH = 4, the CCC drops sharply to 7.2×10^{-6} *M* and remains almost constant up to pH = 7. This drop implies the formation of a hydrolyzed species with a charge greater than +3. A variety of titration and spectroscopic experiments suggest the equilibrium

$$8Al^{3+} + 20\ OH^- = Al_8(OH)_{20}^{4+} \qquad (4.6.10)$$

resulting in a complex with a +4 charge. When recalculated in terms of the normality of this complex, a CCC of 3.6×10^{-6} *M* agrees with the CCC of 4×10^{-6} *M* obtained for other tetravalent ions with negative AgI sols.

3. When the pH increases above 7, the CCC rises strongly, indicating the formation of a species with a lower charge. This behavior is consistent with the equilibrium

$$Al_8(OH)_{20}^{4+} + 2OH^- = Al_8(OH)_{22}^{2+} \qquad (4.6.11)$$

   although evidence for the existence of this structure is not well established.

4. Between pH 5 and 6 and at a sufficiently high concentration of $Al(NO_3)_3$, zeta potential measurements show that the AgI becomes positively charged. This charge reversal results from the preferential adsorption of hydrolyzed metal complexes onto the AgI particles rather than the hydrated metal ions themselves. The hydrolyzed metal complexes are *charge-reversing adsorbed species* on the silver iodide particles. Preferential adsorption of other species, such as $Th(OH)^{3+}$, over Th^{4+}, in such systems indicates this effect is not merely the result of higher valence.

Thus preferential adsorption of complex ionic species can lead to reversal in particle charge. Even at low concentrations, surface-active materials, such as surfactants and some polymers, can have pronounced effects on colloidal stability.

4.6.4 Heterocoagulation between Dissimilar Colloidal Particles Is More Complex Than Homocoagulation between Identical Particles

We conclude this discussion of chemical effects on electrostatic stabilization by noting that we have been preoccupied with the chemistry of the electrolyte, yet the particles themselves may have different chemistries. In many applications we encounter mixtures of colloidal sols with different Hamaker constants and surface potentials.

In Section 2.6.3, eq. 2.6.21, we saw that the Hamaker constant for particles 1 and 3 separated by a solvent 2 is

$$H_{123} = (H_{11}^{1/2} - H_{22}^{1/2}) (H_{33}^{1/2} - H_{22}^{1/2}) \qquad (4.6.12)$$

H_{123} can be either positive or negative, depending on the relative magnitudes of H_{11}, H_{22}, and H_{33}. So far, we have considered identical particles in a solvent, and they always attract. But in some solvents, the London dispersion interaction between unlike particles can be repulsive, and in these instances the two sols would prefer to separate.

A more significant result occurs when the surface potentials on the two particles differ; then we must exchange the symmetrical variation of Φ with h, shown in Figure 4.10, for the asymmetrical condition shown in Figure 4.21. A reformulation of the Gouy–Chapman model using appropriate boundary conditions leads to more complex expressions. However, for two flat particles we can write an approximate solution of the form

$$V_{rep} \approx \frac{64 \, c_{io}^* \, kT \, \Gamma_1 \Gamma_2}{\kappa} \exp(-\kappa h) \qquad (4.6.13)$$

Figure 4.21
When the potential on two particles (1) and (2) differs and $\Phi_{01} > \Phi_{02}$, the change in potential Φ between the surfaces of the two particles ceases to be symmetrical and the DLVO theory must be reformulated. At $z = z_o$, $d\Phi/dz = 0$ and both Φ and σ have their minimal values. (D. F. Evans and H. Wennerström, *The Colloidal Domain: Where Physics, Chemistry, Biology, and Technology Meet*, VCH Publishers, New York, 1994, p. 205. Reprinted with permission by VCH Publishers, © 1994.)

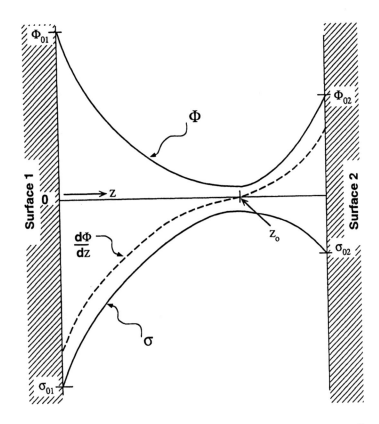

This equation should be compared with eq. 4.4.9. We find that Γ_o^2 from eq. 4.4.9 is replaced by the geometric mean, $\Gamma_1 \Gamma_2$.

Under conditions in which both surfaces have fixed surface potentials of different value ($\Phi_{01} \neq \Phi_{02}$ = constant), electrostatic interactions can switch from repulsive to attractive as the two surfaces approach one another.

4.7 Applications Involving Colloidal Interfacial Systems

4.7.1 Paints Exemplify Complex Interfacial Colloidal Systems

Annual sales of paints and related coating materials in the United States exceed $8 billion dollars. Paints serve two main functions: decoration and protection against corrosion and/or weathering. The first commercial paints were oil based and upon drying released up to 50% of their wet volume as organic vapors into the atmosphere. Environmental restrictions and economic constraints have led to the increasingly widespread use of water based paints. To function effectively, these complex colloidal dispersions must meet a variety of criteria at the same time. In the sections to follow, we describe the main functions that a paint coating must meet, provide examples of systems that fulfill these requirements, and discuss some failure mechanisms and how they can be minimized by proper formulation of the paint.

4.7.1.1 Desirable Properties of a Paint Coating

One important property of a paint is its hiding power. If we paint over a dark wall with white paint, we do not want to see the dark surface through our newly finished surface. How do we achieve high-hiding power?

Nature produces a high hiding coating that illustrates the general principle very nicely—snow. A thin layer of snow on a highway completely obliterates the blacktop underneath. Snow consists of small ice crystals—snowflakes—separated by small microvoids of air. The microvoids cover a size range comparable to the wavelength of light and consequently scatter light rather than transmit it. The intensity of scattered light depends on the difference in refractive index between the snowflakes (n_{water} = 1.33) and the microvoids (n_{air} = 1.0) raised to the second power; that is, $I_{scattered} \propto (n_{air} - n_{water})^2$. In fact, all white materials that occur in nature, such as birchbark, feathers, clouds, and so forth, contain microvoids. Because air is a rather inexpensive ingredient, we might consider using microvoids containing air to achieve high hiding power in paints, and indeed such paints have been produced. However, they have limited utility because whenever the microvoids become filled with water by capillary condensation, the hiding power is diminished or obliterated.

More typically, colloidal-sized titanium dioxide (TiO_2) particles that possess a high refractive index are used as scattering centers in paints. Such materials are called *pigments*. Because they are one of the most expensive ingredients, a major goal in designing a paint is to minimize the required amount of pigment.

To be effective, the pigment must adhere firmly to the surface being painted, and materials called *binders* are used for this purpose. In many water based systems, the binder consists of tiny latex particles. Upon drying, the latex particles coalesce to encapsulate the pigment particles in a continuous film. The difference in the refractive index between the pigment particles and the binder determines the hiding power of the coating.

Pigment and binder are the two main functional components of a paint coating. However, if we attempted to formulate a simple paint by dispersing TiO_2 into a latex, we would not be very satisfied with the results for a variety of reasons: (1) Colloidal stability, which is a major determinant of the paint's shelf life, would be too limited. (2) Hiding power would be poor because we could not disperse enough pigment into the latex without making the paint too viscous to manage. (3) Even if we found a way to control the viscosity, the final paint film would be unacceptable because we would find it almost impossible to eliminate brush marks. (4) We would probably encounter problems with the coat's long-term durability. (5) If the paint were used as an exterior coating, sunlight would rapidly deteriorate the binder as a result of UV photochemical reactions catalyzed by the TiO_2. (6) We might also encounter peeling and blistering associated with low adhesion between the coating and its substrate, or (7) cracking of the film as the substrate changed dimensions with temperature or moisture (which would in turn compromise its aesthetic and protective functions), or (8) mildew formation under humid conditions.

4.7.1.2 Basic Ingredients of a Paint

To show how these challenges are met in formulating a modern paint, Table 4.4 lists specific examples of ingredients used in making two paints: one a high-quality paint usable for both exterior and interior applications and the other a cheaper, less versatile paint. The ingredients also are categorized under generic headings on the left-hand side of the table used to describe the various components in latex paints.

Dispersants used to impart colloidal stability to the TiO_2 sol permit higher pigment content and lower viscosity. In some applications, this stability makes it possible to replace some of the expensive pigment by less costly extenders. Additives, such as coalescing solvents, promote coalescence of the latex particles.

TABLE 4.4 Typical Formulations for Latex-Based Paints

Ingredients matte latex paints	A Interior/exterior (wt %)	B Cheap interior (wt %)
Pigment—Tioxide RXL (TiO_2)	20.0	11.0
Extenders—Snowcal 6ML (Whiting)	11.1	29.0
China clay M100 (Calcined China Clay)	5.0	—
Speswhite (China Clay)	—	5.0
Dispersants—10% Calgon PT solution (polyphosphate)	—	1.0
5 percent Tetron solution (polyphosphate)	2.0	—
Dispex N40 (sodium polyacrylate)	—	0.5
Orotan 850	0.5	—
Defoamer—Bevaloid 691	0.2	0.1
Polymers—3% Methocel J12MS solution (hydroxypropyl-methyl cellulose)	—	16.0
3% Bermacoll E270G solution (ethylhydroxy cellulose)	18.0	—
Ammonia (0.910)	0.1	0.1
Coalescents—butyl diglycol acetate	1.0	—
Texanol	—	0.5
Dibutyl phthalate	—	0.25
Preservative—Proxel CRL	0.1	0.1
Binder—Latex, Emultex AC430 (TSC 55%)	24.9	10.0
Water	17.1	26.45
Total	100	100
Solids content	51%	51%
Volume solids	23.4%	29.4%
Pigment volume concentration	50% (CPVC 47%)	75%
Specific gravity	1.4	1.43

Polymers are used to control paint rheology, particularly during application. Many of these polymers inevitably introduce surface-active compounds that promote foam formation; so defoamers are added to counteract this effect. Some of the ingredients, particularly the polymers, are liable to degradation by microorganisms; so a biocide is added to preserve them.

Before we describe the role of the various ingredients in more detail, we must define two fundamental concepts: pigment volume concentration and critical pigment volume concentration. Pigment volume concentration (PVC) is the volume percent of pigment and extender in the dried paint film

$$PVC = \frac{\text{volume of (pigment + extender)}}{\text{volume of (pigment + extender + polymer + other nonvolatile ingredients)}} \qquad (4.7.1)$$

Generally, gloss and semigloss paints have PVCs in the range of 15 to 25% and mainly use TiO_2 as the pigment. High-quality matte paints, like Formulation A illustrated in Table 4.4, have PVCs ranging from 40 to 65%, while cheaper paints, like Formulation B, have much higher PVCs achieved by adding two or three times as much extender as TiO_2.

The critical pigment volume concentration (CPVC) is defined as the point at which just sufficient binder is present to coat the pigment particles and fill the void spaces between them. CPCVs can be estimated by determining the oil required to fill the void space of a packed bed of pigments.

A major goal in the design of latex paints is to have the PVC significantly larger than the CPVC. Under these conditions, air voids entrained in the paint film produce increased opacity. This strategy—known as dry hiding—provides one way to minimize the amount of pigment required and lower the cost. However, it makes the paint film more porous, which can lead to increased permeability and compromised mechanical properties.

4.7.1.2A Pigments and Extenders. Titanium dioxide is the most widely used pigment. Normally, TiO_2 pigments are not used in pure form because they disperse poorly in most solvents. Usually, they are surface coated with small amounts of alumina, silica, or both. The active adsorption sites they provide for polymers help deagglomerate, disperse, and stabilize the pigments. In addition, the alumina/silica coating provides a protective barrier between the pigments and binder, which reduces photoinduced decomposition of the binder. Virtually all inorganic pigments (such as iron oxides, cadmium sulfide, vermilion, and so forth) and organic pigments (such as copper phthalocyanine blue and quinacridone red) are also surface coated.

The refractive indexes of extenders are very close to those of the binders and consequently do not directly influence the opacity below the CPVC. However, the use of extenders raises the PVC and thus provides an economically attractive way to minimize the use of pigments. In most countries, the cheapest extenders are locally mined and graded whitings, but China clay and dolomite are also used because of their whiteness.

4.7.1.2B Dispersants. Adding anionic dispersants (such as polyphosphates, which preferentially adsorb onto the pigment and extender particles) increases the stability of a paint. They impart a negative surface charge to the colloidal particles, and the resulting electrostatic repulsion stabilizes the paint against flocculation. Because many latex particles are also negatively charged, introducing dispersants also minimizes heterocoagulation between pigments, extenders, and latexes.

4.7.1.2C Latexes. A variety of anionically and sterically stabilized latexes are used as binders. Most latexes used in paints exhibit a Gaussian distribution of particle sizes ranging from 0.1 to 1.0 μm with a peak particle size below 0.6 μm. Particle size directly affects many of the properties of a paint system, as illustrated in Figure 4.22. Smaller latexes provide higher gloss because they reduce light scattering at the surface and at the same time improve pigment binding simply because the mixture contains more latex particles per pigment particle. On the other hand, larger latexes provide higher opacity because they facilitate entrainment of air voids above the CPVC. Improved flow is also associated with larger particle size. Clearly there are tradeoffs in finding the best compromise for optimizing performance of a coating.

A second key variable is the composition of the latex particle, which directly affects its ability to coalesce and form a continuous film. The minimum film-forming temperatures (MFTs) for typical latexes used in paints range from +1 to 16°C, that is, below to just around room temperature. The MFTs are related directly to the glass transition temperature T_g, of the latex polymer, the temperature at which it transforms from a brittle glass to a soft plastic material. In general, latexes possessing higher MFTs yield harder final films, a desirable feature in many applications. However, their use in unmodified form is severely limited in nontropical climates. One way of modifying the latex is to incorporate a small amount of a co-monomer to lower the MFT; another way is to add a coalescent, which acts as a plasticizer (see below). In many applications, such as primers for wood and paints for roofs, flexibility is an essential property of the coating. The range of latexes now available makes it possible to choose a composition to optimize final performance.

4.7.1.2D Coalescents. Coalescents are high-boiling sol-

Figure 4.22
The particle size of pigments and binders has a pronounced effect on the properties of a paint system.

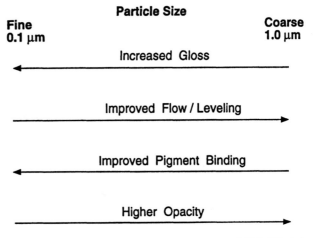

vents, such as diethyl phthalate or diethyleneglycol monobutyl ether, which preferentially partition into the latex phase and act as a plasticizer to facilitate coalescence. By judicious incorporation of coalescents, a latex paint with a MFT of 12°C can be used at temperatures as low as 5°C. Coalescents are not required in tropical countries where 20°C temperatures are normal.

4.7.1.2E Polymers. So far, we have focused mainly on issues of colloidal stability that are important in determining shelf life and optimizing the properties that determine the performance of the finished coat. Now we consider the issues involved in actually applying the paint to the substrate. A coating is applied to a surface by mechanical force using a brush, roller, or sprayer that leaves surface disturbances (such as the brush marks shown schematically in Figure 4.23). The goal is for these surface disturbances to disappear before the coating dries.

Surface leveling needed to achieve a continuous smooth film is driven by surface tension and resisted by viscosity. Mathematical models of the leveling process provide estimates of the leveling half-time

$$t_{1/2} = \eta_L \lambda^4 / \gamma h^3 \qquad (4.7.2)$$

where η_L is the viscosity (at low shear rate), γ is the surface tension, and λ and h are defined in Figure 4.23. Clearly, decreasing the distance between striations or brush marks and increasing the thickness of the film are both effective ways to accelerate leveling. However, there are limits to how much the viscosity of the paint can be reduced. For example, if the viscosity becomes too low, the wet paint will run down vertical walls, creating sags.

One way to circumvent the dilemma of achieving a coating that possesses sufficiently low viscosity so that it levels well and yet sufficiently high viscosity so that it does not sag is to add a suitable polymer to make the paint thixotropic. In thixotropic fluids, viscosity varies with time. Coatings exist that retain a low viscosity for a short period after shearing, thus allowing good leveling, but thicken fast enough to prevent sagging.

Hydrophobically modified hydroxyethylcellulose (HMHEC) exemplifies a polymer used to control the rheological properties of paint. It is a nonionic, water-soluble polymer that contains both hydroxyethyl groups (which hydrogen bond to water and thereby increase the viscosity of the aqueous phase) and long-chain alkyl groups (which associate with the latexes and pigments to create an interconnected network). By disrupting this network, shearing lowers the viscosity, but the network reestablishes itself when shearing ceases.

Figure 4.23
Idealized profile of coating striation marks that result from application with a brush or roller.

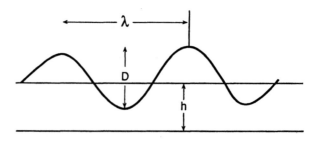

Many natural products employed as thickeners in paints are subject to degradation by microorganisms. Infection at the manufacturing site, after the paint is opened or during subsequent storage, results in an unusable paint, and biocides are added to combat such degradation.

4.7.1.3 Summary

Paints are complex colloidal systems that must fulfill a number of stringent and often conflicting requirements. In this brief discussion, we have highlighted many of the key properties and characteristics of a modern paint. For the sake of brevity, however, we have omitted a large number of more subtle issues.

4.7.2 Protective Latex Coatings Are Formed Using Diffusiophoresis—Movement of Colloidal Particles in the Gradient of a Chemical Solute

Although corrosion of steel costs the United States over $7 billion per year, applying continuous protective coatings to objects with complex shapes constitutes a major technological challenge. One method commonly used for this purpose is electrodeposition. The electrodeposition process applies an electrical field of up to 400 V dc to a metal object immersed in an aqueous bath containing 5 wt % polymer resin and pigment. The electrical field attracts charged polymer to the metal, and after a few minutes, a film 30 μm thick (30 g/m^2) accumulates on the surface.

This section focuses on an alternative metal coating process known as *diffusiophoresis*. It employs a chemical reaction to create an ionic concentration gradient and as a consequence an electric field normal to the metal's surface. The field attracts charged latex particles and leads to their accumulation at the surface. Although no external electrical field is applied, the exposure time required to accumulate 30 g/m^2 of deposit is about the same as with electrodeposition. After the object is removed from the deposition bath, drying causes the latex particles to coalesce and form a continuous protective coating. Sections 4.7.2.1 and 4.7.2.2 describe experiments that reveal the origin of diffusiophoresis.

The main advantages of diffusiophoresis over electrodeposition are lower capital costs and lower operating costs, particularly because latex polymers replace more expensive polymer resins. Automotive manufacturers use the process to protect frames and other steel parts that do not require a second coat. In addition, the technique is used to form rubber gloves and other rubberized articles. We discuss this latter application in Section 4.7.2.3.

4.7.2.1 Forming a Latex Film

A laboratory version of the steel coating process employs a $2 \times 3 \text{ cm}^2$ steel test panel immersed in a stirred beaker containing 5 wt % polymer latex, $0.075M$ HF and $0.044M$ H_2O_2. It is presumed that an anionic surfactant is added to the latex to impart both electrostatic and steric stabilization (see Section 6.5.3). The H_2O_2–HF oxidizes the steel to form ferrous (Fe^{2+}) ions

Figure 4.24
Cross section of wet coating composed largely of uncoalesced latex particles. At the metal surface, dissolution occurs according to: Fe(s) + 2H⁺ + H₂O₂ → Fe²⁺ + 2H₂O. Reactants and product diffuse through the film at different rates, generating an electric field that attracts the negatively charged latex particles to the metal surface. [D. C. Prieve, *Advances in Colloid and Interface Science* **16**, 321 (1982).]

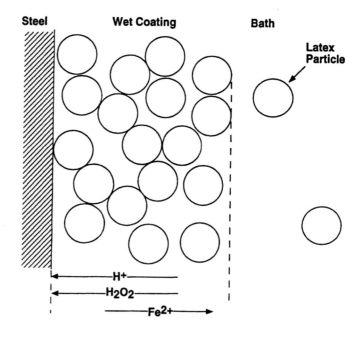

$$Fe(solid) + 2H^+ + H_2O_2 \rightarrow Fe^{2+} + 2H_2O \qquad (4.7.3)$$

without the evolution of gas. The Fe^{2+} ions produced by heterogeneous reaction can be further oxidized by hydrogen peroxide to form ferric (Fe^{+3}) ions. Their subsequent reaction with fluoride ions leads to the formation of a variety of ferrous and ferric complexes. Figure 4.24 shows the direction of counterflowing ion and molecular fluxes in a cross section of the wet coating. After the steel panel has been immersed in the coating bath for several minutes, it is removed, dried, and weighed. The increase in its weight equals the *difference* between deposited latex and dissolved iron. The amount of dissolved iron is determined independently by running duplicate experiments and scraping the latex off the panel before drying.

Figure 4.25 shows typical results of iron dissolution and polymer deposition with time. During a 3-minute period, 25 g/m² of latex are deposited. This amount corresponds to a film thickness of 0.001 inches (25 μm). Deposition and dissolution rates (as given by the slopes of the curves in Figure 4.25) decrease with time. A series of experiments indicates that diffusion of H_2O_2 through the porous polymer film controls the deposition rate, and it decreases with increasing film thickness.

Based on the discussion of critical coagulation concentrations given in Section 4.5.1, we might expect coagulation of the negatively charged latex spheres to result from the local concentration of multivalent Fe^{2+} and Fe^{3+} ions adjacent to the metal surface. However, two additional experiments make it clear that this is not the case.

First, if we repeat the dissolution experiment in the presence of an electrolyte, such as KCl, latex deposition onto the metallic surface is almost entirely *suppressed*. We can add up to 0.1*M* KCl without causing the latex to coagulate because the latex is sterically stabilized. Nor does the addition of KCl appreciably affect the CCC for ferric ions even though the concentration of KCl

Figure 4.25
Typical evolution of the coating mass that accumulates on the dissolving metal. As the polymer coating grows, metal dissolution continues, but the rate of growth, which is controlled by diffusion of H_2O_2 through the coating, is reduced. Coating bath contains 50 g/liter of latex solids, $0.075M$ of HF (pH = 2.1), and $0.044M$ of H_2O_2. [D. C. Prieve, *Advances in Colloid and Interface Science* **16**, 321 (1982).]

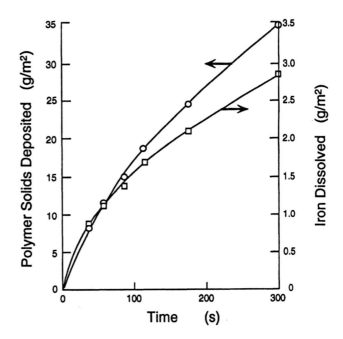

is a hundredfold greater than that of the ferric salt (similar to effects discussed in Section 4.6.3). Furthermore, if we start a dissolution experiment without added KCl and then add it after a film has formed, latex deposition ceases, but the dissolution of iron continues.

Second, if we repeat the dissolution experiment using a zinc panel, we obtain dissolution results almost identical to those we obtain with steel—except that zinc ions, unlike iron ions, do not cause the latex to coagulate even at $0.2M$ HF.

These two additional experiments clearly illustrate that explanations of latex particle accumulation on metal surfaces based solely on electrostatic coagulation phenomena are insufficient, and we must seek other reasons for latex aggregation at the metal surface.

4.7.2.2 A Simple Experiment Illustrates Diffusiophoresis

We can understand the mechanism leading to the formation of latex films by performing a simple experiment in which salt diffuses through a porous membrane in contact with a solution containing latex particles (see Figure 4.26). In this experiment, an $0.1M$ salt solution is added to a cell C capped by a detachable Nuclepore membrane M. Then the cell assembly is lowered into a beaker containing latex diluted to 0.5% with distilled water. The membrane serves as a collector for the deposited latex particles and as a barrier to prevent convective mixing of the two solutions.

By inserting electrodes (Ag/AgCl electrodes with KCl salt bridges) into each compartment, we can measure a potential across the membrane during the deposition experiment. This potential is called the *diffusion potential*.

A number of different $0.1M$ nonflocculating salts, bases, and acids are added to the cell, and we can correlate the diffusion potential that is generated with the mass of polymer solid depos-

Figure 4.26
Schematic of experiment. Porous membrane M separates salt solution held inside the cell C and agitated by magnetic stirrer S from latex dispersion outside the cell. Diffusion of salt through the membrane generates an electric field that can attract charged latex particles to the membrane. [D. C. Prieve, *Advances in Colloid and Interface Science* **16**, 321 (1982).]

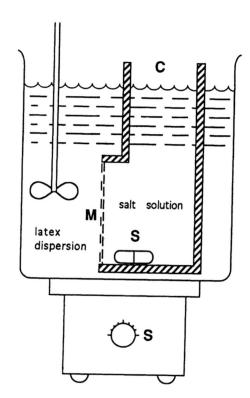

ited in a fixed time interval, say, 4 minutes, for both anionic and cationic latexes. Figure 4.27 plots the latex deposition rates versus the diffusion potential. With anionic latexes, the amount of deposited material decreases as we transverse the series of salts from NaOH to KBr, and no deposition is observed with the series KClO₄

Figure 4.27
Cross correlation of the mass of polymer accumulating next to the membrane in Figure 4.26 during an exposure of 4 min with the corresponding diffusion potential when 0.1M solutions of different electrolytes are placed in the cell. Latex particles that have a surface charge of the same sign as the diffusion potential are attracted to the membrane and tend to accumulate, while particles of opposite sign are repelled. [D. C. Prieve, *Advances in Colloid and Interface Science* **16**, 321 (1982).]

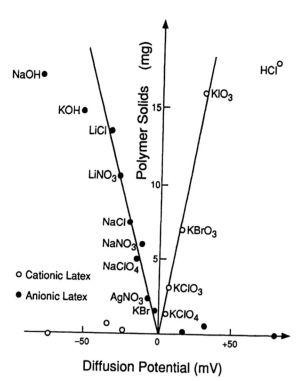

to HCl. With cationic latexes, we observe the opposite behavior, the amount deposited decreases from HCl to KClO$_4$, and no deposition occurs from KBr to NaOH.

We can explain the specific salt and acid effects by examining the origin of the diffusion potential. In the diffusion experiment in Figure 4.26, we start with a sharp interface between a solution containing two solute ions and an adjacent solvent (water, containing the latex particles). No convective mixing takes place during the experiment. We consider first what happens to two uncharged, noninteracting solutes (1 and 2), in which solute 1 has a diffusion coefficient five times larger than solute 2. Due to Brownian motion, solute 1 will move faster and farther than solute 2, and solute 1 will begin to separate from solute 2, generating a concentration difference between them. Because the solutes are noninteracting, they continue to separate with time.

Now if the two solutes are *charged* ions (from the salt), their separation will generate an electrical field. For example, if the solution contains H$^+$Cl$^-$, and solute 1 is the cation, H$^+$, and solute 2 the anion, Cl$^-$, the intrinsic diffusion velocities of these two ions differ by a factor of 5. As the proton races ahead of the chloride ion, charge separation occurs, and as we saw in Chapter 2, this separation generates an electric field. The field continues to increase, eventually causing the diffusion of protons to slow down, while diffusion of chlorides speeds up until the field reaches a magnitude such that no net flow of current remains. If the solution contains Na$^+$OH$^-$, the hydroxide ion diffuses three times faster than the sodium ion, and the electric field generated polarizes in the opposite direction. Indeed, the sign of the electric field and the magnitude of the electrical potential are determined by the relative diffusivity of the ions—hence the term *diffusion potential*.

The magnitude of the field E relates to the diffusion coefficients of a symmetrical electrolyte by

$$E = \frac{kT}{ze} \frac{D_+ - D_-}{D_+ + D_-} \frac{d(\ln c_i)}{dz} \tag{4.7.4}$$

where c_i stands for the local ion concentration and D_+ and D_- are the respective diffusion coefficients of cations and anions estimated from ionic conductivities.

Figure 4.28 compares the diffusion potential measured in the deposition experiment (abscissa of Figure 4.27) with the diffusion coefficient ratios. The linear correlation agrees with the prediction of eq. 4.7.4.

Taken with the observations of Figure 4.27, these results suggest that the relative difference in cation and anion mobilities sets up an electrical field that drives polymer deposition. The diffusion potential causes the charged latex particles to move. Depending on the direction of the field and the sign of the charge on the latex particles, we can understand why the particles are sometimes attracted to and sometimes repelled from the membrane when we add different salts. These experiments also provide the basis for understanding the differences between deposition on steel and zinc panels in various electrolytes described in the previous section.

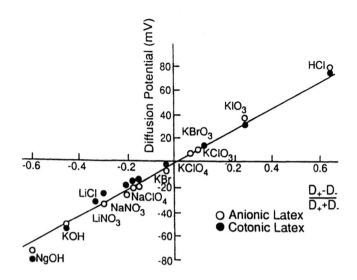

Figure 4.28
Diffusion potential measured between one of a dozen electrolytes and each of two latexes. The linear correlation with the relative difference in cation and anion mobility is suggested by eq. 4.7.4. Concentrations of latex and electrolyte are the same as in Figure 4.27 [M. M. Lin and D.C. Prieve, *J. of Colloid and Interface Science* **95**, 327 (1983).]

The diffusion potential generated spontaneously when ions diffuse at different rates represents only one cause of diffusiophoretic migration. More generally, diffusiophoresis refers to any migration of rigid colloidal particles caused by a macroscopic gradient in (electro)chemical potential of some molecular solute that adsorbs to form a *diffuse* and *mobile* layer at the surface of the particles. Historically the "diffusio" in diffusiophoresis refers to the diffuseness of the adsorption layer and not to the diffusion of ions or to the associated diffusion potential.

4.7.2.3 Using Latex Dipping to Produce Surgical Gloves and Other Rubberized Items

Many rubberized items, such as gloves, balloons, prophylactics, catheters, seamless football bladders, Wellington boots, and overshoes, are fabricated by dipping a mold called a "former" into a latex solution. All these irregularly shaped items require a continuous defect-free film to function properly. Although much of the discussion in the literature and some of the terminology used to describe this process refer to coagulation, clearly diffusiophoresis plays the major role in forming such rubberized items.

We will focus on the production of surgical gloves because they illustrate many key features of the dipping process. Some applications, such as neuro- or ophthalmic surgery, require thin, smooth gloves. Other applications employ roughened or textured finger and palm areas to improve the grip of the glove. Textured items can made by using a suitably roughened mold.

Surgical gloves are made by a two-step dipping process. The first step immerses the former in an aqueous or ethanol "coagulant" bath containing 15–20 wt % hydrated calcium nitrate, 1–2 wt % parting aid (an inert powder), and 0.05–0.25 wt % surfactant. When the former is removed from the bath, evaporation leaves a surface residue consisting mostly of calcium nitrate. A parting aid serves two purposes: It decreases fluid flow during evaporation, ensuring a uniform residue over the former, and it minimizes latex film adhesion when the finished item is stripped from the former. The surfactant also serves two purposes: it stabilizes the parting aid powder, maintaining its uniform dispersion in the fluid, and promotes effective wetting of the former during immersion.

In the second step, the former is immersed in an aqueous latex solution. Dissolution of the calcium nitrate residue creates an electrolyte concentration gradient, and the resulting diffusion potential concentrates the latex at the former's surface. Total immersion time is 5 to 20 s—depending on the thickness of the glove. The rate of immersion and withdrawal of the former is important; it is slower over the finger region than over less intricate parts of the mold.

After the latex–former assembly is removed from the latex bath, an initial drying makes it possible to trim and roll the top cuff area to form a bead. Next, the assembly is leached in 80°C water for 10 min to remove water-soluble ingredients, such as calcium nitrate, that could irritate the skin if they remained in the final product. Finally, the latex is dried at 80—120°C for 30 min before the glove is stripped off the mold.

Thicker houseware and electricians' gloves are made by the same procedure except they require several latex baths. Fabric-lined gloves are made by first placing a cotton liner on the former and adjusting processing conditions in the "coagulant" and latex baths so that it bonds with the latex coat.

4.7.3 Modern Ceramics Place Stringent Demands on Homogeneity and Quality Achievable through Colloidal Particle Processing

Ceramic materials typically have very high melting points; so methods that employ melting and resolidification, such as the casting techniques used for metals, are not useful for processing them. Most ceramics are fabricated by powder processing techniques.

The art of making ceramic pottery is over 10,000 years old. The basic raw materials—blends of clay (hydrous aluminum silicates), flint (silica), and potash feldspar (anhydrous potassium aluminosilicate)—are taken from the earth. By adding varying proportions of water, the suspension can be adjusted to the right consistency for throwing on a potter's wheel or pouring into a mold. Small additions of electrolyte in the form of sodium silicate (water glass) also produce remarkable alterations in the fluidity of the system. Thus ceramic manufacturing technology has used the principles of colloid science for many years. Old and proven processes are still followed in the manufacture of bricks and refractories, sanitary "whitewares," tableware (cups, saucers, and plates), and fine porcelain china and figurines, but new processes are needed for advanced ceramics, as will be discussed later.

Traditional ceramic powder processing can be broken down into the four basic stages illustrated in the upper sequence of Figure 4.29:

1. Synthesis—the raw material may have the desired chemical composition (typically an oxide) or a chemical composition, such as the hydroxide or carbonate, that can readily be converted to an oxide by heating (calcination).

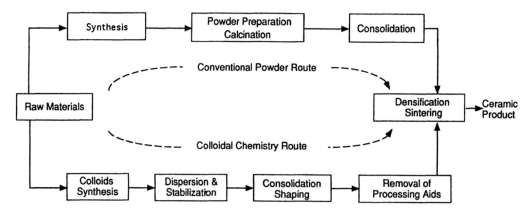

Figure 4.29
Schematic of the processing sequence from materials synthesis to final product following the powder route used to make conventional ceramics and the colloidal chemistry route used to fabricate "advanced" ceramic materials. (After I. A. Aksay).

2. Powder preparation—the starting material is ground or milled to a fine powder, sorted according to powder size and size distribution, and treated to prevent agglomeration. Surfactants are added to separate the particles and maintain a state of suspension in solution.

3. Powder consolidation—the powder is compacted and preformed to a shape close to the desired final shape. It may be poured as a slurry into a mold, handled as a clay on a potter's wheel, pressed into a block, or injected into a die, depending on the product.

4. Densification/microstructure development—the final compacted powder is fired ("sintered") so that it shrinks to the desired shape, density, and microstructure.

Many alternative paths lead from the raw material to the final product; the route chosen depends on the application. Selecting the appropriate set of variables for each stage constitutes the proprietary "art" of ceramics processing.

Until the 1900s carefully selected raw materials, taken straight from the earth, ground, washed, and filtered to improve their quality, and treated to enhance their workability, met manufacturing needs. But since the 1950s, ceramics have undergone a technological revolution, and a new set of standards is in place. Ceramics have gone beyond passive applications (such as furnace liners and insulators) to active applications (such as sensors, superconductors, electro-optic shutters, capacitors, high-performance machine tools, and components in high-temperature gas turbine engines). These new "advanced" ceramics place stringent demands on purity, homogeneity, and quality.

New electrical ceramics must be pure and homogeneous. An electro-optic ceramic, such as lanthanum-doped lead zirconate titanate [9 at. % La in $(PbZrO_3)_{0.35}(PbTiO_3)_{0.65}$], must be pure and maintain the proper stoichiometric ratios for all size domains down to molecular dimensions. Otherwise minute variations in composition will cause local changes in the electro-optic coefficient and refractive index and lead to undesirable light scattering. New optical ceramics must be fully dense. For example, alumina for high-intensity street lamp lenses or for radar domes must be completely transparent and mechanically strong.

New structural ceramics must be free from flaws and exhibit high strength over a wide range of temperatures. For example, silicon nitride rotors for use in high-performance gas turbine engines must be reliable up to 1000°C. Ceramics are known to break in a brittle manner due to microscopic flaws or cracks introduced during the processing sequence, and great care must be exercised to avoid flaws. In particular, diamond machining often is used to shape a ceramic part precisely, and the impact abrasion introduces flaws. This problem has led to the idea of "net-shape" forming in which the ceramic is fired to the desired final shape with no need for subsequent shaping. Net-shape forming places great demands on uniform compaction of powders and reproducible behavior during the firing operation.

This evolving part of the ceramics industry is projected to reach a value of $10 billion per year by the year 2000, and considerable investment is being made into new processing methods. One approach employing colloid science is illustrated in the lower sequence in Figure 4.29. We will examine each of the critical steps in this process sequence in more detail.

4.7.3.1 Powder Synthesis

The ultimate goal of powder synthesis is the production of extremely fine powders of submicrometer dimensions that possess high purity and homogeneity. Two processing approaches are available: precipitation from chemical solutions or controlled vapor-phase reactions to form small micromolecular clusters. We will concentrate on the former, the so-called *sol–gel process.*

The first step in the process is to prepare the sol. As we saw in Section 4.2, most metal oxides and hydroxides can be precipitated to form a colloidal suspension or sol, which can be prepared from an aqueous or an alcohol solution. The aqueous solution uses classical wet chemistry reactions to provide insoluble precipitates and controls the nucleation and growth sequence by introducing impurities and by pH adjustments. The alcohol solution uses metal alkoxides such as

Titanium tetramethyloxide	$-Ti(CH_3O)_4$
Silicon tetraethyloxide	$-Si(C_2H_5O)_4$
Zirconium tetraethyloxide	$-Zr(C_2H_5O)_4$
Barium isopropyloxide	$-Ba(C_3H_7O)_2$

as the starting liquids. With the addition of water, the alkoxides hydrolyze to an intermediate hydroxide complex and then condense to give a net reaction like

$$Ti(C_2H_5O)_4 + 2H_2O \rightarrow TiO_2 + 4C_2H_5OH \qquad (4.7.5)$$

Under controlled conditions, the metal oxide precipitates as a monodisperse spherical sol, as discussed in Section 4.2.1 and illustrated in Figures 4.2 and 4.32a.

Mixing alkoxide solutions in the appropriate chemical ratios presents an opportunity to blend the precipitate on a molecular scale. For example, titanium tetra n-butyl and zirconium tetra n-butyl oxides are mixed and coprecipitated onto fine lead oxide particles to serve as powder precursors for processing lead zirconate–titanate piezoelectric ceramics. Barium isopropyl oxide and titanium amyl oxide are mixed and hydrolyzed directly to form powders for barium titanate dielectrics. Monodisperse spherical sols can be prepared for most modern mixed-oxide ceramic materials, including the yttrium barium copper oxide ($YBa_2Cu_3O_7$) superconductors.

4.7.3.2 Powder Stabilization/Powder Conditioning

Colloidal *suspensions* must be prevented from agglomerating or flocculating during the second step in the lower sequence in Figure 4.29. Most oxide or hydroxide particles are positively charged; so we can control the stability of the sol by adjusting the pH of the solution. Figure 4.30 shows typical measurements of the zeta potential (the charge carried by the particle in the electrolyte, as described in Section 4.6.2) of aqueous alumina and zirconia sols as a function of pH. For the three samples of zirconia the zeta potential varies from +75 mV for pH < 4 to –50 mV for pH > 9 and passes through zero between pH 5.5 and 7.5. For alumina, any pH above 8 is undesirable. Thus preparation of a stable mixed alumina–zirconia sol requires an acidic solution with a pH less than 4.

A very low-viscosity aqueous sol is desirable for some applications where the sol, or liquid slurry, is poured into a porous mold. Capillary forces suck the liquid into the fine pores of the mold, leaving the powder behind to conform to the mold shape. This procedure is not commonly used for advanced ceramics because particles may aggregate unevenly, residual capillary stresses may develop, and sintering to full density without flaws is more difficult to achieve.

Most advanced ceramic applications require removal of the solvent. One method is by spray drying. The dispersed sol, or liquid slurry, is forced through a spray nozzle into a drying chamber, and then the resulting fine powder is collected as a free-flowing dry powder for subsequent compaction or can be reformulated with additives to form a paste. In this stage, keeping the particles from aggregating is a challenge.

A veritable "cookbook" of *additives* is used to condition fine

Figure 4.30
Zeta potentials of zirconia and alumina sols as a function of pH. For the zirconia sols, the isoelectric point (the zero charge point), at which the particle bears no net charge, occurs in the pH range 6.0 to 7.5, while for the alumina sol the value lies above a pH of 9. [E. M. DeLiso, W. R. Cannon, and A. S. Rao, *Defect Properties and Processing of High-Technology Nonmetallic Materials, Materials Research Society Symposia Proceedings* **60**, 43 (1986).]

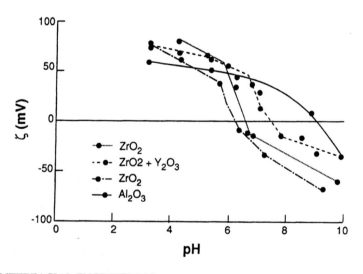

TABLE 4.5 Deflocculants, Binders, and Plasticizers Used in Ceramics Processing

Organic	Inorganic
(a) Common deflocculants	
Sodium polyacrylate	Sodium carbonate
Ammonium polyacrylate	Sodium silicate
Sodium citrate	Sodium borate
Sodium succinate	Tetrasodium pyrophosphate
(b) Common binders	
Cellulose	Clays: kaolin, bentonite
Natural gums: gum arabic	Soluble silicates
Polysaccharides: refined starch	Organic silicates: ethyl silicate
Lignin extracts: paper waste	Soluble phosphates, aluminates
(c) Common plasticizers	
Water	
Glycols: ethylene, diethylene, triethylene, tetraethylene glycol	
Glycerol	
Phthalates: dibutyl, dimethyl phthalate	

J. S. Reed, *Introduction to the Principles of Ceramic Processing,* © 1988, John Wiley & Sons, pp. 139, 153, 177. Reprinted by permission of John Wiley & Sons, Inc.

ceramic powders prior to compaction. To prevent aggregation, the powder must be stabilized in the absence of an electrolyte. *Deflocculants* (dispersants), such as the sodium salts listed in Table 4.5a or fish oil, are added for this purpose. For some handling operations, *binders* listed in Table 4.5b are added to bind the particles together. Finally *plasticizers* (e.g., glycols and phthalates) similar to those used in paints are added in varying quantities to control the mixture's viscoelastic properties.

4.7.3.3 Consolidation and Shaping

To obtain a prescribed shape, the ceramic paste or slurry (depending on the application) is injected into a die (injection molding), spread out as a thin sheet (tape casting), or loaded directly into a chamber for pressing.

We consider tape casting. As illustrated in Figure 4.31, the colloidal slurry is fed underneath a leveling or "doctor" blade and spreads out on a Mylar sheet as a thin layer of ceramic paste. The slurry's viscosity must be adjusted to give the desired flow characteristics. After passing through a drying oven, the sheet of unfired ceramic powder plus binder and plasticizer has the con-

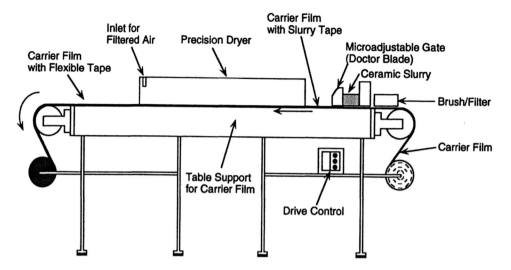

Figure 4.31
Schematic of a continuous tape-casting apparatus for producing ceramic materials. (J. S. Reed, *Introduction to the Principles of Ceramic Processing,* © 1988, John Wiley & Sons, p. 396. Reprinted by permission of John Wiley & Sons, Inc.)

sistency of rubber. It can be picked up and transported to a machine that cuts it into intricate shapes. In particular, fine holes about 0.1 mm in diameter can be punched into the unfired ceramic layer. In the IBM multichip ceramic package 10—20 such layers—each carrying molybdenum metal interconnect patterns deposited by screen printing—are placed on top of one another and cofired to form multilayer ceramic stacks. A similar procedure is commonly used to manufacture barium titanate multilayer capacitors. Obviously to reproduce this process, the ceramic slurry must be properly stabilized, have the correct viscosity, and contain enough solid particles to minimize shrinkage during firing. Table 4.6 gives typical compositions of ceramic slurries used for tape casting thin sheets of alumina microelectronic substrates and barium titanate capacitors.

TABLE 4.6 Compositions of Tape Casting Slurries

Component	Alumina tape	Vol. %	Titanate tape	Vol. %
Ceramic powder	Alumina powder[a]	27	Titanate powder[a]	28
Liquid system	Trichlorethylene	42	Methylethyl ketone	33
	Ethanol	16	Ethanol	16
Deflocculant/ dispersant	Menhaden fish oil	1.8	Menhaden fish oil	1.7
Binder	Polyvinyl butyral	4.4	Acrylic	6.7
Plasticizer	Polyethylene glycol	4.8	Polyethylene glycol	6.7
	Octyl phthalate	4.0	Butyl benzyl phthalate	6.7
Surfactant			Cyclohexanone	1.2

J. S. Reed, *Introduction to the Principles of Ceramic Processing,* John Wiley & Sons, New York, 1988, p. 397.

[a] < 5 μm particle size; includes sintering aids, grain growth inhibitor.

4.7.3.4 Densification

In the final stage of processing a ceramic product, the compacted powder material is fired at a high temperature (1000—1500°C) for a few hours to sinter the solid particles together. However, before raising the kiln temperature to its highest value, it must be held at a lower temperature to evaporate or decompose all the hydrocarbon additives used to stabilize and lubricate the powder, a stage known as binder burnout.

4.7.3.5 Sol–Gel Processing

We can further improve the purity of a ceramic material and enhance its processing if we can avoid some of the steps along the colloidal chemistry route and the additives they require. One obvious approach would be to go directly from the powder synthesis stage to densification. The *sol–gel process* offers this possibility.

If monodisperse sols synthesized by the alkoxide precipitation process described in Section 4.7.3.1 are allowed to stand and dry, evaporation of the water or the volatile alcohol changes their electrolyte concentration and reduces the repulsion between their charged sol particles. Oxide particles agglomerate and interconnect to form a *gel* (see Section 6.3.4.2). The state of agglomeration determines the rigidity of the gel network.

In some instances, allowing a rigid gel to form is desirable. For example, we can prepare complex ceramic thin films by spinning a mixed sol onto a substrate, allowing it to gel in situ, and then firing it to form a solid ceramic thin film.

In other instances, large blocks of rigid gels may be used as precursors for fabricating ceramic blocks. Unfortunately, the rigid blocks crack and break up as they dry due to strong capillary forces (in the same manner as cracks appear at the surface of a drying lake bed). To avoid capillary force and density gradients that cause cracking, we must extract the liquid slowly. Drying under controlled temperature and humidity conditions is a long and expensive process. An interesting alternative is to place the gel in an autoclave and raise the temperature and pressure to the critical point before removing the liquid solvent. Because the liquid and its vapor have the same density at the critical point, the surface tension between them is zero. From eq. 3.3.8, the capillary forces are also zero. Thus the liquid can be extracted quickly and the gel dried without cracking. This process is referred to as *critical point drying*.

Figure 4.32a shows a carefully prepared gel of titanium dioxide in which the particles have agglomerated to an array of tightly packed spheres. Firing this gel at a comparatively low sintering temperature (1100°C) for a short time (1 h) produced the fully dense polycrystalline ceramic shown in Figure 4.32b. Not a single additive was necessary beyond the synthesis stage. The potential for using this process to make high-purity, homogeneous, fully dense, material is clear. Research is continuing to make the sol–gel process more reproducible and less costly. Opportunities to apply our understanding of solid–liquid interface phenomena abound in the area of modern ceramic processing.

Figure 4.32
(A) Titanium dioxide particles prepared by the sol–gel process that have been agglomerated to form a highly packed array.
(B) Dense polycrystalline ceramic produced by sintering the TiO_2 particles. [E. A. Barringer and H. K. Bowen, *J. of the American Ceramic Society* **65**, C-199 (1982). Reprinted by permission of the American Ceramic Society.]

4.7.4 Zeolite Synthesis

4.7.4.1 Zeolites Are Molecular Sieves

The sale of zeolite materials constitutes a business worth several hundred million dollars annually in the United States. Even so, the technological value of these materials should be measured on an even larger scale because they are essential to the processing of virtually all petrochemical products. Use of zeolites is often said to save the refining industry a billion dollars a year.

How are zeolites used? They make extremely efficient "molecular sieves," by which we mean they can admit one type of molecule into their structure yet exclude another based on its size or chemical properties. For example, one application uses zeolite A to purify O_2 from air. Figure 4.33 shows the crystal structure of zeolite A. The channels entering the structure are 4.2 Å wide—large enough to admit either O_2 and N_2. However, N_2 is more

Figure 4.33
Structures of sodalite, zeolite A, and zeolites X and Y.

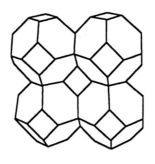

Sodalite

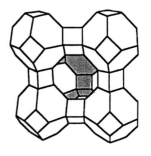

Zeolite A

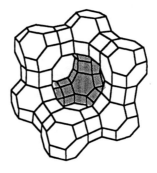

Faujasite (zeolites X and Y)

attracted to this material because of the way its electrons are distributed. In the pressure swing adsorption process, a column filled with the zeolite is first charged with high pressures of air, and then the pressure is released. Because N_2 is held by the zeolite, O_2 escapes much more rapidly. This process provides O_2-rich gas for both industrial processes and medical facilities.

In another application, zeolite Y (Figure 4.33) is used to break large molecules from crude oil into more useful, smaller molecules. Other catalysts can perform this "separation" process, but they are poisoned quickly by large, polymeric molecules that adsorb too strongly, permanently blocking the reactive sites. Because the window entering zeolite Y is only 7.4 Å wide, it can admit medium-sized hydrocarbons while excluding molecules that are too large. This zeolite is used for processing nearly half of all the material going through a modern refinery.

4.7.4.2 Zeolite Synthesis Involves Crystallizing Materials under Controlled Conditions— Quaternary Ammonium Salts Are Key to Success

Zeolites were not widely used in industrial operations until the early 1960s, when Union Carbide researchers commercialized the synthesis of an early version of zeolite catalyst. In fact, synthesis of zeolites was only discovered in the 1940s by Professor R. M. Barrer. He accomplished this by mimicking the geothermal environments where natural (but, sadly, not very useful) zeolites were formed. In nature and in the laboratory, silicate and aluminate

Figure 4.34
Basic building block structures for zeolites that have been identified by ^{29}Si NMR measurements. Vertices represent silicon atoms; lines represent bridging oxygen atoms. (Alon McCormick, Ph.D. Thesis, University of California, Berkeley, 1987.)

solids are dissolved in hot alkaline aqueous solutions, and then zeolites slowly crystallize and precipitate.

In the 1970s and 1980s, solute molecules in the synthesis solutions were found to resemble the structural units making up the zeolites themselves. A wealth of evidence from kinetic, microscopic, and spectroscopic studies showed that zeolites grow from solution by the assembly of these oligomeric intermediate polyanions, known for that reason as *secondary building units* (Figure 4.34).

It was also found that quaternary ammonium cations, R_4N^+, could stabilize cagelike structures in the solution and that zeolites formed in this way were structurally distinct from those formed without the ammonium cations. This startling and promising

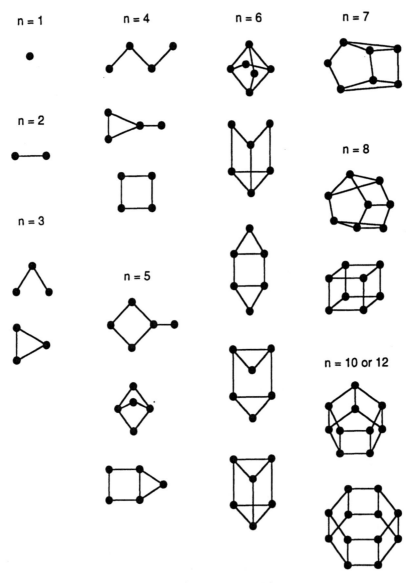

Structures Identified by ^{29}Si NMR

finding showed that controlling zeolite synthesis on a *molecular* scale by the addition of quaternary ammonium cations could result in a change in the yield of a desired zeolite on a *macroscopic* scale.

The predominant view of why this happens is as follows. Suppose we try to assemble the complex cage structure of zeolite A from the individual atoms of a molecular modeling set. At each stage we have to spend some time deciding what to do next to build such a complicated structure. If we simulate the random processes that real molecules undergo, it takes a long time to finish even one unit cell (rather like the proverbial example of a monkey typing a sonnet). Furthermore, by the time the zeolite is finally assembled, the dissolved precursors have already been used up in building much more compact and less ordered structures, and further growth is curtailed.

Next, suppose we are given some preassembled cages. For instance, if we are given cube structures, we can build zeolite A just by connecting these cages together in a simple cubic packing (compare Figures 4.33 and 4.34 for proof). As it turns out, adding a quaternary ammonium cation does just that—it constructs cages, and whichever zeolite is actually formed from these cages can now nucleate and grow much more quickly.

Why do quaternary ammonium cations build such cages? One possibility is that hydrophobic interactions associated with the quaternary ammonium's hydrocarbon chains induce ordering of water molecules and small silicates in the vicinity of the cation.

The following example illustrates how quaternary ammonium salts can be essential for the selective synthesis of a zeolite instead of some other, undesired, aluminosilicate material.

First, we demonstrate the difficulty that might be encountered in achieving a desired zeolite yield without a quaternary ammonium salt. Figure 4.35 shows a flow diagram for the zeolite synthesis reactor system. It consists of a digester where minerals are dissolved, a reactor where aluminates are added to begin zeolite crystallization, and a filter to remove solid products. The filtrate is recycled to the digester.

We can independently control the flow rates of the feed chemicals and the heat provided to the reactor. $SiO_2(s)$, $NaOH(aq)$, and H_2O are fed to the digester (along with some recycled solution). If the SiO_2 completely dissolves according to the reaction:

Figure 4.35
Schematic of a reactor used to synthesize zeolites.

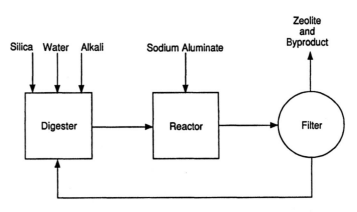

(1)
$$8\,SiO_2(s) + 2\,OH^-(aq) + 2\,H_2O \rightarrow H_6Si_8O_{20}^{2-}\,(aq) \qquad (4.7.6)$$

and the dissolved silicate species equilibrates with other silicate species according to the four reactions:

(2) $\quad H_6Si_8O_{20}^{2-}(aq) + 2\,OH^-(aq) + 2\,H_2O = 2\,H_6Si_4O_{12}^{2-}(aq)$ $\qquad (4.7.7)$

(3) $\quad H_6Si_4O_{12}^{2-}(aq) + 2\,OH^-(aq) = H_6Si_3O_{10}^{2-}(aq) + H_2SiO_4^{2-}(aq)$ $\qquad (4.7.8)$

(4) $\quad H_6Si_3O_{10}^{2-}(aq) + 2\,OH^-(aq) = H_4Si_2O_7^{2-}(aq) + H_2SiO_4^{2-}(aq) + H_2O$ $\qquad (4.7.9)$

(5) $\quad H_4Si_2O_7^{2-}(aq) + 2\,OH^-(aq) = 2\,H_2SiO_4^{2-}(aq) + H_2O$ $\qquad (4.7.10)$

then $K_{(2)} \approx 0.01\ mol^{-3}$, $K_{(3)} \approx 2\ mol^{-1}$, $K_{(4)} \approx 30$, and $K_{(5)} \approx 30$.

After equilibration, the silicate solution is mixed with sodium aluminate and then fed to a continuous stirred tank reactor, which must operate at 100°C. In the reactor all the aluminum is consumed in two reactions that produce either an undesired amorphous byproduct or the desired zeolite crystals:

$$H_6Si_4O_{12}^{2-}(aq) + NaAl(OH)_4 \rightarrow H_8NaSi_3AlO_{12}(s) + H_2SiO_4^{2-}(aq) \qquad (4.7.11)$$
$$\text{dense, amorphous material}$$

$$H_6Si_8O_{20}^{2-}(aq) + NaAl(OH)_4 \rightarrow H_8NaSi_7AlO_{20}(s) + H_2SiO_4^{2-}(aq) \qquad (4.7.12)$$
$$\text{zeolite}$$

The principal operating constraint in this process is that the pH in the digester must be high enough so that SiO_2 remains dissolved. Figure 4.36 shows that if the pH falls too low, the original SiO_2 may reprecipitate.

On the other hand, zeolite formation is favored by relatively low pH values. Reactions 2 though 5 show the reason. These reactions interconvert the different precursors with each other. By Le

Figure 4.36
Solubility of amorphous silica versus pH. (Shawn Thornton, M.S. Thesis, University of California, Berkeley, 1985.)

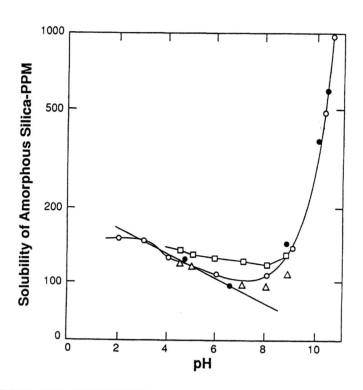

Chatelier's principle it is clear that lowering the concentration of OH⁻, that is, lowering the pH, favors the chief zeolite precursor, the cage achieved by biasing the conditions in favor of reaction (1). A set of mass balances can be performed to show that at the lowest tolerable pH (before SiO_2 would precipitate according to Figure 4.36) we can attain a selectivity of only 50%. This means that in the best case scenario, 50% of the dissolved precipitate would be wasted. Even worse is the lack of any convenient way to separate zeolite from the byproduct, meaning it will have to be sold impure.

Suppose we add a small amount of tetrapropylammonium chloride. This salt stabilizes the cage—that is, it increases the equilibrium constants for the formation of cage structures. Following the same analysis used in the first example now shows that a selectivity of 80% is possible.

4.7.4.3 Applications of Zeolites

Let us finish this section by considering interfacial phenomena that are important in the various fields of zeolite science. We can divide zeolite applications into three classes:

1. Separations: We have already seen an example in which zeolite A specifically adsorbs N_2 rather than O_2. In a variety of other applications, zeolites are used to remove organic molecules from aqueous streams, polar molecules from organic streams, or small molecules from larger molecules. Recently zeolites even have been added to cat litter to remove ammonia. Clearly these applications use the fundamentals of adsorption.

2. Catalysis: Aluminum incorporated into the zeolite lattice introduces a negative charge (because the valence of Al^{3+} is less than that of Si^{4+}). In some zeolites, this charge is balanced by acidic protons. These zeolites can break C-H and C-C bonds to perform cracking and isomerization reactions. When small clusters of noble metals also are added inside the zeolite pores, these materials (known as reforming catalysts) can simultaneously hydrogenate, dehydrogenate, crack, polymerize, isomerize, and cyclize organic molecules.

3. Ion exchange: Up to 50% of the zeolite structure is pore space accessible by small fluid molecules; so the cations (which may be any electropositive element) are readily exchanged. Thus zeolites can be used to remove radioactive, heavy metal, and alkaline earth ions from contaminated streams. The largest single use of zeolites (by volume) is as a water softener in laundry detergents (see Section 5.5.1.3).

In both catalysis and ion exchange, rates are usually limited by the transport of molecules through the very narrow zeolite pores. The diffusion coefficient of a fluid molecule can plunge dramatically as the pore size reaches zeolitic proportions. For this reason, controlling the size of zeolite crystals is very important, and the fundamentals of nucleation and crystal growth play a significant role in doing so.

Bibliography

D. J. Shaw, *Introduction to Colloid and Surface Chemistry,* London: Butterworth & Co. Ltd., 1980.

D. F. Evans and H. Wennerström, *The Colloidal Domain,* New York: VCH Publishers, Inc., 1994

D. H. Everett, *Basic Principles of Colloid Science,* London: Royal Society of Chemistry, 1980.

P. C. Heimenz, *Principles of Colloid and Surface Chemsitry* (2nd edition), New York: Marcel Dekker, Inc., 1986.

R. J. Hunter, *Foundations of Colloid Science,* New York: Oxford University Press, 1987.

H. van Olphen, *An Introduction to Clay Colloid Chemistry,* New York: Interscience, John Wiley & Sons, Inc., 1963.

Exercises

4.1 The potential at a fixed distance from a charged particle increases _ ; decreases _ ; or remains the same _ ; as salt is added to a sol.

4.2 The potential at a fixed distance from a charged particle increases _ ; decreases _ ; or remains the same _ ; as the valence of the ions in solution increases.

4.3 The charge density on the surface of a colloidal particle increases _ ; decreases _ ; or remains the same _ ; as salt is added to a sol.

4.4 Calculate the Debye screening length at 20°C for potassium chloride solutions containing 0.1 and 1.0 g of KCl per liter, respectively.

4.5 Clay particles with a surface charge of 0.12 C/m^2 are immersed in the two electrolytes in Exercise 4.4. What is the potential in millivolts at distances of 30 and 100 Å from the particles in each solution?

4.6 Describe the origins of the forces that keep a colloid dispersed by (a) electrostatic stabilization, and (b) steric (polymer) stabilization.

4.7 Consider two identically charged spherical colloidal particles with a surface potential of 25 mV dispersed in an aqueous medium. The effective Hamaker constant is 7×10^{-20} J. Determine the separation h_o for which the interaction potential goes through zero close to the primary minimum. Assume $ze\Phi_o \approx kT$ and $h_o << \kappa^{-1}$.

4.8 The sedimentation rate of a colloidal suspension increases _ ; decreases _ ; or remains the same _ ; as salt is added.

4.9 What can you say about the sols, A and B, from the following information concerning the relative molar critical coagulation concentrations (CCCs)?

CCC's (Relative Concentrations)		
Electrolyte	Sol A	Sol B
KCl	64	64
$MgCl_2$	32	1
Na_2SO_4	1	32

Suggest a mechanism for this behavior.

4.10 Draw curves that depict the potential energy of interaction V(h) between two colloidal particles as a function of their separation h for the following conditions (express relevant energies in units of **kT**): (a) a stabilized sol that experiences slow coagulation; and (b) a sol that experiences rapid coagulation. (c) How do you convert a sol from condition (a) to condition (b)?

4.11 The coagulation rate of a sol
increases _ ; decreases _ ; or remains the same _ ;
as the size of the spherical particles increases.

4.12 Tuorila [*Kolloid Chem. Beiheft 22*, 191 (1926)] measured the rapid coagulation of a gold sol, radius 51 nm, and obtained the following values:

Time t (s)	Concentration of particles $[P] - (10^8 \, cm^{-3})$
0	20.2
30	14.7
60	11.2
120	7.74
240	4.78
480	2.71

Determine the association rate constant and compare with the theoretical value for a diffusion-controlled reaction.

4.13 Calculate the expected value of the stability ratio W for two spheres of radius 100 nm in a $2 \times 10^{-3} M$ aqueous KCl solution at T = 300 K and with surface potential 35 mV and Hamaker constant 5×10^{-20} J.

4.14 What is the zeta potential for a sol? How is it measured? What is the significance of zero zeta potential?

4.15 Increasing the concentration of I^- ions in a silver iodide sol above $2.5 \times 10^{-12} M$ causes the surface potential on the silver iodide particles to
increase _ ; decrease _ ; or remain the same _ .

4.16 What are the key constituents of a good outdoor paint?

APPENDIX 4A
THE COMPLETE
SOLUTION OF THE
POISSON-BOLTZMANN
EQUATION

We wish to develop the expression for Φ starting with the Poisson-Boltzmann equation, equation 4.3.5

$$\varepsilon_r \varepsilon_o \frac{d^2\Phi}{dz^2} = -\sum_i z_i e\, c_{io}^* \exp\left(-\frac{z_i e\Phi}{kT}\right) \qquad (4A.1)$$

Solving this second order differential equation is made easier by the identity

$$\frac{d}{dx}\left(\frac{df}{dx}\right)^2 = 2\frac{d^2f}{dx^2}\frac{df}{dx} \qquad (4A.2)$$

so that equation 4A.1 can be rewritten as

$$\frac{d}{dz}\left(\frac{d\Phi}{dz}\right)^2 = -2\left(\frac{d\Phi}{dz}\right)\frac{1}{\varepsilon_r \varepsilon_o} \sum_i z_i e\, c_{io}^* \exp\left(-\frac{z_i e\Phi}{kT}\right) \qquad (4A.3)$$

Changing the differential on the right hand side, this becomes

$$\frac{d}{dz}\left(\frac{d\Phi}{dz}\right)^2 = +\frac{2kT}{\varepsilon_r \varepsilon_o} \sum_i c_{io}^* \frac{d}{dz} \exp\left(-\frac{z_i e\Phi}{kT}\right) \qquad (4A.4)$$

The terms on either side can now be integrated. By choosing z and ∞ as the limits of integration we get

$$\left(\frac{d\Phi}{dz}\right)^2 = \frac{2kT}{\varepsilon_r \varepsilon_o} \sum_i c_{io}^* \left(\exp\left(-\frac{z_i e\Phi}{kT}\right) - 1\right) \qquad (4A.5)$$

When we take the square root of this equation the result depends on the sign of the surface charge density. If we choose a negatively charged particle and specialize to a symmetrical electrolyte (such as NaCl, $z_i = \pm1$, or MgSO$_4$, $z_i = \pm2$) at bulk concentration c_o^* then

$$\left(\frac{d\Phi}{dz}\right) = -\left(\frac{8kT\, c_{io}^*}{\varepsilon_r \varepsilon_o}\right)^{1/2} \sinh\left(\frac{|z_i|\, e\Phi}{2kT}\right) \qquad (4A.6)$$

The advantage of selecting a symmetrical electrolyte is that the right hand term of equation 4A.6 contains an even square.

Equation 4A.6 is a non-linear first order differential equation that can be integrated to give

$$\Phi(z) = \frac{2kT}{ze} \ln\left(\frac{1 + \Gamma_o \exp(-\kappa z)}{1 - \Gamma_o \exp(-\kappa z)}\right) \tag{4A.7}$$

where

$$\frac{1}{\kappa} = \left(\frac{\varepsilon_r \varepsilon_o kT}{\sum_i (z_i e)^2 c_{io}^*}\right)^{1/2} \tag{4A.8}$$

and Γ_o relates to the surface potential Φ_o through

$$\Gamma_o = \frac{\exp\left(\dfrac{ze\Phi_o}{2kT}\right) - 1}{\exp\left(\dfrac{ze\Phi_o}{2kT}\right) + 1} = \tanh\left(\frac{ze\Phi_o}{4kT}\right) \tag{4A.9}$$

These three equations represent the complete solution to the Poisson-Boltzmann equation, they describe the drop in electrical potential in an electrolyte solution in the direction, z, normal to the charged surface, reproduced in equation 4.3.6.

To obtain the equation for the surface charge density we note from equation 4.3.17 that

$$\sigma_o = \varepsilon_r \varepsilon_o \frac{d\Phi}{dz}\Big|_o^\infty = -\varepsilon_r \varepsilon_o \left(\frac{d\Phi}{dz}\right)_o \tag{4A.10}$$

Combining equations 4A.10 and 4A.6 we obtain the result

$$\sigma_o = (8\varepsilon_r \varepsilon_o kT c_{io}^*)^{1/2} \sinh\left(\frac{|z_i| e\Phi}{2kT}\right) \tag{4A.11}$$

used in equation 4.3.20.

APPENDIX 4B CALCULATIONS ILLUSTRATING NUMERICAL RESULTS WITH THE GOUY-CHAPMAN AND THE STERN LAYER EQUATIONS

Example 1. We show how to calculate the various quantities given in Table 4B.1 for a negatively charged surface, with

TABLE 4B.1 Gouy-Chapman Calculations

1 Φ_o mV	2 c_{io} M	3 z	4 $1/\kappa$ 10^{-8}m	5 Γ_o	6 σ_o C/m^2	7 $\Phi_{30Å}$ mV	8 $\Phi_{100Å}$ mV	9 $\Phi_{\kappa^{-1}}$ mV
51.4	10^{-4}	1	3.04	0.46	.00138	45.9	35.6	17.5
51.4	10^{-2}	1	0.304	0.46	.0138	17.9	1.8	17.5
51.4	10^{-4}	2	1.54	0.76	.0043	37.7	21.5	14.8
51.4	10^{-4}	3	1.01	0.90	.0118	28.0	12.1	11.9
51.4	10^{-4}	1	0.486#	0.46	.0002#	26.2#	6.08#	17.6#

Results for particles with constant surface potential Φ_o (= −51.4 mV) immersed in different concentrations of salt in water or oil.

= oil (ε_r = 2)

surface potential Φ_o = −51.4 mV, in various aqueous salt solutions containing NaCl and other symmetrical electrolytes, z = 2,3.

First we evaluate the Debye lengths, using the general equation 4.3.13. For aqueous solutions at 25°C we can use equation 4.3.14, and for the $1 \times 10^{-4}M$ NaCl electrolyte

$$1/\kappa = 3.043 \times 10^{-10} \times (1 \times 10^{-4})^{-1/2} = 3.04 \times 10^{-8} \text{ m} = 304 \text{ Å} \tag{4B.1}$$

This result and results for different electrolyte concentrations, valencies and dielectric constant are contained in column 4 of Table 4B.1.

We need to know the value of the ratio $ze\Phi_o/kT$ because it occurs repeatedly in the Gouy-Chapman equations for surface charge density and potential. We can evaluate this ratio by noticing that when $e\Phi/kT = 1$

$$\Phi = \frac{kT}{e} = \frac{(1.38 \times 10^{-23}\text{JK}^{-1})(298 \text{ K})}{1.602 \times 10^{-19}\text{C}} = 2.57 \times 10^{-2} \text{ J/C} = 25.7 \text{ mV} \tag{4B.2}$$

In our example Φ_o = 51.4 mV, and so $e\Phi_o/kT = 2$. Using equation 4.3.7 we can evaluate Γ_o for monovalent electrolytes, z = 1 and obtain

$$\Gamma_o = \frac{\exp(2/2) - 1}{\exp(2/2) + 1} = 0.46 \tag{4B.3}$$

This value for Γ_o and results for other values of z and ε_r are included in column 5 of Table 4B.1.

To evaluate σ_o, we use equation 4.3.20. First we need to convert molar concentration, c_{io} (moles per liter), to particle concentration per unit volume, c_{io}^* (ions per cubic meter). The conversion is as follows

$$c_{io}^* \text{ (particles/m}^3) = (6.02 \times 10^{23} \text{ particles/mole})(10^3 \text{ l/m}^3)(c_{io} \text{ mole/l})$$

$$c_{io}^* = 6.02 \times 10^{26} c_{io} \tag{4B.4}$$

For the 10^{-4} M NaCl solution $c_{io}^* = 6.02 \times 10^{22}$ particles/m^3. Second we need to evaluate $2\varepsilon_r\varepsilon_o kT$. Using ε_r = 78.5 for water and assuming room temperature, we obtain

$$(2\varepsilon_r\varepsilon_0 kT) = (2 \times 78.5 \times 8.85 \times 10^{-12} \text{ C}^2\text{J}^{-1}\text{m}^{-1} \times 4.12 \times 10^{-21} \text{ J})$$

$$(2\varepsilon_r\varepsilon_0 kT) = 5.725 \times 10^{-30} \text{ C}^2/\text{m}. \qquad (4B.5)$$

Inserting these values in equation 4.3.20 gives

$$\sigma_0 = (5.725 \times 10^{-30} \text{ C}^2/\text{m})^{1/2} (6.02 \times 10^{22} /\text{m}^3)^{1/2} [\exp(1) - \exp(-1)]$$

$$= 1.38 \times 10^{-3} \text{ C/m}^2 = 8.62 \times 10^{15} \text{ charges/m}^2 \qquad (4B.6)$$

This value and other results are in column 6 of Table 4B.1

To find the potential away from the surface we use equation 4.3.6. For the $10^{-4}M$ NaCl electrolyte at 30Å

$$\frac{ze\Phi}{2kT} = \ln\left(\frac{1 + 0.46 \exp - (3.29 \times 10^7 \text{m}^{-1})(30 \times 10^{-10}\text{m})}{1 - 0.46 \exp - (3.29 \times 10^7 \text{m}^{-1})(30 \times 10^{-10}\text{m})}\right) = 1.776 \qquad (4B.7)$$

so that $\Phi_{30Å} = 45.9$ mV. At 100Å from the surface $\Phi_{100Å} = 35.6$mV. These values together with results for the potential at the Debye distance, κ^{-1}, and other electrolytes are in columns 7–9 of Table 4B.1.

Example 2. Muscovite mica is a charged clay particle with one charge per every 48 Å^2. Calculate the surface potential, Φ_0, using the Gouy-Chapman equation and calculate Φ_0 and Φ_δ, using the Stern-Gouy-Chapman equation, for 10^{-2}, 10^{-4}, and $10^{-6}M$ electrolytes containing $z = 1$ counter-ions.

First, we calculate Φ_0(GC) using the Gouy-Chapman equation. Inverting equation 4.3.20, we can write

$$\Phi_0(GC) = \left(\frac{2kT}{ze}\right)\sinh^{-1}\left(\frac{\sigma_0}{(8\varepsilon_0\varepsilon_r kT\, c_{io}^*)^{1/2}}\right)$$

$$\Phi_0(GC) = 51.4 \text{ mV} \sinh^{-1}\left(\frac{\sigma_0}{(1.38\times10^{-2}c_{io})^{1/2}}\right) \qquad (4B.8)$$

where the result applies for $z = 1$, in water at 25°C. Inserting the value for σ_0 for mica (from equation 4B.10 below) and the different solution concentrations, we obtain the values for Φ_0(GC) listed in column 3 in Table 4B.2.

Second, we calculate Φ_0(SGC) and Φ_δ using the Stern-Gouy-Chapman equations 4.6.4 and 4.6.6. In order to use these equations, we must specify a number of parameters. Our knowledge of the structure of mica allows us to calculate the number of charged sites per unit area (S_0). One charge per 48 Å^2 means there are

TABLE 4B.2 Stern Gouy-Chapman Calculations

1 c_{io} M	2 z	3 Φ_0(GC) mV	4 Φ_δ mV	5 Φ_0(SGC) mV	6 σ_δ C/m^2	7 σ_d C/m^2	8 σ_δ/σ_d
10^{-2}	1	−208	−20.9	−1530	.324	.009	36
10^{-4}	1	−326	−139.2	−1650	.324	.009	36
10^{-6}	1	−444	−257.6	−1770	.324	.009	36

Results for mica in different concentrations of salt in water

$$S_o = \frac{10^{20} \text{ Å}^2/\text{m}^2}{48 \text{Å}^2/\text{site}} = 2.083 \times 10^{18} \text{ sites/m}^2 \qquad (4B.9)$$

Assuming that each negatively charged site corresponds to an adsorption site for a counter-ion, the surface charge density is

$$\sigma_o = eS_o = -3.333 \times 10^{-1} \text{ C/m}^2 \qquad (4B.10)$$

We have assigned a negative sign to σ_o because we know from mica's structure that its surface will bear a negative charge.

In deciding on a value for δ, we assume that the potassium ions retain their first hydration shell of water, so

$$\delta = (2R_{H2O} + R_{K^+}) \approx 4 \text{ Å} \qquad (4B.11)$$

Although we possess little detailed information on the dielectric constant of a fluid near a charged interface, we know that the high electrical fields ($\sim 10^6$ volts/cm) associated with a charged surface will completely orient dipoles adjacent to the surface. This orientation considerably reduces ε_r from its bulk value. Studies of aqueous systems indicate a value of ε_r in the range of 3 to 10 rather than 78.5. Here we assume a value $\varepsilon_r = 10$.

From equation 4.6.5 we have the general three-part charge balance condition

$$\sigma_o + \sigma_\delta + \sigma_d = 0 \qquad (4B.12)$$

that leads to the complex expression in equation 4.6.6. When $ze\Phi_\delta > kT$, we can simplify equation 4.6.6 for a negatively charged surface to the three parts

$$\sigma_o + \frac{zeS_o \, c_{io}^* V}{N_{Av}} \exp\left(-\frac{ze\Phi_\delta + \Delta G_{ads\,c}}{kT}\right) + (2\varepsilon_r\varepsilon_o kT \, c_{io}^*)^{1/2} \exp\left(-\frac{ze\Phi_\delta}{2kT}\right) = 0 \qquad (4B.13)$$

In the expression for σ_δ we have neglected the right-hand term of equation 4.6.3 because it is so much smaller than the left-hand term; and in the expression for σ_d we have neglected the exponential with the negative quotient in equation 4.3.20 because only the exponential with the positive quotient contributes significantly to the charge in the diffuse layer. Inserting appropriate values for mica in equation 4B.13, we have

$$-3.33 \times 10^{-1} + 6 \times 10^{-3} \, c_{io} \exp\left(\frac{-(\Phi_\delta + \Delta G_{ads\,c}/ze)}{25.7 \text{mV}}\right) + 5.86 \times 10^{-2} \, (c_{io})^{1/2} \exp\left(\frac{-\Phi_\delta}{51.4 \text{ mV}}\right) = 0 \quad (4B.14)$$

Using $\Delta G_{ads\,c}/ze = -200$mV, we can iterate to obtain the values of Φ_δ listed in column 4 of Table 4B.2. Knowing Φ_δ we can then use equation 4.6.4

$$\Phi_o - \Phi_\delta = \frac{\delta\sigma_o}{\varepsilon_o\varepsilon_r} = \frac{(4 \times 10^{-10}\text{m})(-3.33 \times 10^{-1}\text{C/m}^2)}{(10)(8.885 \times 10^{-12} \text{ C}^2/\text{Jm})} = -1.51 \text{ V} \qquad (4B.15)$$

to obtain Φ_o (listed in column 5), equation 4.6.3 to obtain σ_δ (column 6) and equation 4B.12 to obtain σ_d (column 7).

APPENDIX 4C
DIFFUSION-CONTROLLED COAGULATION—LEADING TO FORMATION OF DIMERS

The initial step in the coagulation of colloidal sols is when two isolated particles, P, come together to form a dimer, P_2. Our goal is to calculate the rate constant k_r in equation 4.5.10,

$$-\frac{d}{dt}[P] = k_r [P]^2 \tag{4C.1}$$

Our model system consists of uniform spherical particles of radius R undergoing Brownian motion. We assume they follow the square wave interaction potential curve of Figure 4.14 and adhere on contact to form the dimer.

We focus our attention on one central stationary particle, i, and calculate the number of particles that diffuse towards it. The flux, J, of particles crossing unit area is given by Fick's law

$$J = -D \nabla[P] \tag{4C.2}$$

where D is the diffusion coefficient of the spheres. The total flux of particles crossing a spherical surface of radius r surrounding the central particle, i, is

$$(J\,A)_i = -(4\pi r^2)\, D\frac{d[P]}{dr} \tag{4C.3}$$

At steady state the flux is constant and this equation can be integrated using the boundary conditions $[P] = [P]_b$ at $r = \infty$, where $[P]_b$ is the bulk concentration, and $[P] = O$ at $r = 2R$ to give

$$(J\,A)_i = 8\pi\, R\, D\, [P]_b \tag{4C.4}$$

This equation gives the number of particles that move towards a fixed particle per unit time. However, a given particle is not fixed but undergoes Brownian motion. The effect on the rate of flocculation can be taken into account by replacing the diffusion coefficient for a single particle, D, with the relative diffusion coefficient for two particles, D_{12}. From Einstein's random walk equation

$$D = x^2/2t \tag{4C.5}$$

where x is the distance that a particle moves in time t. Assuming that the Brownian motions of particles 1 and 2 are uncorrelated, their relative diffusion coefficient is

$$D_{12} = \frac{(x_1 - x_2)^2}{2t} = \frac{(x_1^2 - 2x_1 x_2 + x_2^2)}{2t}$$

and since $x_1 x_2 = 0$ for uncorrelated Brownian motion

$$D_{12} = D_1 + D_2 \qquad (4C.6)$$

and the relative diffusion coefficient for two similar but un-correlated particles is

$$D_{11} = 2 D \qquad (4C.7)$$

If one includes hydrodynamic interactions, the analysis of the relative Brownian motion becomes more complex.

Equation 4C.4 gives the flux toward one central particle. There are P such particles in the system, and one should sum over all these and divide by two to avoid counting the association of particle i with particle j twice. Thus,

$$\frac{d[P]}{dt} = \frac{8\pi R D_{11} [P] \cdot [P]}{2} = 8\pi R D [P]^2 \qquad (4C.8)$$

where we have dropped the subscript b denoting the bulk concentration. An identification with equation 4C.1 yields an expression

$$k_r = 8\pi R D \qquad (4C.9)$$

for the binary association rate constant of two similar particles. This is the von Smoluchowski equation 4.5.12. For a hetero-association with two different particles

$$k_r = 4\pi (R_1 + R_2) (D_1 + D_2) \qquad (4C.10)$$

For the case $R_1 \approx R_2$ and $D_1 \approx D_2$ the two equations differ by a factor of two because the two particles are dissimilar.

APPENDIX 4D DIFFUSION-CONTROLLED COAGULATION—LEADING TO FORMATION OF MULTIMERS

Our goal is to obtain an expression describing the kinetics for complete aggregation of a colloidal sol. The model used is identical to that given for the formation of dimers in Appendix 4C. However, we generalize it to include trimers, tetramers, etc., leading to the formation of macroscopic particles which coagulate and precipitate.

The rate constant k_{ij} for the diffusion controlled association of an i-mer, P_i, with a j-mer, P_j, is equation 4C.10

$$k_{ij} = 4\pi (D_i + D_j)(R_i + R_j) \qquad (4D.1)$$

Using Stokes law, $D = kT/6\pi\eta R$, we can express equation 4D.1 as

$$k_{ij} = \frac{2kT}{3\,\eta}(1/R_i + 1/R_j)(R_i + R_j) \qquad (4D.2)$$

The rate constant has only a weak dependence on the size of the particles because the factor $(1/R_i + 1/R_j)(R_i + R_j)$ can only vary between 1 and 4, depending on whether $R_i \gg R_j$, $R_i \approx R_j$ or $R_j \gg R_i$. Thus, it is a reasonable approximation to make all rate constants equal and replace $(1/R_i + 1/R_j)(R_i + R_j)$ by a value in the interval 1 to 4, say 2. Then

$$k_{ij} \approx \frac{4}{3}\frac{kT}{\eta} \qquad (4D.3)$$

is a universal rate constant to insert in the kinetic scheme of the total aggregation process.

For the concentration of species P_m the rate equation is

$$\frac{d[P_m]}{dt} = k\left(\frac{1}{2}\sum_{i<m}[P_i][P_{m-i}] - [P_m]\sum_i[P_i]\right) \qquad (4D.4)$$

As m goes from 1 to $m_{\max}$, equation 4D.4 represents $m_{\max}$ coupled differential equations. In fact it is possible to solve this system of equations.

Let us start by considering the change in the total particle concentration $\sum_m [P_m]$. This quantity is changed by all associations, and since the rate constant is the same, we have in analogy with equation 4C.1

$$\frac{d}{dt}\sum_m [P_m] = \frac{k}{2}\left(\sum [P_m]\right)^2 \qquad (4D.5)$$

where the factor of two is introduced to correct for the double counting of each association.

Given that initially at time $t = 0$ the total particle concentration $[P]_{tot} = 0$, equation 4D.5 has the solution

$$\sum_m [P_m] = \frac{[P]_{tot}}{1 + t/\tau} \qquad (4D.6)$$

where $\tau = 2/(k[P]_{tot})$. This shows that for short times $t < \tau$ the decrease in total number of particles is linear in time, while for long times $t > \tau$ the decrease is inversely proportional to t. This result corresponds to the upper curve in Figure 4.15. The characteristic time τ separating the two regimes decreases with total particle concentration. When $t = \tau$, one half of the total particles exist in some aggregated form P_m.

Now equation 4D.4 can be integrated in an iterative way by noting that $\sum_i [P_i]$ in the last term is given explicitly by equation 4D.6. For $m = 1$ the first term on the right-hand side of equation 4D.4 vanishes and

$$\frac{d[P_1]}{dt} = -k[P_1] \frac{[P]_{tot}}{1 + t/\tau} \qquad (4D.7)$$

where $[P_1]$ is the only unknown. A straightforward integration then yields

$$[P_1] = [P_1]_o (1 + t/\tau)^2 \qquad (4D.8)$$

Thus, the concentration of monomers declines parabolically with time as shown in Figure 4.15. An uncoupled equation for $m = 2$ can now be constructed and so on for $m = 3$, etc. The general solution for the initial condition of only monomers at t = 0 is

$$[P_m] = [P]_{tot} (t/\tau)^{+m-1} (1 + t/\tau)^{-m-1} \qquad (4D.9)$$

which is the result quoted in equation 4.5.15 and illustrated in Figure 4.15.

APPENDIX 4E
KINETICS OF SLOW COAGULATION

Our goal is to calculate the coagulation rate for dimer formation when a potential energy barrier like that shown in Figures 4.1 and 4.12 is present. We employ Fick's equation [4C.2] as our starting point, $J = -D \nabla[P]$, but generalize it to the presence of an external potential. A general form of Fick's first law says that the flux, J, is proportional to a friction factor, D/kT, a concentration term $[P]$ and a driving force, the gradient in chemical potential $\nabla\mu$, so that

$$J = -\frac{D}{kT} [P] \nabla\mu(P) \qquad (4E.1)$$

The chemical potential is in an external potential V(r) so that

$$\mu = \mu^0 + kT \ln[P] + V(r) \qquad (4E.2)$$

Combining equations 4E.1 and 4E.2 leads to

$$J = -D \nabla[P] - \frac{D}{kT} [P] \nabla V \qquad (4E.3)$$

which compared with equation 4C.2 has an extra potential-dependent term.

We can solve equation 4E.3 using the same procedure as in Appendix 4C for rapid coagulation. With the central particle, i, located at the origin of the coordinate system and for a spherically symmetrical particle, the total flux through a shell of radius r is

$$(J\ A)_i = -4\pi r^2 D \left(\frac{d[P]}{dr} + \frac{[P]dV/kT}{dr} \right) = \text{constant} \qquad (4E.4)$$

at steady state. In this equation we have again invoked the approximation that the relative diffusion coefficient is twice the individual D for similar particles.

Equation 4E.4 can be integrated by noting that $\exp(V/kT)$ is an integrating factor and the concentration profile around the central particle is

$$[P] = -\frac{(JA)_i}{4\pi D} \int_r^\infty \exp\left(\frac{V(r)}{kT}\right) r^{-2} dr \cdot \exp\left(-\frac{V}{kT}\right) + [P]_o \exp\left(-\frac{V}{kT}\right) \qquad (4E.5)$$

where we have used the boundary condition that $[P]$ should tend to the bulk value $[P]_o$ for large r. At a radius $r < 2R$, we consider a dimer to have formed. Thus, a second boundary condition is $[P] = 0$ for $r = 2R$ as used earlier in Appendix 4C. We can now solve for the flux $(JA)_i$ in equation 4E.5 and obtain

$$(JA)_i = 4\pi D [P]_o \left(\int_{2R}^\infty \exp\left(\frac{V(r)}{kT}\right) r^{-2} dr \right)^{-1} \qquad (4E.6)$$

With the same argument that leads from equations 4C.4 to 4C.9, the rate constant, k_s, in the presence of a barrier is

$$k_s = \frac{8\pi RD}{2R \int_{2R}^\infty \exp\left(\frac{V(r)}{kT}\right) r^{-2} dr} \qquad (4E.7)$$

or

$$k_s = \frac{k_r}{2R \int_{2R}^\infty \exp\left(\frac{V(r)}{kT}\right) r^{-2} dr} = \frac{k_r}{W} \qquad (4E.8)$$

where W is the stability ratio, equation 4.5.18. Thus, an energy barrier decreases the rate of flocculation by an amount determined by the stability ratio, W, given by

$$W = 2R \int_{2R}^\infty \exp\left(\frac{V(r)}{kT}\right) r^{-2} dr \qquad (4E.9)$$

If we are able to express the interaction potential energy $V(r)$ as a function of particle separation, that is $V(h)$ where h is the particle separation distance, we can obtain W. The expression is usually so complex that the integral must be evaluated numerically.

When we use equation 4E.9 to calculate W for typical values of Φ_o and h_{max} in the DLVO theory, the relationship between W and the electrolyte concentration becomes

$$\log_{10} W = -k_s \log c + k_r \qquad (4E.10)$$

In other work Reerink and Overbeek showed that

$$k_s = 2.15 \times 10^9 \frac{Rz^2}{y^2} \qquad (4E.11)$$

where $y = ze\Phi_o/kT$ and R is the particle radius.

5

AMPHIPHILIC SYSTEMS— LIQUID–LIQUID INTERFACES

with H. Wennerström

LIST OF SYMBOLS

a	area per molecule at surface
a_{hg}	effective area of amphiphile headgroup (HG) at micelle surface (eq. 5.1.6)
a_{hc}	cross-sectional area of amphiphile hydrocarbon tail (hc) ($=v_{hc}/l_{hc}$)
c	molar concentration
CMC	critical surfactant concentration for micelle formation
ΔG_{em}	free energy of emulsification (eq. 5.4.1)
ΔG_{mic}	free energy of micelle formation (eq. 5.2.6)
$\Delta G_{(HP)}$	free energy associated with hydrophobic (HP) interactions (eq. 5.2.10)
$\Delta G_{(HG)}$	free energy associated with headgroup (HG) interactions (eq. 5.2.11)
ΔG_{el}	free energy associated with electrostatic headgroup interactions (eq. 5.2.11)
h	distance between bilayers
H	curvature
H_{ij}	Hamaker constant (eq. 2.6.6)
ΔH_{mic}	enthalpy of micelle formation (eq. 5.2.7)
k_{No}^{+}	association rate constant for micelle formation
k_{No}^{-}	dissociation rate constant for micelle disintegration
K_{No}	equilibrium reaction constant for micelle formation (eq. 5.2.3)
l_{hc}	length of fully extended hydrocarbon chain (hc) in a macromolecule (eq. 5.1.5)
n_c	number of carbon atoms along a hydrocarbon chain
N_o	aggregation number; number of amphiphile molecules per micelle (eq. 5.2.2)
N_s	surfactant number (eq. 5.3.2)
R_{mic}	spherical micelle radius

$[S_N]$	concentration of N aggregated surfactant molecules (S)
ΔS_{mic}	entropy of micelle formation (eq. 5.2.8)
T_{Kr}	Krafft temperature; T_c cloud point
v_{hc}	volume of hydrocarbon chain (hc) (eq. 5.1.3)
γ_{12}	surface (interface) tension between phases 1 and 2
Γ_i	excess concentration of i at surface (eqs. 3.5.13 and 5.1.2)
κ^{-1}	Debye screening length (eq. 4.3.13)
Π_{osm}	osmotic pressure (eq. 4.4.1)
τ_1	relaxation time for exchange of monomers between micelles (eq. 5.2.13)
τ_2	relaxation time for disintegration of micelles (eq. 5.2.14)

CONCEPT MAP

Amphiphilic Molecules and Self-Assembly Processes

- Amphiphilic molecules (surfactants) possess polar headgroups and nonpolar chains (Table 5.1) and in order to minimize unfavorable solvophobic (solvent-hating) interactions they spontaneously aggregate to form a variety of microstructures (Figure 5.1). To self-assemble, the hydrocarbon chains must be in a "liquid" state, that is, above the Krafft temperature.

- The surfactant number, $N_s = a_{hc}/a_{hg}$ (eq. 5.3.2), relates amphiphilic molecular structure (Table 5.1) to aggregate architecture (Figure 5.1). a_{hc} is the cross-sectional area of the hydrocarbon chain and a_{hg} is the effective area per headgroup. When N_s is <1, interface curvature is high, and spherical structures (micelles) form; when N_s =1, planar structures (bilayers) are favored; when N_s >1, inverted structures (inverted micelles) form.

- Amphiphilic molecules are associated physically, not chemically, and aggregate size and shape change in response to variations in solution conditions, such as temperature, pH, salt concentration, etc.

Spherical Micelles

- Spherical micelles occur above a critical surfactant concentration, the CMC, and are characterized by an optimum aggregation number, N_o, with a small standard deviation σ (Figure 5.4). N_o depends on a_{hg} and l_{hc}, the length of the hydrocarbon chain. For SDS, $N_o = 58 \pm 5$.

- In water, a_{hg} is set by a compromise between solvophobic and electrostatic interactions. Hydrocarbon chain exposed at the micelle's surface seeks to miminize contact with water and drives the headgroups together; electrostatic or steric headgroup repulsion drives them apart.

Thermodynamics

- Surfactant monomers (S) aggregate by a stepwise process, $S_{N-1} + S \rightarrow S_N$ (eq. 5.2.1) to form monodisperse micelles, i.e., $N_0 S = S_{N_0}$.

- The free energy of micellization is $\Delta G_{mic} = -RT \ln K = N_0 RT \ln CMC$ (eq. 5.2.6).

- We can write $\Delta G_{mic} = \Delta G_{(HP)} + \Delta G_{(HG)}$. $\Delta G_{(HP)}$ and $\Delta G_{(HG)}$ are the free energies of transferring the surfactant hydrocarbon chain out of the solvent into the micelle and of the headgroup interactions, respectively.

Dynamics

- Two relaxation times characterize molecular processes in micelles.

- τ_1 (10^{-6} to 10^{-3} s^{-1}) measures the rate of exchange of monomers between micelles, while the number of micelles remains constant.

- τ_2 (10^{-3} to 1 s^{-1}) measures the rate of micelle formation and disintegration. Because micelles must pass through a kinetic bottleneck ($N_0 \rightarrow N \rightarrow 1$, Figure 5.7) as they disintegrate, τ_2 is slower than τ_1.

Comparison of Ionic and Nonionic Micelles

Ionic Micelles

- Adding salt lowers the CMC and increases N_0. Raising the temperature increases the CMC and lowers N_0.

- The Poisson–Boltzmann equation provides a qualitative expression for $\Delta G_{(HG)}$ that relates changes in CMCs and N_0s with chain length, temperature, and salt.

Nonionic Micelles

- Adding salt has relatively little effect on either the CMC or N_0. Heating causes N_0 to increase until phase separation occurs at the cloud point temperature. For a given hydrocarbon chain, the CMC is lower for a nonionic surfactant.

- Nonionic micelles are more difficult to model than ionic micelles.

Aggregation in More Concentrated Solutions

- With increasing surfactant concentration spherical micelles become unstable and new aggregate structures form. With varying concentration amphiphilic aggregates traverse the manifold structures shown in Figure 5.1, going from micelles to inverted micelles.

- Phase diagrams (Figure 5.12) for amphiphiles are complex, particularly in the regions where bilayers, bicontinuous, and neat phases coexist because differences in free energy between phases are very small and $\approx RT$.

Bilayers, Vesicles, and Bicontinuous Structures

- When $N_s \approx 1$, planar structures, such as bilayers, form. The thickness of a bilayer is set by the length of the amphiphilic chains, but growth in the lateral direction has no structural limits.

- Solubility of bilayer-forming amphiphiles is very small ($\sim 10^{-5}$ to 10^{-13} M). Transformations between bilayer structures via monomer diffusion is very slow compared to micellar systems.

- Because bilayers are so large, they interact via colloidal forces. Two additional interactions are important in bilayer stability. Short-ranged hydration forces and thermally induced undulatory motions lead to repulsion between bilayers.

- Vesicles are closed bilayers (molecular containers) formed by sonic agitation of lamellar structures in excess water. Most vesicles are metastable and revert back to bilayers. However, vesicular lifetimes are sufficiently long to enable vesicles to be characterized and for them to be useful in a number of applications.

- Bicontinuous structures consist of two continuous, coexisting phases, separated by a well-defined amphiphile interface; a sponge filled with water is a macroscopic structural analogy.

Emulsions

- Emulsions are homogeneous mixtures of oil and water stabilized by emulsifiers, which are often amphiphilic molecules.

- Because emulsions are liquid, they display the flexibility of amphiphilic systems

Microemulsions

- Microemulsions are thermodynamically stable, clear, or slightly opaque solutions.

- By adding oil or water to the structures shown in Figure 5.1, we can generate all the microemulsion structures included in Figure 5.15.

- Because of their small microstructural size, microemulsions contain enormous oil–water interfacial area and require large concentrations of emulsifier.

Macroemulsions

- Macroemulsions are thermodynamically unstable and eventually separate into oil-rich and water-rich bulk phases.

- Usually emulsion formation requires injection of mechanical energy to generate water in oil (W/O) or oil in water (O/W) droplets that range in size from 100 to 1000 nm.

- Because emulsions have such large droplet sizes, colloidal forces determine their stability. Under the influence of attractive forces, emulsion droplets, like colloidal particles, flocculate or coagulate. But, unlike colloidal particles, they also coalesce. Two droplets fuse to form a larger droplet (Figure 5.16). Repetition of this process leads to separate oil-rich and water-rich bulk phases.

5

A major challenge in interfacial engineering is to devise new strategies for the production of sophisticated interfacial systems. Interfacial systems typically consist of structures with nanometer dimensions and extremely high surface-to-volume ratios. The usual manufacturing approach is to use brute force. For example, we form colloidal particles by grinding bulk material, or we form sols by exceeding an equilibrium condition, such as a solubility limit. In the case of electronic devices, we repeat a sequence of processing steps over and over again: (1) deposit a thin layer onto a substrate; (2) coat with photoresist; (3) expose to imprint a pattern; (4) dissolve unreacted photoresist; (5) etch away unprotected regions to produce a desired shape; (6) remove the remaining photoresist material; then go back to the beginning and repeat the sequence for the next layer. Deposition, coating, exposing, dissolving, and so forth are all achieved by imposing external forces on the system to produce thermodynamically irreversible processing paths.

An alternative approach to the manufacture of interfacial systems is to design them so that they spontaneously self-assemble to the desired microstructures. The basic idea is to synthesize molecules that contain multiple functional groups that can interact with one another and with other molecules in their vicinity to form specific structures either upon mixing or in response to slight changes in concentration, pH, temperature, and so forth. *The hallmark of a self-assembling process is that small departures from equilibrium reversibly drive molecular assembly.*

Where do we go for inspiration and useful prototypes for self-assembly systems? We can just look at ourselves. Our bodies contain numerous molecules that self-assemble to form a variety of microstructures that direct and guide living processes. Phospholipids are the building blocks of our cell membranes; lipoproteins shuttle water-insoluble materials, such as fats, cholesterol, and hormones, through our bodies; bile salts, the body's detergent, permit us to ingest water-insoluble fats that provide much of the energy for metabolic processes. Increasing our understanding of these self-assembling molecules provides models we can emulate in designing interfacial systems. In fact, a new field called *mimetic chemistry* exploits these insights to fabricate devices consisting of organic and inorganic nanostructures. Specific examples of such devices include sensors containing macromolecules that respond selectively to molecules in their immediate vicinity, and controlled drug delivery systems that target specific tissues or meter out drugs, such as insulin, in response to need.

This chapter focuses on self-assembling systems involving *amphiphilic molecules,* also known as amphiphiles. The word amphiphilic comes from the Greek *amphi,* which means equivalent, and *philic,* which means loving. Amphiphilic molecules have an equivalent affinity for other molecules in their vicinity. We can translate this abstraction into a more meaningful form by considering some specific examples from the list in Table 5.1.

TABLE 5.1 Typical Amphiphilic Molecules

Chemical structure	Name	CMC at 25°C (mM)	Agg. number N_o
Anionic			
$C_8H_{17}OSO_3^- \cdot Na^+$	Sodium octylsulfate (SOS)	13.4	24
$C_{12}H_{25}OSO_3^- \cdot Na^+$	Sodium dodecylsulfate (SDS)	8.1	58
$C_{16}H_{35}OSO_3^- \cdot Na^+$	Sodium hexadecylsulfate (SHS)		
$C_{12}H_{35}CO_2^- \cdot Na^+$	Sodium dodecanoate	2.44	
$C_8F_{17}CO_2^- \cdot Na^+$	Sodium perfluorooctanate		
Cationic			
– single chained			
$C_{12}H_{25}NH_3^+ \cdot Br^-$	Dodecylammonium bromide	12	
$C_{12}H_{25}N(CH_3)_3^+ \cdot Br^-$	Dodecyltrimethylammonium bromide (DTAB)	14.5	54
$C_{12}H_{25}N(CH_3)_3^+ \cdot OH^-$	Dodecyltrimethylammonium hydroxide (DTAH)	34	29
$C_{14}H_{29}N(CH_3)_3^+ \cdot Br^-$	Tetradecyltrimethylammonium bromide (TTAB)	3.0	70
– double chained			
$(C_{12}H_{25})_2N(CH_3)_2^+ \cdot Br^-$	Didodecyldimethylammonium bromide (di-DDAB)		
$(C_{16}H_{33})_2N(CH_3)_2^+ \cdot CH_3CO_2^-$	Dihexadecyldimethylammonium acetate		
Nonionic			
$C_{10}H_{21}N(CH_3)_2O$	Decyldimethylamine oxide	1.9	
$C_{12}H_{25}(OCH_2CH_2)_5OH$	Dodecylpentaoxyleneglycol monoether $(C_{12}E_5)$		
$C_{12}H_{25}(OCH_2CH_2)_6OH$	Dodecylhexaoxyleneglycol monoether $(C_{12}E_6)$	0.068	
$C_{12}H_{25}(OCH_2CH_2)_8OH$	Dodecyloctaoxyleneglycol monoether $(C_{12}E_8)$	0.071	
Zwitterionic			
$C_{12}H_{25}N^+(CH_3)_2(CH_2)_3SO_3^-$	Dodecyldimethyl propane sultaine	3.6	
	Dipalmitoyl phosphatidyl choline	4.7×10^{-7}	

Amphiphilic molecules are either anionic (negatively charged) or cationic (positively charged) molecules in solution and possess well-defined polar head groups [$-OSO_3^-$ or $-N(CH_3)_3^+$, for example] and well-defined nonpolar tail groups—usually hydrocarbons or fluorocarbons ($C_{12}H_{25}-$ or $C_8F_{17}-$, for example). The amphiphilic molecule's headgroup experiences *solvophilic* (solvent-loving) interactions with polar solvents like water, while the nonpolar group experiences *solvophobic* (solvent-hating) interactions with water. The headgroups want to stay in the water, but the nonpolar tail groups want to escape. In nonpolar solvents like oil, the roles of polar and nonpolar groups are reversed. Some amphiphilic molecules are nonionic, possessing no net charge, but they do possess dipolar or zwitterionic head groups that are attracted to the water.

To minimize unfavorable solvophobic interactions, amphiphilic molecules spontaneously aggregate, or self-assemble, to form a variety of structures like those shown in Figure 5.1. In water, amphiphilic molecules form spherical and cylindrical configurations called *micelles,* arranged so that the polar head groups point out into the solvent (water) and the nonpolar tails are sequestered in the aggregate to escape from the unfavorable polar environment. *Inverted micelles* with the opposite orientation are found in oil; there the nonpolar tails point out into the oil and the polar head groups aggregate. In systems containing both oil and water, amphiphilic molecules naturally adsorb at the oil–water interface because the polar headgroup points into the water while the hydrocarbon tail points into the oil. This adsorption lowers the interfacial tension considerably and permits us to form *macroemulsions,* which are classical liquid–liquid colloidal dispersions consisting of water or oil droplets suspended in an oil or water continuum, with the amphiphile at the interface. Useful and characteristic properties associated with emulsions are employed in many interfacial processes and products.

Amphiphilic molecules are attracted to each other by physical, not chemical, interactions. For this reason, amphiphilic aggregates can change their size and shape significantly in response to small changes in concentration, pH, temperature, and pressure. Such sensitive response contrasts markedly with that of the colloidal particles discussed in Chapter 4.

In this chapter, we analyze self-assembly processes, relate molecular structure to aggregate architecture, and delineate specific properties of the various interfacial structures shown in Figure 5.1.

5.1 Aggregation of Amphiphilic Molecules to Form Spherical Micelles Illustrates Many Features of Self-Organizing Systems

As we have noted, amphiphilic molecules in solution assemble to form a variety of microstructures. We are interested in knowing what factors determine the exact geometric form, size, and stability of these self-assembled structures. In particular we will ex-

Figure 5.1
Examples of self-organizing structures formed by the association of amphiphiles: (a) spherical micelle; (b) cylindrical micelle; (c) bilayer; (d) bicontinuous structure; (e) inverted micelle; (f) spherical vesicle (see Section 5.3.4). The curvature of the aggregate surface is inversely related to the surfactant number ($N_s = a_{chains}/a_{hg}$, eq. 5.3.2). Spherical micelles possess high curvature and small values of $N_s \approx 1/3$, bilayers and bicontinuous structures have zero curvature and $N_s \approx 1$, and inverted structures are characterized by negative curvature and $N_s > 1$. (D. F. Evans and H. Wennerström, *The Colloidal Domain: Where Physics, Chemistry, Biology, and Technology Meet*, VCH Publishers, New York, 1994, pp. 14–15. Reprinted with permission of VCH Publishers, © 1994.)

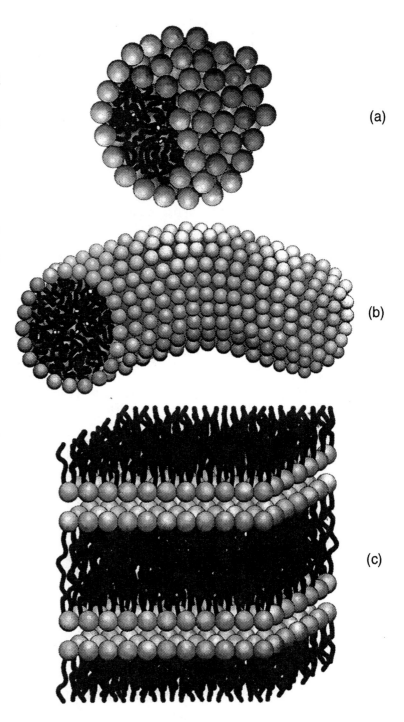

(a)

(b)

(c)

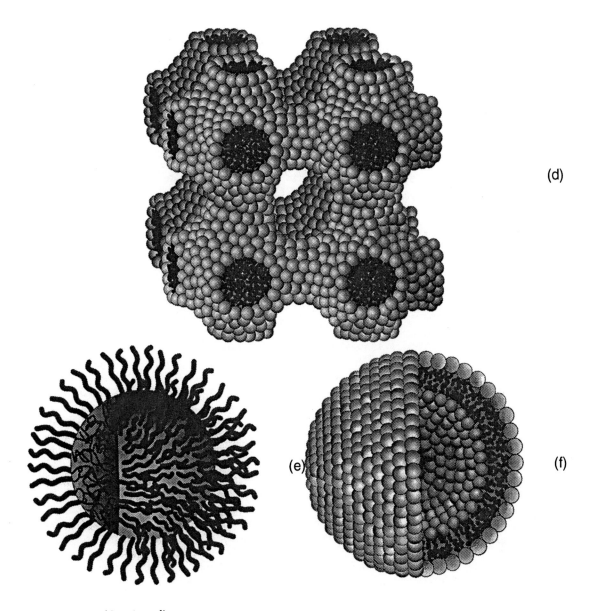

(d)

(e)

(f)

Figure 5.1 *(Continued).*

amine the behavior of a well-studied amphiphile, sodium dodecylsulfate, or SDS.[1] Each SDS molecule has the chemical composition $C_{12}H_{25}OSO_3^- \cdot Na^+$ and the structural formula

$$H - \overset{\overset{\displaystyle H}{|}}{\underset{\underset{\displaystyle H}{|}}{C}} - \overset{\overset{\displaystyle H}{|}}{\underset{\underset{\displaystyle H}{|}}{C}} - \overset{\overset{\displaystyle H}{|}}{\underset{\underset{\displaystyle H}{|}}{C}} - \overset{\overset{\displaystyle H}{|}}{\underset{\underset{\displaystyle H}{|}}{C}} - \overset{\overset{\displaystyle H}{|}}{\underset{\underset{\displaystyle H}{|}}{C}} - \overset{\overset{\displaystyle H}{|}}{\underset{\underset{\displaystyle H}{|}}{C}} - \overset{\overset{\displaystyle H}{|}}{\underset{\underset{\displaystyle H}{|}}{C}} - (O\text{-}SO_3)^- \cdot Na^+$$

---

[1] *As an aid to relating names of amphiphilic molecules to chemical structures, all compounds discussed in this chapter are listed in Table 5.1.*

The molecule consists of a long, eleven-unit methylene (CH$_2$) chain, terminated on the left-hand end with a methyl (CH$_3$) group. This chain is the *nonpolar tail*. The right-hand end terminates with the sulfate *polar headgroup*. In polar solvents like water this amphiphilic molecule ionizes to form an anionic amphiphile and a sodium cation. The negatively charged hydrophilic polar headgroup (O-SO$_3$)$^-$ is attracted to the water molecules while the hydrophobic tail is repelled.

In reality at low temperatures the carbon bonds must satisfy the covalent bonding angle of 109.5°; so the chain exhibits a zigzag (all-trans) configuration. For the aggregate structures of interest to us the bonds are less rigid at ambient temperatures and the chain can twist and bend. This flexibilty leads to the graphical representation of an amphiphile as a wiggly line, representing the tail, joined to a circle, representing the headgroup. The length of the tail may be 5–10 times the diameter of the head.

5.1.1 Concentration of Amphiphilic Molecules in Solution Must Exceed a Critical Value Before Micelles Will Form

To understand aggregation processes, we consider what happens when we add sodium dodecylsulfate (SDS) to water above 22°C. Figure 5.2 shows the changes that occur in the physical properties of an SDS solution as a function of concentration. Initially the SDS solution displays many properties similar to those we observe for any typical electrolyte. However, at 0.008M (8 mM) virtually all the solution's physical properties show a pronounced change. We designate this concentration as the *critical micelle concentration* (CMC).

At the CMC and above it is energetically favorable for SDS molecules to aggregate in the form of spherical micelles (illustrated in Figure 5.1a). The negatively charged polar headgroups are arrayed on a roughly spherical surface, and the hydrocarbon chains are contained within the sphere. Spherical micelles have a characteristic size, so that as we add more and more SDS beyond the CMC, the number of micelles increases, but their size remains constant.

Figure 5.2
Effect of surfactant concentration on various physical properties. Virtually all physical solution properties display a pronounced change in slope over a fairly narrow concentration range that can be used to define the critical micelle concentration (CMC). The small variations in CMC determined from different measurements remind us that micellization is a cooperative process that occurs over a finite concentration range, not an abrupt phase transition. (B. Lindman and H. Wennerström, *Topics in Current Chemistry*, Vol. 87, © Springer-Verlag, Berlin, 1980, Fig. 2.1, p. 6.)

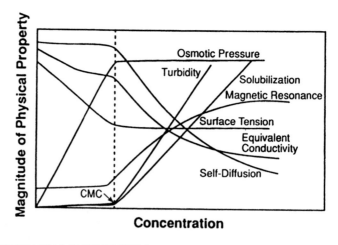

Figure 5.3
Change in the surface tension γ of water with the addition of sodium dodecylsulfate (SDS). The initial lowering of γ results from concentration of surfactant at the air–water surface; the value then becomes almost constant (note the logarithmic scale); the CMC is determined as the intersection of the two linearly extrapolated solid lines A and B. The dashed line shows the effect of a small addition of dodecanol, illustrating how the presence of small amounts of surface active impurities can dramatically perturb aggregation processes. (D. F. Evans and H. Wennerström, *The Colloidal Domain: Where Physics, Chemistry, Biology, and Technology Meet*, VCH Publishers, New York, 1994, p. 6. Reprinted with permission of VCH Publishers, © 1994.)

Before proceeding further, let us examine how we determine the CMC using the change in surface tension with added SDS as our example. Figure 5.3 shows that surface tension decreases rapidly from 72.5 mJ/m^2 (dyn/cm) for pure water to a value of 38 mJ/m^2 at $8 \times 10^{-3} M$ SDS and then decreases slowly at higher surfactant concentrations. The intersection resulting from linear extrapolation of the two lines shown in Figure 5.3 provides the CMC value. Amphiphilic molecules that lower the surface tension, like SDS, are called surface-active amphiphiles or *surfactants*.

We can learn more about the surfactant SDS from curves like Figure 5.3. From the change in surface tension γ with log concentration for concentrations less than the CMC, we can determine the excess adsorption of SDS at the air—water interface by using the Gibbs adsorption equation (eq. 3.5.13) in the form

$$-d\gamma = RT[\Gamma_{DS^-}\, d(\ln c_{DS^-}) + \Gamma_{Na^+}\, d(\ln c_{Na^+})] \qquad (5.1.1)$$

Electroneutrality at the surface requires that $\Gamma_{DS^-} = \Gamma_{Na^+}$, and if the water is pure and no other electrolyte is added, then $c_{DS^-} = c_{Na^+}$ and

$$\Gamma_{DS^-} = -\frac{1}{2RT}\frac{d\gamma}{d(\ln c_{DS^-})} \qquad (5.1.2)$$

Thus, from the slope of the left-hand portion of the solid curve in Figure 5.3, we can estimate the excess concentration of surface

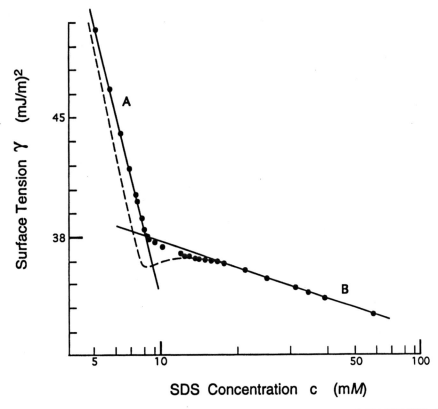

SDS Concentration c (m*M*)

molecules per unit area, Γ_{DS^-}. The inverse of Γ_{DS^-} gives the *effective* area a of a surfactant molecule at the air–water interface. The data in Figure 5.3 give a value for a equal to 60 Å^2 for SDS. This relatively large effective area, about five times the actual area, reflects the electrostatic repulsion of the negatively charged polar headgroups arrayed in close proximity at the air–water interface.

To understand the significance of the effective surface area, measurements of the surface tension versus SDS concentration can be repeated in the presence of a "swamping" electrolyte. When sufficient salt (NaCl) is added to the water, electrostatic effects become unimportant. Na$^+$ need not adsorb at the surface for electroneutrality and so $\Gamma_{Na^+} = 0$ and eq. 5.1.1 becomes $\Gamma_{DS^-} = -d\gamma/RT \, d(\ln c_{DS^-})$. We find the value for a reduces to 38 Å^2 for SDS in 0.1M NaCl. The decrease in a between these two situations reflects a reduction in electrostatic repulsion of the negatively charged headgroups at the air–water interface of the NaCl electrolyte solution.

Measurements of surface tension constitute a sensitive test for surfactant purity. For example, 0.01% impurity of 1-decanol in SDS results in the curve shown as the dotted line in Figure 5.3. Although both compounds adsorb at the air–water interface, the alcohol preferentially concentrates there because it is even more surface active than SDS. The surface tension curve falls below that observed for pure SDS because γ(alcohol) $< \gamma$(SDS). In the bulk solution, the presence of alcohol initiates SDS micelle formation at a value below the CMC for the pure surfactant. The aggregation process also extends over a much broader concentration range. As we add more SDS surfactant above the CMC, alcohol desorbs from the interface and solubilizes back into the micelles as the micellar concentration increases. That is, alcohol molecules leave the surface and become "encapsulated" within the micelles to avoid interaction with the water molecules. Encapsulation of alcohol results in a gradual increase in γ up to the level characteristic of SDS above the CMC. Therefore, a small amount of alcohol impurity leads to a pronounced dip in the surface tension curve, such as we find in the dotted line in Figure 5.3. Impurity also is responsible for the very shallow dip in the solid curve in Figure 5.3 and explains why we must extrapolate the linear portions of the curves to their intersection to find the true CMC value.

Small amounts of impurities present in amphiphiles initially constituted a major obstacle in the development of our understanding of this subject. In research, we take elaborate precautions to assure ourselves that we have pure samples. Commercial applications, on the other hand, often use extremely impure samples or mixtures, as we will see in Section 5.5.

5.1.2 The Radius of Spherical Micelles Is Determined by the Energy Balance between Unfavorable Hydrocarbon–Water Interaction and Polar Headgroup Coulombic Repulsion

Now we return to our discussion of the characteristics of amphiphilic aggregation processes. We want to know how many amphiphilic molecules make up a spherical micelle (the *aggrega-*

tion number N_o) and how closely they are packed together. Light-scattering measurements on SDS solutions above the CMC show that these micelles contain an average of 58 dodecylsulfate ions. How broad or how narrow is the distribution that defines this average? Kinetic measurements (to be described in more detail in Section 5.2.6 and in particular in eq. 5.2.13) yield a very low standard deviation σ for the aggregation number N_o. In the case of SDS micelles, $N_o = 58$ ± 5, and to a very good approximation, we can treat spherical micelles as if they are monodisperse. Calculated micelle size distributions for SDS are reproduced in Figure 5.4.

To determine the packing density, we can carry out a simple but detailed geometric analysis of the SDS micellar structure. First, we calculate the volume v_{hc} of the hydrocarbon tail on the amphiphilic molecule. The hydrocarbon chain volume is

$$v_{hc} = [27.4 + 26.9\ n_c]\ \text{Å}^3 \qquad (5.1.3)$$

where $27.4\ \text{Å}^3$ equals the volume of the CH_3 methyl cap at the end of the chain, $26.9\ \text{Å}^3$ the volume of each methylene (H-C-H) group along the chain, and n_c is the number of carbon atoms in the chain—12 for SDS. Then we multiply this volume by the aggregation number N_o to estimate the volume of the micelle core. For SDS this calculation gives a micellar hydrocarbon volume of $2.03 \times 10^4\ \text{Å}^3$. By assuming

$$N_o v_{hc} = \frac{4\pi R_{mic}^3}{3} \qquad (5.1.4)$$

we can calculate a spherical radius for the micelle hydrocarbon core of $R_{mic} = 16.9\ \text{Å}$.

We also can estimate the size of a spherical micelle if we assume the radius is set by the length of a fully extended hydrocarbon chain. The center of the micelle must contain at least one terminal methyl group (otherwise a vacuum would exist in the micelle center, and nature abhors emptiness), and we expect the polar headgroups to be located on the micelle's surface, not buried in its oily interior. The maximum length of a fully extended hydrocarbon chain is

$$l_{hc\ max} = (1.5 + 1.265\ n_c)\ \text{Å} \qquad (5.1.5)$$

In this equation, $1.5\ \text{Å}$ comes from the van der Waals radius of the terminal methyl group and $1.265\ \text{Å}$ from the carbon–carbon bond length projected onto the direction of the chain—assuming the all-trans (zigzag) configuration. For SDS, $l_{hc\ max} = 16.7\ \text{Å}$. Taking the diameter of the sulfate headgroup to be about $5\ \text{Å}$, the total radius for the spherical SDS micelle is therefore $16.7 + 5 \approx 22\ \text{Å}$.

The fact that the hydrocarbon core radius from eq. 5.1.4 and the chain length from eq. 5.1.5 are almost equal means that the radius of a spherical micelle is dictated by the maximum length of the hydrocarbon tail and the aggregation number is dictated by the number of other hydrocarbon tails needed to fill the volume of a sphere defined by that radius.

We can determine the configuration of the headgroups on the surface of the spherical micelle in the following way. The total

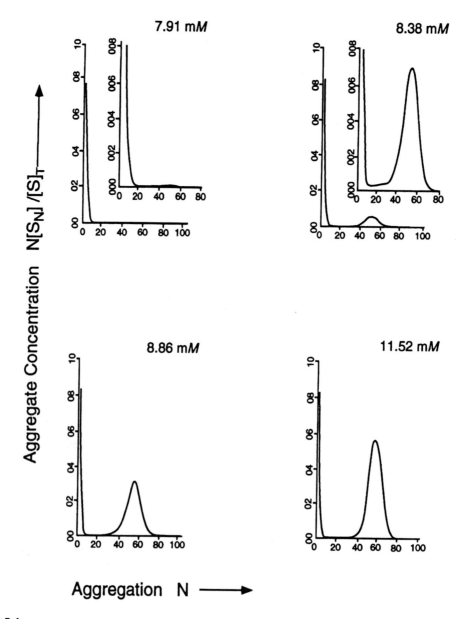

Figure 5.4
Calculated micelle size distributions for SDS with a CMC of 8.3 mM. Curves show the aggregate concentration, defined by $N[S_N]/[S]_T$ (see Section 5.2.1) versus state of aggregation, N, for different surfactant concentrations. Above the CMC, the distribution peaks around the most probable aggregation number N_{o_2}. The peaks themselves can be well described by a Gaussian distribution $P(N) \propto \exp\text{-}[(N_o\text{-}N)^2/2\sigma^2]$ with a standard deviation σ of approximately 7. Even below the formal CMC (top left panel), calculations show that aggregation exists in very low concentration (G. Gunnarsson, Ph.D. thesis, University of Lund, Lund, Sweden, 1981)

area per amphiphilic molecule headgroup (a_{hg}) at the surface of the hydrocarbon core is

$$a_{hg} = \frac{4\pi R_{mic}^2}{N_o} \qquad (5.1.6)$$

Taking $R_{mic} = 16.9$ Å and $N_o = 58$, a_{hg} is 60 Å^2 (a value that is close to the effective surface area of the SDS surfactant molecule at the air–water interface determined from Figure 5.3). By comparison, the actual cross-sectional area of a sulfate headgroup is only 27 Å^2, while that of a hydrocarbon chain is 21 Å^2. Thus our geometric analysis leads us to conclude that *less than half* of the micelle's hydrocarbon core surface actually is covered by polar headgroups, which means that the hydrocarbon chains still experience considerable contact with water.

This conclusion may seem surprising. After all, the micelle forms to minimize unfavorable hydrocarbon–water interactions. Transfer of the hydrocarbon chain from water into the oil-like interior of the micelle drives micellization. However, forming the micelle also forces the ionic headgroups into close proximity at the micelle surface and creates an enormous Coulombic repulsion. Therefore, two opposing interactions are taking place in micelle formation: solvophobic hydrocarbon–water interactions drive the amphiphiles together, while electrostatic repulsion drives their headgroups apart. The equilibrium configuration of the SDS micelle represents a compromise between these two competing interactions. It represents a structure in which the hydrocarbon tails are bunched together as compactly as possible to minimize hydrocarbon–water interactions, while the polar headgroups are separated as far as possible to minimize Coulombic interactions. The representation of the spherical micelle in Figure 5.1a attempts to depict these features, although it is difficult to capture the full three-dimensional character in a two-dimensional drawing.

This energy balance interpretation of micellization explains the existence of well-defined CMCs. At the CMC, the balance of two almost equal opposing forces leads to a cooperative transformation and results in micelle formation. The energy balance interpretation also explains why micelle size deviates so little from the average aggregation number, $N_o = 58$. If $N = 50$ much more of the hydrocarbon in an SDS micelle is exposed to water, while if $N = 70$, the closely packed headgroups have increasing difficulty in maintaining $R_{mic} = l_{hc\,max}$ as more hydrocarbon chains are stuffed into the interior.

Above the CMC adding surfactant to a micellar solution simply produces more micelles rather than leading to growth of existing micelles. For SDS this behavior persists over a considerable concentration range, from $8 \times 10^{-3}M$ up to $4 \times 10^{-1}M$. The CMC is relatively insensitive to temperature. As the horizontal right-hand curve of Figure 5.5 illustrates, for anionic surfactants like SDS the CMC increases very slightly as the temperature is raised to 40°C.

Figure 5.5
The Krafft temperature T_{Kr} is defined by the temperature at which the surfactant solubility equals the critical micelle concentration. At temperatures below T_{Kr}, hydrated surfactant crystals precipitate out of solution before micelles form; above T_{Kr}, the solubility increases dramatically and monomers assemble to form micelles above the CMC. Data for SDS. (B. Lindman and H. Wennerström, *Topics in Current Chemistry*, Vol. 87, © Springer-Verlag, Berlin, 1980, Fig. 2.8, p. 13.)

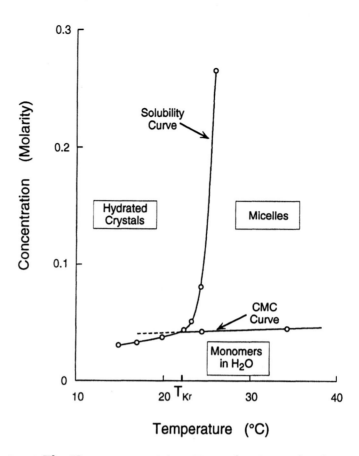

5.1.3 The Temperature Must Exceed a Critical Value Before Micelles Will Form

SDS illustrates another important property of amphiphilic molecules. If we lower the temperature of an SDS micellar solution below 22°C, the surfactant precipitates out in the form of hydrated crystals. In other words, below 22°C the solubility limit of SDS is less than the critical micelle concentration, as shown in the left-hand portion of the curve in Figure 5.5. The point at which the solubility and the CMC curves intersect is called the *Krafft temperature*, T_{Kr}. T_{Kr} = 22°C for SDS. It corresponds to the temperature at which all the hydrocarbon chains solidify, forming the inflexible, all-trans configuration. At this point, it is energetically unfavorable for the molecules to assemble into spherical micelles; instead they prefer to align and form hydrated crystals.

For an amphiphilic system to self-organize as a micelle, its hydrocarbon chains must be sufficiently flexible to undergo the contortions demanded by the micelle shape. The temperature must be high enough for the chains to be in a melted or liquid state, so that the bonds ar not aligned rigidly. NMR measurements show that local chain motions of hydrocarbons in micelles and other amphiphilic aggregates are as rapid and flexible as in the corresponding liquid hydrocarbons.

In large measure the stability of the hydrated crystals determine the Krafft temperature. Consequently, it varies from one surfactant to another. For example, T_{Kr} for potassium dodecylsulfate is 34°C, while for the corresponding magnesium and calcium salts it occurs above 100°C. With carboxylic surfactants the T_{Kr} with divalent counterions also is above 100°C, and the interaction between soap (sodium long-chain carboxylates) and hard water containing soluble divalent calcium salts precipitates the hydrated crystals to cause the infamous ring around the bathtub.

5.2 We Can Quantitatively Model Micelle Formation and Stability

5.2.1 Amphiphile Assembly Can Be Modeled As a Reaction; the Equilibrium Constant Relates Exponentially to the Free Energy of Micellization

Our goal in this section is to develop models of the self-assembly process and show how we can relate CMCs to free energies of micelle formation. We begin by writing micelle formation in terms of a stepwise association process. Two amphiphile monomers S combine to form a "dimer," which goes on to form a "trimer" and eventually an aggregate with N surfactant molecules. Each intermediate association step can be written as

$$S_{N-1} + S \leftrightarrow S_N \tag{5.2.1}$$

This sequence describes the "start" of the aggregation process, but not the "stop" that results in well-defined micellar aggregation numbers.

We can incorporate both "start" and "stop" features by assuming that one aggregation number, N_o, dominates. For a solution containing only molecules S and aggregates S_{N_o}, of fixed aggregation number N_o, we can write an equilibrium reaction for aggregate formation in the form

$$N_o S \leftrightarrow S_{N_o} \tag{5.2.2}$$

for which the equilibrium reaction constant is

$$K_{N_o} = \frac{[S_{N_o}]}{[S]^{N_o}} = \exp\left(\frac{-N_o \, \Delta G_{mic}}{RT}\right) \tag{5.2.3}$$

where ΔG_{mic} is the free energy of micelle formation *per mole of surfactant monomer.*

The following analysis helps us understand how eq. 5.2.3 leads to a well-defined CMC. We write the total surfactant concentration, $[S]_T$, expressed in terms of moles of surfactant per unit volume, as the sum of the concentration in micelle form, $N_o[S_{N_o}]$, and the concentration still in solution [S], so that

$$[S]_T = N_o[S_{N_o}] + [S] \tag{5.2.4}$$

The derivative, $d(N_o[S_{N_o}])/d[S]_T$, describes the fraction of added surfactant that enters into micelles. Figure 5.6 plots a straight-

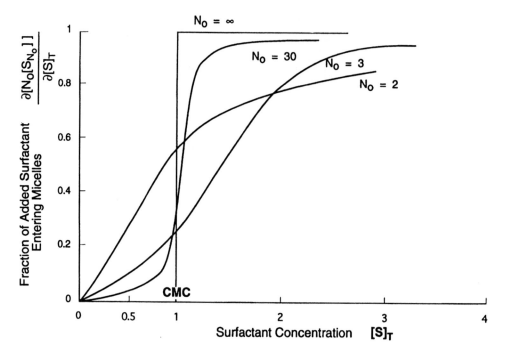

Figure 5.6
As the aggregation number N_o increases, the change in micelle concentration $N_o[S_{N_o}]$, with total concentration, $[S]_T$, as measured by $d[N_o[S_{N_o}]]/d[S]_T$ (see eq. 5.2.4), becomes increasingly sharp. In the limiting case in which the aggregation number becomes infinite, the transition becomes a step function that unambiguously defines the critical micelle concentration, CMC. Smaller micelle aggregation numbers lead to less well-defined CMCs.
(R. J. Hunter, *Foundations of Colloid Science*, Oxford University Press, Oxford, 1989, p. 576. By permission of Oxford University Press.)

forward evaluation of the derivative. It shows a number of curves of $d(N_o[S_{N_o}])/d[S]_T$ versus $[S]_T$ with varying values of N_o. The larger the N_o value, the more abruptly the derivative changes from the low-concentration value of zero to the high-concentration value of unity. When $N_o = 60$ the transition is sharp, in agreement with the experimental observation. In the limit $N_o \rightarrow \infty$, the concentration of micelles follows the step function shown in Figure 5.6, and we obtain the condition for phase separation sometimes used to model micelle formation.

Now we are in a position to relate CMCs to the free energy of micelle formation. Using eq. 5.2.3, we can write

$$-\frac{N_o \Delta G_{mic}}{RT} = \ln K_{N_o} = \ln[S_{N_o}] - N_o \ln[S] \qquad (5.2.5)$$

We divide both sides of the equation by N_o and ignore $(1/N_o) \ln[S_{N_o}]$ because when N_o is large (> 50 to 100), $(1/N_o)\ln[S_{N_o}]$ is small compared to $\ln[S]$. We then identify $[S]$, the monomer concentration, as being equal to the CMC and obtain

$$\Delta G_{mic} = RT \ln CMC \qquad (5.2.6)$$

While we can express the CMC in any concentration units, it is usually quoted as a mole fraction, that is $c_{surfactant}/(c_{surfactant} + c_{water})$, where $c_{water} = 55.5$ mol/liter.

Furthermore, combining the Gibbs equation

$$\Delta G_{mic} = \Delta H_{mic} - T \Delta S_{mic} \qquad (5.2.7)$$

and the Gibbs–Helmholtz equation

$$\frac{\partial(\Delta G_{mic})}{\partial T} = -\Delta S_{mic} \qquad (5.2.8)$$

allows us to calculate the enthalpies and entropies per mole monomer of micellization as well.

Thus we have developed a model that explains a well-defined CMC and can relate it to the thermodynamics of micellization.

5.2.2 The Free Energy of Micellization for Ionic Micelles Contains Two Components—Solvophobic Effects and Headgroup Repulsion

In Section 5.1.2 we argued that micellization is driven by the solvophobic effect and opposed by the headgroup repulsion. This argument suggests that we can write the free energy of micellization in two parts

$$\Delta G_{mic} = \Delta G_{(HP)} + \Delta G_{(HG)} \qquad (5.2.9)$$

$\Delta G_{(HP)}$ represents the free energy per mole monomer associated with transferring the hydrophobic (HP) hydrocarbon chains out of the solvent and into the oil-like interior of the micelle. It is generally negative (attractive). $\Delta G_{(HG)}$ is the free energy per mole monomer associated with the headgroup (HG) interactions and it is generally positive (repulsive).

We can obtain $\Delta G_{(HP)}$ using

$$\Delta G_{(HP)} = -[(n_c - 1) \, 3.0 + 9.6] \text{ kJ/mol} \qquad (5.2.10)$$

where n_c equals the number of carbon atoms in the alkyl chain of the surfactant monomer. 3.0 kJ/mol is the measured value of the free energy change for each methylene group along the chain, and 9.6 kJ/mol is the value for the methyl end group. By substituting $\Delta G_{(HP)}$ in eq. 5.2.9 and inserting the value for ΔG_{mic} from eq. 5.2.6, we can obtain $\Delta G_{(HG)}$.

5.2.3 The Contribution of Headgroup Interaction to the Free Energy of Micellization Can Be Modeled Using the Gouy–Chapman Theory

We can also calculate $\Delta G_{(HG)}$ directly for ionic micelles by expressing $\Delta G_{(HG)}$ in terms of two contributions

$$\Delta G_{(HG)} = \Delta G_{el} + N_o a_{hg} \, \gamma_{mic,w} \qquad (5.2.11)$$

ΔG_{el} accounts for the purely electrostatic interactions arising from headgroup repulsions and counterion effects that tend to push headgroups apart. $\gamma_{mic,w}$ can be viewed as the *micelle–water* interfacial tension per unit area and $N_o a_{hg}$ is the headgroup area per micelle; so $N_o a_{hg} \gamma_{mic,w}$ accounts for solvophobic interactions that tend to push headgroups together.

The Gouy–Chapman theory of Section 4.3.1 can be used to compute ΔG_{el} except that now we must solve the Poisson–Boltzmann equation for charged spheres rather than flat surfaces. The solution is called the *dressed micelle model*. Here we simply note two results from the solution:

1. $\Delta G_{el} = f(\kappa, N_o, R_{mic})$ (κ, the inverse Debye length), and can be evaluated in terms of measurable micellar properties; and

2. $\gamma_{mic,w}$ can be evaluated indirectly from ΔG_{el}.

To obtain the latter we note that the equilibrium configuration of a micelle corresponds to a minimum in ΔG_{mic} as a function of size. We can therefore differentiate equation 5.2.11 with respect to $N_o a_{hg}$ and equate it to zero to obtain

$$\frac{d(\Delta G_{(HG)})}{d(N_o a_{hg})} = \frac{d(\Delta G_{el})}{d(N_o a_{hg})} + \gamma_{mic,w} = 0 \qquad (5.2.12)$$

and thus a value for $\gamma_{mic,w}$.

Table 5.2 displays values of quantities calculated using eqs. 5.2.6 and 5.2.9–5.2.12. Columns 1–3 give experimental values used in the calculations. Column 4 derives the value for ΔG_{mic} per mole monomer using eq. 5.2.6. The dressed micelle theory provides an estimate for ΔG_{el} in column 5 and $\gamma_{mic,w}$ in column 6, and thence $N_{Av} a_{hg} \gamma_{mic,w}$ in column 7. We can add columns 5 and 7 to get a value for $\Delta G_{(HG)}$ in column 8, using eq. 5.2.11. We can subtract $\Delta G_{(HG)}$, column 8, from ΔG_{mic}, column 4, to obtain an

TABLE 5.2 Quantitative Analysis of the Free Energy Contributions in the Formation of Ionic Micelles: Calculations Based on the Dressed Micelle Model

Surfactant property	1 N_o	2 a (Å^2)	3 CMC (mM)	4 $RT \ln CMC^a$ (kJ/mol)	5 ΔG_{el} (kJ/mol)	6 γ_{mic,w_2} (mJ/m²)	7 $N_{Av} a \gamma$ (kJ/mol)	8 $\Delta G_{(HG)}$ (kJ/mol)	9 $\Delta G_{(HP)}$ (kJ/mol)
NaCl in SDS									
$c_{salt}(M)$									
0	58	62	8.3	−22.00	12.75	16.15	6.03	18.78	−40.77
0.01	64	60	5.7	−22.94	11.67	15.8	5.70	17.36	−40.30
0.03	71	58	3.1	−24.44	10.44	15.2	5.30	15.76	−40.20
0.10	93	53	1.47	−26.28	8.76	15.15	4.84	13.60	−39.88
NaCl in SOS									
$c_{salt}(M)$									
0	24	65		−14.71	6.72	11.2	4.38	11.10	−25.81
0.1	29	61		−15.37	6.23	11.95	4.39	10.62	−25.99
Temp. of TTAB									
25°C	70	64	3.8	−23.7	14.0	16.1	6.2	20.2	−43.9
76°C	55	70	6.7	−26.2	14.7	16.6	7.0	21.7	−47.9
114°C	35	81	13.5	−26.8	13.3	15.2	7.4	20.7	−47.5
166°C	20	100	39	−26.5	10.9	12.45	7.5	18.4	−44.9

D. F. Evans and H. Wennerström, *The Colloidal Domain: Where Physics, Chemistry, Biology, and Technology Meet,* VCH Publishers, New York, 1994, p. 163. Reprinted with permission of VCH Publishers © 1994.

^{a}CMC expressed in mole fractions.

estimate of $\Delta G_{(HP)}$, column 9, using eq. 5.2.9. The derived values of $\Delta G_{(HP)}$ per mole monomer can be compared with values estimated from eq. 5.2.10; that is, −42.6 kJ/mol for SDS, −30.6 kJ/mol for SOS, and −48.6 kJ/mol for TTAB. The results are very self-consistent and support the models of micelle structure and formation presented here.

5.2.4 The Behavior of Ionic Micelles Can Be Explained by Hydrocarbon Tail and Headgroup Characteristics

So far this chapter has focused on the behavior of one surfactant, SDS. But as Table 5.1 demonstrates, there are many other possibilities. Amphiphiles can have one, two, or even three nonpolar chains, that is to say, one, two, or three, hydrocarbon tails attached to the headgroup. For example, didodecyldimethylammonium bromide (di-DDAB) has two dodecyl nonpolar hydrocarbon tails attached to the dimethylammonium headgroup. Also there are a variety of polar headgroups. They can fall into one of four categories, depending on the charge the molecules carry in solution: anionic, cationic, nonionic (uncharged but possessing a permanent dipole moment), and zwitterionic (in which both a cationic and an anionic charge exist on the molecule).

Now we are in a position quantitatively to explain many features of these other ionic amphiphiles.

- With a given headgroup, increasing the length of the hydrocarbon tail, n_c, makes the value of $\Delta G_{(HP)}$ (eq. 5.2.10) more negative and thus lowers ΔG_{mic} and the CMC (eq. 5.2.6). The aggregation number N_0 also increases with a longer hydrocarbon tail. A comparison of the data for sodium octylsulfate (SOS) and sodium dodecylsulfate (SDS) given in Table 5.1 illustrates this point.

- Increasing the temperature raises the CMC slightly and decreases N_0. Using data for tetradecyltrimethylammonium bromide (TTAB) given in Table 5.2 reveals that from 25 to 166°C the CMC increases by a factor of eight. Since $\Delta G_{(HP)}$ remains almost constant over this temperature range, most of the change must originate in $\Delta G_{(HG)}$. One effect of increasing the temperature is that the double layer (see Section 4.3.2) becomes more diffuse, decreasing κ and leading to greater headgroup repulsion, and thus a higher ΔG_{el} (eq. 5.2.11), a higher ΔG_{mic}, and an increase in CMC.

- Adding salt increases κ and decreases the electrostatic repulsion between headgroups. This in turn decreases $\Delta G_{(HG)}$, but produces considerably less effect on $\Delta G_{(HP)}$, as illustrated in Table 5.2 for SDS with added NaCl. Adding salt lowers the CMC and increases the aggregation number because more surfactant molecules can pack together. If enough salt is added, the packing changes to such an extent that another shape with less curvature is preferred, and spherical micelles transform into cylindrical ones (see Section 5.3.1).

- For a given surfactant, differences between counterions reflect specific chemical interactions at the micellar surface (see the discussion on chemical effects at charged interfaces in Section 4.6). To illustrate this effect, we can compare the properties of the cationic surfactants dodecyltrimethyl-ammonium hydroxide (DTAH) and bromide (DTAB) in Table 5.1. Since OH⁻ is more highly solvated, it is effectively larger than Br⁻. As a consequence, the Br⁻ counterions can move closer into the micelle's headgroup region and decrease headgroup repulsion more effectively than the OH-complex. Thus the bromide has a lower CMC (by a factor of ten) and a greater N_o (by a factor of two).

- A measure of the interaction between micelles in dilute solution is given by the second virial coefficient. For ionic micelles like SDS the second virial coefficient B is approximately 40 times larger than the micelle's volume. Adding salt to the solution screens electrostatic repulsion and substantially reduces the value of B.

We conclude this section by returning to one of our major themes: amphiphilic aggregation is a physical process driven by differences in solvophobicity. Therefore, we should expect the aggregation process also to be sensitive to the polar properties of the *solvent*. To emphasize this point, we show in Table 5.3 CMCs and N_os for micelles in two solvents other than water, hydrazine (H_4N_2, a hydrogen-bonded solvent with many properties similar to those of water) and ethylammonium nitrate (EtNH$_3$NO$_3$, an $11M$ fused salt that becomes a liquid at room temperature). The free energy for transferring a methylene group, $\Delta G_{(CH_2)}$, out of these solvents and into the micelle provides a measure of solvophobicity. $\Delta G_{(CH_2)}$ is 3.0 kJ/mol for water, 2.8 J/mol for hydrazine, and 1.7 J/mol for ethylammonium nitrate. As $\Delta G_{(CH_2)}$ decreases, the CMCs increase and the aggregation numbers become smaller. Smaller aggregation numbers mean the CMC becomes less well defined (Figure 5.6), until eventually amphiphiles do not aggregate at all in solvents such as ethanol or acetone because distinct interactions do not exist between the solvents and headgroups and hydrocarbon tails.

TABLE 5.3 Comparison of Micellar Properties in Water, Hydrazine (H_4N_2), and EthylAmmonium Nitrate (EAN)

Solvent	Surfactant	$T\,(°C)$	CMC (10^3M)	Agg. no. N_o
H_2O	$C_{10}H_{21}OSO_3Na$	25		
	$C_{12}H_{25}OSO_3Na$	25	14.6	58
H_4N_2	$C_{10}H_{21}OSO_3Na$	35	2.6	
	$C_{12}H_{25}OSO_3Na$	35	6.4	
EAN	$C_{12}H_{25}N(CH_3)_3Br$	50	23	
	$C_{14}H_{29}N(C_5H_5)_3Br$	25	7.2	17
	$C_{16}H_{33}N(C_5H_5)_3Br$	30	1.8	26

5.2.5 Many Nonionic Micelles Separate into Two Phases with Increasing Temperature

Nonionic micelles are more difficult to model in a quantitative way than ionic micelles. However, we can obtain a qualitative understanding of nonionic micelle properties by noting how they differ from ionic micelles.

Nonionic surfactants possessing dipolar or zwitterionic headgroups exhibit substantially lower CMCs than corresponding ionic surfactants. For example, CMC(SDS) = $8 \times 10^{-3}M$, while CMC($C_{12}E_8$) = $9.7 \times 10^{-5}M$ (see Table 5.1). Adding electrolyte to the solvent produces only a marginal effect on micellization. Such low values for the CMCs and insensitivity to salt concentration simply reflect lower headgroup repulsion.

Increasing the temperature creates opposite effects for ionic and nonionic surfactants. For ionic micelles the aggregation number usually decreases with increasing temperature, whereas for nonionic micelles the aggregation number increases significantly with temperature. We can understand this increase if we study the headgroups of typical uncharged surfactants, such as polyethylene oxide chains, sugars, or amine oxides. They all interact with water via hydrogen bonding and/or dipole–dipole interactions that decrease in magnitude with increasing temperature. As water becomes less and less a good solvent for the polar headgroups, they repel less and therefore larger aggregates are easier to form.

When we heat aqueous solutions of nonionic surfactants of the polyethylene oxide type, they become turbid at a temperature called the *cloud point*. At a somewhat higher temperature, they separate into two phases. One of these phases contains a low concentration of surfactant approximately equal to the CMC. The other phase is surfactant-rich. The cloud point varies slightly with surfactant concentration; it is not a unique temperature like the Krafft point.

Adding NaCl to nonionic micelles lowers the cloud point. This effect mainly results from lowering the activity of water in the bulk, which makes the solvents less prone to hydrate the headgroup. Expressed in another way, the electrolyte solution can be seen as a less good solvent for the headgroups than pure water. If the ions can interact specifically with the headgroup, this trend is reversed.

5.2.6 Relaxation Measurements Provide Information on Micellar Lifetimes and Dynamics

Earlier in this chapter we stated that amphiphilic assemblies are associated physically, not chemically, and thus respond to changes in temperature. pressure, and concentration by changing their aggregation number or structure. How fast can these transformations occur? For micelles we can determine the time scales using fast-relaxation measurements. The idea is to take a system that is initially at equilibrium, subject it to a small but rapid change in an intensive parameter (such as a temperature jump, pressure jump, or ultrasonic agitation) and then follow the relaxation to the new equilibrium state by tracking changes in scattering or spectroscopic properties.

Measurements on a variety of surfactant solutions yield two well-defined relaxation times: τ_1, typically in the range 10^{-6}–10^{-3} s, and τ_2, in the range 10^{-3}–1 s, as shown in Table 5.4a. Since the perturbation associated with a relaxation measurement causes only a small, negligible shift in the optimal size distribution, the two relaxation times indicate that two different mechanisms are involved in the return to normalcy. To understand the kinetics involved, we need to think about the micellar system in terms of multiple equilibria (eq. 5.2.1) and a narrow aggregate size distribution, as indicated in Figure 5.4.

Detailed analysis of the fast relaxation time, τ_1, shows that it corresponds to the rate of exchange of surfactant monomers between micelles, as illustrated in Figure 5.7a. τ_1 is related to the "dissociation" rate constant, $k_{N_0}^-$. This constant defines the rate at which a monomer leaves a micelle (the rate at which the micelle dissociates) and is related to τ_1 in the following way

$$\frac{1}{\tau_1} = \frac{k_{N_0}^-}{\sigma^2} + \frac{k_{N_0}^-}{N_0} \frac{(c - CMC)}{CMC} \qquad (5.2.13)$$

c is the amphiphile concentration in the solution, N_0 represents the most probable aggregation number, and σ stands for the standard deviation of the micelle size-distribution function (see Figure 5.4). We can determine $k_{N_0}^-$ and σ from the slope and intercept of a $1/\tau_1$ versus $(c - CMC)$ plot if we know the aggregation number, N_0, independently.

Table 5.4b summarizes the parameters obtained from an analysis of kinetic data for a series of sodium alkyl sulfates. The dissociation rate constant, $k_{N_0}^-$, decreases by a factor of 1400 in going from the C_6 to the C_{14} compound. This large variation in $k_{N_0}^-$ occurs because in the process of leaving a micelle, the surfactant's hydrocarbon chain moves out of the oily environment and into the water phase. As we can see from eq. 5.2.10, the longer the hydrocarbon chain, the higher the free energy cost and thus the slower the rate.

TABLE 5.4a Relaxation Times, τ_1 and τ_2, for Some Sodium Alkyl Sulfates[a]

Surfactant	Temperature ($°C$)	Concentration (M)	τ_1 (μs)	τ_2 (ms)
$C_{12}SO_4Na(SDS)$	20	1×10^{-2}	15	1.8
	20	5×10^{-2}		50
$C_{14}SO_4Na(STS)$	25	2.1×10^{-3}	320	41
	30	2.1×10^{-3}	245	19
	35	2.1×10^{-3}	155	7
	25	3×10^{-3}	125	34
$C_{16}SO_4Na(SHS)$	30	1×10^{-3}	760	350

TABLE 5.4b Kinetic Parameters of Dissociation and Association of Alkyl Sulfates from Their Micelles[a]

Surfactant	N	CMC (M)	$k_{N_o}^-$ (s^{-1})	$k_{N_o}^+$ $(M^{-1} s^{-1})$
C_6SO_4Na	17	0.42	1.32×10^9	3.2×10^9
C_7SO_4Na	22	0.22	7.3×10^8	3.3×10^9
C_8SO_4Na	27	0.13	1.0×10^8	7.7×10^9
C_9SO_4Na	33	6×10^{-2}	1.4×10^8	2.3×10^9
$C_{11}SO_4Na$	52	1.6×10^{-2}	4×10^7	2.6×10^9
$C_{12}SO_4Na(SDS)$	64	8.2×10^{-3}	1×10^7	1.2×10^9
$C_{14}SO_4Na$ (STS)	80	2.05×10^{-3}	9.6×10^5	4.7×10^8

[a]E. A. G. Aniansson, S. N. Wall, M. Almgren, H. Hoffmann, I. Kielmann, W. Ulbricht, R. Zana, J. Lang, and C. Tondre, *J. of Physical Chemistry* 80, 905 (1976). Reprinted with permission from *J. of Physical Chemistry*. Copyright 1976 American Chemical Society.

We can evaluate the rate at which a monomer arrives at a micelle, the "association" rate constant, $k_{N_o}^+$, by noting that the ratio of the two rate constants for arrival and departure are equal to the equilibrium constant (equation 5.2.3), that is, $K_{N_o} = k_{N_o}^+/k_{N_o}^-$. If we use the relation between $\ln K_{N_o}$ and ΔG_{mic} in eq. 5.2.5 and between ΔG_{mic} and $\ln$ CMC in eq. 5.2.6 (per surfactant molecule), we obtain $k_{N_o}^+ = k_{N_o}^-/CMC$.[2] Table 5.4b shows that the association rate constant, $k_{N_o}^+$, changes only by a factor of 7 in going from C_6 to C_{14}. In fact, $k_{N_o}^+$ is close to the diffusion-limited rate, the rate set by diffusion of the monomer through the solution. This closeness reflects the fact that a micelle is essentially a liquidlike drop, and the monomer association reaction involves no structural reorganization or bond formation. The only barrier felt by a monomer entering a micelle arises from long-range electrostatic repulsion due to the micellar charge.

Thus τ_1 really corresponds to the rate at which surfactant monomers redistribute themselves between existing micelles, leaving one micelle and being accepted by another, as depicted in Figure 5.7a. This process involves continuous ongoing dynamic exchange. It does not change the number of micelles in solution nor the average aggregation number.

The slower relaxation time, τ_2, is associated with the formation or disintegration of entire micelles. In contrast to the τ_1 process, it does change the number of micelles. By analyzing Figure 5.7b we can understand why this process occurs more slowly than the one described by τ_1. If we take a particular micelle with an aggregation number N_o, it disintegrates by losing one

---

[2]*Note in Table 5.4b that CMC is expressed in molar units M so that $K_{N_o}^+$ has dimensions $M^{-1} s^{-1}$.*

(a)

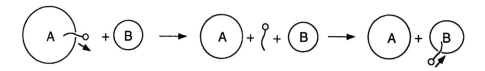

(b)

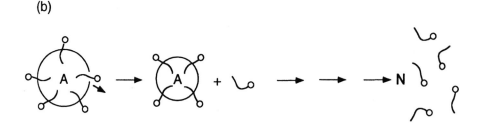

Figure 5.7
Schematic of the two dynamic micellar reorganization processes characterized by relaxation time measurements. (a) Transfer process corresponding to fast relaxation time τ_1. Transfer of monomers between two micelles A and B is a three-step process. Transfer changes the distribution of micellar sizes but not their total number. (b) Disintegration process corresponding to slow relaxation time τ_2. Disintegration of micelle A by sequential departure of monomers is a multistep process. Disintegration occurs more slowly than monomer exchange because it involves micelles passing through a kinetic bottle neck as $N_o \to N \to 1$. Disintegration changes the number of micelles. Reversal of the process leads to formation of micelles.

monomer at a time and in so doing passes through a stage involving very unfavorable aggregation numbers. This kinetic bottleneck involves formation of the least probable aggregation numbers between N_o and unity.

A detailed analysis yields

$$\frac{1}{\tau_2} = \frac{N_o^2}{CMC} \frac{1}{R_1} \left(1 + \frac{\sigma^2}{N_o} \frac{(c - CMC)}{CMC} \right)^{-1} \qquad (5.2.14)$$

where the equilibrium and the kinetic properties of micelles with the least probable aggregation numbers determine the residence time. R_1 is a "resistance" term describing the constriction of the bottleneck and is inversely proportional to micelle concentration and the micelle dissociation rate constant $k_{N_o}^-$.

Thus from relaxation measurements we learn that micelles are highly dynamic entities. The average residence time for a monomer in a micelle ranges from 10^{-6} s for short-chain surfactants, such as octylsulphanate, to 10^{-3} s for longer chain surfactants. The average lifetime of a micelle is on the order of milliseconds to a second.

Using insights gained from relaxation measurements, we also can estimate the average excursion of a surfactant headgroup in and out of the micelle surface due to thermal motion. As the surfactant headgroup begins to move out away from the micelle surface, it draws its attached hydrocarbon tail into the aqueous phase, but this process is unfavorable and so the tail is pulled back into the micelle. The monomer molecule oscillates in the surface of the micelle. Fluctuations in the distribution of thermal energy occasionally give a surfactant molecule enough energy to escape from the micelle, and the rate constant, $k_{N_o}^-$, measures this effect.

On average, however, the headgroup simply oscillates around its equilibrium location. At room temperature, the average excursion exposes two carbon atoms along the chain, corresponding to a movement of 4 Å. Thus Figure 5.1a depicts the surface of the micelle to be smoother than it is in reality.

In fact, all the aggregate structures shown in Figure 5.1 are misleading in that they depict a static average appearance and do not convey any sense of structural dynamics. Most of the information we have about molecular structure comes from x-ray measurements (in which we immobilize molecules so that we can obtain Bragg diffraction patterns), electron microscopy (in which we absorb molecules onto grids in a high vacuum chamber), or light and neutron scattering (which yield time-averaged information on structures). We translate this information into static pictures or sketches of molecules and in the process condition ourselves to think statically instead of dynamically. By so doing, we oversimplify the true state of matter!

5.2.7 Micelles Can Solubilize Other Hydrocarbon Molecules

So far we have focused on the dynamic properties of micelles containing only surfactant. However, a micelle's hydrocarbon core provides a small oil-like reservoir that can be used to *solubilize* other nonpolar molecules. This characteristic of micellar solutions finds wide application in interfacial processes.

We can measure the residence time of a molecule solubilized in a micelle by choosing molecules that can be probed by their phosphorescence. Residence times shown in Table 5.5 correlate with the molecule's solubility in water. As the solubility increases down the table, the residence time decreases. We can attribute the variation in residence times directly to differences in the distribution coefficients of the probe molecule between the aqueous phase and the micelle. The association rate, or the rate of solubilization, is diffusion limited just as it is for surfactant monomers. On the time scales required for most industrial processes, such exchanges result in a uniform distribution of many

TABLE 5.5 Residence Times in Micelles

	Residence times (μs)		
Probe molecule	In SDS (micelles)	In $C_{16}TAB$ (micelles)	Solubility in water (M)
Anthracene	59	303	2.2×10^{-7}
Pyrene	243	588	6.0×10^{-7}
Biphenyl	10	62	4.1×10^{-5}
Naphthalene	4	13	2.2×10^{-4}
Benzene	0.23	1.3	2.3×10^{-2}

D. F. Evans and H. Wennerström, *The Colloidal Domain: Where Physics, Chemistry, Biology, and Technology Meet,* VCH Publishers, New York, 1994, p. 158. Reprinted with permission of VCH Publishers, © 1994.

solubilized molecules. As we discuss in Section 5.5.1, solubilization of oil particles is an important function of a detergent.

5.3 Many Amphiphilic Molecules Form Bilayers, Liquid Crystals, Vesicles, and Bicontinuous Structures

5.3.1 Surfactant Numbers Provide Useful Guides for Predicting Different Aggregate Structures— Spherical Micelles, Cylindrical Micelles, and Bilayer Structures

If we increase the concentration of SDS in water by fiftyfold beyond the CMC to $0.4M$, the concentration of spherical micelles is so great that intermicellar interactions become important. Spherical micelles become increasingly unstable, and the molecules aggregate in the form of oblate spheres and eventually *cylindrical micelles,* as illustrated in Figure 5.1b. If we examine the behavior of the longer-chain homologue of SDS, sodium hexadecylsulfate, we find that cylindrical micelles form at concentrations just above the CMC. If we use a double-chain amphiphile, such as didodecyldimethylammonium bromide (di-DDAB), or a phospholipid (phospholipids are examples of zwitterionic amphiphiles), the molecules never form micelles; instead they assemble into *planar bilayers,* illustrated in Figure 5.1c. Bilayers consist of sheets of amphiphilic molecules, two molecules thick, with the polar headgroups on the surface facing the solvent. The layers can be organized in the form of a *lamellar* structure composed of a stack of parallel sheets whose separation is dictated by the surfactant concentrations. There are a rich variety of other self-assembled microstructures, as displayed in Figure 5.1. Observations suggest a direct relationship between the structure of amphiphilic molecules and their aggregate architecture.

One way of classifying the structures in Figure 5.1 is in terms of their curvature. Spherical micelles are highly curved; bilayers are flat. We can define the mean curvature H by

$$H = \frac{1}{2}\left(\frac{1}{R_1} + \frac{1}{R_2}\right) \tag{5.3.1}$$

where R_1 and R_2 are the principle radii of curvature. For a sphere $R_1 = R_2$ and $H = 1/R$; for a cylinder $R_1 = R$, $R_2 = \infty$, and $H = 1/2R$; and for a bilayer $R_1 = R_2 = \infty$ and $H = 0$. Another structure that exhibits zero curvature is the continuous saddle-shaped surface of Figure 5.1d, for which $R_1 = -R_2$. To assign a sign to a radius of curvature, we must specify a direction, $\vec{n}$, normal to the surface. By convention $\vec{n}$ points in a direction from the hydrocarbon tail toward the headgroup. Therefore, by convention the surface curvature of normal aggregates is positive and that of inverted aggregates is negative.

We can relate molecular structure to aggregate curvature through the *surfactant number*

$$N_s = \frac{v_{hc}}{l_{hc\,av} a_{hg}} = \frac{a_{hc}}{a_{hg}} \qquad (5.3.2)$$

where v_{hc} stands for the volume of the hydrocarbon chain (or the hydrophobic tail) of the surfactant molecule (eq. 5.1.3), $l_{hc\,av}$ represents the *average* length of the tail, and a_{hc} is the cross-sectional area of the hydrocarbon tail. The ratio a_{hc}/a_{hg} emphasizes that the surfactant number compares the cross-sectional area of the hydrocarbon chain to the *effective* area of the headgroup. When headgroup repulsion is strong and the effective headgroup area is much greater than the hydrocarbon cross section, that is when N_s is small, the aggregate surface *must be curved*.

Smaller values of N_s correspond to greater curvature of the aggregated structure. When N_s is small, the surface-area term is large compared to the volume term, highly curved surfaces are preferred, and spherical micelles usually form. When N_s is close to unity, a balance is implied between the surface-area and volume features and bilayers usually form. Comparison of different aggregates reveals that optimal shape depends on surfactant numbers in the following sequence:

Spherical micelles	$N_s = 1/3$	H = 1/R
Infinite cylinders	$N_s = 1/2$	H = 1/2R
Planar bilayers	$N_s = 1$	H = 0
Inverted aggregates	$N_s > 1$	H < 0

So long as the solutions are dilute and we can ignore interactions between aggregates, eq. 5.3.2 predicts both the surfactant number and the preferred aggregated structure. The most problematic quantity in the estimation of the surfactant number is the effective area per polar group, a_{hg}. For typical ionic micelles, such as the SDS micelle in Section 5.1.1, a_{hg} equals 60 Å^2 at the CMC in water, and upon addition of salt at 0.1M, a_{hg} decreases to 40 Å^2. We calculate N_s for the SDS micelle. Using the values for v_{hc} and l_{hc} given in Section 5.1.2, we obtain $N_s = 323/(16.7 \times 60) = 0.32$ in water, which is consistent with the observation of spherical micelles in this system. With the addition of salt, N_s approaches 0.5, and we anticipate a transition to cylindrical micelles.

Either spherical or cylindrical micelles can reach a radius $R_{mic} \approx l_{hc\,max}$ (eq. 5.1.5) because only a few of the chains in the aggregate need to extend to their full length to establish the radius. On the other hand, a bilayer's thickness never reaches $2l_{hc\,max}$ and seldom exceeds 1.6 $l_{hc\,max}$. The average length, $l_{hc\,av}$, of the chains in a bilayer is always less than $l_{hc\,max}$ because the chains are in a liquid state and rarely fully extended.

While the radius of the cylindrical micelle R_{mic} is set by the length of the hydrocarbon chain, the cylinder can grow indefinitely in length with little effect on headgroup interactions. As a consequence, cylindrical micelles tend to be more polydisperse than spherical micelles. When the cylindrical micellar length becomes sufficiently long, the micelles become entangled like spaghetti and show behavior more typically associated with polymer solutions (to be discussed in Chapter 6).

5.3.2 Bilayers Are the Basic Building Blocks of a Number of Amphiphilic Structures

Bilayers are the basic building blocks of a number of amphiphilic structures. They are formed in dilute solution by amphiphilic molecules with surfactant numbers between $\frac{1}{2}$ and 1. So they are favored by amphiphiles with large v_{hc} or small a_{hg}, such as those with double hydrocarbon chains or noncharged headgroups. Examples include the dialkyldimethylammonium compounds and the phospholipids shown in Table 5.1. The thickness of a bilayer is set at twice the average lengths of the hydrocarbon chains, $2l_{hc\,av}$. However, its growth has no structural limit in the lateral directions. In fact, the unfavorable hydrocarbon–water (solvophobic) interaction that occurs at the edges of a bilayer sheet encourages lateral growth.

Surfactants that form bilayers typically exhibit low monomer solubility in water ($10^{-5}M$ for didodecyldimethylammonium bromide, di-DDAB, and $10^{-10}M$ for dipalmitoyl phosphatidyl choline). Therefore, any transformation from one aggregated structure to another or attainment of equilibrium that proceeds via monomer exchange through the solvent takes place very slowly. For these systems relaxation times τ_1 and τ_2 are very large; days, weeks, or even years may pass before a bilayer system attains equilibrium. These sluggish transformations contrast markedly with those in micellar solutions where the time scale is on the order of milliseconds.

An important property of bilayers is their low permeability to polar molecules. Even though the hydrocarbon core of a bilayer may be only 30–40 Å thick, it provides an effective barrier to ions and other polar compounds. This feature plays an important role in biological phospholipid membranes, which isolate the interior of a cell from its environment.

5.3.3 Hydration and Thermal Undulatory Interaction Forces Play an Important Role in Stabilizing Bilayer Systems

The fact that bilayers grow to form large extended structures with micrometers dimensions means that colloidal forces on the scale discussed in Chapter 4 begin to play an important role in determining their aggregated structures. Bilayers will move together under the influence of van der Waals forces and stack up to form lamellae. At equilibrium, repulsive forces must balance these attractive forces.

With respect to the attractive forces between two bilayers, we are reminded that the distance dependence of the van der Waals attractive forces depends on geometry as we saw in Chapter 2. In fact, for two bilayers whose thickness is small relative to their separation (h), summing over all the pairwise interactions produces an attractive force per unit area of bilayer that varies as H_{121}/h^5. H_{121} is the Hamaker constant for the bilayer amphiphiles (1) in the solvent (2) discussed in Section 2.6.3. The h^{-5} dependence is shorter ranged than the h^{-3} attraction between flat colloidal particles (eq. 2.6.7).

With respect to the repulsive forces between two bilayers, we note that zwitterionic phospholipid bilayers have an equilibrium spacing of about 20 Å. This result is contrary to what we would expect from our discussion of classical colloidal interactions. Since nonionic or zwitterionic amphiphiles carry no net electrical charge on the headgroup, they exhibit no double layer and no electrostatic stabilization. Bilayers should move into contact under the van der Waals attraction. But they do not. In fact, if we try to force them together, we encounter a large, short-ranged repulsive force. This repulsive force, which is called the solvation force (or if the solvent is water, the *hydration force*), constitutes a new type of interaction in addition to the overlap core repulsion and electrostatic interactions already discussed in Chapters 2 and 4. It is one of a number of additional interparticle interactions that have been characterized during the past decade, thanks to the availability of new instrumentation.

To characterize the hydration force, we pause to examine its measurement and properties in more detail. We can measure the force required to bring bilayers together by placing them in an osmotic pressure apparatus similar to that shown in Figure 4.12b. A membrane permeable only to water and simple electrolytes separates the bilayer system from an adjacent polymer solution. By adding more polymer to the adjacent solution, we lower the activity of water, which in turn increases the osmotic pressure of the polymer solution ($\Pi_{osm} = \mathbf{R}Tc$) with respect to the bilayer system. Adding water decreases the osmotic pressure. When the osmotic pressure on the polymer side of the membrane becomes higher than that on the bilayer side, water flows through the membrane from the bilayer to the polymer solution, diluting the polymer solution and concentrating the bilayer system by driving its bilayers closer together. When the osmotic pressure on both sides of the membrane becomes equal, the flow of water stops. Then we can use x-ray scattering to measure the separation of the bilayers in the lamellae under equilibrium conditions. Repeating the measurements with different polymer concentrations in the adjacent phase, we obtain a relationship between osmotic pressure or force per unit area versus bilayer separation. A series of independent osmotic pressure measurements taken in the absence of bilayers yield a calibration curve relating the polymer's osmotic pressure to concentration. Results for di-HDAA above and below the Krafft temperature are plotted in Figure 5.8.

The steep rise in the force curve at about 20 Å displayed in Figure 5.8 marks the onset of interactions dominated by hydration forces. The hydration force is extremely short ranged, and the force between bilayers per unit area is found to change exponentially with their distance apart h

$$F = F_o \exp\left(-\frac{h}{\lambda}\right) \qquad (5.3.3)$$

In many systems, the decay length λ corresponds to the diameter of the water molecule, 2.8 Å. Hydration forces control short-range interactions in many bilayers and in some polymers, such as DNA. Explanation of the hydration force is a subject of lively debate, and we simply do not yet understand its molecular origin.

Figure 5.8
Separation force curves for dihexadecyldimethyl-ammonium acetate (di-HDAA) bilayers. Osmotic pressure measurements are plotted versus average bilayer separation at temperatures below and above the surfactant Krafft temperature of 34°C. The steep rise at small distances, < 20 Å, marks the onset of repulsive hydration forces. Above the Krafft temperature (40°C), as the osmotic pressure is decreased and becomes sufficiently small, the bilayers separate spontaneously at ≈ 115 Å and form vesicles and micelles. This transformation is driven by undulatory forces induced by thermal fluctuations. Below the Krafft temperature (23°C) there is no such discontinuity.

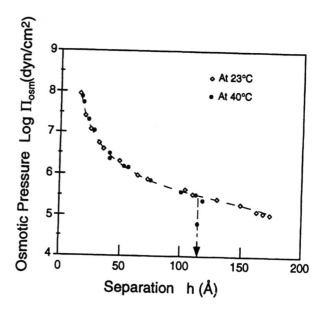

We also encounter a second interaction force in bilayer systems that is not seen in classical colloidal systems called the *undulatory force*. In Section 5.1.3 we emphasized that amphiphilic assemblies form under conditions in which the hydrocarbon chains are in a liquid or melted state. Bilayers and lamellae are extremely flexible, and thermal fluctuations induce undulatory motions in them. In fact, undulatory motions contribute a lowering of the free energy of the system through the increase in entropy associated with the random undulations (in the same way that coil fluctuations in long polymer molecules contribute to the entropy of polymer solutions, to be described in Section 6.3.3.2). The larger the separation, the more significant the undulating motion and the higher the entropy, and so the free energy decreases with increasing bilayer separation, which is equivalent to saying undulations contribute an effective repulsive force of entropic origin. Theoretical estimates of the undulatory force per unit area of bilayer show that it varies as the inverse cube of the separation h

$$F = \frac{3\pi^2}{64} \frac{(kT)^2}{\kappa_b} h^{-3} \qquad (5.3.4)$$

where κ_b represents the so-called bending rigidity that measures the stiffness of the bilayer. κ_b has dimensions of energy and typically lies in the range $(1-20)kT$ for an amphiphilic bilayer.

The repulsive undulation force ($\sim h^{-3}$) is more long ranged than the attractive van der Waals force ($\sim h^{-5}$). Whether attractive or repulsive forces dominate depends on the temperature, the bilayer rigidity κ_b, and the Hamaker constant H_{121}. When attractive forces dominate, the bilayers stack up in lamellae at the equilibrium separation; when repulsive forces dominate, the bilayers move apart and the disintegrating lamellar system is replaced by new, larger microstructures called *vesicles* (to be described in the next section).

We can illustrate the interplay between the van der Waals attractive force and the electrostatic, hydration, and undulatory repulsive forces in a liquid bilayer system by considering the force curve shown in Figure 5.8. The open diamonds plot the change in separation with applied osmotic pressure for a bilayer containing a charged amphiphile, dihexadecyldimethylammonium acetate (di-HDAA), above its Krafft temperature. Starting from the left-hand side of the curve, we observe a rapid decrease in force as the separation increases up to 20 Å. In this region, repulsive hydration forces dominate. Beyond 20 Å, the net force results from a combination of repulsive electrostatic and undulation forces and attractive van der Waals forces.

As we reduce the applied force even further and increase the water content of the lamellar system, the bilayers continue to move apart until at a critical value indicated by the arrow in Figure 5.8 (separation distance 125 Å, $\log \Pi_{osm} = 5.5$), the opaque lamellar dispersion disappears and the solution becomes clear. If we examine this disintegration under a video-enhanced microscope, we can witness individual bilayers peeling off the lamellar stack to form vesicles.

Viewed from another perspective, this experiment can provide additional insight. The di-HDAA bilayers do not want to stack to form a lamellar structure, but we force them to do so by manipulating the osmotic pressure. Below the critical osmotic pressure, the bilayers are driven apart because the electrostatic and undulatory forces become larger than the attractive forces. The short-ranged hydration and van der Waals forces have ceased to be important, undulatory forces dominate, and the larger vesicle microstucture is preferred. This sequence demonstrates how the interplay of a variety of forces—each with its own magnitude and distance dependence—combine to dictate the interaction between bilayers.

As we lower the temperature, the bilayers stiffen, and the lamellar phase undergoes a phase change that can lead either to the formation of crystalline phases or sometimes to a stable or metastable "gel" phase. Biophysicists call the temperature at which this transformation occurs the *chain transition temperature,* but it merely represents another manifestation of the Krafft temperature T_{Kr}, described in Section 5.1.3. Figure 5.9 illustrates some possible bilayer "gel"-phase structures below the Krafft temperature. The characteristic feature is that the alkyl chains are ordered and aligned, but liquidlike solvent still exists between the bilayers. The actual configuration depends on the mismatch between the cross-sectional areas per chain(s) and per headgroup. Depending on the degree of mismatch, the chains may interdigitate to accommodate the mismatch (shown in Figures 5.9a and 5.9b) or they may tilt relative to the bilayer plane (Figure 5.9c) or the bilayer may develop a ripple (Figure 5.9d).

Below the chain transition temperature, where the hydrocarbon chains are solid, undulatory motions and undulatory forces disappear. If we repeat the experiment displayed in Figure 5.8 on dihexadecyldimethylammonium acetate below its Krafft point of 34°C (results represented by solid diamonds), we cannot

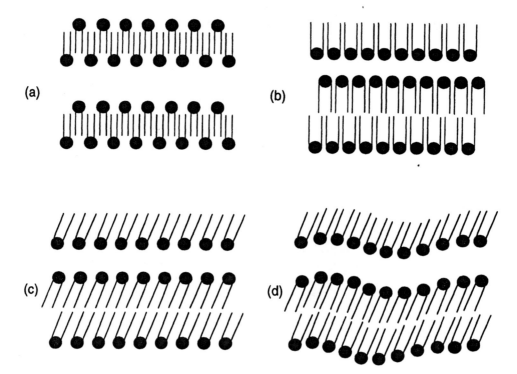

Figure 5.9
Various gel-phase structures formed by bilayers when the amiphiphles' hydrocarbon chains are frozen in an ordered state while the headgroups remain solvated by solvent. The hydrocarbon chains can become (a) interdigitated, (b) offset, or (c) and (d) inclined, depending on the relative headgroup to hydrocarbon chain cross-sectional dimensions. (D. F. Evans and H. Wennerström, *The Colloidal Domain: Where Physics, Chemistry, Biology, and Technology Meet*, VCH Publishers, New York, 1994, p. 247. Reprinted with permission of VCH Publishers, © 1994.)

detect a critical osmotic pressure. The lamellar system continues to exist even if we decrease the applied osmotic pressure to its minimum value by placing pure water in the adjacent phase across the membrane. The difference between the force curves in these two instances demonstrates the role that undulatory forces can play in bilayer systems. It also illustrates how changes in temperature and other variables can turn on or off some interaction forces.

5.3.4 Vesicles Are Molecular-Sized Containers Formed by the Breakup of Bilayers

Our discussion has considered bilayer systems principally in terms of lamellar structures consisting of planar bilayers packed in large stacks separated by solvent. We have assumed that the bilayers extend indefinitely in the lateral direction. In reality, bilayers tend to avoid interaction between the hydrocarbon chains and solvent at their edges by closing up on themselves to form enclosed regions. If the wall separating the inside and outside regions is one bilayer thick, the simple enclosed structure is known as a *vesicle*. A simple spherical vesicle is illustrated in Figure 5.1f. However, aggregate structures are rarely that simple.

If we take bilayers in equilibrium with excess water and agitate the solution by sonification, the bilayers break up and reassemble in the form of closed tubes or multilayered tubes, or closed spheres or multilayered spheres. Multilayered shells having onionlike structures are called *liposomes*.

Figure 5.10 shows many liposome features in a video-enhanced optical microscope image of the surfactant TEXAS I in water. Figure 5.11 shows a cryogenically frozen transmission

Figure 5.10
Video-enhanced micro-
scope images of liposomes
formed by the amphiphile
Texas I illustrating the
variety of structures and
defects that occur in
bilayer structures.
Bar = 10 μm. Regions A
and B are large liposomes
that appear birefringent
under crossed polars. C is
a large vesicle containing
entrapped smaller vesi-
cles. D is a vesicle with a
"dust storm" appearance
in real time, indicating
that it is filled with small,
unresolvable particles.
(D. D. Miller, J. R. Bellare,
D. F. Evans, Y. Talmon,
and B. W. Ninham, *J. of
Physical Chemistry* **91**, 674
(1987). Reprinted with permission from *J. of Physical
Chemistry.* Copyright
1987, American Chemical
Society.)

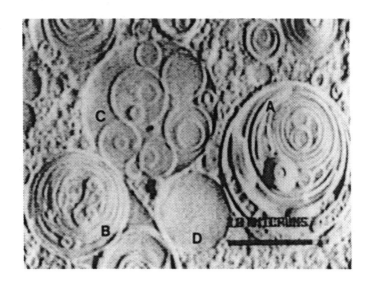

electron microscope (TEM) image of the same surfactant, revealing even more subordination of detail: vesicles within vesicles,
tubes within tubes, liposomes within liposomes. Depending on
the nature of the surfactant and the vesicles, we usually distinguish between small unilamellar vesicles (SUVs), large unilamellar vesicles (LUVs), and multilamellar liposomes. Figures 5.10
and 5.11 provide examples of each type. Vesicles constitute a
particularly interesting amphiphilic aggregate because like living
cell walls they provide a barrier between an internal compartment
and the external world.

The properties of a vesicular solution typically depend on
how it has been prepared. Most vesicles are thermodynamically
unstable, but reaching an equilibrium state can be a slow process
because monomer solubility in the solvent (water) is so low.
Sometimes vesicles are metastable enough to allow detailed physical and chemical studies within a reasonable time frame. For
example, they have proved valuable in *in vitro* studies of membrane proteins. Generally, a vesicle's instability depends on the
rate at which it and other vesicles fuse to form a larger one. Vesicle
fusion is really another manifestation of the general phenomenon
of colloidal coagulation discussed in Section 4.5.

Because vesicles are molecular-sized containers, they have
many exciting potential applications and have been extensively
studied. Methods devised to prepare vesicles so that solutes are
isolated within them have made it possible to study the permeability of different solutes through vesicular walls. One method
widely used for measuring permeability first forms the vesicle in
a solution containing a suitable solute probe, such as a radioac-
tively tagged tracer or a fluorescence molecule; then the vesicular
solution is passed through a gel permeation column to achieve
separation; the largest particles are elevated first and the smallest—containing both small solutes and solvent—last. We can use
standard analytical techniques to follow the emergence of the
probe solute into an external solution. Alternatively, we can
measure permeability by following the flux of solute from the
external solution into the internal one. For example, we can

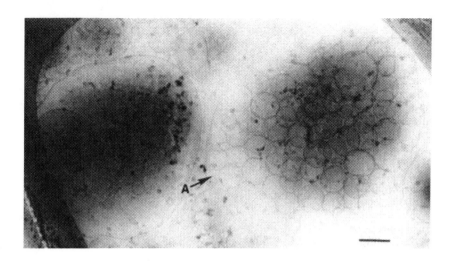

Figure 5.11
Cryo-transmission electron micrograph of the same sample of Texas I shown in Figure 5.10. Bar = 0.25 μm. Higher magnification reveals large vesicles containing smaller vesicles and microtubules. (D. D. Miller, J. R. Bellare, D. F. Evans, Y. Talmon, and B. W. Ninham, *J. of Physical Chemistry* **91**, 674 (1987). Reprinted with permission from *J. of Physical Chemistry*. Copyright 1987, American Chemical Society.)

follow H^+ permeation into a vesicle solubilized with a pH indicator. Ability to prepare vesicles loaded with a desired reagent provides a key step in exploiting their potential application capabilities.

5.3.5 Amphiphilic Bicontinuous Structures Contain Two Coexisting Phases

Figure 5.1d shows a particularly fascinating bicontinuous phase of cubic symmetry in which a sheet of amphiphilic molecules is draped on a saddle interface of zero mean curvature. This bicontinuous structure separates space into two continuous coexisting phases, one "outside" the interface and the other "inside" it. In a truly zero curvature bicontinuous structure the "outside" and "inside" phases both have the same three-dimensional shape and size. A sponge is an everyday example of a macroscopic bicontinuous structure. If we get inside a sponge, we can access every other part of it without ever going back into the aqueous phase contained in its interconnected channels. Similarly, we can access every part of the sponge's channels without ever entering the sponge itself. Macroscopic bicontinuous structures also are found in certain corals, where they obviously provide a very efficient means for extracting nutrients from the sea.

In our drawing the "outside" of the bicontinuous structure consists of the headgroups and so the "outside" phase compatible with the headgroups is a polar solvent like water. The "inside" of the bicontinuous structure consists of the hydrocarbon tails and so the "inside" phase is made up of the amphiphile molecules themselves or, in the case of some microemulsions (to be discussed in Section 5.4), other hydrocarbons like oil. In such a bicontinuous structure, solvent molecules and amphiphiles each diffuse in three dimensions within their own medium with an effective diffusion coefficient only slightly smaller ($\approx 2/3$) than its value in a homogeneous solution. As a consequence the relaxation times for microstructural reorganization of bicontinuous structures are expected to be short.

5.3.6 Temperature Versus Concentration Phase Diagrams for Amphiphilic Molecules in Solution Are Complex

In Sections 5.1.3 and 5.3.1 we indicated that aggregate structure favored by a given amphiphile varies with amphiphile concentration and temperature. In Section 5.2.1 we found that the onset of spherical micelle formation can be very sharp and exhibits many of the characteristics of a phase transformation. These results may be summarized in the form of a phase diagram. Figure 5.12 shows a typical temperature versus composition phase diagram for the nonionic surfactant polyethylene oxide, $C_{12}H_{25}(CH_2\text{-}O\text{-}CH_2)_5OH$ ($C_{12}E_5$). The diagram displays a rich variety of phases. In general, as we increase the surfactant concentration in a given solution at a given temperature, we traverse the manifold structures shown in Figure 5.1. For example, going along A–B in Figure 5.12, we pass from spherical micelles (phase L_1) to bilayers (phase L_3) to lamellar liquid crystal structures (phase L_α) to inverted micelles (phase L_2). The sequence of phases encountered varies considerably from one surfactant to another, and for some systems, such as SDS, the phase diagram is particularly complex in regions involving bicontinuous (cubic) configurations and bilayers. This complexity arises because the difference in free energy between these phases often is only $\approx RT$ and the system is sensitive to small changes in temperature or composition, causing one structure readily to transform into another.

Figure 5.12
Effect of temperature and surfactant concentration on the phase behavior of the polyethylene oxide nonionic surfactant $C_{12}H_{25}O(CH_2OCH_2)_5OH$. Regions L_1 and L_2 are normal and inverted micellar phases, respectively; L_α is a lamellar liquid crystal; L_3 is an isotropic solution containing bilayers. Horizontal tie lines denote regions where the two phases at the ends of the tie line coexist. Region (L_1' + L_1'') is a region of phase separation where micellar-rich and micellar-lean phases coexist above the cloud point (T_c). The solid line below (L_1' + L_1'') defines the cloud point curve. (R. Strey, R. Schomäcker, D. Roux, F. Nallet, and U. Olsson, *J. of the Chemical Society, Faraday Transactions* **86**, 2253 (1990).)

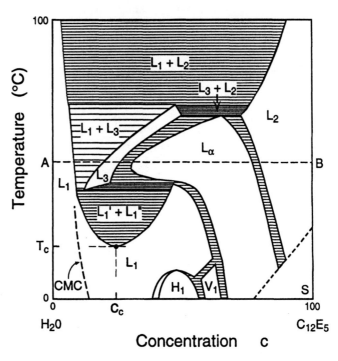

Figure 5.12 introduces another feature of nonionic surfactant solutions, the phase separation of a micellar solution into two coexisting micellar phases, one micellar lean, L_1', the other micellar rich, L_1''. When we heat an aqueous solution of $C_{12}E_5$ into the region where the two phases coexist ($L_1' + L_1''$ in the figure), the solution suddenly becomes turbid above a temperature called the *cloud point*. The critical temperature below which there is no demixing is T_c. The solid curve defining the boundary of $L_1' + L_1''$ is the cloud point curve.

5.4 Micro- and Macro-Emulsions Are Homogeneous Systems Containing Oil and Water Stabilized by Amphiphilic Molecules

So far we have been concerned with the properties of amphiphiles and their aggregate structures in water or water solutions. Now we add a third component to the system and examine the behavior of amphiphiles in liquids containing water and oil.

Oil and water do not mix. However, many situations in the food and health industry, oil recovery, and machining fluids require them to do so. If the oil and water phases are separated by surfactant molecules, they can coexist more readily and form an *emulsion*. Emulsification is a widely used and important interfacial process. From a thermodynamic point of view, we can distinguish two types of emulsion. Systems in a thermodynamically stable state are called *microemulsions,* while metastable (or unstable) systems are known as *macroemulsions*. Both types of system share a number of common structural features, although, as their prefixes suggest, microemulsions typically involve a smaller characteristic structural length scale (< 100 nm) than macroemulsions.

The simplest emulsion structures are droplets of oil in water, O/W, or water in oil, W/O, with amphiphiles or surfactant molecules lining the interface, as illustrated in Figure 5.13. In dilute solution the droplets are isolated and spherical, but packing constraints at higher volume fractions result first in the formation of distorted spheres and finally—when they are packed as densely as possible—in rhombic dodecahedryl shapes. When the composition is nondilute, microemulsions form bicontinuous structures like Figure 5.1d as already discussed in Section 5.3.5. Bicontinuous structures do not occur in macroemulsions because they are inherently unstable, and any thermal perturbation in the system immediately destroys them.

5.4.1 The Curvature of the Oil–Water Interface Can Be Manipulated

To create an O/W system, Figure 5.13b, we need an interface with a positive curvature, remembering the convention that the surface normal points from the hydrocarbon tail toward the polar headgroup for positive curvature. Figure 5.14 shows a schematic

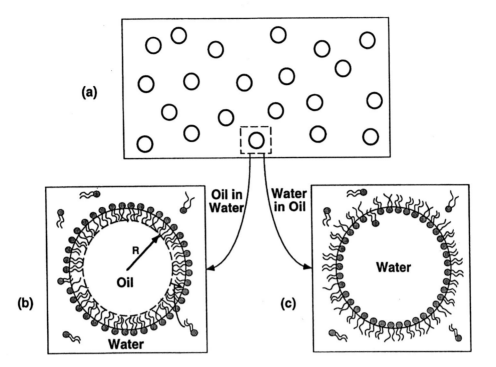

Figure 5.13
A droplet microemulsion phase (a) can be of the oil-in-water (O/W) type shown in (b) or the water-in-oil type (W/O) shown in (c). Typically the drops are nearly spherical and have a small size polydispersity. The radii R of the drops are usually one to four times the thickness of the surfactant film. The excess surfactant generally resides in the continuous phase. (D. Evans and H. Wennerström, *The Colloidal Domain: Where Physics, Chemistry, Biology, and Technology Meet,* VCH Publishers, New York, 1994, p. 462. Reprinted with permission of VCH Publishers © 1994.)

picture of a surfactant film at a curved oil–water interface. At the most primitive level, we can envision the surfactant behaving as a wedge at the oil–water interface as depicted in Figure 5.14a. Obvious surfactant choices to enhance the curvature are single-chain ionic surfactants with a large "wedge angle," such as SDS with its large cross-sectional ratio between headgroup ($60 \, \text{Å}^2$) and hydrocarbon tail ($19 \, \text{Å}^2$), or nonionic surfactants, such as $C_{12}E_6$, which undergo substantial hydration to swell the area of the headgroup at low temperatures. Using the definition of Section 5.3.1, we can define the degree and sense of curvature of the interface by the effective surfactant number, $N_s = a_{hc}/a_{hg}$, eq. 5.3.2.

Depending on the application (foodstuffs, medicine, lubricants, oil recovery), we may wish to increase or decrease the curvature at the interface. Figure 5.14 summarizes a number of strategies for decreasing the curvature in an emulsion system. We can understand the effectiveness of these different methods in terms of the effective surfactant number and can use it to distinguish the different emulsion structures accessible in any given system.

We can decrease the curvature (and increase N_s) of an O/W system by adding electrolyte to decrease headgroup repulsion with ionic surfactants, as depicted in Figure 5.14b, or we can raise the temperature with nonionic surfactants.

We can decrease the curvature (increase N_s) of an O/W system by adding double-chain surfactants as depicted in Figure 5.14c. In practice, mixtures of surfactants quite often are used for this purpose.

We can decrease the curvature (increase N_s) of an O/W system by adding an uncharged cosurfactant, such as an alcohol, that also adsorbs at the oil–water interface, as depicted in Fig-

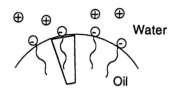

(a) Normal oil/water interface in an emulsion

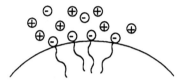

(b) Adding salt decreases head group repulsion and decreases the curvature at the oil/water interface

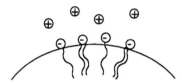

(c) Adding a double chain surfactant to a single chain surfactant decreases the curvature at the oil/water interface

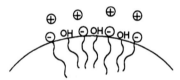

(d) Adding a long chain alcohol decreases the curvature at the oil/water interface. The hydroxyl groups are uncharged and generally do not change the effective area per surfactant head group

(e) Oil penetration can decrease curvature at the oil/water interface when the oil chain is shorter than the surfactant chain (right) because it increases the effective hydrocarbon volume (V_{hc}) in the surfactant number

Figure 5.14
Schematic diagram showing how various additives to the water and to the oil can affect the curvature of the oil–water interface in an O/W emulsion system.

ure 5.14d. Adding alcohol produces little change in the effective headgroup area, a_{hg}, of an ionic surfactant because the OH is uncharged; the OH simply is incorporated into the oil–water interface. However, the hydrocarbon chain of the alcohol is tethered to the OH; so the total volume of hydrocarbon chain at the interface is increased.

Interaction between the oil molecules and the surfactant chains also affects curvature at the oil–water interface. Penetration of the oil molecules into the surface region between surfactant chains decreases curvature (increases N_s) because the molecules contribute additional hydrocarbon volume, as shown in Figure 5.14e. The extent of penetration depends on the relative lengths of the oil chain and the surfactant chain. When the chain length of the oil exceeds that of the surfactant, oil penetration is negligible (left-hand drawing). When the oil chain is shorter than the surfactant chain (right-hand drawing), penetration increases and the curvature decreases.

The above additions all act to *decrease* the curvature of an O/W emulsion. Obviously, similar additions will serve to *increase* the curvature of a W/O emulsion, and opposite strategies must be employed to force the curvature in the opposite directions.

From this discussion we can appreciate that selecting the surfactant in the design of an emulsion system involves a number of complex issues. At present we understand some of these issues only at an empirical level.

5.4.2 Microemulsions Are Thermodynamically Stable Systems Possessing Microstructures with Characteristic Dimensions Less Than 100 nm

Microemulsions are thermodynamically stable and have a long lifetime, which makes them particularly suitable for detailed study. Typically they are clear or slightly opaque because their structural dimensions are less than ¼ the wavelength of light (4000 Å), and special optical enhancement or nonoptical characterization techniques are needed in their study.

Microemulsions show a rich range of stable microstructures. Figure 5.15b presents a ternary phase diagram for a three-component system—oil–water–surfactant. The triangular diagram is a succint way of summarizing the spectrum of structures formed at a given temperature and pressure with varying compositions. Figure 5.15a briefly explains how to interpret this type of representation.

Many of the structures shown in Figure 5.15b have the same basic origin as those shown in Figure 5.1. For example, if we start with a surfactant solution that consists of micelles in water and add oil, the oil solubilizes in the interior of the micelle, and we produce an O/W microemulsion droplet. On the other hand, if we start with an inverted micelle in oil and solubilize water in its interior, we produce a W/O microemulsion droplet. Figure 5.15b shows that if we start with spherical micelles at A and add surfactant, we can adjust the composition to B. By adding oil, we can move continuously along the tie line from B to B' and then by removing surfactant end up at A' with inverted micelles. In the process we never cross a phase boundary, but we do pass through irregular bicontinuous structures.

For a thermodynamically stable microemulsion, the oil–water interface must be covered by surfactant, and in many microemulsions coverage requires 10 to 20 wt % surfactant. Surfactants are expensive, and commercial applications of microemulsions are relatively few. (See Section 5.5.2 for one example.) While scientific studies have contributed enormously to our

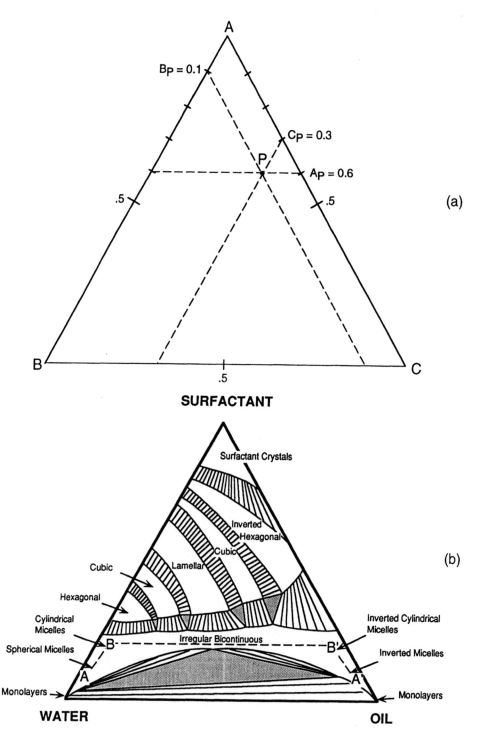

Figure 5.15
Triangular phase diagrams summarize the phases in equilibrium in a three-component system *at a given temperature*. (a) Evaluating the composition of the point P on the triangular diagram is obtained from lines parallel to the sides of the triangle. The relative amounts of A, B, and C in this example are 0.6, 0.1, and 0.3, respectively. (b) Triangular phase diagram illustrating the rich variety of microstructures possible in a ternary oil–water–surfactant microemulsion system. (Redrawn from H. T. Davis et al., 1987).

knowledge of emulsion behavior, cost places a limit on their commercial application. So we will not delve further into the behavior of microemulsions; instead we move on to more economically viable macroemulsions.

5.4.3 Macroemulsions Are Thermodynamically Unstable Colloidal Systems Containing Oil (Water) Droplets Dispersed in Water (Oil) with Microstructure Dimensions Greater Than 100 nm

Macroemulsions are thermodynamically unstable dispersions containing two immiscible liquids (usually oil and water) and an emulsifier to separate them. The emulsifier can be an amphiphilic surfactant as previously discussed, or a polymer, or finely dispersed particles. Typically, macroemulsions consist of droplets larger than 100 nm (1000 Å). Macroemulsions find wide commercial use because we often want to process a macroscopically homogeneous system containing both water and oil.

Macroemulsions exhibit properties that combine features associated with colloidal sols and amphiphilic aggregates. Both macroemulsions and sols are thermodynamically unstable, and we use similar strategies to prepare them. Both macroemulsions and sols show kinetic instability and tend to coalesce. However, emulsion droplets are liquid and can transform their size, shape, and number like amphiphilic systems. This combination of characteristics makes emulsions unique and particularly challenging and complex to deal with in practical applications.

We can state general goals for forming, stabilizing, and "breaking" a macroemulsion in a simple way. Forming an emulsion requires the creation of a metastable state containing either O/W or W/O droplets, stabilization demands that we keep the droplet intact as long as possible, while "breaking" an emulsion requires opening a pathway for destabilization and eventual coalesence.

The final fate of an emulsion is clear: it will separate by coalesence into two or more distinct equilibrium phases, as shown in Figure 5.16. In large part, the way we think about emulsion stability depends on our final goal. If long shelf life or durability is necessary, we want to stabilize the emulsion; if the emulsion is

Figure 5.16
Schematic illustration of the steps involved in the evolution from a freshly prepared emulsion (left) to final separation into two phases (right). When two emulsion droplets approach one another, they can maintain their identify by flocculation or coagulation into the secondary or primary interaction energy minimum (see Figure 4.1). Eventually the liquid droplets coalesce to form larger droplets until separation is complete. (D. F. Evans and H. Wennerström, *The Colloidal Domain: Where Physics, Chemistry, Biology, and Technology Meet,* VCH Publishers, New York, 1994, p. 487. Reprinted with permission of VCH Publishers, © 1994.)

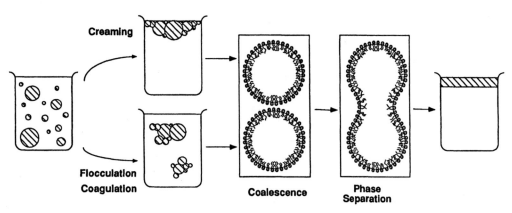

Creaming

Flocculation Coagulation

Coalescence

Phase Separation

an unwanted byproduct or if the success of a manufacturing process requires destabilization, we want to get rid of the emulsion. Stability and instability represent two sides of a Janus mask.

5.4.3.1 Formation of Macroemulsions

We employ two basic schemes for preparing emulsions just as we do for colloidal sols. Mechanical energy can be used to mix all the emulsion ingredients (mechanical dispersion), or chemical energy can be stored in the emulsion ingredients and liberated upon contact.

A number of commercial emulsion mills mix and shear oil, water, and emulsifier to produce emulsions. The free energy ΔG_{em} needed to disperse a liquid of volume V into drops of radius R in a solvent is

$$\Delta G_{em} = \gamma \frac{3V}{R} \qquad (5.4.1)$$

where γ equals the *interfacial* tension. For example, dispersing water in oil ($\gamma = 50 \times 10^{-3}$ J/m^2) to form drops of R = 100 nm requires a free energy of 27 J/mol. This amount of energy is modest from a chemical point of view. Making the drops larger and reducing γ by adding a surfactant reduces ΔG_{em} to even smaller values.

In practice, we must expend considerably more energy to form an emulsion than the thermodynamic value required to reach the dispersed state for two reasons. First, we cannot direct the input energy solely to forming droplets. Mechanical agitation converts much of the input energy to heat through viscous dissipation, and if we start from a chemical nonequilibrium state, the process of diffusional and convective mixing converts much of the free energy to heat or dissipates it. Second, the energy barrier against coalescence that stabilizes an emulsion implies the existence of a barrier against forming two droplets from a bigger one. The external energy must be large enough to overcome this barrier.

Mechanical agitation generally forms both oil and water droplets. What determines whether the final product is a W/O or O/W emulsion? The fate of a droplet depends on two competing processes: migration of the emulsifier to the droplet's surface to stabilize it versus coalescence with another droplet to destroy it. The phase that coalesces most rapidly—for whatever reason— becomes the continuous phase. Obviously composition is a major factor determining the continuous phase. Adding water to an oil-continuous W/O emulsion eventually will result in its *inversion* to a water-continuous O/W emulsion.

Another factor that determines the final form of the emulsion is the order in which we mix the ingredients. We can dissolve the emulsifier directly in either the water or oil phase, or we can dissolve components of the emulsifier in both phases. We can carry out emulsification by simply pouring all the ingredients into a container and mixing them, or we can mix them by adding the aqueous phase bit by bit to the oil–emulsifier phase, or vice versa. Hollandaise sauce is an oil-continuous macroemulsion formed when the egg yolk plus lemon juice is mixed with butter.

Processes that use chemical energy stored in the ingredients to form emulsions are sometimes called "spontaneous" emulsification processes because they do not require mechanical energy. This terminology is misleading. Such processes involve going from a state of higher free energy (associated with the initial form of the ingredients) to a metastable one (associated with the emulsion) rather than proceeding directly to the final equilibrium state involving separate phases.

We can demonstrate emulsification driven by chemical reaction and/or local surface tension gradients by contacting Nujol mineral oil containing oleic acid with either an aqueous alkaline solution or an aqueous sodium oleate solution. Bringing Nujol into contact with the aqueous alkaline solution produces emulsification in the aqueous phase. However, bringing Nujol into contact with the sodium oleate solution does not produce emulsification. Presumably, neutralization of the oleic acid by the base in the first instance provides the energy needed for the emulsification process.

5.4.3.2 Stabilization of Macroemulsions

A freshly prepared macroemulsion changes its properties with time due to a series of events that occur on a microscopic scale. Depending on the particular molecular characteristics involved, these events either lead to a metastable state (a "stable emulsion") or to the final equilibrium state consisting of macroscopically separated phases (an "unstable emulsion"). The distinction between a stable and an unstable emulsion is qualitative and depends partly on the time scale and context of the application.

A number of effects contribute to stabilization of an emulsion. As with the colloidal sols discussed in Chapter 4, the existence of a repulsive range of interaction between droplets forms the basis for stability. We can use the general DLVO approach and terminology from Chapter 4 to describe the qualitative features of droplet–droplet interactions, particularly for O/W emulsions. However, we must recognize that for liquid droplets repulsion does not arise solely from electrostatic forces; it also can originate from hydration or undulatory forces.

Figure 5.16 summarizes the steps leading to destabilization, or "breaking," of an emulsion. Emulsion droplets rather easily flocculate into the secondary minimum (see Figure 4.1). Whether flocculated or not, drops can migrate in the gravitational field to increase the concentration at either the top or the bottom of the vessel. The best known manifestation of this phenomenon in milk accounts for its name—*creaming*.

Once droplets have been brought into close proximity by flocculation and/or by creaming, they can coagulate into the primary minimum while still keeping their identity as separate droplets, as shown in Figure 5.16. This stage of destabilization is final for colloidal sols. For emulsions, however, a further process occurs—*coalescence*—in which two droplets merge into a bigger one, as shown in Figure 5.16. Coalescence involves reorganization of the stabilizing surface layer surrounding the droplet. The surfactant film changes its topology, and this change usually is connected with passing a substantial free energy barrier. Coalesence constitutes the final stage in the life of an emulsion droplet.

Emulsifiers simultaneously serve three functions in stabilizing an emulsion. They play an essential role in determining whether oil or water is the continuous phase. They control many of the attractive and repulsive colloidal interactions that govern flocculation and coagulation. They determine the mechanical properties of the interfacial film that controls coalescence rates. In the literature, emulsifiers often are classified on the basis of an empirical scale called the hydrophobic–lipophilic balance (HLB) number. Typical W/O emulsifiers have an HLB in the range 4–8, and O/W emulsifiers lie in the range 12–16. In fact, for ionic surfactants, the HLB number correlates well with the surfactant number, N_s, using a headgroup area $a_{hg} = 50 \text{ Å}^2$.

5.5 Applications Involving Amphiphilic Interfacial Systems Illustrate the Rich Diversity of This Topic

5.5.1 Detergency Is a Complex, Everyday Process Involving Surfactants

The world is a dirty place, and we spend an enormous amount of energy and resources cleaning up. In the United States, the domestic cleaning agents with which we wash ourselves, our clothes, our utensils, and the interiors of our homes is a multi-billion dollar industry. Many manufacturing operations, such as metal surface preparation prior to coating, also require cleaning processes.

Cleaning agents must be formulated to meet a wide variety of needs, and literally thousands of different systems for this purpose are produced commercially. Our discussion focuses mainly on detergents and their use in removing soils from fabrics. Modern detergents contain a number of components tailored to meet widely different washing conditions, such as variations in soil, fabric, and water hardness, as well as variations in laundering conditions that differ from one culture to another. In addition, the inescapable impact on the environment of widespread detergent use has played a decisive role in the reformulation of commercial detergents over the past several decades.

When we wear clothes, we get them dirty on the outside by accumulation of grime from the environment and on the inside from body secretions. Dirty clothes, therefore, contain a complex mixture of adsorbed greases, oils, and particulates, such as clays, dead skin, and so forth. The way dirt sticks to clothes depends in part on the nature of the fabric; so the challenge in cleaning a natural fabric, such as cotton or wool, differs from that encountered with synthetic materials, such as rayon or polyester. At the end of the cleaning process, we expect to see the grime removed, not evenly redistributed over the article of clothing. If the article is white, we are reassured if it appears "dazzling" white. The next section presents a brief history of detergency, describes the key components of soils and detergents, and discusses the detergency process.

5.5.1.1 The Development of New Surfactants Has Spawned a Variety of New Technologies

A Sumerian tablet first mentions soap as a cleaning product for woolen clothing in 2500 B.C. This tablet also identifies soap's two starting ingredients, wood ash and oil. In modern terms these ingredients react to produce long-chain carboxylates or surfactants

$$KOH \text{ (ash)} + \text{triglycolides (oil)} \rightarrow CH_3(CH_2)_nCOOK \text{ (potassium carboxylate)}$$
$$+ 3H_2O + glycerol \qquad (5.5.1)$$

This Sumerian tablet may constitute the oldest known record of a chemical reaction.

Over the ensuing centuries, soaps were increasingly used as cosmetic and cleaning agents, although no one except the very wealthy washed a great deal. However, beginning in the late 1700s, more widespread use revealed a major shortcoming of soaps: they became inactive and curdled in hard waters, due to precipitation of Mg and Ca carboxylates. This limitation stimulated twentieth-century chemists to develop improved synthetic replacements, such as the alcohol sulfates. Initially synthetic products met with limited success. For example, in 1940 the sale of soaps in the United States exceeded 1.4 million tons, while that of synthetic detergents was only 4.5 thousand tons. However, by 1960 sale of synthetics outpaced soap sales by a factor of three, and in 1982 synthetic detergent sales reached 5.1 million tons, while soap sales declined to 0.5 million tons.

During the period between 1940 and 1960, the composition of the entire surfactant industry changed, resulting in new sources of raw materials, new synthetic surfactant molecules, new applications, and new additives. The effort that began with the aim to improve the detergent properties of soap resulted in specialized amphiphilic molecules tailored to meet specific needs, such as low- or high-foaming properties, defoaming, oxidation resistance, and salt tolerance. Molecules also were required to be caustic-stable, biodegradable, bacteriocidal, tolerant of hard water, edible, effective at low concentrations, and easily removable. Clearly, no single molecule or type of molecular structure could meet all these requirements simultaneously; different combinations are required to satisfy different needs, and thousands of alternative synthetic surfactants are now available. While this development in detergents was taking place, entirely new enterprises were stimulated and spun off: more effective herbicides and insecticides, new emulsion polymers, latex paints, and adhesives, more effective mining and flotation processes, new food products, and improved bituminous and concrete products for heavy construction. It was a very productive era.

Detergency involves mass transfer in which undesirable substances adsorbed onto a fabric are forced out and become solubilized or suspended in the aqueous phase during the washing cycle. Key variables in the detergent process are soil type, detergent formulation, fabric makeup, and mechanical action during washing. The next two sections focus on types of soil and the composition of detergents.

5.5.1.2 Composition of Soils Is Complex—Soils Are Classified into Four Main Categories

We can divide soils into two broad categories based on their solubility in water. Soluble soils, such as salts and short-chain polar organics, usually are removed simply by immersing the fabric in water. Insoluble soils require detergents. We classify insoluble soils in terms of the properties they exhibit during the washing process.

5.5.1.2A Oily Soils. Oily soils usually contain mixtures of hydrocarbons and long-chain fatty acids, esters, and alcohols. Used motor oil is a specific example. Under most laundering conditions, removal of oily soils is conditioned by their liquidlike properties, although oily soils also may contain some particulate solids, such as the carbonaceous matter in used motor oils.

5.5.1.2B Particulate Soils. These soils contain mixtures of solid particles. Street dust and vacuum cleaner dirt are examples. A recipe for formulating "street dust" in the laboratory requires mixing carbon black (1.75%), iron oxide (0.50%), peat moss (38%), cement (17%), silica (17%), kaolin clay (17%), and mineral oil (Nujol, 8.75%). Particulate soils often contain small amounts of lipid materials (greases, fatty acids), but their properties during washing are dominated by the particulate nature of the soil.

5.5.1.2C Fatty Soils. These soils usually derive from a biological source and contain mixtures of proteins from abraded skin, pigments, and various lipids secreted from sweat glands. A typical composition of secreted lipids includes triglycerides (30–50%), mono- and diglycerides (5–10%), free fatty acids (15–30%), cholesterol, sterols and their esters (2–6%), fatty acid esters of aliphatic alcohols (12–16%), and squalene (10–12%). Depending on their composition, fatty soils may be found either in a liquidlike or solidlike state; so detergent removal of these materials presents a complex problem.

5.5.1.2D Stains. Stains are intensively colored substances that cause noticeable soiling even when present in small amounts. Common stains include tea, coffee, chocolate, wine, mustard, grass, rust, blood, ink, and natural colorants in general. Usually detergent action cannot completely eliminate these stains, and successful stain removal requires the use of bleaches or enzymes.

5.5.1.3 Composition of Detergents Is Complex— Modern Detergents Contain a Variety of Components, Each of Which Accomplishes a Specific Task during Washing

Like the paints described in Section 4.7.1, detergents are complex mixtures that must meet conflicting constraints encountered in a wide variety of processing conditions. Modern detergents contain much more than just surfactants; they also contain builders, bleaches, antideposition agents, and enzymes.

5.5.1.3A Surfactants. Of the four classes of surfactants categorized in Table 5.1, anionics and nonionics are most widely used in detergents. Zwitterionics are used to a much lesser extent. Cationics are almost never used, because most naturally occurring materials, including some of the constituents of grime, are negatively charged and precipitate in the presence of cationic surfactants.

The most widely used *anionic surfactants* contain sulfate and sulfonate headgroups and either hydrocarbon or alkylbenzene tails. Because of their wide pH tolerance, lower sensitivity towards hardness ions (Ca^{2+} and Mg^{2+}), and ease of synthesis, branched-chain alkylbenzene sulfonates gained popularity during the 1950s. However, their widespread use was accompanied by increased foaming in a number of rivers. In some cases even ground water came from the tap with a head of foam. This environmental problem was traced back to the slow biodegradation of the branched alkylbenzene sulfonate. Switching to linear alkylbenzene sulfonates improved the rate of biodegradation and reduced levels of surfactants in the effluents to acceptable levels.

Nonionic surfactants used in laundry applications are generally linear *alcohol ethoxylates* synthesized by reacting mixtures of C_{10}, C_{12}, and C_{14} alcohols with ethyleneoxide to give

$$CH_3(CH_2)_{10-14}OH + n(CH_2OCH_2) \xrightarrow{KOH} CH_3(CH_2)_{10-14} \text{-O-} (CH_2 \text{-} CH_2 \text{-O})_{7-10} H$$

Ethoxylates react slightly faster with ethylene oxide than does the alcohol itself; so the final product is a complex mixture containing significant amounts of unreacted alcohols, low-molecular-weight ethoxylates, and a tailing distribution of high-molecular-weight ethoxylates where $n \gg 10$. Because the various homologs can partition or adsorb to different degrees, the distribution of ethyleneoxide surfactants found at a water–substrate interface may be substantially different from that in the bulk solution. This distribution in turn can have a major impact on adsorption, wetting, oil penetration, and other surface processes.

Because nonionic surfactants are uncharged, they are less sensitive to water conditions, such as hardness, than anionic surfactants. However, they are much more sensitive to temperature changes. Near their cloud points nonionic surfactants exhibit optimum cleaning ability, and their foam properties change markedly. Above the cloud point they become low foaming or even defoaming, and surface adsorption is greatly enhanced. For this reason, they make particularly good surfactants for removing oily soils from polyester and polyester–cotton blends. In addition, they can be formulated into highly active aqueous solutions, which makes them an important component in the liquid laundry detergents that have gained widespread use.

Many nonionic surfactants that are available now contain hydrophobes from both natural and synthetic sources. A particularly interesting new class of surfactants called Pluronics™ contains two ethyleneoxide headgroups attached to the ends of a more hydrophobic polyoxypropylene tail. By controlling the ratio of hydrophobe (ethylene oxide) to hydrophile (propylene oxide) and varying the total molecular weight of the polymer, we can control the cloud points, surface tensions, wetting properties, and foaming of Pluronics™ in an orderly way. These new materials possess surprisingly low critical micelle concentrations ($3-11 \times 10^{-6}M$ compared to $1-2 \times 10^{-3}M$ for normal nonionics or $1-8 \times 10^{-3}M$ for typical anionic surfactants). Because they possess a low CMC and inherent polydispersity, Pluronics™ are much more difficult to characterize than other surfactants, and we do not know a great deal about their aggregate structure. Their unusual molecular structure suggests that Pluronics™ self-assemble to form unusual micellar aggregates.

One interesting aspect of the solubilizing properties of the Pluronics™ is their ability to combine with a number of surface-active biocides without affecting the biocidal properties. In this respect Pluronics™ differ from the more typical surfactants based on alcohol ethoxylates (AEs) or nonylphenolethoxylates (NPEs). AEs and NPEs form classical micelles that tend to incorporate or solubilize the biocide and thus diminish its biocidal activity.

Although *cationic surfactants* are not used for detergents in the washing cycle, they do play an important role as antistatic agents in the cleaning cycle. Long-chain dialkyldimethylammonium surfactants, such as $(CH_3(CH_2)_{17})_2N(CH_3)_2Cl$, are dispersed onto a small sheet of paper that is tumbled with clothes during drying. Small amounts of cationic surfactant absorbed onto fabrics reduces static cling. (Single-chain cationic surfactants make good biocides and are found in many patent medicine cough syrups.)

5.5.1.3B Builders. Builders enhance the cleaning effectiveness of surfactants by two main mechanisms: (1) chelating divalent cations and thereby reducing the hardness of water; and (2) raising the pH so that the wash solution becomes alkaline. Alkalinity suffices to promote dissolution of many soils, and proteins and milk soils are solubilized more readily by the surfactant following the formation of soluble salts when their carboxylate groups are neutralized by base. Fats or oils often require ester hydrolysis, a slower process promoted by elevated temperatures. These varying conditions are not available in all cleaning operations and may require more specialized machinery.

Sodium triphosphate is an optimum builder because it simultaneously controls alkalinity at a useful level by its buffering action and acts as an excellent chelant for both calcium and magnesium ions. In addition, it has the ability to strip divalent metal carbonates or carboxylates that may have precipitated on surfaces. Thus it effectively prevents incrustation on fabrics and minimizes deposits in washing machines that otherwise can lead to unsightly and nonhygienic conditions. As we saw in Section 4.7.1, polyphosphates are used as dispersants in paints because they impart a negative charge to the surface of the colloidal pigment particles to stabilize them electrostatically. These same properties also make sodium triphosphate a good agent for stabilizing dirt particles once they have been removed from a fabric's surface.

Desirable features of sodium triphosphate led to tremendous growth in its use in detergents from 1945 until detergent phosphates were targeted as major polluters responsible for rapid algae growth in lakes in the mid-1970s. Although fertilizers and human and animal wastes are more significant sources of phosphate pollution than detergents, detergent phosphates are now banned in many states.

Finding a single builder to replace sodium triphosphate has proved a difficult and elusive goal. Instead, combinations of components, each of which fulfills some of the builder requirements, are employed. In domestic applications, sodium carbonate and silicates are used to control alkalinity. Unfortunately they form insoluble precipitates with divalent ions, and therefore must be used in conjunction with a compatible chelating agent. Alternative chelating agents include ethylene diamine tetraacetic acid, nitrile triacetic acid, citric acid polyacrylates, and zeolites (discussed in Section 4.7.4). Incorporation of zeolites constitutes a shift from soluble to insoluble builders.

Zeolites have gained widespread use in Europe and Japan because laundering conditions in those countries are more compatible with the limitations of zeolites as builders. In Europe, wash water that is initially cold is heated during the course of the wash cycle; so the ion exchanger has a longer time to work. However, zeolite ion exchangers are ineffective in removing calcium and magnesium from soils or from surfaces, and a soluble cobuilder must be added to strip the hardness ions from the fabric and give them up to the zeolite. In this way zeolites have replaced over 50% of the detergent phosphates formerly used in Europe.

In Japan, low washing temperatures are used throughout the wash cycle, and the cycle time is extended. Under these conditions zeolites perform acceptably without added builders. Almost all laundry detergents sold in Japan are phosphate free and based on zeolite A.

5.5.1.3C Bleaches. Not all stains can be removed by the action of surfactants and builders alone; often oxidative bleaching is required. From a chemical view, bleaching saturates or destroys conjugated double bonds in complex molecules, a process that often leads to formation of more water-soluble subunits.

The most widely used bleaches are *peroxides* and *hypochlorites*. Hypochlorites dominate in the United States and are usually added after the wash cycle. Recently a number of products have been introduced that contain peroxycarboxylic acid–based systems.

In Europe little hypochlorite is used. Instead, bleaching involves using *sodium perborate tetrahydrate* to initiate the bleaching agent during the wash cycle. This compound is stable in dry detergent and only begins to generate hydrogen peroxide as the wash temperature increases. In alkaline washes, the peroxide anion is the active bleaching species. It obtains the best results at the higher temperatures commonly used in European machines. At low temperatures, reactive activators generate organic peroxycarboxylic acids. Most peroxyacids possess pK_As around 8.0–8.5; thus the anion is the active bleaching agent under typical wash conditions. Normally peroxyacetic acid is used for this purpose, but more recently, longer-chain peroxyacids have been extensively used. These peracids are surface active and can deposit onto the fabric where their action is required.

5.5.1.3D Antiredeposition Agents. Added to detergent systems, antiredeposition agents prevent the soils that have been removed from the fabric from being deposited again. They are especially important to inhibit graying or accumulation of dispersed soil after numerous wash cycles.

Carboxymethylcellulose was the first truly effective antiredeposition agent. Its mode of action depends on its ability to adsorb preferentially onto the fabric surface. Carboxymethylcellulose is especially active on cotton, but it is ineffective on synthetic fabrics, such as nylon or polyester. Nonionic derivatives of cellulose, such as hydroxyethylcellulose or hydroxypropylcellulose, are used to prevent redeposition on those materials.

5.5.1.3E Enzymes. Incorporating enzymes into cleaning systems constituted a significant change in the detergent industry in recent years. As early as 1913, a German patent was granted for the use of an enzyme to remove protein soils from fabrics, but alkaline and temperature-stable *proteases* were not developed until the 1960s. As a result of advances in biotechnology, proteases stable at pH 9–11 and temperatures of 50–70°C have become available and are used in laundry, dishwashing, dairy, and food-processing plant applications.

Amylases, which hydrolyze starch-type soils, also have been developed. These have found an application niche in potato and pasta plants and in products for automatic dishwashing where rice or pasta are frequently encountered. *Lipases,* which hydrolyze oils and greases, have more recently come on the market, but they have not found as widespread application.

Enzymes are most effective when they have time to hydrolyze the proteins, starches, or fats during presoaks or longer wash times. Because longer wash times are used in Europe and Japan, these countries generally lead the United States in the use of enzyme materials.

Clearly, detergent systems contain a high proportion of reactive components, and we might anticipate that they would deactivate one another during storage or use. More recent sophisticated detergents use encapsulation, controlled release, and triggered release to achieve detergent stability and to release components at the appropriate times during the wash cycle.

When we go to the supermarket, we encounter row upon row of laundry products, bearing witness to the size and diversity of the detergent industry. Our mail contains a continual stream of coupons offering rebates for the detergent of the week. Television advertisements thrust upon us mythical figures symbolizing the magical powers of a particular product or distraught housewives contemplating the woes of junior's stained and soiled garments who become transformed into radiant caretakers at the end of a washing cycle. Such inducements are part of a commodity-driven market.

As consumers, how do we make our way through this thicket of advertising to make rational purchasing decisions? Short of consulting *Consumer Reports,* we probably cannot. However, in a commodity-driven market, perceptible deficiencies in a product's performance can mean potential economic famine for the manufacturer, and product failure can lead to almost certain death. Consequently, evaluating the performance of a particular detergent formulation as well as that of the competitors has become an important activity in the detergency industry. Standardization of soils, designated application methods, set washing procedures and equipment, protocols for evaluation, and even the existence of American Society for Testing and Materials (ASTM) protocols testify to the seriousness of this enterprise.

5.5.1.4 During the Cleaning Process, Detergents Operate by a Number of Mechanisms

So far, we have focused on the composition of soils and detergents. In this section we discuss what happens when soil and detergent encounter one another in the washing process. We can describe the removal of insoluble soils by a four-step mass transfer process involving:

1. Diffusion and adsorption of detergent components onto the soil and fabric.

2. A chemical or physical process (or a combination of both) that leads to desorption and release of the soil from the fabric.

3. Transport of soil components away from the fabric and into the bulk wash solution.

4. Stabilization of soil components in the solution so as to minimize redeposition.

Detergency is a kinetic process, and in most cases the wash time is so short that equilibrium considerations are largely irrelevant. We can use equilibrium arguments to decide whether a step in the process can occur, but we cannot tell whether or not it will take place at a useful rate.

5.5.1.4A Detergency Mechanisms for Oily Soils.

The specific mechanistic steps in removing oily soil are:

1. An induction period during which surfactant diffuses to the soil/fabric interfaces;

2. A soil-removal period during which the soil is separated from the fabric, mainly by the rollup mechanism described in the next paragraph; and

3. A final period when soil removal becomes negligible, signaling that any residual soil cannot be removed under the particular washing conditions.

In general, redeposition of oily soils is not a major problem.

Rollup is the most important detergency mechanism in the wash sequence for oily soils. In this step, changes in surface tension resulting from surfactant adsorption cause the oily material to roll up to form a sphere that detaches from the fiber, as shown in Figure 5.17. By analogy with the spreading coefficient (Section 3.2.5), we can define a rollup coefficient f as

$$f = \gamma_{OF} - (\gamma_{FW} + \gamma_{OW} \cos \theta) \tag{5.5.2}$$

Rollup occurs when f is positive, that is, when the soil–fabric interfacial tension γ_{OF} is large, and the fiber–water γ_{FW} and soil–

Figure 5.17
Sequence illustrating the
rollup mechanism.
Changes in surface tension
accompanying adsorption
of surfactant induce oily
soil to roll up into spheres
and detach from the fabric.
(A. W. Adamson, *Physical
Chemistry of Surfaces*, 2nd
ed., Copyright © 1967,
John Wiley & Sons, p. 496.
Reprinted by permission
of John Wiley & Sons, Inc.)

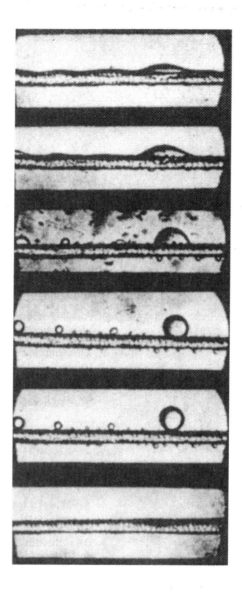

water γ_{OW} interfacial tensions are small. If the soil viscosity is high, interfacial forces may not suffice to drive rollup, and external forces like mechanical agitation may be needed to dislodge the oil.

Other processes may reduce the effectiveness of the detergent. Penetration of the soil by the surfactant may form viscous liquid crystalline layers that inhibit rollup. Surfactant-induced solubilization and emulsification also occur, but they are usually less important with oily soils than rollup processes.

5.5.1.4B Detergency Mechanisms for Particulate Soils. Particles adhere to fabric by a combination of attractive forces and mechanical interlock into the fiber. In large measure, the relative importance of these two interactions depends on the method of soil deposition. With a light dusting of soil, attractive forces are most important, but when particles have been ground into the surface, mechanical inclusion dominates.

When attractive forces dominate, we can use a two-step process to analyze the energetics of particle removal within the framework of the DLVO theory. In the first step, wash liquid penetrates between the fabric and the particle surfaces, separating them by a distance h. In general this distance is so small that attractive forces still dominate. In the second step, the particle is transported by fluid shear forces to distances at which attractive interactions become negligible.

We can express the total work for particle removal as the sum of the work involved in the first two steps less the net change in surface–interface energy

$$w_T = w_1 + w_2 - \sum \gamma \qquad (5.5.3)$$

where $\sum \gamma$ represents the sum of the wetting tensions of the surfaces involved. The kinetics of particulate removal is first order, but the rate coefficient is not constant because soil particles are attached to fabrics with a spectrum of interaction energies. Mechanically imbedded particles are difficult if not impossible to remove.

5.5.1.4C Detergency Mechanisms for Fatty Acid Soils. Multiple mechanisms operate in the removal of fatty acid soils. They include rollup, emulsification, formation of liquid crystalline phases, and mechanical and chemical breakup of the oil. Several mechanisms often operate simultaneously on a given fatty soil deposit, and their relative importance can change during the wash cycle.

5.5.1.5 We Still Do Not Understand the Complex Interfacial Process of Detergency at a Fundamental Level

We have shown how various detergent components combine to remove a variety of soils and indicated some of the main mechanistic features of the detergency process. For the sake of brevity, we have omitted many subtle aspects. Nevertheless, detergency exemplifies an enormously complex, widely used process that has evolved into a highly developed empirical industrial and domestic art without being well understood at a fundamental level. For example, we have mentioned the important role that fluid convection plays in detergency. If we had to depend on diffusion-controlled detergency, our clothes would probably spend more time in the washing machine than on our bodies. Useful models to describe the action of a washing machine's agitator simply do not exist.

5.5.2 Pressure-Sensitive Adhesives Illustrate the Use of Emulsions in Interfacial Processes

As a participant at a professional meeting, you frequently receive an adhesive name tag. After removing the backing, you attach it to your garment and at the end of the day you peel it off again with the firm expectation that it will come off without taking part of your expensive jacket with it. You have used a pressure-sensitive adhesive, one of the products of a 70×10^6 lb per year business involved in sealing, reinforcing, insulating, masking, and protecting surfaces. Other applications include 3M Post It™ notes, masking tapes, protective films used on printed circuit boards or on plexiglass, and bandages. The adhesive used in these products is a sticky, viscoelastic material that would be a disaster as a permanent glue, yet represents a triumph of technological engineering for purposes of easy removal.

The following sections consider how pressure-sensitive adhesive materials are formulated to display varying adhesive properties. Section 8.3 describes how they are placed on a variety of substrates using a continuous coating process. The technological challenge is to find the right combination of adhesion and substrate for the application. A bandage with the characteristics of a 3M Post It™ note would not be very useful. One way the challenge is being met is through the use of oil–water macroemulsions initially formed as water–oil microemulsions and then inverted by adding water. Using macroemulsions enables us to obtain a fluid system whose viscosity is low enough that it can be coated on a substrate to achieve the desired final product.

5.5.2.1 Pressure Sensitive Adhesives Are Prepared by Mixing Latexes, Tackifiers, and Plasticizers

Pressure-sensitive adhesives (PSAs) contain two main components:

- Latexes prepared from either natural rubber, butadiene–styrene (at, for example, a 75:25 weight ratio), or certain acrylic polymers that are not sufficiently "tacky" in themselves to serve as pressure-sensitive adhesives; and
- "Tackifiers" and plasticizers that can be combined with the latexes to produce the "permanently and aggressively tacky" properties that are the hallmark of a pressure-sensitive adhesive.

"Tackiness" has been defined as the property that enables an adhesive to form a bond with the surface of another material on brief contact under light pressure. A large number of different tackifiers and plasticizers are available for use in formulating adhesives. Our discussion focuses on natural-product-derived *rosins*. Other chemical classes of tackifiers are available via synthetic approaches. A rosin is a solid resinous material found in pine trees that is harvested by three methods:

1. Removing sap from the tree (like harvesting maple syrup);

2. Extracting the rosin from ground-up pine tree stumps; and

3. Concentrating byproducts from paper mills.

Rosin consists of a small amount of neutral material and at least 15 different rosin acids, each of which possesses varying degrees of unsaturation and a distribution of methyl groups on the steroidlike ring. The relative amount of each compound the rosin contains depends upon the source of the rosin, the time of year, and the chemical treatment it has received. Conjugated double bonds make some rosin compounds susceptible to air oxidation, but if partial hydrogenation removes one of the conjugated double bonds, the rosin acids become more stable and resistant to air oxidation, and therefore more suitable for use in commercial adhesive systems.

Figure 5.18a indicates the three types of rosin esters that are combined to formulate a PSA. Tackifying compounds, such as the fully hydrogenated rosins Foral 85® and Foral® 105, are glycol and pentaerythritol esters, respectively.[1] They are solids with nominal softening points of 85 and 105 C. The plasticizer compound, such as the hydrogenated ester of rosin Hercolyn® D, is a methyl ester. It is a viscous Newtonian liquid (η = 2300 cp at 25 C). None of these compounds achieves complete esterification, and the final products also contain free rosin acid and, in the case of the Foral compounds, residual hydroxyl groups.

Mixtures of tackifier and plasticizer such as Foral® 85 and 105 and Hercolyn® D form highly viscous fluids or solid materials that are almost impossible to apply by the continuous coating processes of Chapter 8. They become even more unmanageable when they are compounded with latex spheres. To overcome this limitation, the rosin mixture is transformed into a water-continuous emulsion before it is mixed with the latexes. An emulsion contains three main components: oil, water, and surfactant. In this system, the oil components are the rosin esters, and the surfactant is prepared by neutralizing unreacted rosin acids with concentrated potassium hydroxide to form the anionic surfactant shown in Figure 5.18b. (Note that producing this surfactant is particularly economical; if the initial oil does not contain sufficient free rosin acid to form it, a little more can be added.) The traditional approach to accomplish emulsification is to add water to the oil–surfactant mixture and to agitate the mixture vigorously to create the oil–surfactant–water interface. A different sequence— used to achieve the desired results with, for example, Hercolyn® D—is described in the next section.

5.5.2.2 Emulsification of the Plasticizer and Tackifier Involves an Unusual Mechanism

The process used to emulsify the resins is unusual, and we use it to illustrate some of the principles established earlier in this and prior chapters. We focus on emulsification of Hercolyn® D alone because the process is easier to follow with a liquid, but the same mechanism applies to fluid mixtures of Foral® and Hercolyn® D at higher temperatures.

Adding small amounts of water to Hercolyn® D (plus surfactant) results in a W/O microemulsion. We can determine this by placing two electrodes in the mixture and measuring the electrical conductivity. As long as the water content is below 2% we observe high resistance, consistent with water-solubilized inverted micelles in an oil continuum.

If we bring the Hercolyn® D W/O microemulsion containing 1–1.8% water into contact with a bulk aqueous phase, the micro-

---

[3]*FORAL and HERCOLYN are registered trademarks of Hercules Incorporated.*

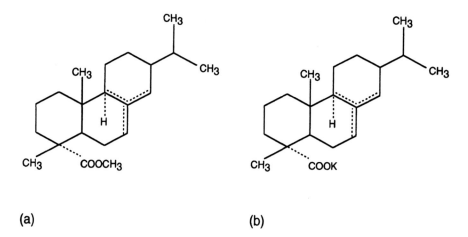

(a)   (b)

Figure 5.18
Structures of (a) rosin acid
ester oil and (b) surfactant
used in the formulation of
pressure-sensitive
adhesives.

emulsion undergoes spontaneous inversion to form a water-continuous O/W macroemulsion, and the manner by which it does
so is interesting. Figure 5.19 shows four frames from a series of
video-enhanced images taken of the process. Oil droplets are seen
streaming away from the microemulsion–water interface into the
water continuum. At room temperature, the resulting oil spheres
have an average diameter of 225 nm, and form an O/W macroemulsion so stable that it can remain unchanged for years. The
size of the emulsion droplets increases as we raise the temperature. In fact, the droplet size correlates directly with the viscosity
of the microemulsion. When the temperature is above 60 °C, a
coarse, unstable, macroemulsion forms.

Observations suggest the following emulsification mechanism. Due to the high concentration of counterions within the
inverted micelles contained in the microemulsion, the osmotic
pressure inside the micelles is considerably lower than the osmotic pressure of the contacting water. Consequently, the inverted
micelles imbibe water and grow. Because the microemulsion is
highly viscous, this process is confined to the interfacial region
between the microemulsion and the water. The inverted micelles
remain fixed in place, until they grow so large that they interconnect, and eventually invert to normal micelles. Immediately after
the inversion process, the oil droplets—stabilized by the anionic
surfactant—behave like a concentrated colloidal dispersion.
Electrostatic repulsion drives them apart and repels them into
the adjacent water phase. The size of the resulting oil droplets
is determined by the constraints of the initial microemulsion
structure.

Commercial production of the macroemulsions involves
a continuous process in which Foral®–Hercolyn® D microemulsion formation and subsequent contact with water is stimulated with vigorous stirring so as to drive the inversion
process more rapidly. Typically, emulsions contain 50 vol %
of discontinuous resin phase and are adjusted to give viscosities of 20 cp at 25 °C. The resulting emulsion is so stable it can
be shipped as a fluid in railroad tank cars with no fear of
coagulation or coalesence.

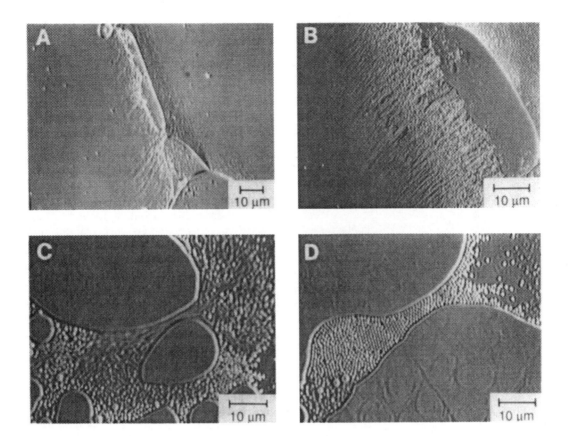

Figure 5.19
Video-enhanced microscope images of spontaneous emulsification of Hercolyn® D–potassium rosinate–water microemulsion with water:
(A) approximately 25 s after contact; (B) approximately 40 s after contact;
(C) 3—5 min after contact, explosive injection of oil phase droplet into water;
(D) 3—5 min after contact, showing nearly uniform spheres of oil phase before injection into water.
(R. W. Greiner and D. F. Evans, *Langmuir* **6**, 1793 (1990). Reprinted with permission from *Langmuir*. Copyright 1990, American Chemical Society.)

5.5.2.3 Emulsification Can Achieve the Low Viscosity Essential for Coating Pressure-Sensitive Adhesives

Pressure-sensitive adhesive is prepared by blending the rosin emulsion with an aqueous dispersion of latex. As a rule of thumb, the size of the emulsion droplets should be comparable to that of the latex spheres. Water concentration is adjusted so that the viscosity of the final material ranges from 100 to 2000 cp, suitable for the coating process.

PSA is deposited on the substrate using the continuous coating method analyzed in more detail in Section 8.1. Upon leaving the coater, the substrate passes through a dryer, where 99.8% of the water is evaporated. As a result of intermixing between the latex polymer and the resin components, the film becomes homogeneous. Dried PSA film has a viscosity of 10^6 to 10^8 cp and a glass-transition temperature T_g usually well below room temperature when measured by the differential scanning calorimetry (DSC) method.

5.5.2.4 Pressure-Sensitive Adhesives Must Satisfy Various Requirements

What criteria do we use to decide whether we have a good product? Some general guidelines for PSAs are that they:

- Adhere to the surface of another material (the adherent) upon brief contact and with application of light pressure;

- Separate from the adherent without leaving a visible residue; and

- Meet the ASTM specification that the bond between the PSA and the adherent be of measurable strength. A number of standardized tests are available for evaluating PSA materials.

We can analyze these guidelines in terms of two general principles. The first focuses on interaction energies, the second on kinetic properties.

The most critical factor is the *adhesion* determined by the molecular contact between the adhesive and the adherent. This contact relates directly to the energy or work of adhesion (eq. 3.2.2) and constitutes yet another manifestation of the pervasiveness of van der Waals attractive forces. Forming a "bond" of measurable strength requires the adhesive to spread onto the adherent, a process that can be related to the spreading coefficient (eq. 3.2.6).

Another critical factor is the time required for the adhesive to respond to mechanical forces. When we press our name tag onto the fabric of our clothes, we create shear forces. If PSAs were purely elastic, the adhesive would momentarily deform, but the deformation would be recoverable and the name tag would pop off again when we stopped pressing on it. If PSAs were purely viscous, the adhesive would deform, but the name tag would slowly slip off the fabric. For the tag to adhere to the irregularities of the fabric, the adhesive must respond to the shear forces by viscous flow but the distortion must be retained; in other words, the adhesive must be both viscous and elastic. PSAs are viscoelastic materials (see Section 6.6.4) with relaxation times about 1 s, comparable to the usual bonding times. The right combination of latex, tackifier, and plasticizer can tune the viscoelastic behavior to be responsive to the need.

Another critical factor is the force required to break the adhesive bond. Adhesive force measures the material's tack. The adhesive must display quite different properties when we peel the backing off the tag and when we peel the tag off our jacket. Studies of the interface between a label and its backing show that during removal long strands of adhesive extend between the two surfaces. For easy removal of the backing we want these strands to break and not stretch forever like chewing gum. During rupturing, the elastic modulus (Appendix 11A.1) of the adhesive at the strain rates and elongation magnitudes imposed on it must be high enough so that it does not stretch too much. Each of the components of a PSA makes different contributions to the elasticity and plasticity toward meeting these needs. "Tack" and "bond" strengths increase if stress can be distributed so that more of the adhesive actively resists bond rupture. High-molecular-weight and flexible polymers can provide this stress-distributing property.

Thus we can see that designing a material that possesses the physical and chemical properties required of a PSA involves simultaneously meeting a number of stringent requirements.

5.5.3 Low-Density Foams Can Be Fabricated by Emulsion Polymerization

Developing new materials for specific and demanding applications constitutes a major challenge in materials science. One example is the low-density foam required for laser inertial-confinement fusion (ICF) techniques. The following paragraphs describe the ICF process, the characteristics required of an ICF foam, and how such foams can be fabricated by polymerizing a high-internal-phase-volume polystyrene emulsion.

5.5.3.1 The Inertial Confinement Fusion Process Duplicates Nuclear Fusion Processes That Occur in the Stars

If we could replicate on earth the sustained nuclear fusion that generates the sun's energy, we would have a virtually inexhaustible energy supply. One approach to imitating this process involves rapidly heating the surface of a capsule containing a deuterium–tritium (DT) fuel mixture with many high-powered laser beams focused onto the capsule target from all directions. Simultaneous laser pulses generate a plasma envelope surrounding the target's surface, and the rocketlike blowoff of plasma material forces the capsule to compress. As a consequence of this enormous hydrostatic pressure, the fuel in the core of the capsule reaches a density about 10^3 times that of liquid DT and a temperature of 10^8 K. Ignition leads to a thermonuclear burn that spreads radially throughout the compressed core.

5.5.3.2 Design Characteristics of ICF Targets Place Stringent Limits on Low-Density Foams

Figure 5.20 shows a schematic diagram and a prototype ICF target. A very low-density sponge in the shape of a spherical shell holds liquid DT. The interior contains the DT fuel in a vapor form, and the shell is coated with a thin protective hydrocarbon membrane to prevent the liquid DT from evaporating at the outer surface. Compressed vapor provides the central ignition region that initiates the thermonuclear reaction.

The physics of the ICF process places severe restrictions on the design and material characteristics of the capsule, particularly the properties of the sponge or foam that contains the liquid DT. To function effectively, the foam must possess the following properties:

- Low density. The goal is to achieve a foam density of 50 mg/cm^3, which gives a 0.95 volume fraction of DT. The aim is to maximize the volume fraction of DT and to minimize the volume fraction of support material that ends up diluting the fuel.

- Porosity. To be able to fill the foam with liquid DT, the foam must consist of cells that are interconnected (a bicontinuous structure). Any isolated or closed cells result in inhomogeneities in density that may grow catastrophically during implosion.

Figure 5.20
(a) Schematic of an inertial confinement fusion (ICF) structure consisting of a thin protective membrane, a low-density foam to contain liquid deuterium-tritium (DT) fuel, and a hollow core filled with DT vapor. (b) Protype polystyrene hemishell.

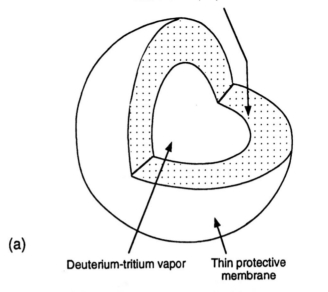

Low-density, low-atomic-number foam saturated with liquid deuterium-tritium

(a)

Deuterium-tritium vapor   Thin protective membrane

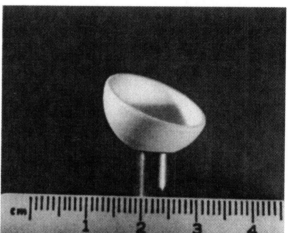

(b)

- Wettability. The liquid DT must readily wet the foam.

- Mechanical stability. During implosion, the capsule is subjected to an acceleration of ≈ 1000 g and local inward velocities of ≈ 1000 m/s. The foam must be strong enough to withstand the inertial forces, and the internal structure of the foam must be sufficiently small to confine the liquid DT by capillary forces. This restriction sets an upper limit of 1 μm on the diameter of the foam's cells.

- Machinability. The foam must be robust enough to withstand the machining operations required to form a smooth, spherical shell geometry. Simultaneous requirements of low density and machinability are incompatible in many ways because mechanical robustness diminishes with decreasing density.

- A composition consisting mostly of carbon and hydrogen. To optimize energy output and reduce ignition temperature, the foam must be made of elements possessing low atomic number. The design limits require less than 10 at. % of elements with atomic number between nitrogen and silicon and 1 at. % of elements with higher atomic numbers. Thus hydrocarbon foam is the most desirable material.

- Resistance to radiation damage. During the time needed for the target to be filled and stabilized before implosion is initiated, the foam must tolerate exposure to tritium radiation without changing its dimensions.

5.5.3.3 Low-Density Foams Can Be Fabricated Using Emulsion Polymerization Techniques

No commercial polymer foam meets all the ICF design requirements (see Figure 5.21). Most foams have cell sizes in the range of hundreds of micrometers and densities that are too large. Although inorganic silica aerogels meet many of the physical requirements, their atomic number is too high.

One strategy for forming an ICF foam involves making an oil-continuous emulsion in which the oil phase can be polymerized to lock in the foam framework, removing the aqueous phase to produce the foam, and then backwashing the foam to remove any high-atomic-number contaminants.

To achieve the low-density goal, we need to prepare an emulsion in which the discontinuous water phase constitutes 92 to 95 vol % of the system. Such a system cannot be constructed from spheres of equal size because packing constraints limit the volume fraction to 0.74. Figure 5.22 shows two alternative structures for high-internal-phase-volume (HIPV) emulsions. The first

Figure 5.21
Diagram showing the foam cell size and density goals for an ICF foam. While SiO_2 aerogel meets the size–density specifications, its atomic number is too high; no commercially available foams are suitable.

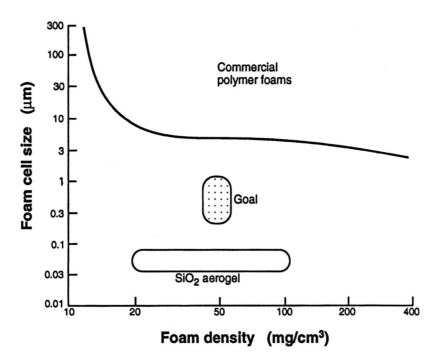

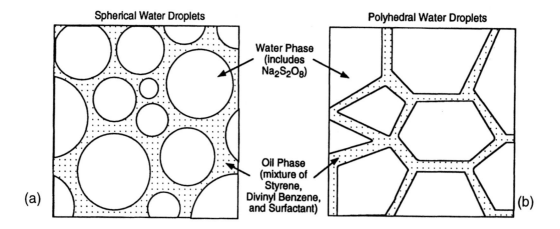

Figure 5.22
Two limiting structures for a high-internal-phase-volume water-in-oil emulsion. (a) A heterogeneous mix of water spheres; (b) water polyhedra in an oil continuum.

involves spheres of varying size that can be packed close together. The second involves water-filled polyhedra separated by thin oil-filled membranes interconnected to form the continuous phase. The exact structure of the emulsions used in forming the ICF foams is unknown, but it is probably a compromise between these two extremes.

Usually HIPV emulsions are quite viscous. Consequently, many commonly used emulsification techniques cause air bubbles to be irreversibly trapped in the emulsion. These bubbles create unacceptable variations in foam density. A satisfactory method for producing HIPV emulsions involves connecting two syringes (one containing the water phase and the other the oil phase) via a small orifice and forcing the liquid through the orifice from one syringe to another until the emulsion forms.

The continuous oil phase contains a mixture of two monomers, styrene and divinylbenzene, and an oil-soluble surfactant, sorbitan monoleate. The internal aqueous phase contains a free radical initiator, sodium persulfate, ($Na_2S_2O_8$). Heating the emulsion to 50°C for 24 h results in polymerization (see Section 6.1). By itself, styrene polymerizes to form linear chains, but combining it with divinylbenzene results in chemically cross-linked chains that form a mechanically stable network. After polymerization, the aqueous phase is removed by drying the foam in a vacuum oven for several days. The resulting foam has densities ranging from 50 to 80 mg/cm^3 and the cells are interconnected. Foams with densities of 30 mg/cm^3 can be prepared by adding a nonpolymerizable oil, such as ethylbenzene, to the initial oil phase. However, these foams lack the necessary mechanical strength.

Drying involves removing water vapor from the foam. This process leaves a residue of surfactant and initiator that contributes extra mass and undesirable high-atomic-number material. However, virtually all this residue can be removed by backwashing with isopropanol and water to yield a foam composed almost entirely of hydrocarbon.

Scanning electron micrographs reproduced in Figure 5.23 show foams possessing densities of 50 and 30 mg/cm^3. Open cells with diameters of 2–3 µm are interconnected by pores of 1 µm

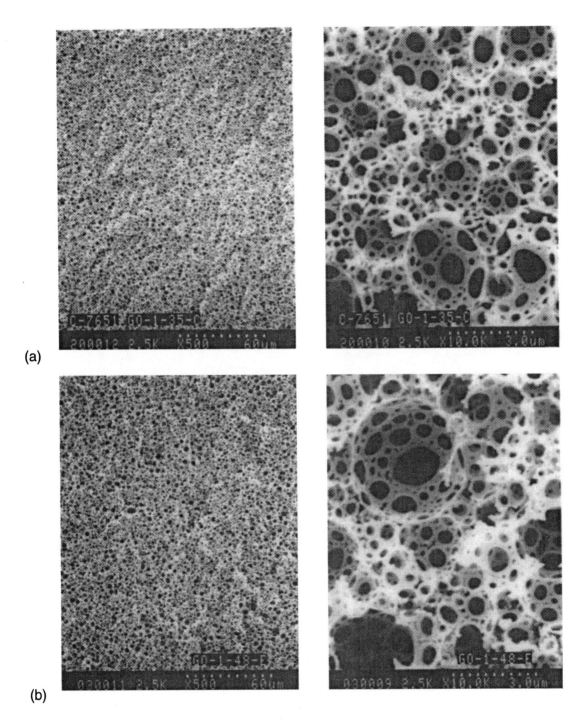

Figure 5.23
Scanning electron micrographs of polystyrene foams. (a) Standard formulation, 50 mg/cm^3.
Magnification; left 500×, right 10,000×. (b) Foam formed by adding ethylbezene to the polymerization mixture leading to a lower density of 30 mg/cm^3 but lacking the necessary mechanical
strength for subsequent processing. Magnification; left 500x, right 10,000×.

diameter. Because wetting properties are determined mainly by pore size, the foam shows acceptable wetting properties for DT. Nevertheless, the size of the cells is a little larger than desired because the scale length is larger than the very fine (1 µm) surface finishes required. The foam is sufficiently resistant to radiation damage.

Bibliography

D. J. Shaw, *Introduction to Colloid and Surface Chemistry*, London: Butterworth & Co. Ltd., 1980.

D. F. Evans and H. Wennerström, *The Colloidal Domain*, New York: VCH Publishers, Inc., 1994

D. H. Everett, *Basic Principles of Colloid Science*, London: Royal Society of Chemistry, 1980.

P. C. Heimenz, *Principles of Colloid and Surface Chemsitry* (2nd edition), New York: Marcel Dekker, Inc., 1986.

R. J. Hunter, *Foundations of Colloid Science*, New York: Oxford University Press, 1987.

Exercises

5.1 What is the origin of the favorable interaction between the polar headgroup of an amphiphilic molecule and water? Why does the hydrocarbon tail interact unfavorably with water?

5.2 The surface tensions of aqueous solutions of sodium dodecyl sulfate (SDS) at 25°C at different concentrations c are:

c (mM)	0	2	4	5	6	7	8	9	10	12
γ (mN/m)	72.0	62.3	52.4	48.5	45.2	42.0	40.0	39.8	39.6	39.5

Estimate the surface excess and the area per surface SDS molecule at three concentrations: 3, 7, and 10 mM.

5.3 The surface tensions of aqueous solutions of the nonionic surfactant $CH_3(CH_2)_9(OCH_2CH_2)_5OH$ at 25°C at different concentrations c are:

c (mM)	0	0.01	0.03	0.1	0.2	0.5	0.8	1	2	3
γ (mN/m)	72.0	63.9	56.2	47.2	41.6	34.0	30.3	29.8	29.6	29.5

Determine the critical micelle concentration (CMC). What is the surface area per surfactant molecule at the CMC?

5.4 (a) The CMC for the nonionic surfactant $C_{12}E_8$ is $10^{-4}M$. If each micelle contains 100 surfactant molecules, estimate the concentration of micelles in micelle moles/liter when the total surfactant concentration is 10^{-3} and $10^{-1} M$.

(b) Osmotic pressure ($\Pi = RTc$) provides a way to count the total number of particles in solution. Calculate the osmotic pressure at 298K for the surfactant solution in (a) at the CMC, when $c_{total} = 10^{-3} M$ and $c_{total} = 10^{-1} M$. ($1 Jm^{-3} = 1 Pa$)

5.5 CMCs and aggregation numbers for the nonionic surfactant dodecyldimethylamine oxide ($n\text{-}C_{12}H_{25}(Me)_2\, N^+ \cdot O^-$) (M.W. ~ 229) are given in the following table.

T(°C)	CMC (g/l)	N_o
1	0.0284	77
27	0.0210	76
40	0.0183	78
50	0.0175	73

Data from *J. Phys. Chem.* **66**, 295, 1962.

(a) Calculate the area per head group assuming that the micellar radius equals the length of the fully extended hydrocarbon chain. (b) Using the equilbrium model, calculate the values of ΔG_{mic}, ΔH_{mic}, and ΔS_{mic} for micellization at 25 and 40°C. (c) Estimate the vaules for $\Delta G_{(HG)}$ at 25 and 40°C using the value of $\Delta G_{(HP)}$ from eq. 5.2.10. (To a good approximation and over this narrow temperature range, you can assume that $\Delta G_{(HP)}$ is almost constant.)

5.6 The effective area of the amphiphile headgroup at the surface of a micelle
increases _ ; decreases _ ; or remains the same _ ;
as salt is added to a surfactant solution. Why?

5.7 The surfactant number
increases _ ; decreases _ ; or remains the same _ ;
as salt is added to a surfactant solution. Why?

5.8 The following table gives the measured critical micelle concentration (CMC) of an SDS aqueous solution as a function of the concentration of sodium chloride in the solution. The table also gives the measured mean aggregation numbers.

c_{NaCl} (M)	CMC (mM)	N_o
0.00	8.1	58
0.01	5.7	64
0.03	3.1	71
0.10	1.5	93
0.30	0.71	123

Assuming spherical micelle geometry, calculate for all five salt concentrations: (a) the volume of hydrocarbon core inside a micelle; (b) the effective radius of the hydrocarbon core; (c) the cross-sectional area per chain at the aggregate surface; (d) the area per charge assuming it is located 0.3 nm outside the hydrocarbon core.

5.9 Micelles are dynamic entities—always changing. (a) What exactly changes with time? (b) What are the characteristic time scales for these changes? (c) How are these relaxation times measured?

5.10 In pressure-jump experiments on micellar solutions of sodium hexadecyl sulfate (SHS) at 45°C, one observes a fast relaxation process with a time constant that varies with SHS concentration as follows:

Concentration SHS (mM)	Relaxation time τ_1 (10^{-6} secs)
0.58	807
0.66	665
0.70	603
0.90	430
1.0	365
1.1	325
2.0	114
5.0	38

The CMC is 0.52 mM and the aggregation number N = 100. Determine the rate constant for a monomer leaving the micelle, $k_{N_0}^-$. Estimate the standard deviation σ of the micelle size distribution. What is the value of the association rate constant $k_{N_0}^+$? Calculate the average lifetime of a monomer in the micelle.

5.11 Not all amphiphilic molecules in solution form spherical micelles. (a) What other self-assembled structures are there? (b) What are the key features of amphiphilic molecules that dictate their choice of self-assembled structure? (c) Define the surfactant number N_s. What are the typical values of N_s for different self-assembled structures?

5.12 Describe the physical differences between spherical micelles and spherical vesicles.

5.13 A bilayer of amphiphilic molecules separates two aqueous electrolyte solutions. The bilayer acts like a capacitor separating two conducting media. What is the capacitance per unit area if the bilayer is 4 nm thick and has a relative dielectric permeability of $\varepsilon_r = 3$? When charge is transferred from one of the solutions to the other, an electrical potential develops across the bilayer. Calculate how many potassium ions per unit area must be transported across a bilayer to develop a potential difference of 0.1 V.

5.14 Tetradecane is added to a micellar solution of 10 wt % $C_{12}E_5$ surfactant to form an O/W emulsion. The tetradecane and the $C_{12}E_5$ are present in equal amounts by weight. The density of the oil and surfactant are 0.84×10^3 and 1.0×10^3 kg/m^3, respectively. (a) Calculate the resulting emulsion droplet size assuming spheres covered by a monolayer of surfactant 3.5 nm thick. (b) Calculate the free energy of the emulsion per unit volume assuming the energy of the oil–surfactant–water interface is 10 mJ/m^2. (c) Calculate the creaming rise rate of the emulsion droplets, neglecting droplet–droplet interactions.

6

POLYMERS

with H. Wennerström

LIST OF SYMBOLS

c_m monomer concentration

c^* weight fraction of polymer

C^* polymer overlap concentration (eqs. 6.3.11 and 6.3.12)

D diffusion coefficient (eq. 3.8.26)

ΔE_{mix} internal energy of mixing

f friction factor (eq. 3.8.22)

ΔG_{mix} free energy of mixing for dilute (eq. 6.3.14) and semidilute (eq. 6.3.17) polymer solutions

l_p persistence length of polymer molecule (eq. 6.3.5)

L length of fully extended polymer molecule

$\langle L_r^2 \rangle$ mean square length of polymer molecule

L_t length of tube in reptation model

M molecular weight

$\langle M \rangle_n$ number-averaged molecular weight of polymer molecules (eq. 6.1.6)

$\langle M \rangle_w$ weight-averaged molecular weight of polymer molecules (eq. 6.1.7)

n_c number of carbon atoms along a hydrocarbon chain

n_i number of moles of component i

N_i number of molecules of component i

N_p degree of polymerization (number of monomer repeats in a polymer)

$\vec{r_i}$ position of segment i in a polymer molecule

R_g radius of gyration of a polymer molecule (eq. 6.3.2)

R_{1N} average end-to-end length of a polymer molecule (eq. 6.3.3)

R_h hydrodynamic radius (eq. 6.6.9)

ΔS_{mix} entropy of mixing for dilute (eq. 6.3.13) and semidilute (eq. 6.3.15) polymer solutions

T_g glass transition temperature

w effective molar interaction parameter (eq. 3.1.8)

W effective molecular interaction parameter (eq. 3.1.8)

W molecular pair interaction energy in liquid or solid phase

z distance from surface

γ_{12} surface (interface) tension between phases 1 and 2

Γ_i surface excess concentration of i (eq. 3.5.13)

ζ friction factor per segment of polymer molecule (eq. 6.6.5)

η viscosity (eq. 3.8.1); $[\eta]$ intrinsic viscosity (eqs. 3.8.19 and 6.6.1)

κ^{-1} Debye screening length (eq. 4.3.13)

μ_i chemical potential of component i

τ time for molecule to escape tube in reptation model (eq. 6.6.15)

τ_r mechanical relaxation time (eq. 6.6.20)

ϕ_p volume concentration of polymer p in semidilute polymer solution

χ interaction parameter for semidilute polymer solutions; $= w/\mathbf{R}T$ (eq. 6.3.20)

CONCEPT MAP

Basic Molecular Structural Units

- Polymer molecules contain N_p covalently bonded monomers whose collective properties are unique. N_p is the degree of polymerization.

Homopolymers

- N_p identical monomers linked with uniform interactions along a linear chain. Chains feature random coils. Molecular weight $M_{polymer} = N_p M_{monomer}$.

Block Copolymers

- B_m-C_n: m units of monomer B linked to n units of monomer C. If B and C display different solvent interactions, block copolymer molecules may self-organize to form micelles.

Heteropolymers

- Different monomers joined in a nonregular pattern. Crosslinking between chains is a common feature. Protein molecules, for example, contain 23 different monomers.

Molecular Size Distribution—Polydispersity

- Polymers are synthesized by free radical chain reactions or condensation reactions that join monomers together to form a macromolecule. Usually such reactions produce a range of N_ps, which makes the polymer polydisperse. Polydispersity is measured by the ratio of $\langle M \rangle_w / \langle M \rangle_n$—the weight- and number-average molecular weights (eqs. 6.1.6 and 6.1.7). Many biopolymers are monodisperse.

States of Matter

Solid Polymers

- Most solid polymers (particularly those containing structural irregularities and networks) are amorphous at all temperatures. Transition from solid to liquidlike behavior occurs at the glass transition temperature T_g.

- Some solid polymers form ordered crystalline solids that transform to an amorphous state on melting.

Polymer Solutions

- Three concentration regimes dominate (Figure 6.7):

  1. Very dilute—polymer molecules dispersed, interactions unimportant
  2. Semidilute—polymer chain overlap becomes important
  3. Concentrated—chains entangle.
- Transition between (1) and (2) occurs at the overlap concentration C^* ($\propto M$) $\approx 0.1 - 5$ wt %.

Homopolymer Solutions— Characteristic Properties

- The size of a random coil in solution is characterized by the average radius of gyration, R_g (eq. 6.3.2), and the average end-to-end distance, R_{1N} (eq. 6.3.3).

- In dilute solution, polymers can fold (Figure 6.8) in many ways because of their many internal degrees of freedom. In the compact state $R_g \propto M^{1/3}$; in the linear configuration $R_g \propto M$; in the random coil state $R_g \propto M^a$, where $1/3 < a < 1$.

- Actual conformity depends on polymer–solvent interaction parameter W.
  When $W > \frac{1}{2}kT$ solvent is "bad" and polymer molecule shrinks to compact form.
  When $W < \frac{1}{2}kT$ solvent is "good" and polymer molecule expands to random coil or linear form.

Thermodynamic Properties

- A polymer's internal degrees of freedom lead to significant entropy of mixing, ΔS_{mix} (eq. 6.3.13 for dilute, eq. 6.3.15 for semidilute solution). ΔS_{mix} has the same form as for an ideal solution (eq. 3.7.6).

- ΔG_{mix} (eq. 6.3.14) has the same form as for regular solution theory. It incorporates the effective interaction parameter W.

- W provides a basis for classifying polymer–solvent interactions. $W > \frac{1}{2}kT$, solvent is "bad"; $W < \frac{1}{2}kT$, solvent is "good." $W = \frac{1}{2}kT$ defines the "theta condition."

Transport Properties

- In dilute polymer solutions:
  Viscosity $\eta \propto M^{1/2}$ (eq. 6.6.4) if polymer behaves as a rigid body, and $\eta \propto M$ (eq. 6.6.7) if polymer behaves as a free draining coil.
  Diffusion coefficient $D \propto M^{-3/5}$ (eq. 6.6.10) for rigid body and $D \propto M^{-1}$ (eq. 6.6.11) for free draining coil model.

- In concentrated polymer solutions and melts:
  Chain entanglements dominate transport behavior. de Gennes' reptation model predicts viscosity, $\eta \propto M^3$, and diffusion coefficient, $D \propto M^{-2}$.

Other Polymer Structures

- When individual monomers possess amphiphilic properties, helices and other ordered structures form at low temperatures. Transformation between ordered forms and random coils is a cooperative process occurring over a narrow temperature range.

- Polyelectrolytes are charged polymer molecules. Repulsive electrostatic interactions extend chains, leading to low overlap concentrations.

- Gels result from physical association or chemical cross-linking of polymer molecules in solution. On a macroscopic level, gels are solidlike, but on a microscopic level they retain fluidlike properties. For example, D for solvent in gel $\propto$ D for solvent in bulk.

Polymer–Surfactant Interactions

- Surfactant micelles self-assemble in a polymer solution at a critical aggregation concentration (CAC) that is usually less than the surfactant's CMC. Combined polymer—micelle structures are pictured as coiled strings (polymer chains) of beads (surfactant micelles).

- A given micelle can solubilize hydrocarbon chains from two different polymer molecules. This linking process dramatically increases the viscosity of the solution.

Polymers on Surfaces— Adsorption Processes

Thermodynamics

- Physical adsorption leads to many polymer–surface contacts (Figure 6.13). If $E_{abs} > kT$, the process is usually irreversible.

- Configuration of adsorbed polymer depends on W and surface coverage. In a good solvent, $W < \frac{1}{2}kT$, adsorbed polymer molecules coil out into the solvent. When grafting density $> R_g^{-2}$, adjacent chains interact and extend away from the surface. At very high surface coverage, they form a dense polymer bush.

- In a bad solvent, $W > (1/2)kT$, polymer molecules concentrate at the surface to minimize contact with the solvent.

Kinetics

- Rate of adsorption is limited by the slow rate of polymer diffusion through the solution and onto the surface.

- Polydispersity can affect the adsorption process. Short-length polymer molecules adsorb first because they diffuse faster than large ones, but the final equilibrium favors adsorption of large polymer molecules.

- Achieving equilibrium can take a very long time and may never be reached.

- With amphiphilic block copolymers, one block selectively adsorbs onto the surface, while the other interacts with the solvent.

- Polyelectrolytes adsorb strongly onto oppositely charged surfaces. Adding electrolyte can desorb them.

- Polymers can be permanently grafted onto a surface by covalent bonding.

Forces between Surfaces with Adsorbed Polymers

- In a good solvent, two surfaces covered with polymer begin to interact when their separation $\approx 2R_g$ (10–100 nm). A repulsive force arises because the configurational entropy of the interacting chains is reduced as they interpenetrate (steric stabilization).

- Colloidal particles covered with a polymer coagulate when the solvent changes from "good" to "bad," with a drop in temperature, for example. This aggregation process is usually reversible.

- When polymers do not completely cover the surface of a colloid particle, bridging between particles can lead to the formation of loosely bound flocs.

6

Our discussion of fluid interfacial systems began in Chapter 4 by focusing on colloidal particles. On a molecular scale, colloidal particles are large immutable objects whose properties are dominated by their bulk and their surface chemistry. Internal motions within the particle are not very important. Colloidal dispersions are thermodynamically unstable and must be stabilized kinetically by electrostatic or steric repulsive forces to overcome ubiquitous attractive forces if they are to remain in solution. Colloidal particles resemble a row of dominoes (or an autocratic society) in which discrete entities all fall down together.

In Chapter 5 we turned our attention to amphiphilic systems whose properties are dictated by schizophrenic molecules containing solvophilic and solvophobic moieties. To minimize solvophobic interactions, amphiphilic molecules self-assemble to form a variety of aggregates that can change their size and shape in response to their environment. Specific properties associated with each aggregate structure, such as micelles, bilayers, and vesicles, find widespread use in commerce and biology. Amphiphilic aggregates are like cooperative societies in which free individuals (molecules) join together to build larger, more effective structures.

We now move on to consider polymer systems, which comprise the third important fluid interfacial system. The word *polymer* derives from the Greek words *poly* and *meros* meaning many and parts. Polymers are formed by permanently joining together many smaller *monomer* units to form a large macromolecule. Although the effect of each monomer may be small, the sum of their collective contributions leads to unique and highly useful properties. Because polymer molecules are generally quite flexible, intramolecular motions play an important role in determining their behavior. Polymers made up of identical monomers are like ideal communist communities in which the members of the proletariat (monomers) are indistinguishable and make equal contributions to enhance the performance of the whole (polymer).

In interfacial engineering, we encounter several different types of polymers that we can classify in terms of their structure. A linear *homopolymer* consists of a small monomer unit repeated N_p times. N_p is called the *degree of polymerization*. In linear homopolymers, chemical interactions take place along the chain in a uniform way, and we often observe random coil conformations. In *block copolymers*, the polymer chain consists of blocks of one repeating monomer unit followed by one or more blocks of other repeating unit(s). Because block copolymers have two parts, they can display properties similar to amphiphiles and form self-assembly structures analogous to those found with surfactants. In *heteropolymers,* many different monomer units are joined in nonregular patterns. Some natural polymers are heteropolymers, and the different units typically are composed of proteins, nucleic acids, and biopolymers. Proteins can contain up to 23 different amino acid monomers, which form specific conformational structures determined by detailed sequences of amino acids. Polymers with charged monomers are called *polyelectrolytes,* whether they are homo-, hetero-, or block copolymers.

Another way to classify polymers is in terms of their origin. Many are natural products. These include cotton and wool fibers, rubber elastomers, and proteins and carbohydrates derived from a range of sources such as wood, cereals, algae, insect shells, and even rooster combs. Many of these materials have been used since antiquity. Natural polymers can be processed to yield useful materials, such as leather, adhesives, or explosives. The biological molecules in our bodies that carry the genetic code, enable us to move, act as catalysts for physiologic processes, and provide the connective tissues that hold us together are all polymers.

Since the 1920s, synthetic polymers have played an increasingly important role in manufacturing processes. They have provided alternatives to natural products; for example, nylon and dacron have become substitutes for cotton. Increasingly sophisticated understanding of macromolecules has enabled us to tailor new polymers for specific, demanding applications. Examples of more recent polymers include inert fluorocarbons, silicone adhesives, and block copolymers that possess both structural rigidity and high impact resistance.

Clearly, polymer systems show a rich range of behavior. This chapter considers the general properties of the different polymers in bulk and in solution form, examines the behavior of polymers confined to interfaces, and analyzes their interaction with surfactants. The chapter concludes with some examples of polymer systems in which interfacial behavior is important in processing and application.

6.1 Polymer Synthesis Produces Macromolecules That Are Generally Polydisperse

Understanding how polymers are synthesized provides us with insight into several of their characteristic properties. The two major kinetic schemes for synthesis of polymers involve chain and stepwise polymerization; these two schemes are sometimes known as *addition* and *condensation* polymerization, respectively. We can clarify the differences between them by analyzing specific examples.

6.1.1 Chain (Addition) Polymerization

We illustrate chain polymerization by the polymerization of styrene to form polystyrene. Free radical polymerization of styrene induced by a benzoyl peroxide initiator illustrates the three mechanistic steps that constitute chain polymerization: initiation, propagation, and termination.

1. Initiation: Upon heating, benzoyl peroxide decomposes to form two free radicals ($\bullet$)

$$\phi-\overset{\overset{\text{O}}{\|}}{\text{C}}-\text{O}-\text{O}-\overset{\overset{\text{O}}{\|}}{\text{C}}-\phi \quad \overset{\Delta}{\rightarrow} \quad 2 \ \phi-\overset{\overset{\text{O}}{\|}}{\text{C}}-\text{O}\bullet \tag{6.1.1}$$

where ϕ represents the phenyl- group.

2. Propagation: Each of the free radicals from reaction 6.1.1 can add a styrene monomer molecule to form a new free radical

$$\phi-\overset{\overset{\text{O}}{\|}}{\text{C}}-\text{O}\bullet + \text{CH}_2=\underset{\underset{\phi}{|}}{\text{CH}} \ \rightarrow \ \phi-\overset{\overset{\text{O}}{\|}}{\text{C}}-\text{O}-\text{CH}_2-\overset{\bullet}{\underset{\underset{\phi}{|}}{\text{CH}}} \tag{6.1.2}$$

The free radical product of reaction 6.1.2 rapidly adds N_p other styrene monomer molecules to give

$$\phi-\overset{\overset{\text{O}}{\|}}{\text{C}}-\text{O}-\text{CH}_2-\overset{\bullet}{\underset{\underset{\phi}{|}}{\text{CH}}} + N_p \ (\text{CH}_2=\underset{\underset{\phi}{|}}{\text{CH}}) \rightarrow \phi-\overset{\overset{\text{O}}{\|}}{\text{C}}-\text{O}\left\{\text{CH}_2-\underset{\underset{\phi}{|}}{\text{CH}}\right\}_{N_p}-\text{CH}_2-\overset{\bullet}{\underset{\underset{\phi}{|}}{\text{CH}}} \tag{6.1.3}$$

As each monomer is added "head to tail," the free radical unit moves to the right, always to remain at the tail end of the chain. In this case chain polymerization proceeds by opening a double bond (the one between CH_2 and $\text{C}\phi\text{H}$) to form a link to the next monomer, sequentially developing a linear molecule containing N_p monomer units.

3. Termination: Two growing chains, R and R', can react with each other to eliminate the free radical by a "tail-to-tail" combination

$$\text{R}-\text{CH}_2-\overset{\bullet}{\underset{\underset{\phi}{|}}{\text{CH}}} + \overset{\bullet}{\underset{\underset{\phi}{|}}{\text{CH}}}-\text{CH}_2-\text{R}' \rightarrow \text{R}-\text{CH}_2-\underset{\underset{\phi}{|}}{\text{CH}}-\underset{\underset{\phi}{|}}{\text{CH}}-\text{CH}_2-\text{R}' \tag{6.1.4}$$

This combination stops further reaction and completes the synthesis of the polymer molecule. Termination also can occur by disproportionation, in which a hydrogen atom is transferred from one chain to another.

Table 6.1 gives examples of polymers synthesized by chain polymerization. Many homopolymers are named simply by adding the prefix *poly* to the monomer, as in polystyrene.

TABLE 6.1 Selected Chain Polymer Structures and Nomenclature

Structure	Name	Where used
$\left[\text{CH}_2 - \text{CHR}\right]_n$ **Vinyls**		
R : –H	Polyethylene	Plastic
R : –CH$_3$	Polypropylene	Rope
R : –ϕ (Phenyl Group)	Polystyrene	Drinking cups
R : –Cl	Poly(vinyl chloride)	"Vinyl"
R : –O–CO–CH$_3$	Poly(vinyl acetate)	Latex paints
R : –OH	Poly(vinyl alcohol)	Fiber

$$\left[\text{CH}_2 - \overset{\overset{\text{X}}{|}}{\underset{\underset{\text{OCOR}}{|}}{\text{C}}}\right]_n$$

X : –H **Acrylics**
X : –CH$_3$ **Methacrylics**

X : –H R : –C$_2$H$_5$ Poly(ethyl acrylate)		Latex Paints
X : –CH$_3$ R : –CH$_3$ Poly(methyl methacrylate)		Plexiglass
X : –CH$_3$ R : –C$_2$H$_5$ Poly(ethyl methacrylate)		Adhesives

$$\left[\text{CH}_2 - \overset{}{\underset{\underset{\text{R}}{|}}{\text{C}}} = \text{CH} - \text{CH}_2\right]_n \quad \textbf{Dienes}$$

R : –H	Polybutadiene	Tires
R : –CH$_3$	Polyisoprene	Natural rubber
R : –Cl	Polychloroprene	Neoprene
$\left[\text{CX}_2 - \text{CR}_2\right]_n$ **Vinylidenes**		
X : –H R : –F	Poly(vinylidene fluoride)	Plastic
X : –F R : –F	Polytetrafluoroethylene	Teflon
X : –H R : –CH$_3$	Polyisobutene[a]	Elastomer
	Common copolymers	
EPDM	Ethylene-propylene-diene-monomer	Elastomer
SBR	Styrene-butadiene-rubber Poly(styrene-*stat*-butadiene)[b]	Tire rubber
NBR	Acrylonitrile-butadiene-rubber Poly(acrylonitrile-*stat*-butadiene)	Elastomer
ABS	Acrylonitrile-butadiene-styrene	Plastic

L. H. Sperling, *Introduction to Physical Polymer Science*, 2nd ed., Copyright © 1992, John Wiley & Sons, p. 12. Reprinted by permission of John Wiley & Sons, Inc.

[a]Also called polyisobutylene. The 2% copolymer with isoprene, after vulcanization, is called butyl rubber.

[b]The term *stat* (statistical) indicates a random sequence of monomers.

Crosslinked polymers can be formed by polymerizing a mixture of difunctional monomers, such as 1,4-divinyl-benzene and styrene, to form a crosslinked network like that shown in Figure 6.1. Many crosslinked polymer networks are produced by chemical reactions triggered by heating. The resulting polymers are "set" by the reaction and are known as *thermosets*. Once solidified, thermoset polymers do not soften or flow on subsequent heating.

Allowing monomers and initiators to flow into a mold before reaction takes place presents one strategy for processing solid polymers into intricate shapes—reaction injection molding (RIM). During synthesis, the reactant mixture in the mold consists of monomer, a small amount of rapidly reacting free radical species, and terminated polymer molecules that usually do not undergo further reaction after the termination step.

6.1.2 Step (Condensation) Polymerization

We illustrate step polymerization by the formation of nylon 4 through the creation of an amide linkage

$$N_p \; H_2N\text{-}(CH_2)_3\text{-}\overset{\overset{\displaystyle O}{\displaystyle \|}}{C}OH \rightarrow H\text{-}\left[\,NH\text{-}(CH_2)_3\text{-}\overset{\overset{\displaystyle O}{\displaystyle \|}}{C}\,\right]_{N_p} OH + (N_p - 1)H_2O \tag{6.1.5}$$

where the 4 in nylon 4 stands for the number of carbon atoms in the monomer unit.

A different type of nylon (designated as nylon 4,8) can be prepared by reacting a diamine with a dicarboxylic acid, usually in the form of acidchloride. Figure 6.2 illustrates a simple scheme for synthesizing nylon 4,8. The acid chloride dissolved in carbon tetrachloride and the diamine dissolved in water are brought into contact at the CCl_4–water interface where polymerization occurs. By reaching into the interface with a pair of tweezers, we can draw a nylon fiber out of solution. As the polymerization proceeds, the reactants at the interface are replenished by diffusion from the adjacent bulk solutions. This reaction forms the basis for nylon spinning.

Figure 6.1
Crosslinking, at •, by difunctional or polyfunctional monomers leads to the formation of networklike structures.

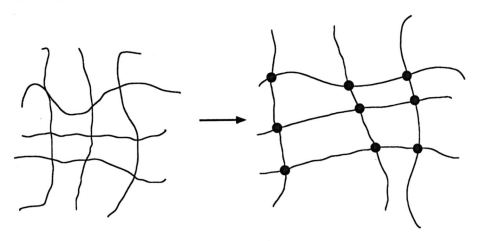

Figure 6.2
When a water-soluble diamine and an oil-soluble dicarboxylic acid chloride are brought into contact at an oil–water interface, polymerization leads to the formation of nylon. Using a pair of tweezers, the polymer can be withdrawn as a continuous string from the interface as reactants are replenished by diffusion. (P. W. Morgan and S. L. Kwolek, *J. of Chemical Education* **36**, 182, 1959)

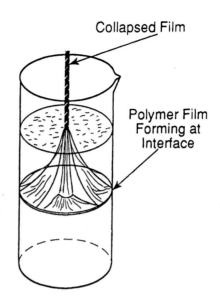

Collapsed Film

Polymer Film Forming at Interface

6.1.3 Both Chain and Step Polymerization Produce Polydisperse Polymers

Because of the variation in N_p, both chain polymerization and step polymerization produce macromolecules that possess a distribution of molecular weights. We characterize the average molecular weight of a polymer by either the *number-average molecular weight,*

$$\langle M \rangle_n = \frac{\Sigma N_i M_i}{\Sigma N_i} \tag{6.1.6}$$

where N_i represents the number of molecules of molecular weight M_i, or the *weight-average molecular weight,*

$$\langle M \rangle_w = \frac{\Sigma N_i M_i^2}{\Sigma N_i M_i} \tag{6.1.7}$$

Different experimental methods for determining molecular weight yield either the number-average or the weight-average value. Techniques that involve colligative solution properties, such as osmometry, count individual molecules and yield $\langle M \rangle_n$. Other methods, such as light, x-ray, and neutron scattering, give values of $\langle M \rangle_w$.

The degree of polydispersity is given by the ratio $\langle M \rangle_w / \langle M \rangle_n$, called the *polydispersity index.* For many typical polymers the polydispersity index ranges from 1.5 to 2.0. To obtain a more homogeneous sample, we must use more specialized synthetic procedures, such as anionic polymerization (for which $\langle M \rangle_w / \langle M \rangle_n \leq 1.01$) or some sort of fractionation after synthesis, for example, chromatography. Many biopolymers, particularly proteins, have $\langle M \rangle_w / \langle M \rangle_n = 1.0$; that is, all the polymer molecules have the same polymerization number, N_p. They are said to be *monodisperse.*

6.1.4 Polymers Are Used in Both Bulk and Solution Form

Polymers find application both in bulk form, in which the polymer molecules aggregate together in the absence of any solvent to form crystalline and amorphous arrays, and in solution form, in which the polymer molecules are dispersed in a solvent. To understand the nature of surfaces and interfaces involving polymers or polymer solutions, we must first understand the basic characteristics of both forms. We discuss the structure of polymers in their bulk form in Section 6.2 and in their solution form in Section 6.3.

6.2 Bulk Solid Polymers Exist in Both the Amorphous and Crystalline State

Polymer macromolecules assemble in many different ways and thereby demonstrate a wide variety of physical properties in bulk form. For example, epoxies form when monomers are mixed together to initiate a polymerization reaction, and they harden when the macromolecules become multiply connected, as depicted in the crosslinked heteropolymer structure in Figure 6.1. The resulting *bulk* polymer structure is completely random and lacking in long-range order; it is *amorphous*. Crosslinking in other polymers, particularly homopolymers, may be less extensive, but they are still amorphous because individually the molecules usually possess some structural irregularity, such as random coils, along their chains. They are like a spaghetti composed of randomly interwoven coiled strands of unequal length. A more subtle effect exists for vinyl polymers $+CH_2-CHR+_N{}_p$. As the chain grows, the stereochemistry for the asymmetric carbon is random along the chain. This is called an *atactic polymer*. When such a polymer is cooled to form a solid, the heterogeneity along the chain precludes the formation of a perfect crystal. Depending on the degree of heterogeneity, the solid may contain crystalline domains separated by amorphous regions, or it may fail to form ordered structures altogether, resulting in a completely amorphous material. Atactic vinyl polymers, such as polystyrene and polyvinylchloride (see Table 6.1), are examples of polymers with this structure. In some polymers, large regions exist where the macromolecules are ordered in *crystalline* form, and the resulting polymer structure is a mixture of crystalline and amorphous material. Polyethylene is an example of a polymer that can be made in crystalline form.

Varying amounts of amorphous, crosslinked, and crystalline macromolecules exist in a typical polymer. Synthetic polymers show a wide size distribution because the chain termination point varies. The proportion, distribution, and orientation of amorphous and crystalline regions are determined by the processing. Together they constitute the microstructure of the bulk polymer that dictates its physical behavior.

6.2.1 The Decrease in Young's Modulus with Increasing Temperature Marks Transitions from Crystalline Solid to Amorphous Glass to Rubbery Material to Viscous Fluid

A useful parameter to characterize a polymeric material is its Young's modulus of elasticity. (See Appendix 11A for definitions and a brief introduction to the mechanical properties of solids). Figure 6.3 shows how Young's modulus for a linear *amorphous* polymer changes with temperature. At low temperatures, only vibrational motions along the chain are possible and the polymer is glasslike—hard and brittle (region I). In this state the material is rigid, and the Young's modulus is high (10 GPa). As the temperature is raised, the polymer undergoes a transition from a glasslike to rubberlike consistency at the *glass transition temperature*, T_g (region II). (Glass transition temperatures lie somewhere between −100 and +150°C for most polymers.) This change marks the onset of long-range coordinated molecular motion. During the transition, the polymer softens and the Young's modulus decreases by three to four orders of magnitude. The material transforms to a rubbery material (region III), displaying rubbery flow with low elastic modulus (1 MPa). At higher temperatures, the material "melts" to become a viscous fluid, a bulk *polymer melt.* Then it becomes incapable of supporting a stress; so the elastic modulus drops to zero (region IV–V).

Crystalline polymers also generally contain appreciable amounts of amorphous material. They too are hard and brittle at low temperatures. When heated, the crystalline regions transform to an amorphous liquidlike state, and the decline in Young's modulus at the glass transition temperature is even more sharply defined than it is for totally amorphous polymers. An example is

Figure 6.3
Schematic of the change in Young's modulus with temperature for a linear polymer as its transforms from a glass-like material to a viscous fluid. The transformations labeled in the figure are described in the text. (L. H. Sperling, *Introduction to Physical Polymer Science,* 2nd ed., Copyright © 1992, John Wiley & Sons, p. 8. Reprinted by permission of John Wiley & Sons, Inc.)

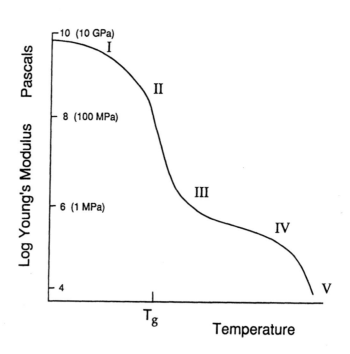

the simplest polymer, polyethylene, which we write as $+CH_2+_{N_p}$.
This tough, flexible crystalline polymer is widely used as tubing, milk bottles, fibers, piping, and packaging material. It is hard at room temperature, but it becomes fluid at high temperatures.

Solid polymers that become fluid at high temperatures are called *thermoplastics*. Most linear polymers included in Table 6.1 are thermoplastics. The sensitivity of their flow behavior to temperature presents another strategy (see Section 6.1.1) for processing polymers—thermoplastic injection molding (TIM). While fluid, they can be forced (injected) into molds at high temperatures and then cooled to form intricate solid shapes.

6.2.2 Many Crystalline Polymers Consist of Lamellar Structures That Form Spherulites

We can understand the structure of many crystalline polymers by first noting the characteristics of solid waxes. Waxes are normal alkanes with N_p = 25–50, and at room temperature they are predominantly crystalline solids. Their melting point increases with chain length, approaching an asymptotic limit of about 140°C as $N_p \to \infty$. The molecules are ordered by van der Waals attraction forces to form lamellar structures like those depicted in Figure 6.4a. The length of the lamellae is defined by the total length of the hydrocarbon chain in the all-trans (zigzag, see Section 5.1) configuration. Such materials are brittle—as anyone who has tried to bend a candle knows.

Polyethylene is also an alkane, with N_p increased orders of magnitude up to 10^3–10^5. Crystallized from dilute solution, polyethylene consists of ordered layers of lamellae, but in this instance many of the lamellae form when the long-chain molecules fold back upon themselves, as depicted in Figure 6.4b. Then the individual lamellae may be one-tenth to one-hundredth as long as the fully extended polymer chain. Furthermore, different parts of long-chain molecules frequently cross from one lamella to another, thereby providing covalently bonded links between lamellae. Such materials are tough, and they can be flexed many times without breaking.

When crystallized from concentrated solution or by solidification from the melt, polyethylene and many other polymers form *spherulites*. In these crystalline structures—shown in Figure 6.5—individual lamellar crystalline plates grow by radiating out from a single crystalline nucleation site at their center. Figure 6.6 shows schematically the development of a spherulite from a chain-folded lamellar precursor. As growth initiates, lamellae develop on either side of the nucleating structure. These lamellae deviate from the original plane of formation as the structure begins to grow. Gradually they diverge and fan out, and as they do so, new lamellar structures are initiated between them. The process repeats so that the structure eventually develops the characteristic spherical shape. If nucleation occurs simultaneously from many nuclei throughout the melt, equal-sized spherulites form with flat planar boundaries between them; otherwise, the boundaries are hyperboloids.

Figure 6.4
(a) Lamellar structures formed by solid crystalline normal alkanes (N_p = 25–50) under the influence of van der Waals attractive forces. (b) Lamellar structures formed by polyethylene (N_p = 10^3 to 10^5) in which chains fold back upon themselves and also cross from one lamella to another. (L. H. Sperling, *Introduction to Physical Polymer Science,* 2nd ed., Copyright © 1992, John Wiley & Sons, p. 4. Reprinted by permission of John Wiley & Sons, Inc.)

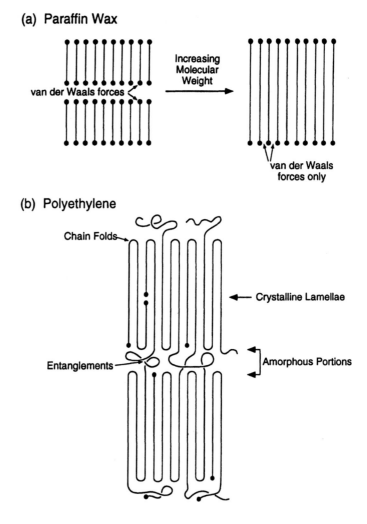

(a) Paraffin Wax

Increasing Molecular Weight

van der Waals forces

van der Waals forces only

(b) Polyethylene

Chain Folds

Crystalline Lamellae

Entanglements

Amorphous Portions

The folded chains in spherulites are more irregular in length and alignment than in the single crystals of Figure 6.4b. Between the lamellar structures lie amorphous unaligned chains that provide the intercrystalline links. These links tie the entire structure together with covalent bonds and give *semicrystalline* polymers a high mechanical toughness.

As we shall see in Sections 6.7.2 and 6.7.3, controlling the degree of crystallinity, spherulite size, and relative orientation in bulk polymer processing dictates the final physical properties of many polymers.

6.2.3 Elastomers Are Crosslinked Polymer Systems above Their Glass Transition Temperature

Elastomers provide a further illustration of how microstructure influences the elasticity and mechanical properties of bulk polymer materials. *Elastomers* are materials whose ambient temperatures lie in region IV in Figure 6.3, well above the glass transition temperature. Rubber and related polymer systems are elastomers.

Figure 6.5
Electron micrographs of
spherulite replicas formed
in a melt-crystallized thin
film of poly(4-methyl-
pentene-1). (a) An edge-on
view of a spherulite showing the distinctly lamellar
character and "sheaflike"
arrangement of the lamel-
lae. (b) A flat-on view of a
spherulite. (F. Khoury and
E. Passaglia, in *Treatise on
Solid State Chemistry,*
Vol. 3 of *Crystalline and
Noncrystalline Solids,*
N. B. Hannay, Ed., Plenum
Press, New York, 1976,
p. 470.)

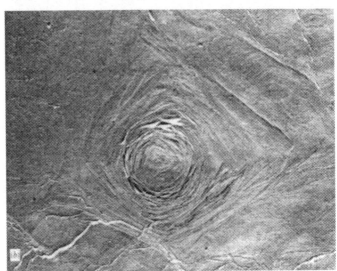

Figure 6.6
Schematic development of
a spherulite from a chain-
folded precursor crystal.
Row (a) shows the edge-on
view of the spherulite,
while row (b) gives the flat-
on view. (F. Khoury and
E. Passaglia, in *Treatise on
Solid State Chemistry,*
Vol. 3 of *Crystalline and
Noncrystalline Solids,*
N. B. Hannay, Ed., Plenum
Press, New York, 1976,
p. 468.)

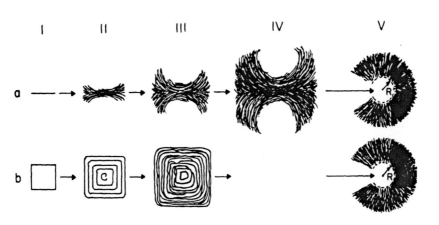

As harvested, rubber is a linear molecule (molecular weight $M_W \approx 1 \times 10^5$ g/mol) with a cis 1,4 polyisoprene structure. Rubber molecules can be crosslinked—using sulfur (vulcanization), radiation, or a multifunctional monomer—to form a randomly connected network of the kind illustrated in Figure 6.1. Typically 10 to 20 crosslinks are formed per molecule, so that one huge interconnected macromolecule results. In between the crosslinked sections, the polyisoprene forms an amorphous structure with $T_g = 201$ K ($-72°C$).

Under mechanical stress at room temperature, the polymer molecule segments between the crosslinked points respond to the deformation forces by uncoiling and rotating along the chain backbone to straighten out the kinks in the amphorous chains. These changes occur reversibly, and the elastomer elongates elastically but with a very low modulus of elasticity. The actual value of the modulus is sensitive to the spacing of the crosslinks. As the spacing becomes shorter, the modulus goes up. The modulus, and therefore the rigidity, of rubber increases with the degree of vulcanization or time of radiation during the crosslinking process. As noted in Section 6.7.3, many synthetic analogs of rubber have been developed.

6.2.4 Mechanical Behavior of Bulk Polymer Solids

This section concludes by examining the mechanical behavior of bulk polymers when the stress is raised to such a level that the elongation is no longer purely elastic and reversible. Although the behavior may be roughly categorized into three types, brittle, tough and elastomeric, the particular behavior depends on the microstructure of the polymer and the temperature.

For *brittle* polymers the stress–strain relationship is linear and reversible up to about 1–2% elongation. At this point the material breaks suddenly and catastrophically. [Appendix 11A, Figure 11A.2, curve (b), illustrates this behavior.] The breaking stress defines the polymer's *tensile strength*. Stresses at the breaking point are typically 60 MPa. Brittle polymers elongate elastically by the stretching of bond lengths and the bending of bond angles. Polymers like polystyrene behave this way at temperatures well below their glass transition temperature (T_g for polystyrene is 120°C).

The stress–strain relationship for *tough* polymers is linear up to a certain critical stress, after which elongation continues at almost constant stress. In the plateau region, the material undergoes extensive plastic flow. After an elongation of a few percent (~10%) strain, the polymer finally hardens due to molecular alignment (crystallization) and then ruptures in a brittle manner. [Appendix 11A, Figure 11A.2, curve (d), illustrates this behavior.] The stress at the onset of the plastic region defines the *yield strength* of the polymer. Typical stresses at the yield point are 20–50 MPa. Plastic flow occurs when individual molecules slide past each other or become unravelled in an irreversible manner. Polyethylene (in its semicrystalline state and at temperatures above the $T_g \approx 0°C$ of its amorphous portions) exemplifies a polymer system that behaves this way. One-half the integrated area under the stress–strain curve equals the mechanical energy consumed per unit volume of material. Therefore, the greater the region of plastic flow, the more energy is needed to fracture the polymer. For this reason, the area under the stress–strain curve is an indication of the polymer's *toughness*.

Elastomers typify the third stress–strain relationship. Their distinguishing feature is that they can be stretched elastically to several times their original length (several hundred percent strain) before they finally harden and fracture. [Appendix 11A, Figure 11A.2, curve (e), illustrates this behavior.]

Several criteria are used to describe the mechanical stability of a polymer. Tensile strength describes the stress required to break the material. Toughness describes the amount of energy expended in deforming the material to the breaking point. In engineering practice polymers are rarely stressed to their breaking point, and toughness is usually a more useful measure of mechanical performance.

6.3 A Complete Description of Polymer Solutions Involves Integrating Information from Many Fields

6.3.1 The Character of a Polymer Solution Changes with Concentration; There Are Three Different Concentration Regimes—Very Dilute, Semidilute, and Concentrated

We move to the characterization of polymers in *solution*. When discussing polymer solutions, we must distinguish three very different concentration regimes—very dilute, semidilute, and concentrated. They are illustrated in Figure 6.7. The distinction among them will become clear as our discussion progresses from very dilute to semidilute to concentrated solution behavior.

6.3.2. Macromolecule Conformation in Very Dilute Solution Depends on Molecular Weight and Solution Interaction

When a polymer is dissolved in an excess of solvent, the individual macromolecules are so far apart that we can neglect intermolecular interactions. These conditions prevail in *very dilute solutions* (Figure 6.7a).

In a very dilute solution, the polymer molecule can fold in so many different ways and with so many internal degrees of freedom that it is difficult to describe the actual state of an individual macromolecule. Statistical averages must be used instead. In this section we define parameters used to describe the conformation and the effective "size," shape, and form of polymer molecules in solution. *We shall find that the effective size of a molecule depends both on its molecular weight and on its interaction with the solution.*

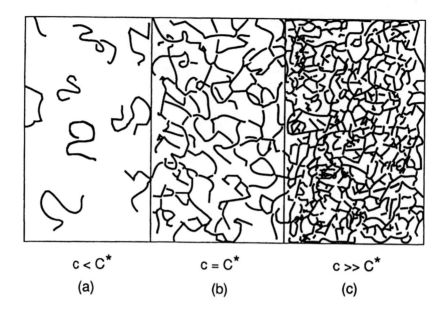

$$c < C^*$$
(a)

$$c = C^*$$
(b)

$$c \gg C^*$$
(c)

Figure 6.7
The three concentration regimes of a polymer solution can be described most conveniently in terms of the overlap concentration C^*, as defined in eq. 6.3.11. (a) Very dilute solution, $c < C^*$—intermolecular interactions are unimportant; (b) semidilute solution, $c \geq C^*$—chain overlap becomes important; (c) concentrated solution, $c \gg C^*$—chain entanglements dominate. (D. F. Evans and H. Wennerström, *The Colloidal Domain: Where Physics, Chemistry, Biology, and Technology Meet,* VCH Publishers, New York, 1994, p. 295. Reprinted with permission VCH Publishers © 1994.)

6.3.2.1 Macromolecule Size Is Described in Two Ways: Radius of Gyration and Average End-to-End Distance

How can we characterize polymer chain conformation in solution? Figure 6.8 shows three possible types of chain folding. In two extreme cases the chain is either completely compacted or completely stretched out. In the compact state, Figure 6.8a, the chain behaves somewhat like a micelle: it folds back on itself to minimize polymer–solvent contact. The "size" of the polymer molecule is related to the radius of the globule R, and therefore increases as $N_p^{1/3}$, where N_p is the degree of polymerization. Many soluble proteins in their native state show this size dependence on molecular weight. Other polymers in very dilute solution stretch out and adopt a linear configuration (Figure 6.8b). In this case, chain "size" is related to the longest linear dimension L, and increases linearly with N_p. This configuration often is dictated by helical structures like the DNA double helix.

With most synthetic polymers in dilute solution, the chain is very flexible on a length scale that is large compared to the size of a monomer unit. This property leads to formation of a randomly configured coil (Figure 6.8c). In this case, the characteristic "size," $2R_g$ increases with the degree of polymerization as some power, $R_g \sim N_p^{\alpha}$, where α is larger than 1/3 and less than 1.

The most commonly used measure of coil dimension is the *radius of gyration*. For a fixed set of masses, we define the radius of gyration by

$$R_g^2 = \frac{\sum_i m_i (r_i - r_{cM})^2}{\sum_i m_i} \tag{6.3.1}$$

Figure 6.8
Schematic illustrating the
three extreme types of
polymer configuration in
solution: (a) Compact globule in which $R \sim N_p^{1/3}$;
(b) stiff rod in which
$L \sim N_p$; (c) coil in which
$R_g \sim N_p^{\alpha}$ with $1/3 < \alpha < 1$.
(D. F. Evans and H. Wen-
nerström, *The Colloidal
Domain: Where Physics,
Chemistry, Biology, and
Technology Meet*, VCH
Publishers, New York,
1994, p. 292. Reprinted
with permission of VCH
Publishers © 1994.)

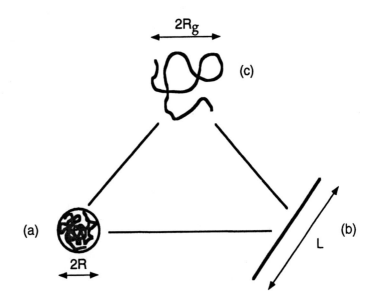

where m_i represents the mass, r_i the position of the i segment, and r_{cM} the location of the center of mass (essentially the moment of inertia divided by the total mass). R_g describes the radius of a thin spherical shell with mass equal to the molecular weight of the polymer such that the shell and actual polymer molecule have the same moment of intertia. The radius of gyration characterizes the particle size without specifying its shape. To relate R_g to the actual dimension of a particle, we must specify its geometry. For example, $R_g^2 = (3/5)R^2$ for a solid sphere of radius R, $R_g^2 = L^2/12$ for a thin rod of length L, and $R_g^2 = (1/2)R^2$ for a disk of radius R.

Because polymer chains adopt many different configurations, we can only characterize their averaged properties. The radius of gyration for a homopolymer containing N_p monomers of equal mass is

$$\overline{R}_g^2 = \left\langle \sum_{i=1}^{N_p} | \vec{r_i} - \vec{r_{cM}} |^2 \right\rangle / N_p \qquad (6.3.2)$$

A second way to characterize a polymer chain is in terms of the *average end-to-end distance*, R_{1N}, from monomer 1 to monomer N, where $N = N_p$

$$\overline{R}_{1N}^2 = \left\langle | \vec{r}_{N_p} - \vec{r_1} |^2 \right\rangle \qquad (6.3.3)$$

Angular brackets denote an average over all allowed coil conformations.

Even though we characterize the averaged chain conformation of a polymer molecule by one length in eqs. 6.3.2 or 6.3.3, we should not picture a typical polymer molecule as a loose, uniform spherical ball. The density of monomer units decreases with distance from the center of the coil. Furthermore, for most configurations, the decrease is more rapid in some directions than in others, so that although the coil is spherical on average, it may exhibit a nonspherical distribution of monomers and a corresponding distinct asymmetry.

6.3.2.2 Macromolecule Size Depends on Solution Interaction; Persistence Length Is a Key Parameter in Characterizing Homopolymers in Solution

How much the chain extends from its center of mass into the solvent depends on the interactions between the monomer units and the solvent. The interaction parameter W between the monomer unit of a polymer molecule and a solvent molecule (discussed in Appendix 6A, eq. 6A.5) provides a convenient way to classify polymer solvent interactions. When W is positive and greater than $\frac{1}{2}kT$, monomer–solvent contact is unfavorable, and the chain contracts to the compact state of Figure 6.8a. When W is less than $\frac{1}{2}kT$ or negative, R_g increases as the polymer seeks to increase its contact with the solvent and for negative values, it stretches to the fully extended state of Figure 6.8b. The difference between $W > \frac{1}{2}kT$ and $W < \frac{1}{2}kT$ distinguishes a "bad" solvent from a "good" solvent for a particular polymer molecule. In a detailed mean-field derivation, Flory showed that in a good solvent both R_g or R_{1N}, scale as $N_p^{0.6}$; that is, $R_g \sim M^{0.6}$ providing N_p, and thus M, the polymer molecular weight, is large.

The situation when W is positive and equal to $\frac{1}{2}kT$ is a particularly important one in helping us to understand polymer conformations in solution. It is called the *theta (θ) condition*. Under theta conditions, the slightly unfavorable monomer–solvent interactions that tend to compact the polymer are exactly compensated by thermally activated monomer–monomer (mutual) size exclusions that tend to expand the polymer. As a consequence, the polymer extends in space in a way that can be described by a random walk process in which each step lays down a link and the direction of the next step or link bears no relation to its neighbor. This leads to the chain conformation depicted in Figure 6.8c. Using random walk statistics to characterize polymer chain conformations, we shall find later in eq. 6.3.8 that for the theta (θ) condition $R_g \propto M^{0.5}$.

The mean square length $\langle L_r^2 \rangle$ of a random walk of N_r steps with step size l_r is

$$\langle L_r^2 \rangle = N_r l_r^2 \qquad (6.3.4)$$

When applying this equation to a polymer, it is tempting to equate each step length l_r with monomer size and the total number of steps N_r with the degree of polymerization N_p. However, this correlation does not properly account for local stiffness due to specific bond angles along the polymer chain.

To accomodate these geometrical constraints, an appropriate step distance, the so-called *persistence length* (or Kuhn length), l_p, is assigned to each monomer in a particular chain so that we can assume random walk properties. Using eq. 6.3.4, the value for the chain's mean square length becomes

$$\langle L_r^2 \rangle = N_p l_p^2$$

This value is the same as the square of the end-to-end distance, so that

$$R_{1N}^2 = N_p l_p^2 \tag{6.3.5}$$

We can relate N_p to the Kuhn length by

$$N_p = L/l_p \tag{6.3.6}$$

where L equals the length of a fully stretched chain.

For a purely random coil, we can also use random walk statistics to relate the end-to-end distance to the radius of gyration and obtain

$$R_{1N}^2 = 6R_g^2 \tag{6.3.7}$$

Substituting eq. 6.3.7 in 6.3.5 gives

$$R_g = l_p \sqrt{\frac{N_p}{6}} \tag{6.3.8}$$

and because R_g and N_p can be measured, l_p is an experimentally accessible number. Equation 6.3.8 shows R_g scales as $N_p^{0.5}$ and therefore $R_g \sim M^{0.5}$ for the theta condition. This analysis leads to the conclusion that $R_g/M^{0.5}$ is constant. Table 6.2 demonstrates that results for a wide range of polymers of different molecular weights are consistent with this conclusion (values of $R_g/M^{0.5}$ for all the dilute θ-solutions in the fourth column are similar).

Persistence lengths vary from several tenths of a nanometer for very flexible polymers, such as ethylene $+CH_2\text{-}CH_2+$, di-

TABLE 6.2 Molecular Dimensions in Bulk Polymer Samples

| | | $\left(R_g^2/M\right)^{1/2} \dfrac{\text{Å mole}^{1/2}}{g^{1/2}}$ | | |
| | | | Light scattering dilue solution | |
Polymer	State of bulk	SANS[a] bulk	θ-Solvent	SAXS[b]
Polystyrene	Glass	0.275	0.275	0.27
Polystyrene	Glass	0.28	0.275	—
Polyethylene	Melt	0.46	0.45	—
Polyethylene	Melt	0.45	0.45	—
Poly(methyl methacrylate)	Glass	0.31	0.30	—
Poly(ethylene oxide)	Melt	0.45	—	—
Poly(vinyl chloride)	Glass	0.30	0.37	—
Polycarbonate	Glass	0.457	—	—

L. H. Sperling, *Introduction to Physical Polymer Science*, 2nd ed., Copyright © 1992, John Wiley & Sons, p. 169. Reprinted by permission of John Wiley & Sons, Inc.

[a]SANS, small-angle neutron scattering.

[b]SAXS, small angle x-ray scattering.

methyl siloxane $+$O$-$Si(CH$_3$)$_2$$+$, and ethylene oxide $+$O$-$C$_2$H$_4$$+$, to over 100 nm for some polyelectrolytes in low salt concentration solutions.

For two different polymers that possess the same chain length L, the effective size of the coil will increase with increasing persistence length l_p. We can see this by substituting eq. 6.3.8 into 6.3.6 and rearranging to give

$$R_g = \sqrt{\frac{Ll_p}{6}} \qquad (6.3.9)$$

Although this result applies strictly only to the theta condition, it is a useful guide for other practical applications.

Comparing the third column with the fourth column in Table 6.2, it is interesting to note that values of R_g for bulk solid polymers in either the amorphous glassy or molten state, the third column, equal those for the corresponding polymers in dilute solution in the theta condition, the fourth column.

6.3.3 The Crossover from Very Dilute to Semidilute to Concentrated Solution Behavior Depends on Chain Conformation

So far, our discussion has dealt with very dilute solutions, where the polymer molecules are so far apart that polymer–polymer interactions are not an issue. As the polymer concentration increases, polymer–polymer interactions become more important, and two distinct concentration regimes (semidilute and concentrated) appear in which polymer–polymer interactions play somewhat different roles. We remember that the sphere of influence of a polymer molecule with a characteristic radius R_g contains mostly solvent and only a small amount of polymer in its random coil. Therefore, we can expect that polymer chains will start to interact even at low polymer concentrations.

For many applications polymer solutions are used in the *semidilute solution* regime, where the solvent remains in excess of the polymer on a weight-by-weight basis, but coils of individual molecules overlap so that a monomer on one chain is likely to make contact with monomers of other chains. Figure 6.7b depicts the semidilute solution. In semidilute solutions, the chains form a disconnected network whose mesh size is equal to or smaller than the radius of gyration. Mesh size decreases as concentration increases because the coils are pushed even more into one another. Eventually they become entangled.

Crossover from the very dilute solution to the semidilute solution regime is described by the characteristic concentration, C*. Clearly, C* does not have a precise value, but we can estimate it by assigning a volume, v_{coil}, to each macromolecule and calculating the total number of coils to fill the volume of the solution. If we assume

$$v_{coil} = \frac{4}{3}\pi R_g^3 \qquad (6.3.10)$$

then the overlap concentration C* is either

$$C^* = \frac{M}{N_{Av}} \frac{3}{4\pi R_g^3} \qquad (6.3.11)$$

where M is the polymer molecular weight when C^* is expressed in units of g/cm^3 or

$$C^* = \frac{N_p 10^{-3}}{N_{Av}} \frac{3}{4\pi R_g^3} \qquad (6.3.12)$$

where N_p is the degree of polymerization (the number of monomers in the polymer molecule) when C^* is expressed in units of molar monomers.

Because R_g varies with M to the power 0.5 in a theta solvent and 0.6 in a good solvent, $C^* \sim N_p^{-0.5}$ for a theta solvent and $C^* \sim N_p^{-0.8}$ for a good solvent. Thus the higher the molecular weight of the polymer, the lower the overlap concentration. Estimates of C^* based on measured R_g values show that the transition to a semidilute solution typically lies in the range 0.1–5.0 wt %.

At higher concentrations of polymer, where solvent and polymer are present in approximately the same amount on a weight basis (50 wt %), we find yet another concentration regime, the *concentrated solution*. The picture of a loose polymer network is no longer relevant in such concentrated solutions. Instead the system consists of highly entangled chains that exhibit properties similar to those of a bulk polymer melt. Figure 6.7c depicts the concentrated solution. We return to the behavior of concentrated solutions later in the chapter, Sections 6.4.5 and 6.6.3, but in the meantime focus on the characteristics of semidilute polymer solutions.

6.3.4 The Regular Solution Model Provides the Basis for the Flory–Huggins Theory of Semidilute Polymer Solutions

A thermodynamic description of semidilute polymer solutions is essential before we can understand their stability and behavior in interfacial applications. The Flory–Huggins theory of semidilute polymer solutions provides such a description. As an aid in understanding the Flory–Huggins theory, we refer first to the basic equations used for the regular solution model and then see how they are modified by the Flory–Huggins model. Regular solution theory assumes that the entropy of mixing is the same as that of an ideal mixture, as defined in eq. 3.7.6. That is

$$\Delta S_{mix} = -k(N_A \ln X_A + N_B \ln X_B) \qquad (6.3.13)$$

where X_A and X_B represent the *mole fractions* of molecules A and B, respectively. The Gibbs free energy of mixing is $\Delta G_{mix} = \Delta E_{mix} - T \Delta S_{mix}$. The analysis presented in Appendix 6A gives the result, expressed in molar units,

$$\Delta G_{mix} = (n_A + n_B) \, wX_A X_B + RT \, (X_A \ln X_A + X_B \ln X_B) \qquad (6.3.14)$$

where n_A and n_B are the number of moles of A and B. This free energy expression forms the basis of *regular solution theory*. It contains only one unknown parameter, w, the effective interaction parameter, already introduced in Section 3.1.1, where $w = z\,N_{Av}[W_{AB} - (W_{AA} + W_{BB})/2]$ is a measure of the relative interaction energy between a solute and a solvent molecule, W_{AB}, compared to the average interaction of solvent and solute molecules with their own kind, W_{AA} and W_{BB}.

We return to semidilute polymer solutions, where we will find two important parameters determine the stability of a solution: the entropy of mixing ΔS_{mix}, and the polymer–solution interaction parameter w.

The seminal feature of the Flory–Huggins theory of semidilute polymer solutions is the recognition that two entropy effects exist when a polymer and solvent are mixed together. As with any binary mixture, simply mixing the two different molecular species increases the entropy. However, this entropy of mixing contribution is small because a semidilute polymer solution contains such a low molar concentration of polymer molecules. A much more important contribution comes from the entropy associated with the *multiple configurations* of the individual polymer molecules when they are in solution. Having small solvent molecules as neighbors allows the molecular chain much more flexibility than it has in the bulk or a concentrated solution, where it is surrounded by other polymer molecules. The contribution to the configurational entropy increases with the space available in the solution and therefore is large in a semidilute solution. Configurational entropy decreases for a concentrated solution or *for any other system that forces polymer molecules to share the same volume defined by their radius of gyration*. (We shall return to this important point many times in the course of this chapter.)

By accounting for the connectedness between monomers in a polymer chain, Flory and Huggins independently succeeded in 1953 in generalizing regular solution theory and adapting it to semidilute polymer solutions. They used a lattice model of the kind shown in Figure 6.9. Assuming a random distribution of monomer units on the lattice sites, they introduced the additional constraint that the monomers must be connected for the polymer chains to remain intact. (Figure 6.9 should be compared with Figure 3.16; the only difference is the connectedness.) Under these conditions, the entropy of mixing adopts the deceptively simple form

$$\Delta S_{mix} = -k\,(N_s \ln \phi_s + N_p \ln \phi_p) \qquad (6.3.15)$$

where N_s and N_p are the number of solvent and polymer molecules, respectively, and ϕ_s and ϕ_p are their corresponding *volume fractions*. In deriving eq. 6.3.15, Flory and Huggins assumed that solvent molecules and monomer units occupy the same volume, so that

$$\phi_s = N_s/(N_s + N_p N_p) \text{ and } \phi_p = N_p N_p / (N_s + N_p N_p) \qquad (6.3.16)$$

Although the form of eq. 6.3.15 is remarkably similar to the ideal entropy of mixing of two liquids described by eqs. 3.7.6 and

Figure 6.9
Placement of polymer monomers (+) and solvent (O) molecules on two-dimensional lattice model according to precepts of Flory–Huggins theory (compare with **Figure 3.16**).

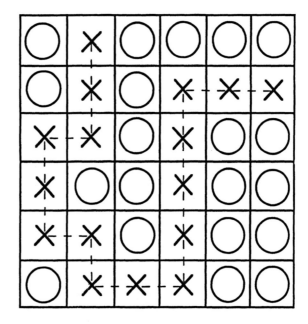

6.3.13, its simplicity masks the fact that the main contribution to ΔS_{mix} for a semidilute polymer solution comes from the change in configurational entropy of the polymer chain.

We can derive the interaction contribution to the free energy of mixing exactly as we do for the regular solution theory in Appendix 6A. The only difference is that the number of nearest neighbors z in the expression for W (and its molar equivalent w = $N_{Av}W$) is reduced by two because the monomer units are covalently linked to two other monomers in the chain.

In molar units, the resulting expression for the free energy of mixing derived with the *Flory–Huggins dilute solution theory* is

$$\Delta G_{mix} = (n_s + n_p N_p)w\phi_s\phi_p + RT(n_s \ln \phi_s + n_p \ln \phi_p) \qquad (6.3.17)$$

where n_s and n_p are the number of moles of solvent and polymer, respectively. Note again the close analogy with eq. 6.3.14 for a regular solution.

If we hold the volume concentration of polymer ϕ_p constant, ΔG_{mix} varies only slightly with the degree of polymerization N_p, because only the small term, $n_p \ln \phi_p$, changes with N_p at constant ϕ_p. This slight variation of ΔG_{mix} with N_p makes it difficult to determine the molecular weight of a polymer in the semidilute regime by measuring a thermodynamic quantity. However, as we will see in Section 6.6, a number of transport properties (viscosity, diffusion) are extremely sensitive to the size of the molecules.

We obtain the chemical potential of the solvent by substituting the values of ϕ_s and ϕ_p from eq. 6.3.16 into 6.3.17 and differentiating with respect to n_s

$$\mu_s = \mu_s^0 + w\phi_p^2 + RT \ln \phi_s + RT[\phi_p - (\phi_p/N_p)] \qquad (6.3.18)$$

where μ_s^0 is the chemical potential of the pure solvent. When the volume fraction of polymer ϕ_p is small, we can write

$$\ln \phi_s = \ln (1 - \phi_p) \approx -\phi_p - \frac{1}{2}\phi_p^2 \qquad (6.3.19)$$

so that the chemical potential, μ_s, equals

$$\mu_s = \mu_s^o - RT(\phi_p/N_p) + \left(w - \frac{1}{2}RT\right)\phi_p^2$$

$$= \mu_s^o - RT\left[(\phi_p/N_p) - \left(\chi - \frac{1}{2}\right)\phi_p^2\right] \qquad (6.3.20)$$

where the second equation introduces the χ parameter, where $\chi = w/RT$. Note the linear term in ϕ_p vanishes when N_p becomes large and that the term quadratic in ϕ_p becomes positive when $\chi \leq \frac{1}{2}$.

The interaction parameter w, or the more commonly used χ parameter, serves to classify the character of a solvent for a particular polymer. If $\chi > \frac{1}{2}$, the polymer is hardly soluble, and as $N_p \to \infty$, it becomes totally insoluble. In this case we have a bad solvent. For $\chi < \frac{1}{2}$, the polymer is soluble over all N_p, and we have a good solvent. The borderline case, $\chi = \frac{1}{2}$, corresponds to the theta condition, a theta solvent. Providing w does not change as the temperature is increased, a polymer solvent goes from being a bad solvent below the theta temperature (= 2w/R) to a good solvent above it.

This theory enables us to understand how the magnitude of the molecular interaction parameter and temperature affect phase stability and lead to phase separation. When the effective interaction parameter w is positive, small, and less than the thermal energy, w < 2RT—as occurs at high temperatures—the mixture forms one stable liquid solution at all compositions. When the effective interaction parameter is positive (repulsive) and large compared to 2RT, w > 2RT, two phases, one polymer rich the other solvent rich, coexist for some of the average compositions, and a miscibility gap occurs in the phase diagram, as shown in Figure 6.10. When w >> 2RT—as occurs at low temperatures—only a small amount of polymer can dissolve in the solvent and vice versa.

6.4 Many Additional Features Affect Macromolecule Conformation in Solution

So far, we have been concerned with the solution properties of homopolymers made up of simple single-unit monomers characterized by a single solvent interaction parameter w. However, this assumption oversimplifies the complexity of many macromolecular structures. Among the complicating factors are the following. In some homopolymers the chemistry of the monomer units is so complex that the chemical affinity for the solvent varies *within the monomer unit;* in some heteropolymers the different monomer units may interact very differently with the solvent; and in polyelectrolytes the monomer unit carries localized charges along

Figure 6.10
Phase diagram calculated for two liquids, A and B, using the regular solution theory; X_A and X_B are mole fractions of A and B, respectively. A similar diagram is derivable from Flory–Huggins theory for semidilute solutions of polymer, p, in solvent, s; ϕ_p and ϕ_s are then volume fractions of polymer and solvent, respectively. A phase quenched below the spinodal line is unstable, and the components segregate spontaneously without requiring a nucleation process. (D. F. Evans and H. Wennerström, *The Colloidal Domain: Where Physics, Chemistry, Biology, and Technology Meet*, VCH Publishers, New York, 1994, p. 434. Reprinted with permission by VCH Publishers © 1994.)

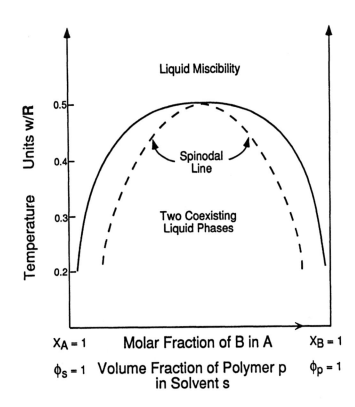

its chain, again leading to localized interactions with the solvent that differ from those with the rest of the monomer. For concentrated solutions of these polymers, the complex monomer units become entangled and interact with each other in novel ways to produce new structures. This section examines the conformation of some of these more complex monomers in dilute and concentrated solutions.

6.4.1 Homopolymers with Amphiphilic Monomer Units Often Form Ordered Helical Structures— The DNA Double Helix

We often encounter homopolymers in which one part of the monomer interacts favorably with the solvent (solvophilic) while the rest of the monomer dislikes the solvent (solvophobic). This situation occurs with an *amphiphilic monomer* unit.

What type of polymer structure minimizes solvophobic interactions yet maximizes solvophilic interactions at the same time? Commonly encountered structures that meet this dual need include single or double helices in which the monomers' solvophobic parts are located inside the helix, while its solvophilic parts are exposed to the solvent. The DNA double helix is the best known example. Helix formation in DNA basically is driven by the favorable interaction between the hydrophobic base pairs that prefer to face each other within the spiral. This attraction is opposed by electrostatic repulsion between charged phosphate groups and by repulsion due to loss of solvation energy between water and the polar hydroxy and nitrogens on the purine and pyrimidine rings along the helix. Repulsion due to loss of solvation energy is partly offset by the formation of hydrogen bonds between the bases in the helix. The net result is a strongly organized helix.

Like micellization, helix formation is a cooperative process, and the change from random coil to helical conformation on cooling takes place over a relatively narrow temperature range. When a polymer changes its conformation from a random coil to a helix, it loses most of its chain-configurational entropy.

In Section 6.3.3.2, we concluded that an increase in configurational entropy plays a significant role in the dissolution of a polymer in a good solvent; consequently, we might expect that helix formation even in dilute solutions would result in precipitation as a polymer solution is cooled. However, other factors come into play. Precipitation does not occur for DNA because the helices are charged and they repel one another. Other polymers, such as gelatin or the polysaccharides, exhibit less effective charge stabilization and a form of precipitation does occur. Rather than forming a compact crystalline solid, however, precipitation of the helices results in gel formation, a structure to be described in Section 6.4.5, and illustrated in Figure 6.11.

6.4.2 Heteropolymers Often Form Folded Structures—Protein Chain Folding

Another way for polymer molecules to satisfy a dual need for hydrophobic and hydrophilic interaction is through chain folding. Many protein molecules demonstrate this behavior. Proteins are heteropolymers assembled from 23 different amino acid monomer units in a specific sequence dictated by DNA. The amino acids are joined by amide linkages similar to those observed on the right-hand side of eq. 6.1.5 in the step polymerization of nylon 4. As Table 6.3 indicates, amino acids contain a variety of side groups, including hydrocarbon chains, sulfides, and positive or negative charged moieties (molecular entities). Consequently, the different units along the protein heteropolymer chain display a wide spectrum of interaction parameters, χ in eq. 6.3.19, with respect to water. Depending on the amino acid sequencing, very different molecular structures result, ranging from long fibrous structures that occur in silk, hair, or tendons to catalysts for specific reactions, such as enzymes.

Usually protein structure is discussed in terms of three levels. The sequence of amino acids along the chains constitutes the primary structure; formation of disulfide bonds between amino acids to link different chains at prescribed points constitutes the secondary structure; and folding of the protein polymer chain to hide the hydrophobic parts and form a globular unit constitutes the tertiary structure. The tendency for a protein chain to fold results from a delicate balance between the hydrophobic

Figure 6.11
Double-stranded helices
form gel networks by two
mechanisms: (a) chemical
crosslinking. Single-strand
sections join one helix to
the next; each molecule
participates in two or more
helices. (b) Physical crosslinking. Individual helices
associate to form irregular
patterns; each molecule is
involved in only one helix.
(D. F. Evans and H. Wen-
nerström, *The Colloidal
Domain: Where Physics,
Chemistry, Biology, and
Technology Meet*, VCH
Publishers, New York,
1994, p. 314. Reprinted
with permission of VCH
Publishers © 1994.)

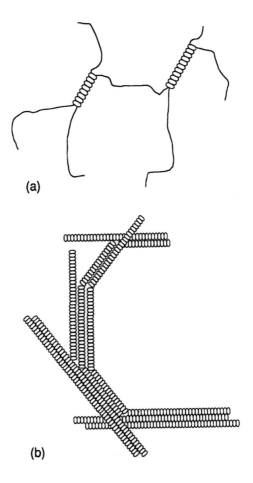

(a)

(b)

and hydrophilic interactions of different monomer units along the
chain and the configurational entropy.

Hydrophobic interactions involving hydrocarbon groups
located along the polymer chain mainly drive protein folding, and
they are opposed by the loss of configurational entropy. The
protein chain strives to adopt a conformation that hides the apolar
side chains from the aqueous solvent while it retains the polar
side chains at the surface of the folded molecule. In the folded
state, some compromises are necessary because some polar
groups, such as the amide linkages, must enter the apolar interior.
In response to an intrinsically unfavorable environment, these
amides tend to link up via hydrogen bonds to form helices or
sheets. The resulting folded conformation of a *globular protein
molecule* in its native state is relatively unique and exhibits little
flexibility for change.

On heating, globular proteins *denature;* that is, they transform from their native folded state into a more or less random coil
with a large number of accessible conformations. Denaturation
occurs upon raising the temperature because of the importance of
the chain conformation entropy, the **TS** contribution, to the free
energy. Usually denaturation occurs over a relatively narrow
(~10 K) temperature range, demonstrating that it is also a cooperative process. Denaturation can also be induced chemi-

TABLE 6.3 Name, Abbreviation, and R Group for Some Common Amino Acids

Name	Abbreviation	R group
Alanine[a]	Ala	$-CH_3$
Arginine	Arg	$-CH_2\,CH_2\,CH_2\,NH\overset{\overset{\displaystyle NH}{\|\|}}{C}NH_2$
Aspartic acid	Asp	$-CH_2\,COOH$
Cysteine	Cys	$-CH_2\,SH$
Glutamic acid	Glu	$-CH_2\,CH_2COOH$
Glycine	Gly	$-H$
Histidine	His	
Isoleucine[a]	Ile	$-CH\,(CH_3)\,CH_2\,CH_3$
Leucine[a]	Leu	$-CH_2\,CH\,(CH_3)_2$
Lysine	Lys	$-CH_2\,CH_2\,CH_2\,CH_2\,NH_2$
Methionine[a]	Met	$-CH_2\,CH_2\,SCH_3$
Phenylalanine[a]	Phe	
Serine	Ser	$-CH_2\,OH$
Threonine	Thr	$-CHOHCH_3$
Tryptophan[a]	Trp	
Tyrosine	Tyr	
Valine	Val	$-CH\,(CH_3)_2$

[a]Nonpolar R groups.

P. C. Hiemenz, *Polymer Chemistry: The Basic Concepts,* Marcel Dekker, New York, 1984, p. 20.

cally by changing the properties of the solvent by, for example, adding urea.

6.4.3 Block Copolymers Can Be Amphiphilic and Show the Same Self-Assembly Properties as Surfactants

In some heteropolymers the different monomers are segregated into blocks of monomer units. A diblock copolymer is made up of one block of monomer units B followed by a second block of monomer units C. The resulting molecular chain has the general formula $B \cdots B - C \cdots C$, or, $B_m - C_n$. In a solvent, the conformation of a single, isolated block copolymer chain depends strongly on the respective solvent quality of the two monomer units. A good solvent ($\chi < 1/2$) for both units will dissolve the block copolymer; in dilute solutions it will result in an open coil configuration similar to that of a homopolymer in a good solvent. A bad solvent ($\chi > 1/2$) for both units will not dissolve the block copolymer molecule. If the solvent is good for one of the monomers, say B, and bad for the other, say C, then the molecule displays an amphiphilic character. In this case, the C part of the single molecule in the chain is in a bad solvent, so the C chain adopts a compact structure to which a random coil B chain is attached (see, for example, Figure 6.12a).

Increasing the block copolymer concentration leads to strong attraction between the compact C parts from different molecules and results in their association. This association process is closely analogous to the self-assembly of amphiphiles to form micelles (Chapter 5). The process exhibits a well-defined critical micellization concentration (CMC), although it occurs at such low concentrations that it is difficult to measure. The self-assembly process starts because of the attraction between the C parts and stops because the random coils from different B parts overlap more and more as the aggregation number increases. The decrease in their configurational entropy results in an effective repulsion. When the repulsion of the B parts balances the attraction between the C parts, aggregation stops and the structure resembles that shown in Figure 6.12b.

The geometry of the resulting aggregate depends on the relative length of the two blocks as well as on the persistence length of the B chain in the solvent. We can make the C parts longer and thus increase the ratio n/m between C and B units, which causes a decrease in the curvature at the interface between the self-assembled aggregate and the solution. As described in Section 5.3.1, the preferred self-assembly structure changes progressively from spherical micelles to ellipsoidal micelles and then to hexagonal, cubic, and lamellar phases. When the C part dominates, we obtain inverted aggregates—hexagonal and spherical.

Figure 6.12
(a) When a single block copolymer molecule, BC, is in a solvent that is good for B but bad for C, the C segment (thick line) adopts a compact configuration, while the B segment (thin line) forms a solvent-expanded coil. (b) When many block copolymer molecules, BC, are in a solvent that is good for B but bad for C, attraction between C segments leads to self-assembly and the formation of micellelike structures. The B chains are crowded by the aggregation and stretch to contact the solvent.
(D. F. Evans and H. Wennerström, *The Colloidal Domain: Where Physics, Chemistry, Biology, and Technology Meet,* VCH Publishers, New York, 1994, p. 304. Reprinted with permission of VCH Publishers © 1994.)

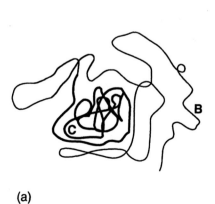

(a)

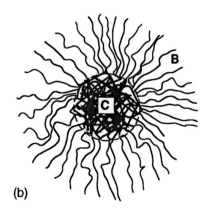

(b)

6.4.4 Polyelectrolyte Chains Have a More Extended Conformation Because They Are Charged

Another feature that determines polymer conformation is the existence of charged species along the polymer chain. Charged polymers, or polyelectrolytes, play important roles in many interfacial applications. Synthetic polyelectrolytes include polyacrylic acid, polystyrene sulfonate, and polyamines. Biological materials include DNA and hyaluronic acid, which is one of many charged polysaccharides. In general, polyelectrolytes are soluble in water and other highly polar solvents but not in apolar solvents.

Both intra- and intermolecular polymer interactions are influenced by charged groups. Intramolecular repulsions between charged groups lead to more extended chain conformation and a *rodlike macromolecule* configuration (as shown in Figure 6.8b). Charged polymers have a larger R_g than we observe with corresponding neutral polymers. In fact, chain persistence lengths (and thus R_g) are determined mainly by electrostatic effects. Odijk derived an approximate relation for the persistence length

$$l_p = l_p^o + l_p^{el} = l_p^o + \frac{e^2}{16\pi\varepsilon_r\varepsilon_o kT\kappa^2\alpha^2} \qquad (6.4.1)$$

Here, l_p^o stands for the persistence length of the chain in the absence of the long-range electrostatic interactions, l_p^{el} represents the electrostatic contribution, and α is the distance between charges (e) along the chain.

As we might expect, the persistence length of a polyelectrolyte and its R_g depend on the presence of salt in solution. Increasing the salt concentration decreases the Debye screening length κ^{-1}, and the persistence length reduces to l_p^o.

Intermolecular interactions also are affected. Owing to their long persistence length, polyelectrolyte coils or rods are very extended and consequently reach the overlap concentration C* at very low concentrations. When the rods overlap, viscosity increases. This property has numerous uses. For example, hyaluronic acid is used in eye surgery primarily for its ability to increase the viscosity of the aqueous medium at very low concentrations.

6.4.5 Polymers Can Form Gels through Chemical Crosslinking and Self-Association

In discussing the nature of semidilute polymer solutions above the overlap concentration C* in Section 6.3.3, we noted that polymer chains become entangled and the mesh size of these entanglements decreases as the concentration increases. However, we did not specifically discuss the nature of the polymer–polymer interactions at the points of contact. For some polymers, and particularly for the complex polymers that are the subject of this section, these interactions may be chemical or physical in nature, resulting in a crosslinked structure similar to that shown for bulk polymers in Figure 6.1. With crosslinking in polymer solutions, polymer dynamics, such as chain diffusion (to be discussed in Section 6.6), are markedly slowed down, while the diffusion coefficients of the liquid solvent and small solute ions or molecules within the solvent are virtually unaffected. Systems that retain the transport properties of the liquid solvent while providing a semirigid polymer framework are called *polymer gels.*

In a *chemically* crosslinked gel, overlapping polymer coils are covalently connected at contact points along the polymer chain. As the degree of crosslinking increases, larger and larger polymer molecules are created. Eventually high-molecular-weight polymers extend throughout the whole macroscopic sample. Once a chemically crosslinked polymer gel is formed, the solvent can be removed by evaporation. An interesting aspect of this process is that the dried polymer retains a memory of its macroscopic shape, which can be recovered when water is added.

Chemically crosslinked polymer gels find many practical uses. Gel electrophoresis, in which the larger solute molecules are moved through a gel by the application of an electric field, is routinely used to analyze protein solutions or to fractionate DNA. One main role of the semirigid gel framework during this process is to prevent liquid convection, which otherwise destroys resolution of the fractionated parts. Another use involves formation of fibrous membranes for purification and separation processes. For example, the active component of a reverse osmosis membrane employs a chemically crosslinked polyelectrolyte gel whose small pore size allows only water to pass through the membrane.

In a *physically* crosslinked gel, overlapping polymer coils are connected by physical association at contact points along the polymer chain. As we might expect, physical interaction is weaker and physically associated gels are more sensitive to changes in ambient conditions. Gelation can be induced by lowering the temperature, increasing the salt concentration in the solvent, or adding specific chemicals that adsorb onto the polymer chains.

A useful aspect of physical gels is that gel formation is almost reversible. Gels formed by cooling can be dissolved by heating. However, the process is not thermodynamically reversible. The slight displacement of heating and cooling curves relative to one another demonstrates that physical gels are not equilibrium structures. In some cases, we can observe aging processes, in which the gel contracts with time and leaves solvent behind. A gel undergoing this aging process is said to undergo *syneresis.*

In Section 6.4.1 we found that helical polymers have intrinsically low solubility because their configurational entropy is low and that precipitation is expected to accompany helix formation. However, polydispersity and chemical monomer heterogeneity preclude the ordered packing of helices; instead they precipitate to form a gel. Figure 6.11 shows two different ways of generating helix polymer gel networks. Polymer strands can cross from one helix to the next, leaving coil segments between them, as illustrated in Figure 6.11a. When this happens, the gel network structures become chemically connected. Helices also may be physically associated as illustrated in Figure 6.11b.

Physically associated polymer gels and helical polymer gels find many practical uses, particularly in the food and pharmaceutical industries. Gelatin, a protein obtained from collagen (a constituent of ligaments), is the best-known gel former. A number of polysaccharides, including agarose (which is widely used in chromatography), pectin (the basic gelling component of natural fruit jelly), and carageenan (a common food additive obtained from seaweed) also form gels. The role of gelatins in photographic emulsions is highlighted in Section 8.4.2.

6.5 Polymer Adsorption at Surfaces Plays an Important Role in Stabilizing Interfacial Systems

6.5.1 Polymers Can Be Attached to a Surface by Spontaneous Adsorption or Grafting

A polymer molecule in solution will adsorb to a surface provided the total adsorption energy is substantially larger than kT. When a polymer adsorbs, it typically makes many contacts with the surface (see Figure 6.13). Thus, even if the energy per individual monomer contact with the surface is relatively small, the cumulative effect of multiple contacts results in a high adsorption energy. If a polymer has any affinity for a surface, therefore, it usually adsorbs strongly and in practice often irreversibly. This feature accounts for the fact that many polymers show excellent surface adhesion, making them suitable for *adhesives* and coatings.

A noncharged homopolymer that adsorbs from a good solvent onto a surface will not change its coil conformation dramat-

Figure 6.13
When a homopolymer adsorbs onto a surface from a good solvent, it attaches to form loops, trains, and tails. The polymer density decreases with distance away from the surface, z. (J. N. Israelachvili, *Intermolecular and Surface Forces,* 2nd ed., Academic Press, London, 1992, p. 291.)

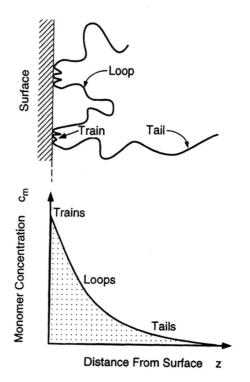

ically. Typically, the thickness of the adsorbed layer, as well as its lateral extension, is of the same magnitude as the radius of gyration R_g. We can describe the adsorbed chain conformation in terms of the *trains, loops,* and *tails* illustrated in Figure 6.13. A particular adsorbed polymer chain configuration depends on: (1) the number of trains or sequences in which the chain is in contact with the surface; (2) the number of loops or sequences in which the chain makes long excursions into the solution between two trains; and (3) the length of the two tails, the two ends of the chain sticking out from the surface.

We can promote polymer adsorption through specific molecular selection and design. Polyelectrolytes adsorb strongly onto oppositely charged surfaces. Amphiphilic block copolymers are very surface-active; the solvophobic block has a strong tendency to adsorb on the surface, while the solvophilic block extends into the solution. In some instances polymer chains can be permanently attached, or *grafted,* to a surface by covalent bonds. Grafting is enhanced by activating the surface chemically with a coupling agent that bonds covalently to the termination group of the polymer chain.

The actual chain configuration at the surface depends on the adsorption forces that bring the polymer down onto the surface. For grafted polymers, if the monomer has no affinity for the surface, the chain will adopt a coil configuration stretching out into the solution away from the grafting points (see Figure 6.14). Neighboring chains repel one another, and this repulsion influences the coil conformation when the grafting density exceeds R_g^{-2}. When many polymer molecules are attached to a surface in this way, the resulting structure is referred to as a

Figure 6.14
When a polymer chain in a good solvent possesses no affinity for the surface, the polymer tries to avoid the surface. Avoidance leads to a maximum in polymer density at some distance z_o in the solution adjacent to the surface. The polymer can be attached to the surface by grafting or by possessing a copolymer block that has a high affinity for the surface. (J. N. Israelachvili, *Intermolecular and Surface Forces,* 2nd ed., Academic Press, London, 1992, p. 291.)

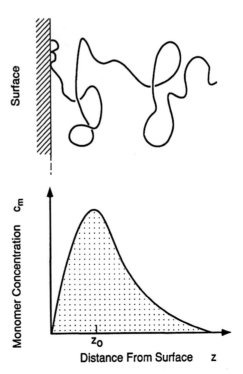

polymer brush. For a dense polymer brush, the monomer concentration, c_m, decays parabolically out from the surface

$$c_m (z) = constant \ (z - z_0)^2 \qquad (6.5.1)$$

where z_0 is the distance from the surface where the monomer density reaches a maximum, as shown in Figure 6.14.

With a block copolymer, B_m–C_n, in which the C part adsorbs strongly on the surface, the conformation of the B coil in the adjacent solution depends on the surface density of the adsorbed C-chain components. If we make a series of block copolymers with increasing m and constant n, chain—chain repulsion of the B coils in the solution becomes important as m increases. The reason for this is that for small m values, the polymer density is determined by the packing of the C chains on the surface; the B coil extends out into the solution on a scale of R_g in both the parallel and the perpendicular direction. When m is much greater than n, the chain conformation becomes similar to that of a grafted polymer, and the distribution function of eq. 6.5.1 for the dense polymer brush will apply. When m is much smaller than n, the average density of C chains adsorbed onto the surface exceeds the value of R_g^{-2} for the B-polymer coils in solution and they overlap; and when m is very much smaller than n, some of the C chain must break away from the surface into the solution to optimize conditions.

Polyelectrolytes adsorb onto oppositely charged surfaces by the attractive action of electrostatic forces. The charged polymer acts as a giant counterion at the surface, and the adsorption of the polyelectrolyte is favored energetically over individual ions because the entropy of the configuration increases greatly when the single polyelectrolyte molecule replaces the small counterions at the surface and simultaneously releases its own counterions. Because adsorption and conformation of the polyelectrolyte in solution are governed mainly by electrostatic effects, we might expect surface conformation to depend strongly on aqueous salt concentration. Pure water encourages strong adsorption, and the polyelectrolyte adopts a conformation that provides an optimal match between polymer and surface charges. Strong electrostatic interaction leads to a chain that essentially lies flat on the surface. The surface density of chains will be close to $\sigma_0 / Z_p e$, where σ_0 represents the surface charge density of the bare surface and $Z_p e$ the total charge of the polyelectrolyte. As the salt concentration increases, the effective strength of the electrostatic interactions decreases. The polymer gains some conformational entropy by extending slightly into the solution, and, to some extent, the ions of the electrolyte act as counterions. The electrostatic binding gradually loosens with increasing salt concentration. Unless other chemical forces influence the adsorption behavior, the polymer eventually ceases to bind to the surface. For this reason rinsing a surface with pure water will never remove polyelectrolyte from the surface, whereas rinsing with a concentrated electrolyte solution enables the polyelectrolyte to come off the surface more readily.

6.5.2 Kinetics Often Determine the Outcome of a Polymer Adsorption Process

If we expose a clean surface to a solution containing an adsorbing polymer, we immediately create a nonequilibrium situation. Adsorption usually is controlled by diffusion of the polymer to the surface through an adjacent, unstirred layer of solution. Because the polymer diffusion coefficients are very small, the adsorbed layer forms over a period of seconds or minutes.

Polymer polydispersity also influences the adsorption process. Because the largest polymers make more surface contacts per molecule, they have the highest thermodynamic affinity for the surface. However, smaller polymers have a larger diffusion coefficient; so they reach the surface first. Thus, in the initial stages, shorter polymer molecules are preferentially concentrated at the surfaces, while longer polymers are favored at equilibrium adsorption.

To reach equilibrium, the larger polymers must replace the smaller ones. This is a slow process because it requires desorption of the shorter chains. Depending on the circumstances, changes may take place over hours or days, or the exchange may be so slow that for all practical purposes the initial state remains stable. Changes made in the solution may have little or no effect on the configuration of adsorbed polymer surfaces.

As a consequence of their slow kinetics, polymer adsorption processes are complex events that are delicate and difficult to study in the laboratory. The exact sequence and timing of steps become important issues. Obviously these slow kinetic effects must be taken into account in practical applications.

6.5.3 Adsorbed Polymers Drastically Change Forces between Surfaces—They Can Lead to Colloidal Stabilization or Flocculation

Surfaces covered with polymers "see" one another through the outer part of the adsorbed polymer coil, as illustrated in Figure 6.15. These coils begin to overlap at surface–surface separations on the order of $2R_g$. In practice, they range from 10 to 1000 nm. These distances are so large that they render the van der Waals attraction and the double-layer repulsion between the bare surfaces negligible, so that polymer–polymer overlap provides the dominant contribution to the force between particles.

In a good solvent, that is, $\chi < \frac{1}{2}$, the polymer coil expands away from the surface to gain configurational entropy. When it encounters a coil emanating from the second surface, reduction of the conformational entropy of the coils creates a corresponding increase in free energy and thus a repulsive force between the surfaces. (Compare this with the undulatory force described in Section 5.3.3.) The dimensions of the polymer coil determine the range of the force, and configurations that extend farthest from the surface, indicated by the arrows in Figure 6.15, are particularly important in determining the onset of the repulsive force. Although this repulsive force is basically entropic in origin, it is commonly referred to as *steric stabilization*.

Figure 6.15
Schematic illustrating
how the configuration of
polymers located on two
surfaces oppose one
another. Where they
attempt to overlap there is
a decrease in configura-
tional entropy producing a
repulsive interaction force.
(D. F. Evans and H. Wen-
nerström, *The Colloidal
Domain: Where Physics,
Chemistry, Biology, and
Technology Meet,* VCH
Publishers, New York,
1994, p. 318. Reprinted
with permission of VCH
Publishers © 1994.)

Surface

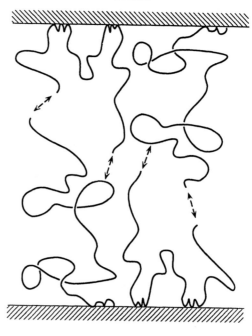

Surface

Figure 6.16 shows an experimentally measured force curve
for mica surfaces covered with polyethyleneoxide (PEO) in water
(which is a good solvent for PEO) at 20°C. The stabilizing force is
always repulsive (positive) and strong enough to be measurable
out to several multiples of R_g. It increases with molecular weight
in the same manner as R_g does.

In a bad solvent, that is, $\chi > \frac{1}{2}$, the polymer concentrates at
the surface to minimize its contact with the solvent. When two
such polymer-covered surfaces come into close proximity, the
chains anchored on one surface can make favorable excursions
toward chains anchored on the second surface without too much
exposure to the solvent. These extensions lead to an increase in
configurational entropy and an *attractive* (negative) force as
shown in the right-hand portions of the curves in Figure 6.17.
However, the attraction rapidly turns into repulsion as the
polymer chains start to overlap and confine each other. The net
result of this interaction leads to a minimum in the force
separation curve at about R_g, as can be seen in Figure 6.17. (A
similar secondary minimum for electrostatic stabilization was
discussed in Section 4.5.) For polymers of similar molecular
weight, repulsion in a good solvent is sensed at a greater
separation (about 100 nm) than attraction in a bad solvent
(about 30 nm), as we can see by comparing the distances in
Figure 6.16 with Figure 6.17.

Table 6.4 demonstrates the crucial role played by solvent
quality in the steric stabilization of colloids. Here the kinetic
stability of suspensions of colloidal particles covered by adsorbed
polymers at different temperatures are compared with the theta

Figure 6.16
Normalized interaction force F between two curved mica surfaces, radius R, with adsorbed polyethylene oxide (PEO) in water at room temperature, which is above the θ temperature. Water is a *good* solvent for PEO. Results shown for two samples of PEO of different molecular weight. A repulsion (positive force) is observed for all separations. The range of the force scales with the radius of gyration, R_g. Note that measurable forces are obtained at distances clearly larger than $2R_g$, showing the influence of tails sticking far out into the solution. [J. Klein and P. Luckham, *Nature* **300**, 429 (1982). Reprinted with permission from *Nature*. Copyright 1982 Macmillan Magazines Limited.]

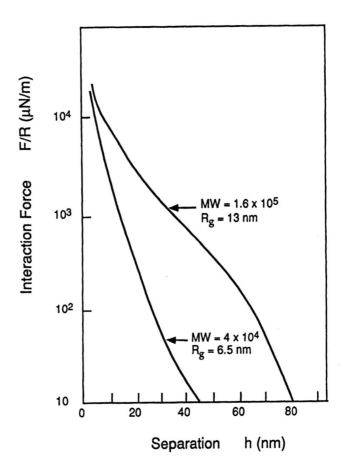

temperature (= $2W$/k) for the polymer–solvent pair. Below the theta temperature (where the polymer–solvent interaction is "bad"), the colloidal particles attract and flocculate; above it (where the polymer–solvent interaction is "good"), the particles are stabilized. As Table 6.4 shows, the critical flocculation temperature (CFT) is almost identical to the theta temperature.

If we create a situation in which the system lacks sufficient polymer to saturate the surfaces, an attraction can occur even in a good solvent. In this case, a tethered polymer coil extending into the solution from one surface can make molecular contact with another uncovered surface and also become absorbed on it. The attractive interaction that results is usually called *bridging*.

A similar attractive mechanism can operate in a nonequilibrium state in which two surfaces have been pushed toward one another and left to equilibrate for some time. In the compressed state, the polymer coil from one side can adsorb on the opposing surface or become entangled in loops of the polymer adsorbed on the opposing surface, like Velcro™. If one tries to separate the surface at this stage, an attractive nonequilibrium force appears. This force is one of many sources of hysteresis effects with polymer-covered surfaces.

Figure 6.17
Normalized interaction force F between two curved mica surfaces, radius R, with adsorbed polystyrene in cyclohexane at 24°C, which is below the θ temperature. Cyclohexane is a *bad* solvent for polystyrene. Results shown for two samples of PS of different molecular weight. A strong attraction (negative force) is observed as the surfaces approach but switches to repulsion as they close. Note that maximum attraction occurs at R_g and attraction is measurable out to ~$2R_g$. (J. N. Israelachvili, *Intermolecular and Surface Forces*, 2nd ed., Academic Press, London, 1992, p. 300.)

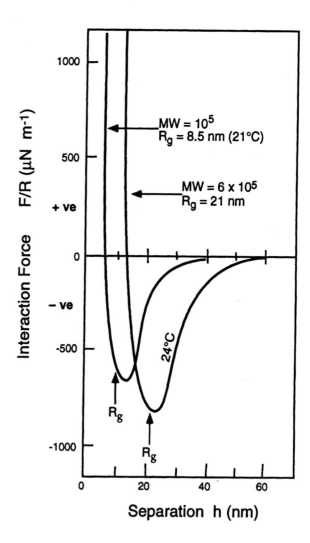

6.5.4 Polyelectrolytes Can Be Used to Flocculate Charge-Stabilized Colloidal Dispersions

Polyelectrolytes dramatically affect the interaction between charged colloidal particles by acting as giant counterions with a potentially large spatial extension. Charged surfaces equilibrated with an excess of polyelectrolyte become neutralized, and the long-range double-layer electrostatic repulsion force disappears. The particles are drawn closer together and flocculate. At shorter separations, the force curve depends on the configurational properties of the polyelectrolyte chain and contains both repulsive and attractive components.

We observe more dramatic flocculation effects when only small amounts of the polyelectrolyte are added to an electrostatically stabilized colloidal system. One polyelectrolyte molecule easily can end up with one segment attached to one particle and the remainder to another, thus providing a bridge tying the two surfaces together. In practice, such interactions are very useful because by adding small amounts of polyelectrolyte, we can produce major changes in the stability of colloidal sols. For

TABLE 6.4 Comparison of Polymer–Solvent Theta Temperatures with Critical Flocculation Temperature (CFT) of Steric-Stabilized Colloidal Dispersions

Stabilizer	Molecular weight	Dispersion medium	CFT (K)	θ temp. (K)
Poly(acrylic acid)	9,800	0.2M HCl	287±2	287±5
	51,900	0.2M HCl	283±2	287±5
	89,700	0.2M HCl	281±2	287±5
Polyacrylamide	18,000	2.1M $(NH_4)_2SO_4$	292±3	—
	60,000	2.1M $(NH_4)_2SO_4$	295±5	—
	180,000	2.1M $(NH_4)_2SO_4$	280±7	—
Poly(vinyl alcohol)	26,000	2M NaCl	320±3	300±3
	57,000	2M NaCl	301±3	300±3
	270,000	2M NaCl	312±3	300±3
Poly(ethylene oxide)	10,000	0.39M $MgSO_4$	318±2	319±3
	96,000	0.39M $MgSO_4$	316±2	315±3
	1,000,000	0.39M $MgSO_4$	317±2	315±3
Polyisobutylene	23,000	2-methylbutane	325±1	325±2
	150,000	2-methylbutane	325±1	325±2

D. H. Napper, *Polymeric Stabilization of Colloidal Dispersions,* Academic Press, London, 1983, pp. 116–17.

example, positively charged polyelectrolytes are used extensively in papermaking to counteract negatively charged clay particles used as fillers and other negatively charged additives. A small amount of polyelectrolyte aids flocculation of cellulose fibers in the papermaking process. Practical experience indicates that the order in which the different components are mixed alters the end result. This fact reinforces the important role nonequilibrium states play in polymer adsorption.

6.5.5 Polymers Can Facilitate the Self-Assembly of Surfactants

When a surfactant is added to a dilute polymer solution, we often observe a cooperative self-assembly process at a concentration less than the CMC of the pure surfactant. A micellelike aggregate forms that incorporates the polymer coil. These polymer-bound micelles form at a *critical association concentration* (CAC) that depends on the nature of both the amphiphile and the polymer. Table 6.5 shows CACs for a series of amphiphile–polymer pairs and compares them with the CMCs of the amphiphile. Typically, the CAC is smaller than the CMC by a factor between 3 and 10 for simple systems and between 10 and 1000 for more complex long-chain polymers.

The structure of polymer-bound micelles is often visualized as a coiled string of beads; the polymer chain provides the string, and the micelles are the beads. Usually micelle–micelle repulsion limits the number of micelles attached to a given polymer chain,

TABLE 6.5 The Critical Association Concentration (CAC) for Some Surfactant Polymer Systems Compared with the Critical Micelle Concentration (CMC) for the Surfactant

Surfactant	Polymer	CAC (mM)	CMC (mM)
$C_{12}H_{25}OSO_3Na$ (SDS)	Polyethylene oxide	5.7	8.3
	Polvinylpyrrolidone	2.5	8.3
$C_{10}H_{21}OSO_3Na$	Polvinylpyrrolidone	10	32
$C_{12}H_{25}N(CH_3)_3Br$ (DTAB)	Sodium hyaluronate	7	16
	Sodium alginate	0.4	16
	Sodium polyacrylate	0.03	16
$C_{14}H_{29}N(CH_3)_3Br$ (TTAB)	Sodium hyaluronate	0.4	3.8
	Sodium alginate	0.03	3.8
	Sodium polyacrylate	0.0025	3.8

D. F. Evans and H. Wennerström, *The Colloidal Domain: Where Physics, Chemistry, Biology, and Technology Meet,* VCH Publishers, New York, 1994, p. 313. Reprinted with permission of VCH Publishers © 1994.

Figure 6.18
Schematic illustration of the variation of the surfactant chemical potential with increasing surfactant concentration in a solution containing a constant amount of polymer. At the critical association concentration, CAC, the surfactant self-assembles onto the polymer. At the critical micelle concentration, CMC_o, obtained in the absence of the polymer, aggregation continues on the polymer. At still higher concentrations, the surfactant chemical potential finally reaches the value that permits free micelles to form. This concentration is denoted as the "CMC" although its value depends on the polymer concentration. (D. F. Evans and H. Wennerström, *The Colloidal Domain: Where Physics, Chemistry, Biology, and Technology Meet,* VCH Publishers, New York, 1994, p. 312. Reprinted with permission of VCH Publishers © 1994.)

particularly when the surfactants are charged. When the coils become saturated with micelles, any additional surfactant enters into the solution as monomer until the CMC is reached. Figure 6.18 shows a schematic plot of the amphiphile chemical potential versus concentration that illustrates this point. The "CMC" of the surfactant now depends on the polymer concentration.

The specific nature of the interaction between polymer and surfactant varies from system to system. With SDS and polyethylene oxide, the fairly polar polymer adsorbs on the micelle surface primarily to reduce hydrocarbon–water contact. Polyelectrolytes electrostatically bind micelles of opposite charge. For a polyelectrolyte, such as polystyrene sulfonate, an added hydrophobic interaction occurs between the styrene moiety and the interior of the micelle.

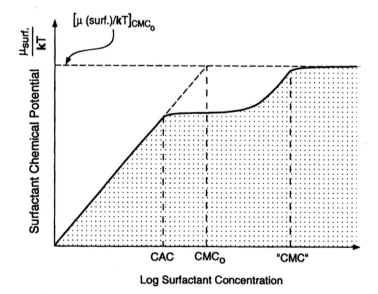

More specific micelle–polymer interactions can be obtained by adding a few long alkali side chains to the polymer main chain. These side chains promote stronger polymer–polymer interactions and create strong binding sites for micelles. Figure 6.19 shows the pronounced changes in viscosity that occur in solutions of such hydrophobically modified polymers. These changes result when the micelles link hydrophobic groups from two polymer chains, creating a gel-like polymer network. As we saw in our discussion of paints in Chapter 4, hydrophobically modified polymers are used to thicken paint and other colloidal dispersions when high viscosity is desirable.

6.6 Transport Properties of Homopolymer Solutions Can Change by Orders of Magnitude with Variation in Molecular Weight, Concentration, and Temperature

In many applications involving polymer solutions and polymer melts, we encounter enormous variation in transport properties. Mass transfer and fluid flow change by orders of magnitude with

Figure 6.19
The viscosity of hydrophobically modified polyacrylate (PAA) solutions (1% by weight in water) as a function of added surfactant sodium dodecylsulfate (SDS). The polyacrylate polymer has either 150 or 500 monomer units and is hydrophobically modified by substituting C_{18} alkyl side chains to levels of 1% and 3% (as distinguished in the specimen identifcation numbers). Note that the viscosity increases in the vicinity of the CMC. For N_p = 500 and 3% substitution (PAA-500-3-C18) the viscosity increases by at least four orders of magnitude, consistent with the formation of a gel-like polymer network. [I. Iliopoulos, T. K. Wang, and R. Audebert, *Langmuir* 7, 617 (1991). Reprinted with permission from *Langmuir.* Copyright 1991 American Chemical Society.]

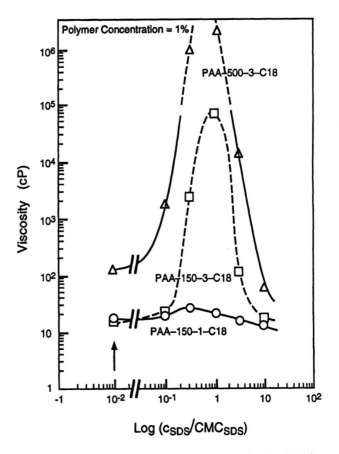

polymer molecular weight and concentration as well as with temperature. In this section, we consider polymer diffusion and rheology in two regimes: dilute solutions, in which we can ignore intermolecular interactions between polymer molecules, and concentrated solutions and melts, in which polymer entanglements dominate transport. For simplicity we restrict the discussion to homopolymer solutions.

Our approach to this topic differs somewhat from the one we used in previous sections. Rather than attempt to develop explicit relationships between variables, we focus on how diffusion coefficients and viscosities scale, that is, how they depend on quantities like molecular weight. Consequently, we write most equations as scaling relationships (~) rather than equalities (=).

6.6.1 In Dilute Polymer Solutions, Viscosity Scales with Molecular Weight to a Power between 0.5 and 1.0

In dilute solution, polymer viscosity increases with molecular weight M in a manner described by the Mark–Houwink equation

$$[\eta] = KM^a \tag{6.6.1}$$

where $[\eta] = (1/c^*)(\eta/\eta_o - 1)$ is the intrinsic viscosity, the relative change in viscosity per unit *weight* fraction of polymer c^*. K and a are characteristic constants for a given polymer–solvent system at a specified temperature. Table 6.6 gives typical examples in which the values for a range from 0.5 to 1.0. We can understand this variation by considering two limiting models for the way a polymer molecule affects fluid flow. Because we are interested in the scaling of η and M, we generally ignore numerical coefficients in this discussion.

TABLE 6.6 Parameters in the Intrinsic Viscosity–Molecular Weight Relationship $[\eta]$ = KM^a for Different Dilute Polymer Solutions

Polymer	Solvent	T (°C)	K × 10³ (ml/g)	a
cis-polybutadiene	Benzene	30	33.7	0.715
it-polypropylene	1-Chloronaphthalene	139	21.5	0.67
Poly(ethyl acrylate)	Acetone	25	51	0.59
Poly(methyl methacrylate)	Acetone	20	5.5	0.73
Poly(vinyl acetate)	Benzene	30	22	0.65
Polystyrene	Butanone	25	39	0.58
Polystyrene	Cyclohexane (θ solvent)	34.5	84.6	0.50
Polytetrahydrofuran	Toluene	28	25.1	0.78
Polytetrahydrofuran	Ethyl acetate hexane (θ solvent)	31.8	206	0.49
Cellulose trinitrate	Acetone	25	6.93	0.91

L. H. Sperling, *Introduction to Physical Polymer Science*, 2nd ed., Copyright © 1992, John Wiley & Sons, p. 104. Reprinted by permission of John Wiley & Sons, Inc.

In the first limiting model, we assume that the entire domain of the polymer coil is unperturbed during flow, so that it behaves like a rigid spherical body of constant radius equal to the radius of gyration R_g. We further assume that the polymer is in a theta solvent, an assumption that assures us that interaction between the polymer molecule and solvent within the excluded volume (the spherical volume defined by the polymer molecule) does not affect the conclusions. With these assumptions, the viscosity of the solution can be described by the Einstein equation 3.8.17, which relates the viscosity of a fluid η containing spherical particles to the viscosity of the solvent η_o

$$\left(\frac{\eta}{\eta_o} - 1\right) \sim \phi_p \qquad (6.6.2)$$

where ϕ_p is the *volume* fraction of polymer in solution, equivalent to the volume fraction of 'spheres' in equation 3.8.17.

Now the volume of each molecule is proportional to $(R_g^2)^{3/2}$ and the number of polymer molecules is proportional to c^*/M, so that the volume fraction is related to the weight fraction by

$$\phi_p \sim (R_g^2)^{3/2} \frac{c^*}{M} \qquad (6.6.3)$$

We know from eq. 6.3.8 that for the theta solvent condition $R_g^2 \sim N_p$, and N_p, the degree of polymerization, is equal to M/M_o. So we can write

$$\phi_p \sim N_p^{3/2} \frac{c^*}{M}$$

which inserted into eq. 6.6.2 gives

$$[\eta] \sim M^{1/2} \qquad (6.6.4)$$

thus providing a direct comparison to the Mark–Houwink equation with $a = 1/2$.

In the second limiting model, we assume that the polymer coil is open to perturbation during flow; more energy is dissipated by hydrodynamic work with a corresponding increase in viscosity. This situation is the basis for the *"free draining" chain model*. We focus on how a velocity gradient in a fluid affects the motion of a nonentangled polymer chain. Similar to Figure 3.23, we set up a coordinate system located at the molecule's center of mass, coincident with the layer whose velocity is labeled v_p in Figure 6.20a. By subtracting v_p from each velocity vector of Figure 6.20a, we obtain the relative motion of the different parts of the polymer molecule indicated in Figure 6.20b and can see that the polymer molecule tumbles with a clockwise rotation. Induced rotation and accompanying deformation of the polymer molecule consume energy intended to produce translational motion, and the additional dissipative loss contributes to an increase in viscosity.

By approximating the free draining polymer chain to be comprised of N_p segments of length l_p, Debye showed that the

(a) **(b)**

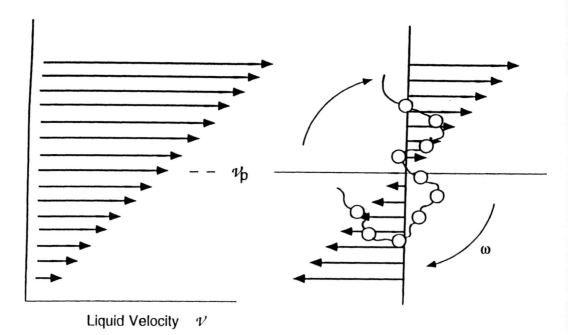

Liquid Velocity $\;\nu$

Figure 6.20
When the velocity gradient of a flowing liquid acts on a nonentangled polymer, it causes the polymer to deform and rotate. This effect can be seen most readily in (b) by setting up a coordinate system located at the molecule's center of mass (where the velocity of the liquid $v_l = v_p$, the velocity of the polymer molecule) to emphasize the relative motion of different parts of the polymer in the liquid. Rotation (ω) contributes a dissipative energy loss and a corresponding increase in viscosity. (P. C. Hiemenz, *Polymer Chemistry: The Basic Concepts,* Marcel Dekker, New York, 1984, p. 107.)

energy of dissipation per unit volume, dU/dV, due to rotation of an individual molecule can be expressed as

$$\frac{dU}{dV} \sim (N_p \zeta R_g^2)\left(\frac{dv}{dx}\right)^2 \tag{6.6.5}$$

In this equation ζ is the friction factor *per segment* and dv/dx is the velocity gradient. Remembering from Section 3.8.1.1 that viscosity is equal to the energy dissipated per unit volume per velocity gradient squared, then $N_p \zeta R_g^2$ represents the additional contribution to the viscosity made by a single rotating polymer molecule.

A solution containing $N_{Av}\, c^*/M$ molecules has an excess viscosity of

$$\eta - \eta_o \sim N_p\, \zeta R_g^2\, \frac{N_{Av}\, c^*}{M} \tag{6.6.6}$$

Using the proportionalities $R_g^2 \sim N_p$, and $N_p \sim M$, we obtain

$$\left(\frac{\eta - \eta_o}{\eta_o}\right) \sim Mc^*$$

or

$$[\eta] \sim M^1 \tag{6.6.7}$$

thus providing a direct comparison to the Mark–Houwink equation with $a = 1$.

Equations 6.6.4 and 6.6.7 show that the range of a values found in the Mark–Houwink equation can be explained by differences in the permeability of the polymer coil to the flow streamlines: when it is impervious, $a = 1/2$, and when it is completely pervious, $a = 1$. Typically, some flow occurs through the "excluded" volume of the molecule, and the value of a lies somewhere between these two extremes. In addition, as we found in Section 6.3.2.1, the radius of a polymer coil increases with molecular weight as $R_g \sim M^\alpha$, where α varies from 1/3 to 1 depending on the polymer–solvent interaction. This also contributes to the dependence of viscosity on molecular weight for polymer coils. Experimentally values for a range from 0.5 for a theta solvent to 0.8 for a good solvent as seen in Table 6.6.

Due to the interaction between the fluid flow field and the polymer, the average molecular weight, M_V, determined in a viscosity experiment varies with the value of a and has the form

$$M_V = \left(\frac{\Sigma_i\, N_i M_i^{1+a}}{\Sigma_i\, N_i M_i} \right)^{1/a} \tag{6.6.8}$$

When $a = 1$, $M_V = \langle M \rangle_w$ (see eq. 6.1.7). In general, $a \neq 1$, and M_V lies between $\langle M \rangle_n$ and $\langle M \rangle_w$.

6.6.2 In Dilute Polymer Solutions, Diffusion Scales Inversely with Molecular Weight to a Power between 0.5 and 1.0

The same two limiting models also provide a basis for understanding the diffusion of polymer molecules in the dilute solution regime. When we assume that the diffusing polymer behaves as a rigid sphere, we can relate the diffusion coefficient to the hydrodynamic radius R_h using the Stokes equation 3.8.36

$$D = kT/6\pi \eta\, R_h \tag{6.6.9}$$

For a rigid sphere model, generally $R_h > R_g$ because solvation increases the size of the diffusing sphere, as discussed in Section 3.8.5.3. Typically $R_h/R_g \approx 1.2 - 1.5$.

For a good solvent $R_h \sim M^{3/5}$, and for a theta solvent $R_h \sim M^{1/2}$. The dependence of D on the polymer's molecular weight for the rigid sphere model ranges therefore from

$$D_{\text{rigid sphere}} \sim M^{-0.6} \text{ to } M^{-0.5} \tag{6.6.10}$$

depending on the polymer–solvent interaction.

With free draining polymer coils diffusion involves hydrodynamic coupling between the internal motion of the polymer chains and the solvent. Consequently, obtaining a simple relation between diffusion coefficient, radius of gyration, and molecular weight is more difficult for the free draining chain model. However, using eq. 3.8.35, $D = kT/f$, and replacing the friction factor f by $N_p\, \zeta$ gives

$$D_{\text{free draining}} \sim M^{-1.0} \tag{6.6.11}$$

The typical diffusion coefficient for a polymer in dilute solution $(10^{-10} - 10^{-11} \text{ m}^2/\text{s})$ is one to two orders of magnitude less than the diffusion coefficient for the pure solvent $(10^{-9} \text{ m}^2/\text{s})$. For example, the diffusion coefficient for polystyrene $(M \approx 6.7 \times 10^5)$ in cyclohexane at 35°C is $1.8 \times 10^{-11} \text{ m}^2/\text{s}$, compared with $10^{-9} \text{ m}^2/\text{s}$ for pure liquid cyclohexane.

It is useful to compare the average displacement of a solute polymer molecule to that of a solvent molecule. Using the Einstein equation ($l^2 = 2Dt$, eq. 3.8.25), we estimate that in the course of 1 s the average displacement of the polymer molecule is 10^4 nm, while the average displacement of the solvent molecule is 3×10^5 nm. As we noted in Section 6.5.2, the slow diffusion coefficients of polymers in dilute solution play a decisive role in many processes, such as adsorption onto surfaces.

6.6.3 The de Gennes Reptation Model Describes Viscosity and Diffusion in Concentrated Polymer Solutions and Polymer Melts

Now we consider transport in concentrated polymer solutions and melts, the regime in which chain entanglements dominate behavior. Scaling factors as a function of molecular weight are higher in this regime than for dilute solutions. Solid sphere and free draining chain models are no longer appropriate, and new insights are necessary to account for the transport behavior.

First, the viscosity is extremely sensitive to molecular weight. Figure 6.21 plots log viscosity versus log degree of polymerization (and thus log molecular weight) for several polymers in their melted state. The curves consist of two straight lines with a change in slope in the vicinity of log $N_p = 2.7 - 3.0$ ($N_p \approx 1000$). Each segment can be expressed as

$$\eta = \text{Constant } N_p^a = \text{constant } M^a \qquad (6.6.12)$$

Below the break, a equals 1 and the viscosity increases linearly with molecular weight as it does for the free draining model. The linear relationship holds because below the break—where N_p is small—the polymer chain is too short for entanglement to take effect and the melt behaves like a dilute solution. Above the break, however, N_p is so large that entanglements do occur, a equals 3.4, and the viscosity increases by about six orders of magnitude when the degree of polymerization is increased 100-fold.

Second, the diffusion coefficients of polymer molecules in highly viscous entangled polymer solutions and melts are also very sensitive to molecular weight and temperature. The numerical values of the diffusion coefficients are small; for polymer melts they are typically 10^{-15}–$10^{-19} \text{ m}^2/\text{s}$ or 4–7 orders of magnitude less than in dilute solutions.

How can we possibly model polymer transport in such entangled systems? The *de Gennes reptation model* provides a successful approach that is easy to visualize. In this model, a single polymer chain is considered to be trapped inside a three-dimensional network formed by other entangled polymers (like reptiles in a box). The network can be reduced to a set of fixed obstacles, $O_1, O_2, \ldots O_n$, as indicated in Figure 6.22a. Because the

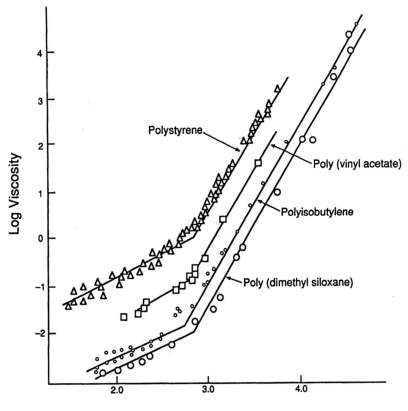

Log Degree of Polymerization Log N_p

Figure 6.21
Plots of log viscosity versus log degree of polymerization, N_p (~molecular weight), for polymers in their melted state show a distinct break in the region of log N_p, 2.7–3.0, where chain entanglement becomes important. The curves are displaced vertically for display purposes. (T. G. Fox and V. R. Allen, *J. of Chemical Physics* **41**, 344, 1964).

polymer chain cannot cross any of the obstacles, they can be used to define a tube containing the polymer P. Figure 6.22b indicates a possible configuration of an actual polymer chain and an averaged chain that defines the total length of the tube, $L_t = N_s l_s$, where l_s is a segment of the tube. l_s scales similar to the persistence length l_p.

Constructing this tube helps us to distinguish between two types of polymer motion: conformational and reptational. Conformational changes occur within the confines of the tube. Conformational changes such as reorientation of a coil produce no net translation of the polymer with respect to the tube. Reptational changes, on the other hand, are snakelike motions that permit the polymer to escape from the tube. Comparing the length of the actual polymer chain with its averaged length shows that it possesses stored distortions or stored lengths, like b in Figure 6.22c, that can move along the chain. When a stored length moves from A toward C as in Figure 6.22d, the part of the chain containing point B translates to the right. Repetition of this motion leads to net movement of the polymer chain and permits it to escape from the tube.

To determine the sensitivity of the reptation transport mechanism to the overall chain length and thus to the molecular weight of the polymer, we first ask what is the characteristic time τ required for the polymer chain to relax or escape by sliding along

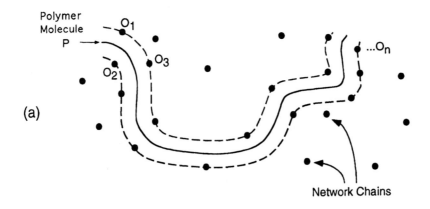

(a)

Polymer Molecule P

O_1

O_2

O_3

...O_n

Network Chains

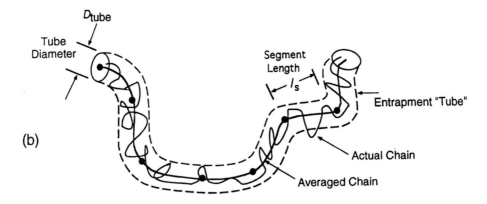

(b)

D_{tube}

Tube Diameter

Segment Length

l_s

Entrapment "Tube"

Actual Chain

Averaged Chain

Figure 6.22
The de Gennes reptation model for polymer molecule transport in concentrated solutions and polymer melts. (a) An entangled coiled polymer chain is trapped inside a three-dimensional network that is defined by a set of fixed obstacles, $O_1, O_2, \ldots, O_n$. (b) These points define a tube whose length can be specified by $L_t = N_s l_s$, and whose diameter equals D_{tube}.

the contour of the tube? Clearly, the polymer molecule must move a distance equal to the tube's length, L_t, so we can use the Einstein equation

$$\tau = \frac{L_t^2}{2D_{tube}} \qquad (6.6.13)$$

to estimate τ, where D_{tube} is the diffusion coefficient for the polymer molecule inside the tube. Using $D = kT/f$ (eq. 3.8.35), we can replace D_{tube} with a friction factor, f_{tube}, for the chain in the tube so that

$$\tau = \frac{f_{tube}L_t^2}{2kT} \qquad (6.6.14)$$

Figure 6.22d suggests that the polymer moves one segment at a time, so we can equate $f_{tube} = N_s \zeta$ remembering that ζ stands for the friction factor per segment) to obtain

$$\tau = \frac{N_s \zeta L_t^2}{2kT} \qquad (6.6.15)$$

Figure 6.22 (*continued*).
(c) and (d) The polymer
moves through the tube
when a stored length b
moves from A toward C
along the coiled chain. As
the stored length configuration moves along the chain,
the chain is displaced to the
right by the amount b. Repetition of this motion leads to
the polymer chain escap-
ing from the tube. (P. G. de
Gennes, *J. of Chemical
Physics* **55**, 572, 1971.)

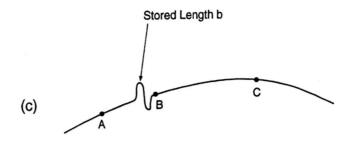

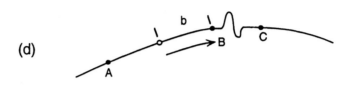

Because we are mainly interested in the scaling properties of the system, we can insert the proportionalities $L_t = N_s l_s \sim N_p l_p$ into eq. 6.6.15 to give

$$\tau \sim \frac{l_s^2 \zeta}{kT} N_s^3 \sim \tau_o N_p^3 \sim M^3 \qquad (6.6.16)$$

which identifies the relaxation time with the cube of the molecular weight. By setting $N_p = 1$, we can identify τ_o with $(l_s^2 \zeta / 2kT)$, the segmental relaxation time.

As we saw in Chapter 3, viscosity involves the movement of one molecule past another to relieve applied shear stress. From a molecular point of view, viscosity directly relates to a fluid's molecular relaxation times. So the de Gennes reptation model predicts that the viscosity of entangled polymer molecules is proportional to the molecular weight cubed

$$\eta \sim M^3 \qquad (6.6.17)$$

This proportionality should be compared with the experimental data in Figure 6.21, which shows $\eta \sim M^{3.4}$, a discrepancy that is a topic of continuing debate.

In discussing translational diffusion, usually we take the polymer's end-to-end distance R_{1N} as a characteristic distance. Substitution into the Einstein equation yields

$$D \sim \frac{R_{1N}^2}{\tau} \qquad (6.6.18)$$

Since $R_{1N}^2 = N_p l_p^2$ from eq. 6.3.5, and $N_p \sim M$, and $\tau \sim M^3$, we obtain

$$D \sim M^{-2} \qquad (6.6.19)$$

which shows that the polymer's diffusion coefficient scales inversely to the molecular weight squared, in accordance with measurements.

In our discussion of transport in polymer solutions, we focused on two limiting situations: the dilute solution, in which we ignore polymer–polymer interactions, and the concentrated solution, in which they are dominated by entanglements. In traversing a polymer concentration range, we move from one limiting situation to the other. As a result, we can expect transport properties to show enormous variation and complexity.

6.6.4 The Mechanical Response of Polymer Systems Is Time Dependent—Polymers Are Viscoelastic Materials

Another aspect of the transport of polymer systems involves their response to an applied mechanical force. In our previous discussions of mechanical response we have dealt with two extremes; either the material is a pure liquid and obeys the characteristics of Newtonian viscosity as described in Section 3.8.1.1, or the material is a pure solid and obeys the characteristics of Hookean elasticity as described in Appendix 11A.1. In reality, concentrated polymer solutions, polymer melts, and bulk polymers simultaneously display characteristics of both a Newtonian viscous fluid and a Hookean elastic solid; they are *viscoelastic* materials (discussed in more detail in Appendix 11A.4). When a viscoelastic polymer system is subjected to a sudden distortion, at first the stress causing the distortion rises to a maximum value dictated by the elastic moduli, but then it relaxes to an equilibrium value. The mechanical response is time dependent. The stress σ drops gradually with time t following an inverse exponential relationship, as illustrated in Figure 11A.3d. The stress and (by implication) the elastic moduli are time dependent, that is

$$\sigma(t) \propto \exp - \left(\frac{t}{\tau_r} \right) \qquad (6.6.20)$$

We can associate this time-dependent behavior with a finite *mechanical relaxation time* τ_r.

Mechanical behavior becomes most revealing when an oscillating strain is applied to a polymer material. The mechanical behavior then depends on the frequency of the oscillations. At very low frequencies the material has plenty of time to respond and assumes relaxed states as the load rises and falls; the elastic modulus measured at low frequencies has a low value. At very high frequencies there is no time for relaxation, and the elastic modulus has a high value. By gradually increasing the frequency we find the modulus goes through a transition from the low to the high value. Coincident with the transition there is a dissipation of mechanical energy, giving rise to a *mechanical loss relaxation peak* of the kind shown in Figure 11A.5.

The relaxation processes that contribute to the viscoelastic behavior of bulk polymers depend on the temperature. We can obtain information about the structure of polymers by probing the frequency dependence of moduli at different temperatures (as discussed in Appendix 11A.4) and by probing the temperature dependence at a given frequency (as shown for example in Figure 6.29). A bulk polymer well below its glass transition temperature is hard and brittle and tends toward Hookean behavior; a bulk polymer well above the glass transition temperature is a viscous liquid and tends toward Newtonian fluid behavior. For temperatures in between the properties change from those of a viscoelastic solid to those of a viscoelastic liquid. Charting the frequency dependence of the moduli through the transition provides a rich source of information concerning the nature of the molecular changes taking place in the polymer system. Developing the models that relate viscoelastic behavior to molecular properties would take us well beyond the scope of this book but can be pursued in reference texts on rheology.

6.7 Processing of Polymeric Systems Illustrates How a Wide Variety of Materials Are Produced

6.7.1 Production of Polyurethane Foams Involves Simultaneous Control of Polymerization and Foam Formation

We enjoy the benefits of polyurethane foams every day in products such as padding in furniture and bedding or insulation in refrigerators and freezers. Other uses for foam products include molded parts on automobiles, flame-retarding insulation layers in composite construction materials, and components in artificial limbs.

Apart from their versatility, polyurethane foams find widespread use because they are efficient to manufacture. Chemical efficiency arises because the polymerization reactions consume all raw materials and leave no byproducts. In addition, selective catalysts promote high production rates and provide control over the physical properties of the final product. Economic efficiency occurs because processing is carried out at low pressures. Consequently, capital costs are relatively low, and improving existing processes or developing new products does not require extensive capital reinvestment. Versatility arises because additives, such as fillers, plasticizers, dyes, and so forth, produce products with a wide range of mechanical properties, surface textures, and colors.

Foams are organized into three major categories: high-density flexible foams, low-density flexible foams, and low-density rigid foams. All are produced by the same strategy. An exothermic reaction between polyisocyanates and polyols generates the polymer; the heat liberated during the reaction initiates the formation of gas bubbles; the polymerization process stabilizes foam growth. To control the shape, the reacting mixture is either injected into a mold or formed as a slab on a conveyer belt and cut to size after curing. To demonstrate the essential interfacial features in this industry, we focus on the production of low-density rigid polyurethane foams.

6.7.1.1 A Sequence of Processing Steps Describe the Formation of Rigid Polyurethane Foams

6.7.1.1A Selection of the Reactants. What reactants are involved in the manufacture of polyurethane foam? The key chemical reaction is between an isocyanate group, R-N=C=O, and an alcohol, R'-OH, to form a urethane RNHCOOR'. Polymerization occurs when bifunctional isocyanates (diisocyanates) and alcohols combine to form polyurethane in what is called the *gelling reaction.*

$$n \, O=C=N-R-N=C=O \; + \; n \, HO-R'-OH \;\rightarrow\; \left[O=C=N \left\{ RN \underset{H}{\overset{O}{\overset{\|}{|}}} C \, OR' \right\} OH \right]_n \qquad (6.7.1)$$

Further reactions take place between urethane and isocyanate to form allophanates,

$$RNHCOR' + R''-N=C=O \;\rightarrow\; RNH\overset{O}{\overset{\|}{C}}-\underset{R''}{N}-\overset{O}{\overset{\|}{C}}R' \qquad (6.7.2)$$

and between urea and isocyanate to form biurets

$$R'NHCNHR' + R''-N=C=O \;\rightarrow\; R'NH\overset{O}{\overset{\|}{C}}N\underset{R''}{\overset{O}{\overset{\|}{C}}}NHR' \qquad (6.7.3)$$

which can lead to crosslinking. The degree of crosslinking can be increased by adding polyamines or polyols, such as $RC(CH_2OH)_3$, with trifunctional groups. In addition, isocyanates may self-condense to form dimers and trimers.

These competing reactions can be controlled by introducing catalysts, such as metal carboxylates that promote the gelling reaction or quaternary amines that catalyze the trimerization of isocyanates. Small (parts per million) additions of inhibitors, such as HCl or benzoylchloride, impede proton transfer to isocyanate groups.

A distinguishing feature of polyurethane foam processes is the inclusion of "blowing" agents that generate vapor bubbles during the polymerization process. Chemical reactions involving monomers generate gases. For example, the reaction of isocyanates with water forms carbon dioxide.

$$2R-N=C=O + H_2O \rightarrow RNHCONHR + CO_2 \qquad (6.7.4.)$$

To promote this reaction, microdroplets of water can be mixed into the reaction mixture. For most applications, however, CO_2 generation does not provide sufficient foaming. Instead almost every case uses halogenated alkanes, such as Cl_3CF or Cl_2CF_2, to stimulate foaming. Concern about the environmental impact of halogenated alkanes has led to a search for alternatives, and organic foaming agents, such as methylene chloride, nitroalkanes, and amides, are finding increasingly widespread use.

Heat from the exothermic polymerization reaction initiates foam production. This process involves: (1) nucleation of the vapor phases (often aided by injection of microbubbles of air into the reaction mixture); and (2) generation of vapor–liquid interfaces as the foam grows. As we saw in Chapter 3, adding surfactants can reduce the energy required to generate new interfaces. In the production of low-density foams, PDMS-PE (polydimethylsiloxane-polyether graft copolymer) is added as a surfactant. Besides reducing the interfacial tension in the growing foam, surfactants serve as emulsifiers and stabilizers in the initial polyol–isocyanate mixture. Adding catalysts and surfactants helps to give higher cell count by controlling the rate of nucleation bubble loss during the induction period. Other additives also influence foaming. Combinations of stabilizers and destabilizers can be added to tweak the process. Because additives used to control the density and integrity of the final product often are discovered by trial and error, they constitute highly prized trade secrets.

Depending on the application, other chemicals, such as flame retardants, fillers, antiaging agents, coloring materials, adhesive agents, and plasticizers, may be added to the reaction mixture during synthesis.

6.7.1.1B Mixing the Reactants.
Manufacture of rigid polyurethane foams may be separated into two major process steps. As indicated in Figure 6.23, the first is the storage, conditioning, and mixing of the reactants, and the second is shaping the foam into the desired final rigid form. We will discuss these two steps in turn.

Figure 6.23
Schematic illustrating the major processing steps involved in the manufacture of polyurethane foams. (H. Boden, K. Schulte, and H. Wirtz, in *Polyurethane Handbook,* 2nd ed., G. Oertel, Ed., Carl Hanser Verlag, Munich, 1994, p. 131.)

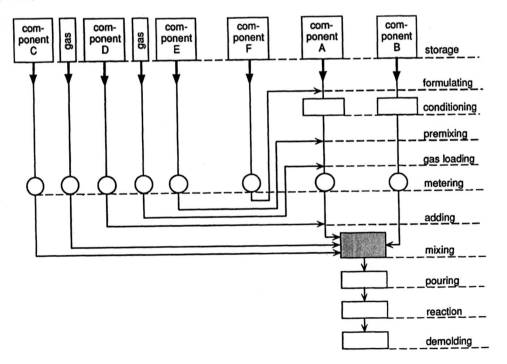

The first step encompasses the sequence from storage to mixing. All the reactants are conditioned, combined, and metered into a mixhead before being injected into a mold or extruded onto a conveyor belt. Careful control of temperature is critical during this process because the viscosity, density, and chemical reactivity of polyols and isocyanates varies with temperature. Chemical components must be maintained within ±2°C to obtain reproducible products. Conditioning is achieved by placing reactants in well-stirred storage tanks in temperature-controlled rooms and delivering them to mixing and metering machines when thermal equilibrium has been achieved.

The metering and mixing machines perform two key functions: (1) they ensure that proper ratios of reactants are maintained throughout the mixing process; and (2) they achieve homogeneous mixing of materials with as short a residence time as possible. A major task is to design mixers that accommodate the changing viscosity, temperature, and sticking coefficients (polyurethanes are very good adhesives) that accompany initiation of polymerization in the mixing chamber.

6.7.1.1C Forming Polyurethane Foam—Production of Construction Slabstock and Refrigerator Insulation. Shaping, curing, and trimming the reacting foam constitutes the second process step. The mixed and nucleated reaction mixture is deposited onto a continuous belt or into a mold. Deposition generally takes place less than 1 s after the start of the mixing process. An induction period occurs before the mixture becomes opaque due to the nucleation and growth of visible bubbles—the "cream" time.

In slab production, illustrated in Figure 6.24, the reaction mixture is poured onto a lower belt by a mixhead that oscillates back and forth. The foaming mixture rises to the surface of an upper belt that controls the thickness of the slab. Then the reaction mixture reacts, cures, and cools. The process time varies according to the thickness, but generally requires about 3–5 m

Figure 6.24
Schematic diagram of a polyurethane foam slab laminator. a, storage; b, metering; c, mixing head; d, upper and lower belts; e, coiled substrates; f, cutter. (H. Boden, K. Schulte, and H. Wirtz, in *Polyurethane Handbook*, 2nd ed., G. Oertel, Ed., Carl Hanser Verlag, Munich, 1994, p. 137.)

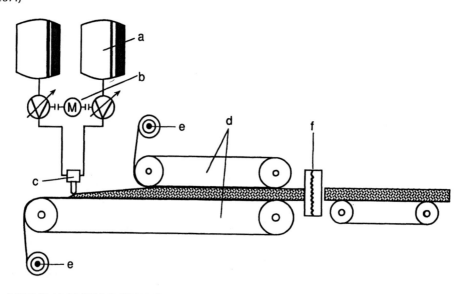

from mixing to curing. Typical volume rates are about 5 m^3/min. Then the stabilized polyurethane foam is sliced for packaging.

In most slabstock applications, one or both of the belts is surfaced with paper or some other material to which the foam adheres. For example, in creating roofing material, the foam forms on the rigid surface that faces outside when it is in use. Slabstock production is used worldwide to produce millions of tons of construction materials, such as insulation, roofing, and blocks, each year.

In mold production, the molds may be stationary or placed on conveyer belts. Often they are the most expensive component used in the production line and must be designed to minimize turnaround time. Their complicated shapes usually are made with cast aluminum, although plywood panels are also used. Because polyurethanes are excellent adhesives, molds are usually coated with a release agent before being used. Accumulation of release agents requires cleaning of the mold after repeated use so that part definition and surface appearance are not impaired. Because temperature is a significant variable, the molds are pre-heated to 40–45°C before injection of the reactants. Careful metering and control of foaming conditions prevents the foam from overflowing the mold vents. After demolding, the foam cores are sent to curing ovens.

Successful production of polyurethane foams requires careful balancing of the polymerization and foaming processes. Usually, process variables are defined in terms of the time lapse from the beginning of the mixing process until a visual change or rise in the viscosity occurs (cream time), the time from the beginning of the mixing to the end of the rise of the foam (rise time), the time after which the product is not longer flowable (setting time), and the mold release time or final curing time.

6.7.2 High-Impact Polystyrene (HIPS) Is a Microphase Polymer Composite Consisting of Rubbery Particles Encased in a Brittle Thermoplastic

In 1985, three million tons of high impact polystyrene (HIPS) were used for packaging, appliance casings, and toys. HIPS is useful in these applications for its strength, flexibility, and good impact strength.

Polystyrene (PS) by itself is a brittle thermoplastic that catastrophically fractures under impact. When a second rubber phase, polybutadiene (PBD), is added to the polymer, it becomes stronger, more flexible, and more fracture resistant than the single-component system. The exact properties of this composite micro-phase system depend on the manufacturing method.

Two methods have industrial importance in producing HIPS: discontinuous and continuous polymerization. In discontinuous polymerization, styrene monomers bind onto preformed rubber particles in an emulsion. The advantage of the discontinuous, or batch, production method is that it permits greater control over the properties of the rubber phase during processing. In continuous polymerization, styrene monomer is polymerized in the presence of dissolved polybutadiene; so the rubber is incorporated into the final crosslinked product. The advantage of the more widely used continuous production method is that the rate of output is much higher.

6.7.2.1 Continuous Processing of HIPS Illustrates How Controlling Phase Separation and Inversion Determines the Final Properties of HIPS

In continuous processing, formation of HIPS evolves as materials are channeled through three tower reactors (linear flow reactors) connected in series, as illustrated in Figure 6.25. Polymerization is initiated, and the rubber particle size is fixed in reactor 1, polystyrene conversion is continued to the desired level in reactor 2, crosslinking of the rubber particles—followed by removal of solvent and unreacted styrene—occurs in reactor 3. The paragraphs to follow examine each process in some detail.

We feed into reactor 1 a mixture containing styrene monomer (65%) and polybutadiene (3–10 wt %) dissolved in ethylbenzene (20 wt %) that serves to reduce the viscosity. As a specific example, we consider a mixture containing an 8 wt % ratio of polybutadiene to styrene; point A in the ternary phase diagram (see Figure 5.15a) shown in Figure 6.26.

Adding a free radical initiator, such as benzoyl peroxide, to the mixture leads to the formation of polystyrene through the reactions of eqs. 6.1.1 and 6.1.2. It also leads to the occasional formation of polybutadiene–polystyrene grafts, as in the reactions of eqs. 6.7.5 and 6.7.6. Grafting is initiated by hydrogen abstraction from some point along the polybutadiene chain. Benzoyl peroxide or styrene free radicals (R″•) react with the polybutadiene to give

$$R\text{-}CH_2CH=CHCH_2\text{-}R' + R'' \bullet \rightarrow R\overset{\bullet}{C}HCH=CHCH_2R' + R''H \quad (6.7.5)$$

that then reacts with styrene

Figure 6.25
Three sequential tower reactors are used in the synthesis and processing of high-impact polystyrene (HIPS). Polymerization is initiated in reactor 1, and a polystyrene–polybutadiene (O/O) emulsion forms. At 35% conversion the emulsion inverts to a polybutadiene–polystyrene (O/O) emulsion. Careful control of agitation and temperature in reactors 2 and 3 result in the desired distribution of spherical rubber droplets in a polystyrene matrix. (A. E. Platt, in *Encyclopedia of Polymer Science and Technology,* Vol. 13, N. M. Bikales, Ed., Copyright © 1970, John Wiley & Sons, p. 199. Reprinted by permission of John Wiley & Sons, Inc.)

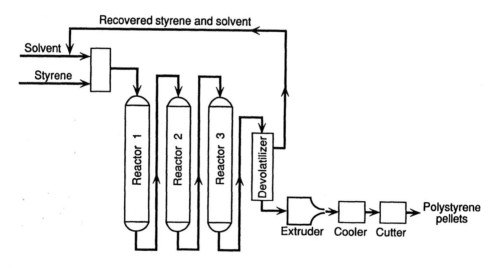

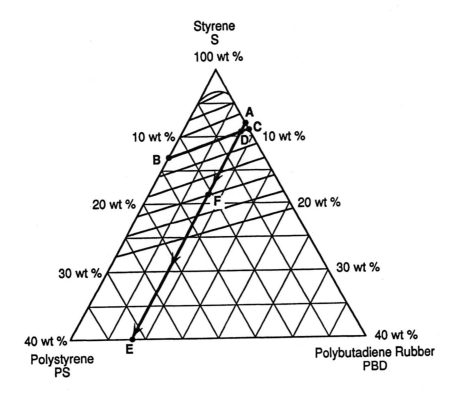

Figure 6.26
Ternary phase diagram for
the system styrene–poly-
styrene–polybutadiene
rubber. (M. Grayson, Ed.,
*Kirk–Othmer Concise
Encyclopedia of Chemical
Technology,* John Wiley &
Sons, New York, 1985,
p. 1116. Reprinted by per-
mission of John Wiley &
Sons, Inc.)

$$RCHCH=CHCH_2R' + n\,CH_2=CH \;\rightarrow\; RCHCH=CHCH_2R'$$

to form either *grafted copolymers,* poly(butadiene-g-styrene), when
the free radical attacks some point along the backbone of the polybutadiene chain, or *block copolymers,* poly(butadiene-b-styrene),
when the free radical attacks the end of the polybutadiene chain.

Since the amount of polybutadiene is constant, the composition of the polymerizing mixture follows along the tie-line A →
E in the ternary phase diagram. As styrene polymerization proceeds, the volume fraction of polystyrene increases until a point
is reached where the two polymers (polystyrene and polybutadiene) are no longer mutually soluble and phase separation occurs
(point D in the phase diagram, the relative volume fractions of
PBD and PS are given by the ratio BD/DC). In an unstirred
solution, two bulk phases would form; however, if rapid stirring
is introduced to avoid phase separation, an oil-in-oil, O/O, emulsion forms instead (see Section 5.4.3).

At first, the O/O emulsion consists of polystyrene droplets
in a continuous styrene–polybutadiene oil phase. As the polystyrene chains continue to grow in reactor 1, the volume fraction of
polystyrene droplets increases until the volume fractions of polybudatiene and polystyrene become comparable (point F on the
phase diagram). Then the emulsion undergoes inversion and

converts to polybutadiene droplets, stabilized by the amphiphilic poly(styrene-b-butadiene) block copolymers, dispersed in a polystyrene–styrene continuum. In the inversion process some small spheres of polystyrene become permanently entrapped in the larger polybudatiene emulsion droplets, and so at point F we actually have an oil (polystyrene) in oil (polybutadiene) in oil (polystyrene–styrene) emulsion.

As more sytrene is converted to polystyrene, the viscosity of the mixture increases, until at 30–35% conversion (close to E in the ternary diagram), the solution becomes so viscous that the polybutadiene droplets can no longer be broken up by agitation. The size of the polybutadiene droplets with their entrapped polystyrene spheres is fixed and remains unchanged during the remaining polymerization process. Because agitation in reactor 1 during the first 35% polymerization determines the size of the polybutadiene emulsion droplets after the inversion process, it plays a crucial role in determining the physical properties of the final product. Discussion below indicates how the toughness of HIPS directly relates to the size of the polybutadiene rubber particles.

Reaction temperature also plays an important role in determining the properties of the final product. The polymerization reactions are highly exothermic, generating 700 kJ/kg. Heat exchangers are used in reactors 1 and 2 to remove heat and maintain the temperature between 100 and 180°C. Processing temperatures at the low end (100°C) reduce free radical transfer between chains and thus increase the degree of polymerization and the molecular weight of the polystyrene. High-molecular-weight polystyrene macromolecules become oriented during injection molding, leading to products with anisotropic properties. For some applications preferred orientation is desirable or acceptable in the final product. The molecular weight of the polystyrene also can be controlled by adding chain transfer promoters like ter-dodecylmercaptan into reactor 1.

In reactor 2, the solution is gently agitated to ensure even temperature distribution as the polymerization continues from 35% to 80–85% conversion. Just before reactor 3, the temperature is increased to 200°C. The higher temperature promotes crosslinking of the polybutadiene chains and thereby stimulates the formation of robust polybutadiene particles possessing elastomeric properties that can withstand the high shear rates encountered later, during injection molding or other fabrication processes.

Also in reactor 3 the ethylbenzene solvent and unreacted styrene are removed by vacuum distillation and recycled back to the feed. After the molten polybutadiene–polystyrene emulsion leaves the vacuum chamber, it is cooled and chopped into HIPS pellets for shipping. HIPS is a thermoplastic that can be remelted and injected into a mold to make the final product.

6.7.2.2 The Size and Separation of the Rubber Particles Plays a Critical Role in Determining the Impact Strength of HIPS

The impact resistance of HIPS results from dispersion of discrete polybutadiene rubber particles in a rigid polystyrene matrix. For optimum performance the polybutadiene particle diameters must be in the range 2–4 μm for most applications. Many appliance and packaging applications require glossy surfaces, and larger particles (5–10 μm) create pits and bumps on the surface of the finished product that scatter light and produce a matte finish.

Particle diameter is fixed by the shear rate during agitation in the reactor 1. The drop break diameter, D, of an emulsion droplet is given by

$$D = \left(\frac{k'\gamma}{\tau_{xy}}\right)\left(\frac{\eta_d}{\eta_c}\right)^{\beta-1} \quad (6.7.7)$$

where γ is the interfacial tension between the emulsion droplet and the continuous phase, τ_{xy} is the shear stress, η_d and η_c are the viscosities of the dispersed and continuous phases, and β and k′ are constants.

Figure 6.27 reproduces optical micrographs illustrating the relationship between agitation and final microstructure. With agitation the microstructure consists of fine polybutadiene spheres distributed uniformly in a polystyrene matrix, as shown in the lower right-hand picture. In the absence of agitation ($\tau_{xy} = 0$), polystyrene and polybutadiene form bicontinuous structures shown in the lower left-hand picture that result in a material with no impact resistance at all.

As well as particle size, interparticle spacing is important in determining the optimum material toughness. The critical interparticle distance, z_c, is given by

$$z_c = D_r[(\pi/6\phi_r)^{0.33} - 1] \quad (6.7.8)$$

where D_r is the rubber particle diameter and ϕ_r is the rubber volume fraction. Thus the optimum particle diameter that exists for a fixed volume fraction re-emphasizes the need for tight process control during the emulsification and inversion stages.

Figure 6.27
Phase-contrast photomicrographs of high-impact polystyrene (HIPS) showing how agitation during polymerization and emulsification affects the particle formation via phase inversion. Required microstructure for high impact resistance consists of spherical polybutadiene particles in polystyrene matrix, lower right. (A. E. Platt, in *Encyclopedia of Polymer Science and Technology,* Vol. 13, N. M. Bikales, Ed., Copyright © 1970, John Wiley & Sons, p. 194. Reprinted by permission of John Wiley & Sons, Inc.)

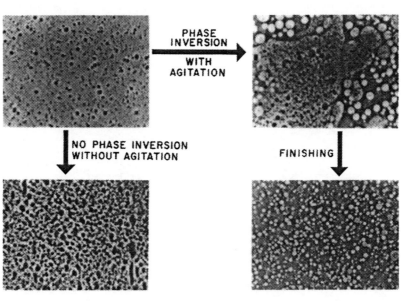

PHASE INVERSION WITH AGITATION

NO PHASE INVERSION WITHOUT AGITATION

FINISHING

6.7.3 Thermoplastic Elastomers (TPEs) Are Block Copolymers That Are Thermoplastic at High Temperatures but Form Rubbery and Rigid Microphases on Cooling

TPEs were originally devised to replace natural rubbers in applications where flexibility and elasticity rather than stiffness were the major requirements. Thermoplastic elastomers (TPEs) are synthetic materials that contain polymer networks analogous to those observed in natural rubber (Section 6.2.3). TPEs are synthesized by a variety of standard polymerization reactions. A major advantage of thermoplastic elastomers is that they do not require vulcanization or chemical crosslinking for strength and durability, and as a consequence they may be remelted and the scraps recycled.

The most prevalent form of TPEs are multiblock copolymers containing "hard" and "soft" segments (Figure 6.28a). Above their melting points, TPEs behave as thermoplastics, and thus can be processed by injection, extrusion, or blow molding. Upon cooling, the hard segments aggregate to form crystalline regions that provide junction points tying the soft segments together to form a network as shown in Figure 6.28b (rather like Figure 6.1). The soft segments are amorphous and well above their glass transition temperature, and they contribute rubbery elasticity to the system.

Figure 6.28
Microstructure of an ideal thermoplastic elastomer (TPE). (a) Schematic of a multiblock copolymer containing "soft" and "hard" segments. Above its melting point, it behaves as a thermoplastic. (b) Upon cooling, the hard segments associate to form crystalline regions, while the soft parts remain in an amorphous state. The crystalline regions tie the soft segments together to produce a rubberlike elastomer. (C. S. Schollenberger and K. Dinbergs, *J. of Elastomers and Plastics* **7**, 65 1975.)

(a)

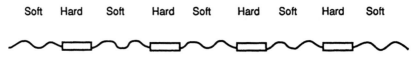

Soft   Hard   Soft   Hard   Soft   Hard   Soft   Hard   Soft

Multi-block Copolymer Primary Chains

Cooling

(b)

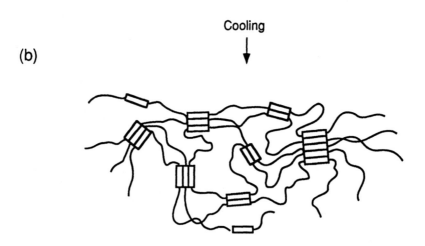

Crosslinked/Extended Network of Polymer Primary Chains

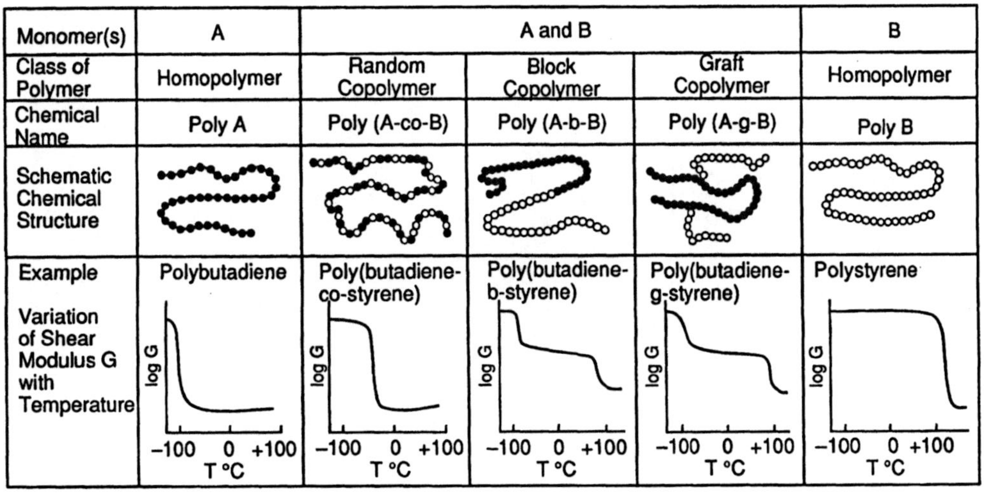

Figure 6.29
Change in shear modulus (G) with temperature for different blends in the polystyrene–polybutadiene system; homopolymer polybutadiene displays single transition at low T_g; homopolymer polystyrene displays single transition at high T_g; random copolymer displays single transition at intermediate T_g; block copolymer and graft copolymer show two T_gs that correspond to the glass transitions of each of the constituents. The block copolymer structure is ideal for a thermoplastic elastomer (TPE).
(N. G. McCrum, C. P. Buckley, and C. B. Bucknall, *Principles of Polymer Engineering,* Oxford University Press, Oxford, 1988, p. 161. By permission of Oxford University Press.)

The elastomeric properties of the TPE material are determined by the relative size and distribution of the hard and soft microdomains. When the volume fraction of the soft material is low, rubbery domains form in a glassy solid. Then we have hard and high-impact-strength solids like HIPS discussed in Section 6.7.2. When the volume fraction of the soft material is high, glassy domains form in a rubbery solid, and we have elastomeric properties. Depending on the processing conditions, the monomers blend to form either a grafted copolymer, a block copolymer, or a random copolymer (the same as a heteropolymer). Figure 6.29 compares the change in shear modulus with temperature for the different blends in the polystyrene–polybutadiene system. At the extremes, 100% polystyrene and 100% polybutadiene homopolymers show high and low T_g, respectively. The random copolymer shows an intermediate value of T_g, which is not desirable for a TPE. Block and graft copolymers show two transition temperatures consistent with the idea of two-phase polystyrene and polybutadiene segments aggregating into separate domains. Block copolymer structures are ideal for a thermoplastic elastomer.

Major types of thermoplastic elastomers are styrenic [for example, styrene–butadiene–styrene (SBS), styrene–isoprene–styrene (SIS)], polyether–ester block copolymers, thermoplastic polyurethanes (TPU), polyether–amide block copolymers, and thermoplastic polyolefin rubbers (TPO).

In large part, the choice of material for an application depends on the conditions the product will experience during its lifetime. For example, butadiene can be copolymerized with acrylonitrile to form a nitrile—butyl rubber (NBR). The advantage of this material over a homopolymer is that it can be synthesized with a range of co-monomer (acrylonitrile) content that allows NBR to be tailored for specific applications. When sufficient acrylonitrile is present in the block copolymer, NBR is highly resistant to hydrocarbons (oils). Increasing the content of one

Monomer(s)	A	A and B			B
Class of Polymer	Homopolymer	Random Copolymer	Block Copolymer	Graft Copolymer	Homopolymer
Chemical Name	Poly A	Poly (A-co-B)	Poly (A-b-B)	Poly (A-g-B)	Poly B
Schematic Chemical Structure					
Example	Polybutadiene	Poly(butadiene-co-styrene)	Poly(butadiene-b-styrene)	Poly(butadiene-g-styrene)	Polystyrene
Variation of Shear Modulus G with Temperature	log G −100 0 +100 T °C	log G −100 0 +100 T °C	log G −100 0 +100 T °C	log G −100 0 +100 T °C	log G −100 0 +100 T °C

polymer component may enhance one property but degrade another. In NBR, increasing the acrylonitrile content beyond 40% results in improved hydrocarbon resistance but a significant increase in the glass transition temperature T_g of the copolymer. Because the rubbery nature of the material is tied directly to the T_g, raising the T_g of NBR decreases its toughness.

6.7.3.1 Fabrication of Athletic Footwear and Modification of Asphalt Illustrate How TPEs Are Used

We consider two widely different specific processing examples involving TPEs—athletic footwear and asphalt.

6.7.3.1A Fabrication of Athletic Footwear. TPEs find widespread use in the manufacture of casual and athletic footwear. Typical polymers for this purpose are styrenic and thermoplastic urethanes. Injection molding is used to mold soles, heels, and combination sole–heels directly onto preshaped shoe tops.

Figure 6.30 shows an injection molding machine in which TPE pellets are fed from a hopper into a heated mixing chamber. The mixing chamber consists of a reciprocating screw that melts, homogenizes, and injects molten polymer into the mold. During the molding cycle, three separate hydraulic systems control the injection process. The first rotates the screw, homogenizing and pressurizing the melt. The second hydraulic system rams the screw forward by a precisely metered amount, forcing melt into the mold. The third closes and opens the mold before and after injection. Typical molding process conditions employ pressures of 30–50 MPa (4000–7000 psi), and molding cycle times, from melt to finished part, of 1 or 2 min.

Advantages include economical processing by injection molding, low compound cost, no vulcanization, recyclable scrap, ease of decorating and painting, and flexibility in shoe design, including an endless variety of shapes, textures, and patterns. Favorable physical characteristics include good low-temperature

Figure 6.30
A reciprocating screw injection molding machine used for processing thermoplastic elastomers (TPEs), such as polystyrene–polybutadiene, in the manufacture of athletic shoes. (N. G. McCrum, C. P. Buckley, and C. B. Bucknall, *Principles of Polymer Engineering,* Oxford University Press, Oxford, 1988, p. 295. By permission of Oxford University Press.)

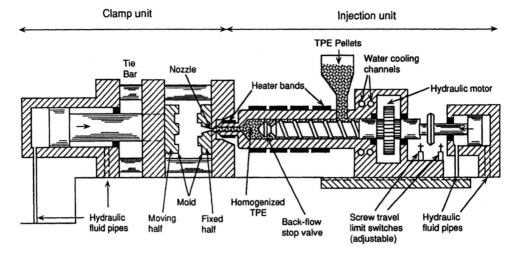

flexibility and high hardness at ordinary temperatures. Disadvantages include poor high-temperature performance, poor wear in athletic shoes, poor oil resistance, and high mold costs for short runs.

6.7.3.1B Modification of Asphalt.

Asphalt finds widespread use as a sealant, adhesive, and binder. However, it tends to become brittle and crack under stresses induced by thermal expansion and contraction and/or mechanical vibration. These undesirable properties can be minimized by adding TPEs. TPE-modified asphalts have found application in automobile body undercoating, mastics, and high-quality roofing membranes, but their high cost precludes widespread use in paving asphalts.

The useful features of asphalt–block copolymer blends can be attributed to their unique composite morphology. At low concentrations, the TPE polymer is the disperse phase and only enhances the asphalt's properties to a slight degree. At higher concentrations (> 10%), the TPE polymer forms a continuous phase. As depicted in Figure 6.31, the continuous phase consists of block copolymer containing light asphalt fractions, and the discontinuous phase consists of asphalt microparticles enriched in heavier asphalt fraction.

When poly(styrene–butadiene–styrene) tri-block copolymers are used, the polystyrene chains become incorporated into the asphalt microparticles, thereby binding them into the total network. The polymer increases the compliance of asphalt by furnishing the polybutadiene phase of greatly lowered T_g. This lower T_g in turn increases plasticity and resistance to cracking. Figure 6.32 illustrates typical compliance creep curves of an asphalt modified with increasing amounts of styrene–butadiene–styrene tri-block copolymer. Unmodified asphalt exhibits viscous creep over very long times, but upon addition of the block copolymer, an equilibrium compliance is attained more rapidly, particularly when the continuous network microstructure is formed.

In roofing compounds, the addition of TPEs produces a material that has excellent strength, elasticity, elongation, and cold-temperature flexibility. When used as crack sealant in roads,

Figure 6.31
Thermoplastic elastomer (TPE) modified asphalt. Schematic of the structure of an asphalt–block copolymer blend. The discontinuous (black) phase consists of asphalt microparticles that have incorporated some polystyrene segments; the continuous (white) phase consists of polystyrene–polybutadiene–polystyrene (SBS) elastomer containing lighter asphalt fractions.

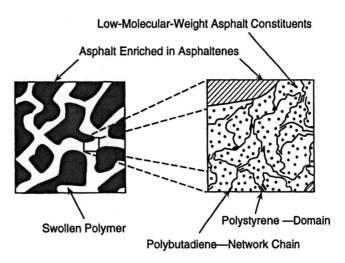

Low-Molecular-Weight Asphalt Constituents

Asphalt Enriched in Asphaltenes

Swollen Polymer

Polystyrene —Domain

Polybutadiene—Network Chain

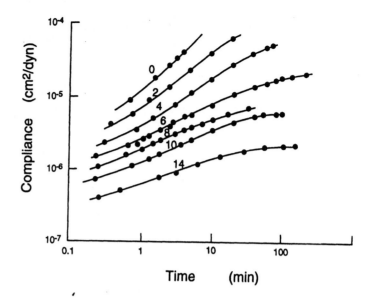

Figure 6.32
Increasing the amount of polystyrene–poly-butadiene (SBS) block copolymer from 0 to 14 wt% increases the creep compliance curve for asphalt at 22°C.

asphalt modified with TPE reduces the damage caused by temperature-induced expansion and vehicle vibration, thereby increasing the durability of the surface. Modified asphalt also provides a balance of adhesive and cohesive properties that prevents stones and water from entering road cracks and destroying the pavement.

6.7.4 Disposable Diapers Exemplify a Highly Engineered, Complex Polymer System

As any parent can tell you, there is no more determined individual than a baby with a wet diaper. However, much of the discomfort and traditional diaper rash has disappeared with the introduction of disposable diapers. When first introduced in the 1970s, disposable diapers consisted of a single layer of absorbent on a waterproof backing fastened with safety pins. Nowadays disposable diapers involve the sophisticated polymer system illustrated in Figure 6.33. The inner liner contains a wettable, nonwoven fabric that draws fluid away from the skin by capillary action. The middle section contains a crosslinked polyelectroyle that absorbs and retains fluid. The outside is a nonpermeable polyethylene film to which fastening tabs and elastic are attached in the leg and belt regions. Patents issued at the rate of about 100 per year provide some indication of the technological advances associated with disposable diapers. The following sections describe how disposable diapers are fabricated and discuss some of the environmental issues.

6.7.4.1 Production of the Inner Liner of a Disposable Diaper Illustrates Fabrication of a Nonwoven Fabric

Most diaper liners are made of polypropylene fibers coated with a surfactant like Triton X 102 to render their surfaces hydrophilic and wettable by water. A widely used method for producing the

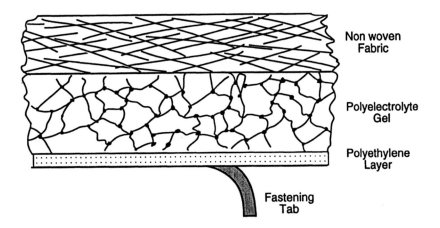

Non woven
Fabric

Polyelectrolyte
Gel

Polyethylene
Layer

Fastening
Tab

Figure 6.33
Schematic of a disposable
diaper containing an inner
liner fabricated from a nonwoven fabric that draws
fluid away from the skin
by capillary forces, a middle section containing
crosslinked polyelectrolyte
that adsorbs and retains
fluid, and a protective
polyethylene outer layer to
which fastening tabs are
attached.

fibers involves the spunbound process. Polypropylene pellets and
surfactant (1–2 wt%) are fed into a heated chamber where a single
screw extruder melts, mixes, and homogenizes the polymer and
surfactant. Then the melt is passed through a metering pump into
a die containing thousands of holes, each 0.015 inches in diameter, and extruded by compressed air into the ambient atmosphere.
The fibers quench as they move downward, until about 6 feet
below the die, they enter an air aspirator where they are gathered
together and drawn. Drawing leads to strain-induced alignment and
crystallization of the polypropylene and results in high fiber strength.
The drawing process actually translates back up the fibers so that
much of the strain occurs as they emerge from the die. Typical fiber
velocities are 12,000–15,000 ft/s, but the velocity and temperature
must be closely monitored because too much drawing leads to
fiber breakage. The surfactant adsorbs at the molten surface and
thereby ends up as a coating on the solidified fiber.

After leaving the aspirator, the fibers are separated by electrostatic charging or directed air flows and uniformly deposited onto a
horizontal screen that is 10 ft wide and moves with a velocity of 300
ft/min. A vacuum on the back side of the screen compacts the fiber
mat. After traveling on the screen for 10 to 12 ft, the fiber mat feeds
into heated steel rollers, one of which has an engraved pattern. The
combination of pressure ($\approx$2000 lbs/in.^{2} exerted by the roller onto
the mat) plus heating ($\approx$140°C) leads to the creation of a bound
fiber mat resembling an inverted waffle in which approximately
15% of the surface area is compressed. This mat is then slit into
appropriate widths and wound onto rolls before being transported
to the diaper fabrication assembly line.

Optimum performance of the mat as a fluid transfer material
occurs when the average fiber size is 20–30 μm. Key variables are
the diameter of the fibers, mat thickness, and weight of material
per area. Some diapers now contain two nonwoven fiber mats
with varying fiber diameters.

6.7.4.2 Superabsorbers Are Crosslinked
Polyelectrolytes

The middle section of the disposable diaper is the superabsorber.
It consists of a crosslinked polyelectroyle that absorbs and retains

fluid. As we saw in Section 6.4.4, intramolecular repulsions between charged groups lead to more extended chain conformations in polyelectrolytes than with corresponding neutral polymers. When charged polymers are crosslinked, intramolecular repulsions lead to swelling of the polymer network and gel formation. These crosslinked polyelectrolyte gels are called *superabsorbers*. The actual amount of water uptake is influenced mainly by the degree of crosslinking, pH, and electrolyte concentration.

Figure 6.34 illustrates the effect of crosslinking on water uptake in a poly-sodium acrylate. At extremely low levels of crosslinking, the polymer simply dissolves in water. Once crosslinking stabilizes the gel, water absorption takes place and the amount absorbed initially increases and then declines with increasing crosslinking. If the cross linking is too dense, other criteria such as the rate of absorbtion and retention may be compromised.

The rate at which a crosslinked polyelectrolyte swells as it absorbs water depends on the granular size, porosity, and surface energy of the particles as well as on the distribution and density of crosslinking. Transport of water into a superabsorbent involves both capillary and diffusion transport mechanisms. Initial transport primarily is the result of capillary flow. However, the initial absorption leads to the formation of a surface layer of swollen gel that impedes the movement of water into the unswollen interior. Known as "gel blocking," this phenomenon becomes important in designing absorbent materials for diapers. Crosslinking reduces gel blocking by resisting excessive surface swelling.

As water enters the gel, counterions dissociate from the polymer backbone and dissolve in the water to form an electrolyte. The counterions are confined to the gel by electrostatic attraction to compensate for the charges attached to the polymer. This local counterion excess generates an osmotic pressure difference between the interior of the gel and water in the bulk. (This

Figure 6.34
Effect of crosslinking on the water absorbency of the polyelectrolyte sodium polyacrylate. (C. Chen, J. C. Vassallo and P. K. Chatterjee, in *Absorbency*, P. K. Chatterjee, Ed., Elsevier, Amsterdam, 1985, p. 205.)

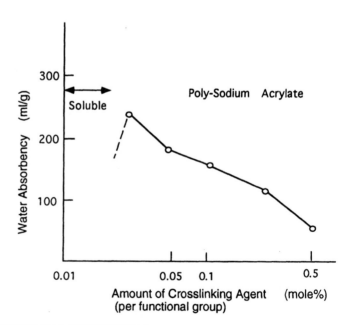

situation is analogous to that encountered in our discussion of colloids in Section 4.4.3 in which the interaction between charged surfaces and counterions restrained to the region between plates had both electrostatic and osmotic origins and was responsible for stabilizing the dispersion.) Electrostatic repulsion and osmotic pressure differences lead to gel swelling, but the swelling process is mediated by the pH and the concentration of bulk electrolyte. Figure 6.35 shows how the addition of salt decreases water adsorbency because electrostatic effects are screened and osmotic pressure differences are reduced. Note also that divalent calcium chloride has a larger effect than monovalent sodium chloride. Urine typically contains 0.8% NaCl, 0.15% $MgSO_4$, 0.06% $CaCl_2$, and 1.9% urea, so pH and salt effects are important in diaper adsorbency behavior.

A second important feature for diapers and related absorbents is the "water retention value," which is determined by flooding a test specimen with water and subjecting it to a high gravitational field, generally above 1000 G in a centrifuge. The water retained is expressed in weight percent based on the weight of dry fiber. An absorbent that upon application of pressure just squirts the water back out like a sponge is not very useful for diapers.

Superabsorbers for diapers are made from a blend of cross-linked synthetic sodium polyacrylates and modified cellulosic fibers. The polyacrylates can adsorb 35–40 grams of saline solution per gram of dry polymer, but because of their high cost they are blended with natural cellulose fiber. An added benefit of the blend is that the composition of the polymer–cellulose mixture can be adjusted to minimize gel blocking.

Processing begins by using hammer mills to transform cellulose boards into fiber, mixing the fibers with the polyelectrolyte granules with diameters of 20–100 μm. The mixed fibers and granular particles are transferred to an air gun and sprayed onto a screen to form an absorbent pad shaped in the form of an hourglass. Typically 300–500 pads are formed per minute.

Figure 6.35
Adding salt decreases water absorbency of a polyelectrolyte because it diminishes osmotic pressure differences and increases electrostatic repulsion between the fixed charges on the polymer chain. (C. Chen, J. C. Vassallo and P. K. Chatterjee, in *Absorbency*, P. K. Chatterjee, Ed., Elsevier, Amsterdam, 1985, p. 202.)

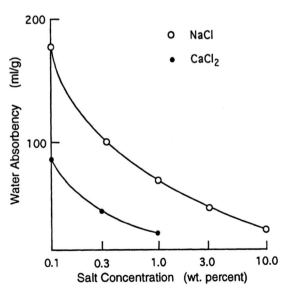

A recent "innovation" has been the production of gender-specific pads with more superabsorbant material in the front for boys and more in the middle for girls. Any difference in product performance remains to be documented. However, packaging gender-specific pads with gender-specific covers (for example, blue for boys and pink for girls) along with designed tabs has made a significant impact on sales. In a commodity-driven market where the product has been optimized and even possibly over-designed, appearances can be the deciding factor in market share!

Production of gender-specific pads involves an air spray machine that contains two nozzles: one delivers the cellulose wood fibers, the other the polyelectrolyte superabsorbent. Wood fibers are spread over the entire hourglass area, while the super-absorbers are concentrated in the regions where most absorption is needed. The screen on which the particles are deposited moves at 300 feet per minute, the cycle time for each pad is 0.2 s, and the required accuracy for targeted delivery of particles is measured in fractions of an inch.

6.7.4.3 Diaper Fabrication Is a High-Speed Multistep Process

As the absorbent pads roll off of the air forming machine, they are directly transferred to the final assembly line to avoid storage costs. In the final assembly, the absorption pad first is joined to the outside polyethylene cover sheet using a hot-melt adhesive such as a polystyrene–polybutadiene–polystyrene tri-block co-polymer and then joined to the polypropylene nonwoven inner-liner using a similar hot-melt adhesive. Then the trilayered assembly is trimmed, and fastening tabs, elastic leg bands, and waistbands are added. Sophisticated assembly lines contain up to nine hot-melt adhesive stations.

6.7.4.4 Biodegradability Is an Important Environmental Issue for Disposable Diapers

Disposal of used diapers presents a growing concern. The average level of biodegradability of disposable diapers now approaches 80%. Considerable effort is being devoted to devel-oping materials and methods that will lead to significantly higher levels of biodegradability. However, unresolved public technology issues (such as whether land fills, composting, or other forms of disposal are most appropriate) make it difficult to decide where to direct the technological efforts. A number of methods have been tried. For example, adding starch to the polymers aids in microbial consumption. Components that lead to polymer degradation upon composting and materials of such low molecular weight (< 1000 g/mol) that they can be metabolized by microbes also are being explored.

A second area of development involves making diapers more compact. Compactness benefits both the consumer end— where higher-density products are easier to ship, store, and display in stores and handle in the home—and the disposal process because compact materials can be handled more readily.

Bibliography

P. Heimenz. *Polymer Chemistry,* New York: Marcel Dekker, Inc., 1984.

C. Macosko. *Rheology: Principles, Measurements, and Applications,* New York: VCH Publishers, Inc., 1994.

N. G. McCrum, C. P. Buckley, and C. B. Bucknall. *Principles of Polymer Engineering,* New York: Oxford University Press, 1988.

L. H. Sperling. *Introduction to Physical Polymer Science,* New York: John Wiley & Sons, Inc., 1986.

D. H. Napper, *Polymeric Stabilization of Colloidal Dispersions,* London: Academic Press Ltd., 1983.

Exercises

6.1 A polyvinyl chloride $+CH_2-CHCl+_{N_p}$ mixture has an average molecular weight of 45,000 g. What is the degree of polymerization of an average chain?

6.2 The sharpness of the glass transition temperature of a semi-crystalline polymer
increases _ ; decreases _ ; or remains the same _
as the crystallinity of the polymer decreases.

6.3 Crosslinking a polymer by radiation causes the elastic modulus to
increase _ ; decrease _ ; or remain the same _ .

6.4 Derive the radius of gyration for a sphere, a disk, and an extended rod in terms of their actual physical dimensions as given in the text.

6.5 In a good solvent the radius of gyration of a given polymer molecule is
less than _ ; equal to _ ; or greater than _
in a bad solvent.

6.6 Estimate the overlap concentration of an aqueous DNA solution with a persistence length of 15 nm and a contour length of the double helix of 300 nm.

6.7 Polytetrafluoroethylene, PTFE, $+CF_2-CF_2+_{N_p}$, coatings on cookware are highly desired because they are insoluble, resistant to heat and chemical attack, and they have the lowest coefficient of friction of any solid. (a) Given that the carbon–carbon bond length is 1.54 Å and the bond angle is 109.5°, what is the persistence length of the PTFE molecule? (b) PTFE molecules with an average degree of polymerization of 10^4 and 10^6 are dissolved in a theta solution. Estimate the radius of gyration and the overlap concentration for both cases. (c) What are the scaling relationships between R_g and N_p, and C^* and N_p.

6.8 Outline the fundamental assumptions made in the Flory–Huggins theory of dilute polymer solutions.

6.9 Discuss the differences between an elastomer and a gel and identify two different applications for each kind of polymer structure.

6.10 Polystyrene can be deposited onto Pyrex glass by adsorption from benzene or toluene solutions. Discuss how the adhesion of the polystyrene film might be influenced by the polymer synthesis, the concentration and temperature of the solution, the time of deposition, and the surface condition of the glass substrate.

6.11 Forcing two random coils into the same volume of a solution causes (a) the entropy to
increase _ ; decrease _ ; or remain the same _;
(b) the free energy to
increase _ ; decrease _ ; or remain the same _;
(c) the repulsive force between the two particles to
increase _ ; decrease _ ; or remain the same _ .

6.12 For a concentrated polymer solution, derive the "scaling relationship" between the coefficient of viscosity and the relaxation time required for a polymer molecule to move the length of its defining tube.

6.13 What are the key requirements and constituents of a good pressure sensitive adhesive?

APPENDIX 6A   REGULAR SOLUTION THEORY

The regular solution model provides the simplest description of solution equilibrium and phase separation for nonpolar or weakly polar liquids. In its most basic form, the theory deals with the thermodynamic equilibrium of an incompressible mixture of two molecular species, A and B, of equal size. These molecules interact only with their nearest neighbors, whose number is fixed at the coordination number z. As a result, the interaction energy, which is assumed to be pairwise additive, depends exclusively on the three different types of near-neighbor pairs, AA, BB, and AB.

With N_A and N_B molecules of the two species, the energy E_o of the unmixed liquids is

$$E_o = \frac{1}{2}zN_AW_{AA} + \frac{1}{2}zN_BW_{BB} \qquad (6A.1)$$

where W_{AA} and W_{BB} represent the microscopic intermolecular interaction energies of the AA and BB pairs, respectively. (The factor ½ compensates for double counting the interaction between any given AA or BB pair.) We can estimate the energy of the pure liquids from their heat of vaporization.

The energy of the mixture E_{mix} is given by

$$E_{mix} = N_{AA} W_{AA} + N_{BB} W_{BB} + N_{AB} W_{AB} \qquad (6A.2)$$

where the number of pairs N_{AA}, N_{BB}, and N_{AB} are related to the total number of molecules by

$$2N_{AA} + N_{AB} = zN_A$$

$$2N_{BB} + N_{AB} = zN_B \qquad (6A.3)$$

The first equality, for example, follows by realizing that a given A molecule is engaged in z pairs, which may consist of AA or AB combinations. When we sum over all the A molecules to form zN_A pairs, we double count the AA pairs but not the AB ones; hence the factor of 2 in eq. 6A.3 applies only to the N_{AA} (and N_{BB}) interactions.

The energy of mixing, ΔE_{mix}, represents the difference between E_{mix} and E_o. Substituting equation 6A.3 in 6A.1, subtracting 6A.2, and rearranging gives

$$\Delta E_{mix} = E_{mix} - E_o = N_{AB} (W_{AB} - W_{AA} - W_{BB}) \qquad (6A.4)$$

If we define an effective *molecular* interaction parameter W

$$W = zE\left(W_{AB} - \frac{(W_{AA} + W_{BB})}{2}\right) \qquad (6A.5)$$

where $W = w/N_{Av}$ (w is the effective *molar* interaction parameter introduced in eq. 3.1.8), we obtain the energy of mixing

$$\Delta E_{mix} = N_{AB} \frac{W}{z} \qquad (6A.6)$$

All the individual interaction energies (W_{AA}, W_{BB}) are negative; so W easily can be positive. In fact, this is normally the case because it requires only that the heterogeneous interaction, W_{AB}, be weaker than the average of the interactions in the respective pure phases.

Because the free energy rather than the energy determines the thermodynamic properties of the solution, we also need to estimate the entropy of mixing. Regular solution theory makes the simplifying assumption that at constant density, mixing affects only the positional order. Furthermore, it assumes that the molecules mix in a fully random way. (In a *real* solution, entropy involves changes in internal conformations, density, and orientational correlations in addition to the mixing of the two species.)

These assumptions produce two consequences. First, the entropy of mixing is the same as that of an ideal mixture, as defined in eq. 3.7.6.

$$S = -k(N_A \ln X_A + N_B \ln X_B) \qquad (6A.7)$$

Second, we can calculate the number of A–B pairs, N_{AB}, to be inserted in eq. 6A.4 using the assumed random distribution. Of the z near neighbors to an A molecule, zX_B are B molecules. Because the number of A molecules is $N_A = X_A(N_A + N_B)$, the number of AB pairs is

$$N_{AB} = z\,(N_A + N_B)\,X_A\,X_B \qquad (6A.8)$$

For an incompressible system, energy and enthalpy are equal except for the trivial constant, PV. Similarly the Gibbs and Helmholtz free energies are equivalent. Substituting eq. 6A.8 in 6A.6 for ΔE_{mix}, and using eq. 6A.7 for ΔS_{mix}, permits us to calculate the Gibbs free energy of mixing $\Delta G_{mix} = \Delta E_{mix} - T\,\Delta S_{mix}$. Expressed in molar units, the result is

$$\Delta G_{mix} = (n_A + n_B)\,X_A\,X_B\,w + RT(X_A \ln X_A + X_B \ln X_B) \qquad (6A.9)$$

where n_A and n_B are the number of moles of A and B. This free energy expression forms the basis of *regular solution theory*. It contains only one unknown parameter, w.

We can derive chemical potentials, $\mu_i = (\delta G/\delta n_i)$, consistent with this free energy by substituting $X_A = n_A/(n_A + n_B)$ and $X_B = n_B/(n_A + n_B)$ in eq. 6A.9 and differentiating with respect to either n_A or n_B. For molecule A

$$\mu_A = \mu_A^o + X_B^2 w + RT \ln X_A \qquad (6A.10)$$

and interchanging indices A and B gives the chemical potential for molecule B. In eq. 6A.10 the pure liquid (represented by the superscript o) is chosen as the standard state.

Equation 6A.1 gives the internal energy (and enthalpy) of a pure liquid, and to obtain the free energy, we must add a term due to the entropy, S_{liq}^o. This entropy arises from molecular disorder in the liquid (which we assumed to be unaffected in the mixing of two liquids). If we include the entropy term, the standard chemical potential per mole, μ^o, becomes

$$\mu_A^o = \frac{z}{2} W_{AA} N_{Av} - T S_{liq}^o \qquad (6A.11)$$

This theory enables us to understand how the magnitude of the molecular interaction parameter and temperature affects phase stability and leads to phase separation. When the effective interaction parameter w is positive and small—and less than the thermal energy, 2RT, (w < 2RT), as occurs at high temperatures—the mixture forms one stable liquid phase at all compositions. When the effective interaction parameter is positive (repulsive) (w > 2RT) and large relative to 2RT, two phases coexist for some of the average compositions and a miscibility gap occurs in the phase diagram, as shown in Figures 6.10 and 11.21. When w >> 2RT, as occurs at low temperatures, only a small amount of B can dissolve in A and vice versa.

7

POLYMER COMPOSITES

with Gibson Batch

LIST OF SYMBOLS

C	number of principal axes of curvature at liquid resin–fiber interface
D_p; D_f	diameter of particles (p), fibers (f) in a polymer composite
E_m; E_f; E_c	static elastic tensile modulus—Young's modulus— of polymer matrix (m), fiber (f), or composite (c)
E_{cL}	Young's modulus of composite containing fibers in longitudinal (L) orientation (eqs. 7.5.4 and 7.5.12)
E_{cT}	Young's modulus of composite containing fibers in transverse (T) orientation (eqs. 7.5.5 and 7.5.13)
G_m	static elastic shear modulus of matrix (m)
K	Kozeny constant (eqs. 7.2.4 and 7.2.5)
l_t	load transfer length (eq. 7.5.7); l_c critical fiber length (eq. 7.5.21)
L	length from liquid polymer resin injection point to flow front
L_{eff}	actual path length for polymer resin liquid through packed fiber bed
$N_{ca\ loc}$	local capillary number (eq. 7.3.15)
P'_c	dimensionless capillary pressure (eq. 7.3.10)
ΔP_e	excess external pressure forcing liquid polymer resin through fiber bed
ΔP_c	capillary pressure acting on liquid resin inside fiber composite (eq. 7.3.3)
Q	liquid polymer resin input rate
R_c	radius of curvature of liquid polymer resin–air surface; along fibers (eq. 7.3.1) and across fibers (eq. 7.3.9)
V	volume fraction of spheres (s), particles (p), fibers (f) in composite
v_{sup}	superficial velocity of liquid through a porous bed (eq. 7.2.1)
v_{int}	average velocity of liquid polymer resin inside the fiber bed (eq. 7.2.3)
v_{tot}	total velocity of liquid polymer front during impregnation (eq. 7.3.11) due to external pressure (eq. 7.2.6) and capillary pressure (eq. 7.3.5)
γ_{lv}	surface tension of liquid polymer resin
γ_f	fracture surface energy

δ	effective diameter of pores between fibers in composite (eq. 7.3.2)
ε_{xx}	tensile strain
$\theta_s; \theta_d$	contact angles, static (s) (eq. 3.2.4) and dynamic (d) (Appendix 8A)
κ	permeability of a porous bed (eq. 7.2.1)
λ	spacing between fibers in a polymer composite
ξ_{tot}	parameter defining liquid diffusivity through a packed fiber bed (7.3.13) due to external pressure (eq. 7.2.9) and capillary pressure (eq. 7.3.8)
σ_{int}	interfacial tensile strength of the matrix–fiber interface
$\sigma_m; \sigma_f$	tensile stress in the matrix (m) or a fiber (f)
$\sigma_{um};$	ultimate tensile strength—fracture strength—of polymer
$\sigma_{uf}; \sigma_{uc}$	matrix (m), fiber (f), or composite (c)
$\sigma_{uc,L}$	ultimate tensile strength of composite containing fibers in longitudinal (L) orientation (eqs. 7.5.17 and 7.5.19)
$\sigma_{uc,T}$	ultimate tensile strength of composite containing fibers in transverse (T) orientation (eqs. 7.5.18 and 7.5.23)
τ_{int}	interfacial shear strength of the matrix–fiber interface (eq. 7.4.3)

CONCEPT MAP

Synergistic Properties of Polymer Composites

- Polymer composites combine high strength of rigid reinforcements with high toughness of solid polymer matrix. Reinforcements may be particulate, short fibers, or long fibers. Most common composites are fiberglass-reinforced thermoset polyesters bonded with a silane coupling agent.

- Polymer matrix composites reinforced with glass, carbon, or Kevlar fibers display elastic moduli 20–1000 times larger than unreinforced solid polymers.

Fabrication Processes

- Fabrication involves impregnating reinforcements with liquids—either unpolymerized monomers and additives that react to form solid polymer (thermoset process) or liquid polymers above their melting point (thermoplastic process). Main manufacturing processes are (Figure 7.4):

Mix–shape–set—short fibers and polymer matrix precursors are mixed and sprayed directly onto an open face mold (spray-up molding) or pulled through a heated die (pultrusion).

Shape–inject–set—a liquid thermoset resin is injected into a mold containing dry fiber mats (resin transfer molding).

Prepreg–shape–set—fiber mats are first impregnated with matrix material; later the "prepreg" sheets are placed in a mold to be shaped and set by heat and pressure.

Polymer Liquid Impregnation of Fiber Reinforcement

Under External Pressure (ΔP_e)

- Fluid velocity through fiber bed is governed by Darcy's Law (eq. 7.2.1)
  $v_e = (\kappa/\eta)(\Delta P_e/L)$

- Permeability, κ, depends on diameter, volume fraction, and orientation of fibers according to Kozeny–Carman analysis (eqs. 7.2.4 and 7.2.5).

- Since $v_{av} \propto L^{-1}$, impregnation depth (L) depends on (time)$^{1/2}$.

- In asymmetrical reinforcements, permeability is direction dependent.

Under Capillary Pressure (ΔP_c)

- Fluid velocity through fiber bed is governed by Hagen–Poiseuille equation (eq. 7.3.4)
  $v_c = (\delta^2/32\eta)(\Delta P_c/L)$

- For flow along fibers, capillary pressure, ΔP_c, remains constant. ΔP_c is determined by liquid surface tension, contact angle, and pore size δ. δ depends on fiber diameter and volume fraction (eq. 7.3.2).

- For flow across fibers, ΔP_c varies and can change from positive (wetting) to negative (nonwetting) value (Figure 7.12). Negative ΔP_c leads to inclusion of air voids.

- Dynamic contact angle is larger than static value, enhancing possibility of air entrapment.

- For a given fiber diameter and volume fraction, key processing variables are external pressure, temperature (viscosity), and time.

- When both external pressure and capillary pressure operate, impregnation of nonuniformly spaced fibers can lead to inclusion of voids (Figure 7.14).

Fiber Surface Pretreaments

- Pretreatments of fiber surfaces are designed to:
  control wetting during impregnation—to minimize voids that lead to failure
  control adhesion—to manipulate bond strength that controls strength and toughness.
- Pretreatments include chemical or plasma etching (Figure 7.18) and addition of chemical coupling agents (amiphiphilic molecules) (Figure 7.19).

Elastic Modulus of Fiber-Reinforced Polymer Composites

- Objective is to increase stiffness of composite without compromising toughness.
- Elastic modulus of continuous fiber composites follows the law of mixtures. Maximum modulus values are achieved when stress is applied parallel to the fibers (longitudinal modulus—eq. 7.5.4) and minimum values when stress is perpendicular (transverse modulus—eq. 7.5.5).
- Elastic modulus of discontinuous fiber composites is given by the Halpin–Tsai equations (longitudinal modulus—eq. 7.5.12; transverse modulus—eq. 7.5.13). Maximum stiffening is achieved with high-modulus fibers, high fiber aspect ratio, and high fiber volume fraction.

Tensile Strength of Fiber-Reinforced Polymer Composites

- Objective is to increase strength of composite without compromising toughness.
- Tensile strength of continuous fiber composites follows the law of mixtures. Maximum composite strength values are obtained when stress is applied parallel to the fibers. Strength is then limited by the fracture strength of the fibers and the strength of the matrix.
- Tensile strength of discontinuous fiber composites depends on the fiber aspect ratio and the fiber–matrix interface shear strength τ_{int}. Maximum strength values are obtained when the fiber length exceeds the critical load transfer length, $l_c = \sigma_{uf} D_f / 2\tau_{int}$ (eq. 7.5.21); then the fiber fails before the matrix or interface. When the fiber length is less than l_c, failure occurs either at the interface or in the matrix itself.

Fracture Toughness of Fiber-Reinforced Polymer Composites

- Fracture resistance of nonreinforced isotropic polymers is generally low; stress-induced cracks propagate in a plane normal to the applied stress, resulting in catastrophic failure (Figure 7.16a).

- Fracture resistance of polymer composites depends on fiber orientation and interfacial shear strength. When a crack runs into weakly bonded fibers, debonding occurs, and the crack is deflected to run parallel to the fibers. Fracture is delayed until the fibers have pulled out of the composite (Figure 7.16b). This process absorbs energy and leads to high fracture toughness.

- For optimal strength and toughness, crack propagation should be confined to the polymer matrix slightly away from the fiber–matrix interface. This is obtained by optimizing the interfacial adhesion between matrix and fibers.

7

A composite material generally is made of a high-modulus reinforcement held together by a weaker binder material. This two-phase structure can be seen in natural materials, such as feathers, leaves, bone, and wood. By combining rigid and flexible materials in a similar manner, we can design products with lighter weight, better physical properties, and greater resiliency. Composites owe their superior performance over unreinforced materials to the synergistic combination of matrix and reinforcement.

Polymer matrix composites emerged as new materials for aircraft construction in the mid-1940s. Today, sales of polymer composites with applications in the aerospace, automotive, marine, and construction industries have grown to 2.5 billion pounds per year (Table 7.1). Their corrosion-resistant properties have attracted applications in gratings and handrails in chemical plants and in underground gasoline storage tanks. Because composites have a greater strength-to-weight ratio than steel, applications are emerging for lightweight automotive drive shafts and truck springs. The impact resistance of composites is utilized in bowling alleys, truck beds, and bus panels. Because composites have low thermal and electrical conductivities, they are replacing aluminum in window moldings and ladder rails for utility companies. The largest volume of composites is used in marine applications, such as boat hulls, barge covers, and other structural members.

This chapter applies the principles of interfacial engineering to the processing and mechanical behavior of composites. First we overview the classification, processing, and performance of composite materials in Section 7.1 and highlight areas where interfacial engineering fundamentals are important. Sections 7.2 and 7.3 explore the behavior of fluids during composite processing, including both basic concepts and modeling equations. Section 7.4 highlights the importance of matrix–reinforcement interface bond strength in the mechanical behavior of composites and describes experimental techniques used to characterize the fiber–polymer interface. Section 7.5 analyzes specific mechanical properties. Section 7.6 discusses some special issues in the processing and application of reinforced composites.

TABLE 7.1 Reinforced Plastic Shipments by Market in 1987

Market	Shipments (millions of pounds)
Aerospace/military	39
Appliances	147
Construction	493
Consumer products	151
Corrosion-resistant equipment	310
Electrical	209
Marine	393
Land transportation and automotive	630
Other	82
Total	2455

Source: Kates, Plastics Engineering, April 1988, p. 37.

7.1 The Composition, Processing, and Properties of Polymer Composites Are Unique

7.1.1 The Typical Matrix Is a Thermoset or a Thermoplastic Polymer

The matrix material used in composites can be polymer, ceramic, or metal. Polymer matrices, which make up by far the largest volume in industrial applications, are our major concern here. The polymers are either thermosets (described in Section 6.1.1) or thermoplastics (Section 6.2.1).

Thermoset polymer composites are prepared by comixing monomer reactants that polymerize *in situ* after they have flowed through and "impregnated" the reinforcements. The most common thermoset is unsaturated polyester, which copolymerizes with styrene by free radical (chain) polymerization to form a crosslinked polymer network. Due to the low viscosity of the reactants before polymerization, thermosets are easy to impregnate into fibers; however, the monomers are very reactive and require special temperature-controlled storage and handling prior to processing. Thermoset epoxies are used where high-temperature properties are important in the application.

Thermoplastics are fully polymerized before impregnation and become fluid only at temperatures elevated well above their glass transition temperature. Because thermoplastics are inert, they are generally easier to store and handle than thermosets. However, their higher fluid viscosity (100,000—500,000 cP for thermoplastics versus 50—2000 cP for thermosets) and high processing temperatures (200—380°C) make impregnation of the reinforcing fibers more difficult.

7.1.2 Reinforcements Are Either Particulates or Fibers

Many combinations of matrix and reinforcements are available. Figure 7.1 shows the classification of composites according to reinforcement type, composition, and orientation. The selection for a particular application is based on property requirements and economic constraints.

Particulate composites are made by mixing powder fillers into a polymer to modify the thermal expansion coefficient and shrinkage, reduce friction and wear, improve machinability, increase surface hardness, or simply reduce material cost. Typically, they have low impact strength due to stress concentrations at the particle–matrix interface. However, some particulate composites are designed specifically for enhanced toughness, and Section 6.7.2 describes impact toughening that occurs when rubberlike particles are dispersed in a brittle matrix. Because rubber particles absorb energy, they arrest crack propagation, such as in rubber-toughened epoxies and the polystyrene–polybutadiene copolymer combinations.

Fiber-reinforced composites are known for their high strength and stiffness. However, the properties of fiber-reinforced composites are highly anisotropic; in the fiber direction the fibers dominate the properties, while in the transverse direction the properties approach those of the matrix. Most common fiber reinforcements are made from glass, carbon, and aramid (Kevlar), with *filament* diameters ranging from 8 to 25 µm and elastic moduli 20–100 times larger than those of typical polymers. The high strength and stiffness of fibers result from the fiber-making process, which removes many of the microscopic defects that otherwise weaken fiber materials (such as glass) in their bulk form. Composite properties can be "engineered" to a specific application by controlling fiber length, volume fraction, and orientation.

Figure 7.2 illustrates some common types of fiber reinforcements used in composites. The basic commercial unit is the *strand;* each strand contains between 200 and 4000 filaments. *Rovings* are aligned bundles of fibers composed of 8–120 strands, for a total yield of 100–400 yards of roving per pound. Because aligned filaments pack tightly together, high fiber volume frac-

Figure 7.1
Classification of composite materials. (B. D. Agarwal and L. J. Broutman, *Analysis and Performance of Fiber Composites,* Copyright © 1980, John Wiley & Sons, p. 4. Reprinted by permission of John Wiley & Sons, Inc.)

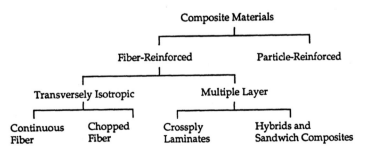

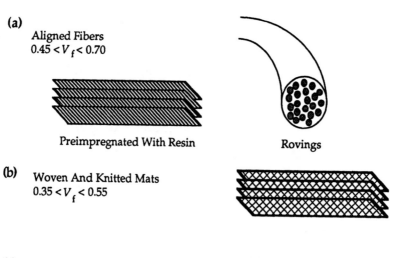

(a) Aligned Fibers
$0.45 < V_f < 0.70$

Preimpregnated With Resin   Rovings

(b) Woven And Knitted Mats
$0.35 < V_f < 0.55$

(c) Random Fiber Mats
$0.05 < V_f < 0.46$

Figure 7.2
Types of reinforcements
and fiber volume fractions
V_f used in composites.

tions (up to 0.70) are possible. Rovings also can be used in woven or knitted *mats* to provide strength in two directions. Continuous random fiber mats, which cost less to manufacture than woven mats, are made by binding randomly oriented strands together. However, due to poor packing, random fibers have lower volume fractions in the composite. To enhance performance for structural applications, multiple layers of reinforcements are used across the thickness of the composite, as illustrated in Figure 7.3. To achieve better appearance and corrosion resistance, a surface veil or mat may be used to create a smooth resin-rich surface.

7.1.3 Impregnation of the Reinforcements by the Matrix Can Be Complex—Good Wetting Is Critical

Composite processing involves three steps: complete wetting of the fibers by the liquid matrix material, molding the fibers and matrix into the desired orientation and shape, and solidifying or setting the matrix material (either by heat cure of a thermoset or cooling of a thermoplastic) into a solid article. Based on these three unit operations—"wet," "shape," and "set"—we can classify composite processing into the three broad strategies depicted in Figure 7.4.

Mix–shape–set (Figure 7.4a). The resin matrix material (reactive monomers and additives) and fibers are mixed shortly before shaping and setting. The resin must wet the fibers before setting into the final shape.

- In *spray-up molding,* resin and chopped fiber segments 1/8–1/2 inch in length are sprayed directly into an open-face mold and compacted by rollers. This spray-up technique is used to fabricate large parts, such as the 10,000 pound barge covers seen on the Mississippi river.

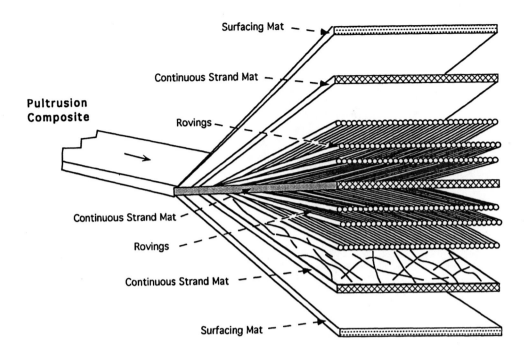

Pultrusion Composite

Surfacing Mat

Continuous Strand Mat

Rovings

Continuous Strand Mat

Rovings

Continuous Strand Mat

Surfacing Mat

Figure 7.3
Exploded view of
pultruded composite
layer stacking sequence.
(R.W. Meyer, *Handbook of
Pultrusion Technology,*
Chapman and Hall, New
York, 1985, p. 2.)

•In *pultrusion,* fibers are wetted in a resin tank before being shaped. They are "set" in a heated die. The thermoset polymer composite then pulls the prewetted fibers and resin through the die. Although pultrusion is restricted to geometries of constant cross section, such as sheets or beams, it is one of the few continuous processes for making polymer composites.

Shape–inject–set (Figure 7.4b). The reinforcements are preformed into their near final shape before being placed in the mold. The resin matrix material is injected into the mold. The resin must wet the fibers before setting into the final shape.

•In resin *transfer molding,* a liquid thermoset resin is injected into a mold that already contains dry reinforcements.

Prepreg–shape–set (Figure 7.4c). Fiber mats and rovings are wetted and impregnated with matrix material by roll coating (see Section 8.2.1.1), dip coating, crosshead extrusion, or continuous belt pressing and then stored until needed. These "prepreg" sheets are cut into sections, transferred into a mold, and shaped and set by pressure and heat.

•Common shaping processes using prepreg include autoclave molding, hand layup, injection molding, *compression molding,* and transfer molding.

Problems can arise in processing due to air entrainment during impregnation, particularly when fibers are packed closely together. Because these voids create stress concentrations that lower ductility, strength, and weatherability, minimizing voids is a major goal in composite processing. Sections 7.2 and 7.3 examine wetting mechanisms for reinforcements and describe models for impregnation and void formation during processing. The flow

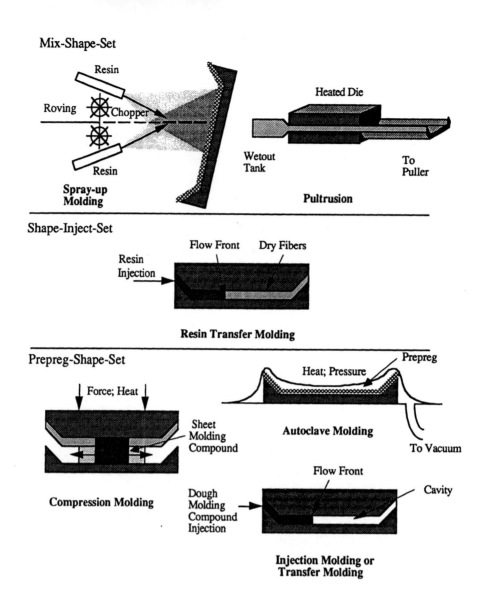

Figure 7.4
Manufacturing processes for polymer composites.

of liquid matrix material into fiber mat is driven by two mechanisms: first, as analyzed in Section 7.2, the fluid is *pushed* against viscous forces by an external pressure; second, as analyzed in Section 7.3, the fluid is *pulled* by capillary forces. We bring both mechanisms together at the conclusion of Section 7.3.

7.1.4 Fiber–Matrix Interfacial Bonding Is Optimized for Strength and Toughness

Pound for pound, the performance of polymer composites exceeds that of metals and unreinforced plastics. Figure 7.5 and Table 7.2 compare the strengths and stiffness for different materials, compensating for density differences. The enhanced mechanical properties of composites result from the synergy between the inherent stiffness of the reinforcements and the light weight of the matrix materials. To achieve that synergy, however,

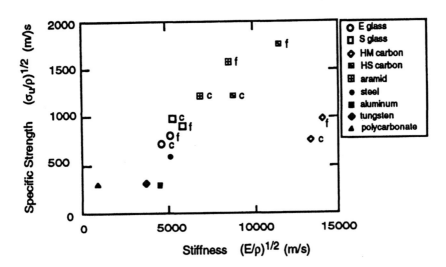

Figure 7.5
Comparison of specific strengths and stiffness of materials listed in Table 7.2. Symbols indicate fiber properties with "f" and epoxy composites of these fibers with "c."

we must optimize the mechanical coupling between the matrix and the reinforcement.

To illustrate the role of interfacial coupling in the reinforcing mechanism of composites, we consider the breakdown of a graphite–epoxy composite in which the graphite fibers are preferentially aligned in one direction, as shown in Figure 7.6. We can monitor the breakdown by following the decline in the dynamic (acoustic) value of Young's modulus (Appendix 11A). Subjecting the specimen to increasing tensile strains in the fiber direction results in a significant decrease in the modulus E relative to the initial value, E_o.

We can understand this loss of modulus as follows. For strains less than 0.1 %, the composite carries the load and both matrix and fibers elongate without breaking any fibers. At strains in excess of 0.1%, some misaligned fibers break as they attempt to straighten. Transfer of the portion of the load, previously carried by the broken fibers generates a shear stress that is redistributed through the matrix to surrounding fibers. Depending on the strength of the interfacial bond and the individual fiber and matrix moduli, full fiber tension is regained 2–200 fiber diameters away from the broken end. Thus the broken fiber continues to carry load over most of its length, and the modulus falls only by a small amount. As the strain increases, additional fibers break (including those aligned with the tensile axis), and the stiffness of the composite deteriorates more rapidly. Eventually, so many fibers have broken that cracks in the polymer matrix start to propagate across the sample and E/E_o drops sharply. To maintain the high strength and stiffness of a polymer composite under load, the interfacial bond between the matrix and the reinforcement should be as strong as possible.

High interfacial bond strength is not necessarily an advantage always for mechanical properties of polymer composites. Take fracture resistance, for example. If the stress concentrations at the tip of an advancing crack can cause the fibers to debond from the matrix, the fibers act as crack arrestors and deflectors. A substantial increase occurs in the energy absorbed in fracture (see

TABLE 7.2 Typical Properties of Composites and Reinforcements Used in Composites

Fiber/composite	Density ρ (g/cm^3)	Tensile modulus E (GPa)	Tensile strength σ_u (GPa)	Ultimate strain ε_b (%)	Fiber diameter D_f (μm)
E Glass	2.56	69–72	1.7–3.5	3	5–25
in epoxy composite	2.10	45	1.1		
S Glass	2.48	85	2.1–4.8	5.3	5–15
in epoxy composite	2.00	55	2.0		
High Modulus Carbon	1.96	390–520	1.9	0.38	8.4
in epoxy composite	1.63	290	1.0		
High Strength Carbon	1.80	240–300	5.6	1.8	5.5
in epoxy composite	1.60	145	2.3		
Aramid	1.44	124	3.6		
in epoxy composite	1.38	80	2.0		
Conventional Materials					
Steel	7.85	210	2.75	1.8–2.5	200
Aluminum alloys	2.70	70	0.14–0.62		
Glass	2.50	70	0.7–2.1		
Tungsten	19.30	350	1.1–4.1		
Polyethylene	0.95	1.07	0.0224	13	
Nylon 12	1.01	1.25	0.0427	12	
Polycarbonate	1.20	1.93	0.0615	6.2	
PMMA	1.19	2.63	0.0731	4.6	

Section 7.4.1), and fracture toughness is increased. This behavior implies that weakly bonded fibers yield composites with higher fracture toughness.

For most applications polymer composites require a high impact strength, a combination of both high tensile strength and high fracture toughness. Furthermore, in structural applications, composites are subjected to compression and shear as well as tension, and good interfacial adhesion is necessary to resist fiber buckling. Thus one must optimize bond strength to balance the conflicting requirements for structural stiffness and strength and fracture resistance—bonding between the matrix and reinforcement must be neither too strong nor too weak. Section 7.4.2 discusses fiber surface treatments to adjust the interface adhesion for several common types of composites. Section 7.5 addresses the consequences on mechanical behavior.

7.2 Liquid/Solid Interfaces and Tortuosity Resist Flow between Fiber Reinforcements during Impregnation

In the wet processing of composites, a liquid polymer resin penetrates dry fiber beds under some externally applied pressure. Our goal here is to understand the effect of processing variables on the rate of impregnation. We model the effect of pressure on

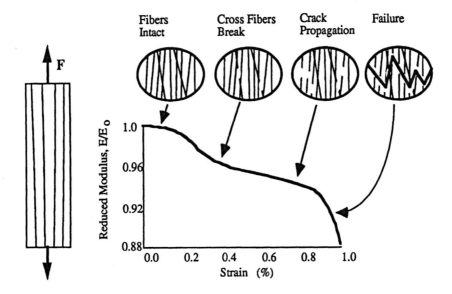

Figure 7.6
Reduction of elastic modulus of a graphite–epoxy composite due to fiber breakage under dynamic loading in the nominal fiber direction.

filling rate through equations used to describe flow through tubes (Section 3.8.1.2). Obviously, fiber content and fiber diameters are important factors in determining filling rate because they affect the size of the pores available for flow.

The spaces between solid reinforcements are randomly interconnected pores of variable cross-sectional shape, as shown in Figure 7.7a. We can estimate the size of these pores by assuming that the fibers (the filaments within a strand) form a regular hexagonal array, as in Figure 7.7b. Table 7.3 gives expressions for the gap distance λ and interstitial pore diameter δ for composites with fibers (cylinders) and particles (spheres) as a function of reinforcement size D (fiber or sphere diameter) and volume fraction V. For a typical fiber diameter of 10 µm and a volume fraction of 0.62, λ equals 2.1 µm and δ is 4.0 µm. Because the gaps and pores are usually much smaller than the size of the composite particles (fibers or spheres), we model impregnation as flow through a porous medium.

Figure 7.7
Arrangement of circular fibers: (a) with random spacing ($V = 0.42$), (b) with hexagonal close-packed spacing ($V = 0.42$), (c) with hexagonal close-packing ($V = 0.907$).

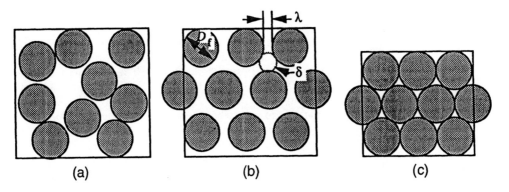

(a) (b) (c)

TABLE 7.3 Spacing of Cylinders and Spheres in Hexagonal Close-Packed Arrays

Geometrical dimension	Cylinders (fibers)	Spheres (particles)
Surface area to volume ratio	$= \dfrac{4}{D_f}$	$= \dfrac{6}{D_p}$
Particle-to-particle spacing λ	$\lambda_f = D_f\left[\left(\dfrac{\pi}{2\sqrt{3}\,V_f}\right)^{1/2} - 1\right]$	$\lambda_p = D_p\left[\left(\dfrac{\pi\sqrt{2}}{6V_p}\right)^{1/3} - 1\right]$
Interstitial pore diameter δ	$\delta_f = D_f\left[\left(\dfrac{2\pi}{3\sqrt{3}\,V_f}\right)^{1/2} - 1\right]$ between 3 fibers	$\delta_p = D_p\left[\left(\dfrac{2}{3}\right)^{5/6}\left(\dfrac{\pi}{V_p}\right)^{1/3} - 1\right]$ between 3 particles $\delta_p = D_p\left[\left(\dfrac{\pi\sqrt{3}}{4V_p}\right)^{1/3} - 1\right]$ between 4 particles
Maximum packing fraction V_{max}	$V_{fmax} = \dfrac{\pi}{2\sqrt{3}} = 0.907$	$V_{pmax} = \dfrac{\pi\sqrt{2}}{6} = 0.740$
Minimum pore diameter at maximum packing fraction δ_{min}	$\delta_{min} = D_f\left(\dfrac{2}{\sqrt{3}} - 1\right) = 0.155 D_f$	$\delta_{pmin} = D_p\left(\dfrac{2}{\sqrt{3}} - 1\right) = 0.155 D_p$ between 3 particles $\delta_{pmin} = D_p\left[\left(\dfrac{3}{2}\right)^{1/2} - 1\right] = 0.225 D_p$ between 4 particles

7.2.1 Impregnation under External Pressure Can Be Modeled as Flow through a Porous Medium—Darcy's Law

The first studies of flow through porous media were carried out by Henry Darcy in 1856. He measured the flow rate of water through beds of sand as a function of column length and pressure head (Figure 7.8) and obtained an empirical relationship

$$v_{sup} = \frac{\kappa}{\eta}\frac{\Delta P_e}{L} \tag{7.2.1}$$

In this equation, v_{sup} is called the *superficial velocity*. It equals the volume of fluid emerging per unit time divided by the total cross-sectional area of the channel. It represents the average velocity of the emerging fluid based on an empty channel. η is the viscosity of the fluid, ΔP_e is the external driving pressure, L is the length of the packed bed, and κ is its *permeability*. The permeability constant κ (with units of length squared) describes the flow conductance through packed beds. Porous media with low permeability constants are more difficult to fill; so they require higher pressures and/or longer fill times.

Figure 7.8
Conceptual description of Darcy's law for vertical drainage flow through a sand bed.

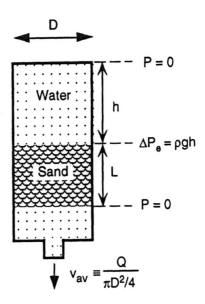

In developing our understanding of permeability, it is useful to compare Darcy's law, eq. 7.2.1, to the Hagen–Poiseuille relation for flow through an open tube of diameter δ and length L that we introduced in Section 3.8.1.2, eq. 3.8.8

$$v_{av} = \frac{\delta^2}{32\eta} \frac{\Delta P_e}{L} \tag{7.2.2}$$

From this comparison we see that the upper limit for the permeability is $\kappa = \delta^2/32$, a value that obviously must be reduced by factors representing the volume fraction V of the particles or the fibers in the tube, and their shape.

7.2.1.1 Tortuosity Reduces the Permeability of the Porous Medium—The Kozeny–Carman Equation

It is useful to recall that the external pressure drop, ΔP_e, in eq. 7.2.2 derives from the energy dissipation at the walls of the tube associated with the irreversible flux of momentum during fluid flow. In a packed bed, the energy dissipation occurs at the particle–fluid interface, and we expect that ΔP_e will need to be larger (or κ will be less than $\delta^2/32$) for several reasons:

1. The actual distance traveled by the fluid follows a tortuous path of distance L_{eff} as it flows around particles. L_{eff}/L is defined as the *tortuosity;* it is always greater than 1.

2. The average fluid velocity v_{int} inside the bed is larger than the superficial velocity, and can be derived as

$$v_{int} = v_{sup} \frac{(L_{eff}/L)}{(1 - V)} \tag{7.2.3}$$

where $(1 - V)$ is the volume fraction of fluid.

Taking these factors into account, the Kozeny–Carman analysis for the permeability of a packed bed containing spherical particles gives

$$\kappa = \frac{D_p^2(1 - V_p)^3}{36K\, V_p^2} \qquad (7.2.4)$$

where D_p is the particle diameter and V_p is the particle volume fraction. For beds containing fibers the Kozeny–Carman analysis gives

$$\kappa = \frac{D_f^2(1 - V_f)^3}{16K\, V_f^2} \qquad (7.2.5)$$

where D_f is the fiber diameter and V_f is the fiber volume fraction.

K is the *Kozeny constant*. It is equal to $2(L_{eff}/L)^2$. Larger values for K yield smaller values for the permeability κ. For flow of Newtonian fluids moving in a direction *parallel* to the length of fibers, $L_{eff} = L$ and K has a theoretical value of 2; *perpendicular* to the length of the fibers, $L_{eff} = \pi L/2$ and K has a theoretical value of $2(\pi/2)^2 \approx 5$. Reinforcements for which the tortuosity is more severe lead to even larger theoretical values of K. For example, K is 14–18 for flow around flat disks (as in clay) and 100–2000 for flow around flat sheets (as in mica).

Experimentally determined values of K shown in Table 7.4 often differ significantly from theory. For example, for longitudinal flow through unidirectional rovings, K is found to be 0.70, much smaller than the theoretical value of 2. This difference is attributed to channeling of the fluid through the more open regions with less resistance to flow. Conversely, for transverse flow perpendicular to fibers, K is found to be 18, much larger than the theoretical value of 5. This difference results from constriction of the flow by "log jamming" of fibers to block flow channels. Here the path for flow is restricted by narrow spaces between fibers and $\lambda \approx 0$.

7.2.2 Darcy's Law Predicts Impregnation Length as a Function of External Pressure

Once K has been determined experimentally for a given configuration of fiber reinforcements, a value for κ can be calculated using eq. 7.2.5. Then we can use Darcy's relationship to predict the rate of flow front advancement during impregnation. The velocity of the fluid front due to external pressure, v_e, is equal to $v_{sup}/(1 - V_f)$; so inserting the permeability (eq. 7.2.5) into eq. 7.2.1 and dividing

TABLE 7.4 Typical Experimental Values of Permeability (κ) for a Given Fiber Volume Fraction V_f, and the Corresponding Kozeny Constant (K)

Reinforcement type	V_f	κ (μm^2)	K
Continuous random fiber mat	0.28	73	0.29
Woven mat	0.56	10	6.9
Aligned fibers (longitudinal or parallel flow)	0.5	2	0.7
Aligned fibers (transverse or perpendicular flow)	0.5	0.2	18

by $(1 - V_f)$ gives

$$v_e = \frac{\Delta P_e}{\eta L} \frac{D_f^2}{16K} \frac{(1 - V_f)^2}{V_f^2} \tag{7.2.6}$$

where L is the distance from the injection point to the flow front. Note that the flow-front velocity is proportional to $1/L$, so that the rate of impregnation decreases with time as the composite is filled by fluid under a constant external pressure.

To determine L at a given time t we insert $v_e = dL/dt$ and integrate eq. 7.2.6 to get

$$L = \left(\frac{\Delta P_e}{\eta} \frac{D_f^2}{8K} \frac{(1 - V_f)^2}{V_f^2} t \right)^{1/2} \tag{7.2.7}$$

or

$$L = (2\xi_e t)^{1/2} \tag{7.2.8}$$

where

$$\xi_e = \left(\frac{\Delta P_e}{\eta} \frac{D_f^2}{16K} \frac{(1 - V_f)^2}{V_f^2} \right) \tag{7.2.9}$$

ξ_e is a parameter that measures the spreadability of the liquid under external pressure, and it has the same dimensions as a diffusion coefficient.

Because L is proportional to $t^{1/2}$, doubling the size of a part will require a fourfold increase in time to fill. Equation 7.2.7 provides a good practical guide for composite impregnation. It predicts the dependence of impregnation length on process variables such as pressure, viscosity (determined in turn by the processing temperature), fiber diameter, and volume fraction.

Table 7.5 presents solutions of Darcy's equation for a variety of flow geometries. These solutions assume different boundary conditions, either constant external pressure ΔP_e (P in the table) or constant fluid input flow rate Q.

In many molding processes, the location of the fluid source is such that impregnation of reinforcements involves filling in two or three directions. In its one-dimensional form, Darcy's equation still provides a useful approximation when the permeability is isotropic (as is the case for a random fiber mat) or when flow is strongly favored in one direction (as with aligned fiber reinforcements). As depicted in Figure 7.9, the permeability for a given fiber configuration is found by injecting a thermosetting polymer into the center of a bundle of rovings and viewing the advancing flow front. The flow front in a random fiber mat appears circular, indicating that permeability is isotropic in the plane of the mold. For an oriented fiber mat the flow front is ellipsoidal. Based on the longitudinal-to-transverse ratio of the resulting pattern, flow parallel to fibers typically is found to be 15–25 times faster than flow across them.

The simple relationships given in Table 7.5 are not appropriate when the permeability is anisotropic. For modeling flow

TABLE 7.5 Solution of Darcy's Law for Simple Geometries

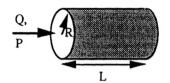 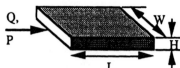 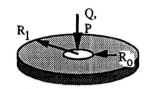

	At constant pressure P	At constant flow rate Q

Tube flow:

Flow length in time t; L

$$L = \left(\frac{2\kappa P}{\eta(1 - V)}\right)^{1/2}\sqrt{t}$$

$$L = \frac{Qt}{\pi R^2(1 - V)}$$

Time to fill length L; t_{fill}

$$t_{fill} = \frac{L^2\eta(1 - V)}{2\kappa P}$$

$$t_{fill} = \frac{L\pi R^2(1 - V)}{Q}$$

Transient condition; Flow rate Q, pressure P

$$Q = \pi R^2\left(\frac{\kappa P(1 - V)}{2\eta}\right)^{1/2}\frac{1}{\sqrt{t}}$$

$$P = \frac{\eta}{\kappa(1 - V)}\left(\frac{Q}{\pi R^2}\right)^2 t$$

Steady-state condition; Flow rate Q, pressure P

$$Q = \frac{\pi R^2\kappa}{\eta L}P$$

$$P = \frac{\eta L}{\pi R^2\kappa}Q$$

Rectilinear flow:

Flow length in time t; L

$$L = \left(\frac{2\kappa P}{\eta(1 - V)}\right)^{1/2}\sqrt{t}$$

$$L = \frac{Qt}{HW(1 - V)}$$

Time to fill length L; t_{fill}

$$t_{fill} = \frac{L^2\eta(1 - V)}{2\kappa P}$$

$$t_{fill} = \frac{LHW(1 - V)}{Q}$$

Transient condition; Flow rate Q, pressure P

$$Q = HW\left(\frac{\kappa P(1 - V)}{2\eta}\right)^{1/2}\frac{1}{\sqrt{t}}$$

$$P = \frac{\eta}{\kappa(1 - V)}\left(\frac{Q}{HW}\right)^2 t$$

Steady-state condition; Flow rate Q, pressure P

$$Q = \frac{HW\kappa}{\eta L}P$$

$$P = \frac{\eta L}{HW\kappa}Q$$

Radial flow:

Flow length in time t; R

$$\frac{R^2}{2}\left\{\ln\frac{R}{R_o} - \frac{1}{2}\left[1 - \left(\frac{R_o}{R}\right)^2\right]\right\} = \frac{\kappa Pt}{\eta(1 - V)}$$

$$R = \left(\frac{Qt}{\pi H(1 - V)} + R_o^2\right)^{1/2}$$

Time to fill radius R_1; t_{fill}

$$t_{fill} = \frac{\eta(1 - V)}{\kappa P}\frac{R_1^2}{2}\left\{\ln\frac{R_1}{R_o} - \frac{1}{2}\left[1 - \left(\frac{R_o}{R_1}\right)^2\right]\right\}$$

$$t_{fill} = \frac{\pi(R_1^2 - R_o^2)(1 - V)}{Q}$$

Transient condition; Flow rate Q, pressure P

$$Q = \frac{2\pi H\kappa}{\eta\ln(R_1/R_o)}P$$

$$P = \frac{\eta Q}{4\pi H\kappa}\ln\left(\frac{Qt}{\pi H(1 - V)R_o^2} + 1\right)$$

Steady-state condition; Flow rate Q, pressure P

$$Q = \frac{2\pi H\kappa}{\eta\ln(R_1/R_o)}P$$

$$P = \frac{\eta Q}{4\pi H\kappa}\ln\left(\frac{R_1^2}{R_o^2}\right)$$

Figure 7.9
Shape of the advancing resin front. Top view of symmetrically and asymmetrically woven mats.

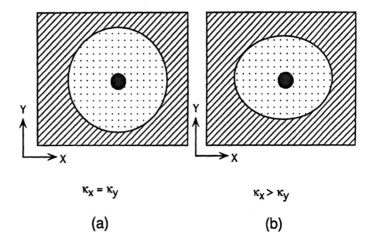

$\kappa_x = \kappa_y$

(a)

$\kappa_x > \kappa_y$

(b)

through anisotropic reinforcements, Darcy's equation must be solved with the permeability expressed as a tensor. When the woven mats have a greater concentration of strands in one direction than the other, the flow front is ellipsoidal. Further complications arise when mats of different fiber volume fractions or orientation or texture are layered together. The flow rate through one layer is not the same as the adjacent layer. Visualization of the flow front in the plane of a three-layered composite in Figure 7.10 shows a strong preference for flow in the most permeable layer, in this case the random fiber mat. Because flow from the most permeable layer feeds fluid transversely into the least permeable layers, simulation of multilayered filling requires simultaneous knowledge of the longitudinal and transverse permeability in each layer. The pressure to fill a cavity reduces when transverse flow occurs out of the most permeable layer into less permeable layers.

7.3 Capillary Action Also Assists Impregnation

When the gaps between reinforcements become sufficiently small, capillary forces assist impregnation by spontaneously drawing fluid into the mat. Flows generated by external pressure and capillary pressure play complementary roles in impregnation. External pressure drives the fluid to fill the macroscopic voids between aligned fiber rovings, while capillary pressure draws the fluid along the microscopic voids between individual filaments. Without capillary forces, it would be very difficult to impregnate fibers without forming voids.

The key variables determining impregnation by capillary flow are surface tension, solid–liquid wetting angle, viscosity, and the microstructure of the pores contained in the reinforcement material. In order to understand the interplay between the key variables, we first consider the static situation using the Young–Laplace relation, eq. 3.3.10, and then generalize the results to the dynamic situations encountered during impregnation. We find it

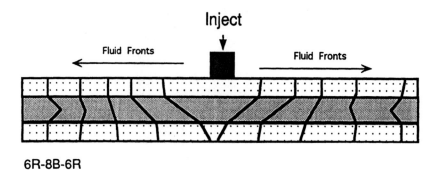

6R-8B-6R

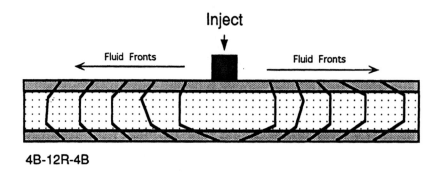

4B-12R-4B

Figure 7.10
Side view of flow front as a function of filling time. Computations for two sandwich laminates consisting of different thicknesses of continuous random fiber mat (R) and biaxial woven fabric (B). (W. B. Young, K. Rupel, K. Han, L. J. Lee, and M. J. Liou, *Polymer Composites* **12**, 30, 1991.)

necessary to distinguish capillary flow *along* the fibers and *across* the fibers in the next two subsections.

7.3.1 Capillary Flow Rate along Fibers Is Based on the Young–Laplace and Hagen–Poiseuille Equations

Figure 7.11a shows a liquid meniscus with a radius of curvature R_c formed between a bundle of five fibers with an effective pore diameter δ. Our model assumes that the curvature is dictated by the contact angle between the fluid and the fibers and that the five fibers constitute a "capillary tube" of diameter δ. From Section 3.3.1, eq. 3.3.10, R_c is related to the contact angle θ and pore diameter by $R_c \cos \theta = \delta/2$. We can write the Young–Laplace equation for capillary pressure as

$$\Delta P_c = \frac{C\gamma_{lv}}{R_c} = \frac{2C\gamma_{lv}}{\delta} \cos\theta \qquad (7.3.1)$$

where γ_{lv} is interfacial tension and C equals the number of principal axes of curvature in the surface of the fluid. For transverse flow perpendicular to fibers in fiber beds, the liquid front is cylindrical and $C = 1$; for longitudinal flow parallel to fibers (and for flow through particle beds), the liquid front is spherical or ellipsoidal and $C = 2$.

We can express δ in terms of fiber diameter D_f and volume fraction V_f by equalizing fiber–pore diameter ratios with volume fraction ratios

Figure 7.11
Schematic of a liquid meniscus advancing through fiber reinforcements: (a) flow along (parallel to) fiber direction, (b) flow across (perpendicular to) fibers. (Figure 7.11b from D. B. Clark and B. Miller, *Textile Research Journal* **48**, 256, 1978.)

(a)

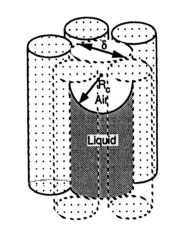

(b)

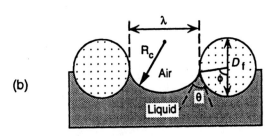

$$\frac{D_f}{\delta} = \frac{V_f}{1 - V_f} \tag{7.3.2}$$

Inserting δ from eq. 7.3.2 into eq. 7.3.1, we obtain ΔP_c as a function of V_f and D_f

$$\Delta P_c = \frac{2C\gamma_{lv} \cos \theta}{D_f} \frac{V_f}{1 - V_f} \tag{7.3.3}$$

We can obtain an approximate expression for the velocity v_c of the fluid front drawn along the fibers due to capillary pressure by inserting values for ΔP_c and δ in the Hagen–Poiseuille equation 7.2.2

$$v_c = \frac{\Delta P_c \delta^2}{32\eta L} \tag{7.3.4}$$

Substituting for δ and ΔP_c from eqs. 7.3.2 and 7.3.3, respectively, and assuming $C = 2$, the capillary flow velocity along the fibers is

$$v_c = \frac{\gamma_{lv} \cos \theta \; D_f}{8\eta L} \frac{1 - V_f}{V_f} \tag{7.3.5}$$

Again, we integrate eq. 7.3.5 to obtain

$$L = \left(\frac{\gamma_{lv} \cos \theta \; D_f}{4\eta} \frac{1 - V_f}{V_f} t \right)^{1/2} \tag{7.3.6}$$

or

$$L = (2\xi_c t)^{1/2} \tag{7.3.7}$$

where

$$\xi_c = \left(\frac{\gamma_{lv} \cos \theta}{8\eta} \frac{D_f}{1} \frac{1 - V_f}{V_f} \right) \tag{7.3.8}$$

ξ_c is a parameter that measures the spreadability of the liquid front under capillary pressure.

In general, we note that the impregnation velocity increases when the contact angle between fluid and fiber is small, fluid surface tension is high, and viscosity is low, and when the volume fraction of fibers is large and their diameter is small. If the contact angle is greater than 90°, ΔP_c is negative and capillary pressure opposes impregnation.

7.3.2 Capillary Flow across Fibers Fluctuates Due to Varying Meniscus Curvature

For capillary flow in the direction parallel to the bundle of fibers discussed in the previous section, the geometry of the interface does not change as the liquid is drawn forward. However, when the fluid moves across the fibers, the curvature of the meniscus, R_c, fluctuates and the model for transverse capillary flow must take this into account.

Figure 7.11b shows the profile of the meniscus of a fluid passing between two long fibers. Assuming that the contact angle θ remains constant, the radius of curvature of the meniscus, R_c, is determined geometrically from the filament diameter D_f, spacing λ, and the angle subtended from the center of the fiber to the contact point ϕ by

$$R_c = \frac{D_f}{2} \frac{(\lambda/D_f) + 1 - \sin \phi}{\sin(\phi + \theta)} \tag{7.3.9}$$

As the fluid comes into contact with the fiber, ϕ increases from 0 at initial contact to 180° when the fluid surrounds the fiber. Substituting eq. 7.3.9 into eq. 7.3.1 with $C = 1$ and rearranging gives

$$P'_c = \frac{\Delta P_c D_f}{2\gamma_{lv}} = \frac{\sin(\phi + \theta)}{(\lambda/D_f) + 1 - \sin \theta} \tag{7.3.10}$$

where P'_c is the dimensionless capillary pressure $\Delta P_c D_f / 2\gamma_{lv}$. Figure 7.12 shows the dependence of P'_c on ϕ and θ. By convention, positive pressure is associated with capillary forces drawing the liquid into the pore space, while negative pressure is associated with a resistance to impregnation.

When the liquid–air contact angle θ is close to zero, the capillary pressure is always positive and the liquid is drawn forward because it spontaneously wets around the entire fiber profile as ϕ goes from 0 to 180°. As the contact angle increases above zero, however, the meniscus curvature changes from posi-

Figure 7.12
Dimensionless pressure difference as a function of angle ϕ (see Figure 7.11) for various values of contact angle θ. (From D. B. Clark and B. Miller, *Textile Research Journal* **48**, 256, 1978.)

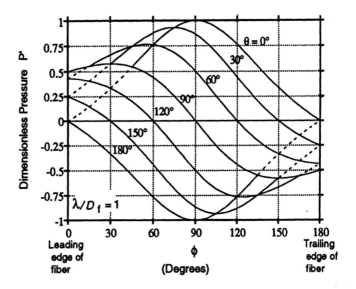

tive to negative somewhere at the back surface of the fiber. Then the liquid is drawn forward until it reaches the equilibrium contact point, where the curvature reverses. Keeping the fluid moving forward requires external pressure to overcome the increasingly negative capillary pressure. Under some conditions, flow hesitation at the equilibrium contact point may result in air entrapment when the fluid front bypasses the local resistance and moves through larger gaps.

The maximum and minimum values of the capillary pressure as the liquid progresses across the fibers depend on wetting angle and fiber spacing, as depicted in Figure 7.13. Capillary pressure is greatest when the fluid wets the fiber (θ close to zero) and when the fibers are close together ($\lambda/D_f \ll 1$). Adding

Figure 7.13
Dimensionless pressure difference as a function of contact angle θ for different fiber spacings $\lambda/D_f =$ 0.125, 0.25, 0.5, 1.0 corresponding to fiber volume fractions of 0.71, 0.58, 0.40, and 0.23, respectively. Dotted lines indicate unobtainable regions where the meniscus cross section is an arc greater than 180° with R_c greater than $\lambda/2$. (From D. B. Clark and B. Miller, *Textile Research Journal* **48**, 256, 1978.)

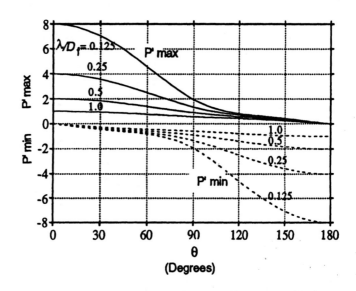

surfactants to the fluid to reduce the contact angle improves impregnation into tightly packed fibers. Capillary resistance to impregnation across the fibers is greatest for large fiber spacing (generally between fiber rovings), and here flow must be driven by external pressure to maintain the overall flow rate.

7.3.3 Flow Rates Due to External Pressure and Capillary Pressure Are Complementary

Assuming that no air entrapment occurs, then the rate of wetting of a fiber mat is given by combined contributions from external pressure, ΔP_e, and capillary pressure, ΔP_c,

$$V_{tot} = V_e + V_c = \frac{\Delta P_e \delta_e^2 + \Delta P_c \delta_c^2}{32 \eta L} \tag{7.3.11}$$

where the effective pore diameter for external pressure is $\delta_e = \sqrt{32 \kappa /(1 - V_f)}$ (obtained by solving eqs. 7.2.2, 7.2.5, and 7.2.6), and for capillary pressure is $\delta_c = D_f(1 - V_f)/V_f$ from eq. 7.3.2. Integrating eq. 7.3.11, we find that the impregnation length increases with $t^{1/2}$ in the same manner as eq. 7.2.8 and 7.3.7

$$L = (2\, \xi_{tot}\, t)^{1/2} \tag{7.3.12}$$

where ξ_{tot} now has contributions from both Darcy's law and capillary flow.

$$\xi_{tot} = \frac{\Delta P_e\, D_f^2}{16 K \eta}\, \frac{(1 - V_f)^2}{V_f^2} + \frac{C \gamma_{lv} \cos \theta\, D_f(1 - V_f)}{16 \eta V_f} \tag{7.3.13}$$

7.3.4 Unequal Flow Rates and Changes in Dynamic Contact Angle Promote Voids

Different rates of impregnation through large and small pores can lead to trapped air. As shown in Figure 7.14, we model this using two adjacent and interconnected pores where $\delta_1 < \delta_2$. Because capillary pressure increases with $1/\delta$, when ΔP_e is small, the smaller pore fills faster than the larger one, leaving air trapped in the larger pore. Conversely, when ΔP_e is large, the larger pore fills before the smaller one, leaving air trapped in the smaller pore. Trapping of air can be avoided only by selecting an external pressure such that the flow velocity of each pore is equal. Reinforcements contain a range of pore sizes; so the optimum

Figure 7.14
Dual-pore model for parallel flow paths with different effective diameters and kinematic viscosities. (R. Davé and S. Houle, "The Role of Permeability during Resin Transfer Molding," in *Proceedings of the American Society for Composites, Fifth Technical Conference,* East Lansing, MI, June 12–14, 1990, p. 539.)

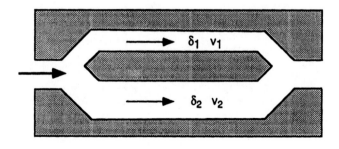

external pressure for one pore size does not suit another, and air entrapment inevitably occurs during impregnation. Trapped air is undesirable during the fluid fill process because it constricts flow and leads to time-dependent permeability as the trapped air becomes dissolved in the fluid.

Another important effect contributing to air entrapment is that the dynamic contact angle θ_d associated with a moving meniscus is larger than the equilibrium or static contact angle θ_s (see Appendix 8A). Simulations and measurements of dynamic contact angle of polypropylene have yielded a semiempirical relationship

$$\theta_d^3 - \theta_s^3 = 53 \, N_{ca \, loc} \qquad (7.3.14)$$

where $N_{ca \, loc}$ represents the capillary number based on the local flow velocity v_{loc} through the porous medium.

$$N_{ca \, loc} = \frac{\eta \, v_{loc}}{\gamma_{lv}} \qquad (7.3.15)$$

Thus the contact angle increases with injection speed, and the dynamic contact angle in narrow passages with high local velocities is larger than it is in large passages. As we have seen, larger contact angles can result in air entrapment when flow fronts join behind reinforcements.

A number of processing steps have been developed to reduce air entrapment. First, air can be removed by a combination of compression, flushing, and dissolution. An increase in cavity pressure compresses trapped bubbles as the flow front extends more and more beyond them. Once bubbles are below a critical size, they are swept away by the flow or rapidly dissolve into the resin. Removal of trapped air in this way has been used to explain increases in permeability with increasing flow rate, as shown in Figure 7.15. The resin solubility for air can be improved by preheating to remove predissolved air. Second, we can reduce air

Figure 7.15
Relative permeability as a function of flow rate for transient flow through 3, 4, and 5 layers of continuous random fiber mat. (R. Davé and S. Houle, "The Role of Permeability During Resin Transfer Molding," in *Proceedings of the American Society for Composites, Fifth Technical Conference,* East Lansing, MI, June 12–14, 1990, p. 539.)

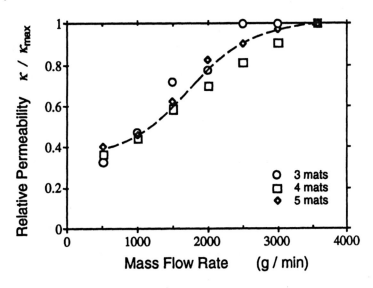

entrainment by lowering the contact angle between the polymer resin and the reinforcement. Surfactants often are added to the polymer or applied to the fiber surfaces to reduce surface tension. Third, we can decrease the flow rate or decrease the viscosity. Preheating reinforcements before processing is known to improve impregnation, probably through the combined reduction of surface tension and viscosity. Fourth, reinforcements can be impregnated under a vacuum to reduce the amount of trapped air.

7.4 Interfacial Bond Strength Can Be Manipulated by Chemical and Mechanical Treatments

As noted in Section 7.1, different mechanical applications have conflicting requirements on the interfacial bond strength, and it is essential to be able to manipulate the bonding between matrix and fiber by chemical and mechanical treatments. Before we describe the alternative treatments, we explore the mechanical behavior of composite systems in more detail to understand why interfacial bond strength is so important.

7.4.1 Interfacial Bond Strength Is Key to the Mechanical Behavior of Polymer Composites

Typically, reinforcing fibers have an elastic modulus 20–100 times larger than that of the polymer matrix. When an aligned fiber composite is under tension in the fiber direction, fibers that are under less stress elongate less than fibers that are under greater stress, creating a shear stress in the matrix. This shear stress transfers load to the fibers that have less elongation. In equilibrium, the displacements of all fibers are nearly the same. Hence, the interfacial bond is instrumental in redistributing the tensile stress among the fibers. For maximum tensile strength the interfacial bond strength should be as high as possible.

Another important physical property is fracture toughness. Fracture toughness is the amount of energy absorbed during fracture. The fracture toughness of a polymer composite may be quite different from that of its polymer matrix. For example, if the matrix is a brittle isotropic polymer (see Section 6.2.4), instantaneous fracture occurs at the tensile strength. A crack propagates rapidly over the plane perpendicular to the stress, as shown in Figure 7.16a; failure is catastrophic, and no energy is absorbed in the process. In this case the fracture toughness is minimal. On the other hand, Figure 7.16b shows that if a composite is made from the same brittle polymer matrix reinforced with fibers, cracks can be deflected and arrested by the fibers. In this case, failure is delayed, considerable mechanical energy is expended to pull the material apart, and the composite has a much higher fracture toughness.

We can understand this difference in fracture toughness from a simple argument based on the classical Griffith theory (see Appendix 11B). The Griffith theory was developed to account for

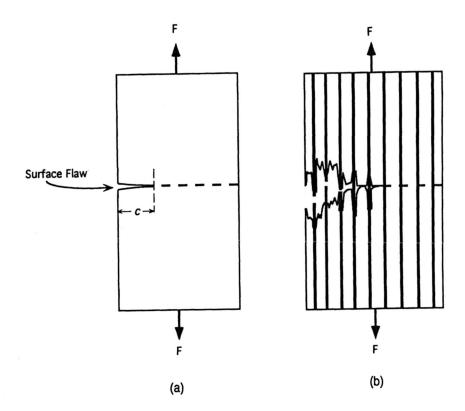

F

Surface Flaw

$\leftarrow c \rightarrow$

F

(a)

F

F

(b)

Figure 7.16
Fracture path in brittle polymer composites. (a) In an isotropic brittle polymer the fracture propagates across the flat plane normal to the applied stress. (b) In a fibrous composite the fracture path is deflected and forced to propagate parallel to the fibers until the fibers pull out or break.

the fact that while the theoretical cohesive tensile strength of a solid is E/15, brittle materials like glass or polystyrene below T_g break at values about E/1000. Griffith proposed that their reduced strength was due to stress concentrations at flaws at the surface or voids inside the materials. In the presence of surface flaws of depth c, the fracture strength, or ultimate tensile strength σ_u, is given by the Griffith equation,

$$\sigma_u = \left(\frac{2E\gamma_f}{\pi c}\right)^{1/2} \tag{7.4.1}$$

The controlling parameter in this fracture process is γ_f, the fracture surface energy—the energy required to fracture the material per unit area of fracture surface. It is given by the sum of the all the energetic processes involved in fracture: (1) the intrinsic surface energy required to create two new surfaces, γ_s; (2) the plastic deformation energy to tear the material apart, γ_p; and (3) energy dissipation by other mechanisms, such as crack deflection at the matrix–fiber interface, γ_d,

$$\gamma_f = \gamma_s + \gamma_p + \gamma_d + \ldots \tag{7.4.2}$$

The material property $\sqrt{2E\gamma_f}$ is defined as the *fracture toughness*. It is the parameter that relates fracture strength to flaw size in a brittle or semibrittle material.

The fracture toughness of different materials is dominated by different components of γ_f. In brittle isotropic polymers the only energy needed to fracture the material is γ_s, and they have almost zero fracture toughness. When the matrix is ductile, as with plastic polymers and metals, γ_p is high and the material is very tough. In a fibrous composite, γ_d can be high if the crack is deflected to run parallel to the applied stress, as shown in Figure 7.16b. The composite can then be much tougher than the matrix material. For optimum fracture toughness the interfacial bond strength must be manipulated, as discussed below.

7.4.2 Impact Strength Depends on Characteristics of the Matrix and the Interface

An important physical property exploited in polymer composites is their impact strength. High impact resistance requires a combination of both high tensile strength and high fracture toughness. The impact strength of polymer composites can be quite different from that of the isotropic matrix materials and may decrease or increase with fiber volume fraction, depending on the nature of the matrix, the fiber direction, and the interfacial bond strength as follows:

- If the matrix is a ductile polymer, the presence of fibers will increase the tensile strength but may decrease the fracture toughness, because they decrease the volume of ductile material and constrain plastic deformation, thereby reducing γ_p.

- If the matrix is a brittle polymer, fibers will increase the strength and act as crack arrestors to increase the amount of energy γ_d absorbed by crack deflection. This enhancement of fracture toughness can be particularly effective when the applied stress is parallel to the fiber direction. In that case, cracks can run either parallel or perpendicular to the applied stress, depending on the strength of the fiber and the interfacial bond strength. The increase in γ_d depends on the strength of the matrix–fiber bond.

  —If matrix–fiber bonding is weak, stress concentration at the tip of a crack causes the fibers to debond from the matrix. The advancing crack is deflected and forced to grow parallel to the stress along the fibers, as shown in Figure 7.16b. Failure is delayed and more energy is consumed in the fracture process. For this condition strength is lost as the fibers break under increasing loads, but fracture toughness is greatly enhanced.

  —If matrix–fiber bonding is strong, then fibers break before debonding from the matrix, causing brittle failure perpendicular to the stress direction, similar to isotropic materials. For this condition little strength is lost because the load carried by the fiber is transferred to neighboring fibers, but the fracture toughness is not enhanced.

Interfacial bond strength requirements for tensile strength may run counter to requirements for toughness. Optimal bonding conditions for both tensile strength and toughness are achieved when the bond strength causes fracture to follow a path in the polymer slightly above the fiber surface, as illustrated in Figure 7.17b. Then plastic deformation in the polymer absorbs a large amount of energy, while a thin layer of polymer adhering to the fiber restrains crack propagation during impact. An optimally bonded interface absorbs energy yet maintains good interfacial stress transfer in tension.

A very weak interfacial bond is undesirable because it results in poor weatherability due to void formation followed by penetration of water or other solvents into the cracks. In this case, the fibers debond from the matrix, resulting in progressive failure along the interface.

Table 7.6 summarizes these points. Clearly, achieving composites with high strength, toughness, and weatherability requires an optimal degree of interfacial bonding. Much effort has gone into the treatment of fiber surfaces and the selection of additives to modify and control the matrix–fiber interfacial bond strength.

7.4.3 Fiber–Matrix Interface Bond Strength Is Controlled by Various Surface Treatments

We can manipulate the bonding of the reinforcement to the matrix in composites by modifying the physical, mechanical, or chemical properties of the interface:

Physical interactions—such as Lewis acid–base interactions, hydrogen bonding

Mechanical interlocking—surface etching followed by matrix shrinkage onto fibers after cure

Figure 7.17
Failure conditions in glass–polypropylene composites depend on chemical coupling. (a) With no chemical coupling, poorly bonded fibers are completely removed from the composite, leaving holes in the polymer matrix. (b) With silane coupling, excellent adhesion leads to rupture in the polymer matrix immediately adjacent to the glass–polymer interface. Fiber diameter ≈10 μm. (*Guide to Silane Coupling Agents,* Dow Corning Corporation, Midland, Michigan.)

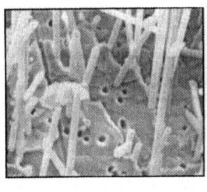
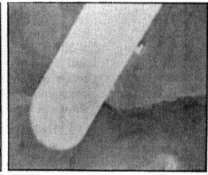

(a)

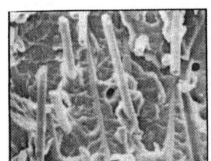

(b)

TABLE 7.6 Qualitative Effects of Interfacial Shear Strength on Composite Properties

	Low	*Intermediate*	*High*
Crack propagation	At interface or under fiber surface[a]	Slightly above interface	Through matrix or transverse to fiber
Weatherability	Poor	Good	Good
Longitudinal strength	Good	Good	Excellent
Transverse strength	Poor	Good	Excellent
Fracture toughness	Good	Good	Poor

[a]For untreated carbon fibers with weakly bonded surface layers. Weak surface layers are usually removed by plasma treatment for acceptable bonding.

Chemical bonding—covalent and polar bonds.

Carbon fibers have weakly bound layers of graphite on the surface that cause premature fracture *under* the fiber surface. To remove these weak surface layers, the fibers are etched. On the new surface, polar oxygen- or nitrogen-containing groups (such as acids, amines, and esters) lattice bond to the matrix by physical interactions. As illustrated in Figure 7.18, etching also creates surface roughness for mechanical interlocking. Etching agents include oxygen–nitrogen mixtures below 1100°C, other oxidizing agents (CO_2, Cl_2, NO_2-NO, NH_3) at higher temperatures, ionized inert gases (plasma), or aqueous solutions of nitric or sulfuric acid. Not surprisingly, excessive amounts of surface treatment with these corrosive environments can reduce the fiber tensile strength.

For chemical bonding, coupling agents are used. Coupling agents are bifunctional (amphiphilic) molecules that react with both organic and inorganic phases. Silanes used for glass fibers have the general structure $Y(CH_2)_nSiX_3$, where n = 0–3, Y is an organofunctional group that bonds with the matrix, and X is a hydrolyzable group (usually methoxy, $-OCH_3$) that reacts with water to form a silanol, $-Si(OH)_3$. Table 7.7 lists commercially available silanes. The attachment mechanism shown in Figure 7.19 includes:

1. Hydrolysis—hydrolyzing the silane to form a trihydroxy silanol.

2. Chemisorption—subsequent reaction of silanol with pendant hydroxy groups on the surface of the glass fiber.

3. Condensation—occurs with other surface sites or with other silanol molecules.

The reaction with the interface is reversible in the presence of water. Hence, the dynamic equilibrium of the silanol–hydroxy bonds provides a means to move silane molecules over the surface, relaxing shear stresses without loss of adhesion. The advantages of silane coupling agents applied to fillers and reinforcements are: (1) improved tensile, flexural, and compressive strengths; (2) improved property retention in wet environments; (3) less stress cracking of the matrix; and (4) improved wetting and dispersion of fillers throughout the resin matrix.

Figure 7.18
The effect of plasma treatment on monofilaments of polyethylene with a draw ratio of 30:1: (a) before treatment, (b) after treatment. (N. H. Ladizesky and I. M. Ward, *J. of Materials Science* **18**, 533, 1983.)

(a)

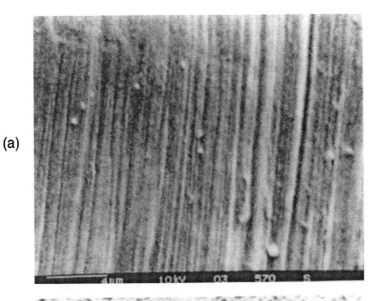

(b)

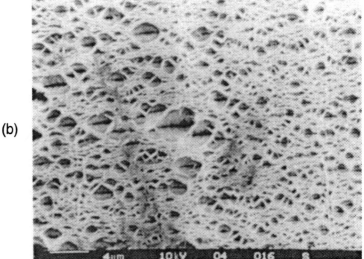

TABLE 7.7 Common Coupling Agents Used for Glass Fiber–Polymer Matrix Composites

Chemical type	Chemical formulation	Applications
Diamino	$NH_2CH_2CH_2NH(CH_2)_3Si(OCH_3)_3$	Epoxies, phenolics, melamines, nylons, PVC, acrylics, polyolefins, polyurethanes, nitrile rubbers
Methacrylate	$CH_2=C(CH_3)COO(CH_3)_3Si(OCH_3)_3$	Unsaturated polyester, rubber, polyolefins, styrenics, acrylics
Epoxy	$CH_2OCHCH_2O(CH_2)_3Si(OCH_3)_3$	Epoxies, urethane acrylate, and polysulfide sealants
Vinyl	$CH_2=CHSi(OCOCH_3)_3$	Polyesters, polyolefins, EPDM, EPM
Chloroalkyl	$Cl(CH_2)_3Si(OCH_3)_3$	Epoxy, styrenics, nylon

Source: Guide to Silane Coupling Agents, Dow Corning Corporation, Midland, Michigan.

Hydrolysis:

$$R\text{-}Si(OR)_3 + 3H_2O \longrightarrow R\text{-}Si(OH)_3 + 3\,HOR$$

Chemisorption:

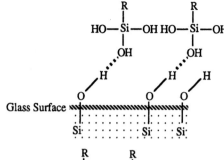

Drying, Self-Condensation and Coupling:

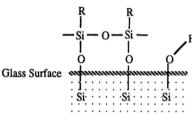

Figure 7.19
Silane bonding to silica glass.

7.4.4 Fiber–Matrix Interfacial Bond Strength Is Measured by Direct or Indirect Techniques

We determine the interfacial bond strength, τ_{int}, by adhesion measurements that are either *direct* or *indirect*. They are summarized in Table 7.8 and Figure 7.20. In each case, we assume the fibers to be unidirectional and uniformly wetted by the matrix to avoid local stress concentrations. We also assume that the stress is constant along the length of the fibers. This assumption may or may not be true, depending on sample geometry and the shear and tensile modulus of the fibers and matrix.

TABLE 7.8 Summary of Methods for Determining Interfacial Fiber–Matrix Interaction

Method	Specimen	Deficiency	τ_{crit} (MPa)[a]
Fiber pull-out	Single fiber coupons	Difficult sample preparation Large data scatter Nonuniform shear stress distribution	50.3 (microdrop)
Critical fiber length	Single fiber coupons	Neglects plastic matrix deformation Distribution of fiber fragment lengths	68.3
Microindentation	Actual composites	Fiber splitting enhances fiber debonding Variable spacing between neighboring fibers Debonding can be difficult to detect	65.8

M. Narkis, E. J. H. Chen, and R. B. Pipes, *Polymer Composites* **9**, 245 (1988).

[a]Values measured for AS-4 carbon fibers in Epon 828 epoxy resin. (L. T. Drzal and P. J. Herrera-Franco, in *Adhesives and Sealants*, H. F. Brinson, Technical Chairman, Vol. 3 of *Engineered Materials Handbook*, ASM International, Materials Park, Ohio, 1990, p. 403.)

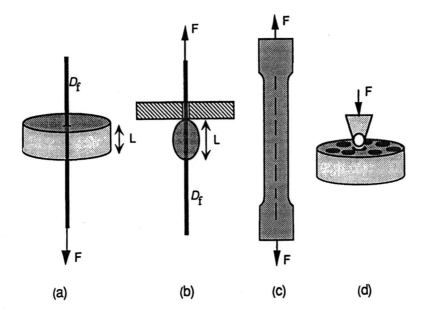

F F F F

D_f D_f L L F

(a) (b) (c) (d)

Figure 7.20
Techniques to measure interfacial shear strength of fiber-reinforced plastics: (a) single fiber pull-out test, (b) single fiber microdroplet test, (c) single fiber fracture test, (d) fiber micro-indentation test.

Direct methods for measuring τ_{int} involve a single filament embedded into a resin disk (Figure 7.20a) or a microdroplet (Figure 7.20b). We relate the force required to pull the fiber from the resin to the interfacial bond strength using a simple force balance,

$$\tau_{int} = \frac{\text{pull–out force}}{\text{interfacial area}} = \frac{F}{\pi D_f l} \qquad (7.4.3)$$

Direct measurement methods are difficult to perform because the fiber often breaks before the interface. Furthermore, preparing and conditioning the test specimens requires substantial time. The microdroplet method (Figure 7.20b) may give low values of τ_{int} due to evaporation of volatile polymer precursors while the microdroplet is being cured. More important, the fiber–matrix interface has different matrix geometry and residual curing stresses in single fiber test samples than in composite parts. As a consequence, direct techniques may not provide quantitatively accurate values of τ_{int}, although they do provide a viable comparison of the relative interfacial bond strengths after various fiber surface treatments.

A particularly useful *indirect* test involves elongating a tensile specimen having a single filament embedded in the resin matrix (Figure 7.20c). (Stress analysis relevant to this technique is discussed in more detail in Sections 7.5.1.2 and 7.5.2.2). Because the matrix elongates more than the fiber, the fiber is forced either to break or to delaminate from the matrix along the interface. If the fiber is sufficiently long, the tensile stress on the fiber reaches its breaking stress σ_{uf} (u, ultimate strength; f, fiber) at its midpoint before interfacial delamination. Then the fiber breaks in two. As more load is applied to the specimen, the fiber breaks into successively shorter half lengths. Ultimately, the fiber lengths are shorter than the critical length l_c, and the interfacial bond shear strength, τ_{int}, cannot transfer enough stress to break the fiber at its midpoint. At this juncture, the tensile force at the midpoint of the fiber due to the interfacial shear stress ($\tau_{int} \pi D_f l_c/2$) must be less than the force needed to break the fiber ($\sigma_{uf} \pi D_f^2/4$). Equating these two forces gives τ_{int} as follows

$$\tau_{int} = \frac{\sigma_{uf} D_f}{2l_c} \tag{7.4.4}$$

Assuming that fibers slightly longer than l_c break in half while fibers slightly shorter than l_c do not break, then broken fiber segments after this test should vary in length from just above $l_c/2$ to just below l_c. To measure the broken lengths, the matrix should either be transparent (so that the segment ends can be located) or x-rays might be used to reveal them. Alternatively, segments can be retrieved after the resin has been dissolved or burned away. Experimental uncertainties arise when this technique is used with very brittle matrix materials because cracks may initiate in the matrix instead of the fiber. As in the fiber pullout test, stress fields in the test specimen may differ from those in the composite; so direct application to composite parts involves some uncertainty.

Fiber microindentation measurements illustrated in Figure 7.20d provide another indirect way to measure τ_{int} in composites. Parts are cut perpendicular to the fiber direction, and single filaments are compressed axially with a rounded tip probe. The force F needed to debond the fiber from the surrounding matrix is proportional to τ_{int},

$$\tau_{int} = A\left(\frac{4F}{\pi D_f^2}\right)\left(\frac{G_m}{E_f}\right)^{1/2} \tag{7.4.5}$$

where the first factor in parentheses is the nominal fiber compressive stress, G_m is the shear modulus of the matrix, E_f is the tensile modulus of the fiber, and the constant A is determined by finite element modeling as a function of the fiber nearest-neighbor spacing. Unlike other techniques we have described, in this test, interfacial shear stress is not assumed to be uniform along the fiber, but rather to reach a maximum at some fraction of a fiber diameter away from the surface, as predicted in a finite element analysis. In practice, this test inolves experimental uncertainties because of difficulties in detecting the actual debonding load, variations in the debonding mechanism (for example, fracture through the fibers or matrix rather than at the interface), and assumptions used in determining A with the finite element structural analysis.

7.5 Polymer Composite Mechanical Properties Depend on Fiber Reinforcement, Matrix Properties, and Interfacial Bond Strength

The principle of stress transfer between the matrix and the reinforcements provides guidelines for using individual constituent fiber and matrix properties to predict the mechanical behavior of fiber-reinforced composites. Mechanical properties of major interest are the composite stiffness, defined by the elastic modulus (E_c where the subscript c indicates composite), and the composite ultimate tensile strength (σ_{uc}). To simplify the analysis, we first derive the mechanical behavior of idealized composites in which the fibers are assumed to be long and *continuous*, running from end to end through the composite. Then we extend the analysis to more typical composites in which the fibers are *discontinuous* or "chopped." In our discussion, we will distinguish between properties in the *longitudinal* direction (distinguished by the subscript L), measured parallel to the fibers, and the *transverse* direction (distinguished by the subscript T), measured perpendicular to them.

7.5.1 The Elastic Properties of Fiber-Reinforced Composites

7.5.1.1 Elastic Modulus of Continuous Fiber-Reinforced Composites Is Determined by the Rule of Mixtures

Figure 7.21 depicts the idealized structure of a continuous fiber-reinforced composite. We distinguish the mechanical behavior for two extreme orientations: when the composite is loaded parallel to the fibers (longitudinal L properties) and when the composite is loaded perpendicular to them (transverse T properties).

Figure 7.21
An idealized continuous fiber composite loaded (a) longitudinally and (b) transversely.

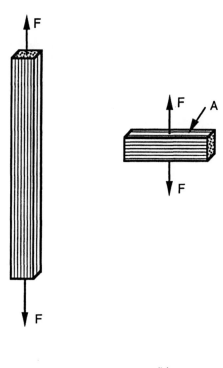

(a)   (b)

For *longitudinal* loading, Figure 7.21a, the tensile *strain* ε_{xx} must be the same for the fibers (subscript f) and matrix (subscript m) that constitute the composite, so that

$$\varepsilon_{xx} = \frac{\sigma_f}{E_f} = \frac{\sigma_m}{E_m} = \frac{\sigma_{cL}}{E_{cL}} \qquad (7.5.1)$$

where stress σ is tensile load per unit area, F/A, and E_f and E_m are the moduli of the fibers and matrix, respectively. For unidirectional composites, the fibers and matrix cross-sectional areas are proportional to their respective volume fractions, and so eq. 7.5.1 becomes

$$\frac{F_f}{E_f V_f} = \frac{F_m}{E_m(1 - V_f)} = \frac{F_m + F_f}{E_{cL}} \qquad (7.5.2)$$

where V_f is fiber volume fraction and $1-V_f$ is the volume fraction of the matrix. The ratio of the loads carried by the fibers and matrix is given by rearrangement of eq. 7.5.2,

$$\frac{F_f}{F_m} = \frac{E_f V_f}{E_m(1 - V_f)} \qquad (7.5.3)$$

and plotted in Figure 7.22. For a typical composite material with E_f/E_m of 40 and V_f of 0.6, $F_f/F_m = 60$, and the fibers carry over 98% of the load.

Manipulating eq. 7.5.2, we can compute the longitudinal composite modulus E_{cL} (c, composite; L, longitudinal) and obtain

$$E_{cL} = E_f V_f + E_m (1 - V_f) \qquad (7.5.4)$$

This result is the typical expression for composite properties by the linear rule of mixtures.

Figure 7.22
Ratio of load carried by fibers and matrix as a function of the ratio of respective moduli and fiber volume fraction.

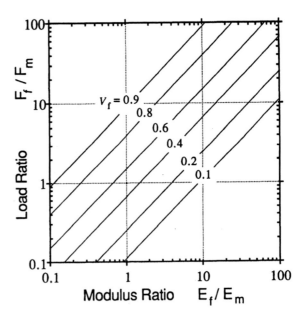

For *transverse* loading, Figure 7.21b, we assume that the *stress* is the same in both fibers and matrix throughout the composite. Using a method analogous to that used to derive eq. 7.5.4, the rule of mixtures then gives

$$\frac{1}{E_{cT}} = \frac{V_f}{E_f} + \frac{(1 - V_f)}{E_m}$$

or

$$E_{cT} = \frac{E_m E_f}{E_m V_f + E_f(1 - V_f)} \tag{7.5.5}$$

7.5.1.2 End Effects Decrease the Elastic Modulus in Discontinuous Fiber Composites—The Longitudinal Elastic Modulus Depends on the Fiber Aspect Ratio

For many composite applications, chopped fiber reinforcements are easier to process than continuous fibers. For example, fibers from 0.05 to 0.5 inches long commonly are mixed with thermoplastics for injection molding complex shapes. For reasons that we will describe later, the longitudinal modulus E_{cL} is less for short-fiber composites than for long-fiber composites.

Differences in modulus arise because the tensile stress in a discontinuous fiber (σ_f) must decline to zero at the unrestrained ends, as shown in Figure 7.23. There is no way to hold the very ends of a fiber for it to be loaded there. Instead, the tensile load is transferred to the fiber through the shear stress at the interface between the fiber and the matrix. (The situation resembles trying to stretch a piece of spaghetti by holding each end between your thumb and forefingers.) A simple analysis shows that the tensile stress in the discontinuous fiber increases linearly with distance z along the fiber according to the relation

Figure 7.23
Profiles of fiber stress σ_f and interfacial shear stress τ along discontinuous fibers in composites under tensile loading.

$$\sigma_f = \frac{4\tau_{int}}{D_f} z \tag{7.5.6}$$

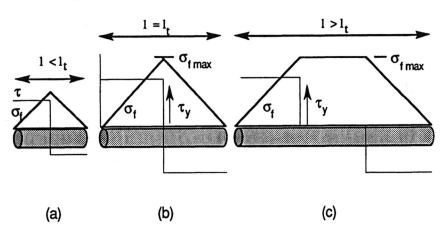

(a) (b) (c)

reaching the highest value at the midlength, $z = l/2$, as depicted in Figure 7.23a and b. τ_{int} is the interfacial shear strength determined by the adhesion between the reinforcement and the matrix. Note that the slope of the curve, and hence the rate at which the tensile stress increases along the fiber, is proportional to τ_{int}. τ_{int} dictates the *efficiency* of load transfer from the matrix to the fiber.

The strain in the fiber cannot exceed the strain in the composite; so for a given load and stress on the composite, the highest stress reached in a fiber, $\sigma_{f\,max}$, is the value it would experience if it were a continuous fiber, given by σ_f in eq. 7.5.1. A very short fiber never reaches this stress at its midpoint, Figure 7.23a. A very long fiber reaches this stress over a portion of its length in the midsection, Figure 7.23c. The fiber that just reaches the maximum stress exactly at its midpoint has a length equal to the *load transfer length* l_t, Figure 7.23b. From eq. 7.5.6

$$l_t = \frac{\sigma_{f\,max}D_f}{2\tau_{int}} \tag{7.5.7}$$

Substituting values for $\sigma_{f\,max}$ into eq. 7.5.7 obtained by solving eqs. 7.5.1. and 7.5.4, we obtain an expression for l_t in terms of the stress on the composite σ_{cL} and the moduli and volume fractions of the matrix and reinforcement

$$l_t = \frac{\sigma_{cL}D_fE_f}{2\tau_{int}[E_fV_f + E_m(1 - V_f)]} \tag{7.5.8}$$

Equation 7.5.8 shows that increasing the load and thus the longitudinal stress σ_{cL} on the composite also increases the load transfer length. Thus, as the load increases in a composite that possesses a range of fiber lengths, fewer and fewer fibers have the required transfer length. As a consequence, a declining fraction of the fibers are able to be fully stressed.

These end effects decrease the *average* fiber tensile stress, which in turn decreases the composite tensile modulus. The magnitude of the decrease depends on the fiber geometry.

1. When fibers extend over the entire length of the composite, they create the continuous fiber configuration analyzed in Section 7.5.1. The average fiber tensile stress $\bar{\sigma}_f$ is uniform throughout the composite

$$\bar{\sigma}_f = \sigma_{f\,max}, \quad \text{when } l \gg l_t \tag{7.5.9}$$

2. With long fiber composites, the fiber lengths always exceed l_t. The average fiber tensile stress decreases slightly due to the end effects

$$\bar{\sigma}_f = \sigma_{f\,max}\left(1 - \frac{l_t}{2l}\right), \quad \text{when } l > l_t \tag{7.5.10}$$

3. With short fiber composites, the fiber lengths are less than l_t, and the maximum fiber stress is never attained in any of them. The average fiber stress depends on the interfacial shear stress

$$\overline{\sigma}_f = \frac{\tau_{int} l}{D_f}, \quad \text{when } l < l_t \qquad (7.5.11)$$

Due to end effects, the resultant longitudinal tensile modulus for discontinuous fibers is obviously more complex than for continuous fibers. It is given by the Halpin–Tsai equation,

$$\frac{E_{cL}}{E_m} = \frac{1 + \beta_L \chi_L V_f}{1 - \chi_L V_f} \qquad (7.5.12)$$

where $\chi_L = (E_f/E_m - 1)/(E_f/E_m + \beta_L)$ and $\beta_L = 2l/D_f$. l/D_f is the *fiber aspect ratio*. Note that χ_L is always less than unity and decreases as the fibers become longer. For very long fibers, β_L becomes a large number, and eq. 7.5.12 approaches the relationship for continuous fibers (eq. 7.5.4) as we would expect.

For long fibers, the ends have a negligible effect on composite physical properties. For example, using eq. 7.5.10, we can see that the average fiber stress in composites whose fiber lengths equal $5l_t$ is 90% of that in continuous fiber composites. Hence, the modulus of long-fiber composites approaches that of continuous fibers, a factor that may be important for processes, such as injection molding, that cannot handle continuous fiber composites.

Figure 7.24 plots changes in longitudinal modulus with fiber aspect ratio and volume fraction according to the predictions of the Halpin–Tsai equation.

7.5.1.3 End Effects Have Less Influence on the Transverse Elastic Modulus and the Modulus of Randomly Oriented Short Fiber Composites

End effects exert less influence on the transverse modulus because the load is applied normal to the fiber direction and E_{cT} is not affected by the presence of fiber ends in discontinuous fiber composites. However, in practice, the simple expression for E_{cT}, eq. 7.5.5, fails because it neglects nonuniform fiber packing. The Halpin–Tsai equation takes packing inefficiency into account and gives the following result for the transverse composite modulus, E_{cT}

$$\frac{E_{cT}}{E_m} = \frac{1 + \beta_T \chi_T V_f}{1 - \chi_T V_f} \qquad (7.5.13)$$

where

$$\chi_T = \frac{E_f/E_m - 1}{E_f/E_m + \beta_T}$$

β_T is a constant that takes account of the shape and alignment of the fibers. For transverse properties of unidirectional fibers, $\beta_T = 2$.

When discontinuous fibers are randomly oriented, an empirical equation relates the resultant modulus to the theoretical longitudinal and transverse values from eqs. 7.5.12 and 7.5.13, respectively,

Figure 7.24
Predictions of the Halpin–Tsai equation for longitudinal modulus of short-fiber composites as a function of fiber length and fiber volume fraction.

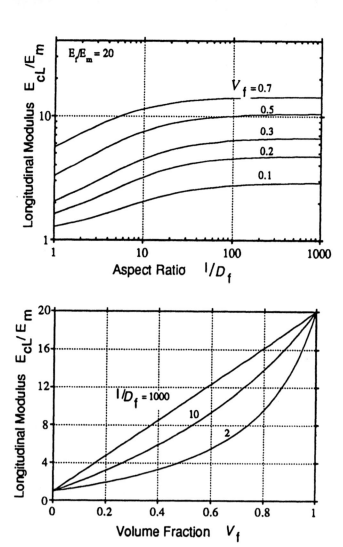

$$E_{random} = \frac{3}{8} E_{cL} + \frac{5}{8} E_{cT} \qquad (7.5.14)$$

7.5.2 The Tensile Strength of Fiber-Reinforced Composites

7.5.2.1 Tensile Strength of Continuous Fiber-Reinforced Composites Is Determined by the Rule of Mixtures

The ultimate tensile strength of a composite, σ_{uc} (u, ultimate; c, composite), made of continuous fibers depends on the stress direction and the failure mechanism. Tensile strength parallel to the fibers ($\sigma_{uc,L}$; L, longitudinal) is limited by failure of either the fiber or the matrix; tensile strength perpendicular to the fibers ($\sigma_{uc,T}$; T, transverse) is limited by the failure of either the matrix or the fiber–matrix interface. We have three failure modes to consider.

When the ultimate breaking strength of the fibers (σ_{uf} in Table 7.2) limits the *longitudinal* strength of the continuous fiber composite, eq. 7.5.1 gives

$$\sigma_{uc,L} = \frac{E_{cL}}{E_f} \sigma_{uf} \qquad (7.5.15)$$

Substituting E_{cL} from eq. 7.5.4 into eq. 7.5.15 gives

$$\sigma_{uc,L} = \frac{E_f V_f + E_m(1 - V_f)}{E_f} \sigma_{uf} \qquad (7.5.16)$$

When the strength of the matrix limits the longitudinal strength of the continuous fiber composite, the rule of mixtures gives

$$\sigma_{uc,L} = \sigma_f V_f + \sigma_m{}^* (1 - V_f) \qquad (7.5.17)$$

where $\sigma_m{}^*$ represents the matrix stress at failure.

The *transverse* strength of the composite ($\sigma_{uc,T}$) is determined when the stress exceeds the tensile adhesion strength of the fiber–matrix interface, σ_{int}, and the matrix between them fails. Using the rule of mixtures, the tensile strength of the composite then becomes

$$\sigma_{uc,T} = \sigma_{int} V_f + \sigma_m{}^* (1 - V_f) \qquad (7.5.18)$$

In reality, the longitudinal tensile failure of a continuous fiber composite is not a precipitous event; it occurs gradually, as we see in Figure 7.6. Failure typically starts when one of the continuous fibers breaks and we observe a momentary load drop as the load it was carrying is transferred to the neighbors. With a further increase in load, another fiber breaks and then another until ultimately all the fibers break in quick succession at the ultimate tensile strength. When all the fibers have broken, the continuous fiber composite is much weaker. It behaves like a discontinuous fiber composite, and its strength continuously declines until the matrix finally gives way.

Equations 7.5.16, 7.5.17, and 7.5.18 predict ultimate tensile strengths generally larger than the measured value. In practice, the longitudinal strength of a continuous fiber composite is reduced by misalignment of the fibers, nonuniform loading, and nonuniform fiber strength. Transverse strength is reduced because the interface is not necessarily uniformly bonded along the complete length of the fiber due to the variable nature of the surface treatments used to enhance bond strength.

7.5.2.2 The Longituduinal Tensile Strength of Discontinuous Fiber Composites Depends on the Interfacial Shear Strength

The ultimate tensile strength of a longitudinally loaded discontinuous fiber composite is determined by three possible sources of failure. The shear strength of the interface between fiber and matrix may be the weakest link, the polymer matrix may rupture, or the fiber may break. We consider each type of failure in turn.

First, if the fiber–matrix adhesion is weak, then as we increase the load, the shear stress at the interface soon exceeds the interfacial shear strength γ_{int}, and the interface gives way. The shear stress along the major length of the fiber–matrix interface drops to zero and the fiber is loaded from both ends by only the short segments, of length $l_t/2$ (eq. 7.5.8) that are still adherent, and is never stressed to the maximum value. Eventually the fiber pulls out of the composite. Under these circumstances, the ultimate strength of the composite is limited by the strength of the matrix

$$\sigma_{uc,L} = \sigma_m{}^* \, (1 - V_f) \qquad (7.5.19)$$

Second, if the shear stress in the vicinity of the interface exceeds the shear strength of the matrix γ_m, the matrix yields plastically and deforms with no increase in resistance. Using the average stress equations already described, we can show that then the strength of unidirectional short fiber composites is

$$\sigma_{uc,L} = \frac{\tau_m V_f}{D_f} + \sigma_m{}^* \, (1 - V_f) \qquad (7.5.20)$$

assuming that failure occurs right at the fiber–matrix interface.

Third, if the interfacial bond strength and the matrix shear strength are both very strong, the tensile stress at the midlength of the fiber can be pushed to exceed the ultimate strength of the fiber (σ_{uf}). The ultimate strength of the composite is then given by eq. 7.5.16. The minimum fiber length needed to accomplish this is the critical load transfer length l_c, where

$$l_c = \frac{\sigma_{uf} D_f}{2 \tau_{int}} \qquad (7.5.21)$$

For a given diameter and interfacial bond strength, fibers much longer than l_c become stressed above their ultimate strength and break in two at about their midpoint, where the stress reaches a maximum. If the new "half-length" still exceeds l_c, then the fiber will break in two again as the load increases. This process continues until all the discontinuous fibers have a length of l_c or less. The ratio l_c/D_f is the critical aspect ratio. As described in Section 7.4.3, this process can provide one means to determine τ_{int}.

To achieve the highest longitudinal strength in composites with discontinuous fibers, it is important, therefore, to use fibers of high intrinsic strength that possess high aspect ratios and have been treated for maximum interface adhesion. Using the average stress equations, we can show that under these circumstances the strength of unidirectional long fiber composites is

$$\sigma_{uc,L} = \sigma_{uf} \left(1 - \frac{l_c}{2l} \right) V_f + \sigma_m{}^* \, (1 - V_f) \qquad (7.5.22)$$

which approaches the value in eq. 7.5.17 for continuous fiber composites when $l \gg l_c$.

A commonly neglected yet important feature of the stress state of a composite is the residual stress exerted by the matrix on the fibers after the resin has set. Residual compressive stresses due to shrinkage during processing may change the composite's response to applied loads through Poisson's ratio effects, but more significantly they affect the tensile strength by mechanical interlocking and by raising the friction between the matrix and reinforcements. Even though debonding may develop along an interface during loading in the fiber direction, stress still can be transferred between the fiber and matrix by mechanical interlocking and friction. The presence of residual and compressive stresses is one reason why τ_{int} measured by the single-fiber pull test described in the Section 7.4.4 may differ from the value encountered in composites. Another reason is that fibers are rarely completely continuous and end effects need to be taken into account.

7.5.2.3 The Transverse Tensile Strength of Discontinuous Fiber Composites Depends on the Interfacial Adhesive Strength

Tensile failure at the interface can occur under transverse loading (Figure 7.20b), where the stress exceeds the tensile adhesive strength of the fiber/matrix interface $\sigma_{int.}$ Hence, the ultimate transverse strength of the discontinuous fiber composite $\sigma_{uc,T}$ is defined by a rule of mixtures equation

$$\sigma_{uc,T} = \sigma_{int} V_f + \sigma_m{}^* (1 - V_f) \qquad (7.5.23)$$

the same as eq. 7.5.18.

7.6 Special Issues in the Processing and Application of Reinforced Composites

7.6.1. Electric Cable Trays for the Eurotunnel Are Processed by the Pultrusion Technique

7.6.1.1 Different Combinations of Resin and Initiator Are Used for Pultrusion

The particular combination of resin and initiator (to initiate the thermoset reaction) used in a pultrusion machine (see Figure 7.4a) affects the processing and performance of the polymer composite part. Careful selection of an appropriate resin can alleviate processing problems such as sloughing (the buildup of resin inside the die), trapped voids between fibers, and rough surface finish. The choice of initiator affects the maximum temperatures permitted inside the pultrusion die and the pultrusion rate. In turn, the pultrusion conditions affect the possibility for cracking due to residual stresses in thick sections. To overcome restrictions imposed by any one initiator, different combinations of initiators are used to promote curing over a wide range of temperatures. For example, Percadox 16N initiator is active at low temperatures (80°C) and is used to begin resin polymerization quickly inside the pultrusion die, but it decomposes at the higher temperatures encountered during the exothermic thermoset reaction, and to complete the cure a high-temperature initiator such as t-butyl perbenzoate is added. Unfortunately, limited data are available to help select the most appropriate resin–initiator formulation, and success in pultrusion is largely dictated by the experience of the operator. A good review of the issues involved in pultrusion is provided by Meyer (1985).

7.6.1.2 Other Components Are Added to Improve Processing, Properties, and Aesthetics—Inorganic Fillers Reduce Cost and Improve Properties

A polymer resin–fiber mixture generally contains other components added for their specific properties. To help reduce cost, inorganic fillers are sometimes added in place of glass fiber. Inorganic fillers improve heat transfer during cure; they reduce peak exothermic reaction temperatures, and thereby reduce the potential for residual stress cracking and shrinkage. Mold release agents, toughening agents, and flame retardants may also be added to the resin mixture. Some pigments can accelerate or decelerate cure. The amount of filler that can be added depends on filler type; it may be as small as 0.5 wt % for certain initiators and as high as 200 wt % for flame retardants in low-viscosity resins.

Ultimately, the viscosity of the mixture limits the amount of filler that can be added to the resin. The maximum useful viscosity for pultrusion is roughly 1000 cP for fiberglass rovings, and 3000 cP with fiberglass mat. Adding clay fillers at a 1:1 ratio to a typical polyester resin, for example, results in a mixture too viscous for pultrusion, but with calcium carbonate powder at a 1:1 ratio the viscosity is about 1000 cP, which is more acceptable.

Fillers also make fiber wetting by the polymer resin more difficult. Wetting is crucial in electrical applications because moisture can penetrate along the fiber–matrix interface if voids are present, leaving a path for electrical current leakage. Water penetration can also lead to premature failure of the composite by delamination. In pultrusion, the process variables that most affect fiber wetout (wettability) are the resin viscosity, roving size and the amount of twist in the roving, and residence time in the wetout tank. Viscosity can be reduced through use of viscosity depressants, and wetting can be improved by special surface treatments of filler particles. Wetout can be improved by reducing the amount of the glass reinforcements, but this results in some loss of strength. The composition of a pultruded composite for a typical electrical application may have approximately 30 percent polyester resin, 25 percent clay filler, and 45 percent fiberglass roving.

7.6.1.3 Methacrylate-Based Resins Have a Processing Advantage over Phenolics for the Electric Cable Trays in the Eurotunnel

The Eurotunnel, commonly known as the Chunnel, is a 32-mile-long tunnel under the English Channel, which for the first time has allowed electric trains to pass directly between England and France. Electrical cables pass through the tunnel for light control, ventilation, and communications. High-voltage cables are also present to power the trains. Cable support trays, which carry these lines along the walls of the tunnel, are required to withstand conditions of 100 percent relative humidity, temperatures ranging from 41 to 104°F, constant exposure to salt water, and exposure to wind speeds of over 220 mph produced by each passing train. In the event of a fire, cable tray materials must pass stringent fire, smoke, and toxicity requirements. Long material life and low maintenance cost are also important considerations in selecting a material.

Conventional steels such as stainless, galvanized, and coated steels do not have the desired corrosion resistance. Structural composites made from phenol and methacrylate resins were considered. Phenolics have excellent properties and flame retardancy, but they were rejected owing to slow processing rates. Methacrylate-based composites were ultimately selected because they have the best processing and performance characteristics. Over 8 million pounds of pultruded reinforced shapes have been processed for installation of the Eurotunnel cable trays, making it the largest single application of pultrusion products in the world. Accelerated testing procedures indicate that the lifetime of these composite parts is at least 20 years.

Methacrylate resins were introduced to the composites industry in the mid-1980s and have gained acceptance because of their low viscosity, rapid cure, and smooth surface appearance. Their resin viscosity is low (15 cP compared to 800 cP for standard polyester resins), and so methacrylate resin can be loaded with greater amounts of fillers to reduce smoke generation and improve flame retardancy. Usually, hydrated alumina and antimony trioxide are used to provide flame retardancy and low smoke evolution with halogenated polyester resins. The methacrylate resin used in the Eurotunnel application, however, relies solely on hydrated alumina because of the higher obtainable concentration. Up to 60 percent by weight of hydrated alumina can be added to the low-viscosity methacrylate resins without severely hindering processing.

7.6.2 Composite Field Repair

This section describes two applications of polymer composites where *in situ* repair is important. Four issues come together in composite repair—cure chemistry, adhesive bonding, shelf life of raw materials, and processing.

7.6.2.1 Composite Armor Repair Utilizes a Light-Cured Polyester Resin Preimpregnated into a Woven Glass Mat

Polymer composite materials for military armor has attracted attention recently, but an issue that still challenges composite armor is one of adequate repair on site. Holes in military vehicles or aircraft must be repaired quickly to protect personnel inside from chemical, biological, or radioactive weapons. If possible, the structural integrity of the vehicle must not be compromised. Even more challenging, the repair should be performable by untrained personnel in extreme weather and battlefield conditions.

Requirements for a field repair system include ease of use, less than 10 min repair time, no external heat or energy source other than the vehicle battery, minimal noxious fumes, formability for any shape hole or crack, sufficient bonding to effect temporary structural repair, compact and light weight, high tolerance to moisture and temperature variation, and a shelf life of at least 6 months.

One polymer resin–initiator system for armor repair meeting many of these requirements uses a photoinitiator activated by sunlight (N. Smith, M. Livesay, and E. Castaneda, "Rapid Composite Armor Field Repair Patch/Kit Cured by the Sun or UV," in *How to Apply Advanced Composites Technology,* ASM Publishers, p. 113, 1988.) The resin selected for this application is an unsaturated isophthalate ester that has good clarity (for light penetration) and tensile strength. As with all commercial polyester resins, styrene is added as a comonomer and diluent to improve wetting. Also for this application a free-radical cure inhibitor is added to prevent premature gelation during shipping and handling. Finally a novel photoinitiator is added. The photoinitiator is activated by short-wavelength radiation (ultraviolet, 360–400 nm), like the light available from a tanning lamp or ambient sunlight. This photocurable polymer resin–initiator combination is sufficiently latent for prolonged storage before application, yet it provides a rapid cure on demand. Photocured systems are generally less temperature sensitive than thermal-cured systems. Thermal insensitivity is an important feature because the existing armor constitutes a large heat sink, making thermal cure difficult.

The penetration of light through the full repair layer is crucial to rapid cure. Fiber mats preimpregnated with the polymer resin–initiator combination (prepregs) that are 1/4 inch thick require 8 min to cure under controlled conditions, while 1/2 inch thick specimens require 25 minutes. Clearly, it is advantageous to cure layers of the thin material sequentially rather than a single thick layer all at once. Fortunately oxygen temporarily inhibits cure of the exposed resin surface. This inhibition provides a tacky surface layer that allows interlaminar bonding of successive layers. By adding each new layer only after the preceding layer is tacky, thick composite repair sections are prepared that show no evidence of preferential failure between the layers. Indeed they fail across the layers, as desired for greater strength.

Use of this unique chemistry and curing technique posed several unexpected challenges. The standard coupling agent used on the glass fibers was found to inhibit the photoinitiator cure completely; so the fiber mats had to be leached with solvent and retreated with a special coupling agent. Fibers other than glass, such as Kevlar, Spectra, or carbon, also significantly inhibited cure.

Composite prepregs are fabricated by squeezing liquid resin into woven glass fabric with a flexible blade. The optimal resin content is about 50 percent by weight. Less resin results in dry patches of fibers that do not bond to the armor; more resin creates weak unreinforced resin pockets. To thicken the resin to prevent loss during application, 1 percent magnesium oxide is added as an oil paste and prereacted with the resin at 250°F for 15 min. The prepreg material is encased in a polyvinylacetate film on both sides and stored in a sealed opaque plastic bag to prevent accidental cure and loss of styrene monomer. The kits are stable for at least 6 months at 72°F (20°C), but shelf life is reduced at higher temperatures.

7.6.2.2 Composite Sleeve Repair inside Sewer Lines Uses Low-Temperature-Cure Polyester Resins with Heavy Polyester Nonwoven Mats

The United States has over 800,000 miles of underground municipal sewer lines, much of which will become damaged with shifting earth or simply age. Replacement of these lines can be very costly and time consuming, and can lead to major disruptions in power, telecommunications, and surface traffic. One process for the repair of damaged sewer lines without actually replacing them involves the installation of a polymer composite liner. The repair process occurs while the polymer matrix is uncured and the composite is still flexible and easily manipulated. Repair is conducted through manholes and requires little or no excavation (D. Johnson, D. Rice, R. Owens, A. H. Horner, "The Insituform Process—A Composite Solution to the Problem of Sewer Line Rehabilitation," 43rd Ann. Conf., Composite Institute, Soc. Plastics Ind., 8-C, 1988.)

The liner material must be flexible to conform to irregular pipe sections during installation and yet after curing it must remain strong and corrosion resistant for several decades. The reinforcement needs to be very flexible to allow handling during the installation procedure, which as discussed below literally involves turning a preformed tube inside out. A polyester nonwoven feltlike fibrous material is selected as the reinforcement because of its high elongation and formability. An unsaturated polyester resin is selected as the matrix, though epoxy and vinyl ester resins are sometimes used in chemical effluent lines where greater corrosion resistance is required. The resin is mixed with a thermally activated initiator that initiates polymerization at a temperature slightly above room temperature, say, 45°C. The resin–initiator formulation is impregnated into the reinforcement mat that has been stitched in the shape of a tube of diameter less than the inner diameter of the pipe to be repaired. Once the liquid polymer resin has impregnated the reinforcement, care must be taken to prevent curing during storage and handling. Refrigeration may be necessary if storage is longer than 24 hours.

The most novel aspect of this composite repair system is the installation. The composite liner must be installed through long sections of pipe where access frequently is available from only one end. After installation the resin cure must be complete to ensure that only very low levels of residual monomer leave the pipe as effluent into the environment. Finally, the composite sleeve must be compatible with unions and tees in the line, which may involve changes in pipe diameter.

A schematic of the installation process is included in Figure 7.25. After the line has been cleaned and inspected, the preimpregnated nonwoven mat is lowered through a manhole at one end of the section to be repaired and attached. Water is then

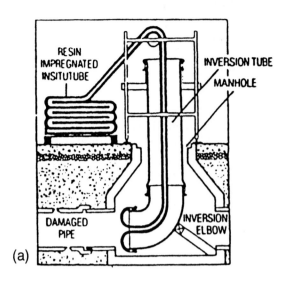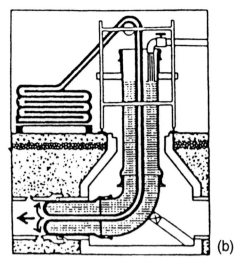

Figure 7.25
Repair of damaged underground sewer pipe. In (a), resin-impregnated fiber "Insitutube"™ is attached to the inversion tube. In (b), the tube is turned inside out with water pressure. After curing the damaged pipe has a new fiber-reinforced polymer lining.

pumped to invert the composite liner while simultaneously pressing it against the inner surface of the damaged sewer pipe. As the nonwoven sleeve enters the pipe, it stretches and literally turns inside out. A water-impermeable flexible coating surrounding the prepreg material protects the uncured resin from washout by the water. After the tube inverts, this coating becomes the new inner pipe surface. On the other side the resin impregnated into the fiber mat bonds with the old pipe surface. By circulating hot water in the tube over a period of several hours, the resin is thermal cured. After cure the ends of the liner are cut using remotely operated equipment, and the repair is complete.

This composite repair process has occurred in pipelines in Detroit, Pittsburgh, Chicago, and Houston. In one instance, repair occurred on a 420-foot section of 23-inch-diameter pipe approximately 33 feet beneath a busy city street in less than a month.

Bibliography

B. D. Agarwal and L. J. Broutman, *Analysis and Performance of Fiber Composites,* New York: John Wiley & Sons, Inc., 1980.

A. R. Bunsell, *Fibre Reinforcements for Composite Materials,* Amsterdam: Elsevier, 1988.

L. T. Drzal and P. J. Herrera-Franco, "Composite Fiber-Matrix Bond Tests," in *Engineering Materials Handbook,* Materials Park, OH: ASM International, 1991, Vol. 3, pp. 391–405.

L. T. Manzione, *Applications of Compter-Aided Engineering in Injection Molding,* Munich: Hanser, 1987.

P. K. Mallick, *Fiber Reinforced Composites,* New York: Marcel Dekker, 1988.

R. W. Meyer, *Handbook of Pultrusion Technology,* New York: Chapman-Hall, 1985.

M. Narkis, E. J. H. Chen, and R. B. Pipes, "Review of Methods for Characterization of Interfacial Fiber-Matrix Interactions," *Poly. Composites* **9**, 245 (1986).

J. A. Rolston, "Fiberglass Composite Materials and Fabrication Processes," *Chemical Eng.*, 96–110, Jan. 28 (1980).

Exercises

7.1 Describe the differences between a thermoset polymer and a thermoplastic polymer. List the relative advantages and disadvantages of these two types of polymer in the processing and application of polymer composites.

7.2 Derive the combined flow equation (eq. 7.3.11) for packed spheres instead of fibers.

7.3 Thermoplastic composites are made by preimpregnating fiber tows with polymer powder. The problem is to apply the proper amount of polymer to the fibers to obtain the correct fiber volume fraction after air is squeezed out. Carbon filaments 9 μm in diameter (density = 1.8 g/cm^3) are coated with PMMA powder spherical particles 25 μm in diameter (density = 1.2 g/cm^3). (a) Find the maximum and minimum fiber spacing between fibers to obtain a final fiber volume fraction of 62 percent. (b) Find the weight of polymer necessary to coat one pound of fiber. (c) Calculate the yield (yards per pound) of 12,000 (12k) filament tows to carbon only. (d) Calculate the yield of the same tow after coating with PMMA. (e) If 25 μm polymer fibers were used in the tow instead of particles, find the number of filaments in a 12k tow.

7.4 Find the maximum rate of immersion of fibers into a bath without entrainment.

7.5 (a) Calculate the longitudinal modulus for a composite made of 60% volume fraction high-modulus carbon fiber and polycarbonate. (b) Calculate the maximum tensile stress this composite is capable of supporting before a fiber breaks. The ultimate fiber strength is 1.9 GPa.

7.6 The value of the interface shear strength τ_c for a high-modulus carbon fiber–polycarbonate composite with an untreated carbon fiber surface is 37 MPa and for a treated surface is 68 MPa. The ultimate fiber strength is 1.9 GPa. (a) Calculate the minimum average lengths of chopped fibers, 8.4 μm in diameter, if the treated and untreated composites are to support the maximum possible tensile stress. Calculate the minimum average lengths if the fiber diameter is increased to 1 mm. (b) What happens when the average lengths of 1-mm-diameter chopped fibers are 10 mm and you try to load the composite to the "maximum possible" stress reached in (a)? (Assume constant volume fraction of fibers.) (c) What happens when the average lengths of 1-

mm-diameter chopped fibers are 40 mm and you load the composite to the "maximum possible" stress reached in (a)?

7.7 Illustrate the stress–strain curves you might expect for the two composites in Exercise 7.5(c) above.

7.8 Outline various fiber surface treatments used to increase the value of the interface shear strength. What are the disadvantages of having the interface shear strength too high? In some instances the interface shear strength is made deliberately low. What purpose is served by a weak interface?

7.9 Show that the Halpin–Tsai equation approaches the rule of mixtures for the longitudinal modulus of continuous fiber-reinforced composites.

7.10 Pultrusion is the continuous lamination of reinforcements with a resin binder as they are pulled through a shaping die. Fillers in the resin can affect pulling force, resin viscosity, and surface finish. Recent data with 30 parts alumina trihydrate fillers in 100 parts methacrylate resin are shown below (R. D. Howard et al., SPI Comp. Inst., 43rd Ann. Meeting, Feb. 1–5, 1988).

Filler	Viscosity (cP)	Pulling Force (kg)	Filler Dia. (μm)	Surface finish
OL104	12.4	70	0.8	poor
ON310	3.0	530	8.0	excellent

(a) Explain why smaller particles have higher viscosity. (b) Compare mean particle diameters to spacing between fibers for a composite with 49% volume fraction of 20-μm diameter glass fibers. (c) Why is the pulling force larger for ON310? (d) Surface finish is related to internal pressure during cure. How else can one obtain high pressures and gloss in pultrusion?

8

LIQUID COATING PROCESSES

with Mario Errico

LIST OF SYMBOLS

d	gap distance in slot coater
h	thickness of wet coated layer
L	length along the die face in slot coater
N_{ca}	capillary number (eqs. 7.3.15 and 8.1.18)
P_d; P_u	pressure *outside* the liquid bead at the downstream (d) and upstream (u) locations
P_1; P_2	pressure *inside* the liquid bead at the downstream (1) and upstream (2) extremities
ΔP	pressure difference across the liquid bead ($\Delta P = P_d - P_u$)
Q	liquid flow rate per unit width in slot coating
R	radius of liquid-free surface
S	spreading coefficient
U	web velocity
v	liquid velocity
γ_{12}	surface (interface) tension between phases 1 and 2
η	viscosity of coating liquid (eq. 3.8.1)
θ_s; θ_d	contact angles, static (s) (eq. 3.2.4) and dynamic (d) (Appendix 8A)

CONCEPT MAP

Coating Processes

- A great variety of coating methods are available for transfer of liquid from a reservoir onto a solid substrate (Figure 8.16). The chosen method depends on rheology of the coating liquid, type of substrate and desired coating thickness, uniformity, and coating speed.

- Nonpremetered methods, such as roll coating, rely on liquid uptake on rolls to deliver liquid to a flexible substrate (the web). Thickness of the delivered layer depends on fluid dynamic behavior in the uptake zone (liquid rheology as a function of roll surface velocity) and the delivery zone (dynamic wettability of the substrate by the liquid).

- Premetered methods, such as slot coating, force liquid onto a substrate (web) at a prescribed rate dictated by liquid flow rate and web velocity. Slide and curtain coating apply multilayer (up to 20 layers) coatings in one pass.

Analysis of Slot Coating Process

- Under equilibrium conditions the shape of the liquid bead between the die face and the web is constant (Figure 8.5). Thickness of the coated liquid layer h depends on web velocity U and liquid flow rate through the bead, Q; h = Q/U.

- Liquid bead stability can be analyzed in two parts, one emphasizing viscous flow, the other capillary flow.

Viscous Flow Model (Figure 8.6)

- Liquid flow rate is determined by viscous drag of the moving substrate on the stationary liquid (Couette flow) and by external pressure difference (ΔP) between upstream and downstream locations (Poiseuille flow).

- For constant coating thickness, ΔP has a range of values—the coating process window—determined by the die geometry (die face lengths and gap dimensions), liquid viscosity η, and substrate velocity (eq. 8.1.6 and Figure 8.7).

Capillary Flow Model (Figure 8.8)

- Liquid flow rate is determined by the pressure difference (ΔP) due to different radii of curvature at upstream and downstream liquid-air surfaces (Laplace flow).

- For constant coating thickness, ΔP has a range of values—the coating process window—determined by die geometry (die face lengths and gap dimensions), liquid surface tension, γ_{vl}, and dynamic contact angle of liquid to web (eq. 8.1.15 and Figure 8.9).

Visco-Capillary Flow Model

- Visco-capillary model combines features from viscous and capillary flow models (eqs. 8.1.16, 8.1.17, and Figure 8.11).

- In practice the most legitimate model is determined by the capillary number, $N_{ca} = \eta U / \gamma_{vl}$. When N_{ca} is large (>10), viscous effects dominate; when N_{ca} is small (<1), capillary effects dominate.

- Coating thickness is also extremely sensitive to small fluctuations in slot region. Uniformity is 20 times more sensitive to pressure change in slot than in bead region (eq. 8.1.22). Careful control of slot gap across coater is essential for uniform deposition. Fine-scale tuning of slot gap is often incorporated in total die design (Figure 8.15).

Coating Defects

- Periodic coating thickness variations arise due to hydrodynamic instability causing ribbing (down-web variations) and barring (cross-web variations).

- Nonperiodic defects arise due to unfiltered impurity particles or particles resulting from stagnant recirculation in the coating bead, localized dewetting of the substrate, nucleation of bubbles at localized hot spots or during accelerated drying.

- Defect-free coatings essential for many applications leads to incorporation of real-time monitoring during manufacture.

8

Coating processes are employed in a number of widely different industries. Clay-based coatings give paper a smooth finish suitable for high-quality printing; adhesives coated onto paper or film produce self-adhesive labels or tapes; photographic films are made of a thin polyester (polyethylene terphthalate) substrate coated with as many as 18 layers of a variety of chemicals in a gelatin–water base; aluminum disks coated with suspensions of iron oxide and other materials are used to make computer magnetic memory products ("hard" disks).

Defined in the most general sense, a coating process replaces the original air–substrate interface with a new layer of material applied onto the substrate's surface. The new material may be delivered via the vapor phase (as in the manufacture of thin-film reflective coatings for sunglasses), via the liquid phase (as in the use of an evaporative solution to deposit a thin layer of adhesive), or via a liquid phase carrying colloidal solid particles (as in the spray painting of automobiles). In this chapter, we restrict our attention to coating processes involving liquid phases. Vapor-phase deposition is treated separately in Chapter 10.

An individual coating liquid contains many ingredients. We already emphasized this fact when we described modern paints in Section 4.7.1. Besides the "active" component—pigment particles in paint, microscopic silver halide crystals in photographic film, or iron oxide particles in magnetic recording tape—the ingredients may also include solvents added to dissolve or suspend the active component, surfactants, binders, plasticizers, cross-linking agents added to aid the spreading and drying process, and chemicals such as biocides added for special purposes. Coating liquids are spread on substrates that may be flexible (like a paper or a polyester film base) or rigid (like glass plates or aluminum disks). The liquid layer is generally thin and continuous. After the liquid coating has been applied, drying, cooling, or curing processes are used to solidify it to obtain the desired end product. Generally, the final coating needs to be of uniform thickness and free from any defects.

With so many variables, it is not surprising that researchers have developed a myriad of methods to deliver coating liquids to different substrates for different product configurations. The first part of this chapter describes in detail the slot coating process used to coat a thin layer of pressure-sensitive adhesive on adhesive labels. The slot coating process highlights some of the issues that must be considered in maintaining a coating process production line. The remainder of the chapter presents an overview of other widely used liquid coating processes, the criteria for selecting which coating process to use for a given application, and the interfacial phenomena that underlie them. We also review common types of defects and flaws encountered in coating processes and the use of real-time imaging to monitor and identify flaws and adaptively control the coating process. Finally, the chapter presents some examples of coated products chosen from real manufacturing operations.

8.1 Making Self-Adhesive Paper Labels: A Tutorial on the Fundamental Aspects of Coating Processes

This section discusses the fundamental aspects common to most coating processes by going through the steps used to make pressure-sensitive self-adhesive paper labels. The product we wish to "manufacture" is a commodity in the modern office, where it may be used in sheet form to produce printed address labels or the pressure-sensitive name tags discussed in Section 5.5.2. Figure 8.1 shows a typical cross section of the label product. Labels are lifted from the sheet of backing material that has been coated with a thin layer of low-surface-energy 'release' agent to allow easy removal of the label. Because the backing material plus release agent is discarded after the label has been removed, manufacturers must deal with obvious environmental implications. The adhesive layer (described in Section 5.5.2) is anchored to the "face stock" or label material by a very thin layer of a "primer" agent. Adhesive and facestock materials are selected with a high work of adhesion (see Section 3.2.1) between them. For release, on the other hand, adhesive and backing materials are selected with a low work of adhesion between them.

The complete self-adhesive label shown in Figure 8.1 generally is produced by passing a flexible and continuous substrate,

Figure 8.1
Cross section of a typical pressure-sensitive adhesive label.

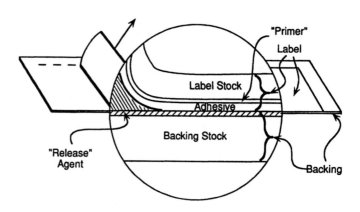

called the *web,* that may be a sheet or strip of paper or plastic, through a multistation "web coater" or "web coating line" of the kind illustrated in Figure 8.2. At the first station (1), the backing stock material is unwound from a large roll and coated with the release agent. After the release agent is solidified by "drying" or "curing," the backing material plus release agent is moved on to the next station (2), where it is coated with the adhesive and cured or dried. At station (3), the label stock or facestock material is unwound from a large roll, coated with a primer agent, cured or dried, and then pressed onto the adhesive-coated backing material at the "laminator" station (4). The final sandwich is rewound as a roll for transportation to a "finishing/converting" machine. Here cuts made into just the facestock produce labels of the desired shape. Finally, the roll is cut into sheets for shipping.

Adhesive used for this application is a water-based, oil-in-water (O/W) macroemulsion that looks and feels like the white "carpenter" glue used for bonding paper and wood. Its discontinuous phase consists of the tackifier, plasticizer, and latex mixture described in Section 5.5.2.1 plus other compounds, while the continuous water phase contains surfactant to stabilize the emulsion and enhance the wettability of the fluid to the facestock substrate. Primer agent coated on the facestock further enhances the wettability of the adhesive fluid to the substrate.

We want to achieve paper coated with 30 g/m^2 (grams per square meter or GSM—a typical industrial unit) of *dry* adhesive. To get some idea of the volume of liquid going through a coating line, we can consider what happens if we feed facestock and backing paper (both 1 m or 40 inches wide) at a machine rate of 100 m/min. Assuming that 50% of the adhesive emulsion is water, we need to coat 60 GSM of emulsion, which corresponds to a

Figure 8.2
Schematic of a multi-station coating line to make pressure-sensitive adhesive products. The coated backing stock and label stock materials are brought together at the laminator

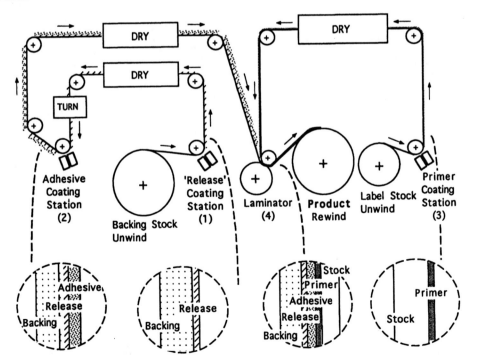

thickness of about 60 μm (since its density is close to water, 1 g/m²
of emulsion liquid equals a thickness of 1 μm). Therefore, we are
feeding the coater with 6000 grams (about 1½ gallons) of emulsion per minute and removing 3000 grams (about ¾ gallon) of
water per minute in the drier.

8.1.1 Adhesive Is Applied to the Backing Using
Slot Coating

The goal of any coating process is to apply a uniform, defect-free
liquid film to a substrate. For simplicity's sake, our discussion
starts with an "ideal process" and then shows how real-world
conditions affect the uniformity and quality of the coating—thus
providing a steady income for a fairly large number of coating
engineers.

Our example is the *slot coating* process (also called *die
coating* and *extrusion coating*), which is illustrated in detail in
Figure 8.3. In this process, adhesive is fed under pressure from a
pipe to a slot designed to distribute the liquid as uniformly as
possible across the whole width of the web. The adhesive exits
the die through the slot and is applied to a moving web supported
by a steel or rubber-covered steel backing roll. Liquid flowing
between the slot and the substrate establishes a *liquid coating
bead*. The equipment is designed to control the stability of the
bead by maintaining a slight vacuum on the upstream side of the
die, so that P_u is generally *less than* P_d in Figure 8.3 and subsequent figures.

Figure 8.4 illustrates the overall flow features of the liquid
bead. This image was taken from a video recording of a two-slot

Figure 8.3
Schematic diagram of a
slot coating head showing
feed of the liquid into the
die and internal flow
regions.

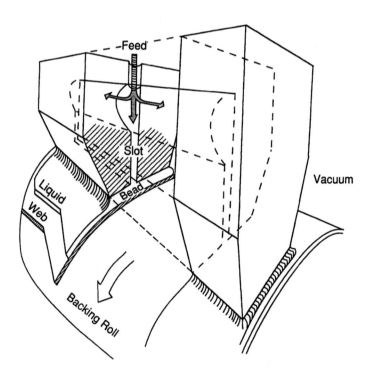

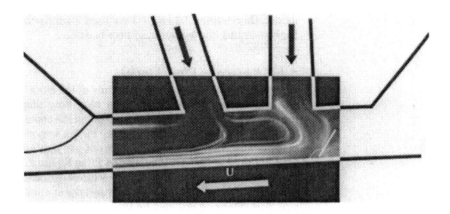

Figure 8.4
Hydrogen-bubble visualization of liquid flow. This example illustrates laminar flow in the liquid bead of a two-slot die. (Courtesy of Dr. Mario Errico and Dr. Wieslaw Suszynski.)

die where small hydrogen bubbles were injected into the liquid to serve as tracers for flow visualization. Due to the long exposure time, the bubble images have blended together into streaks. Notice that part of the liquid exiting the downstream slot moves directly downstream. Another part moves upstream, where it turns around and then is pulled downstream by the web.

8.1.2 Analysis of the Stability of the Slot Coating Process

The analysis of the slot coating process presented in this section is based on research by L. E. Scriven and coauthors. Their work has provided detailed insight into the factors contributing to the stability of the liquid bead in slot coating. Figure 8.5 is a cross section of the die, showing the feed and bead regions. It represents a frozen picture of the liquid bead as it emerges from the die and is carried away downstream on the web. We want to know what factors control the thickness of the wet-coated layer and what operating conditions ensure a film of uniform thickness. The analysis assumes uniform flow across the width of the slot and neglects edge effects.

We can conduct the analysis in three parts: the first part considers viscous flow due solely to pressure gradients and leads to the viscous flow model; the second part considers flow due solely to capillary effects and leads to the capillary flow model; the third part combines the two models into the visco-capillary

Figure 8.5
Cross section of the liquid bead in a single-slot coater used for viscous flow analysis.

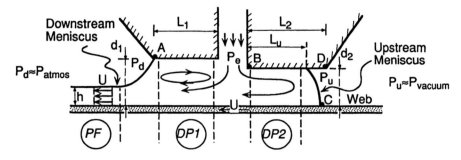

model. (In Sections 7.2 and 7.3 we used a similar breakdown to analyze liquid impregnation of fiber beds)

8.1.2.1 Viscous Flow Model

We can describe viscous flow in terms of the three simple components shown in Figure 8.6. The plug flow diagram of Figure 8.6a represents the liquid carried out of the coater on the web. Liquid of thickness h is carried away on the web moving with a velocity U so that the quantity removed per unit width per unit time in the plug flow region (labeled *PF* in Figure 8.5) is $Q = Uh$. Flow in the bead region is governed by the drag due to the velocity difference between the moving web and the stationary die (Couette flow, Figure 8.6b) and by the pressure gradient (Poiseuille flow, Figure 8.6c). Figure 8.6d illustrates how we can combine the drag and pressure flow profiles to obtain the flow velocity profile in the downstream and upstream drag-pressure regions labeled *DP1* and *DP2*, respectively in Figure 8.5.

In the first part of our analysis, we neglect regions where the flow is turning or readjusting. We also neglect capillary effects at the free surfaces.

The pressure difference ΔP across the bead is given by $\Delta P = (P_d - P_u)$ where P_d and P_u are the pressures acting on the liquid bead at the downstream (d) and upstream (u) extremities, respectively. (When a vacuum is pulled on the uspstream side, $P_d > P_u$.) For a given web speed U, flow rate equations for the three regions give

Figure 8.6
Simple flow models used in the viscous flow analysis.

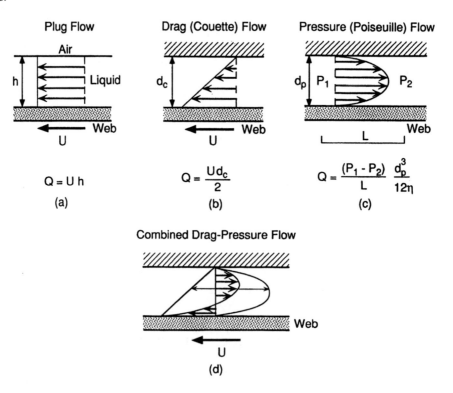

Plug Flow	Drag (Couette) Flow	Pressure (Poiseuille) Flow

$$Q = Uh$$

(a)

$$Q = \frac{Ud_c}{2}$$

(b)

$$Q = \frac{(P_1 - P_2)}{L} \frac{d_p^3}{12\eta}$$

(c)

Combined Drag-Pressure Flow

(d)

$$PF: \qquad\qquad Q_1 = Uh \qquad\qquad (8.1.1)$$

$$DP1: \qquad Q_2 = \frac{Ud_{cd}}{2} + \left(\frac{P_e - P_d}{L_1}\right)\frac{d_{pd}^3}{12\eta} \qquad (8.1.2)$$

$$DP2: \qquad Q_3 = \frac{Ud_{cu}}{2} + \left(\frac{P_u - P_e}{L_u}\right)\frac{d_{pd}^3}{12\eta} = 0 \qquad (8.1.3)$$

P_e is the pressure on the liquid in the slot, L_1 and L_2 are the total lengths of the die face downstream and upstream, respectively, and L_u is the distance the bead moves over the die face upstream. In these equations, d_{cd} and d_{cu} are the downstream and upstream "equivalent" Couette gaps; d_{pd} and d_{pu} are the "equivalent" Poiseuille gaps. They are defined by

$$d_c = \frac{\int_0^L [d(x)]^{-2}\, dx}{\int_0^L [d(x)]^{-3}\, dx}; \qquad d_p = \left(\frac{L}{\int_0^L [d(x)]^{-3}\, dx}\right)^{1/3}$$

These definitions indicate that if the gaps between the die and the web are constant along the length of the bead as drawn in Figures 8.5 and 8.6 (a good approximation if the backup roll diameter is large compared to $L_1 + L_2$), then the equivalent gaps are identical to the actual gaps, d_1 downstream and d_2 upstream.

Net flow in and out of region $DP2$, Q_3 in eq. 8.1.3, is zero because the liquid turns around at the upstream surface. The net flow rate from region $DP1$ must equal the emerging plug flow, that is, $Q_2 = Q_1$. Substituting Q_1 from eq. 8.1.1 and P_e from eq. 8.1.3 in 8.1.2, we obtain

$$\Delta P = P_d - P_u = \frac{12\eta UL_1}{d_{pd}^3}\left(\frac{d_{cd}}{2} - h\right) + \frac{6\eta UL_u}{d_{pu}^3}d_{cu} \qquad (8.1.4)$$

or by rearranging

$$\Delta P = \frac{6\eta UL_1 d_{cd}}{d_{pd}^3}\left\{\left[1 + \frac{L_u}{L_1}\frac{d_{cu}}{d_{cd}}\left(\frac{d_{pd}}{d_{pu}}\right)^3\right] - \frac{2h}{d_{cd}}\right\} \qquad (8.1.5)$$

Note that even when the web speed U and the flow rate Q, and thus the coating thickness h = Q/U, are kept constant, this equation provides a *range of values* for ΔP. This flexibility arises because ΔP varies with L_u, and the upstream meniscus that defines the value for L_u is free to move its location along the die face between the slot exit (where $L_u = 0$) and the die edge (where $L_u = L_2$). The range of pressures defines the process window, or in this case the *coating process window*. In theory, the coating flow should be operable within this window.

Substituting the two extreme values for L_u, we can write the coating window as:

$$\frac{6\eta UL_1 d_{cd}}{d_{pd}^3}\left(1 - \frac{2h}{d_{cd}}\right) < \Delta P < \frac{6\eta UL_1 d_{cd}}{d_{pd}^3}\left\{\left[1 + \frac{L_2}{L_1}\frac{d_{cu}}{d_{cd}}\left(\frac{d_{pd}}{d_{pu}}\right)^3\right] - \frac{2h}{d_{cd}}\right\} \qquad (8.1.6)$$

Figure 8.7 plots ΔP versus $1/h$ for the two extreme values of L_u. The coating window is qualitatively the region between the curves. Equation 8.1.6 tells us that the minimum coating thickness that we can maintain with no applied pressure differential is $h = d_{cd}/2$ ($\approx d_1/2$), corresponding to $\Delta P = 0$ for the lower $L_u = 0$ curve in Figure 8.7. In principle, we can sustain a thinner coating h by applying a vacuum to the upstream meniscus ($\Delta P > 0$) and speeding up the web, or we can maintain the same thickness by applying a vacuum and increasing the gap d_1.

8.1.2.2 Capillary Flow Model

In this second part of our analysis, we neglect contributions from viscous effects and assume that capillary (surface tension) effects dominate. Figure 8.8a highlights the curvature of the free surfaces of the bead cross section. The downstream free surface is an air–liquid interface that has a *static* contact line located at A (in our two-dimensional representation, the line is normal to the page and collapses to a point). The upstream surface is characterized by a *static* contact line located at point B and by a *dynamic* contact line located at point C, where the free surface contacts the web with the dynamic contact angle θ_d. (For a comparison of dynamic contact angle θ_d and static contact angle θ_s, see Appendix 8A.1).

To simplify the analysis, we assume that the upstream free surface is pinned at point B ($L_u = L_2$) and that the shapes of the free surfaces are arcs of circles of radius R. Under these conditions, the following geometric constraints are imposed on the radii of the free surfaces at A and B, respectively:

$$\frac{d_1}{2} \leq R_d \leq \infty, \quad \text{at A (downstream)} \quad (8.1.7)$$

$$\frac{d_2}{1 - \cos \theta_d} \leq R_u \leq \frac{d_2}{1 + \cos \theta_d}, \quad \text{at B (upstream)} \quad (8.1.8)$$

The latter inequality expresses the fact that the upstream radius of curvature at B can never be $< d_2/2$ as the dynamic contact angle at C changes from 0 to 180°; that is, for a given θ_d the upstream free surface is bounded by the two extreme configurations

Figure 8.7
The slot coating operation region ("coating process window") based on the viscous flow model. The operational parameters lie between the curves. (Reprinted from *Chemical Engineering Science*, Vol. 35, B. G. Higgins and L. E. Scriven, p. 673, Copyright 1980, with kind permission from Elsevier Science Ltd., The Boulevard, Langford Lane, Kidlington 0X5 1GB, UK.)

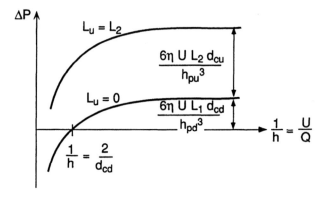

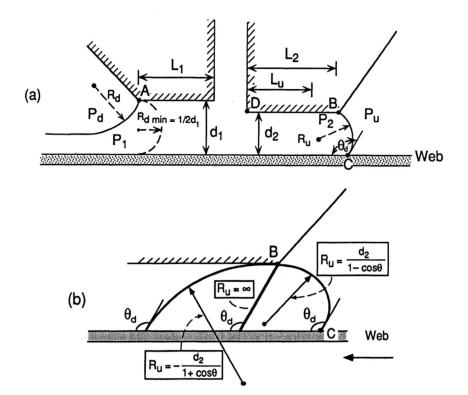

Figure 8.8
(a) Cross section of the liquid bead in a single-slot coater used for capillary flow analysis. (b) The two limiting upstream free surface profiles for a given θ_d. For the right-hand limit (at C) the liquid air interface is convex; for the left-hand limit the interface is concave.

illustrated in Figure 8.8b. The Young–Laplace equation for a cylindrical surface (eq. 3.3.7) gives the pressure difference across the downstream free surface

$$P_d - P_1 = \frac{\gamma_{lvd}}{R_d} \tag{8.1.9}$$

and for the upstream surface

$$P_2 - P_u = \frac{\gamma_{lvu}}{R_u} \tag{8.1.10}$$

where P_1 and P_2 represent the pressure inside the liquid, and γ_{lv} is the surface tension of the coating liquid–vapor interface. Thus the geometric constraints on the radii become constraints on the capillary pressures, and substituting for R gives

$$2\frac{\gamma_{lvd}}{d_1} > P_d - P_1 > 0, \quad \text{downstream} \tag{8.1.11}$$

$$\frac{\gamma_{lvu}(1 - \cos\theta_d)}{d_2} \geq P_2 - P_u \geq -\frac{\gamma_{lvu}(1 + \cos\theta_d)}{d_2}, \quad \text{upstream} \tag{8.1.12}$$

For small web velocities it can be shown that

$$P_d - P_1 = 1.34\left(\frac{\eta U}{\gamma_{lvd}}\right)^{2/3}\frac{\gamma_{lvd}}{h} \tag{8.1.13}$$

where $\eta U/\gamma_{lvd}$ is the capillary number N_{ca}.
Substituting eq. 8.1.13 in eq. 8.1.11, we get

$$0 < 1.34\left(\frac{\eta U}{\gamma_{lvd}}\right)^{2/3}\frac{\gamma_{lvd}}{h} < 2\frac{\gamma_{lvd}}{d_1}, \qquad \text{downstream} \qquad (8.1.14)$$

Assuming $P_1 = P_2$ (a correct assumption in the absence of viscous effects) in eq. 8.1.13, and then substituting eq. 8.1.13 in eq. 8.1.12, remembering $\Delta P = P_d - P_u$, we get

$$-\frac{\gamma_{lvu}(1 + \cos\theta_d)}{d_2} + 1.34\left(\frac{\eta U}{\gamma_{lvd}}\right)^{2/3}\frac{\gamma_{lvd}}{h}$$

$$< \Delta P < \frac{\gamma_{lvu}(1 - \cos\theta_d)}{d_2} + 1.34\left(\frac{\eta U}{\gamma_{lvd}}\right)^{2/3}\frac{\gamma_{lvd}}{h}, \qquad \text{upstream} \qquad (8.1.15)$$

Equations 8.1.14 and 8.1.15 represent another form of coating process window, shown qualitatively in Figure 8.9 for a given set of values for d_1, d_2, and θ. Notice that eq. 8.1.14 defines an upper limit for $1/h$, represented by the vertical line at $1/h_{min}$ in Figure 8.9. Figure 8.10 attempts to visualize the effects of varying d_1, d_2, and θ_d on the coating window that arise from this capillary analysis. Increasing d_1 narrows the coating window horizontally (smaller range of admissible h), while increasing d_2 narrows it vertically (smaller range of ΔP). Increasing the dynamic contact angle θ_d (moving along the diagonal in Figure 8.10) shifts the coating window towards increasing positive ΔP. Therefore, we need more vacuum on the upstream meniscus to increase ΔP. By the way, note that this figure, on a two-dimensional page, shows the effect of five variables (ΔP, $1/h$, d_1, d_2, θ_d).

8.1.2.3 Visco-Capillary Model
In the third part of our analysis, we examine the case in which capillary forces or viscous forces are comparable so that we cannot neglect either one.

Figure 8.9
The slot coating operation region based on the capillary flow model. The coating process window is qualitatively the region between the curves. (Reprinted from *Chemical Engineering Science,* Vol. 35, B. G. Higgins and L. E. Scriven, p. 673, Copyright 1980, with kind permission from Elsevier Science Ltd., The Boulevard, Langford Lane, Kidlington OX5 1GB, UK.)

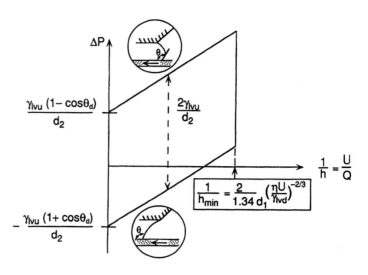

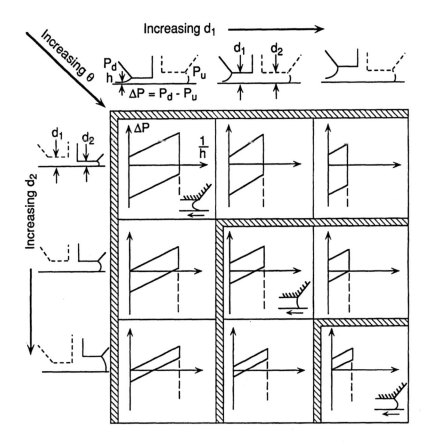

Figure 8.10
Diagram summarizing the effects of slot configuration (d_1 and d_2) and contact angle (θ) on the coating process window for the capillary flow model.

For the sake of simplicity, we assume that the upstream meniscus is pinned at B, as shown in Figure 8.8. The overall pressure difference across the bead can be written as $\Delta P = P_d - P_u = (P_d - P_1) + (P_1 - P_2) + (P_2 - P_u)$. If we replace P_d with P_1 and P_u with P_2, we can use eq. 8.1.5 here. We simply add equations 8.1.13 and 8.1.5 (with $L_u = L_2$) everywhere in eq. 8.1.12 to get for the upstream constraint

$$\Delta P < \frac{\gamma_{lvu}(1 - \cos\theta_d)}{d_2} + 1.34\left(\frac{\eta U}{\gamma_{lvd}}\right)^{2/3}\frac{\gamma_{lvd}}{h}$$

$$+ \frac{6\eta U L_1 d_{cd}}{d_{pd}^3}\left\{\left[1 + \frac{L_2}{L_1}\frac{d_{cu}}{d_{cd}}\left(\frac{d_{pd}}{d_{pu}}\right)^3\right] - \frac{2h}{d_{cd}}\right\} \quad (8.1.16)$$

$$\Delta P > -\frac{\gamma_{lvu}(1 + \cos\theta_d)}{d_2} + 1.34\left(\frac{\eta U}{\gamma_{lvd}}\right)^{2/3}\frac{\gamma_{lvd}}{h}$$

$$+ \frac{6\eta U L_1 d_{cd}}{d_{pd}^3}\left\{\left[1 + \frac{L_2}{L_1}\frac{d_{cu}}{d_{cd}}\left(\frac{d_{pd}}{d_{pu}}\right)^3\right] - \frac{2h}{d_{cd}}\right\} \quad (8.1.17)$$

while the downstream constraint, eq. 8.1.14, is still valid. Figure 8.11 shows the modified coating process window. The vis-

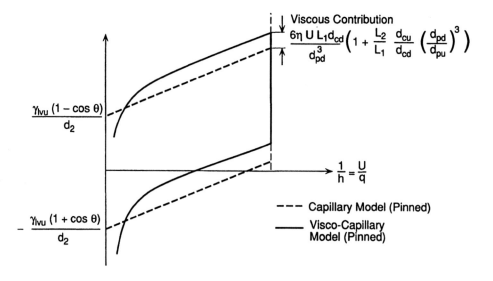

$$\text{Viscous Contribution}$$
$$\frac{6\eta\,U\,L_1 d_{cd}}{d_{pd}^3}\left(1+\frac{L_2}{L_1}\frac{d_{cu}}{d_{cd}}\left(\frac{d_{pd}}{d_{pu}}\right)^3\right)$$

$\dfrac{\gamma_{lvu}(1-\cos\theta)}{d_2}$

$\dfrac{1}{h}=\dfrac{U}{q}$

$-\dfrac{\gamma_{lvu}(1+\cos\theta)}{d_2}$

– – – Capillary Model (Pinned)
—— Visco-Capillary
      Model (Pinned)

Figure 8.11
The slot coating operation region based on the combined visco-capillary flow model. The coating process window is qualitatively the region between the curves. (Reprinted from *Chemical Engineering Science,* Vol. 35, B. G. Higgins and L. E. Scriven, p. 673, Copyright 1980, with kind permission from Elsevier Science Ltd., The Boulevard, Langford Lane, Kidlington 0X5 1GB, UK.)

cous terms shifts it upward relative to Figure 8.7, and more upstream vacuum is needed to maintain stability.

8.1.2.4 The Coating Process Window Depends on Web Speed, Liquid Properties, and Die Design

It is legitimate to ask when can we use the simpler viscous model (eq. 8.1.6) or capillary model (eqs. 8.1.14 and 8.1.15) instead of the full visco-capillary model just described. We can ascertain this from the ratio of the orders of magnitude O of the viscous-originated first term and the capillary-originated third term on the right-hand side of eq. 8.1.16, which gives

$$\frac{O_{viscous}}{O_{capillary}}=\left(\frac{\eta U}{\gamma_{lv}}\right)\frac{L_1}{h} \qquad (8.1.18)$$

We recognize $(\eta U/\gamma_{lv})$ as the capillary number N_{ca}, the ratio of viscous to surface tension forces, and L_1/h is the relative length of the die lip to coating thickness. From eq. 8.1.18 we learn that

1. For large values of the capillary number N_{ca} and L_1/h viscous effects dominate. This occurs at high web speed and with long die lips. In this case, we can define the coating window by eq. 8.1.6 as we allow the upstream meniscus to move between D and B, Figure 8.5.

2. If N_{ca} is small or near 1 and L_1/h is small, then surface tension or capillary effects dominate. This is the case for low web speed and short die lips. Under these circumstances, we can define the coating window by eqs. 8.1.14 and 8.1.15.

Using reasonable parameters for coating our pressure-sensitive adhesive labels, we have U = 100 m/min, η = 10 poise, γ_{lv} = 50 dyn/cm, L_1 = 1 mm, h = 60 μm, and we calculate

$$N_{ca} = 33 \quad \text{and} \quad \frac{L_1}{h} = 16$$

conditions under which viscous forces clearly dominate.

Therefore, the slot-coating process for self-adhesive labels is governed by the variables included in eq. 8.1.6 and Figure 8.7. The coatings engineer can operate within the coating process window by varying the viscosity and surface tension of the liquid. These physical properties are controlled in turn by liquid composition and temperature, the applied pressure, the upstream vacuum, and the geometrical design of the coating head.

8.1.2.5 Real-World Considerations

Some interesting questions lead from these simplified models to real-world situations:

1. What happens if the upstream dynamic contact line moves to the left of D in Figure 8.12a?

2. What happens if the upstream dynamic contact line moves to the right of B in Figure 8.12a?

3. What happens if the downstream static contact line "unpins" and moves to the left of A, up the die face?

4. What happens when the upstream free surface is between D and B? Will the change in ΔP cause the free surface curvature to change, or will the static contact line move first?

5. Is the dynamic contact angle θ_d a constant?

6. How thin can the coated film thickness h be?

Here are some possible answers to each of these questions:

1. Air entrainment will occur as illustrated in Figure 8.12b.

Figure 8.12
Real-world consequences on the stability of the bead due to reduced or excessive feed flow and variations in wetting conditions.

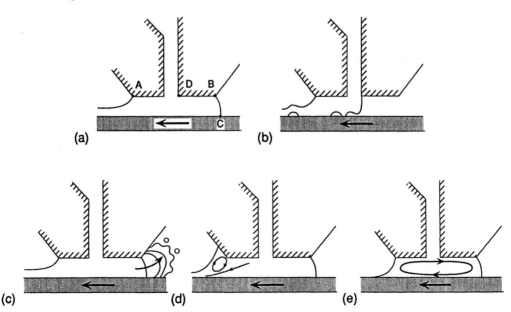

(a)   (b)

(c)   (d)   (e)

2. The upstream meniscus will swell, become unstable, then rupture, as illustrated in Figure 8.12c.

3. A recirculation cell will develop in the vicinity of A, as shown in Figure 8.12d, with possible adverse effects if the liquid is shear sensitive. The liquid might degrade due to long residence times in the recirculation region.

4. The result depends on changes in the wettability of the die face between D and B. Probably the static contact line will be pinned at a corner or where a discontinuity in wettability exists, due to physico-chemical changes that take place there.

5. The dynamic contact angle is not constant; the angle formed by the upstream free surface with the moving web depends on the web speed, as discussed in Appendix 8A.1. It tends toward 180° for high speeds (leading to air entrainment).

6. Every coating method has a minimum wet coverage level, and we cannot coat below it. In slot coating, the viscous model analyzed in eq. 8.1.6 and Figure 8.7 indicates that the minimum thickness with no bead vacuum ($\Delta P = 0$) is half the coating gap. In the presence of a bead vacuum, the wet thickness could be less. The capillary model analyzed in eq. 8.1.15 and Figure 8.9 indicates a minimum thickness, $h_{min} = (h_1/1.34)(N_{ca})^{2/3}$, that decreases linearly with the coating gap for a constant capillary number. If surface tension effects are negligible, it might seem that h could be allowed to become infinitely thin. In reality, however, unstable wetting and/or flow limit the smallest h we can achieve (Figure 8.12e).

Finally, even if we keep ΔP within the coating window, the coating bead can still become unstable due to small environmental disturbances such as external noise or internal molecular fluctuations, leading to three-dimensional phenomena such as ribbing and barring (see Section 8.2.3). Figure 8.13 shows qualitatively an experimental coating window of the kind developed and applied in practice by coating engineers.

Figure 8.13
A qualitative experimental coating process window.

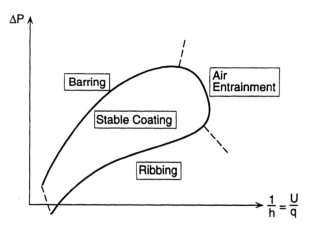

Figure 8.14
Cross section of a single slot coater used to analyze the effects of slot length (L_s) and slot width (d_s) on bead flow and coating uniformity.

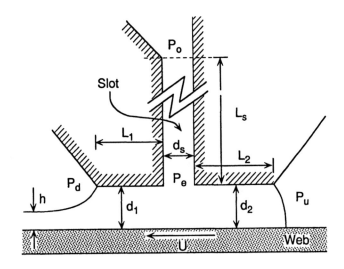

8.1.2.6 Die Design and Slot Configuration Are Critically Important in Slot Coating

So far we have been concerned primarily with flow behavior within the liquid bead. We have neglected flow through the slot, although it can significantly affect flow uniformity, and hence coating thickness uniformity, in the cross-web direction. If coating thickness variations—both along the web (the down-web direction), and across the web (the cross-web direction)—do not lie within prescribed limits, critical properties of the finished product will be adversely affected. Variations of 1–5% might be acceptable for adhesive coatings, whereas photographic films require variations of less than 1%.

Sensitivity to Slot Length. To examine this influence, we initially assume that the liquid is Newtonian and the flow in the slot is laminar and parallel to the slot walls in Figure 8.14. The flow rate Q through the slot is related to the pressure drop by

$$Q = \left(\frac{P_o - P_e}{L_s}\right)\frac{d_s^3}{12\eta} \qquad (8.1.19)$$

where d_s is the slot width and L_s is the slot length.

It is worthwhile to compare the pressure drops in the slot and in the bead. Recall that with the parameters selected above for coating our hypothetical adhesive, viscous forces dominate the flow in the bead. Then eq. 8.1.5 applies. We can simplify the equation by making three assumptions—that the lips lie parallel to the web and are equally distant from it (so that $d_{pd} = d_{pu} = d_{cd} = d_{cu} = d_1 = d_2 = 2h$); that the upstream meniscus is pinned at the edge of the lip; and that both lips have the same length (so that $L_u = L_2 = L_1 = L$). Equation 8.1.5 becomes

$$\Delta P_{bead} = P_d - P_u = \frac{6\eta UL}{d_1^2} \qquad (8.1.20)$$

We can rewrite eq. 8.1.19 after substituting Q = Uh as

$$\Delta P_{slot} = P_o - P_e = \frac{12\eta U L_s h}{d_s^3} \qquad (8.1.21)$$

so that

$$\frac{\Delta P_{slot}}{\Delta P_{bead}} = \frac{L_s}{L}\left(\frac{d_1}{d_s}\right)^3 \qquad (8.1.22)$$

Assuming $d_s = d_1$ for a typical value of slot length $L_s = 20$ mm and lip length $L = 1$ mm, we get

$$\frac{\Delta P_{slot}}{\Delta P_{bead}} = 20$$

In other words, the pressure drop in the slot generally is larger than the pressure drop in the bead. This observation implies that small fluctuations in the slot region are more likely to affect the flow than small upsets in the bead region.

Sensitivity to Slot Width. It is interesting to estimate the flow rate change due to small changes in the slot width d_s about a value d_o. Assuming that the upstream pressure P_o does not change, we can use the Taylor expansion of eq. 8.1.19 about $d = d_o$ to show that the percent flow rate change is given by

$$\frac{Q - Q_o}{Q_o} = 3\frac{d_s - d_o}{d_o} + \text{terms of magnitude } (\Delta d_s)^2$$

or

$$\frac{\Delta Q}{Q} = 3\frac{\Delta d_s}{d_s} \qquad (8.1.23)$$

Thus a 1% error in slot width will produce a 3% variation in flow.

Our emulsion adhesive exhibits shear-thinning non-Newtonian behavior; its viscosity decreases with increasing shear rate. The power-law model given in eq. 3.8.20 provides one simple constitutive model (relating viscosity to shear rate, dv/dy) for shear-thinning liquids. For our purposes we can write it as

$$\eta = \kappa\left(\frac{dv}{dy}\right)^{n-1} \qquad (8.1.24)$$

where κ is the consistency index and n is the flow behavior index with values between 0 and 1 (1 for a Newtonian liquid). Notice that in a log-log plot, eq. 8.1.24 is a straight line with a slope $n-1$. It can be shown that the equivalent of eq. 8.1.23 for a power-law liquid is

$$\frac{\Delta Q}{Q} = \frac{2n + 1}{n}\frac{\Delta d_s}{d_s} \qquad (8.1.25)$$

In many industrial dies, slot width is adjusted by flexing the lip portion of one of the two halves of the die, as shown in Figure 8.15 (hence the name "flex-lip" die). Adjustment bolts are equally spaced (5–10 cm apart) along the die to allow local adjustments by

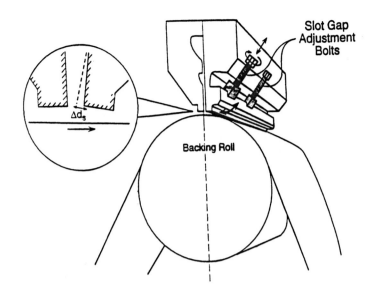

Figure 8.15
A flex-lip slot gap adjustment system allows correction of cross-web coating thickness nonuniformities.

Slot Gap Adjustment Bolts

Δd_s

Backing Roll

flexing the lip along the cross-web direction. Typically, the geometry is set for converging flow, as shown in the inset in Figure 8.15. In this case, it can be shown that the variation in sensitivity of flow rate to slot gap width is one-half that in eq. 8.1.25, that is

$$\frac{\Delta Q}{Q} = \frac{2n+1}{2n}\frac{\Delta d_s}{d_s} \qquad (8.1.26)$$

To summarize, in flow between parallel plates, the flow rate change is three times the gap change for a Newtonian liquid and even larger for a power-law liquid (eq. 8.1.25). With flex-lip dies, the factor is 1.5 for Newtonian and larger for power-law liquids (eq. 8.1.26). Experiments have confirmed these relations.

In this section we have shown that small errors in setting the slot width can result in considerably larger flow rate changes and hence coating thickness nonuniformities. For this reason, die coating heads with massive fixed lips are often used in precision coating applications, such as the multilayer coatings of photographic films described in Section 8.4.2. The liquid distribution manifold (slot included) is designed to achieve optimum cross-web uniformity with a narrow range of liquid rheological properties.

8.2 Overview of Coating Methods

We now turn from our detailed discussion of a specific coating process to a general overview of liquid coating methods.

8.2.1 Coating Methods May Be Classified According to Delivery Procedure

Figure 8.16 schematically illustrates some of the most common coating methods. They can be classified in many ways, but for convenience we have chosen to subdivide them according to the method of delivery to the substrate in the following categories:

Figure 8.16
Different coating methods.
(a) Three-roll pan-fed reverse roll coater. (b) Blade coating and air knife coating. (c) Extrusion coating. (d) Slide coater (after Mercier et al.). (e) Precision curtain coater (after Kistler and Scriven). (*Modern Coating and Drying Technology*, E. D. Cohen and E. B. Gutoff, eds., VCH Publishers, New York, 1992. Reprinted with permission of VCH Publishers © 1992.)

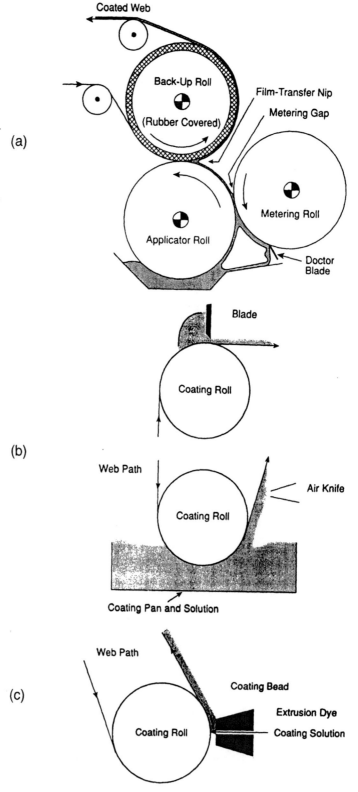

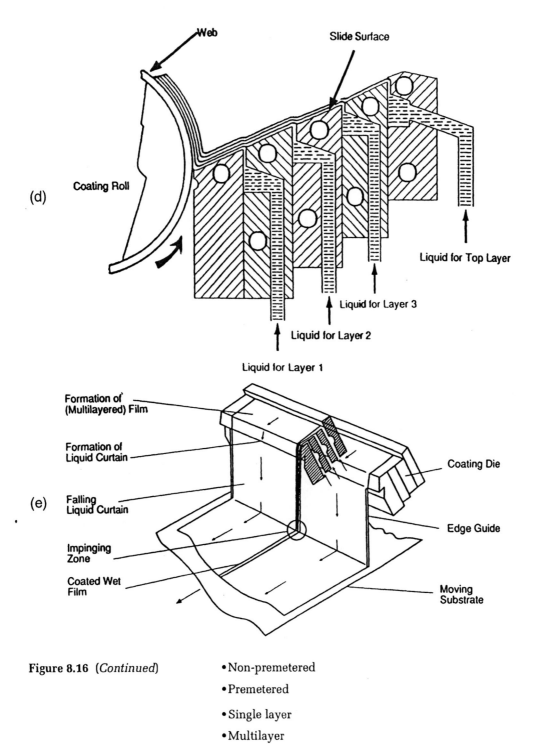

(d)

Web

Slide Surface

Coating Roll

Liquid for Top Layer

Liquid for Layer 3

Liquid for Layer 2

Liquid for Layer 1

(e)

Formation of (Multilayered) Film

Formation of Liquid Curtain

Falling Liquid Curtain

Impinging Zone

Coated Wet Film

Coating Die

Edge Guide

Moving Substrate

Figure 8.16 (*Continued*)

- Non-premetered
- Premetered
- Single layer
- Multilayer

Briefly introducing each method in turn permits us to define some of the terms commonly used in coating technology.

8.2.1.1 Nonpremetered Coating Methods
Nonpremetered coating methods are those in which an excess of coating liquid is available and the amount delivered to the sub-

strate depends on the dynamic liquid uptake and transfer mechanisms.

Roll coating, illustrated in Figure 8.16a, is a typical example of nonpremetered coating. In this instance, the applicator roll (also known as the coating roll) rotates in a large reservoir, picks up some of its liquid, and transfers it to the substrate. Usually this substrate is wrapped around a backup (or backing) roll. A continuous and flexible substrate, like a sheet or strip of paper or plastic, is referred to as the *web*. When the backup roll (and therefore the web surface) is moving in the *same direction* as the applicator roll at their line of contact, the process is known as *forward roll coating*. When they move in *opposite directions,* the process is known as *reverse roll coating*. Figure 8.16a shows a reverse roll coating setup. Additional intermediate rolls may be introduced either to control the amount of liquid carried forward (as in the metering roll shown in Figure 8.16a) or for offset (pattern printing) purposes. In roll coating systems, the surfaces of the applicator and backup rolls may be hard like steel or compliant and covered with an elastomer, depending on the application.

The amount of liquid picked up by the substrate in the region of contact results from fluid dynamic behavior determined by liquid rheology (viscosity of the fluid), wettability of the liquid–substrate interface (characterized by the dynamic contact angle), surface velocites of the rolls, and *nip gaps* (the space filled with fluid between the rolls). Principles of fluid mechanics govern the resultant coating thickness. In the three-roll reverse roll coating method depicted in Figure 8.16a, essentially all the liquid is transferred from the applicator to the web. Coating thickness depends on the applicator roll liquid thickness, which depends on the flow in the nip between the metering and applicator rolls.

Another example of nonpremetered coating methods is *blade coating* (Figure 8.16b), in which liquid contained in the reservoir flows under a blade edge. Because the blade controls, or doctors, the thickness and spreads the liquid layer over the substrate, it is often called the *doctor blade*. Figure 4.32 illustrates the use of this process to control the thickness of a ceramic slurry as it is transferred to a Mylar substrate during the fabrication of thin ceramic sheets. *Air knife coating* is a nonpremetered coating method in which a focused airstream directed at the coated substrate "slices off" the excess liquid.

8.2.1.2 Premetered Coating Methods

Premetered coating methods are those in which all the liquid fed to the coating head is transferred to the substrate. Coaters within this classification differ mainly by how the liquid is fed to the coating head. For example, in *slot coating* (also known as *die coating* and *extrusion coating*) (Figure 8.16c), the liquid is metered to the slot die either by a pump or by a hopper mounted on top of the die. Under stable operating conditions (analyzed in Section 8.1), the flow rate sets the coating thickness for a given web velocity and thickness, and consequently it is independent of the liquid rheology. Other premetered coating methods are *slide coating* (Figure 8.16d) and *curtain coating* (Figure 8.16e), in which a slot die is also used to distribute and apply the liquid. In slide coating, the liquid travels down an inclined plane (the "slide") and is picked up by the web placed very close to the edge of the slide. In curtain coating, the liquid falls freely like a waterfall, and the die may be separated from the web by as much as 10–20 cm.

8.2.1.3 Single-Layer and Multilayer Coating Methods

All the methods illustrated in Figure 8.16—roll coating, blade coating, knife coating, slide coating, and curtain coating—can be used to apply a single layer in one pass. This coating process can be repeated over and over to build up many layers, but doing so requires a number of coating systems.

In multilayer coating, multiple layers are achieved in one pass through the coating head. Slot, slide, and curtain coating can all be used to apply multilayers in one pass. In fact, Figure 8.16d illustrates a multilayer slide coating head in use; in this example four layers are delivered to the substrate sequentially in one pass. As an example of the degree of sophistication reached in multilayer slide coating, the photographic color film industry applies as many as 18 layers in one pass, with a total coating thickness of less than 50 μm!

8.2.2 Many Factors Are Involved in the Selection of a Coating Method

In the end, economics determines the choice of coating method: the best coating method is the one that delivers the desired coating reproducibly and reliably for the lowest cost. Key variables to be considered when choosing a coating method are

- Wet coating thickness
- Rheology of the coating liquid
- Type of substrate (web)
- Coating speed
- Number of layers
- Desired coating uniformity and surface quality

8.2.2.1 Wet Coating Thickness

Certain coating methods are more suitable for delivering very thin or very thick coatings. For example, for thin coatings, *gravure coating* or blade coating can be used to coat wet layers with thickness down to 1 μm. In gravure coating, a cylinder with indentations or "cells" in its surface transfers the liquid from these cells to the substrate. In blade coating, a flexible blade pressed against the web on a backing roll "doctors off" excess liquids. Air knife coating and curtain coating can coat down to several micrometers; reverse roll coating is limited to 5 μm.

Typically the above methods are not suitable for thick coatings. Thick liquid coatings (> 50 μm) are more readily delivered by nonpremetered methods such as roll coating or blade coating. Multilayer coatings where each layer may be less than 5 μm thick can be applied with slide or curtain coating.

For a given dry coating thickness, the thickness of the wet coating to be applied can be varied by diluting or concentrating the coating liquid, possibly allowing the choice of different coating methods. However, changing the dilution/concentration drastically affects the density and rheology of the coating liquid, and this in turn influences the choice of coating method. Also, if we dilute the coating liquid, we must remove more solvent/dispersant during the drying phase, a process that may require a longer dryer with possible increased cost and environmental hazard.

8.2.2.2 Coating Liquid Rheology

In most coating processes, the liquid is subjected to sudden changes in shear at some time in the transfer process. Shear may occur as the liquid is picked up on the coating roll, extruded through the coating head, or squeezed into the nip gaps between rolls. The magnitude of the shear (and the rate at which it occurs) varies with the coating method and the web velocity. As we have seen in Sections 3.8.3 and 6.6.4, respectively, the rheological properties of many liquids are shear-rate dependent and time dependent.

For simple shear flow under steady-state conditions, we can completely characterize the flow by three material parameters: the viscosity coefficient and two normal stress coefficients. A few simple liquids, such as water and organic liquids of low molecular weight, have a viscosity independent of shear rate and very low normal stresses in shear; they are called Newtonian liquids. Most coating liquids, however, are non-Newtonian (see Section 3.8.3.2). Their viscosity may be constant at low shear rates and decrease sharply at higher shear rates ("shear-thinning" fluids), or they may show the opposite behavior ("shear-thickening" or "dilatant" fluids). Other liquids have a "yield stress" below which they behave elastically like a solid ("Bingham" liquids). It is not uncommon for the viscosity of a shear-thinning coating liquid to drop two orders of magnitude (from, say, 5000 to 50 cP) for the shear rates encountered in coating machines.

Most coating liquids are viscoelastic (see Appendix 11.A.4). When they are subjected to a sudden high shear (or elongation)—for example, at the entry to the slot of a slot coater—they exhibit a delay in relaxing the local shear stress and reaching their steady-state viscosity value. Because the residence time of a liquid element in a coating flow is often very short (milliseconds), some liquids, such as polymer melts or concentrated polymer solutions, may never reach their steady-state viscosity value by the time they end up on the web.

Actually, flow in a coating process is more complex than pure shear; it generally involves elongation as well. We can conveniently classify complex flows in terms of simple ones. In die coating, for example, we may encounter simple shear flow (SSF), such as flow between parallel plates, as well as simple elongational flow (SEF), such as stretching a cylinder of material. Coating flows often combine SSF and SEF. Only one parameter, the elongational viscosity, is needed to describe simple elongational (extensional) flow under steady-state conditions. Many coating liquids show an elongational viscosity that varies, either decreasing or increasing with elongation rate.

Frequently low-shear-rate viscosity is the only measured rheological property used to characterize coating liquids. The popular Brookfield viscometer (from which practitioners take the term "Brookfield viscosity") and the Cannon–Fenske glass capillary viscometer, which measure viscosities at shear rates on the order of only 100 s^{-1}, are not much more use than "feeling" the liquid between the thumb and index finger. Shear rates as high as 10^4 to 10^5 s^{-1} are commonplace in most coating flows. For example, in the slot coating process discussed in Section 8.1, we estimate the shear rate in the liquid bead to be 1.7×10^4 s^{-1}. While low-shear-rate viscosities are useful and important in characterizing the reproducibility of the coating liquid, they do not help us understand coating flow behavior of the shear-thinning liquids normally coated. Viscosities at high shear rates are difficult to measure. Modern rheometers can measure viscosities for shear rates up to 2×10^4 s^{-1}. While this capability does not cover the full shear rate range for all flows, it adequately characterizes the shear-thinning behavior of most coating liquids.

8.2.2.3 Type of Substrate (Web)

The variety of coating substrates is as wide as the range of finished coated products. Several substrate properties are relevant to the coating process. The substrate may be *porous* (allowing penetration of the coating liquid), like most kinds of paper, or *nonporous* (impermeable to the coating liquid), like polymer films. Coating methods characterized by higher pressures at the nip force more liquid to enter a porous web. For example, in blade coating of paper with water-based clay suspensions, "dewatering" can be a problem due to the high pressures that develop underneath the blade. In curtain and slide coating, the pressures in the liquid bead are close to atmospheric.

Substrate *wettability* is a measure of the affinity of the coating liquid to the substrate. Other terms also may be used to characterize this phenomenon, such as surface energy. However, what is relevant is the balance between the interfacial tensions of the materials in contact. For example, in slot coating, the balance between the solid–liquid and the liquid–vapor interfacial tension determines how the liquid "wets" the substrate (see Section 8.1.2).

The *roughness* of the substrate can affect the topography of the finished product surface, although the extent of its effect varies with the coating process used. With premetered coating (such as extrusion die coating), the coating surface tends to follow the substrate topography for thin coatings.

8.2.2.4 Coating Speed

Production needs often push coating speeds to the highest levels at which a coating of the desired quality can be obtained. Depending on the coating method, web speeds from 50 to 1000–1500 m/min can be achieved. Some methods are limited inherently to certain ranges of speed. In curtain coating, for example, we cannot maintain the free-falling liquid curtain if the flow rate or web speed is too low. In Section 8.1.2, we found that other parameters, such as pressure, change with coating speed. In practice a "coating process window" which recognizes the interdependence of pressure, film thickness, and coating speed, sets limits for each of these process variables. Operation outside the coating process window results in instability and makes it impossible to obtain a uniform coating.

8.2.2.5 Number of Layers

To coat a substrate with multiple layers, we can either coat one layer at a time and run the web through a multistation coating line or coat all layers simultaneously (multilayer coating). The choice depends on the transport properties of the liquid (viscosity and diffusion) and its drying characteristics. Only three methods allow multilayer coating:

1. Slide coating—the workhorse of the photographic industry, where as many as 18 layers are applied in one pass.

2. Curtain coating—also used for making color photography films.

3. Slot (die) coating—although normally carried out with only one slot, a multislot design can deliver two to three layers or more in one pass.

8.2.2.6 Desired Coating Uniformity and Surface Quality

We must distinguish overall coating uniformity—coating thickness variations across and down the web—from coated surface quality—thickness fluctuations occurring on a smaller scale. Coating thickness normally needs to lie within a specified range for the product to exhibit the desired properties. All coating methods are capable of wide ranges of coating uniformity, depending on the design and operation of the coating station. The parameters responsible for coating nonuniformities are often those that affect resistance to liquid flow in the region of the coating head. We have seen in slot coating that changes in the gap distance across the die face are responsible for coating thickness variations across the web (cross-web), and out-of-roundness of the backing roll causes periodic thickness variations along the length of the web (down-web). In roll coating, uniform and constant clearance between the rolls as they rotate is critical to the quality of the product.

Coating thickness fluctuations on a small scale are often undesirable, although one obvious exception is printing, where we seek to coat isolated dots or lines. The size of the smallest fluctuations and defects that can be tolerated depends on the product. For example, in color photographic film a 5 μm fluctuation is definitely unacceptable because it results in a magnified flaw in the final print or projected image. Local coating thickness fluctuations can originate both at the coating station and during the drying/curing process.

Slot coating, slide coating, precision curtain coating, precision reverse roll coating, and gravure coating give the most uniform coating. Variations less than 2% are readily attainable. In gravure coating, leveling problems can cause local variations.

8.2.3 Defects in Coatings

Inevitably in processes as complex as those outlined above, faults develop and defects appear in the final product. Several steps need to be taken to reduce or eliminate defects: we must detect and identify them, understand their cause, and take the necessary corrective actions. This section discusses different types of defects, the terms used to describe them, and some of their causes.

8.2.3.1 Types of Defects

Ribbing and *barring* (also called *chatter*) are periodic patterns that appear on the wet coating just past the coating station. They originate from hydrodynamic instability during the coating process. Hydrodynamic instability causes the thickness of the liquid coating to vary in sinusoidal fashion across the web, resulting in stripes or ribs along the web (the down-web direction). These thickness variations occur with a small length scale of 1 cm or less. Hydrodynamic instability also causes the liquid volume to build up and fluctuate resulting in ridges or bars across the web (the cross-web direction). Mechanical causes, such as vibrations, out-of-round rolls, or bead vacuum fluctuations, produce bars. These patterns can span the whole web width.

Slide and curtain coating are subject to wave defects rarely seen in other coating methods. Liquids flowing down an incline tend to form *surface waves* at high flow rates. Surfactants can be used to dampen surface waves through the property of surface elasticity. When a wave forms, the surface stretches more in some regions and less in others. Where the surface is stretched, the surface concentration of surfactant decreases and surface tension increases. Higher surface tension attracts liquid and resists the stretching of the surface, thereby damping the magnitude of the wave.

Nonperiodic defects occur more often than periodic ones. Although different evocative terms are used in different industries to describe similar defects, some of the more common terms are *craters, dimples, bubbles, voids,* and *streaks.* The first three are normally micrometer to millimeter size. Voids are uncoated regions of the web (from fractions of a micrometer to centimeters or more in size) caused by breakup of the continuous coated film; they occur most often with substrates of low wettability. Streaks, which normally are found in the down-web direction, can be as short as millimeters—in which case they look like elongated dimples—or as long as meters. With colored coatings (such as those used in the graphics arts industry), *mottling* is a large-scale discoloration often due to variations in coating thickness. In blade coating of paper (see Section 8.4.1 for details), defect terms such as weeping, spitting, and whiskers describe small particles or filaments of dried coating found on the otherwise smooth coated paper (see Figure 8.19).

8.2.3.2 Causes of Defects

We have already stated that instabilities within the die coating bead can cause ribbing and barring on the wet coating. To avoid ribbing, we tend to coat at slower speeds and use less viscous fluids and thicker wet coatings; each of these solutions directly opposes the economic driving forces for higher production rates and lower solvent usage. If the coating liquid is not too viscous and the web speed not too high, surface undulations such as ribbing and barring may level out before the coating reaches the drying unit. Then the dry coated surface is smooth and defect free. This process resembles the leveling of a paint described in Section 4.7.1.2 and Figure 4.23, where we found that the time needed for leveling (eq. 4.7.2) critically depends on both the viscosity and surface tension of the fluid.

Nonperiodic defects arise in many ways. First, attempts to speed the coating process by raising the drying temperature can cause bubble nucleation ("boiling"), resulting in either bubble or crater defects. In multilayer coating, the process of drying and solidifying the coating is particularly tricky because it is often quite thick and the individual layers can have very different properties.

Second, if the wettability of the substrate is low, "dewetting" can occur during coating. Dewetting describes the process in which the coating retracts (or "crawls") from the substrate, leaving uncoated areas or voids.

Third, unfiltered impurities in the coating liquid that become trapped either in the slot (with die, slide, or curtain coating) or on the slide (slide coating) can cause streaks on the coated surface until they break away. Some shear-sensitive liquids, such as emulsions, tend to age and degrade after long residence in stagnant recirculation regions of a coating system. Then they agglomerate to form particles that lead to streaks. Molten liquids (polymer melts or hot melt adhesives) become charred at "hot spots" along the liquid feed system, resulting in carbon impurity particles that can cause streaking or particulate contamination. Scratches or dents on the downstream lip edge in die coating can also cause streaks.

8.2.3.3 Need for On-Line Web Inspection

For some products coating defects must be identified so that faults in the process can be rectified immediately. In these cases, 100% inspection is desirable, which means detecting *all* defects larger than a given threshold and removing the defective portions of the web from the shipped product. Photographic film is an example of a such a product. One way to identify the defects occurring during a coating process is to cut samples from the coated web and use microscopes and any other techniques that emphasize and visualize the defects to analyze them off-line. However, a major drawback of this approach is the length of time required to identify the defect and its cause. It may take minutes to slow down or stop the coating line, and meanwhile the generation of large amounts of defective product results in reduced productivity. Defect identification and on-line analysis with feedback control are attractive for this reason.

Approaches to on-line web inspection vary with the product and desired resolution. Certain coating processes, such as printing, lend themselves to "stroboscopic" image analysis, in which video cameras are synchronized with pulsed light sources to "freeze" the web motion and display on-line stationary images of the coated web. In this situation, the press operator can identify printing problems by observing whether the screen images of the printed web match a "good" template. Other coating processes require continuous monitoring and real-time image analysis to scan the surface of the finished product spot by spot as it emerges from the coater. The next section discusses an approach to real-time imaging applicable to coating systems.

8.2.4 Real-Time Imaging Has Been Applied to On-Line Web Inspection for Defects

High-speed electronic web inspection systems were already in use in industry in the 1970s, and they have continued to evolve rapidly. Million dollar systems for 100% on-line defect detection have existed for several years. They typically consist of one or more laser scanning heads, detectors, and specialized electronics capable of some primitive identification based largely on size and shape factors. They are not so helpful in understanding the causes of the defects. Systems capable of more sophisticated identification are now available. Consequently, more on-line inspection systems are being installed on both pilot and production coating lines.

8.2.4.1 Different Illumination and Scanning Configurations Can Be Combined for Web Inspection

We can conveniently classify real-time imaging sytems in terms of their lighting and sensing configuration, as shown in Figure 8.17.

Scanned illumination and distributed sensing systems use a laser scanner for illumination and a single light sensor for

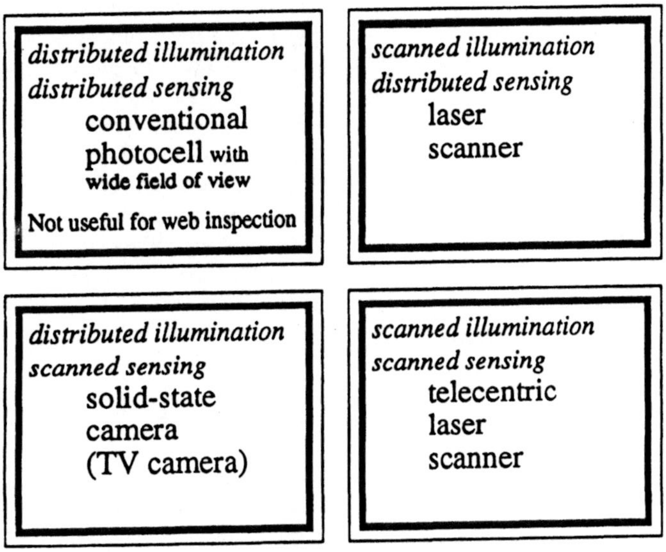

Figure 8.17
Real-time imaging inspection of coated films for defects. Web inspection illumination and sensing configurations.

distributed illumination *distributed sensing* **conventional photocell** with **wide field of view** Not useful for web inspection	*scanned illumination* *distributed sensing* **laser scanner**
distributed illumination *scanned sensing* **solid-state camera (TV camera)**	*scanned illumination* *scanned sensing* **telecentric laser scanner**

detection. A laser beam is reflected off a spinning faceted mirror so that it scans repeatedly across the web at a speed proportional to the rotational speed of the mirror. The better scanners have additional optical components that change the resulting "fan beam" into one with a constant angle of incidence (but not constant intensity). One or more sensors (photomultiplier tubes) collect the reflected and/or transmitted light from the whole web and convert it to a continuous single electrical signal. Fluctuations in the electrical sensor signal are used to identify and locate web defects.

Distributed illumination and scanned sensing systems use a single source of illumination and a solid-state camera to capture the image of the whole scene with a photocell array. The electrical signal from each cell (or pixel) is shifted out sequentially in TV raster-scan format to produce an image that can be displayed to identify and locate web defects. The solid-state camera array can be an area array (for example, 768 columns, 485 rows) or a linear array (for example, 4096 columns, 1 row).

Scanned illumination and synchronously scanned sensing systems use telecentric laser scanners. Light from the laser scanner is reflected from the web and travels back along the same optical path to a beam splitter and sensor.

The main advantage laser scanners have over solid-state cameras is their higher scanning speed and better control of the angle of incidence of the light. In virtually all the more difficult defect detection/analysis applications, the appearance of the defects critically depends on the angle of incidence. Anomalies that are clearly visible at certain angles of incidence can completely disappear at others.

The main disadvantage of laser scanners is their high cost and lack of uniformity of light intensity across the scan. Well-designed modern distributed illumination systems (such as focused lighting—Kohler illumination) can provide good uniformity as well as controlled angle of incidence over an area. A properly designed illumination system combined with a solid-state camera provides a highly effective and economical system. Because camera-based systems have an inherent cost advantage, they continue to displace laser scanner systems for moderate-speed, moderate-resolution applications.

A further challenge is that we can see certain classes of defects only with one kind of illumination (such as near-normal incidence) while others are visible only with a different kind of illumination (such as oblique incidence). Both classes of defect may be present and both must be detected. In such cases, we have no choice but to alternate the two modes of illumination or set up two separate inspection stations. Camera systems are more flexible in such situations because we can easily and rapidly adapt the lighting as well as the inspection algorithm.

8.2.4.2 On-Line Image Analysis Systems Require Ultra-High-Speed Computation

All the web inspection systems we have described produce a data stream that must be processed by the image analysis system. For web coating processes that require 100% inspection (rather than periodic sampling for statistical quality control), the image processing system must keep pace with the data stream and not omit any data from the analysis. Data rates for standard camera systems are typically 10,000,000 samples (or pixels) per second. The newer high-resolution cameras produce data at rates up to 40,000,000 pixels per second, and the data rate is even higher with some laser scanning systems.

Because even an unsophisticated *digital* inspection algorithm must perform 100 or more than 1000 mathematical operations *on each pixel,* the computing requirements (10^{10}–10^{11} per s) for such applications are orders of magnitude beyond the capacity of any affordable general-purpose digital computer system. Therefore, it is not surprising that image analysis systems designed for the highest data rates were once limited to trivially simple algorithms and/or were constructed using analog hardware. Continuing improvements in the price to performance ratio of computing hardware now bring more and more sophisticated applications within the reach of affordable digital systems. Three very different approaches have been followed:

1. As microprocessor and digital signal processing chips have become faster and faster, more and more "low-end" inspection applications are moving to workstations and desktop computers. New chips announced in 1993 and 1994, particularly the reduced instruction set computing (RISC) processors, are at last bringing video-rate speeds to the desktop. We can expect this trend to continue. A few of the software-based image analysis programs on the market are both powerful and easy to use, making them accessible to factory personnel and others who are not experts in image analysis.

2. At the high end of the digital processing range, commercially available image processing boards operating in a "pipelined" mode have become the systems of choice. These systems also benefit from the decreased cost and improved performance of electronic chips and can carry out sophisticated image analysis algorithms at real-time rates (30 images per second) or even faster. Now it is not only possible to monitor high-speed webs but to use the results of the image analysis to *control* the production process. However, these systems are notoriously difficult to program, and operators need extensive training. Wider industrial application must await the packaging of these boards as part of commercial turnkey systems.

3. Some coating processes proceed too fast even for the fastest digital hardware. For these applications, custom-built analog or analog/digital hardware will continue to be used for the foreseeable future.

8.2.4.3 Real-Time Image Analysis Can Be Achieved with Multicamera Systems

Figure 8.18 shows an application in which an eight-camera system has replaced a laser scanner to provide a more sophisticated inspection algorithm at reduced cost. The speed and resolution

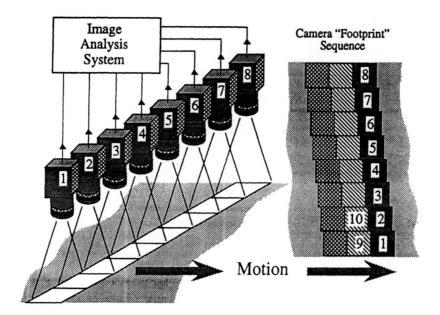

Image Analysis System

Camera "Footprint" Sequence

Motion

Figure 8.18
Real-time imaging inspection of coated films for defects. Web inspection with eight cameras and a single vision system.

requirements for real-time image analysis of the coating process are as follows:

- Required pixel size: 0.008 in. by 0.008 in. = 0.0002 m by 0.0002 m

- Web width: ~32 in. = 0.8 m

- Required number of pixels across web: (0.8 m) ÷ (0.0002 m) = 4000

- Web speed: up to ~72 ft/min = 0.36 m/s

- Required scan rate: (0.36 m/s) ÷ (0.0002 m) = 1800 full-web scan lines/s

A single, standard, solid-state camera (512 columns by 485 rows) cannot meet the 4000-pixel cross-web resolution specification, but *eight* cameras can. This assembly allows for 12 pixels of overlap between adjacent cameras to ensure that no defects on the boundary between two cameras are overlooked. At the standard camera scan rate of 30 frames/s and 485 scans/frame, each camera can scan 30 × 485 = 14,550 lines/s, which is also the processing rate of a standard pipelined image analysis system, but more than eight times the required scan rate of 1800 lines/s. *Therefore, a single image analysis computer is fast enough to handle the data from all eight cameras, one after the other.* The system timing is as follows:

- Pixels cross-web per camera: 512 − 12 for overlap = 500

- Inspection algorithm completed for one frame in real time = 1/30 s

- Inspection time for 8 images = 8 frame times = 8/30 s = 267 msec

- Web motion in 267 msec: 0.36 m/s × 0.267 s = 0.096 m = 480 scan lines

- Overlap in down-web direction: 485 − 480 = 5 scan lines, which is sufficient.

Figure 8.18 also shows the sequence of web areas, or footprints, "grabbed" by the eight cameras. When the system finishes with the area labeled 8, which came from camera 8, it is ready to start on the image labeled 9, which comes from camera 1, and so forth. In this way every part of the web can be analyzed with a little overlap. Because even very good solid-state cameras are inexpensive (< $1000) and the system needs only one image analysis computer system, the total cost for this web inspection application can be less than $50,000. A laser scanner of similar resolution coupled with a much less sophisticated inspection algorithm would cost more than five times as much.

8.2.4.4 Algorithms

To be useful, the imaging system must be programmed to detect anomalies. Assuming that the lighting has been done so that all the important anomalies are actually visible, the "best" algorithm for defect detection will depend in a very subtle way on the nature of the web material and the significant defects. Successful application requires extensive experience and experimentation. To verify that the algorithm works over the range of actual conditions, algorithm developers must have access to a full range of "good" material as well as many samples of each important kind of defect. The most pervasive fallacy in human thinking on this subject is to assume that because people can see a flaw, a machine can see it also. In fact, the machine typically cannot see *anything* it has not been programmed to see.

There is no "standard" or "best" inspection algorithm, and there probably never will be. A good algorithm is "robust"—that is, it is relatively unaffected by changing lighting conditions, web speeds, and other environmental factors or product variations. If the system cannot be made sufficiently robust, it should be self-calibrating. If even this is not economically possible, then as a minimum the system should automatically alert the user or supervisory computer when it is no longer in calibration. Algorithm development for real-time image analysis is a large and complex subject, and space limitations prevent further discussion here.

8.3 Environmental Issues in Coating Processes

Because the media has devoted so much attention to the problem of ozone depletion in the upper atmosphere, everyone is aware of the problem. Although several years ago the ozone "hole" was mostly restricted to the atmosphere above Antarctica, recently alarmingly low ozone levels have been measured in the skies above industrial/urban areas of the United States and other industrialized countries. Thinning of the ozone layer permits increased ultraviolet radiation to reach the earth's surface, which can result in increased risk for skin cancer and cause genetic damage to organisms at the base of the food chain. Studies have shown that chlorinated fluorocarbons (CFCs and HCFCs) are the main culprits in destroying the thin, fragile layer of ozone in the stratosphere. Although proof does not exist yet, it is also widely believed that the "greenhouse" effect, caused by carbon dioxide and other chemicals shielding infrared radiation from the earth's surface, will lead to a steady increase the earth's surface temperature.

8.3.1 Overview of the Clean Air Act Amendments of 1990

In 1990 Clean Air Act Amendments signed into law by the government drastically revised the inadequate Clean Air Act of 1977. The legislation was designed to address three major threats to the nation's environment and the health of the public: acid rain, urban air pollution, and toxic air emissions. Two sections of the Act are particularly relevant to the coating industry: Title III, Air Toxics, and Title VI, Stratospheric Ozone and Global Climate Protection.

Title III requires technology-based controls on sources of 189 toxic air pollutants, such as carcinogens, mutagens, and reproductive toxins. It distinguishes major sources (those emitting more than 10 tons/year of any one or 25 tons/year of any combination of those pollutants) from local sources (small sources, such as dry cleaners). By November 1992, the Environmental Protection Agency (EPA) issued Maximum Achievable Control Technology (MACT) emission standards based on the best demonstrated control technology for many of the listed pollutant sources. Industries that did not reduce emissions according to established guidelines would be penalized in different ways.

Title VI sets out EPA plans to phase out production of substances that deplete the ozone layer. It requires that Class I chemicals, such as CFCs (used in aerosol sprays and as refrigerants), halons, and carbon tetrachloride, be completely phased out by the year 2000 (methyl chloroform by 2002) and that Class II chemicals (HCFCs) be eliminated by the year 2030. It also requires that nonessential products releasing Class I chemicals be banned by November 1992. After 1994 it became unlawful to sell an aerosol that contains an HCFC or a plastic foam that contains or was made with an HCFC. The Title also requests voluntary reductions in the use of many solvents, including carbon tetrachloride, methyl ethyl ketone (MEK), toluene, and xylenes.

8.3.2 Impact on the Coating Industry

The coating industry makes heavy use of many organic chemicals regulated under the new legislation. Industries can follow several approaches in striving to comply with the more stringent regulations:

1. Replace toxic chemicals with others that are less toxic. One approach in the adhesive coating industry is to shift from

organic solvent-based to water-based adhesives. This change is not occurring as fast as it might because the properties of most water-based adhesives are still inferior to those of solvent-based adhesives. Another approach is to replace toxic organic solvents used for cleaning machinery with potent new solvents based on citrus oils that have a pleasant smell. The author will always remember visiting a plant where yard-long tanks filled with methyl ethyl ketone were laid open next to the coating line. Between coughs he asked the line operator for an organic-vapor mask, only to discover that the mask's absorbing cartridge was probably several years old and it was useless. The operator, who had been with the company for 25 years, claimed the choking vapors did not bother him—he had not used a mask for years!

2. Enclose systems or processes. Many coating industries coat polymers dissolved in highly volatile organic solvents such as toluene and xylenes. To reduce air pollution, the coating station can be enclosed and the vapors vented directly to the drying unit, where they can be recovered by condensation.

3. Reduce the use of organic solvents. Within limits, solute concentration can be increased, thus decreasing the amount of solvent. Decreasing the solvent concentration from 50% to 30% would result in a significant downsizing of solvent recovery units.

8.4 Coating Technology Is Widespread in Familiar Products

8.4.1 Printing Is a Selected-Area Coating Process

8.4.1.1 Newspapers

In the past, newspapers were printed using the letterpress method, in which raised characters on the printing plate imprinted the ink into the paper. Most newspapers have changed over to the *offset* or *gravure* printing method, in which the characters are "engraved" in the printing plate. Ink is transferred to the paper by "kiss" printing. The term *offset* derives from the fact that the ink is transferred from the printing plate wrapped around a roller to a rubber-covered applicator roll that finally transfers it to the paper—the printing plate is offset from the paper.

Have you ever wondered why, in this age of technological advances, newspapers still smudge? Both letterpress and offset printing use oil-based inks in which a pigment (often carbon black) is dispersed in an oil similar to lubricating oil. Most of the oil is simply absorbed by the paper fibers, but since it never dries completely, rub-off becomes a problem. The problem is more serious with the offset printing method because less pressure is exerted to drive the ink into the paper than in the letterpress method. Rub-off also depends on how "heavy" the ink is. Some newspapers, such as the *Wall Street Journal,* are printed with very light ink; others, like *The New York Times,* are printed with much heavier ink. Rub-resistant inks do exist. They contain waxes and other compounds to bind the pigment to the paper, but they are more expensive—conventional inks for offset printing cost about 50 cents a pound, whereas rub-resistant inks cost 60 cents and more per pound. An industry trend is replacing offset equipment with flexographic printing presses that use water-based inks. The water actually evaporates during the process, leaving just the pigment and thus reducing the rub-off problem. Moreover, these inks contain latexes that bind the pigment to the paper.

8.4.1.2 Magazines

High-quality glossy paper used in magazines is produced by coating raw paper with a colloidal fluid made by dispersing fine clay powder and calcium carbonate in water containing other constituents. The total solids content is often greater than 50%, giving the liquid the appearance of white glue. Blade coating is traditionally used for coating paper. A thin metal blade wipes off most of the coating that the web picks up from a pond (reservoir), leaving a coating 10–20 μm thick. Web speeds up to 500–600 m/min are used to produce a smooth coated surface. A *calendering* process in which the coated and dried paper passes between two highly polished steel rolls creates the desired high gloss.

Although the photograph reproduced in Figure 8.19 might be reminiscent of some modern sculptural art form, these "statues" are actually only a few millimeters high and represent an unwanted by-product of paper coating. At web speeds above some critical value, coating liquid begins to seep between the blade and the web and liquid build up on the downstream side of the blade, eventually forming the structures shown in the photograph. Sometimes the structures are ejected when still small (100–200 μm) and they end up on the coated paper as small elongated particles (called "whiskers" or "spits"). These defects are unacceptable to the printing industry because they cause inks to smear and can even damage the delicate mechanisms of the printing press.

Figure 8.19
"Stalagmites" grown on the blade edge in high-speed blade coating of paper.

8.4.2 Photographic Films Are Manufactured by Multilayer Coatings of Photographic Emulsions

8.4.2.1 History of Photographic Film/Paper and Photographic Development

Photographic images providing continuous-tone reproductions of scenes were first produced in the early nineteenth century. Some of the oldest images can still be viewed; in fact, daguerreotypes produced by Daguerre in 1837 are on display in the collection of the French Photographic Society in Paris. In spite of impressive technical advances in non-silver halide image recording technologies since Daguerre's time, chemically based silver halide photographic systems continue to occupy a preeminent position. Over 16 billion still photographs are taken by amateurs in the United States each year. This figure corresponds to about one picture per week for every man, woman, and child in the country. Favorable attributes exhibited by silver halide photography include sharp, low-granularity images, superior color reproduction, archival image permanence, and relatively inexpensive cameras, film, and processing. These commercial advantages are a consequence of an intriguing set of physical and chemical properties, a wide variety of interfacial phenomena, and an associated collection of advanced supporting technologies.

The photographic process is an interdisciplinary technology requiring understanding of the fundamentals of chemistry, physics, and interfacial engineering. Production of low-cost photographic products with continuously improving features and quality requires a close partnership among innovators, scientists, engineers, product formulation designers, and manufacturing process technologists.

A typical photographic element is composed of a flexible, yet dimensionally stable, substrate, usually film or paper, over which various chemical-containing layers are coated. Most layers use gelatin (see Section 6.4.5 on gels) as a medium to support and disperse a variety of chemicals, such as colloidal solid particles (photoactive silver halide microcrystals and solid organic dye crystals), colloidal emulsion particles (oil droplets containing organic molecules called couplers for forming color images), and other components used to optimize the physical and image qualities of the final product. Gelatin is not only a dispersing agent but also exerts strong and favorable photographic effects. Black-and-white materials may need only one photographic "emulsion" layer, although it is overcoated with a protective layer. (Note that the term *emulsion* used here in connection with photographic film refers to a gelatin layer or melt containing photoactive particles. It is not a liquid in liquid suspension, as emulsion was specifically defined in Chapter 5.) Modern color films require more than fifteen separate layers—each with a thickness comparable to that of a human hair—to get the desired photographic response. Color films may contain 10 grams of silver per square meter and as many as 100 different chemical compounds to provide the desired sharpness, granularity, color reproduction, image permanence, and light sensitivity. Altogether the color photographic film represents an impressive achievement in interfacial engineering!

8.4.2.2 Production of the Photographic Image Involves Three Stages—Exposure, Development, and Fixation

8.4.2.2A Exposure: The Role and Properties of Silver Halide Microcrystals and Color-Sensitive Dyes.

Silver halide microcrystals are the light-sensitive components within the photographic element. For photographic negatives, silver bromide (AgBr) with a small amount of iodide (AgI) is normally used; for less sensitive photographic papers, silver chlorobromide (AgClBr) is more common. Product needs determine the size of the microcrystals. For high-resolution microfilm products the crystals may be only tens of nanometers in diameter (a fine-grain film); for high-sensitivity photographic film used by amateurs interested in available-light photography, crystals in the micrometer size range (a coarse-grain film) are necessary.

Silver halide is a semiconductor with a room-temperature band gap (energy separation between the top edge of the valence band and the bottom edge of the conduction band) in the range of 2.6 to 3.1 eV (4.2 to 5.0×10^{-19} J). The actual magnitude of the band gap depends on the choice of halide ions (Br, I, Cl). Because the gap lies in the range of 2.6 to 3.1 eV, silver halide absorbs light in the blue range of the visible spectrum (as well as the ultraviolet and beyond) to form a mobile electron in the conduction band and a mobile hole in the valence band. An electron–hole pair is formed for every absorbed photon. Ultimately, the mobile electron may reduce a positive silver ion to yield an elemental atom of silver ($Ag^+ + e^- \rightarrow Ag$), thereby recording the exposure event. If enough ions are reduced at a localized site through iterative processes involving photoelectrons and silver ions, a stable silver center is formed. The smallest stable centers are thought to contain about three silver atoms; stable centers are referred to as *latent image centers*. Latent image centers serve as catalytic sites for subsequent development of larger silver specks through the action of a chemical reducing agent, referred to as a *developer molecule*. The chemical development process can lead to as much as a 10^{10}-fold increase or gain in the number of silver atoms. Gains of this magnitude enable manufacturers to produce high-speed products with available-light capability.

Because photographic images require sensitivity to green and red light (and, for special purposes, infrared light) in addition to the inherent blue sensitivity, the silver halide microcrystals are treated with color-sensitive or *spectral sensitizing organic dyes* that adsorb on the crystal surfaces. On exposure, the dyes absorb the longer-wavelength light (green or red or infrared) to form photoelectrons and positive holes. An electron promoted from the ground state of the dye into a higher-energy electronic state eventually transfers into the conduction band of the silver halide and thereby contributes to the formation of a latent image center that records photons from the "nonblue" visible region of the spectrum. As illustrated in Figure 8.20, all conventional color films require three photosensitive layers or color packs, one for each of the primary colors, blue, green, and red. Sometimes more than one layer is used for one color to obtain better color fidelity over a wider exposure range.

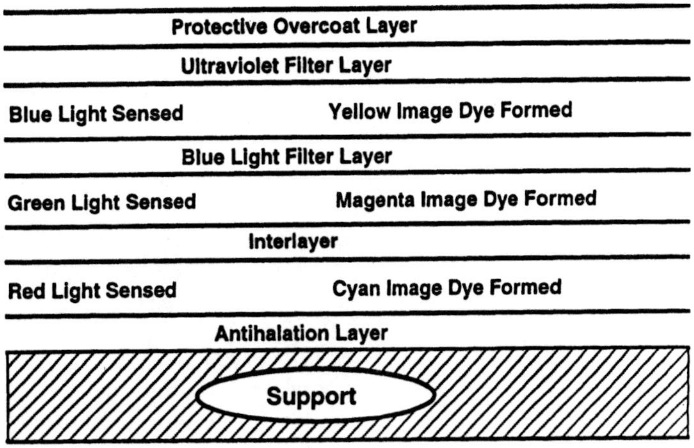

Figure 8.20
Schematic of an integral tri-pack color negative photographic element. In the three image-forming layers blue, green, and red light is absorbed and detected by silver halide microcrysratals held in the gelatin as a colloidal dispersion. Upon development, yellow, magenta, and cyan dyes are formed in the three layers, respectively, when oxidized developer molecules react with coupler molecules dispersed as colloidal emulsions.

Protective Overcoat Layer	
Ultraviolet Filter Layer	
Blue Light Sensed	Yellow Image Dye Formed
Blue Light Filter Layer	
Green Light Sensed	Magenta Image Dye Formed
Interlayer	
Red Light Sensed	Cyan Image Dye Formed
Antihalation Layer	
Support	

In addition to the spectral sensitizing organic dyes, *chemical sensitizers* are also adsorbed onto the surface of the silver halide crystals to improve the efficiency of the latent-image-forming process. Treatments that reduce the number of photons needed to form a latent image by more than a factor of 10 are common and, in fact, essential in astronomical and x-ray films and other systems requiring ultrahigh photographic speeds. Chemical sensitizers are compounds that react with the silver halide surfaces to provide trapping sites that favorably alter the electronic processes occurring during exposure to light. These trapping sites capture photoelectrons from the conduction band and thereby assist in the localized efficient growth of latent image centers. Without such traps, mobile photoelectrons can drift and recombine with the photoholes, annihilating the effect of the original exposure and reducing the photo conversion efficiency. Because these trapping sites are localized on the crystal surfaces, they are readily accessible to subsequent developer molecules.

To avoid exposure to shorter wavelengths, each of the three primary color layers must be separated by a color filter. Because all the silver halide crystals are sensitive to ultraviolet and shorter wavelengths, an ultraviolet filter must first be placed on the outside of the film. Then a blue light filter layer must be placed after the blue-sensitive layer to avoid exposure of the green-sensitive and red-sensitive layers to blue light, and another filter layer must be placed between the green- and red-sensitive layers to avoid exposure of the red-sensitive layer to green light. These filter layers contain dyes that selectively absorb the different colors. The dyes are introduced into the color packs either as dispersed colloidal solid dye particles or as oil-phase droplets. Finally, light that is not absorbed by the three primary color packs may be reflected off the plastic support, causing a halo to appear around exposed areas. To avoid this defect, a grey or black antihalation layer is placed between the base and the image-forming layers. Additional spacer layers and a protective top coat layer may easily bring the total number of separate layers in a color film to 15 or more.

8.4.2.2B How Do We See Color Images? Before we discuss development of the photographic image, it is important to understand our objective. We must realize that what we perceive as a color image actually results from the selective *subtraction* of light—what we see is the light left over after absorption of one, two, or all three of the primary colors.

First, we describe the structure of the *color negative* photographic element. It consists of three separate color-forming packs, each containing chemicals necessary to control the formation and location of color dyes. In a color negative, dyes selectively produced in a given area of the photographic element are *complementary* to the colors we see in the original scene. As demonstrated in the sequence in Figure 8.21, a yellow dye records the location of blue-sensitized crystals, a magenta (blue-red or purple) dye records green-sensitized crystals, and a cyan (blue-green) dye records red-sensitized crystals. The bottom frames of Figure 8.21 show the resulting color negative. Where there is white exposure, all three color dyes—yellow, magenta, and cyan—are found in the negative, giving a black image in transmission. Where blue light exposes the blue-sensitive layer, only yellow dye is found, and so forth.

To obtain a *positive* color print, the image must be reversed by exposing color paper or film to light transmitted through the negative. Then original white exposure leads to an image in which all three color dyes—yellow, magenta, and cyan—are absent and the image appears white. Original blue exposure leads to an image in which the yellow dye is absent so that the image is formed by the remaining magenta and cyan dyes. Since the magenta dye absorbs green light but transmits blue and red light, while the cyan dye absorbs red light but transmits blue and green light, only the blue light is not absorbed in the multilayer system; so the image appears blue. Likewise, original green exposure leads to an image in which the magenta dye is absent, the remaining yellow dye absorbs blue light, and the remaining cyan dye absorbs red light; so only green light; is not absorbed, and the image appears green. Original red exposure leads to an image in which the cyan dye is absent, the remaining yellow dye absorbs blue light, and the remaining magenta dye absorbs green light, so only red light is not absorbed, and the image appears red. In this way we perceive a true color reproduction of the scene in the integrated image.

8.4.2.2C Development: Colloidal Droplets Containing Coupler Molecules Are Critical in the Color Development Process. Immersing the exposed photographic element into an aqueous solution containing chemical reducing agents called developer molecules initiates development. A developer molecule diffuses rapidly through the gelatinous layers and eventually encounters a latent image center on the surface of a microcrystalline silver halide grain. In black-and-white photography, the latent image center behaves like an electrode and receives an electron from the developer molecule. The electron reduces an additional silver ion, causing the latent image center to grow. Iteration of this step is the key mechanism for producing elemental silver specks that constitute black-and-white images.

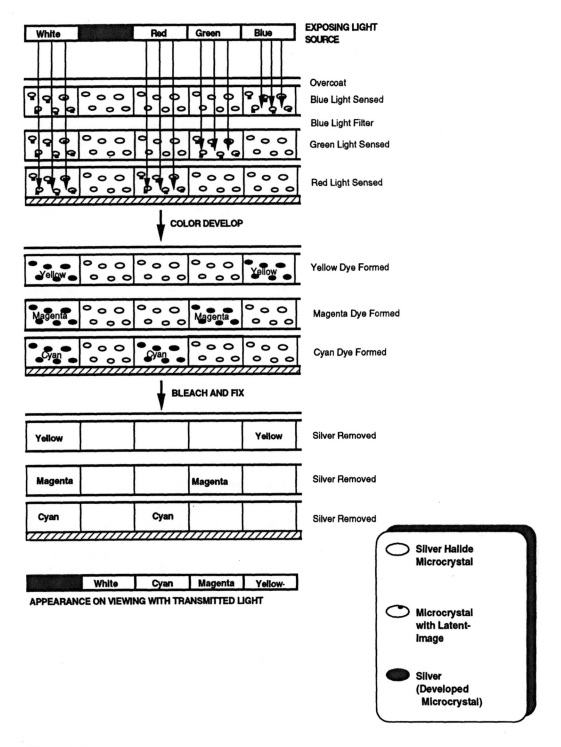

Figure 8.21
A negative-working color photographic image system.

Additional mechanistic steps are involved in color development. After a developer molecule releases an electron to a silver ion near the latent image center, it becomes chemically oxidized and in the oxidized state diffuses away until it encounters the surface of a droplet containing *coupler molecules.* At the interface, a coupling reaction between the developer and the coupler molecule forms an *image dye molecule*. Different coupler molecules produce different dye colors. Coupler molecules are generally introduced into each "emulsion" layer in the color pack as oil emulsion droplets. The coupling reaction occurs at the interface between the aqueous/gelatin gel phase and the oil droplet phase. As is the case with silver halide microcrystals, the size distribution of the coupler droplets is important to the quality of the image. It is important to note the key function of the gel in this process; the gelatin network provides mechanical stability to the structure, while the continuous aqueous phase permits rapid diffusion of developer molecules to the active sites.

8.4.2.2D Fixation. After development, most conventional photographic products are treated with thiosulfate *fixing* baths to dissolve and remove any undeveloped silver halide. Thiosulfate ions dissolve silver ions by forming a soluble silver thiosulfate complex. In color photography, the developed silver specks must first be reoxidized to silver ions via a *bleaching* process and then removed by the fixing process; otherwise the image would contain small silver specks.

8.4.2.3 The Manufacturing Process: Synthesis of a Photographic Element Involves Many Steps

As we have seen, modern color film contains many different components; each of the three color packs contains silver halide microcrystals, color-sensitive organic dyes, chemical sensitizers, and coupler molecules, and a color filter layer is located between each pack. Figure 8.22 provides a flow chart of the complete process from synthesis of the light-sensitive crystals to the production of a finished negative. We follow the different manufacturing steps sequentially.

8.4.2.3A Silver Halide Crystal Growth. The modern preparation of photographic films, papers, and plates begins with the growth of silver halide microcrystals. As noted already in Section 4.2.1, colloidal silver halide crystals nucleate and grow during the controlled reaction of silver nitrate with an alkali halide. The manufacture of a photographic emulsion achieves this result either by injecting aqueous silver nitrate and aqueous alkali halide into a carefully agitated reaction vessel containing a dilute aqueous solution of gelatin (the "double-jet" method) or by injecting silver nitrate into a reaction vessel containing aqueous gelatin and halide (the "single-jet" method). In the photographic industry, nucleation and growth is referred to as *precipitation,* and the resulting microcrystals are referred to as *emulsion grains.* Depending on the flow rates, temperature, agitation, degree of supersaturation, and silver ion concentration in the reactor, silver halide crystals with predetermined sizes and shapes can be precipitated. Depending upon the intended use, crystals can be grown to have a number of different morphologies; tabular cubic, octahedral, and twinned crystals or truncated pyramids are the most common in practical systems. Unwanted flocculation of the microcrystals during growth and subsequent handling is generally not a problem because of Coulombic repulsion and because

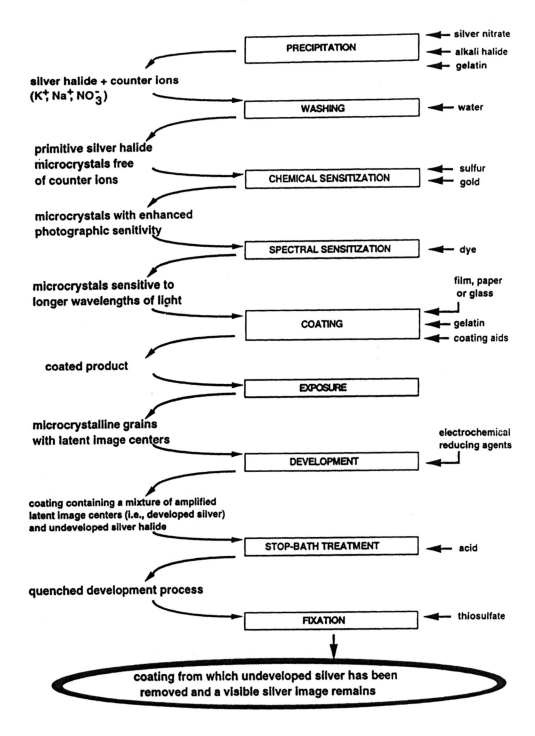

Figure 8.22

Flow chart of the processes involved in the production of a finished negative. Key processing steps are included in the boxes. The first five steps constitute the manufacture of the photographic film. The last three steps constitute development of the negative after exposure. Inputs and raw materials are shown to the right of the boxes. The status of the silver halide emulsion at the completion of each step is shown at the left.

the gelatin provides a protective capsule that adsorbs onto the silver halide grain surfaces.

8.4.2.3B Washing. After precipitation, the emulsion must be washed to remove soluble salts that otherwise crystallize out on drying to desensitize the emulsion's response to light; and then excess water must be removed to concentrate the emulsion. We can use one of three methods for *washing* [acid flocculation (2) is the preferred method].:

1. Chill-setting the emulsion, chopping it into small pieces, and washing the pieces in cold water.

2. Flocculation of the gelatin by adding acid to the solution (which contains gelatin derivatives insoluble in acid) under agitation. The flocculated gelatin contains the silver halide crystals. It is chilled and washed several times through repeated mixing with cold water, gravity settling, and decanting.

3. Separation using ultrafiltration membranes that are impervious to the high-molecular-weight gelatin and silver halide crystals yet permit water and salts to transfer out of the emulsion.

Washing is complete when the conductivity decreases to a desired level or the concentration of halide ions, as measured by a specific ion electrode, reaches a specified level.

8.4.2.3C Response Enhancement: Chemical Sensitization. After washing, the emulsion has very low sensitivity to light, and the microcrystals must be treated to enhance latent image formation. *Chemical sensitization* is achieved by heating the emulsion to ~60°C and adding chemical sensitizers in solution to the agitated liquid. Typical chemical sensitizers are sodium thiosulfate, trivalent gold salts like potassium tetrachloroaurate, and monovalent gold salts like sodium aurous thiocyanate. In particular, sodium thiosulfate diffuses to the silver halide grain surface and chemically reacts to form centers of silver sulfide (Ag_2S) that provide the traps. The role of gold in chemical sensitization appears to be at least twofold; it enhances the catalytic activity of the latent image center and stabilizes a nucleating latent image center by reducing the possibility of its thermal regression during exposure. Proper levels of thiosulfate and gold must be found by experimentation, since adding too much desensitizes the system.

8.4.2.3D Response Enhancement: Spectral Sensitization. The next step in preparing the emulsion is *spectral sensitization*, the process by which organic dyes are added to the silver halide to promote the response to long-wavelength light. Again, the aqueous gelatin is heated slightly above the gelation temperature, appropriate dyes in a methanol solution are added, and the system is agitated. The most widely used spectral sensitizers are cyanines, consisting of heterocyclic moieties such as benzothiazoles linked by a conjugated –CH= chain. The dyes adsorb onto the silver halide crystal surfaces, often aggregating into ordered structures, and the optimal photographic response is usually achieved at near monolayer coverages.

After spectral sensitization is complete, *stabilizers* are added to prevent fogging—the formation of latent image centers without exposure—and to prevent loss of sensitivity with storage.

8.4.2.3E Coupler-Containing Emulsion Droplets.

An oxidized color developer molecule (p-phenylenediamine) reacts with a coupler molecule to form an image dye that may be yellow, magenta, or cyan in color. Photographic coupler molecules are weakly acidic organic compounds that generally have an organic ballast group (for example, a long aliphatic chain). Yellow-forming couplers contain an active methylene group that is not part of a ring; magenta-forming couplers are often pyrazolones or other species that have an active methylene group as part of a conjugated ring; cyan-forming couplers are typically phenols or naphthols. Coupler molecules are suspended in droplets of a high-boiling organic solvent that forms an O/W emulsion of the kind described in Section 5.4. In the industry, these emulsions are referred to as *dispersions*. The ballast groups on the coupler molecules make it favorable for them to remain in the oil droplet phase; and the oil droplets keep the coupler confined so that it does not diffuse through the aqueous phase into neighboring image, forming layers that would lead to poor color reproduction. The colloidal droplets are typically spheres about 0.2 μm in diameter, and they can be prepared by any of a number of different size-reduction techniques. Their small size is necessary so that a large interfacial area is available to facilitate the coupling reaction with oxidized developer molecules. Having a large number of small droplets also enhances image sharpness and reduces granularity.

Coupler emulsions are prepared separately from the photographic "emulsions"; they are blended together immediately prior to the coating process.

8.4.2.4 The Coating Process

8.4.2.4 A Coating Photographic Films Requires a Multilayer Coating Method.

We see that black-and-white films are composed of two or more layers and color films of perhaps 15 or more layers, each of which may contain many chemical constituents. Depending on its function, the thickness of an individual, coated, dried layer is from less than 1 to 30 μm. Rigid constraints apply to the technology of coating photographic materials. The various chemical-containing layers must be uniform in thickness and composition as well as free from streaks and other physical imperfections.

Of the roughly two dozen major coating techniques, only a few are used for producing photographic products. In most of these techniques, the flexible film base (the web) is conveyed to a coating station where specially designed coating dies spread layers of liquids containing chemicals uniformly over the moving web. After coating, the film is conveyed first into a cooling chamber where the aqueous gelatin is transformed into a gelled state and then into a dryer where water is removed. During these latter operations, the layers are reduced by as much as 90% of their original coated thickness.

In almost all cases, a multilayer coating method, such as slide coating or precision curtain coating, is used to coat the many layers of a photographic film simultaneously. In slide coating, Figure 8.16d, the liquids flow out of slots onto an inclined plane and then flow over one another down the slide surface in laminar flow. Laminar flow continues through the coating bead and onto the web. The web is positioned against a backing roll and moves more or less vertically upwards. Throughout this process, the layers remain separate and distinct. In precision curtain coating, Figure 8.16e, the laminar liquids flow over and off a curved section to form a falling curtain that impinges on a horizontally moving web. Here, too, all the layers remain separate and distinct throughout the coating process. The height of the curtain can be from less than a centimeter to 25 cm or more. In both processes, the coating speeds can be hundreds of meters per minute (up to about 1000 ft/min).

We cannot adjust the structure of the slide or curtain coating die. Individual plates feed their layers uniformly across the width of the slide. When the liquid rheology or the flow rate changes greatly, a different plate or slot dimension may have to be used, but once the plates and slots are chosen, the system is fixed. We may be able to adjust either the tilt of the slide surface or the location of the bottom edge of the slide relative to the horizontal plane of the backing roll or the coating gap (the distance between the lip of the slide coating die and the web), but the slot openings remain fixed.

Curtain coating has two distinct advantages over slide coating. First, in slide coating the gap often is under 250 μm (10 mil)—and this may be so narrow as to prevent splices in the plastic base from passing through. Either the coating die or the backing roll must be moved to enlarge the gap to allow splices to pass. Sometimes it is difficult to re-establish the coating bead after the splice has passed through, resulting in a large yield loss. In curtain coating, the gap is much larger, and splices have no problem passing through it. Second, in slide coating, if a bubble gets into a coating liquid it may not be carried out onto the web with the top layer. It simply floats in the coating bead, causing a streak. Because the coating operation is carried out in the dark, the operator is unlikely to notice the streak until a sample is taken at the end of a roll. Then a thin piece of plastic shim must be used to push the bubble to one edge and out of the system. In curtain coating, the bubble flows down the curtain onto the web to cause a single spot defect, which is much less objectionable and causes much less yield loss than a continuous streak.

Slide and curtain coated layers are subject to *interfacial wave defects* that form between layers, causing an unacceptable wavy pattern in the cross section of the finished product. Interfacial waves form only when the physical properties of adjacent layers differ; when the properties are the same, the system behaves as if there were no interface. Physical properties of importance in wave formation are density and viscosity and the thickness of the layers. Density does not vary much; it is always near unity, and usually we cannot change it. Viscosity, on the other hand, can vary by several orders of magnitude. For the sake of stability, the viscosities of adjacent layers should be similar. It has been suggested that the ratio of the viscosity of any layer to the layer below it should be between 0.7 and 1.5. Thicker liquid layers are also more stable. Since the layers on the coating slide need to be thicker as the coating speed increases, interfacial waves are less likely to appear when we increase the speed.

8.4.2.4B Pretreatment of the Film Base. The film base, usually polyester (polyethylene terphthalate), must be treated to be wettable by the liquid photographic layers, even when a surfactant is added to those liquids. The surface energy of the film base, γ_{sv} in Appendix 8A, eqs. 8A.1.1 and 8A.1.2, must be increased to reduce the dynamic contact angle. We can achieve this increase by a number of procedures. The first approach is to coat a thin gelatin layer—known as a subbing layer—on the film base. Gelatin, which is not soluble in alcohol, is coated from an alcohol–water dispersion. With its low surface tension, the alcohol provides good wetting, yet evaporates on drying to leave a high-surface-energy gelatin layer. The second approach is to oxidize the surface of the film base, either by passing the film quickly through an oxidizing gas flame (flame treatment) or by corona treatment in which a silent electrical discharge ionizes the air and the oxygen ions oxidize the surface. For some plastics, the oxidized species appear to diffuse into the surface, making the corona treatment transient; so it is usually done on-line, directly before the coating station.

8.4.3 Magnetic Recording Tapes Are Prepared by High-Speed Coating Processes

The magnetic recording industry spans audio, video, and digital applications in the form of tapes, cassettes, and floppy and hard disks. Almost all of these products consist of magnetic particles coated onto a substrate. In the past, hard disks used a particulate coating, but now thin metal films that provide a higher output and a better signal-to-noise ratio have replaced them (see Section 10.8.1).

Magnetic particles are coated onto flexible substrates from a multicomponent coating liquid that also contains binders to hold the particles onto the substrate, solvents to dissolve the binder, plasticizers to keep the binder flexible, wetting agents to aid in dispersion, lubricants to reduce surface friction, antistatic agents to prevent buildup of static charges, and abrasives to clean the magnetic recording head. Crosslinking agents are added to increase the hardness of the finished coating. Depending on the application, the dried coating is about 0.2–10 μm thick. Tapes for video or audio recording have thicker coatings, and diskettes tend to have thinner coatings.

Flexible magnetic coatings usually are deposited on a polyethylene terphthalate base chosen for its strength, stability, durability, and low cost. The film base for tapes may be 6 to ~25 μm (¼ to 1 mil) thick; diskettes use a heavier base, such as ~75 μm (3 mils).

8.4.3.1 Magnetic Particles with a Needlelike Shape Are Chosen to Achieve High Recording Density and High Magnetic Coercivity
Depending on the application, different types of magnetic particles are used. They include:

Gamma ferric oxide, the most common material because of its low cost, uniformity, and good magnetic properties. The needle-shaped (acicular) particles are about 0.2 to 1 μm long with a width $\frac{1}{10}$ to $\frac{1}{5}$ their length. Small particle size and high aspect ratio increase the magnetic coercivity, and permit a high recording density with low noise. Recently an ellipsoidal-shaped particle was introduced.

Cobalt-modified gamma ferric oxide particles have a higher coercivity and are used in some high-performance tapes. Particles with cobalt adsorbed on the surface have better properties than those with the cobalt dispersed throughout the ferric oxide crystal.

Chromium dioxide particles have a still higher coercivity and give excellent high-density recording performance. These acicular particles are used for superior high-frequency response in audio and video tapes. However, they cost more than the iron oxides, are more abrasive, and raise concerns about chromium toxicity.

Metal particles of iron alloyed with cobalt or nickel have very high coercivities with extremely small size, leading to excellent performance properties with low noise. However, the particles are pyrophoric—they will spontaneously burn when exposed to air—so they must be protected from contact with air at all times. In the dried coating, the binder protects them from contact with air.

8.4.3.2 Magnetic Coating Liquids Contain Other Ingredients That Enhance Coating Flow and Binding to the Substrate

The *binder* is a polymer that holds the magnetic particles together and binds them to the substrate. It must be able to form a continuous film in which the particles are dispersed, and yet the particle concentration must be as high as possible. Binders must have the appropriate strength and elasticity to ensure that the finished coating is hard yet sufficiently flexible to avoid cracking. Polyurethane resins are common binders.

Solvents are chosen that dissolve the binder and act as the carrier for the magnetic particles in the coating process. Solvent concentration plays a strong role in determining the viscosity of the coating liquid. After coating, the solvent is removed in the dryer. The choice of solvent depends on solvent power, evaporation rate, toxicity, and cost. Solvents commonly used in the magnetic tape industry include cyclohexanone, methyl ethyl ketone, methyl isopropyl ketone, tetrahydrofuran, and toluene.

Other ingredients are added to enhance either the coating process or the final qualities of the film. A *plasticizer* may be added to soften the binder film and give it greater flexibility. A *wetting agent* aids in dispersing the magnetic particles and helps keep them dispersed. Frequently lecithin, prepared from soy beans, is used as a surfactant in magnetic coatings. An *antistatic agent* is added to prevent buildup of static electrical charge on the tape surface that would interfere with the equipment and attract dust to the magnetic tape during operation. Conductive carbon black usually is added to lower the electrical resistance of the dried coating. Conductive carbons consist of very small spherical particles agglomerated to form fairly long chains. A *lubricant* is added to the coating to overcome stickiness and scrape-flutter problems. Over time the lubricant should gradually "bloom" to the surface, where its primary function is to reduce friction between the read–write head and the tape surface. An *abrasive* agent is added to remove polymer buildup and clean the read–write head. Very small quantities of fine aluminium oxide powder are included in the coating for this purpose. Sometimes crosslinking *hardening agents* are added to increase the hardness of the finished coating; alternatively the binder may be crosslinked by passing the coated film under an electron beam.

8.4.3.3 Special High-Energy Dispersion Procedures Assure Dispersion of the Magnetic Particles in the Coating Liquid

Because magnetic forces tend to agglomerate magnetic particles, special precautions must be taken to disperse and maintain proper dispersion of these colloidal particles during coating and drying. In the dispersion process, high shear forces are needed to break down agglomerates and disperse the particles throughout the binder–solvent system. However, the shear forces should not be so large as to break or abrade the particles, and if shearing is carried out too long, the magnetic properties suffer.

Optimum dispersion is achieved with a series of treatments. First a very crude dilute dispersion of magnetic particles is obtained in an ordinary mixer by combining all the solvents with some of the binder and with the dispersing agent. This crude dispersion is then transferred to one of several types of mills, such as ball mills, pebble mills, stirred mills, or high-speed agitators. In recent years stirred mills (more commonly called sand mills) have become popular for use with magnetic media. Sand (or metal or ceramic beads) contained in a vertical tank is put in motion by rotating disks or rotating arms. The crude dispersion is added and circulated through the mill and back to a holding tank until the desired degree of dispersion has been reached. A screen prevents the beads from exiting the chamber with the circulating mix. The dispersion process may take hours, and microscopic examination of the mix or measurement of the magnetic properties determines its completion. Viscosity, the temperature, and the ratio of mix to grinding media are important variables to control.

After dispersion is complete, an ordinary mixer or a high-speed agitator is used to incorporate the rest of the ingredients.

8.4.3.4 Special Filtration Procedures Preserve the Quality of the Magnetic Coating Liquid

No large particles can remain in the coating liquid because they cause defects in the magnetic coating and subsequent loss of data or information. Therefore, after dispersion, the coating liquid must be filtered to remove undispersed agglomerates, undissolved binder, and other unwanted particles large enough to interfere with the properties of the final product. Filtration must remove these unwanted particles without removing the magnetic particles. Cartridge filters used for this purpose are formed by winding threads or filaments of synthetic or natural fibers to a depth of about 1 in. (2.5 cm) around a metal sleeve about 1 ft (30 cm) long that contains many holes. The size of the pores in the filter pad depends on the pattern followed in winding the threads or filaments. Particles are trapped inside the filter when they hit a fiber and stick to it. Because the pores in the filter are larger than the particles to be trapped, particles are trapped throughout the depth of the filter. Small particles with their low inertia follow the flow lines of the fluid and pass out of the filter. However, some small particles may be trapped, while some fairly large particles may pass through. There is no sharp cutoff in pore size, and the size rating of the filter is relatively arbitrary. As filtration depends on particles hitting the filaments and then adhering to them, surface forces play a significant role. For this reason, choosing the material the filaments are made of is important.

Cartridge filters have elastomeric gaskets on the top and bottom that must be properly seated. Also, when a filter is first put into service, loose fibers may slough off into the filtrate. Therefore, the first filtrate should be recycled back to the inlet, or the filter should be flushed with solvent.

8.4.3.5 Slot Coating Is the Preferred Process for Coating Magnetic Tapes

The well-dispersed and filtered coating liquid is now ready to be applied to the substrate. Three processes are in use for coating magnetic layers on a flexible base: slot coating, reverse roll coating, and gravure coating. Slot coating is preferred for coating magnetic layers. Because the process is premetered, the wet coverage can be set simply by setting the flow rate. Slot coating gives a uniform coating even for the very thin coverages required by these applications. In reverse roll coating, the wet coverage depends on liquid properties in the nip gap as well as the geometry and roll speeds; in gravure coating, the cell patterns may not level.

For magnetic coatings less than 15 to 25 μm, slot coating must be modified to achieve uniformity and reproducibility. As the coating gets thinner, the gap between the coating die and the web shrinks, and if the die lips are parallel to the film base, as in Figure 8.23a, and if the pressure in the bead region is everywhere atmospheric, then the coating gap will be exactly twice the wet film thickness. But the lips may not be exactly parallel to the web, and pressure in the bead may be above atmospheric (or below as it is when a bead vacuum is used); so the simple one-half relationship does not always hold. However, when thin layers are to be coated, the gap must be narrower, and for very thin layers the gap is very tight. When the gap is very narrow, the eccentricity of the backing roll or the runout of the bearings becomes appreciable relative to the gap, and the gap becomes thicker and thinner as the backing roll rotates. The coating thickness fluctuates in the same manner and approximately to the same extent, causing chatter. A major problem occurs if the gap becomes zero, for then the coating die cuts the web, the web breaks, and coating fluid runs all over the place causing a tremendous mess!

Figure 8.23
(a) Conventional die coater configuration. (b) Adjustments to the tilt and offset make it possible to achieve thinner coatings.

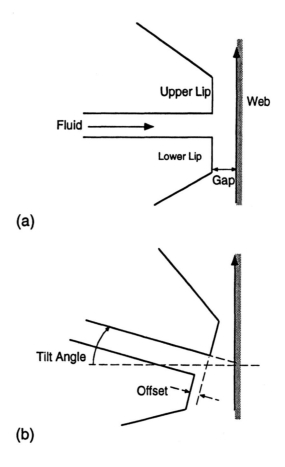

To avoid these problems, we can coat against unsupported web, with the web pressing against the die after it leaves the backing roll. The coating liquid acts as a lubricant and prevents physical contact between the coating die and the film base. In this case, web tension becomes an important variable, and different web tensions should be tried as the die is moved in and out of the web path to define conditions for the best quality coating.

In addition, sometimes it can be advantageous to tilt the die as shown in Figure 8.15 and Figure 8.23b so that the liquid flows in a very slightly converging path in the bead region. Also, we can adjust the lips so that one protrudes very slightly beyond the other. A slight offset, as shown in Figure 8.23b, is helpful for thinner coatings. The coating die can be oriented in any position—vertical (pointed upwards or downwards), horizontal, or any orientation in between.

Sometimes we may find it advantageous to coat two thin layers. A two-layer slot coater (see Figure 8.4) can be used. Each layer can be thinner than could be coated successfully by itself.

8.4.3.6 Magnetic Particles Must Be Oriented after Coating and before Drying

In most cases, magnetic particles are anisotropic. To achieve the best magnetic properties it is desirable to orient them so that all the particles are lined up in one direction, such as perpendicular to the surface or parallel in the direction of the coating. For magnetic tapes strong magnetic field coils are placed immediately after the coating station to orient the magnetic particles as required.

For computer diskettes punched out of a finished coated surface, it is impossible to orient the particles in either the radial direction or the direction of rotation of the disk. For this product all particles should have the same magnetic properties, and they must be completely randomly oriented in the horizontal plane. We can achieve this result with rotating magnet fields, called disorienting magnets.

8.4.3.7 The Coating Is Dried, the Finished Tape Is Passed through a Calender Stack and Cut to Size

The wet coated tape enters the dryers, where recirculating hot air evaporates the solvents. Usually air supplied to both sides of the web supports the plastic base as it travels through the dryer. This air flotation dryer avoids contact with metal idler rolls in the dryer that could scratch the film.

Because the solvents are flammable, explosions are possible unless we prevent conditions in the solvent-laden air from reaching a combustion point. Normally, the maximum solvent concentration is set well below the explosion limit, and concentration sensors in the exhaust air ducts monitor the level. Alternatively, we can use nitrogen in place of air in the dryers. Exhaust gases are sent to a condenser where the solvent vapors are condensed and recovered. Nitrogen from the condenser is recirculated to be reused, with any excess exhausted to the atmosphere. This reduces air pollution to a minimum.

Finally, the dried web travels through a calender stack to compact the coating and make it smooth and flat. Both features are important. Because the upper regions of the coating contain high-frequency recordings, a smooth coating is essential to preserve high-fidelity, high-frequency information.

The finished coating is slit to the required dimensions. For computer disks, the circular areas are punched out. Testing the finished product is an important part of any magnetic coating operation. For precision products such as computer disks, broadcast video tapes, digital audio tapes, instrumentation and computer tapes, 100% testing is carried out to ensure that every unit meets specifications.

8.4.3.8 Special Attention Must Be Paid to Cleanliness at All Stages of Manufacture

High standards of cleanliness are essential at all stages of the coating operation to avoid introducing foreign particles into the finished tape. Earlier we discussed special precautions to take in preparation of the coating liquid. The film base also must be cleaned with a web cleaner situated immediately prior to the coating stand. This cleaner removes almost all dirt or polymeric residues on the film.

Air in the dryer must not blow dirt onto the coating. Normally the air in the dryers will be under a slight vacuum to avoid solvent-laden air reaching operators in the coating alley. As a result, air in the coating alley enters the dryers. To ensure high quality, the coating and drying operation must be carried out under strict cleanroom conditions.

8.4.3.9 Pollution Control

Air quality standards require that solvent removed during the drying process cannot go up the stack to pollute the air. Solvents must be recovered from the dryer exhaust, either for reuse or to be destroyed by combustion. As discussed earlier, solvents can be condensed out of the exhaust stream or extracted with activated charcoal. Extracted solvents are recovered when the charcoal is regenerated.

An alternative approach to the pollution problem is to avoid solvent-based coatings altogether—for example, by using water-based coatings. Although water-based coatings have been reported, we are unaware of any being used in commercial production for magnetic layers. Research and development is also taking place on solvent-free coatings made using liquid monomers and low-molecular-weight liquid polymers as the carrier liquid. These can be polymerized and crosslinked by electron beam radiation.

8.4.3.10 Magnetic Coatings Are Applied on Hard Disks by the Spin-Coating Method

Finally, we note that hard disks used for large archival computer data storage are coated by completely different methods. Typically, aluminum disks (about 50 cm diameter, 1.5 to 6 mm thick) are used because thay are light weight and mechanically stable. These disks are specially cleaned and polished prior to being coated with the magnetic particles. Magnetic coatings are applied by *spraying* or *spin coating*.

In spin coating, a disk rotates at high speed and a small amount of coating liquid is introduced at its center. Centrifugal forces cause the liquid to spread out over the surface. While the liquid is spreading, it is also drying. Final film thickness depends on the centrifugal force, the viscosity at the start and during the course of drying, and the motion of air over the surface that promotes drying. Baffles can be used to control the motion of the air, so that the film thickness is uniform on the disk.

Once the film has spread to the point at which motion stops, the disk is removed and drying is continued in an oven. Then the surface of the disk is polished and burnished, and every disk is individually tested.

Alternative methods for coating metal thin films on hard disks use thermal evaporation, sputtering, and electroplating techniques described in Chapter 10.

Bibliography

E. D. Cohen and E. B. Gutoff, *Modern Coating and Drying Technology,* New York: VCH Publishers, Inc., 1992.

Appendix 8A Dynamics of Wetting and Spreading

8A.1.1 The Dynamic Contact Angle Is Always Greater Than the Static Contact Angle

The basic step in any liquid coating process is the replacement of the substrate–air interface with two interfaces, the liquid–air and substrate–liquid interface. Under static conditions the wetting condition is described by the Young equation 3.2.4

$$\gamma_{sv} = \gamma_{sl} + \gamma_{lv} \cos \theta_s \qquad (8A.1)$$

For complete wetting the contact angle θ_s should be zero, in which case the liquid will spread completely to cover the solid substrate at equilibrium. The spreading coefficient (S) (eq. 3.2.5) is given by the difference in surface energy between the uncovered and the covered surface, that is,

$$S = \gamma_{sv} - (\gamma_{sl} + \gamma_{lv}) \qquad (8A.2)$$

For spreading to occur, S must be positive. The greater the value of S, the more rapidly the liquid will rush to cover the solid.

In a coating process the system is not at rest; it is dynamic, and equilibrium is determined by hydrodynamic behavior. As an example, we examine the situation for a slot coating die. This is part of the practical coating system described in Section 8.1. A cross section in the immediate vicinity of the die head in Figure 8.8a shows how the liquid is forced through the die and into contact with the advancing web.

The situation at the dynamic *contact line* where the substrate and the *liquid bead* make contact is represented by Figure 8A.1a. The substrate or web is moving from right to left at the web speed *v*. The liquid bead is driven to flow from left to right by the spreading coefficient S and by the external pressure P impressed on the fluid in the die. Thermodynamically the liquid wants to spread to cover the web, but the liquid molecules cannot move fast enough to replace the onrushing air molecules adjacent to and attached to the web. Consequently the liquid experiences a viscous drag from right to left. When these three forces (due to spreading, hydraulic pressure, and viscous drag) are in balance, we have dynamic equilibrium, the liquid is in contact with the web, and the liquid contact line appears to be stationary. The liquid contact angle to the web under dynamic conditions, θ_d, is always greater than the static value. For good coating conditions the *dynamic contact angle* must remain smaller than 150°. If the dynamic contact angle exceeds about 150°, the liquid bead will not wet the web; it will roll up and either air will become entrapped under the coating or the substrate will not be coated at all.

Figure 8A.1
(a) Equilibrium configuration for a liquid bead under dynamic conditions. The applied external pressure P and the spreading forces S cause the bead to move left to right; viscous drag due to the moving substrate (web) causes the bead to move right to left. θ_d is the dynamic contact angle along the contact line normal to the page.
(b) Full analysis of dynamic equilibrium requires consideration of concentration gradients, thermal gradients, and hydrodynamic behavior in the contact zone.

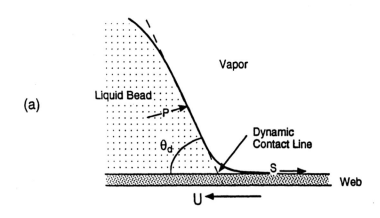

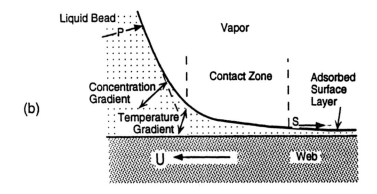

Many expressions both empirical and theoretical have been developed to relate the dynamic contact angle θ_d to the static contact angle θ_s. For liquids with viscosities in the 10 to 100 cP range the dynamic angle θ_d can be described by the Friz equation

$$\tan \theta_d = m \left(\frac{\eta U}{\gamma_{lv}}\right)^n = m \ (N_{ca})^n \qquad (8A.3)$$

where m and n are empirical constants (n has a value typically around 1/3), η is the viscosity and γ_{lv} the surface tension of the fluid, and U is the relative velocity between the fluid and the substrate, the web speed in Figure 8A.1a. The quotient inside the parentheses is the capillary number N_{ca}, a dimensionless quantity that relates the viscous force to the surface force on the droplet. Another relationship due to Joos et al. gives

$$\cos \theta_d = \cos \theta_s - 2 \ (1 + \cos \theta_s) \ (N_{ca})^{1/2} \qquad (8A.4)$$

A semiempirical relationship suitable for molten thermoplastics like polypropylene already mentioned in Section 7.3.4 (eq. 7.3.14) uses

$$\theta_d^3 - \theta_s^3 = 53 \ N_{ca} \qquad (8A.5)$$

The appropriate relationship is the one that works best for the fluids, web velocities, and configurations under development.

8A.1.2 Physical Understanding of the Dynamic Contact Angle Is Complex

Attempts to describe the physical aspects of dynamic wetting are complicated by the number of variables involved. Actually, contact is established over a zone as in Figure 8A.1b, rather than along a line as depicted in Figure 8A.1a. The width of the zone and conditions within the zone depend on the properties of the fluid. Composition gradients and thermal gradients exist at the leading edge of the zone. Surfactants and other impurities contained within the fluid adsorb at the surface to set up chemical gradients. Also, air molecules may become trapped behind the contact line and taken into solution by the coating fluid thus changing its chemistry. Temperature differences between the moving web and the fluid set up temperature gradients. These composition and thermal gradients give rise to local changes in the surface tension and density. They can result in convection cells causing the fluid to recirculate and stagnate within the bead. As noted earlier in connection with Figure 8.12e, stagnation cells generally have adverse affects on the quality of the coating.

Furthermore, conditions are not constant across the width of the web. Evaporation can occur more readily at the edges, causing composition gradients cross-web.

In addition, for coating liquids containing volatile solvents, solvent molecules can be transported much faster by diffusion through the vapor phase than by surface diffusion over the web surface, and so the incoming web is covered with an invisible adsorbed precursor layer of solvent. The interfacial tension betweeen the substrate and the vapor at the contact line, γ_{sv}, is reduced from the normal value, γ_{svo}, by the spreading pressure Π, as described in Section 3.6 (eq. 3.6.1).

$$\gamma_{sv} = \gamma_{svo} - \Pi \qquad (8A.6)$$

Finally, we have assumed in the figures that the substrate is flat and smooth. On a microscopic scale the variation in surface topography means that to maintain a constant contact angle the local contact line must fluctuate and such oscillations may lead to temporary instability and nonuniform coating characteristics.

These complications make it difficult to model dynamic wetting and to describe the equilibrium configuration of a liquid bead in a coating system, although a lot of progress has been made in this direction with modern supercomputer codes. Alternative approaches to the subject make use of flow visualization techniques and real-time imagery to capture the fluctuations in contact angle and flow stability as a function of operational variables. Figure 8.4 is one frame from a flow visualization experiment that illustrates dynamic features at the edge of the liquid bead.

9

GENERAL PROPERTIES OF CRYSTALLINE SOLID SURFACES

LIST OF SYMBOLS

a, b, c	spacing of points along the three axes of a space lattice
b	Burgers vector, slip vector, of dislocation line
B	absolute mobility of an atom in a crystalline solid (eq. 9.5.18)
c	molar concentration of solute in crystalline solid solution
c^*	molecular concentration of solute in crystalline solid solution
D_{self}, D_s, D_v	diffusion coefficients for self-diffusion, solute diffusion (s) (eq. 9.5.11), and vacancy (v) diffusion, respectively
G	elastic shear modulus
ΔG_{ads}	change in free energy due to adsorption/desorption
ΔG_m	maximum change in free energy to move an imperfection
ΔH_f, ΔH_m	change in enthalpy to form (f) or move (m) a point imperfection
ΔH_{fF}, ΔH_{fS}	change in enthalpy to form a Frenkel (F) or Schottky (S) imperfection
J	flux of material through unit cross-sectional area (eq. 9.5.10)
K_{ads}	equilibrium reaction constant for adsorption/desorption (eq. 9.3.3)
n	number of imperfections; N number of atom sites
$[n]_v$, $[n]_i$	fractional concentration ($[n] = n/N$) of vacancies (v), interstitials (i)
P_o	equilibrium vapor pressure of adsorbate
R_{12}	distance between ions 1 and 2
ΔS	change in configurational entropy
T_m	melting temperature
$U_{sublimation}$	sublimation energy
U_{screw}, U_{edge}	strain energy per unit length of dislocation line (eqs. 9.4.9 and 9.4.10)
V_{bond}	interaction potential energy for bonding
v	atom or ion velocity down a potential gradient
$w_{cohesion}$	work of cohesion (eq. 9.2.1)

α, β, γ   angles between axes of a space lattice
α   Madelung constant
γ_f fracture surface energy
γ_{sv} surface energy of a crystalline solid
γ_{sgb} surface energy of a small angle grain boundary (sgb) (eq. 9.4.12)
Γ   excess surface concentration (eq. 9.3.6)
θ   angle of misorientation across a small angle boundary (eq. 9.4.11)
Θ   fractional surface coverage (eq. 9.3.1)
λ   interatomic spacing in a condensed crystalline solid
μ   chemical potential
ν   atomic jump frequency (eq. 9.5.1)
ν_o characteristic surface atom vibration frequency
ν   Poisson's ratio
σ   ionic conductivity (eq. 9.5.29)
σ   tensile stress
Σ   coincident lattice site parameter
τ   shear stress
τ_o characteristic surface atom vibration time ($\tau_o = 1/\nu_o$)
χ   free energy change along a potential gradient

CONCEPT MAP

- Distinctive features of crystalline solids are:
  1. They form a regular array of bonded atoms or ions lying on a space lattice. The crystal structure of the lowest free energy depends on bond character (van der Waals, ionic, covalent, or metallic) and atom or ion sizes.
  2. Solids are mechanically rigid and can support a shear stress.
  3. Kinetic transport is limited by solid state diffusion.

- By analogy with thermodynamic equations for fluids, the excess surface free energy can be written

$$dG = \gamma_{sv}dA + \Sigma\mu_i dn_i$$

$dG = \gamma_{sv}dA$ (Physical Effects)

- Surface energy, γ_{sv}, of a solid depends on character and density of unsatisfied bonds; $\gamma_{covalent} > \gamma_{metallic} > \gamma_{ionic} > \gamma_{van\ der\ Waals}$.
- Surface energy varies with exposed crystalline plane (Wulff plot, Figure 9.7). Closest-packed planes have lowest surface energy.

- Surface atoms (ions) rearrange (Figure 9.8) and/or reconstruct (Figure 9.9) to minimize surface energy.
- Surfaces are not atomically smooth (Figure 9.10). Roughness—atomic terraces, steps, kinks, and adatom sites—is determined by surface orientation and temperature.

dG = $\Sigma\mu_i dn_i$ (Chemical Effects)

- Crystalline interfaces are sharp; interface is typically one atom layer deep.
- Adsorption of impurities from *outside* a solid lowers surface free energy by ΔG_{ads}. ΔG_{ads} varies from 0.05 eV (weak physisorption) to 1.0 eV (chemisorption).
- Coverage of first molecular layer is described by Langmuir adsorption isotherm (eq. 9.3.1); subsequent layers by BET equation (eq. 9.3.5).
- Segregation of impurities to the surface from *within* a solid lowers surface free energy by ΔG_{seg}. Magnitude of ΔG_{seg} depends on chemical potential, mechanical strain, and electrical contributions.

Crystalline Solid Imperfections

- Regular array of atoms or ions is broken by point, linear, and planar imperfections.
- Point imperfections—vacancies and interstitials (Figure 9.12)—exist in thermodynamic equilibrium. Concentration $\propto \exp -\Delta H/kT$. Surfaces act as source and sink for vacancies. Point imperfections in ionic solids are charged. Ionic solid surfaces can become charged. Point defect concentration profile then follows Gouy–Chapman relation (eq. 4.3.6).
- Linear imperfections—dislocations—are introduced by thermal/mechanical stress (Figure 9.15). Edge dislocations lie perpendicular, screw dislocations parallel with their Burgers vector. Dislocation strain energy $\propto Gb^2$ per unit length. Strain energy is relaxed by array formation. Edge-dislocation arrays form pure tilt small angle boundaries; screw dislocation arrays form pure twist small-angle boundaries (Figure 9.16).
- Planar imperfections—small-angle grain boundaries, stacking faults, twins—have surface energy. Small-angle boundary energy varies with angle θ; $\gamma_{sgb} = U_o (A - \ln \theta)\theta$ (eq. 9.4.12).

Crystalline Solid/Crystalline Solid Interfaces

- Three types of solid–solid interfaces are:
  1. Interphase boundary—change in unit cell across interface (Section 11.4)
  2. Epitaxial interface—change in lattice parameter across interface (Section 10.5)
  3. Grain boundary surface—change in crystal orientation across interface (Section 11.2)
- Each type of solid–solid boundary has interfacial free energy, γ_{int}, whose magnitude depends on mechanical strain, angle of misorientation, chemical constituency of boundary region.

Transport Mechanisms

- Solid-state diffusion mechanisms play crucial roles in interfacial transport processes and interface stability.

- Self-diffusion and solute diffusion occur by transfer of atoms to adjacent lattice sites either through the movement of vacancies or through the movement of interstitial atoms. Diffusion coefficient, $D = D_0 \exp[-(\Delta H_f + \Delta H_m)/kT)$, is determined by an activation energy equal to the sum of the energy to form, (ΔH_f), and to move, (ΔH_m), the vacancy or interstitial.

- Diffusion depends exponentially on temperature.

- Diffusion is faster over the free surface than over the grain boundary surface than through the volume of the material.

$$(\Delta H)_{\text{free surface}} : (\Delta H)_{\text{grain boundary surface}} : (\Delta H)_{\text{volume}} \approx 1{:}2{:}4$$

Solid State Diffusion in Chemical, Electrical, and Mechanical Stress Gradients

- Atomic mobility (velocity/force) is given by Nernst–Einstein equation, $B = D/kT$ (eq. 9.5.17). Force driving diffusion may originate from chemical, electrical, or mechanical potential gradient.

Chemical Gradient

- Diffusion leads to interdiffusion between two layers of different composition (eq. 9.5.21).

- Depth of interdiffusion after time t is proportional to $\sqrt{Dt}$.

Electrical Field

- Diffusion leads to ionic conductivity (eq. 9.5.27) or electromigration (eq. 9.5.29).
- Ionic conductivity is proportional to $D(ze)^2/T$, where z is ion valence.

Stress Gradient

- Diffusion due to mechanical stress gradient leads to creep (eq. 9.5.38).
- Creep strain rate, $d\varepsilon/dt$, is proportional to $\sigma D/d^2 T$.

    —Linear dependence on stress, σ, is characteristic of Newtonian flow.

    —For polycrystalline creep, D is grain boundary diffusion coefficient and d corresponds to grain size (see Chapter 11).

9

We now turn our attention to crystalline solid surfaces and interfaces. As indicated in Chapter 1, almost all modern technological applications of solid systems depend in the broadest sense on the properties of interfaces. This is clearly the case for semiconductor devices, metallurgical alloys, and ceramic sensors. Before describing the processing and behavior of different types of solid–solid interfaces in Chapters 10 and 11, this chapter reviews some general features of crystalline solids and the particular characteristics of the free solid surface.

The free solid surface is the *external* surface exposed to the environment, which may be vapor, gas, or vacuum. The solid–vapor interface is important in the processing of many solid–solid interfacial systems; vapor-phase processing will be described in Chapter 10. A unique feature of crystalline solids is that they also can have *internal* surfaces called grain boundaries. (In this text we intend to follow the convention that an interface between two phases of the same material—liquid–vapor, solid–vapor—is known as a surface, while that between two different materials is called an interface. Thus a boundary between two orientations of the same material is a grain boundary *surface*.) The general properties of grain boundaries and their role in processing and properties of multigrain structures will be described in Chapter 11.

This chapter deliberately follows the organization used for fluids in Chapter 3 and extends the principles established there to the solid state. As with fluid systems, the discussion is organized into two parts corresponding to the components of the expression for the surface free energy. Again we distinguish physical effects associated with surface tension γ from effects associated with surface chemistry.

$$dG^\sigma = \gamma\, dA + \left(\sum_i \mu_i^\sigma dn_i^\sigma + \tfrac{1}{2}\int \sigma\varepsilon dV \right)$$

Of course, major differences exist between fluids and solids, and they must modify the effects described previously for fluids. First, solids resist shear stress and can be elastically strained and distorted. As a result, the surface energy dependence on chemistry (the second component in the above expression) is not determined solely by the chemical potential μ, but also by the elastic strain energy of the surface states. The strain energy density per unit volume is equal to one-half the product of stress σ and strain ε, and appears as an additional component on the right-hand side of the surface free energy equation. Second, the solid crystalline state is characterized by long-range order and ideally consists of a regular structural array of atoms (ions) on a lattice. This structural regularity leads to anisotropic behavior, a variation of properties with orientation. Third, for solids ambient temperatures are typically less than one-half the melting point; so the kinetics are slower than for fluids and the solid–solid interface can "freeze" into a metastable state.

In this chapter, Sections 9.1 and 9.2 deal with ideal crystal structures and the origins and consequences of surface energy associated with the first term in the equation; Section 9.3 deals with chemical phenomena, such as adsorption and segregation, associated with the second term in the equation. Section 9.4 recognizes the fact that crystalline solids are not perfect, and defines the nature of crystalline imperfections and their role in the properties of external and internal surfaces. The final section of the chapter, Section 9.5, covers transport processes in solids.

9.1 Crystalline Solids Consist of a Regular Array of Atoms

9.1.1 Each Crystalline Material Is Defined by Its Unit Cell and Lattice Parameters

Long-range order distinguishes a crystalline solid from a liquid or a noncrystalline solid, such as a glass or a bulk amorphous polymer. Crystalline solids are formed by the periodic repetition of a regular array of atoms or ions distributed on a space lattice. A space lattice is an infinite three-dimensional array of points in which every point has surroundings identical with that of every other point. Lattice points can be arranged to fill space in only fourteen different ways, known as the Bravais lattices, which can be further organized into the seven crystal systems of Figure 9.1. The smallest repeat unit of the space lattice is called the *unit cell*.

In a real crystalline solid, the atoms and ions want to pack together as closely as possible to minimize the interaction potential. However, the bond character and relative sizes of different atoms fixes the configuration and spacing, and these features in turn define the unit cells' fundamental lattice angles, α, β, and γ, and its lattice parameters, *a, b,* and *c*. For a perfect single crystal of a given solid material, each atom or molecule falls into its allocated geometric position, and the pattern repeats precisely and indefinitely. The characteristic length over which order is maintained, the *correlation length,* is infinite in the perfect crystalline solid.

For the most part, we restrict our discussion to crystal structures in the cubic crystal system. In this system all three

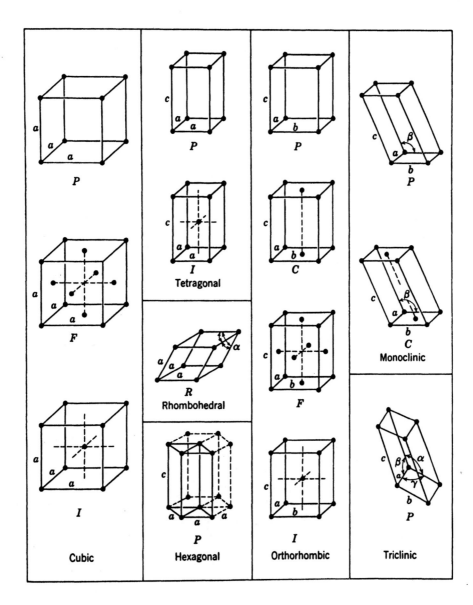

Figure 9.1
Conventional unit cells
of the 14 Bravais space
lattices. (W. Hayden,
W.G. Moffatt, and J. Wulff,
Mechanical Behavior,
Vol. III of *Structure and
Properties of Materials,*
John Wiley & Sons, New
York, 1965, p. 47. Repro-
duced with permission of
the W. G. Moffatt Trust.)

lattice angles are 90° ($\alpha = \beta = \gamma = 90°$), and all three lattice parameters are the same ($a = b = c$).

A single block (or grain) of a crystalline solid is uniquely defined by three features: the unit cell, the lattice parameters, and the orientation of the axes of the unit cell in space. In preparation for Chapters 10 and 11, we note that basically three kinds of solid–solid interface can exist within crystalline solids. We can distinguish each type of interface by features that change across its boundary:

1. *Interphase boundary*—the crystalline phase changes completely across this boundary; thus the unit cell, lattice parameter, and orientation all change (see Chapters 10 and 11).

2. *Epitaxial boundary*—lattice parameter and in some instances the unit cell change across the boundary, but an

orientational relationship always exists between the phases on either side of the boundary. A repeat distance in the crystalline phase on one side of the boundary typically matches a repeat distance in the crystalline phase on the other side (see Chapter 10). (There is a type of interface growth called grapho-epitaxy where a single crystal film can grow on a single crystal substrate with no orientational relationship. An example is the growth of Au on NaCl.)

3. *Grain boundary*—this boundary involves merely a change in orientation with no change in unit cell or lattice parameter (see Chapter 11).

These three boundary types are fundamental to the behavior, processing, and application of thin films, metals, ceramics, and composites in ways to be described in the respective later chapters.

As previewed in Section 2.6.2, the stucture actually chosen by a solid varies with the constituent atoms, ions, or molecules and the nature of the primary bonding forces between them. We now describe a few basic crystal structures associated with different primary bonding mechanisms.

9.1.2 Inert Gas Solids Are Composed of Close-Packed Arrays of Spherical Molecules Bonded by Dispersion Forces

Inert gases, such as argon and neon, form solids at very low temperatures. We can regard the inert atoms as spheres whose radius is defined by the limit of the short-range repulsive interaction due to "electron cloud overlap," interaction '5' in eq. 2.1.1, that contributes the $1/R^{12}$ term in eqs. 2.1.2 and 2.5.2. No electron sharing or transfer takes place between the inert spherical atoms. They are attracted to each other solely by London dispersion forces (one of the three van der Waals attractive interactions discussed in Section 2.5). Van der Waals bonds are *nondirectional;* that is, they are equally exerted in all directions. As already noted in Section 2.5.2, for inert gas solids the free energy in the condensed state is minimized when each atom is surrounded by and contacts the greatest possible number of nearest neighbors (the number of nearest neighbors is called the *coordination number*). When all the atoms are the same size, the highest coordination number is twelve, and this coordination is achieved with either the hexagonal close-packed (hcp), or face-centered cubic (fcc) crystal structures illustrated in Figure 9.2.

These crystal structures derive from two alternative ways of stacking identical spheres. If the spheres in the first layer are positioned to contact each other in locations denoted by **a** in Figure 9.2a and the spheres in the second layer are placed on them in locations denoted by **b**, then two options are available for spheres in the third layer. They can go either into positions that have a spatial location similar to **a**—which results in the ·a·b·a·b·a·b· stacking sequence and gives the hexagonal close-packed crystal structure shown in Figure 9.2c—or they can go into a new positions denoted by **c**—which results in the ·a·b·c·a·b·c·

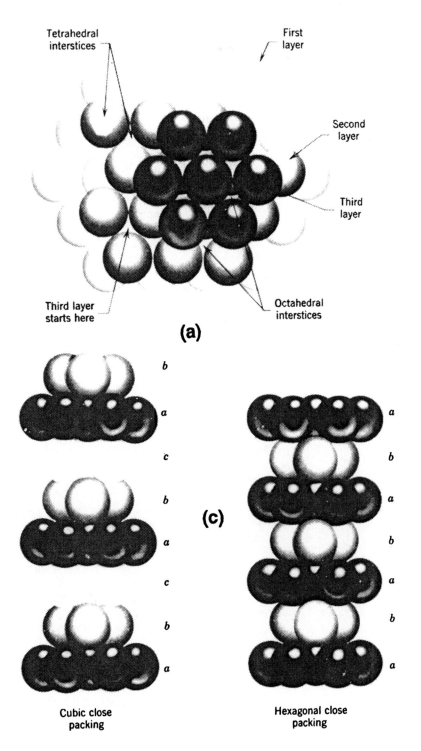

Tetrahedral interstices

First layer

Second layer

Third layer

Third layer starts here

Octahedral interstices

(a)

(c)

b

a

c

b

a

c

b

a

a

b

a

b

a

b

a

Cubic close packing

Hexagonal close packing

Figure 9.2
Alternative stacking sequences for close-packed spheres all of the same diameter. (a) Once the first and second layers have been positioned, spheres in the third layer can be placed in one of two alternative locations, either (b) in a third position to give $\cdot\,\mathbf{a}\cdot\mathbf{b}\cdot\mathbf{c}\cdot\mathbf{a}\cdot\mathbf{b}\cdot\mathbf{c}\cdot$, the face-centered cubic (fcc) stacking sequence, or (c) in the same position as the first layer to give $\cdot\,\mathbf{a}\cdot\mathbf{b}\cdot\mathbf{a}\cdot\mathbf{b}\cdot\mathbf{a}\cdot\mathbf{b}$, the hexagonal close-packed (hcp) stacking sequence. (W. D. Kingery, H. K. Bowen, and D. R. Uhlmann, *Introduction to Ceramics,* 2nd ed., Copyright © 1976, John Wiley & Sons, p. 52. Reprinted by permission of John Wiley & Sons, Inc.)

stacking sequence and gives the face-centered cubic crystal structure shown in Figure 9.2b.

9.1.3 Ionic Solids, Basic to Many Ceramics, Are Composed of Different Sized Positive and Negative Ions Bonded by Coulombic Forces

Ionic solids are formed when two atoms with large differences in electronegativity transfer electrons from one to the other to become charged ions. The atom that gives up the electron(s) becomes a positively charged cation (+ze), while the atom that receives the electron(s) becomes a negatively charged anion (–ze). Alternating cations and anions are held together on crystal lattices by Coulombic forces (Section 2.2.1). The interactions that take place, like those that occur in van der Waals bonds, are nondirectional. Free energy in the condensed state is minimized when each ion is surrounded by and in contact with the highest number of oppositely charged ions.

Because the cations and anions are not the same size, and the relative numbers of anions and cations may not be equal, other requirements, known collectively as Pauling's rules, come into play to decide the minimum energy crystal structure. Cations and anions arrange themselves into configurations that provide the highest possible coordination consistent with the relative size and valence of the ions—that is, the cations are surrounded by the maximum number of anions, while the anions are surrounded by the maximum number of cations. The cations and anions must always remain in contact. Assuming the ions to be spheres leads to the conclusion that the coordination number depends on the cation–anion radius ratio. The highest possible coordination number for ions of slightly different size is 8, and it occurs with the cesium chloride structure shown in Figure 9.3a. The next highest possible coordination number for ions with a greater difference in size (and thus a smaller cation/anion size ratio) is 6, which occurs with the sodium chloride structure shown in Figures 9.3b and 9.4. For an even greater difference in size and smaller size ratio the coordination number drops to 4 with the zinc blende (zinc sulfide) structure as shown in Figures 9.3c and 9.5b.

Pauling's rules also require that the relative number of ions in the crystal structure must equal the stoichiometric composition of the ionic compound. Table 9.1 summarizes the crystal structures for different ion size ratios and stoichiometry. Figure 9.4 reproduces the sodium chloride crystal structure frequently used as a model in this chapter.

To determine the total interaction potential for a given ion in a crystalline solid, we must evaluate its interaction with all other ions in the crystal. For example, for the NaCl crystal structure, each Na^+ ion has six nearest Cl^- neighbors at $R = 0.276$ nm, 12 Na^+ neighbors at $2^{1/2}R$, eight Cl^- at $3^{1/2}R$, and so forth.

From eq. 2.2.3 the bonding potential for a given ion is

$$V_{bond} = \Sigma \pm \frac{z_1 e\, z_2 e}{4\pi\varepsilon_o R_{12}} + \frac{C}{R_{12}^{12}} \qquad (9.1.1)$$

	Coordination Number	Disposition of Ions about Central Ion	Range of – Cation Radius Ratio Anion Radius	
(a)	8	Corners of cube	≥ 0.732	
(b)	6	Corners of octahedron	≥ 0.414	
(c)	4	Corners of tetrahedron	≥ 0.225	

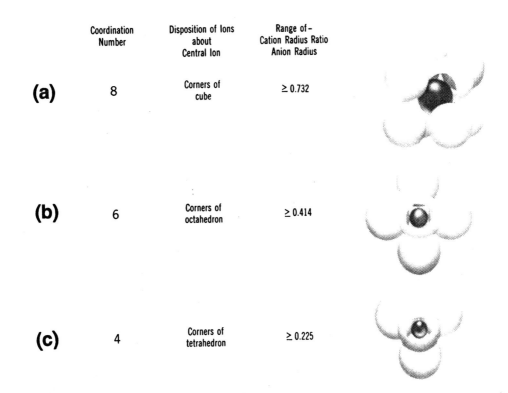

Figure 9.3
Effect of ionic radius ratio on coordination number. (W. D. Kingery, H. K. Bowen, and D. R. Uhlmann, *Introduction to Ceramics*, 2nd ed., Copyright © 1976, John Wiley & Sons, p. 57. Reprinted by permission of John Wiley & Sons, Inc.)

where the summation is over all other possible ions located on the different lattice sites. Those with the opposite sign make a negative contribution to V_{bond}, and those with the same sign make a positive contribution to V_{bond}. The expression in eq. 9.1.1 can be written as

$$V_{bond} = \alpha \frac{|z_1|e|z_2|e}{4\pi\varepsilon_o R} + \frac{C}{R^{12}} \qquad (9.1.2)$$

where the summation is represented by α the *Madelung constant*. R is the minimum separation of opposite ions in the lattice.

For sodium chloride,

$$\alpha = (-6 + \frac{12}{2^{1/2}} - \frac{8}{3^{1/2}} + \frac{6}{4^{1/2}} - \cdots)$$

$$= -(6 - 8.485 + 4.619 - 3.000 + \cdots)$$

$$= -1.748$$

Since $|z_1| = |z_2| = 1$, the first term in eq. 9.1.2, the attractive term, is

$$V_{bond} = -1.748 \frac{e^2}{4\pi\varepsilon_o R} = -1.46 \times 10^{-18} \text{ J per NaCl pair}$$

and the lattice energy per mole is $V_{bond}N_{Av} = 880 \text{ kJ mol}^{-1}$. This estimate of the lattice energy is 15% higher than measured values because the calculation neglects the repulsive C/R^{12} interactions.

Figure 9.4
Crystal structures with six-fold coordination. (a) The sodium chloride crystal structure. (b) Depiction of sodium chloride crystal structure with ions drawn to scale. (W. D. Kingery, H. K. Bowen, and D. R. Uhlmann, *Introduction to Ceramics,* 2nd ed., Copyright © 1976, John Wiley & Sons, p. 42. Reprinted by permission of John Wiley & Sons, Inc.)

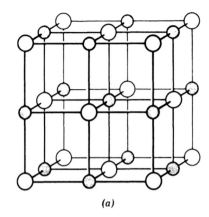

(a)

(b)

Note that the Madelung constant α is a measure of the bonding energy gained by having an ion surrounded by many others in a regular crystalline array compared to the bonding energy of an isolated pair of ions given by eq. 2.2.3. The Madelung constant, whose magnitude is always greater than one, takes on different values for different crystal structures. Consistent with the coordination number ranking, these values range from the highest value of 1.763 for cesium chloride to 1.748 for sodium chloride to 1.638 for zinc blende for the 1:1 stoichiometry ionic solids, as shown in Table 9.1. For 2:1 salts like CaF_2, the higher valence difference increases it to 5.

Due to the strength of the ionic bond, ionic solids generally have high melting points and high latent heats of melting. This is particularly true of divalent ionic solids like magnesium oxide (MgO) or trivalent solids like aluminum oxide (Al_2O_3). Their thermal and chemical stability make them useful ceramics for high-temperature and corrosive environment applications.

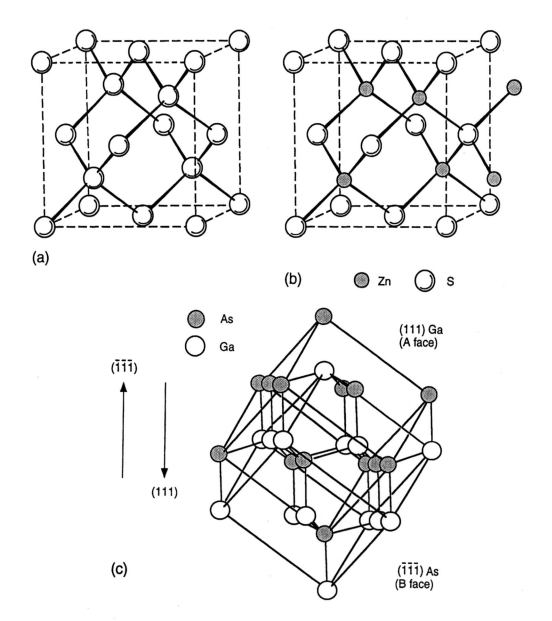

(a)

(b)   ● Zn   ○ S

● As
○ Ga

$(\overline{1}\overline{1}\overline{1})$

(111)

(c)

(111) Ga
(A face)

$(\overline{1}\overline{1}\overline{1})$ As
(B face)

Figure 9.5
Crystal structures with four-fold coordination. (a) The diamond crystal structure for covalently bonded carbon. (b) The zinc blende (ZnS) crystal structure. (c) The zinc blende crystal structure of (b) observed at right angles to the [111] axis and along a [110] direction. This view highlights the alternate Ga and As layer structure in gallium arsenide and the asymmetry in layer spacing.

9.1.4 Covalent Bonding Plays a Key Role in Determining the Structure of Many Materials Used in Electronic Devices, Such as Diamond and the III-IV and II-VI Compounds

Unlike ionic solids, in which the valence electron liberated from one atom attaches itself to and is localized at another atom, covalently bonded solids share the electrons from two or more atoms so that each has the desired outer-shell configuration, if only for part of the time. Covalent bonds between ions are strongly *directional;* that is, they are exerted in specific angular directions. Spatial configuration then dictates the packing arrangement and overrides the desire for unlike ions to have maximum coordination.

TABLE 9.1 Ceramic Crystal Structures

Valency ratio	Coordination Number cation	anion	Crystal structure	Formation	Madelung constant α
1:1	8	8	Cesium chloride	Anions on simple cubic array. Cations in body center sites.	1.763
1:1	6	6	Sodium chloride (rock salt)	Anions on fcc array. Cations in octahedral sites.	1.748
1:1	6	6	Nickel arsenide	Anions on hcp array. Cations in octahedral sites.	
1:1	4	4	Zinc blende (sphalerite)	Anions on fcc array. Cations in 1/2 tetrahedral sites.	1.638
1:1	4	4	Zinc oxide (wurtzite)	Anions on hcp array. Cations in 1/2 tetrahedral sites.	1.641
2:1	8	4	Calcium fluoride (fluorite)	Anions on simple cubic array. Cations in 1/2 body center sites. (*Equivalent to* Cations on fcc. Anions in all tetrahedral sites.)	5.04
2:1	6	3	Titanium dioxide (rutile)	Anions on fcc array. Cations in 1/2 octahedral sites (distorted).	4.82
3:2	6	4	Aluminum oxide (corundum)	Anions on hcp array. Cations in 2/3 octahedral sites.	
			Spinel	Anions on fcc array. Trivalent cations in 1/2 octahedral sites. Divalent cations in 1/8 tetrahedral sites.	

The covalent bond appears in its purest form in the diamond form of carbon. Here the normally spherically symmetrical outer orbitals of the four bonding electrons in the carbon atom ($2s^2$, $2p^2$) become hybridized to form four asymmetrical (sp^3) orbitals. Each hybrid orbital assumes a path that on average corresponds to one of the medians of the tetrahedron formed with the carbon atom at the center. Sharing two electrons along each of the four hybrid orbitals produces the covalent bond and leads to the classical tetrahedral configuration of carbon atoms in the diamond crystal structure of Figure 9.5a. This configuration can be replicated in space; so we can regard a perfect diamond crystal as one continuous covalently bonded carbon macromolecule.

Other materials in which the covalent bond plays a prominent role are the IV-IV compounds (such as silicon carbide), III-V compounds (such as gallium arsenide), and the II-VI compounds (such as cadmium sulfide). Here the ions share different numbers of electrons (3+5 in III-V and 2+6 in II-VI) or the same number of electrons (4+4 in IV-IV) to mutually satisfy the need to have an outer shell of eight. The zinc blende (ZnS) crystal structure, which is characteristic of these compounds (shown in Figure 9.5b) preserves both the fourfold coordination and directional bonding. It is the same crystal system as diamond except that the cube diagonal layers—the {111} planes described by the Miller indexing system in Appendix 9.A—now alternate between one ion and the other. This alternation profoundly affects the crystal structure. It ceases to be centrosymmetric. When viewed normal to the {111} planes down the cube diagonal axis—the ⟨111⟩ direction—the ion stacking sequences are either $\cdots$ III $\cdot$ V $\cdots$ III $\cdot$ V $\cdots$ III $\cdot$ V $\cdots$ or $\cdots$ V $\cdot$ III $\cdots$ V $\cdot$ III $\cdots$ V $\cdot$ III $\cdots$, depending on whether you look down or up. Figure 9.5c shows this stacking sequence. Surface properties of these crystals depend strongly on "which way up" the crystal is oriented, a feature that strongly influences their chemical behavior, as described in Section 9.2.1.2.

Covalent bonds are extremely energetic and are associated with the highest-melting-point materials like diamond, boron nitride, or silicon carbide.

9.1.5 The Special Properties of Metals Arise from the Metallic Bond in Which Valence Electrons Are Shared by All Atoms

Metals display features of both covalent and ionic bonding. They resemble covalent solids in the sense that each ion shares one or more electrons with the total community of other ions to create an electron gas. They resemble ionic solids in the sense that metallic bonding results from the Coulombic attraction of the metal ions to this electron gas. Metallic crystals are pictured as an array of positive ions, embedded in a sea of mobile, negatively charged electrons. The electrons are no longer localized, as they are in the covalent bond, and thus the metallic bond is nondirectional. Because all the ions have the same size in pure metals, they prefer to form close-packed structures similar to those outlined for the inert gas solids. The lowest-energy configurations are those with 12-fold coordination. The result is either close-packed structures formed by the stacking of spheres in the hexagonal close-packed (hcp) array (Figure 9.2c) for metals like Zn, or the face-centered cubic (fcc) array (Figure 9.2b) for metals like Cu, Al, or Au. Alloying mixes metal ions of different size. For certain compositions alloys prefer to form close-packed "intermetallic" structures with features similar to those outlined for ionic solids in Table 9.1 and Figure 9.3.

The strength of the metallic bond is high and similar in magnitude to the electron pair covalent bond. For example, the binding energies of metallic lithium and sodium are 1.7 and 1.13 eV (or 2.7×10^{-18} and 1.8×10^{-18} J) per atom, respectively. Most metals have a high melting point and a high latent heat of evaporation, which makes them suitable for high- and low-temperature structural applications.

9.1.6 Molecular Solids Exhibit Mixed Bonding

Many crystalline materials consist of molecular units, such as the sulfate or carbonate ion, bonded to other molecular units by ionic, covalent, or van der Waals forces.

Ionically bonded. Complex ionic crystalline solids are generated when covalently bonded molecular units, such as SO_4^{2-} or SiO_4^{4-}, carry residual charge and are electrostatically attracted to simpler ions, such as Mg^{2+}, to form crystalline $MgSO_4$ or Mg_2SiO_4.

Polar bonded. If molecular units do not carry residual charge, they are attracted to each other by polar bonding (for example, H_2O molecules form ice crystals with dipole bonding) or by induced polar bonding to form crystalline solids (see Sections 2.2 and 2.3). Some solids, such as $CdCl_2$ and MoS_2, demonstrate mixed bonding, polar bonding between the molecules in the plane of the molecules, and van der Waals bonding across the planes. Graphite consists of covalently bonded carbon layers held together by van der Waals bonding, as illustrated in Figure 9.6a. Such "layered" structures find use as solid lubricants.

Mica, $K(Al_3Si_3O_{10})(OH)_2$, provides an extreme example of multiple bonding; covalently bonded OH^- molecules are attracted to Al, Si, and O by ionic or covalent attraction to form a layered silicate structure in which the layers are held together by electrostatic attraction to potassium ions, as illustrated in Figure 9.6b. (This crystal cleaves easily between the layers to give atomically flat surfaces. Such flat surfaces are useful in the surface forces apparatus, SFA, which measures the forces of attraction between materials deposited on to the mica directly; see, for example, Figures 6.16 and 6.17.) Potassium ions on a cleaved mica surface are easily drawn into an aqueous solution to generate charged mica surfaces. Section 4.3.1 described the key role that charged crystalline particles play in the electrostatic stabilization of colloidal clays, of which mica is a key constituent.

9.2 Characteristics of the Free Solid Surface

9.2.1 The Free Surface Energy Equals One-Half the Cohesive Energy

Section 3.2.1 showed that the energy required to create new surfaces in a liquid equals one-half of the work of cohesion. We can define the surface energy for a solid in exactly the same way

$$W_{\text{cohesion}} = 2\,\gamma_{sv} \tag{9.2.1}$$

where γ_{sv} now denotes the surface energy for the interface between the solid and its vapor. However, important differences exist for solids when compared with liquids. First, the value of γ_{sv} is anisotropic; it depends on the orientation of the crystallographic planes that are exposed as the new surface. Second, surface energy and surface tension are not necessarily equal for a solid. Third, the surface may not be atomically flat, and roughness will affect the value for γ. We explore these differences in the following sections.

9.2.1.1 Surface Energy Depends on Bond Character and Crystal Structure and Is Generally Higher for Crystalline Solids Than for Amorphous Solids and Liquids

In Section 3.1.1, we saw that surface energy originates from differences in atomic or molecular bonding in the bulk compared

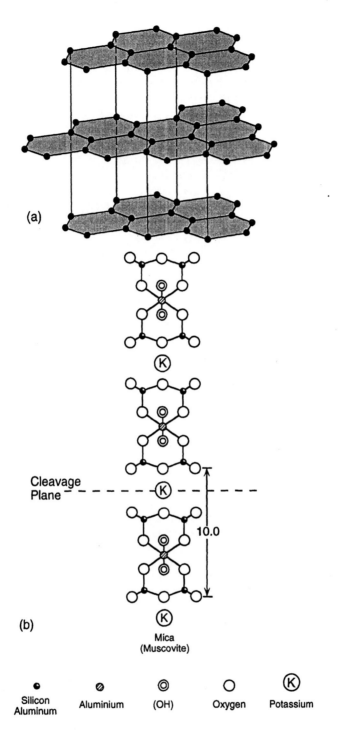

(a)

Cleavage
Plane

10.0

(b)

Mica
(Muscovite)

•	⊘	◎	○	Ⓚ
Silicon Aluminum	Aluminium	(OH)	Oxygen	Potassium

Figure 9.6
Layered crystal structures. (a) Graphite form of carbon. (b) Mica, chemical formula
$Al_2K(Si_{1.5}Al_{0.5}O_5)_2(OH)_2$. Substitution of Al on one of every four silicon sites requires adding a
potassium ion for charge equalization (compare with Figure 4.4). (G. W. Brindley, in *Ceramic
Fabrication Processes*, W. D. Kingery, Ed., MIT Press, Cambridge, MA, 1958, p. 13.)

with the surface. For a solid, the solid–vapor interface may be considered discontinuously sharp. So we can estimate excess free energy per unit area of surface (dG^σ/dA) from the change in interatomic bonding between an atom (or an ion) that is fully coordinated within the crystal and an atom on the free surface that lacks neighbors on one side. The surface energy per atom is higher by an amount that is equal to the missing contributions to its bonding. Often we can estimate interfacial energy in solid systems by measuring the surface density of the incomplete or "dangling" bonds.

For the least complex case, we consider a crystal, such as sodium chloride, consisting of atoms located on a cubic lattice (see Figure 9.4). When the crystal is broken in two through a {100} cube face, each surface atom loses one of its six nearest neighbors, resulting in one dangling bond per surface atom. To estimate the surface energy, we note that when the five remaining bonds are broken, each surface atom escapes into the vapor and the crystal sublimates. Therefore, to a first approximation, the surface energy multiplied by 2, $2\gamma_{sv}$, equals one-fifth of the energy for sublimation. The factor 2 is included because two atoms originally shared the incomplete bonds on each surface. This analysis leads to an expression for the surface energy per unit area

$$2\,\gamma_{sv} = 1 \times \frac{U_{\text{sublimation}}}{5}$$

or

$$\gamma_{sv} = \frac{U_{\text{sublimation}}}{10} \qquad (9.2.2)$$

For a metal in a close packed configuration (see Figure 9.2) each metal ion is surrounded by twelve nearest neighbors, but when the close-packed planes ({0001} planes for hcp and {111} planes for fcc) are exposed as a free surface, three of the nearest neighbors have been split away to leave three incomplete bonds per ion. On sublimation, the remaining nine bonds are broken. Thus for {111} fcc surfaces

$$2\,\gamma_{sv} = 3 \times \frac{U_{\text{sublimation}}}{9}$$

or

$$\gamma_{sv} = \frac{U_{\text{sublimation}}}{6} \qquad (9.2.3)$$

For a strongly bonded crystalline material, such as a metal, with an interatomic separation of approximately 3Å and heat of sublimation 335 kJ/mole, this broken bond relation gives a surface energy γ_{sv} of about 1–2 J/m^2 or 1000–2000 ergs/cm^2. As Table 9.2 indicates, these values are typical for metals. Comparing eqs. 9.2.2 and 9.2.3, we might expect that the surface energies would be greater for metals than for simple ionic materials, a prediction borne out by the values listed in Table 9.2a.

TABLE 9.2a Surface Energy of Selected Solids at Room Temperature $(mJ/m^2 = ergs/cm^2 = dyn/cm)$

	Theoretical	Measured
Covalent solids		
C (diamond) (111)	5650[1]	
(100)	9820[1]	
Metals		
Tungsten (110)	5510[2]	
	1000	>1000
Ionic solids		
MgO (100)	360—924	1200
CaF$_2$	540	450
LiF (100)	480	340[4]
NaCl (100)	212	110
Inert gas solids	50	
Argon (111)	41[3]	

Sources: W. D. Harkins, *Journal Chemical Physics* **10**, 268 (1942); I. N. Stranski and R. Suhrmann, *Annalen der Physik,* **1,** 153 (1947); R. Shuttleworth, *Proceedings of the Physical Society,* **A62,** 167 (1949); J. J. Gilman, *Journal Applied Physics,* **31,** 2208 (1960).

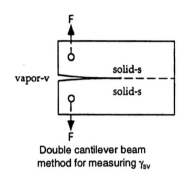

Double cantilever beam method for measuring γ_{sv}

Actually this broken bond approximation produces a valid relationship between the surface energy and the heat of sublimation for solids only if the attractive component of the bonding is short ranged. This is the case for the metallic or van der Waals bond. For Coulombic forces, the attraction is long ranged and the surface energy must be calculated directly. In that case, theoretical calculation of the surface energy involves summing all the attractive interactions between ions located at and beneath the surface of the solid and comparing the result with the bulk lattice energy. For ionic solids, we find that while the Coulombic attraction contributes nearly all the bulk lattice energy, van der Waals attractions contribute nearly 30% of the surface energy. Table 9.2a provides selected theoretical results taking these differences into account.

Table 9.2a also lists values for the surface energy of ionic solids that have been determined by various experimental methods. The double cantilever notched beam method basically measures the stress required to pull apart a crystal into which an atomically sharp notch has been introduced. As the crack extends, the mechanical strain energy stored due to the stress concentration at the tip of the notch is released. When this strain energy is totally converted to surface energy to generate new surface, the method provides a direct way to measure the work of cohesion defined in eq. 9.2.1. Surface energy determined by such "fracture mechanics" techniques is termed *fracture surface energy,* γ_f. (See Chapter 7.4.1 and Appendix 11B.) Actually some of the mechanical strain energy is consumed in other ways, such as by plastic deformation through the motion of dislocations; so the value of γ_f

TABLE 9.2b Measured Surface Energies of Various Materials in Vacuo or Inert Atmosphere $(mJ/m^2)^a$

Material	Temperature ($°C$)	Surface energy γ mJ/m^2
Metals:		
Mercury (liquid)	20	480
Lead (liquid)	350	442
Copper (liquid)	1120	1270
Copper (solid)	1080	1430
Silver (liquid)	1000	920
Silver (solid)	750	1140
Platinum (liquid)	1770	1865
Gold (solid)	1130	1100
Iron (solid)	1450	2300
Ceramics:		
Sodium sulfate (liquid)	884	196
Sodium phosphate, $NaPO_3$ (liquid)	620	209
Sodium silicate (liquid)	1000	250
0.20 Na_2O—0.80 SiO_2	1350	380
0.13 Na_2O—0.13 CaO—0.74 SiO_2 (liquid)	1350	350
B_2O_3 (liquid)	900	80
FeO (liquid)	1420	585
Al_2O_3 (liquid)	2080	700
Al_2O_3 (solid)	1850	905
TiC (solid)	1100	1190
$CaCO_3$ crystal (1010)	25	230

[a]These values should be compared with the surface energy for other substances given in Table 3.1.

is generally higher than the absolute surface energy. For brittle solids (solids in which dislocations do not move easily like the ionic solids included in Table 9.2a) fracture surface energy should differ only a little from surface energy. Indeed, the agreement between experimental and theoretical values for ionic solids in Table 9.2a is quite good. For less brittle solids, fracture does not occur so cleanly, and plastic flow and tearing in the fracture surface increases the work that must be done to achieve decohesion. Then the fracture surface energy rises to a value many orders of magnitude above the absolute surface energy.

The magnitude of the surface energy of a solid depends on the crystal structure (coordination number), the bond character (strength of the broken bond), and the orientation of the surface plane (density of broken bonds). Many of these features can be appreciated from Table 9.2. We might expect a strongly covalent solid like diamond to exhibit a high surface energy, and its theoretical value is the highest in Table 9.2a. High coordination numbers and the higher strength of the metallic bond mean that metals generally have higher surface energies than ionic soids (ceramics). For ionic solids, surface energy increases with the valence of the ion, so that divalent ionic solids like MgO have a higher surface energy than monovalent ionic solids like LiF. The surface energies of crystalline solids are generally higher than those of noncrystalline materials, such as polymers and glasses. A solid's surface energy is higher than that of the equivalent liquid. For example, the surface energy of sodium chloride is estimated to be 212 ergs/cm^2 below its melting point, but 190 ergs/cm^2 above it. Solid and liquid alumina surface energy measures 900 and 700 ergs/cm^2, respectively.

Differences in the relative magnitude of surface energy or surface tension for the various classes of solids make it possible to predict the ease with which a solid of one type covers another. Referring to the Young equation and the spreading coefficient in Section 3.2.2, we can anticipate that a solid with a higher surface energy would prefer to be covered by a solid with a lower surface energy. Thus we can easily coat a metal with a ceramic (porcelainized steel) or a polymer (Teflon-coated pans), whereas coating a ceramic with a higher surface energy metal (copper interconnects on ceramic substrates for example) requires special pretreatments of the ceramic surface.

9.2.1.2 The Surface Energy of a Crystalline Solid Depends on Orientation and Is Anisotropic

Crystallographic orientation of the surface controls the surface energy of a solid. Surface energy generally is lowest for the closest-packed planes in the crystal structure. This is because the number of bonds lying within the planes is at the highest possible value for close-packed planes, and the number of bonds reaching across the planes is at the lowest possible value. Thus the density of broken bonds left dangling on the close-packed surface is at a minimum. For crystallographic planes that are less densely bonded in the plane, the relative density of broken bonds across the plane increases, and as a result, the surface energy is higher.

We expect that while a liquid minimizes its surface energy by forming a spherical droplet, the equilibrium shape of a solid should be terminated in surfaces with the lowest energy. Surfaces such as the close-packed {111} planes of the face-centered cubic metals should be preferred, and surfaces with the largest density of broken bonds should be avoided. Prediction of the equilibrium shape follows directly from the *Wulff plot* of surface energy.

The Wulff plot is a polar plot that presents a spatial representation of the variation of surface energy with orientation, as shown in Figure 9.7. The surface energy of a particular plane is represented by a vector drawn from a point at the center of the unit cell in a direction normal (perpendicular) to that plane. The length of the vector is proportional to the surface energy of the planar surface. For close-packed planes, in which the surface energy is low, the vector is short; for the least densely packed planes, in which the surface energy is high, it is long.

We can derive the equilibrium shape of the solid by placing normal planes at the end of each radius vector and constructing a surface that lies totally within all the planes so placed. For example, if no orientation sensitivity exists, as in the case of a liquid or noncrystalline solid, the vectors are the same in every direction and the equilibrium shape is a sphere, as in a spherical

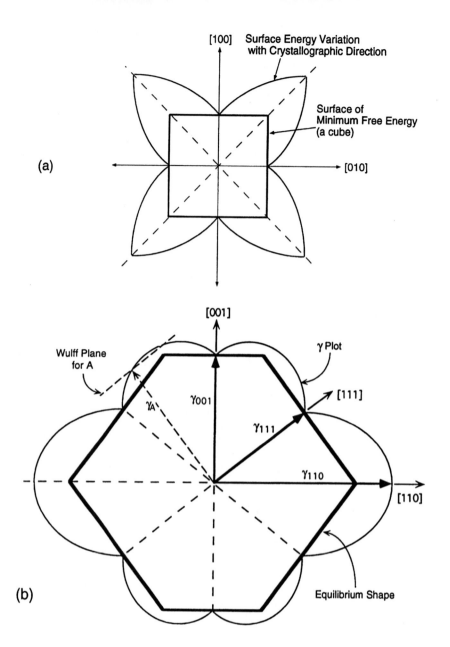

Figure 9.7
Surface energy is anisotropic for crystalline solids. (a) Wulff plot of surface energy for different directions in the NaCl structure, showing construction of the surface with minimum surface energy, in this instance a cube. (b) Wulff plot of surface energy for different directions in the {110} plane of the face-centered cubic crystal structure, showing construction of the surface with minimum surface energy. (C. Herring, *Physical Review* **82**, 87, 1951.) (c) Three-dimensional surface of minimum surface energy for face-centered cubic crystal structure in (b), in this instance a truncated octahedron (almost a sphere).

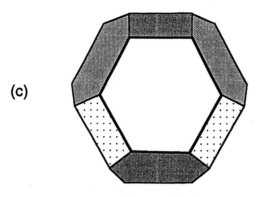

(c)

Figure 9.7 (*Continued*)

liquid droplet. The equilibrium shape for the sixfold-coordinated sodium chloride crystal structure (Figure 9.4) predicted from the Wulff plot shown in Figure 9.7a is a cube. Indeed, the "smoke" particles of magnesium oxide (which has the sodium chloride crystal structure) are tiny cubelets. The equilibrium shape for the close-packed face-centered cubic metal predicted from the Wulff plot shown in Figure 9.7b is a tetrakaidecahedron formed by truncating the octahedron bound by all eight {111} planes with {100} facets, as shown in Figure 9.7c (see also Figure 11.12). This shape is almost spherical. As a general rule, the planes of greatest density—between which the bonding is weakest and, therefore, the density of broken or dangling bonds is lowest—define the external surfaces and the equilibrium shape.

The symmetry of the crystal structure affects the orientation sensitivity of the surface energy. For some materials, such as gallium arsenide, noncentrosymmetry has important consequences. Consider the crystal structure of GaAs illustrated in Figure 9.5c. As we have already noted, cleavage between the cube diagonal {111} planes produces two different kinds of surface; looking one way the surface is gallium terminated, and looking the other way arsenic terminated. The dangling bond configurations of these two surfaces are very different. The gallium ion has three valence electrons, and because it can find three arsenic neighbors immediately beneath the surface to share these electrons, its surface is relatively free from dangling bonds and the {111}$_{Ga}$ surface has a low surface free energy. By contrast, the arsenic ion has five valence electrons, and because it can find only three gallium neighbors at the surface, two bonds are left dangling, unsatisfied. The {111}$_{As}$ surface has a high surface free energy. This difference in surface energy accounts for the different chemical reactivity of the pair of surfaces: acid etches the {111}$_{As}$ arsenic surface more easily than the {111}$_{Ga}$ gallium surface. The arsenic surface is more favorable for rapid vapor deposition of new material than the gallium surface, a characteristic that is important in the processing of gallium arsenide semiconductor devices (see Section 10.5). Two reference edges must be cut on {111} gallium arsenide wafers to define which surface is "up" and to make sure they are properly oriented for processing. (Two edges can be seen, for example, on the GaAs wafer in Figure 10.15.)

9.2.1.3 Surface Energy and Surface Stress Are Not Equivalent in Solids

Section 3.1.1 defined the surface tension of liquids as either the excess free energy needed to form additional surface per unit area or the force per unit length opposing the formation of new surface. For liquids the excess free energy and the force opposing surface formation are the same, and an equivalence exists between surface energy and surface tension. For solids this equivalence no longer necessarily holds true because a solid can sustain a state of stress in the free surface, particularly at low temperature. The major difference between a liquid and a solid is that in a liquid, energy is used to bring atoms from the bulk to form the new surface, whereas in a solid, strain energy is spent stretching the old surface without bringing new material to the surface. Thus, in a solid at normal temperatures, the surface stress or surface tension exceeds the surface energy.

As the temperature approaches the melting point, a solid material begins to deform more like a liquid. The high concentration of vacancies in thermal equilibrium (see eq. 9.3.3) and their rapid diffusion at high temperature (see eq. 9.5.1) enables the solid to change shape in response to the surface stress and it tries to ball up into an equilibrium shape just like a liquid. In fact, we can determine the surface tension of a solid metal near its melting point by measuring the tensile stress required to maintain the shape of a wire or foil so that it neither stretches nor shrinks. The procedure is comparable to the wire loop method depicted in Figure 3.1. Near the melting point, surface tension and surface free energy approach each other in magnitude.

Later in this chapter (and in other chapters) we neglect the significant difference between surface energy and surface stress. For purposes of analyzing an equilibrium configuration, we assume that a solid displays a liquidlike surface tension even at low temperatures, and we assume that the tension can be represented by a vector tangential to the surface, and that the equilibrium configuration can be predicted by balancing tensions. Indeed, these assumptions were already made in the derivation of the Young equation in Chapter 3, eq. 3.2.4. There the surface energy of the solid was represented by a surface tension that was balanced with other tensions to establish the equilibrium contact angle of a liquid on a solid. While this assumption is useful to establish principles, it is not acceptable when detailed experimental observations must be explained or when more precise values are needed for process design.

9.2.2 The Free Solid Surface Has a Structure Other Than That Rendered by Simple Dissection of the Material into Two Halves

9.2.2.1 Surface Atoms Rearrange Themselves to Minimize Free Energy—Surface Relaxation and Surface Reconstruction at Low Temperatures

Many of our diagrams depict the free surface of a solid as a simple dissection of the crystalline material into two perfect halves. This notion is not accurate. Atoms on the surfaces of pure crystalline materials rearrange themselves into new configurations to lower the surface energy by whatever means possible. Bonding, crystal structure, and the magnitude of the driving forces involved as well as kinetics dictate the process of rearrangement.

The ability to reduce surface energy by atomic rearrangement depends on the size and sign of the ions involved. For example, in an ionic crystal of zinc oxide in which the oxide ions are 1.5 times the diameter of the zinc, the smaller zinc ions may retreat to semi-interstitial locations just beneath the surface where they are shielded from complete surface exposure. In this case, *surface relaxation* leads to minor rumpling of an otherwise flat planar surface, as shown in the cross section through the surface in Figure 9.8.

Rearrangement is more pronounced in covalent crystals. The bonds are so strong and directional that major *surface reconstruction* occurs. Atoms on the free {111} surface of silicon completely rearrange their atomic positions so that only every seventh silicon atom in the ⟨110⟩ directions coincides with the bulk position. Surface reconstruction of silicon has been studied theoretically for many years, and various equilibrium configurations have been proposed and analyzed. Recent development of surface analytical tools has made it possible to detect the rearrangement experimentally. One of the triumphs of scanning tunneling microscopy (STM) is its confirmation of the 7×7 surface reconstruction of the {111} surface of silicon. In this nomenclature, $A \times B$ means that A and B "times" the normal lattice parameters in respective directions are required to reregister with the undistorted crystal. The {100} surface of silicon reconstructs to the 2×1 structure depicted in Figure 9.9.

Rearrangement extends only one or two atomic layers beneath the surface because crystallographic structure imposes a strict order in the bulk. Solid–vapor surfaces, therefore, are very abrupt, unlike liquid–vapor surfaces in which the influence of the vapor phase extends about 10 atomic layers deep (Section 3.5). The reconstructed

Figure 9.8
Surface rearrangement in zinc oxide to minimize surface energy. Cross section through the surface shows how smaller zinc ions drop beneath the surface to shield their free bonds. (W. D. Kingery, H. K. Bowen, and D. R. Uhlmann, *Introduction to Ceramics*, 2nd ed., Copyright © 1976, John Wiley & Sons, p. 205. Reprinted by permission of John Wiley & Sons, Inc.)

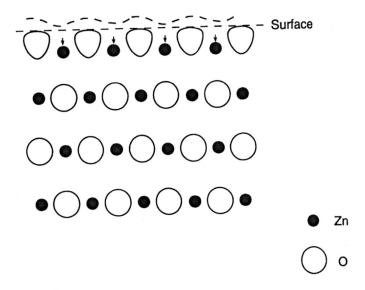

Figure 9.9
Surface reconstruction on the surface of silicon; (a) planar view and (b) cross section of (100) surface. Left-hand portion shows atomic configuration before reconstructed and right-hand after reconstruction. Surface energy is minimized by small movements of Si ions toward each other along [100]. (G. A. Somorjai, *Chemistry in Two Dimensions,* Cornell University Press, Ithaca, NY, 1981, p. 149.)

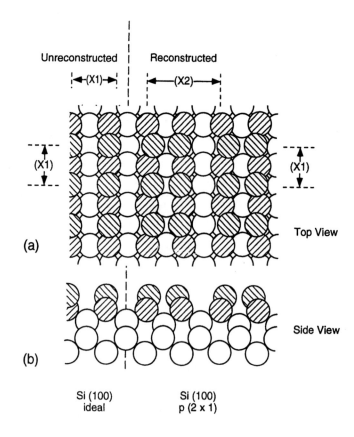

state of the surface is an important consideration in substrate behavior when new layers are to be added to a surface, as in semiconductor device processing (see Section 10.5.3).

9.2.2.2 Surface Roughening at High Temperatures

From our macroscopic viewpoint, we also assumed that solid surfaces are flat, but in practice a clean surface develops a texture or roughness on an atomic scale. At high temperatures, the dynamic exchange of atoms (ions) between the solid and its vapor generates terraces, steps, kinks, adatoms, and vacancies, as illustrated in Figure 9.10.

The energy associated with each surface site differs because the coordination number changes. Coordination decreases, and the excess energy associated with a surface site increases, as we go from A to F in the following sequence:

Location	Coordination (assuming simple cubes)
Atom in the surface (A)	5
Atom in the surface adjacent to a vacancy (B)	4
Atom in a step (C)	4
Atom in a kink (D)	3
Step adatom (E)	2
Surface adatom (F)	1

Atoms in the higher-excess-energy (lower coordination number) locations tend to move to locations of lower energy either by

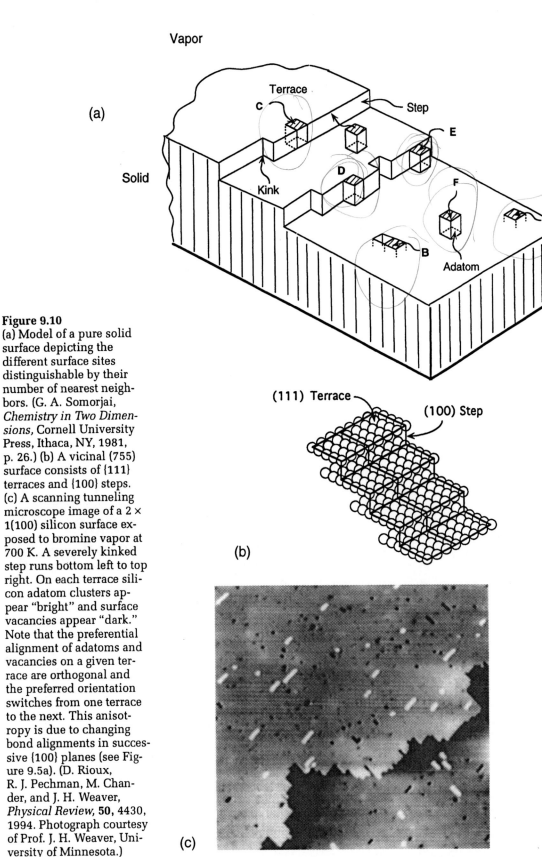

(a)

Vapor

Terrace

C

Step

E

Solid

D

F

Kink

A

B

Adatom

(111) Terrace

(100) Step

(b)

(c)

Figure 9.10
(a) Model of a pure solid surface depicting the different surface sites distinguishable by their number of nearest neighbors. (G. A. Somorjai, *Chemistry in Two Dimensions,* Cornell University Press, Ithaca, NY, 1981, p. 26.) (b) A vicinal (755) surface consists of {111} terraces and {100} steps. (c) A scanning tunneling microscope image of a 2 × 1(100) silicon surface exposed to bromine vapor at 700 K. A severely kinked step runs bottom left to top right. On each terrace silicon adatom clusters appear "bright" and surface vacancies appear "dark." Note that the preferential alignment of adatoms and vacancies on a given terrace are orthogonal and the preferred orientation switches from one terrace to the next. This anisotropy is due to changing bond alignments in successive {100} planes (see Figure 9.5a). (D. Rioux, R. J. Pechman, M. Chander, and J. H. Weaver, *Physical Review,* **50,** 4430, 1994. Photograph courtesy of Prof. J. H. Weaver, University of Minnesota.)

leaving the surface to reenter the vapor or, more likely, by diffusing rapidly over the surface (see Section 9.5.6). The net result is that a free surface in equilibrium with its vapor at high temperatures is atomically rough and continuously changing.

Surface roughness also is sensitive to surface orientation. Close-packed surfaces of high atomic density (and low Miller index; see Appendix 9A) that have the lowest surface energy have the lowest density of surface irregularities. As the macroscopic surface orientation moves away from a low-index orientation, the surface tries to maintain the lowest-energy configuration by introducing terraces of low-index planes terminated by steps. Electron diffraction analysis establishes the fact that a surface with a high Miller index consists of an ordered array of terraces and steps that produces the macroscopic surface orientation. Thus a vicinal surface orientation such as {755} in Figure 9.10b can be obtained by {111} terraces five atoms wide terminating in {100} steps two atoms high $(5\{111\} + 2\{100\} \approx \{755\})$. The steps also can contain kinks, and more complex surface orientations can be constructed by placing kinks every few atoms along the steps. For a random orientation, a "thermally etched" surface may contain terraces, steps, and kinks. In extreme cases, faceted pits terminated in low-index planes exist as well.

In the manufacture of thin-film semiconductor devices, substrates commonly are used slightly off axis. For example, GaAs substrates are available 1° off {100} with the misorientation tilted toward either {110} or {111} to change the step–kink relationship. Before deposition processes begin, they deliberately are annealed at high temperatures (600°C in vacuum) to roughen the surfaces thermally (see Section 10.5.3).

9.3 The High Surface Energy of a Crystalline Solid Is Reduced through a Change in the Chemical Composition of the Surface

In the case of liquid surfaces, we were able to predict the equilibrium chemical composition from thermodynamic principles through the Gibbs adsorption isotherm (Section 3.5). However, for a solid system the situation is complicated by the difference in transport kinetics (see Section 9.5) and the extreme sensitivity of diffusion to temperature. We will find it convenient to distinguish phenomena that occur predominantly at low temperatures from those that occur predominantly at high temperatures. At low temperatures ($<\frac{1}{2}T_m$, T_m the melting temperature) and typically at room temperature, surface chemistry is influenced primarily by direct interaction with the environment *outside* the crystalline solid—*adsorption*. At high temperatures ($>>\frac{1}{2}T_m$), surface chemistry is also influenced by the diffusion of impurities from *inside* the crystalline solid to the surface—*segregation*. As discussed in Chapter 10, adsorption is critical in the deposition of thin films on the free surface. As discussed in Chapter 11, segregation is critical in the behavior of grain boundaries and processing based on their behavior.

9.3.1 Adsorption Reduces Surface Energy at Low Temperatures

At *low* temperatures the freshly exposed surface of a crystalline solid will immediately reduce its surface energy by adsorbing molecular species from the environment. Remnant oxygen, nitrogen, or hydrocarbon molecules are unavoidably present even in the best vacuum systems. As Section 3.7.2 notes, we cannot represent the changes in composition at a solid interface by the Gibbs adsorption isotherm because the solid adsorbent cannot adjust to changes in composition with the speed that a fluid can and surface excess has no meaning. Other isotherms have been derived for solids that describe the change in surface composition. Furthermore, we must distinguish between the processes involved in establishing the first "monolayer" and the growth of subsequent layers.

9.3.1.1 The Langmuir Adsorption Isotherm Describes Adsorption of the First Monolayer

The Langmuir adsorption isotherm describes the coverage of a bare solid surface as a function of its chemical environment. This relation is derived in two other chapters in this book: eq. 3.7.11 in Chapter 3 describes the adsorption of an impurity from a liquid solution onto a solid surface and eq. 10.6.35 in Chapter 10 describes the adsorption from a gas.

Both results give the same form for the Langmuir equation

$$\Theta = \frac{K_{ads}A}{K_{ads}A + 1} \tag{9.3.1}$$

where Θ is the fraction of available sites covered and A is either the molar concentration of solute in solution, c, or the partial pressure of the gas, P. K_{ads} is the equilibrium constant for the adsorption/desorption reaction. For adsorption from the gas phase

$$\Theta = \frac{K_{ads}P}{K_{ads}P + 1} \tag{9.3.2}$$

and

$$K_{ads} = \frac{\alpha \tau_o N_{Av}}{\sqrt{2\pi MRT}} \exp\left(-\frac{\Delta G_{ads}}{RT}\right) \tag{9.3.3}$$

where M is the molecular weight of the adsorbing gas, $1/\tau_o$ equals the characteristic crystalline solid surface atom vibration frequency, and ΔG_{ads} represents the energy of adsorption (and desorption). α is the accomodation coefficient—a measure of the interaction between the gas atom and the surface atom discussed in Section 10.6.2.2. The derivation assumes ideal behavior; that is, gas atoms arriving at the surface are not influenced by the atoms already there.

The Langmuir adsorption isotherm has the form shown in Figure 9.11a. From the Clausius–Clapeyron equation

$$\left[d(\ln P)/d\left(\frac{1}{T}\right) \right]_\Theta = -\frac{\Delta G_{ads}}{R} \tag{9.3.4}$$

Thus, from a set of Θ versus P curves at different temperatures (T_1, $T_2, \ldots$) we can use eq. 9.3.4 to obtain a value for ΔG_{ads}—the energy of adsorption (see Figure 9.11c).

Figure 9.11
Adsorption of nitrogen on rutile (TiO_2) at different temperatures. (a) Low-pressure region; (b) high-pressure region. [(a) and (b) from A. W. Adamson, *Physical Chemistry of Surfaces*, 5th ed., Copyright © 1990, John Wiley & Sons, p. 592. Reprinted by permission of John Wiley & Sons, Inc.] (c) Plot summarizing pressure and temperature dependence as a function of coverage. For a given coverage the activation energy for adsorption (desorption) may be estimated using eq. 9.3.4.

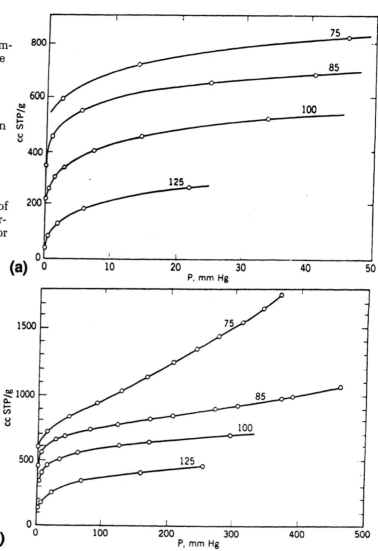

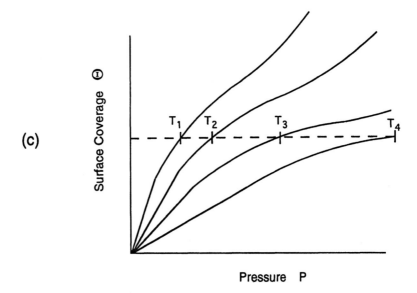

(c)

9.3.1.2 A Distinction between Physisorption and Chemisorption Can Be Based on Adsorption Energy

A range of adsorption behavior depends on the value of ΔG_{ads}. At one extreme lies the weak physical attraction of the gas (the adsorbate) to the crystalline solid surface (the adsorbent) known as physical adsorption (*physisorption*); at the other extreme lies the strong chemical reaction known as chemical adsorption (*chemisorption*). The break between physisorption and chemisorption generally is taken to be $\Delta G_{ads} \approx 60$ kJ/mole ($\cong 0.6$ eV/atom), although additional features are used to distinguish the two phenomena. For example, physisorption is a reversible process. Heating the substrate releases physisorbed gases, and they are chemically unchanged. Thus we can reactivate a charcoal filter by heat, and we can "clean" a semiconductor or a noble metal surface by heating it in a vacuum prior to thin-film deposition to drive off adsorbed gas molecules. Heating a chemisorbed surface may release a molecule of different chemical composition, or it may not be possible to release the chemisorbed material at all. We explore these differences in more detail.

- A low value for ΔG_{ads}, such as 6 kJ/mole (0.06 eV/atom) for inert argon gas atoms adsorbed on a potassium chloride ionic solid surface, characterizes weak physisorption;

- A moderate value for ΔG_{ads}, such as 30 kJ/mole (0.3 eV/molecule) for active ammonia (NH_3) polar molecules on an inert charcoal surface, characterizes strong physisorption;

- A high value for ΔG_{ads}, such as 65 kJ/mole (0.65 eV/molecule) for active carbon monoxide (CO) on a copper metal surface, characterizes chemisorption;

- A value for $\Delta G_{ads} > 100$ kJ/mole (1 eV/atom)—associated with active surface oxidation of pure metals—clearly characterizes a chemical reaction process.

At normal ambient temperatures and pressures, adsorption covers a solid surface with the first monolayer almost instantly. Although adsorption energy varies with location on a clean surface because the value of ΔG_{ads} will be greater at kinks and steps like those shown in Figure 9.10, physisorption still results in an instantaneous and complete monolayer covering everything, including terraces, steps, and kinks. *Even at low vacuum pressures of 10^{-6} torr, physisorption results in complete coverage by residual gas molecules in a millisecond.*

Chemisorption, on the other hand, may not be instantaneous. A chemical reaction and a change in bonding must take place. The end result may be the formation of a surface compound, such as Ag_2S or Al_2O_3, which may or may not be ordered and epitaxial with the substrate. Indeed, chemisorption often takes place in two stages: first the physisorption of the adsorbate due to weak intermolecular van der Waals forces; then a chemical reaction to form a compound bound by strong covalent intermolecular forces.

9.3.1.3 The BET Equation Describes Multilayer Coverage of Solid Surfaces

Once the adsorbed monolayer is complete, the surface chemistry changes. Adsorbate atoms striking the surface experience a quite different interaction with a covered surface. Whereas the first monolayer particles were strongly attracted to the surface in a manner that depended on the chemistry, subsequent particles may experience a weaker attraction or even a repulsion. Thus we can assume that $\Delta G_{ads,1}$, the energy of adsorption for the first layer, will be different and generally greater than the energy of adsorption for all succeeding layers, $\Delta G_{ads,v}$. We associate $\Delta G_{ads,v}$ with the heat of condensation of the adsorbate.

Following an analysis similar to that used to derive the Langmuir adsorption isotherm, a new relationship can be derived to describe the adsorption of subsequent layers on the surface as a function of pressure. The BET equation, due to Brunauer, Emmett, and Teller, is

$$\Theta = \frac{Cx}{(1-x)[1+(C-1)x]} \qquad (9.3.5)$$

where $C \sim \exp[(\Delta G_{ads1} - \Delta G_{adsv})/RT]$ and $x = P/P_o$, where P is the partial pressure and P_o is the equilibrium vapor pressure for the adsorbate.

9.3.2 Segregation Reduces Surface Energy at High Temperatures

At *high* temperatures the kinetics of surface adsorption change so that now the surface chemistry is influenced as much by the internal environment as by the external one. Thermal energy desorbs the gas molecules physisorbed onto the solid surface, and gas atoms move continuously to and from the surface. Inside the solid, rapid diffusion of impurities enables the solid to approach thermodynamic equilibrium. The chemistry of the surface now is determined by the effective reduction in surface energy, $d\gamma_{sv}$, due to the presence of solute elements (molar concentration c) in the same manner as already described for fluids. The Gibbs isotherm, eq. 3.5.13, governs segregation of excess solute (impurity), Γ, to the free surface of a solid,

$$\Gamma = -\frac{1}{RT}\left(\frac{d\gamma_{sv}}{d(\ln c)}\right)_T \qquad (9.3.6)$$

Impurity atoms concentrate at the surface by diffusion from within the bulk. Section 9.5.4 shows that impurities redistribute and come to equilibrium in a matter of seconds as temperatures approach the melting point of a solid. These equilibrium times should be compared with the microsecond timescales for fluid systems to come to equilibrium.

9.4 Crystalline Solids Contain Many Kinds of Imperfections

Our representation in Section 9.1 of the internal structure of a crystalline solid as a perfect array of atoms or ions reproduced indefinitely is not accurate. Many breaks in the perfect periodicity lead to well-defined imperfections.

The regularity in a single-phase, pure, single crystalline solid can be broken by:

1. *Point imperfections,* such as vacancies and interstitials (zero-dimension imperfections);

2. *Linear imperfections,* such as dislocations (one-dimensional imperfections);

3. *Planar imperfections,* such as small-angle boundaries, stacking faults (two-dimensional imperfections);

4. *Volume imperfections,* such as voids (three-dimensional imperfections)

In describing the nature of these imperfections, we highlight their interaction with surfaces and examine their role as constituents of internal surfaces (boundaries) within crystalline solids.

9.4.1 Point Imperfections Exist in Thermodynamic Equilibrium

Our mental picture of a crystalline solid as a uniform, continuous, and perfect array of hard particles flies in the face of the disorder characterized by entropy. Indeed, such perfection can exist only at absolute zero temperature. The total Gibbs free energy of a solid is reduced when point defects are introduced in the form of the *vacancies* and *interstitials* shown in Figure 9.12 (neglecting for the moment differences in size and sign). The total free energy of the solid is then

$$G = G^o + n\,\Delta H_f - T\,\Delta S \qquad (9.4.1)$$

Figure 9.12
Point imperfections in ionic solids. (a) The Frenkel imperfection consists of an interstitial ion and a vacancy. (b) The Schottky imperfection consists of a cation–anion vacancy pair.

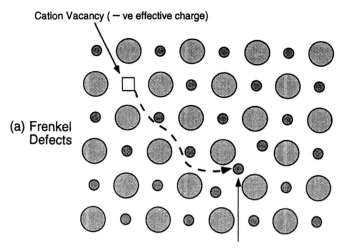

Cation Vacancy (− ve effective charge)

(a) Frenkel Defects

Cation Interstitial (+ ve effective charge)

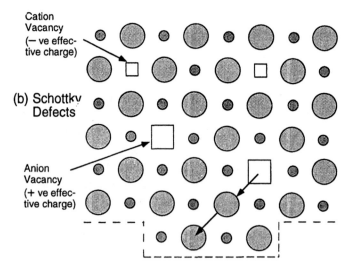

Cation Vacancy (− ve effective charge)

(b) Schottky Defects

Anion Vacancy (+ ve effective charge)

Ions Move to Surface to Create Vacancies

where G^o stands for the standard state free energy of the solid in the absence of any imperfections, n is the number of imperfections, ΔH_f is the enthalpy required to form each imperfection, and ΔS is the configurational entropy due to them.

To calculate the concentration [n] of these imperfections, we must calculate and minimize the free energy change due to their presence. Increase in entropy associated with a random distribution of imperfections is given by eq. 3.7.1, where Ω now represents the number of possible ways of distributing vacancies or interstitials on the lattice sites, N. The computation for the configurational entropy from eqs. 3.7.4 to 3.7.6 gives

$$\Delta S = k \, [N \ln N - (N\text{-}n) \ln (N\text{-}n) - n \ln n] \qquad (9.4.2)$$

Substituting eq. 9.4.2 into 9.4.1, recognizing that at equilibrium the free energy change $\Delta G = G - G^o$ is at a minimum (that is,

$\partial\Delta G/\partial n) = 0$), and assuming $\ln (N-n) \approx \ln N$, gives the result that the equilibrium concentration $[n]$ $(=n/N)$ of point imperfections is

$$[n] = A' \exp\left(-\frac{\Delta H_f}{kT}\right) \qquad (9.4.3)$$

In metals and covalent solids, ΔH_f represents the enthalpy required to *form* the individual interstitial or vacancy, typically 1 eV per imperfection; A' is a term containing the nonconfigurational entropy due to lattice strain and atomic vibrations associated with forming the imperfection, and it has a value ≈ 1.

9.4.1.1 Free Surfaces Act as Both Sources and Sinks for Point Imperfections

Note that 1 eV, the approximate energy to form a vacancy or interstitial, is equivalent to 40kT at room temperature and 13kT at 600°C. Thus, although these imperfections can exist in thermodynamic equilibrium, the energy required to form them is far above the thermal energy ($\approx kT$) even at moderate temperatures. Vacancies do not appear and disappear spontaneously. Bringing a crystalline solid to thermodynamic equilibrium involves a sequence of steps and takes a lot of time. As the temperature rises, additional vacancies must be created, and as the temperature falls, vacancies must be dissipated for the solid to remain in thermodynamic equilibrium. The easiest place for this to occur is at the free surface. For example, a vacancy is generated when an atom jumps from a surface lattice site (such as A in Figure 9.10) onto the surface to form an adatom (such as F), leaving behind a vacancy (such as B). The newly created vacancy diffuses into the bulk of the crystal by exchanging its position with other adjacent atoms (see Section 9.5.1). In this way the surface acts as the *source* for additional vacancies as the temperature rises. As the temperature falls, vacancies must be dissipated, and the surface reverses its role and acts as the *sink* for reducing the concentration of point imperfections.

The time t needed to reach equilibrium is determined by the Einstein relationship (eqs. 3.8.25 and 9.5.21)

$$t = L^2/D_v \qquad (9.4.4)$$

where L is the cross-sectional dimension of the crystalline solid, and D_v is the vacancy diffusion coefficient.

$$D_v = D_o \exp\left(-\frac{\Delta H_m}{kT}\right) \qquad (9.4.5)$$

Note that ΔH_m is the enthalpy required to *move* the vacancy. If a solid is heated and held at a high temperature close to its melting point, it comes to equilibrium in hours, assuming $D_v \approx 10^{-12}$ m²/s and a crystal size of 0.1 mm.

Quenching a crystalline solid by cooling it rapidly from a high temperature freezes in an imperfection concentration that is in excess of the concentration in thermodynamic equilibrium. Point imperfections become trapped inside the crystal and do not have sufficient thermal energy to diffuse out to the surface.

Initially, they simply represent a large nonequilibrium concentration that provides the means for rapid transport of impurity atoms (see Section 9.5.1). Eventually, they may aggregate together and condense inside the crystal; they may form either an internal void or a single planar sheet of vacancies one atom layer thick.

9.4.1.2 Point Imperfections in Ionic Solids Are Charged

For ionic solids the situation is rather more complicated because there are two (or more) ionic species and each point imperfection has an *effective charge* associated with it, as shown in Figure 9.12. The magnitude of the effective charge equals the change in charge distribution within the host crystal caused by removing or adding an ion. Thus a vacancy created by the removal of a positive cation results in an effective negative charge, whereas a vacancy created by the removal of an anion results in an effective positive charge. The interstitial created by the addition of a cation has an effective positive charge, while the interstitial created by the addition of an anion has an effective negative charge. To maintain charge balance, equal numbers of positively charged and negatively charged imperfections must be generated in thermodynamic equilibrium.

As a consequence, ionic solids exhibit point imperfection pairs. The *Frenkel imperfection* (Figure 9.12a) is formed by moving a cation to an interstitial site. The positively charged cation interstitial and the negatively charged cation vacancy left behind are paired to maintain charge balance. The *Schottky imperfection* (Figure 9.12b) is formed by a cation and anion vacancy. The negatively charged cation vacancy and the positively charged anion vacancy are paired to maintain charge balance.

The concentration [n] (= n/N) of point imperfections in ionic solids is now given by

$$[n]_F = A' \exp\left(-\frac{\Delta H_{fF}}{2kT}\right)$$

or

$$[n]_S = A' \exp\left(-\frac{\Delta H_{fS}}{2kT}\right) \tag{9.4.6}$$

Here ΔH_{fF} represents the enthalpy required to form the Frenkel (F) interstitial–vacancy pair and ΔH_{fS} the Schottky (S) vacancy-vacancy pair. Generation in pairs accounts for the factor 2 in the exponential denominator.

The values for ΔH_{fF} and ΔH_{fS} depend on the nature of the imperfection (generally Frenkel imperfections require more formative energy than Schottky imperfections), the valence of the ion being displaced, and the crystal structure. Values range from approximately 10 eV per interstitial–vacancy pair (ΔH_{fF}) in magnesium oxide to approximately 2 eV per vacancy pair (ΔH_{fS}) in sodium chloride. The concentration is extremely sensitive to the particular value for ΔH_f and temperature; for a 2 eV imperfection at 200°C the concentration is 2×10^{-11}; at 600°C it just exceeds 1×10^{-6} (one part per million); and at 1000°C it rises to 1×10^{-4}.

9.4.1.3 Ionic Solid Surfaces Can Become Charged due to Unequal Formation of Cation and Anion Vacancies

The previous section pointed out that point imperfections in ionic solids are effectively charged. For charge balance, we tacitly assumed cation and anion vacancies are generated in equal numbers. However, in actuality, cation vacancies usually have a slightly smaller formation energy than anion vacancies because cations are smaller. More cation vacancies than anion vacancies are created at the surface, and as a consequence the surface becomes *positively* charged when it is held at a high temperature and allowed to come to equilibrium. This surface charge is compensated by an excess concentration of cation vacancies just beneath the surface, forming a negative space charge. (If anion vacancies are easier to produce, the charge pattern is reversed.)

The situation resembles that described in Section 4.3.2 for charged particle surfaces in an electrolyte. The surface of the ionic crystalline solid is positively charged, the solid itself behaves as the electrolyte, and the negatively charged vacancies are equivalent to the counter-ions. Indeed the electrical potential in the space charge beneath the crystal surface follows the Gouy–Chapman relationship (eq. 4.3.6) derived for the concentration of counter-ions in an electrolyte in the vicinity of a charged surface. The vacancy concentration profile can be computed, and we can compare the result shown in Figure 9.13a with Figure 4.8.

Impurity ions in ionic solids also exhibit an effective charge associated with their presence if their valence is different from that of the host, and they too are preferentially attracted to the surface to redress the charge imbalance. They act in the same manner as co-ions and counter-ions in the electrolyte solutions described in Section 4.3.2. As shown in Figure 9.13b, the equilibrium distribution of these impurities also follows the concentra-

Figure 9.13
Difference in cation and anion vacancy concentration leads to surface charge and compensating internal space charge. (a) Pure NaCl; cation vacancies form more readily than anion vacancies. Positively charged surface is compensated by a negative space charge due to excess cation vacancy concentration in surface region. (b) Impure NaCl; containing aliovalent cation impurity (Ca^{2+}) of valence greater than sodium. Negatively charged surface is compensated by excess impurity and excess anion vacancy concentration in surface region. [K. L. Kliewer and J. S. Koehler, *Physical Review* **140** (4A), 1226, 1965.]

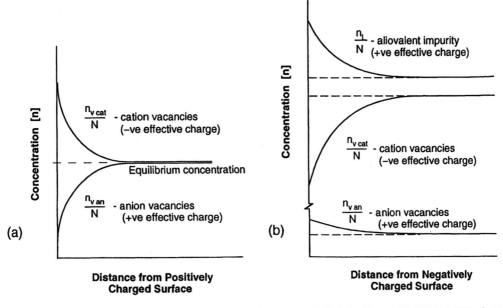

tion profile described by the Gouy–Chapman relation in eqs. 4.3.6 and 4.3.12. In this way electrostatic potential adds to the chemical potential to enhance the driving force for segregation of charged impurites to the surfaces of ionic solids at high temperatures (described in Section 9.3.2 and later in Section 11.3.1).

9.4.2 Linear Imperfections, Known as Dislocation Lines, Are Always Present in Crystalline Solids

Dislocations are present in most crystalline solids. They are easily introduced by thermal/mechanical stresses developed during crystal growth or other solid fabrication processes. Only by exercising extreme caution during solidification can we obtain crystalline solids free from dislocations. Examples of dislocation-free crystals include silicon crystals grown specially for the semiconductor industry.

The next section shows that dislocation lines are important to our discussion of surfaces inside crystalline solids because they can rearrange to form small-angle boundaries, also known as subgrain boundaries. But first we must introduce the concept, configuration, and properties of dislocation lines that are essential for understanding how these subgrain boundaries form.

9.4.2.1 Dislocation Lines are Defined by the Burgers Vector

Dislocation lines are associated with the microscopic aspects of permanent strain or "plastic deformation" of a crystalline solid. We can understand the role of dislocations by considering details of the plastic shear translation of the upper half of the crystalline block over the lower half in Figure 9.14a to cause permanent offset or plastic deformation. Plastic deformation is distinct from elastic deformation. Whereas elastic deformation is reversible and recovered when the load is released, the strain associated with plastic deformation is not recovered when the load is released. (For further information on the microscopic aspects of mechanical behavior, refer to Appendix 11B.)

Plastic shear also differs from elastic shear in that it cannot occur homogeneously over the whole planar surface at one instant. The total energy required to take all the atoms in the plane completely over their neighbors in one coordinated move is too great; moreover the correlation required for all the atoms in the plane to move together at the same instant prohibits the likelihood of instantaneous plastic shear. Instead, we imagine that shear starts along the left-hand edge of the block shown in Figure 9.14a and spreads progressively across the plane. (This movement has been compared to using a fold or a "ruck" to shift a carpet across a floor.) Stopping the displacement part way through its movement across the "slip plane" surface generates a crystalline discontinuity (like leaving the ruck in the middle of the carpet). The discontinuity so produced is equivalent to the insertion of an extra half-plane of material into the crystal. The line where this extra half-plane intersects the slip plane is called the *dislocation line.*

We can also imagine that shear starts along the front edge of the block shown in Figure 9.14b and spreads part way across

Figure 9.14
Movement of line imperfections in crystalline solids leads to permanent displacement (plastic shear strain). (a) Partial plastic shear strain due to the movement of an edge dislocation from the left-hand surface to the center of the crystal. (b) Partial plastic shear strain due to the movement of a screw dislocation from the front surface to the center of the crystal. (c) Partial plastic shear strain due to the movement of a dislocation line. Shear originates at the corner and moves part way over the slip plane. At any instant the dislocation line separates the displaced from the undisplaced portions of the slip plane. The dislocation eventually moves out of the crystal along the back surfaces to complete the displacement.

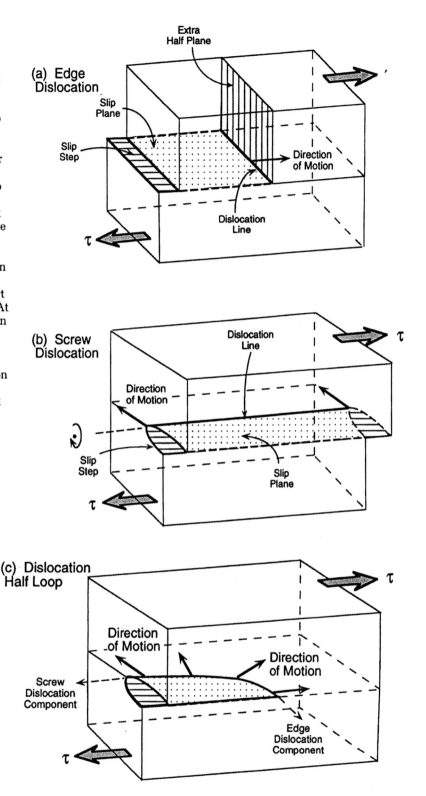

the slip plane to generate another kind of dislocation line. In this instance the crystalline discontinuity is equivalent to a spiral whose axis lies along the dislocation line.

From another perspective, dislocation lines lying in the slip plane separate the shaded portion of the slip plane, which has undergone permanent displacement, from the undisplaced portion.

In reality, we cannot expect shear to initiate instantaneously all along the edge of the crystal either. Figure 9.14c shows the actual sequence by which permanent displacement is achieved. Shear translation starts from a point, such as the corner in Figure 9.14c, then spreads gradually over the slip plane by moving the dislocation line forward in the form of an expanding loop. Proof that plastic deformation occurs by the movement and expansion of dislocation loops over slip planes is now well documented for all kinds of crystalline solids, metals, ceramics, and semiconductors.

Plastic shear in crystalline solids occurs between the closest-packed crystallographic planes for the same reason that they exhibit a low surface energy: interatomic bonding across those planes is weak. Close-packed planes usually define the slip plane of a crystalline solid. However, plastic shear over the plane also must occur in a specific direction and by an amount such that the periodic crystal structure is restored when shear is complete. This combination of direction and magnitude of shear defines the "slip vector." Typically, the slip vector is the shortest vector in the slip plane that connects two identical atoms (or ions) in the crystal structure because it constitutes the smallest shear translation that restores the crystal to its original configuration.

As we have seen at any position, a dislocation line separates the displaced (hatched) portion of the slip plane from the undisplaced (unhatched) portion. As the dislocation moves, it leaves in its wake material that has been permanently displaced by an amount equal to the slip vector, now called the *Burgers vector* of the dislocation line.

Note that the Burgers vector is *invariant* along the length of the dislocation line. As a result the angle between the dislocation line and its Burgers vector varies for different portions of the line. Most important for our present interest is the change in atomic configuration at the core of the dislocation line as the angle varies. We consider the details for two extremes, one where the Burgers vector and the dislocation line are *perpendicular,* giving rise to the *edge dislocation* configuration in Figure 9.14a, and another where the Burgers vector and the dislocation line are *parallel,* giving rise to the *screw dislocation* configuration in Figure 9.14b. These are the two forms that exist at the termination points of the loop on Figure 9.14c.

1. Edge dislocations correspond to that section of the dislocation line in which the Burgers vector b lies perpendicular to the dislocation line, as shown in Figure 9.15a. An edge dislocation is constructed by inserting an extra half-plane of atoms or ions into the crystal. The half-plane terminates on the slip plane and traces the location of the edge dislocation line, as shown along A→D in Figure 9.15a.

2. Screw dislocations correspond to that section of the dislocation line in which the Burgers vector b lies parallel to

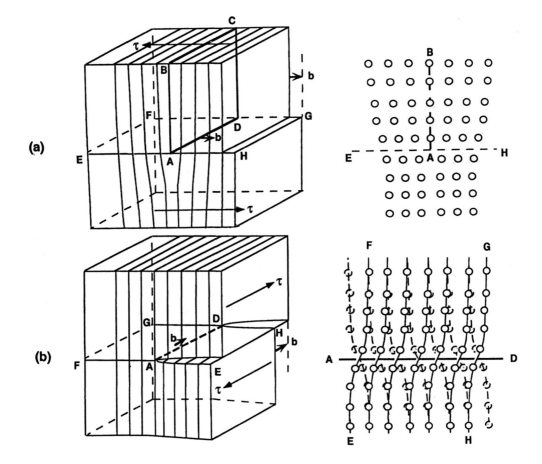

Figure 9.15
Line imperfections in crystalline solids. (a) The configuration of atoms in the vicinity of an edge dislocation. The Burgers vector b is perpendicular to the edge dislocation line AD. Distortion due to the extra half-plane ABCD leads to compression above the slip plane EFGH and extension below EFGH. (b) The configuration of atoms in the vicinity of a screw dislocation. The Burgers vector b is parallel to the screw dislocation line AD. Distortion leads to elastic shear strain about the screw dislocation axis.

the dislocation line, as shown in Figure 9.15b. A screw dislocation is constructed by distorting the crystallographic planes to form a spiral rather than a flat planar surface. The axis of the spiral lies in the slip plane and traces the location of the screw dislocation line, as shown along A→D in Figure 9.15b.

Note that between these two extremes, the character of the dislocation line is mixed and varies from screw to edge as the angle between the Burgers vector and the dislocation line increases from 0 to 90 degrees.

Plastic deformation of crystalline solids, such as metals and ionic solids, is now known to be due to the movement and multiplication of dislocation lines. The transmission electron microscope (TEM) enables us directly to witness dislocation loops moving and spreading over slip planes under the influence of an applied shear stress. We can see the consequences of plastic deformation as a dense dislocation residue of approximately 10^{12} to 10^{16} dislocation lines per square meter. Whereas the reader may have the impression that a metal deforms homogenoeously like a piece of putty, a description that is adequate for many macroscopic engineering purposes, the picture on an atomic scale is very different (see Appendix 11B). The key to manipulating the strength of crystalline metals and alloys lies in the ability to control plastic deformation by dislocation motion. Introducing internal structural interfaces to block dislocation motion is one way to increase the strength of a solid. Chapter 11 features processing techniques used to achieve desirable internal interfaces.

9.4.2.2 Dislocation Lines Possess a High Strain Energy per Unit Length; They Do Not Exist in Thermodynamic Equilibrium

Figure 9.15 shows that dislocations produce elastic distortions in the crystal structure and therefore have a stress field and strain energy associated with them. For the screw dislocation, the shear stress τ is symmetrical about the dislocation line, it is independent of φ. At a distance z from the axis of the screw dislocation line

$$\tau = \frac{Gb}{2\pi z} \tag{9.4.7}$$

where G is the shear modulus and b is the magnitude of the Burgers vector. For the edge dislocation, the stress σ is not symmetrical but varies with the angular position φ about the dislocation line at a distance z in the manner

$$\sigma = \frac{Gbf(\varphi)}{2\pi z(1 - v)} \tag{9.4.8}$$

where v represents Poisson's ratio. The function $f(\varphi)$ changes sign around the edge dislocation. The stress field associated with an edge dislocation varies from a state of compression above the dislocation as drawn in Figure 9.15a to a state of tension beneath it.

We can obtain the strain energy U associated with the dislocation line by integrating the local strain energy (one-half stress times strain per unit volume) for all values of z and φ about the axis of the dislocation line. This procedure gives the strain energy per unit length of a screw dislocation in a crystal of dimension L as

$$U_{screw} = \frac{Gb^2}{4\pi} \ln \frac{L}{L_{core}} + U_{core} \tag{9.4.9}$$

and the strain energy per unit length of an edge dislocation as

$$U_{edge} = \frac{Gb^2}{4\pi(1 - v)} \ln \frac{L}{L_{core}} + U_{core} \tag{9.4.10}$$

L_{core} is the dimension of the dislocation core, where the distortion cannot be described by normal elasticity theory, and the energy required by this distortion is represented by U_{core}. The magnitude of U_{core} is small in comparison with the first term and is often neglected.

Computing the amount of strain energy required to form dislocations yields values >10 eV per atomic length of the line or 4×10^{10} eV/m (64×10^{-10} J/m). Such high values preclude dislocations from existing in thermodynamic equilibrium. As noted earlier, they are introduced into crystals during preparation—they are either "grown in" during the solidification process due to stresses from mechanical vibrations and thermal expansion mismatch, or they are introduced by mechanical forming processes.

9.4.3 Small-Angle Boundaries Are Planar Imperfections; They Are Formed from Arrays of Dislocations

Small-angle boundaries, stacking faults, and twins are terms used to denote the different planar imperfections that break the perfect spatial periodicity of a crystalline solid.

Small-angle boundaries separate small blocks of otherwise perfect crystal. Across the boundary a small angle of misorientation can be resolved into two components: pure *tilt* rotation about an axis *in* the boundary, and pure *twist* rotation about an axis *perpendicular* to the boundary.

We can construct small-angle boundaries from arrays of dislocations. For example, a series of pure edge dislocations, spaced equally one above the other and wedged into the crystal as shown in Figure 9.16a, gives a pure, symmetrical small-angle

Figure 9.16
Planar imperfections in crystalline solid. (a) A small-angle tilt boundary consists of a vertical array of edge dislocations.

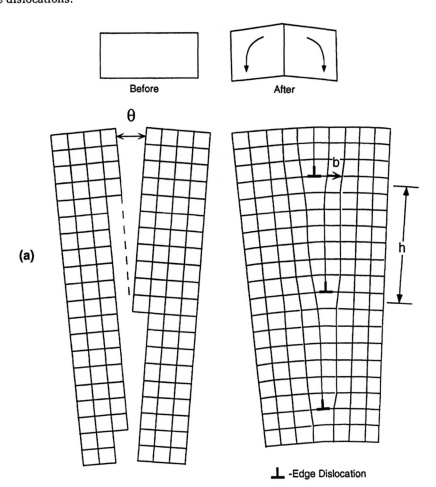

(a)

Before   After

⊥ -Edge Dislocation

Figure 9.16 (*Continued*) (b) A small-angle twist boundary consists of a cross grid of screw dislocations. [(a) and (b) from W. T. Read, Jr., *Dislocations in Crystals*, McGraw-Hill, New York, 1953, pp. 157, 179.] (c) Transmission electron micrograph of a small-angle boundary in a magnesium oxide crystal.

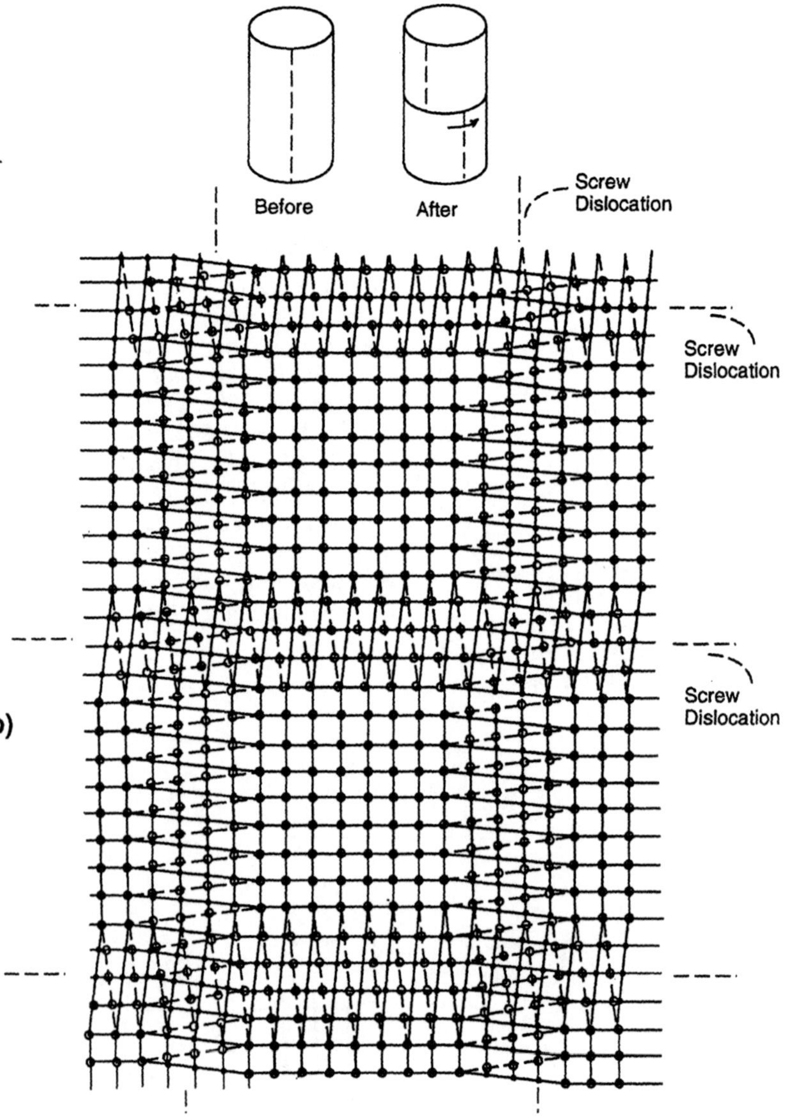

(b)

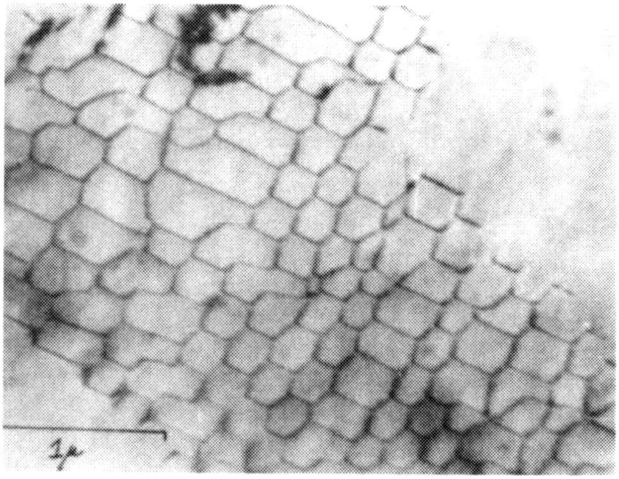

Before

After

Screw Dislocation

Screw Dislocation

Screw Dislocation

tilt boundary. It is symmetrical in the sense that the boundary plane acts as a mirror plane. Aligning edge dislocations one above the other in this way is energetically favorable because the tensile stress field of one dislocation relieves the compressive stress field of the other. The angular misorientation θ is given by

$$\theta = \frac{b}{h} \tag{9.4.11}$$

where b represents the Burgers vector and h the vertical spacing between each edge dislocation.

Figure 9.16b shows that a pure small-angle twist boundary may be constructed with a cross grid array of pure screw dislocations. The twist angle is again related directly to the Burgers vector and the spacing of the screw dislocations through eq. 9.4.11. Ideally, the energy of the boundary is at a minimum when the two sets of dislocations are equally spaced and the Burgers vectors of the screw dislocations are orthogonal, a requirement not easily realized in most crystal structures.

In general, a small-angle boundary consists of a mixture of edge and screw dislocations. The equilibrium configuration is an interconnected hexagonal "chicken wire" mesh that extends through the solid, as illustrated in Figure 9.16c. The sum of the respective tilt and twist components accounts for the small random misorientation across the boundary surface. Present in most single crystals, small-angle boundaries are technologically significant because they scatter electrons and modify the behavior of semiconductor devices, they act as the source or sink for point defects and impurities in ionic solids, and they impede the motion of dislocations in metals and alloys to raise the yield strength.

9.4.3.1 Small-Angle Boundaries Have a Surface Energy Whose Value Derives from the Strain Energy of the Dislocations That Form Them

Small-angle boundaries have an associated surface energy that increases with the misorientation. We can compute the total energy by summing the strain energies of the constituent dislocations, taking into account the interactions between them. The Read–Shockley relationship for the small-angle grain boundary energy per unit area, γ_{sgb}, as a function of the small tilt angle θ is

$$\gamma_{sgb} = U_o(A - \ln \theta)\theta \tag{9.4.12}$$

where $U_o = Gb/4\pi(1-v)$ and $A = 4\pi(1-v)U_{core}/Gb^2$. This equation gives the curve plotted in Figure 9.17. For small misorientations γ_{sgb} increases linearly, but above 15 degrees misorientation the value of γ_{sgb} approaches a constant value asymptotically. The asymptotic levels for boundary surface energy are about 500 mJ/m^2 or 3×10^{18} eV/m^2.

As the misorientation increases, the dislocation spacing decreases until at about 15 degrees, the boundary loses the identity of resolvable dislocation arrays and becomes a fully fledged

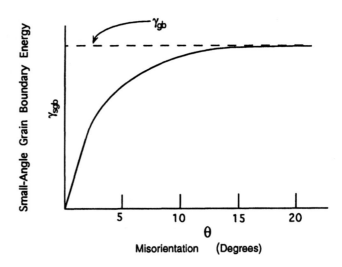

Figure 9.17
Plot of small-angle grain boundary energy as a function of misorientation.

grain boundary. Thus the surface energy of a grain boundary corresponds approximately to the asymptotic value in Figure 9.17. Grain boundaries separate two blocks of crystal with different orientations in space. They have great importance in the behavior, processing, and properties of solid–solid interfacial systems. Chapter 11 treats their formation and behavior.

9.4.3.2 Stacking Faults and Twins Are Other Planar Imperfections—They Have a Low Surface Energy

Stacking faults result when the sequence of planes that make up the crystal structure is disrupted, as illustrated in Figure 9.18b. For example, if the normal stacking sequence for {111} planes in the face-centered cubic crystal structure, $\cdots$ a · b · c · a · b · c $\cdots$ (Figure 9.2a) is faulted through a stacking-sequence change, such as the insertion of b in $\cdots$ a · |b · c · b| a · b · c $\cdots$, then one layer of hexagonal close-packed crystal structure, |b · c · b ·| (Figure 9.2b), appears as a stacking fault in the face-centered cubic crystalline material. Stacking faults commonly result from crystal growth or plastic deformation processes that occur during material processing. As noted in Section 9.4.1.1, they also can form from the condensation of excess vacancies into a planar sheet following a quench. Then one of the layers, such as a in the sequence $\cdots$ a · b · c · a · b · c $\cdots$; is missing to give $\cdots$ a · |b · c · b ·| c $\cdots$; and again one layer of hexagonal close-packed crystal structure appears at |b · c · b ·|.

Planar faults also have an associated energy, the stacking fault energy per unit area γ_{sf}. Stacking fault energies vary from values as high as 160 mJ/m^2 for stacking faults within aluminum to 16 mJ/m^2 within silver (see Table 11.2). These values are about an order of magnitude less than the energy per unit area for small-angle boundaries.

Twins result when the stacking sequence is permanently switched in the manner $\cdots$ a · b · c · | a · c · b · a · c $\cdots$ across the

Figure 9.18
Planar imperfections
in crystalline solids.
(a) Normal plane stacking
sequence in face-centered
cubic crystal structure,
· a · b · c. (b) Stacking fault
sequence · b · c · b ·.
(c) Twin boundary at a.

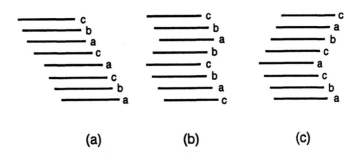

(a)   (b)   (c)

plane ǀa that mirrors the structure. Figure 9.18c depicts this structure. The key feature of a twin boundary is that the atoms (or ions) located at the boundary share in the periodic repetition of atoms (or ions) on both sides, even though the crystallographic orientation changes in a very specific way. Twin boundaries correspond to a special form of grain boundary in which many lattice sites of the grains coincide on either side of the boundary. The *coincidence site lattice* (CSL) across the twin boundary defines the fraction of coincident sites, and the inverse of this fraction is the parameter Σ. For twin boundaries Σ has a low whole number value. In a silicon crystal, for example, the simplest twin configuration is a {111} plane $\Sigma 3$ boundary.

The twin-boundary planar imperfection also has a surface energy associated with it, the twin-boundary surface energy. The twin-boundary energy is one-half the stacking fault energy (see Table 11.2). Chapter 11 discusses how low surface energy distinguishes a twin boundary from a grain boundary.

9.5 Kinetic Transport Mechanisms Are Key to Formation, Processing, and Stability of Interfaces in Solid–Solid Systems

In crystalline solids, the transport of material occurs by diffusion of atoms or ions through the crystal structure. We already have referred to diffusion at high temperatures a number of times in this chapter; now we look in more detail at the mechanisms involved in this phenomenon and at the kinetics, particularly as they pertain to the rate at which material transport and rearrangement can occur. These mechanisms are important in the processing of solid–solid interfaces and their stability during use.

9.5.1 Self-Diffusion and Solute Diffusion in Substitutional Solid Solutions by the Vacancy Mechanism

Our earlier description of a solid as a configuration of atoms or ions rooted at fixed points on a crystal lattice ignored the reality of their highly energized state. Each atom vibrates constantly with a frequency ν_0 of about 10^{13} Hz (oscillations per second). The symmetry and amplitude of the oscillation depends very sensitively on the bonding character of the solid and on the temperature. Close to the melting point the amplitude approaches 1 Å.

In Section 9.4.1 we saw that a solid contains both vacancy and interstitial point imperfections in thermodynamic equilibrium. Their concentration depends on crystal structure, bond character, and temperature. Given so much agitation and so many vacancies, it is not surprising that an atom occasionally jumps from its lattice location into an adjacent vacant site. When such a jump is completed, the atom and the vacancy exchange locations. The atom moves one lattice spacing λ from its starting position, and the vacancy moves one lattice spacing in the opposite direction. Figure 9.19 details such an exchange; we see that the extra free energy required to displace the atom goes through a maximum as the atom squeezes through the saddle point.

Assuming vacancy diffusion to be a thermally activated process, the atomic jump frequency ν is given by

$$\nu = \nu_0 \exp\left(-\frac{\Delta G_m}{kT}\right) \tag{9.5.1}$$

where ΔG_m is the maximum extra free energy required for the vacancy migration mechanism (the subscript m signifies move-

Figure 9.19
Diffusion by vacancy movement. Curve shows the free energy variation as the ion moves through the saddle point in the crystal structure.

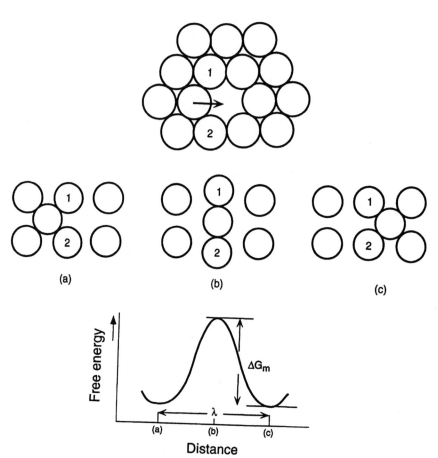

ment of the point imperfection). v_0 is the characteristic atom vibration frequency taken to be 10^{13} per second. Substituting appropriate figures for gold ($\Delta G_m \approx 1$ eV, $T_m = 1336$ K) gives a jump frequency of about 10^9 per second, close to the melting temperature; that means 1 in 10^4 oscillations results in an exchange of positions with the vacancy. The value drops rapidly to only one jump every 10^5 seconds at room temperature, stressing the exponential temperature sensitivity of this and related solid-state transport mechanisms.

Under normal conditions, the atom can equally well jump back to its starting location to give no net displacement. From the vacancy's perspective this corresponds to random movement with no net directional migration. But the presence of a potential gradient changes the situation. Figure 9.20 shows that the change of free energy for the atom moving down the gradient is slightly less than normal, while the change for moving up the gradient is slightly larger than normal. Under these circumstances, migration is no longer totally random; there is a net force that results in a directional flow of atoms down the gradient (and vacancies up the gradient).

The potential gradient may have chemical, electrical, mechanical, or thermal origins. For the moment we will consider the effect of a chemical potential gradient, $d\mu/dx$, on the migration of solute atoms in a substitutional solid solution. Solute atoms are substituted for solvent atoms on the lattice sites in a substitutional solid solution. The net decrease in free energy per solute atom for forward movement by one atomic spacing down the gradient is

$$\chi = \frac{1}{N_{Av}} \lambda \frac{d\mu}{dx} \qquad (9.5.2)$$

(We divide by Avogadro's number, N_{Av}, because the chemical potential is a molar quantity.) The extra free energy needed for the solute atom to pass through the saddle point down the gradient in Figure 9.20 is reduced to $\Delta G_m - \chi/2$, while up the gradient it is increased to $\Delta G_m + \chi/2$. Therefore, the net jump frequency is given by

Figure 9.20
Diffusion in a potential gradient. Curve shows the free energy variation as the ion moves through a saddle point down the potential gradient.

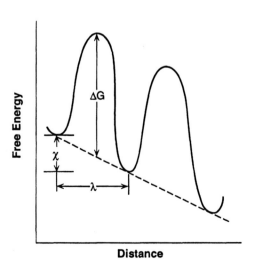

$$v_{net} = v_o \left\{ \exp\left[-\left(\frac{\Delta G_m - \chi/2}{kT} \right) \right] - \exp\left[-\left(\frac{\Delta G_m + \chi/2}{kT} \right) \right] \right\} \qquad (9.5.3)$$

which can be expanded to give

$$v_{net} = 2v_o \exp\left(-\frac{\Delta G_m}{kT} \right) \sinh\left(\frac{\chi}{2kT} \right) \qquad (9.5.4)$$

Since $\chi \ll kT$, $\sinh(\chi/2kT) \approx (\chi/2kT)$. Substituted in eq. 9.5.4 this gives

$$v_{net} = \frac{v_o \chi}{kT} \exp\left(-\frac{\Delta G_m}{kT} \right) \qquad (9.5.5)$$

The net velocity, v_{net}, of solute atoms associated with this jump frequency is

$$v_{net} = v_{net}\lambda$$

and so we obtain

$$v_{net} = \frac{v_o \lambda \chi}{kT} \exp\left(-\frac{\Delta G_m}{kT} \right) \qquad (9.5.6)$$

So far we have assumed that a vacancy always is situated conveniently right next to the solute atom to facilitate the movement. But the limited concentration of vacancies, $[n]_v$ (measured as a fraction), and the low concentration of solute molecules, c^* (measured in molecules per cubic meter), modifies the chance that this will be the case. So the net flux of solute atoms, J, down the composition gradient is given by their velocity multiplied by the concentration of vacancies multiplied by the concentration of solute per unit volume or

$$J = v_{net} [n]_v c^* = v_{net} \lambda [n]_v c^* \qquad (9.5.7)$$

J is measured in units of the number of atoms crossing a unit area per unit time. Substituting χ from eq. 9.5.2 into eq. 9.5.5 and thence v_{net} into eq. 9.5.7 gives

$$J = \frac{v_o \lambda^2 [n]_v c^*}{N_{Av} kT} \exp\left(-\frac{\Delta G_m}{kT} \right) \frac{d\mu}{dx} \qquad (9.5.8)$$

Assuming an ideal dilute solution, we have from eq. 3.7.10

$$\mu = \mu^o + N_{Av} kT \ln c$$

so that

$$\frac{d\mu}{dx} = \frac{N_{Av} kT}{c} \frac{dc}{dx} = \frac{N_{Av} kT}{c^*} \frac{dc^*}{dx}$$

When the latter is substituted into eq. 9.5.8 we obtain

$$J = v_o \lambda^2 [n]_v \exp\left(-\frac{\Delta G_m}{kT} \right) \frac{dc^*}{dx} \qquad (9.5.9)$$

Substituting $\Delta G_m = \Delta H_m + T \Delta S_m$, and remembering from eq. 9.4.3 that $[n]_v = A' \exp(-\Delta H_f/kT)$, we obtain

$$J = v_o \lambda^2 A \exp\left(-\frac{(\Delta H_f + \Delta H_m)}{kT}\right) \frac{dc^*}{dx} \qquad (9.5.10)$$

where A now contains configurational entropy terms associated with both the formation, A', and movement, A'', of the vacancy.

Comparing this result with Fick's first law, eq. 3.8.26, reveals that the diffusion coefficient for the substitutional solute atom, D_s, is given by

$$D_s = v_o \lambda^2 A \exp\left(-\frac{(\Delta H_f + \Delta H_m)}{kT}\right)$$

or

$$D_s = D_o \exp\left(-\frac{\Delta H_d}{kT}\right) \qquad (9.5.11)$$

where $D_o = v_o \lambda^2 A$ and $\Delta H_d = (\Delta H_f + \Delta H_m)$. ΔH_f and ΔH_m are the respective enthalpies for the formation and the movement of the vacancy point imperfection.

If we assume an ideal solid solution in which the size and chemical character of the solute and solvent atoms (ions) are identical, the value of D_s equals that for diffusion of the solvent atom in the solvent itself, or the *self-diffusion coefficient*, D_{self}. So that

$$D_{self} = v_o \lambda^2 A \exp\left(-\frac{(\Delta H_f + \Delta H_m)}{kT}\right) \qquad (9.5.12)$$

For nonideal solid solutions the value of D_s for the solute may be higher or lower than the coefficient for self-diffusion, depending on how the value of ΔH_m is changed by misfit strain, ε_m, and elastic modulus, E, $[d(\Delta H_m) \sim E\varepsilon_m^2/2]$, and chemical interactions.

Since a vacancy always has an adjacent atomic site that it is able to move into, the activation energy for vacancy migration is less than the activation energy for atom (ion) diffusion; the enthalpy is only ΔH_m. Therefore, in a given solid, the vacancy diffusion coefficient D_v is always much larger than the diffusion coefficient for self-diffusion by the vacancy mechanism, D_{self}. In fact

$$D_{self} = [n]_v D_v$$

and so

$$D_v = v_o \lambda^2 A'' \exp\left(-\frac{\Delta H_m}{kT}\right) \qquad (9.5.13)$$

Diffusion in ionic solids and ceramics occurs through the movement of cation and anion vacancies. Solute cations substituted on cation sites generally move into cation vacancies as they migrate. Then the value for ΔH_f in eqs. 9.5.11 and 9.5.12 is equal to $\Delta H_{fs}/2$, where ΔH_{fs} is the energy required to form the Schottky defect. Values for ΔH_m in eqs. 9.5.11 and 9.5.12 are different for the anion and cation species. Because the anion is generally larger than the cation, it requires greater energy to squeeze through the saddle point. Thus different diffusion coefficients exist for cation and anion diffusion. Total transport in an ionic solid is controlled by the *slowest* diffusing species.

Table 9.3 gives typical values for D_o, and ΔH_d and the corresponding coefficients for self-diffusion at different temperatures. Units for diffusion coefficients are typically expressed in m^2/s. Just below the melting point a metal has a self-diffusion coefficient of about $10^{-13}\ m^2/s$, while just above the melting point it increases about four orders of magnitude to $10^{-9}\ m^2/s$, reflecting the increase in free volume of the fluid state. At room temperature self-diffusion coefficients vary between 10^{-18} and $10^{-24}\ m^2/s$, although absolute values are extremely sensitive to the class of solid and the relative melting points. Diffusion is most difficult in covalently bonded solids because of the need to break and remake covalent bonds as the atom moves to an adjacent site.

9.5.2 Self-Diffusion and Solute Diffusion in Interstitial Solid Solutions by the Interstitial Mechanism

An atom may move about within a solid in other ways. If a solute atom is much smaller than the solvent atom (<60%), then the solute atom prefers to locate at interstitial sites to minimize mechanical distortion. Other influences such as solute–solvent interaction, bonding directionality, and ion polarizability also can dictate an interstitial site preference. Solute migration then occurs rather easily by jumping from one interstitial site to another, as depicted in Figure 9.21a. An empty interstitial site can always be found nearby; so in this instance, the enthalpy for solute migra-

TABLE 9.3a Self-Diffusion Data for Pure Cubic Metals[a]

Metal	Crystal structure	ΔH (kJ/mole)	D_o (cm^2/s)	$\Delta H/T_m$ (J/K)
Cu	fcc	197.1	0.20	145.6
Ag	fcc	184.5	0.40	149.8
Ni	fcc	279.5	1.30	161.5
Au	fcc	174.5	0.091	130.5
Pb	fcc	101.3	0.28	169.0
α-Fe	bcc	239.3	1.9	155.6
Nb	bcc	439.3	12.0	160.7
Na	bcc	43.9	0.24	118.4
Ge	diamond cubic	286.6	7.8	232.6

Source: P. G. Shewmon, *Transformations in Metals,* McGraw-Hill, New York, 1969, p. 54.

[a]In metals with noncubic structures, D is not the same in all directions. Thus in hcp metals one needs D in the directions parallel and perpendicular to the c axis to uniquely determine D in all directions. We avoid these considerations by discussing only cubic materials.

TABLE 9.3b Values of the Schottky
Formation Enthalpy ΔH_{fS} and the Cation
Jump Enthalpy ΔH_m of Several Halides

Substance	ΔH_{fS} (eV)[a]	ΔH_m (eV)
LiF	2.34	0.70
LiCl	2.12	0.40
LiBr	1.8	0.39
LiI	1.34, 1.06	0.38, 0.43
NaCl	2.30	0.68
NaBr	1.68	0.80
KCl	2.6	0.71
KBr	2.37	0.67
KI	1.6	0.72
CsCl	1.86	0.60
CsBr	2.0	0.58
CsI	1.9	0.58
TlCl	1.3	0.5
$PbCl_2$	1.56	
$PbBr_2$	1.4	

[a] 1 eV 96.5 kJ per mole.

tion dominates the activation energy, and ΔH_d in eq. 9.5.10 is approximately equal to ΔH_m.

In single-element solids like metals in which all the atoms are the same size, the interstitial mechanism is a more formidable mode of self-diffusion than the vacancy mechanism due to the large strain energy required to form the interstitial. Once formed, an interstitial can move by the *interstitialcy* mechanism illustrated in Figure 9.21b. In this two-step mechanism, an interstitial atom moves into an adjacent lattice site at the same instant the atom located there moves into another interstitial site. If the atom jumping into the lattice site is tagged or if it is a solute atom, the first interstitialcy step produces only half the translation needed to transfer from one interstitial site to the next; the step must be repeated to complete the transfer.

Interstitial and interstitialcy diffusion mechanisms are more common in ionic solids because the two elements are of different size and the crystal structure is less compact. The value for ΔH_f in eq. 9.5.9 now must be replaced by $\Delta H_{fF}/2$, where ΔH_{fF} is the energy to form the Frenkel defect. The value for ΔH_m differs a great deal between cations and anions because they experience different distortions in moving from one interstitial site to another. Thus the total activation energy for diffusion by the interstitial mechanism depends on the diffusing species.

In some ionic solids the preferred diffusion mechanism, whether vacancy or interstitial, is different for cations and anions. In silver chloride, for example, the silver ion moves interstitially (via the interstitialcy mechanism) while the larger chlorine ion moves via the vacancy mechanism. Interstitial migration of silver

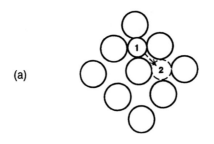

(a)

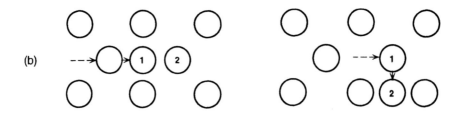

(b)

Figure 9.21
Diffusion by interstitial movement. (a) An impurity atom much smaller than the host diffuses through the solid by jumping from an interstitial site (1) into an adjacent interstitial site (2). (b) Diffusion by the interstitialcy mechanism. An atom moving in from the left displaces atom (1) into an interstitial site. Atom (1) then displaces atom (2) resulting in net diffusion through the crystal structure.

is key to the formation of the silver latent image in the silver halide photographic process (see Section 8.4.2.2) and the operation of fast-ion silver iodide battery electrolytes. In other ceramic materials like zirconium dioxide, interstitial migration of oxygen anions is the preferred mechanism. Rapid oxygen diffusion through zirconia is the key property in the application of this material as a high-temperature electrolyte in fuel cells and gas sensors for automobile engine exhaust monitoring.

This discussion emphasizes the fact that the diffusion coefficient for mass transport in a solid varies significantly with the mechanism involved. We may be concerned with the transport of a solute atom or the self-diffusion of a solvent atom in a metal; we may be concerned with the diffusion—by either the vacancy or the interstitial mechanism—of one or both of the ions in an ionic solid. Descriptions of crystalline solid transport processes, therefore, must state explicitly the carrier and the diffusion mechanism involved.

9.5.3 Diffusion and Absolute Mobility Are Related by the Nernst–Einstein Equation

We also can define material transport by diffusion in terms of a more fundamental property—the absolute mobility of an atom. The product of the absolute mobility B and the force F driving an atom through the lattice gives the velocity v of the atom. That is

$$v = FB \qquad (9.5.14)$$

For diffusion of a solute atom the driving force F_{diff} derives from the gradient in chemical potential. From the analysis already developed in Section 3.8.5.3, the driving force for diffusion is

$$F_{\text{diff}} = -\frac{kT}{c}\frac{dc}{dx} \qquad (9.5.15)$$

Since the resultant flux of material J is given by

$$J = v_{\text{diff}}c \qquad (9.5.16)$$

and

$$J = -D\frac{dc}{dx} \qquad (9.5.17)$$

from Fick's first law, then through substitution for F_{diff} and v_{diff} using eqs. 9.5.15, 9.5.16, and 9.5.17 in eq. 9.5.14, we obtain

$$B = \frac{D}{kT} \qquad (9.5.18)$$

This is the *Nernst–Einstein equation* relating mobility and diffusivity.

The result should be compared with the expression relating the friction factor f and the diffusion coefficient for a particle passing through a fluid given in eq. 3.8.35. We can see that 1/f is equivalent to absolute mobility B. Diffusion can be regarded as the "flow" of the solute atom through the solvent crystal lattice under the influence of the chemical potential gradient.

Equations 9.5.14–9.5.16 provide an example of a linear relation between a potential gradient and the flux J that it generates. This relationship applies in general to all transport phenomena and $J = K \, dX/dz$. K stands for the transport coefficient and dX/dz is the potential gradient. As noted earlier in Section 9.5.1, the potential gradient can be chemical—giving rise to diffusion phenomena discussed in Sections 9.5.1 and 9.5.2; electrical—giving rise to ionic conductivity or electromigration phenomena to be discussed in Section 9.5.7.1; thermal—giving rise to thermal migration phenomena; or mechanical—giving rise to high-temperature creep phenomena to be discussed in Section 9.5.7.2.

9.5.4 Fick's Second Law Describes Changes in Composition with Time

In many applications of interfacial systems we are concerned with the breakdown of a sharp interface due to the interdiffusion of the two solids originally placed on either side of the interface. As already noted in Section 3.8.5.2, Fick's second law states that the change in solute concentration in a given volume with time is equal to the solute flux gradient. This condition is expressed mathematically in the differential equation (eq. 3.8.27)

$$\frac{dc}{dt} = -D\frac{d^2c}{dx^2} \qquad (9.5.19)$$

Various solutions to this equation depend on different boundary conditions.

We are interested in the penetration of a surface coating into the interior of a semi-infinite solid, as shown in Figure 9.22. We assume that the coating material is constantly replenished so that

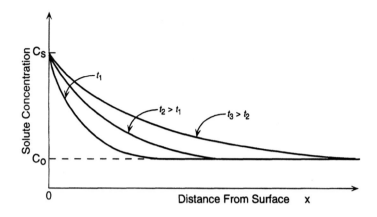

Figure 9.22
Concentration profiles that develop through diffusion when a solute concentration of c_s is maintained at the surface of a specimen of initial composition c_o. Successive curves show effect of increasing time.

its composition remains constant at c_s. The composition of the solid into which the surface layer is diffusing is initially c_o but changes with time t and depth x. So for our boundary conditions we have $c = c_s$ at $x = 0$ for all times $0 < t < \infty$ and $c = c_o$ at $t = 0$ for all depths $0 < x < \infty$. The solution to eq. 9.5.19 under these conditions gives

$$c_x = c_s - (c_s - c_o)\, \mathrm{erf}\left(\frac{x}{2\sqrt{D_s t}}\right) \qquad (9.5.20)$$

for the composition profile as a function of depth and time. erf is the Gaussian error function. D_s is the diffusion coefficient for diffusion of the surface coating element s in the bulk material, a value that depends on the temperature and varies from element to element according to the chemistry, the strain, and the controlling diffusion mechanism. This result presumes that D_s *does not vary* with composition.

To determine the penetration beneath the surface with time, we can track the depth at which the composition is midway between the surface and bulk composition, that is, the depth $x_{0.5}$ where $c_x = (c_s + c_o)/2$. Substitution in eq. 9.5.19 gives $\mathrm{erf}(x_{0.5}/2\sqrt{D_s t}) = \frac{1}{2}$. From the error function table the solution for $x_{0.5}$ is approximately

$$\frac{x_{0.5}}{2\sqrt{D_s t}} = \frac{1}{2} \quad \text{or} \quad x_{0.5} = \sqrt{D_s t} \qquad (9.5.21)$$

Thus the penetration of the surface coating follows a parabolic relationship with time, and for the midcomposition the parabolic rate constant is equal to D_s. Note that this result approximates the average displacement of a particle enjoying a random walk that was given in eq. 3.8.25.

An important guide to remember is that penetration is given by $\sqrt{D_s t}$. With diffusion coefficients 10^{-8}, 10^{-14}, and 10^{-20} m^2/s, the penetrations in 1 sec are approximately 10^{-2} cm, 10^{-1} μm, and 1 Å, respectively. These dimensions highlight two important points in our consideration of interface stability. First, the stability of an interface is extremely sensitive to temperature, as predicated by the high values for ΔH_d in the diffusion coefficient exponential.

Second, whether we consider an interface to be stable or unstable depends on the scale of our concern. On a macroscopic scale, 10^{-2} cm may not represent significant smearing of an interface since it is barely visible to the eye, but on the microscopic scale, movements of even a few angstroms are discernible with the electron microscope. The scale of concern in turn depends on the nature of the interface application.

Note that the time for diffusion to a fixed depth x is proportional to x^2/D_s. With diffusion coefficients of 10^{-8}, 10^{-14}, and 10^{-20} m²/s, movement of 1 μm takes about 10^{-4} s, 100 s, and 10^8 s (3 years), respectively.

We are also interested in the interdiffusion of two solids of compositions c_1 and c_2 joined at an interface as shown in Figure 9.23. The boundary conditions are changed because one solid is the source of solute, and the other is the sink. As already presented in Section 3.8.5.2, under these conditions the concentration profile follows the relationship

$$c_x = \frac{c_1 + c_2}{2} - \frac{c_1 - c_2}{2}\, \text{erf}\left(\frac{x}{2\sqrt{D_s t}}\right) \qquad (9.5.22)$$

reproduced in Figure 9.23. At x = 0 the composition remains constant at $(c_1 + c_2)/2$; so the interface stays in its original location. Compositions above or below this value spread away from the interface. The extent of the interdiffusion is also proportional to $\sqrt{D_s t}$. This result also presumes that D_s *does not vary* with composition.

9.5.5 Large Differences in Diffusivity Lead to Movement of Interfacial Boundaries with Time and the Formation of Interfacial Defects

Next we consider what happens when D does depend on composition. Analysis shows D for a solution or alloy to be

$$D = D_{s2}\, X_1 + D_{s1}\, X_2 \qquad (9.5.23)$$

where D_{s2} is the diffusion coefficient of solute 1 in pure solvent 2 and D_{s1} represents the diffusion coefficient of solute 2 in pure

Figure 9.23
Concentration profiles that develop through diffusion when two solids with compositions c_1 and c_2 are bonded at an interface. Successive curves show effect of increasing time. Results assume D is independent of composition.

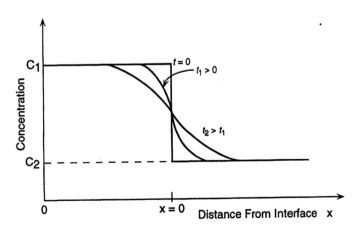

solvent 1. X_1 and X_2 are the atom fractions (and mole fractions) of 1 and 2, respectively, so that $X_2 = 1 - X_1$. Note that when X_1 approaches unity and X_2 approaches zero, D approaches D_2. This somewhat surprising result indicates that in a dilute alloy the diffusion coefficient of the solute controls the interdiffusion behavior of the dilute species.

When the values of D_{s2} and D_{s1} are very different, further complications arise. A net mass transfer occurs across the original interface, and the interface appears to move. The rate of movement is proportional to the difference in the two diffusion coefficients. This movement is the *Kirkendall effect*. Originally, it was found for the interface between brass and copper. The zinc in the brass diffuses by the interstitial mechanism and at a much faster rate than the copper diffuses by the substitutional vacancy mechanism. The number of zinc atoms moving out of the brass far exceed the number of copper atoms moving in to replace them; as a result the brass–copper interface moves into the copper. A net excess of vacancies left behind in the brass must be absorbed within the material somehow. At low concentrations the vacancies can absorb on dislocation lines or agglomerate to form closed internal dislocation loops, as shown in Figure 9.24, but at high concentrations they condense as macroscopic spherical voids in spite of the extra surface energy this process requires.

The formation of voids at interfaces due to differential diffusion presents a possible mechanism for failure of solid–solid interfaces facing prolonged exposure to high temperatures. This possibilty exists in systems in which diffusion involves both interstitial and vacancy mechanisms, such as the interfaces between the noble metals, gold, and platinum, and silicon in semiconductor devices.

9.5.6 Diffusion over Surfaces Is Faster Than through the Bulk of the Material

So far we have been concerned with processes that take place through the bulk or volume of the material. There are other, easier paths for diffusion. For example, the lattice is distorted in the vicinity of a dislocation line, or a small-angle boundary, and the energy required for migration by an interstitial atom or a vacancy along the dislocation line is correspondingly lower. Diffusion along dislocation lines is referred to as *pipe diffusion*. Diffusion at surfaces such as a grain boundary (see Chapter 11) or a free surface will be even easier because vacant lattice sites need not be generated to facilitate movement. Migration from one adatom site to another in Figure 9.10 clearly proceeds much faster over the free surface than through the bulk. Diffusion along grain boundaries is referred to as *grain boundary diffusion* and over free surfaces as *surface diffusion*.

In the general expression for the diffusion coefficient, eq. 9.5.10, both the enthalpies ΔH and the D_o values are changed for the alternative paths. We can summarize the results as follows

$$\Delta H_{lattice} > \Delta H_{pipe} > \Delta H_{gb} > \Delta H_{surf}$$

$$D_{o,lattice} \geq D_{o,pipe} \geq D_{o,gb} \geq D_{o,surf}$$

Figure 9.24
Diffusion in an infinite diffusion couple when D is strongly dependent on the composition of the two solids. Relative diffusion leads to movement of the couple interface to the right in (b) and appearance of vacancies that condense to form vacancy loops or voids in (c). (J. H. Brophy, R. M. Rose, and J. Wulff, *Thermodynamics of Structure*, Vol. II of *The Structure and Properties of Materials,* Copyright © 1964, John Wiley & Sons, p. 87. Reprinted by permission of John Wiley & Sons, Inc.)

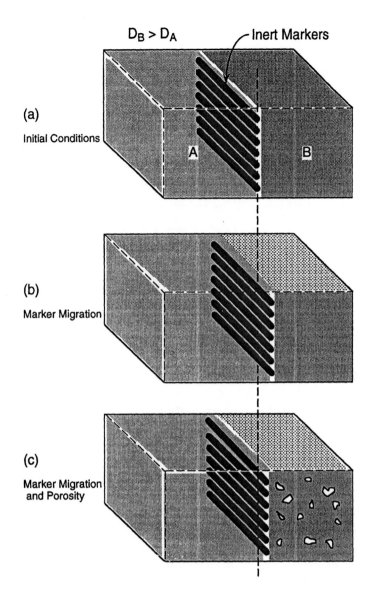

with the net result that

$$D_{lattice} \ll D_{pipe} \ll D_{gb} \ll D_{surf} \tag{9.5.24}$$

Typically, we can assume that

$$\Delta H_{lattice} : \Delta H_{gb} : \Delta H_{surf} \approx 4:2:1 \tag{9.5.25}$$

while the D_os range between 10^{-5} and 10^{-4} m^2/s.

The activation energy for surface diffusion equals only one-quarter of the energy required for diffusion through the crystal lattice. The surface diffusion coefficient, D_{surf}, is approximately five orders of magnitude larger than the lattice diffusion coefficient, $D_{lattice}$, at half the melting temperature. These differences account for the relatively high mobilities of surface adatoms skittering over the roughened surfaces shown in Figure 9.10.

The activation energy for grain boundary diffusion equals only one-half of the energy required for diffusion in the bulk. The grain boundary diffusion coefficient, D_{gb}, can be ten orders of magnitude larger than the lattice diffusion coefficient, $D_{lattice}$, at moderate temperatures below one-third the melting point. These differences lead to penetration five orders of magnitude deeper down a grain boundary than into the surface of a solid.

9.5.7 Material Transport in Solids Also Can Be Driven by Electrical Potential Gradients and Mechanical Strain Gradients

9.5.7.1 Ionic Conductivity and Electromigration Occurs in an Electrical Field

Other potential gradients move atoms through a solid. If the gradient in Figure 9.20 is due to an electrical field E then the force F acting on an ion of charge ze equals Eze Inserting this force in eq. 9.5.14 and using the Nernst–Einstein relationship eq. 9.5.18, we see that the velocity of an individual ion is

$$v_{ion} = \frac{EzeD}{kT}$$
(9.5.26)

Since the ionic conductivity σ is the charge flowing across a unit cross-sectional area per second per unit electrical field, then

$$\sigma = \frac{(ze)[n]v_{ion}}{E}$$
(9.5.27)

where ze is the charge transported and [n] is the concentration of charge carriers. Inserting eq. 9.5.26 into eq. 9.5.27 gives

$$\sigma = \frac{(ze)^2[n]D}{kT}$$
(9.5.28)

When the charge carriers are vacancies, we use values for $[n]_v$ from eq. 9.4.3 and D_v from eq. 9.5.13 to obtain

$$\sigma = \frac{(ze)^2 D_o}{kT} \exp\left(-\frac{(\Delta H_f + \Delta H_m)}{kT}\right)$$
(9.5.29)

for the *ionic conductivity*. Transport of ions in an electrical field has many practical implications. First, ionic conductivity is critical to the operation of solid-state electrolytes for applications in high-temperature batteries, gas sensors, and fuel cells. Second, as we have noted already in Section 9.4.1.3, internal electrical fields develop within ionic solids due to inhomogeneities in vacancy concentration. These internal fields can drive charged impurities to the surface and thereby enhance impurity segregation in these solids. Third, electromigration can be detrimental in the operation of high-power devices and induce failures in microelectronic circuits.

Even in good conductors, such as metals and semiconductors, the passage of a large electronic current causes a net migration of atoms in the direction of electron flow, known as *electromigration*. This effect has been likened to striking a Volkswagen with high-velocity billiard balls to move it slowly in the direction of the balls. Equation 9.5.25 can be used to give the velocity of atoms during electromigration with z replaced by z^*, the "effective charge number" for electromigration (where the magnitude of z^* actually depends on the momentum exchange between the electrons and atoms and experimentally determined values range from -1 to -100), and E, the electrical field, replaced by ρJ, where ρ is the resistivity of the material and J is the current density flowing through the conductor. So the drift velocity for electromigration is

$$v_{ion} = \frac{\rho J z^* e}{kT} D_o \exp\left(-\frac{U_e}{kT}\right) \qquad (9.5.30)$$

where U_e is the activation energy for electromigration. Activation energies for electromigration-enhanced diffusion have been found of the order 0.6 eV for aluminum. These values lead to drift velocities of about 3×10^{-13} ms^{-1} at 300 K for current densities (10^5 A/cm^2) experienced in microcircuit interconnects. Velocities in this range could easily destroy a submicrometer device in 1000 h of continuous operation. Special alloys and interconnect patterns are designed to circumvent this failure mechanism.

9.5.7.2 Creep Occurs in Solids under Mechanical Stress at High Temperatures

When a mechanical stress is applied to a crystalline solid at very high temperatures, the solid deforms continuously at a constant strain rate. This mode of plastic deformation is called *creep*. Often the strain rate increases linearly with the applied stress; the flow is then viscous and the linear relationship is consistent with Newtonian flow—*viscous creep* (see Section 3.8.1). Figure 9.25 shows how diffusion of atoms and counterdiffusion of vacancies along a mechanical stress gradient causes a permanent change in shape by transferring atoms from crystal faces under compression to faces under tension. The faces involved are either the free surfaces of a single crystal, Figure 9.25a, or the grain boundary surfaces in polycrystalline material, Figure 9.25b.

In this transport mechanism, the applied stress affects the energy to *form* a vacancy rather than the energy to move it (as was the case for transport driven by chemical and electrical potential gradients). By examining the migration of the vacancy in Figure 9.19, we can see that the movement of a vacancy per se does not lead to a change of shape. Changing from configuration (a) to (c) in Figure 9.19 has not altered the relative positions of the upper and lower rows of atoms—no shear, and therefore no shear strain, has taken place. A change in shape requires the formation of an excess of vacancies at the surface in tension, which must diffuse to the surface in compression and then dissipate there. We want to compute the strain rate due to this vacancy flux.

At the surface in tension, the force acting on each atom due to the applied stress σ is $\sigma\lambda^2$, where λ represents the atom spacing (about 0.3 nm). The work done by the atom when it jumps by one lattice spacing from the bulk onto the surface, leaving a vacancy

Figure 9.25
Migration of *atoms* by
diffusion at high temp-
eratures due to a stress
gradient results in creep
strain. (a) Creep in a single
crystal. Atom movement
by diffusion from the
surfaces in compression to
the surfaces in tension, as
shown by the arrows,
results in net elongation.
Counterdiffusion of *va-
cancies* occurs in the
opposite direction.
(b) Creep in a polycrystalline solid. Atom move-
ment by diffusion from
grain boundaries in
compression to boundaries
in tension results in
elongation. Figure depicts
bulk diffusion, but grain
boundary diffusion also
contributes to the creep
deformation process.

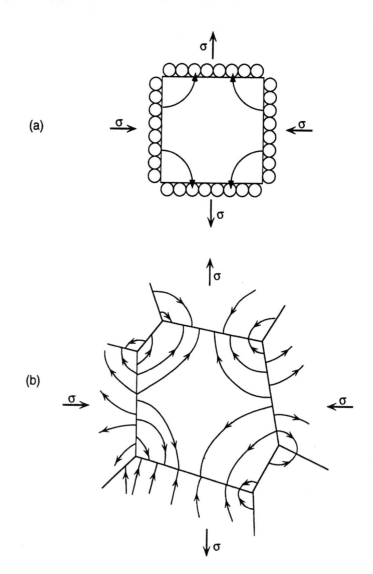

(a)

(b)

behind, is force times distance or $\sigma\lambda^3$. So the energy required to
form the vacancy at the surface in tension is reduced from ΔH_f to
$(\Delta H_f - \sigma\lambda^3)$. At the surface in compression, this energy is increased
to $(\Delta H_f + \sigma\lambda^3)$. The *fractional* concentration of vacancies in the
tension region, $[n]_{vt}$, increases from the thermodynamic equilibrium value $[n]_v = A' \exp(-\Delta H_f/kT)$, given in eq. 9.4.3, to

$$[n]_{vt} = A' \exp\left(-\frac{(\Delta H_f - \sigma\lambda^3)}{kT}\right)$$

$$\approx [n]_v\left(1 + \frac{\sigma\lambda^3}{kT}\right) \quad \text{when } \sigma\lambda^3 \ll kT \qquad (9.5.31)$$

and the *fractional* concentration of vacancies in the compression
region, $[n]_{vc}$, decreases to

$$[n]_{vc} \approx [n]_v\left(1 - \frac{\sigma\lambda^3}{kT}\right) \qquad (9.5.32)$$

The *number* of vacancies per unit volume in equilibrium is $[n]_{vt}/\lambda^3$ and $[n]_{vc}/\lambda^3$ at the respective surfaces. If the surfaces are separated by a distance d, the vacancy concentration gradient $d[n]_v/dx$ is given by

$$\frac{d[n]_v}{dx} = \frac{[n]_{vt} - [n]_{vc}}{\lambda^3 d} \tag{9.5.33}$$

Substituting eqs. 9.5.31 and 9.5.32 into eq. 9.5.33 gives

$$\frac{d[n]_v}{dx} = \frac{[n]_v}{\lambda^3 d}\left[\left(1 + \frac{\sigma\lambda^3}{kT}\right) - \left(1 - \frac{\sigma\lambda^3}{kT}\right)\right]$$

$$= \frac{2[n]_v\sigma}{dkT} \tag{9.5.34}$$

The vacancy flux J_v is the net number of vacancies leaving a unit area of the tensile surface and arriving at the compression surface in unit time. From Fick's first law $J_v = D_v\, d[n]_v/dx$; so eq. 9.5.34 becomes

$$J_v = \frac{2D_v[n]_v\sigma}{dkT} \tag{9.5.35}$$

and substituting eq. 9.5.13 in eq. 9.5.35 gives

$$J_v = \frac{2D_{self}\sigma}{dkT} \tag{9.5.36}$$

Since each vacancy has a volume λ^3, the product $J_v\lambda^3$ represents the net vacancy volume *leaving* a unit area of tensile surface in unit time. This is equal to the net displacement of the tensile surface due to the *arrival* of atoms in unit time, $d\delta/dt$, so that

$$J_v\lambda^3 = \frac{d\delta}{dt} \tag{9.5.37}$$

Since the length over which the displacement occurs is d, the strain rate is

$$\frac{d\varepsilon}{dt} = \frac{1}{d}\frac{d\delta}{dt} \tag{9.5.38}$$

and substituting eqs. 9.5.37 and 9.5.36 gives

$$\dot{\varepsilon} = \frac{2D_{self}\sigma\lambda^3}{d^2 kT} \tag{9.5.39}$$

This is the *Nabarro–Herring relationship* for vacancy-controlled creep (see Appendix 11B.3). The linear relationship between strain rate and stress defines Newtonian viscous behavior (see Figure 3.19). The coefficient of viscosity η (obtained by inserting eq. 9.5.39 in eq. 3.8.1) is equal to $d^2 kT/2D_{self}\lambda^3$.

In the above equations, d becomes identified with the bulk size in the case of a single crystal specimen and with the grain size in the case of a polycrystalline specimen.

Vacancy diffusion creep only occurs at high temperatures, typically above one-half the melting temperature, $T_m/2$, where the concentration and mobility of vacancies is sufficiently high to produce measurable strain. If d = 0.1 cm, T = 1250 K, kT = 1.73 × 10^{-13} ergs, $D_{self} \approx 10^{-16}$ m²/s, and $\lambda^3 \approx 10^{-29}$ m³, then

$$\eta = \frac{(0.01 cm^2)(1.73 \times 10^{-13} ergs)}{2(10^{-16} m^2/s)(10^{-29} m^3)} = 10^{19} P$$

a value too high to be regarded as fluidlike behavior using the criterion established in Section 3.8.1.

Grain boundary surfaces can function almost as efficiently as a free surface in providing a source and sink for vacancies. Furthermore, the flux of vacancies from the tensile to the compression region can occur by the faster grain boundary route rather than through the bulk. Under the right combination of conditions—grain size, temperature, and applied stress—polycrystalline solids can be deformed in a viscous manner by hundreds of percent strain without failure. For fine-grained ceramics with a grain size d = 10μm, loaded at 1500 K, and for conditions under which grain boundary diffusion controls the creep process $(D_{gb} \approx 10^{-14}$ m²/s), the viscosity is equal to 10^{14} P, a value comparable to pitch. Both metals and ceramics are said to demonstrate *superplasticity* under these conditions.

Bibliography

A. H. Cottrell, *The Mechanical Properties of Matter*, New York: John Wiley and Sons, Inc., 1964.

J. P. Hirth and J. Lothe, *Theory of Dislocations, Second Edition*, New York: John Wiley and Sons, Inc., 1982.

W. D. Kingery, H. K. Bowen, and D. R. Uhlmann, *Introduction To Ceramics*, New York: John Wiley and Sons, Inc., 1976.

P. G. Shewmon, *Diffusion in Solids*, New York: McGraw-Hill Book Company, 1963.

G. A. Somorjai, *Chemistry in Two Dimensions: Surfaces*, Ithaca, N.Y.: Cornell University Press, 1981.

K.-N. Tu, J. W. Mayer, and L. C. Feldman, *Electronic Thin Film Science*, New York: Macmillan, 1992.

Exercises

9.1 Aluminum is a face-centered cubic metal with a lattice parameter of 4.05 Å. (a) What is the atom size? (b) What is the spacing of the {111} planes?

9.2 Derive the ion radius ratio limits for the three coordination configurations in Figure 9.3.

9.3 The lattice parameter for a single crystal of cesium chloride is 4.11 Å. Make an estimate of the lattice energy in kilojoules per mole.

9.4 For a face-centered cubic metal crystal determine the number of bonds per atom that are broken when the exposed surface is (a) the {111} crystal plane; (b) the {110} crystal plane; (c) the {100} crystal plane. If surface energy is determined by the density of broken bonds, what are the relative values of γ_{sv} {111}, γ_{sv} {110}, γ_{sv} {100} for an fcc metal?

9.5 For AB compounds of the same crystal structure and lattice parameter, the surface energy for covalently bonded ions is equal to _ ; less than _ ; or greater than _ ; the surface energy for ionically bonded ions.

9.6 In terms of the sublimation energy, estimate γ_{sv} for the {110} surface of a sodium chloride crystal. How does this compare with γ_{sv} {100} and is it consistent with Figure 9.7(b)?

9.7 For silver the heat of sublimation is 345 kJ/mole and the surface energy is 1200 mJ/m^2. Show that these results are consistent with the broken bond approximation for estimating surface energy. (The area per silver atom on the close-packed {111} plane is 7.18×10^{-20}m^2).

9.8 For sodium chloride single crystals the surface energy γ_{sv} of the {100} plane is lower than any other plane and about one third the surface energy of a {110} plane.
(i) Draw the Wulff plot of surface energy for different directions in the {100} plane of the NaCl crystal structure that reflcts this anisotropy.
(ii) What is the shape of the polyhedron with minimum surface energy?
   Assume the surface energy γ_{sv} of the {110} plane is *equal* to the surface energy of the {100} plane (and lower than any other plane).
(iii) Redraw the Wulff plot of surface energy for different directions in the {100} plane of the NaCl crystal structure.
(iv) What now is the shape of the equilibrium polyhedron with minimum surface energy?

9.9 A very fine single crystal copper wire (0.014 cm diameter) is loaded in tension at a temperature just below its melting temperature. Under a load of 0.076 g the wire maintains constant length; for a smaller load it shortens, and for a larger load it lengthens. What is the surface tension of copper?

9.10 Surfaces are often heated in vacuum prior to vapor deposition to clean the surface. At a given temperature the time to clean a perfectly oriented {110} single crystal surface is equal to _ ; less than _ ; or greater than _ ; a surface deliberately misoriented by tilting it a few degrees away from the [110] pole.

9.11 Given that $\Delta H_f = 1.01$ eV and $A' = 1$, calculate the concentration of point defects in thermodynamic equilibrium in a metal crystal at 900, 700, and 25°C.

9.12 The lattice parameter of a sodium chloride single crystal is 5.63 Å. The Burgers vector for dislocations in NaCl is $(a/2)\langle 110 \rangle$. Calculate the dislocation spacing for a symmetrical small-angle pure tilt boundary of 2°. What are the Miller indices of the plane of the boundary?

9.13 Write expressions for the self-diffusion coefficients of cations and anions, respectively, in ionic solids when the predominant defects are (a) Schottky defects; (b) Frenkel defects.

9.14 Use the data of Table 9.3b to calculate the activation energy for cation diffusion in NaCl, LiF, and CsBr.

9.15 If a diffusion anneal time at a given temperature is doubled, the depth of penetration is increased by a factor of ___.

9.16 From radioactive tracer studies on silver we know that the lattice diffusivity is 10^{-21}, 10^{-15}, and 10^{-10} cm^2 s^{-1} and that the grain boundary diffusivity is 10^{-10}, 10^{-8}, and 10^{-6} cm^2 s^{-1} at 200, 400, and 800°C, respectively. Estimate the approximate ratios of grain boundary to lattice penetration at these three temperatures.

9.17 In the fabrication of semiconductor devices a doping element is often introduced by diffusion into surface layers. The activation enthalpy ΔH_d for diffusion of gold in silicon is 1.12 eV, and the diffusion constant D_o is 1.1×10^{-3} cm^2/s. Assuming you have a continuous source of gold available at the surface of a silicon wafer, recommend an annealing time and temperature to control the depth of diffusion to 1 μm. Copper is present as an impurity in the gold; it has $\Delta H_d = 1.0$ eV and $D_o = 4 \times 10^{-2}$ cm^2/s. How deep will copper atoms diffuse during your process? How deep will copper penetrate if the silicon is polycrystalline?

9.18 We set up an interface between two elements A and B and heat them to a temperature at which the diffusion of A into B is a factor of ten times faster than B into A. Relative to an external reference, the boundary between them moves toward A _ or B _ ; and the interface is likely to fail due to voids in A _ or B _ .

APPENDIX 9A
MILLER INDEXING
SYSTEM FOR DENOTING
DIRECTIONS AND PLANES
IN SINGLE CRYSTALS

Directions and planes in single crystals are described relative to the edges of the unit cell. The direction of a line in a crystal lattice is defined by drawing a line through the origin parallel to that line and assigning coordinates to any point on the line. The coordinates are measured in multiples or fractions of the unit cell. Although the coordinates have different values for different points, they always are related by the same ratio. The convention is to represent the direction by the *set of smallest integers*, as shown in Figure 9A.1a. Thus, while a point may have coordinates $2u, 2v, 2w$, the *direction* is specified by the *Miller indices* [uvw]. Square brackets denote a specific direction. If one or more of the coordinates is negative, then the indices are written with a bar, as in [$\bar{u}\bar{v}\bar{w}$]. A set of directions related by symmetry is denoted by ⟨uvw⟩. For example, the cube face diagonals having directions such as [011], [101] and [1$\bar{1}$0] are represented collectively by ⟨110⟩; the specific direction of one cube diagonal may be [11$\bar{1}$], the set of all cube diagonals is denoted by ⟨111⟩. It is interesting to note that opposite directions are obtained by reversing the bars, so that [12$\bar{1}$] is in the opposite direction to [$\bar{1}$2$\bar{1}$], while [1$\bar{1}$1] is in the opposite direction to [$\bar{1}$1$\bar{1}$].

The orientation of a plane in a crystal lattice is defined by the *reciprocals* of the intercepts of the plane on the three crystal axes. These intercepts are measured in multiples or fractions of the unit cell. The *Miller indices* defining the *plane* are the *set of smallest integers* having the same ratio as the reciprocal intercepts. Thus, the intercepts may have values a, b, and c units along the x, y, and z axes, respectively. The reciprocal values $1/a, 1/b, 1/c$, are in the ratio h, k, l, and the plane is specified by the Miller indices (hkl), as shown in Figure 9A.1b. Curved brackets denote a specific plane. If the plane

Figure 9A.1
Miller indices are used to designate (a) crystallographic directions and (b) planes in crystalline solids.

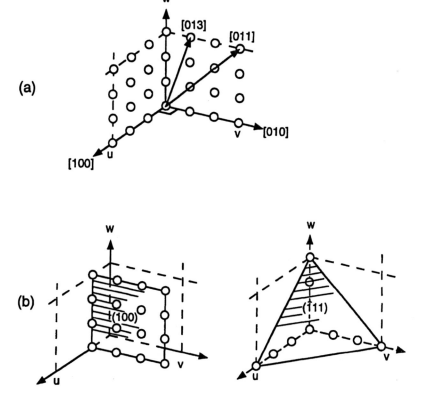

does not intercept one of the axes, then the intercept is taken to be ∞ and the reciprocal is 0. If the plane makes one or more negative intercepts, then the indices are written with a bar, as in (h$\bar{\text{k}}$l). A set of planes related by symmetry is denoted by {hkl}. For example, the specific plane of one cube face may be (010), the set of all cube faces is {100}; the different planes (011), (101) and (1$\bar{1}$0) are represented collectively as {110}.

It is interesting to note that for cubic crystals the direction normal to a plane has the same set of indices as the plane itself. Thus, [010] is the direction normal to the (010) planar surface.

Figure 9A.1 provides examples of simple directions and planes in the cubic crystal sytem using the Miller index notation.

10

THIN FILMS— SOLID–SOLID INTERFACES PROCESSED FROM THE VAPOR PHASE

with W. Gladfelter

A. Franciosi

W. Gerberich

V. Lindberg

LIST OF SYMBOLS

a_s, a_c	lattice parameters for substrate and condensate
b	Burgers vector of epitaxial misfit dislocation lines
d	separation distance between epitaxial misfit dislocations
D_{atom}	atomic diameter
E	Young's modulus of elasticity; G elastic shear modulus
f	flux of gas atoms impinging on a unit area (eq. 10.1.9)
f_e, f_d	flux of vapor atoms emitted by a source (e) deposited on a surface (d)
ΔG_d	change in free energy due to condensation of a circular disk of atoms onto a surface
ΔG_{ads}, ΔG_{des}	change in free energy due to adsorption or desorption
$\Delta G_{surf\ diff}$	activation energy for surface diffusion
h	film thickness; h_c critical thin film thickness (eqs. 10.7.5 and 10.7.6)
ΔH_e	heat of evaporation
K_{ads}	equilibrium reaction constant for adsorption/desorption
L_o	surface step separation
n	number of moles of gas; n_i number of moles of component i
N_v	number of gas atoms per unit volume—volume density

N_s, N_A	number of surface atom sites (s) occupied by adatoms (A)
N_c	number of critical-size nuclei per unit area of substrate
N_i	number of clusters containing i atoms
P_o	equilibrium vapor pressure of adsorbate
P_{gA}	partial pressure of gas A
r_{aA}	rate of adsorption of adatom A on surface (eq. 10.6.16)
r_{dA}	rate of desorption of adatom A from surface (eq. 10.6.19)
r_m	rate of mass transfer
T	deposition rate
T_s	substrate temperature; T_m melting temperature
U_ε	film strain energy per unit area (eq. 10.7.1)
U_d	film energy per unit area due to epitaxial misfit dislocations (eq. 10.7.4)
v_m	molecular volume of molecule m
v	gas molecule velocity
v_{rms}	root-mean-square velocity of a gas molecule (eq. 10.1.3)
W	effective molecular interaction parameter (eq. 3.1.9)
α	thermal expansion coefficient
α	accommodation coefficient
γ_{sc}	interface tension between substrate (s) and condensate (c)
γ_{cv}	surface energy of condensate; γ_{sv} surface energy of substrate
$\varepsilon_{\text{latt. misfit}}$	lattice misfit strain (eq. 10.6.41)
Θ	fractional surface coverage
λ	jump distance between surface sites
Λ	mean free path between gas molecule collisions (eq. 10.1.5)
v_o	characteristic surface atom vibration frequency
v_{diff}	frequency of atomic jumps in diffusion
v_{des}	frequency of desorption (eq. 10.6.17)
v	Poisson's ratio
Π	probability of gas molecule collision (eq. 10.1.7)
σ	sticking coefficient (eq. 10.6.25)
σ_{xx}	tensile stress
τ_s	residence time of an adatom on a surface before desorption (eq. 10.6.18)
τ_o	characteristic surface atom vibration time ($\tau_o = 1/v_o$)
Ω	interface energy parameter
ϕ, θ	angles between source and substrate

CONCEPT MAP

- Thin films are a critical feature of many devices used for data processing, recording, and sensing.

- Films deposited on a substrate constitute either a totally different phase (introducing an interphase interface); or a continuation of the substrate, sharing its crystal orientation, structure, or chemistry (epitaxial interface).

- Thin films are deposited from the vapor phase by physical or chemical means and usually in a vacuum.

- Ideal gas laws and simple kinetic theory describe the behavior of vapor molecules in a vacuum. The number of vapor molecules (molecular weight M) impinging on a unit area per unit time in a system under pressure P is $f = PN_{AV}/\sqrt{2\pi MRT}$ (eq. 10.1.9).

Vapor Deposition Processes

Physical Vapor Deposition (PVD)

- Generally requires high vacuum (10^{-3}–10^{-7} Pa).

- Atom fluxes are produced from sources by heating (evaporation) or by ion beam bombardment (sputtering).

- Deposition is directional (line-of-sight); uniform coverage of steps and valleys is difficult to achieve. Substrate may be rotated during deposition to improve uniformity of coating.

- Substrate may be heated or cooled during deposition.

- Molecular beam epitaxy (MBE) is a precise form of physical deposition requiring ultra-high vacuum (10^{-8}–10^{-10} Pa).

    —Sources are molecular beams directed from effusion cells.

    —Substrates are generally cleaned by ion sputtering or thermal desorption and then roughened to provide growth steps.

    —Multilayer thin films of atomic dimensions can be achieved by sequential switching of sources.

Chemical Vapor Deposition (CVD)

- Generally requires atmospheric to low vacuum (10–10^{-1} Pa).

- Sources are gases that are decomposed or mixed to react in the vicinity of the heated substrate.

- Deposition is nondirectional and all steps and surfaces are covered equally (a conformal coating).

- Substrate is heated.

- Metal-organic CVD (MOCVD) is a precise form of chemical deposition.

    —Sources are metal-organic vapors.

    —Substrates are physically cleaned.

    —Multilayer thin films can be achieved by controlled switching of vapor streams.

Nucleation and Growth of Thin Films

Nucleation

- Heterogeneous nucleation usually dominates under evaporation conditions. Nucleation of a thin film phase occurs at surface steps, kinks, or other energetically favorable sites. Nucleation is controlled experimentally by careful slection of substrate material and surface preparation.

- The energy barrier for heterogeneous nucleation is $\Delta G_{het} = (1/4) \Delta G_c (2 + \cos \theta)(1 - \cos \theta)^2$. ΔG_c is the energy barrier for homogeneous nucleation; $\Delta G_c \propto 1/\ln(P_c/P_o)$, where P_c/P_o is vapor supersaturation. θ is the contact angle. θ is related to the interaction between film molecules and substrate molecules defined by the effective molecular interaction parameter W (eq. 3.1.8).

    When W is small, $\theta = 0$, $\Delta G_{het} = 0$ and nucleation can occur anywhere on the substrate

    When W is large, $\theta = 180°$, $\Delta G_{het} = \Delta G_c$ and the substrate is inactive in the nucleation event.

    When W is moderate, $\theta = 45–90°$, $\Delta G_{het} < \Delta G_c/2$ and nucleation is preferred at surface steps.

Growth

- Growth depends on balance between mass transport to and loss from the substrate.

- Mass transport to the substrate depends on vapor flux f (eq. 10.1.9) in PVD; and on diffusion across boundary layer in CVD.

- Mass loss from the substrate depends on accommodation of incoming atoms and rate of desorption.

- The accommodation coefficient, $0 < \alpha < 1$, accounts for the intermolecular interaction between vapor atoms and substrate atoms and the energy of incoming atoms.

- The coefficient $S = (\lambda^2/L_o^2) \exp[(\Delta G_{des} - \Delta G_{surf\ diff})/kT_s]$ (eq. 10.6.24) represents the statistical probability that an adatom can diffuse to a surface step before it desorbs from the substrate. $\Delta G_{des} - \Delta G_{surf\ diff}$ compares the energy for desorption with the energy for surface diffusion, L_o is the seperation of surface steps, and λ the interatomic spacing. S depends on substrate temperature, T_s, surface diffusitivity, surface roughness, and adsorption/desorption energy.

- Experimental measure of growth conditions is the sticking coefficient σ where

$$\sigma = \frac{\text{rate of mass accumulation}}{\text{rate of mass delivery}}.$$

The sticking coefficient incorporates features of accommodation coefficient α and coefficient S.

Growth Modes

- Depending on the nature of the interface and growth conditions, one of three thin film growth modes prevails (Figure 10.21).

- Volmer–Weber island growth—usually observed for interphase deposits. Islands (clusters of deposited atoms) merge through a sequence of ripening, coalescence, and filling in. Produces polycrystalline films with a columnar structure whose morphology depends on deposition rate, substrate temperature, and surface preparation (Figure 10.27).

- Frank–van der Merwe layer growth—usually observed for epitaxial deposits. Films grow one atom layer at a time by surface diffusion of adatoms to surface steps. Produces a single crystal or highly textured polycrystalline film whose orientation and crystal structure is dictated by the substrate.

- Stranski–Krastanov layer plus island growth. A combination of the two growth modes above.

Stability of Thin Film Systems

- Thin films are produced with residual stress due to energetic deposition processes (such as sputtering) or due to thermal expansion coefficient mismatch. Residual stress can cause separation (delamination or flaking) of interfaces or a gradual decline in physical properties.

- Residual stress due to processing can be relieved by a thermal anneal.

- Residual stress stress due to thermal expansion coefficient mismatch, $\sigma = E \, \Delta\alpha\Delta T/(1 - \nu)$ (eq. 10.7.9), can be reduced by choosing a substrate with a similar thermal expansion, $\Delta\alpha \to 0$, or a lower deposition temperature, $\Delta T \to 0$.

- Epitaxial films become stressed due to lattice misfit strain, $\varepsilon_{\text{latt misfit}}$. Strain energy increases with film thickness until at the critical thickness, h_c, it is energetically favorable for strain to be accommodated by edge dislocation arrays; $h_c = 2b/\varepsilon_{latt\ misfit}$ (eq. 10.7.5). Dislocations can interfere with device performance; eliminating them requires a substrate with lower misfit strain or reduced film thickness.

- Stresses arising from mechanical, thermal, chemical, or electromagnetic origins during device operation may destabilize previously engineered thin film interfaces and adversely affect adhesion and lifetime.

- A variety of other operational phenomena, such as electromigration, creep, and thermal fatigue, may cause failure.

10

A great variety of modern thin-film devices are made by vapor deposition. Applications of this technique include solid-state films for electronic devices that range from simple components (such as rectifying or ohmic semiconductor–semiconductor and metal–semiconductor thin-film junctions) to sophisticated multi-layer structures for quantum-well electro-optic devices. They include optical coatings designed to be either antireflective or totally reflective, depending on applications that range from decorative coatings to ultra-high-performance laser gyroscope mirrors, magnetic films (such as the ferromagnetic ternary-alloy thin films commonly used in the magnetic recording industry), metallurgical thin films designed to improve hardness and resis-tance to chemical attack, or thin-film sensors used to detect gases, mechanical stress, and inertia. Thin films inevitably involve the preparation of solid–solid interfaces through vapor deposition on solid surfaces. The variety of high-vacuum deposition methods developed to meet these varied needs in a production environ-ment constitutes one of the marvels of modern technology.

Within what limits should a film be considered thin? The question is difficult to answer because specific film properties often differ substantially from those of bulk materials and occur in a range of film thicknesses that vary from material to material and from property to property. At one end of the spectrum are polycrystalline metallic films, several micrometers in thickness,

THIN FILMS—SOLID–SOLID INTERFACES / 533

used for magnetic recording media or for coarse interconnects on ceramic substrates. At the other end are semiconductor epitaxial films an exact number of atom layers thick tuned for a very specific electro-optic response. In practice, the physics, chemistry, and technology of thin films deals with films between several tenths of a nanometer (0.1 nm = 1 Å) and several micrometers thick.

Besides overall thickness, other issues important in thin-film technology include thickness uniformity, chemical composition, and sharpness of the interface. Table 10.1 summarizes these broad structural requirements for different devices. The figures are approximate, but they convey the important fact that requirements do vary. For microelectronic devices, thickness control to 1% means that one-micrometer films must be uniform within 100 Å, while 1000 Å films must be atomically flat! Chemical composition and purity must be controlled within set limits. The transition from one chemical composition to the next may need to be very abrupt. Some epitaxial thin films feature transitions within one atomic layer. The more precise the demands on thickness and/or composition control, the more expensive the process. A critical analysis of absolute needs should always precede embarking on the manufacture of a thin-film material system.

This chapter focuses on different methods used to synthesize thin films by physical and chemical vapor deposition, referred to as PVD and CVD, respectively. Unlike the liquid coating processes described in Chapter 8, vapor deposition processes are atomistic, with material transported to the growing film as single atoms or molecules. Film growth is also dry, with no liquids involved. Two types of thin film are produced. Usually the films' crystalline structure and chemistry differ completely from those

TABLE 10.1 Range of Thin-Film Requirements for Different Applications

Material/Application	Composition[a]	Purity[b]	Thickness[c]	Uniformity[d]	Sharpness[e]
Metal/conductors	$1:10^3$	$1:10^5$	10^{-5} m	$\pm10^{-7}$ m	*
Metal/magnetic recording	$1:10^4$	$1:10^5$	10^{-3} m	$\pm10^{-7}$ m	*
Optical/antireflection	$1:10^3$	$1:10^3$	10^{-6} m	$\pm10^{-7}$ m	√
Optical/laser mirrors	$1:10^6$	$1:10^8$	10^{-7} m	$\pm10^{-9}$ m	√√
Semiconductor/ transistors	$1:10^6$	$1:10^7$	10^{-7} m	$\pm10^{-9}$ m	√
Semiconductor/ quantum well devices	$1:10^6$	$1:10^9$ (1 ppb)	10^{-8} m (100 Å)	$\pm10^{-10}$ m (±1 Å)	√√

[a]*Composition* refers to metal alloy composition, or inorganic crystal stoichiometry or semiconductor dopant level.

[b]*Purity* refers to tolerable level of unwanted impurity.

[c]*Thickness* refers to thickness of individual layers in case of multilayer devices.

[d]*Uniformity* refers to uniformity of thickness and/or smoothness.

[e]*Sharpness* refers to need for a sudden change of composition. *—Not important; √—Important for quality of device; √√—Essential for device function.

of the substrate. In this case, an *interphase interface* is formed at the boundary between substrate and film. Other thin films have crystal structure, crystallographic orientation, or chemistry in common with the substrate. In this case, an *epitaxial interface* is formed between the two materials.

Processing issues for both types of interface are basically the same, and for the most part the information contained in this chapter is pertinent to either kind. In Sections 10.1 and 10.2, we begin by reviewing some of the gas kinetics equations that are important for understanding physical and chemical deposition under high-vacuum conditions. In Sections 10.3 and 10.4, we describe the physical and chemical methods used for vacuum deposition of thin films. Section 10.5 treats the special molecular beam processes needed for certain solid–solid epitaxial films. Section 10.6 distinguishes the different modes of thin-film growth by considering the thermodynamics that determine the nucleation event and surface-related reaction kinetics that determine the growth of thin films on a substrate. Key issues for the processing of thin-film devices are the competing growth mechanisms: island growth plus faceting versus smooth atomic layers, columnar grains versus epitaxy. In Section 10.7, we consider the stability of thin films produced by the different techniques and identify the sources of defects and failures in thin-film devices. Finally, in Section 10.8, we present selected examples of solid-state devices whose manufacture illustrates some of the principles involved in thin-film processing.

10.1 Ideal Gas Behavior Can Describe the Properties of the Vapor Phase under Most Vacuum Conditions

Vapor deposition involves the transport of material from a source to a substrate, with deposition occurring atomistically at the substrate. Generally, the deposition occurs in a vacuum chamber. Several aspects of the behavior of gases in vacuum systems are important to vapor-phase processing, and we begin the chapter with a brief review of the gaseous state. Throughout the chapter we use the words *atoms* and *molecules* interchangeably to describe the gaseous species. For example, for general discussion purposes, the behavior of metal atoms or water molecules is qualitatively the same in the vapor phase.

10.1.1 Gas (Vapor)-Phase Equation of State

An ideal gas is a collection of infinitesimally small atoms that have no long-range interaction (attractive or repulsive) and that interact as repulsive hard spheres upon collision. The pressure, volume, and temperature of a sample of ideal gas are interrelated through the simple equation of state

$$PV = nRT = nN_{Av}kT = NkT$$

or

$$P = N_v kT \qquad (10.1.1)$$

where **R** represents the gas constant, **k** is Boltzmann's constant, n is the number of moles of the gas, and N_{Av} is Avogadro's number. Also in this equation, $nN_{Av} = N$, the number of gas molecules, and $N/V = N_v$, the number of gas molecules per unit volume, known as the *number density*.

Several units of pressure are common.[1] The international standard unit, the pascal (Pa), is gaining acceptance, although two other units are still more common. The torr (millimeters of mercury) and millibar are often used. Complicating this issue is the practical fact that vacuum gauges are sensitive to the composition of gas in the system, and this composition is generally unknown. For most purposes, we can use the approximate conversion 1 torr ≈ 1 mbar = 100 Pa.

We will consider five different pressure ranges characteristic of vapor deposition. As listed in Table 10.2, they are: atmospheric (760 torr) for chemical vapor deposition (CVD); rough vacuum of 1 torr for low-pressure CVD (LPCVD); low vacuum (LV) of 10^{-3} torr for sputtering; high vacuum (HV) of 10^{-6} torr for evaporation, and ultra-high vacuum (UHV) of 10^{-10} torr for molecular beam epitaxy (MBE) and surface analysis. (Table 10.3 provides a list of acronyms used in thin-film technology.)

At standard temperature and pressure (T = 0°C, P = 760 torr), one mole of ideal gas has a volume V of 22.4 liters. So the spacing of particles, $(V/N)^{1/3}$, or $(kT/P)^{1/3}$, has ambient values of approximately 33×10^{-10} m (33 Å). These dimensions are about 10 times larger than the interparticle spacing for condensed matter in liquid or solid form. At a HV pressure of 10^{-6} torr, the average spacing increases one-thousand-fold to a few micrometers, and the basic assumption of ideal gas laws—that individual molecules do not interact—becomes more valid.

10.1.2 Kinetic Behavior of Gas Molecules.

Gas (vapor) molecules move with equal probability in random directions with an average velocity *v*. The average kinetic energy of each molecule, mass m, equals $(1/2)mv^2$, and the principle of equipartition of energy states that a molecule possesses an average energy of $(1/2)kT$ for each degree of freedom. Because each isolated particle has three degrees of freedom (x, y, z), the energy per particle is $(3/2)kT$. Therefore

$$\frac{1}{2} mv^2 = \frac{3}{2}kT \qquad (10.1.2)$$

The root mean square (rms) velocity is given by

$$v_{rms} = \sqrt{\frac{3kT}{m}} \qquad (10.1.3)$$

---

[1] *The standard unit of pressure is the pascal (Pa), where 1 pascal = 1 newton/m². 1 bar = 10^5 Pa; 1 m bar = 10^2 Pa. 1 atm = 760 mm of Hg = 760 Torr; 1 Torr = 133.32 Pa; 1 atm = 1.0132×10^5 Pa.*

TABLE 10.2 Residual Gas Properties for Different Vacuum Conditions[a]

	Atmospheric	Rough vacuum	Low vacuum	High vacuum	Ultra-high vacuum
Pressure					
P (Torr)	760	1	1×10^{-3}	1×10^{-6}	1×10^{-10}
Number density					
N_v (molecules/cm^3)	2.5×10^{19}	3.3×10^{16}	3.3×10^{13}	3.3×10^{10}	3.3×10^{6}
Mean free path					
Λ (mm)	6.5×10^{-5}	5.0×10^{-2}	50	5×10^{4}	5×10^{8}
Impingement flux					
f (molecules/cm^2 s)	2.9×10^{23}	3.8×10^{20}	3.8×10^{17}	3.8×10^{14}	3.8×10^{10}
f (molecules/nm^2 s)	2.9×10^{9}	3.8×10^{6}	3.8×10^{3}	3.8	3.8×10^{-4}
Monolayer time					
$t_{monolayer}$ (s)	2.5×10^{-9}	1.9×10^{-6}	1.9×10^{-3}	1.9	1.9×10^{4}

[a]Calculated using the kinetic theory of ideal gases. The gas is "air," 70% nitrogen, 30% oxygen; assumed molecular weight M = 29 g/mole; assumed molecular diameter D_{atom} = 0.372 nm; at room temperature T = 22°C = 295 K. Pressures chosen are typical vacuum chamber pressures for the different processes discussed in Chapter 10.

TABLE 10.3 CVD and PVD Methods

Acronym	Name
CVD	Chemical vapor deposition
LPCVD	Low-pressure chemical vapor deposition
OMCVD	Organometallic chemical vapor deposition
MOCVD	Metal–organic chemical vapor deposition
OMVPE	Organometallic vapor-phase epitaxy
MOVPE	Metal–organic vapor-phase epitaxy
PECVD	Plasma-enhanced chemical vapor deposition
LCVD	Laser-induced chemical vapor deposition
AACVD	Aerosol-assisted chemical vapor deposition
PVD	Physical vapor deposition
VPE	Vapor phase epitaxy
LPE	Liquid phase epitaxy
MBE	Molecular beam epitaxy
GSMBE	Gas source molecular beam epitaxy
MOMBE	Metal–organic molecular beam epitaxy
OMMBE	Organometallic molecular beam epitaxy
CBE	Chemical beam epitaxy

Statistical mechanics arguments lead to the well-known Maxwell–Boltzmann distribution law for the speed of gas molecules shown in Figure 10.1. From this distribution we can determine the most probable speed, the average speed, and the rms velocity. These characteristic speeds are indicated on the diagram, they are similar in value. Finally, the ensemble average of the speed is given by

$$\langle v \rangle = \sqrt{\frac{8kT}{\pi m}} \tag{10.1.4}$$

which is very close in value to the rms velocity.

In a gas at room temperature, these average velocities are of the order 500 m/s for air to 2000 m/s for hydrogen molecules (about the velocity of sound). The average speed increases with temperature and decreases with the mass of the individual molecule.

The kinetic theory of gases uses the Maxwell–Boltzmann distribution and the laws of mechanics to derive several important results concerning an ideal gas.

10.1.3 Mean Free Path between Collisions

The average distance a gas molecule will travel without colliding with another gas molecule is called the *mean free path* Λ. It should depend on the size of the molecule, given by its diameter, D_{atom}, and on the number density of molecules and is

$$\Lambda = \frac{1}{\sqrt{2}\pi D_{atom}^2 N_v} = \frac{kT}{\sqrt{2}\pi D_{atom}^2 P} \tag{10.1.5}$$

For gases such as O_2, N_2, and H_2O, the value of D_{atom} to be used in eq. 10.1.5 ranges from 2 to 5 Å. For air, we use a molecular weight of 29 and $D_{atom} = 3.72$ Å, and at room temperature, the relationship for the mean free path reduces to the convenient form

$$\Lambda \ (\text{air, room temp., in mm}) = \frac{6.6}{P(\text{in pascals})} = \frac{0.05}{P(\text{in torr})} \tag{10.1.6}$$

Figure 10.1
The Maxwell–Boltzmann distribution law for the speed of gas molecules. The curve is plotted for air at 295 K; the markers show, from left to right, the most probable, the average, and the root mean square speed.

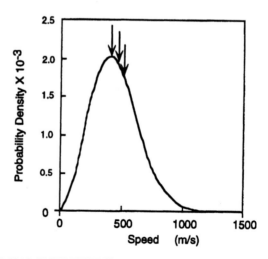

Table 10.2 gives the values of the mean free paths for air at the five pressures we are considering. At 760 torr, the mean free path is only 65 nm, but it rapidly increases as the pressure decreases, becoming 50 m at a pressure of 10^{-6} torr.

The probability Π that a particle in a dilute gas can move a distance x without colliding is

$$\Pi = \exp\left(-\frac{x}{\Lambda}\right) \qquad (10.1.7)$$

and the probability of traversing the distance Λ without collision is 63%. In some thin-film vapor deposition systems, we need to be certain that a vapor molecule will travel from source to target in a vacuum system (whose typical dimensions are 1 m) undeflected by atoms or molecules due to residual gases. This condition requires high vacuum—pressures less than 10^{-4} Pa (7.5×10^{-7} torr). Under high-vacuum conditions, molecules rarely collide with each other; they predominantly collide with the walls of the vacuum chamber or travel in straight paths from source to target: this condition is called *molecular flow*.

10.1.4 Impingement Flux—The Number of Particles Striking a Unit Area of Surface per Unit Time

One of the most important parameters for vacuum technology and vapor deposition is the impingement flux f, which represents the number of molecules striking a unit area of surface per unit time. Kinetic theory shows that flux depends on number density and the average speed of molecules

$$f = \frac{N_v <v>}{4} \qquad (10.1.8)$$

Substituting the value for the velocity from eq. 10.1.4 and the number of gas molecules per unit volume from eq. 10.1.1, the flux is

$$f = \frac{P}{\sqrt{2\pi mkT}} = \frac{PN_{Av}}{\sqrt{2\pi MRT}} \qquad (10.1.9)$$

where M represents the molecular weight, $N_{Av}m$.

In more pragmatic form

$$f = 2.63 \times 10^{20}\frac{P \text{ (in pascals)}}{\sqrt{MT}} = \frac{3.51 \times 10^{22}\, P \text{ (in torr)}}{\sqrt{MT}} \qquad (10.1.10)$$

where f is the number of molecules impinging on one square centimeter per second when we measure the pressure P in pascals or torr, the temperature T in K, and M is the molecular weight.

The hypothetical time required to cover a surface with one layer of molecules, $t_{monolayer}$, relates inversely to the arrival rate and the molecular size. Assuming all the molecules stick and spread out to form a monolayer and that each one occupies an area D_{atom}^2

$$t_{monolayer} = \frac{1}{f D_{atom}^2} \qquad (10.1.11)$$

Table 10.2 gives values of the flux and the monolayer formation time for the five characteristic pressures. It is also of interest to look at the flux from the perspective of atoms already on the surface. Because we typically size atoms in fractions of a nanometer, we are interested in the flux to an area of ≈ 1 nm^2. Under high-vacuum and ultra-high-vacuum conditions, the flux is less than one atom per second impinging onto a surface atom. Impingement flux of impurity atoms becomes important in considering the rate of incorporation of foreign atoms into growing films.

Depositions requiring exceptional purity, such as thin films grown by MBE, or instances where the surface must remain unchanged for a long time, such as surface analysis, require ultra-high vacuum. Under these conditions, monolayers form in hours rather than seconds or fractions of a second.

10.1.5 Other Properties of Gases in Vacuum Are Sensitive to the Pressure

In Section 10.1.3, we defined *molecular flow* as gas flow under such high-vacuum conditions that molecules collide almost exclusively with the walls of the chamber rather than with each other. When this is the case, other physical properties of the gas change. In molecular flow, heat conduction approaches zero, with the result that residual gas in a vacuum chamber has the temperature of the walls of the chamber, generally room temperature. Viscosity also approaches zero in high vacuum. Viscosity is due to momentum transfer between molecules, and when the mean free path exceeds the size of the chamber, molecules cannot interact, and the viscosity drops. At higher pressures, such as the 1 torr typical of CVD, the viscosity of the gas phase is important and, as we shall see later, must be taken into consideration when we establish processing conditions.

10.2 For Practical Purposes, We Can Relate the Choice of Vacuum System to the Kinetic Theory of Gases

Figure 10.2 shows a typical vacuum system arranged for thermal evaporation. A full treatment of vacuum pumps, gauges, and systems would require a separate text such as *A User's Guide to Vacuum Technology* by J. F. O'Hanlon. Modern reviews on ultra-high-vacuum techniques can be found in the shop notes section of the *Journal of Vacuum Science and Technology*. Here we will discuss only the rudiments of vacuum pumping and vacuum chambers selected for physical and chemical vapor deposition.

10.2.1 Vacuum Pumps: Speed and Throughput

A wide variety of vacuum pumps is available. Three factors about vacuum pumps concern us here: cleanliness, gaseous species to be pumped, and speed.

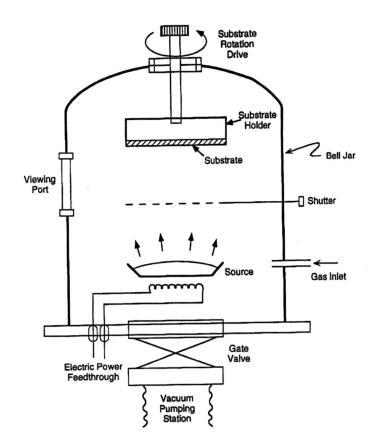

Figure 10.2
The electrically heated source evaporates and deposits on the substrate (and on the walls of the bell jar) in this typical thin-film vacuum processing system.

Applications that do not require ultra-high purity use pumps such as rotary and diffusion pumps. These pumps use oils as a pumping fluid, and some of the oil can become incorporated into the growing film. For ultra-high-purity films, therefore, oil-free turbo-, cryo-, and ion pumps are used.

The effectiveness of any pump varies with the species of gas. Thus, while we may know the composition of air at the start of pumping, that composition will become very different at low pressure. Figure 10.3 shows an analysis of the residual gas in a chamber pumped to ultra-high vacuum. The gas is now predominantly hydrogen, with little or no oxygen or nitrogen remaining.

We define a pump's speed S as the effective volume of gas per unit time removed by that pump; typically it is given in liters per second, $l\,s^{-1}$. The speed will vary with pressure (and with the type of gas), but for most pumps it is constant over a fairly broad range of pressures, as shown in Figure 10.4. At rough vacuum, rotary pumps with speeds of $10\ l\,s^{-1}$ are available, while high-vacuum pump speeds range upwards from $500\ l\,s^{-1}$.

As the pressure decreases, each liter of gas removed by the pump contains fewer molecules. A more meaningful measure of pumping efficiency would be the rate of removal of molecules, dN/dt. We can resolve the practical problem of how to measure N by using the ideal gas law and assuming that the temperature is constant. Then N is proportional to PV, and the rate of removal of molecules is proportional to the throughput Q.

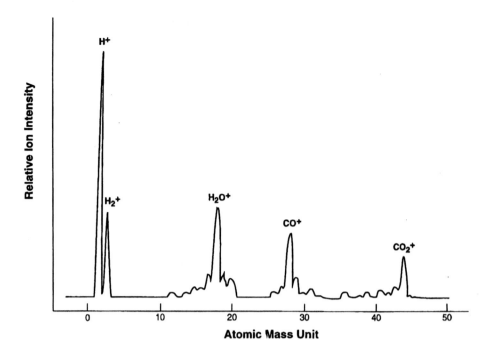

Figure 10.3
Residual gas analysis in a system pumped to an ultra-high vacuum of 2×10^{-10} Pa.

$$Q = d(PV)/dt \qquad (10.2.1)$$

with units such as torr liters/second. Here V equals the volume of gas in the system, a value that can change due to the deliberate introduction of gas or to outgassing and leaks.

If pump speed and gas volume are constant, we equate expression 10.2.1 to the throughput of a pump, $Q_{pump} = -PS$, and

Figure 10.4
A typical pump-down curve for a high-vacuum system. The system is initially pumped by a 25 liter per second rotary pump, followed by a 300 liter per second high-vacuum diffusion pump. Once the pressure has dropped to 10^{-6} torr, the pumping speed is limited by outgassing the system, and the pressure decreases inversely with time over many hours.

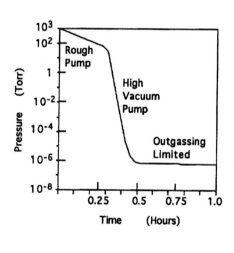

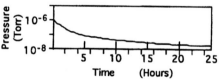

solve the differential equation to find pressure as a function of time.

$$P = P_{initial} \exp\left(-\frac{St}{V}\right) \qquad (10.2.2)$$

where $P_{initial}$ is the starting pressure. One problem remains: the volume of gas in a chamber is not constant. It increases with time as gas desorbs from the surface, a process called *outgassing*. Outgassing of gases adsorbed on the surfaces generally dominates at lower pressures. The outgassing throughput into the chamber, Q, will decrease slowly with time, requiring hours or days if the UHV chamber has been opened to air. Then we must modify eq. 10.2.2 to

$$P = P_{initial} \exp\left(-\frac{St}{V}\right) + \frac{Q}{S} \qquad (10.2.3)$$

We can present the resulting pumpdown curve for a chamber on a semilog plot, as shown in Figure 10.4. Initially the pumps dominate the drop in pressure, but after a few minutes outgassing limits the rate at which pressure can be reduced.

In a clean system that has reached HV or UHV conditions, water vapor remains the main contributor to the residual pressure. We can eliminate water vapor by a 24-hour vacuum "bake-out," achieved by heating the entire vacuum chamber to about 300°C. After bake-out, the system pressure reaches the 10^{-9} Pa ($= 10^{-11}$ torr) range. Then the major contribution to the residual pressure comes from low-molecular-weight gases like hydrogen, which are pumped at comparatively low speed by ion pumps and turbopumps. Figure 10.3 shows a mass spectrum of the residual gas in an ultra-high-vacuum system at a pressure in the 10^{-9} Pa range.

10.2.2 Angular Distribution of Emitted Molecules

When a gas molecule strikes a flat surface, it usually is adsorbed onto the surface for some time. During this time, the molecule loses all memory of its incoming direction. When the molecule is re-emitted, it can go in any direction. However, kinetic theory predicts that the probability of emission will vary with the cosine of the polar angle. If the number of molecules per second emitted from a surface of area A is f_eA (f_e represents the emission flux), then the number of molecules (dN) emitted per second into a small solid angle $d\Omega$ located at polar angle θ is independent of the azimuthal angle (ϕ) and is

$$dN = \frac{(C+1)f_eA}{2\pi} \cos^C \theta \, d\Omega \qquad (10.2.4)$$

Ideally $C = 1$, and we have the simple cosine law for emission. Figure 10.5 is a polar plot of this function. Also shown are more realistic curves where the cosine is raised to higher powers, $C > 1$, for which the vapor flux is more highly directed. These curves approximate emission from evaporation sources in physical vapor deposition systems, as we will discuss later.

Figure 10.5
The number of molecules emitted per second from a flat surface source into a small solid angle at polar angle θ follows a simple cosine law when C = 1. When the cosine is raised to higher powers, C > 1, the vapor flux is more highly directed. (H. K. Pulker, *Coatings on Glass*, Vol. 6 of *Thin Films Science and Technology*, Elsevier, Amsterdam, 1984, p. 175.)

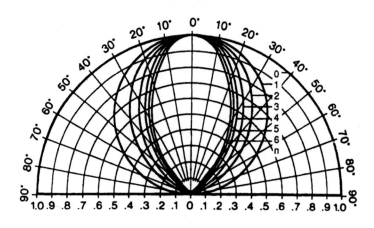

10.2.3 Vacuum Chambers Come in a Variety of Materials and Shapes

The materials predominantly used in vacuum systems are stainless steel and glass. Careful choice of the proper steel can minimize diffusion through the walls of the chamber and eliminate outgassing of some of the elements in the steel, such as sulfur.

We can ensure the vacuum integrity of connections within the chamber by the use of O-rings and gaskets. For LV or HV conditions, elastomeric materials may be used for this purpose, while for UHV conditions soft copper gaskets are used.

Systems can be designed for almost any type of substrate. In applications where the substrates are small, like the half-inch-diameter quartz substrates used for optical mirrors, deposition is done in batch mode. If the substrate is large and flexible, like a three-foot-wide polymer sheet, the sheet is unwound from a roll, coated, and rewound all inside the vacuum system. For large rigid substrates, such as architectural glass, the substrate is continuously fed into the vacuum system through a series of vacuum-tight rollers.

10.3 Physical Vapor Deposition Generally Employs High-Vacuum Techniques

We focus our attention first on processing thin films by physical vapor deposition (PVD) methods. Later sections focus on chemical vapor deposition (CVD) and molecular beam epitaxy (MBE) methods. Physical vapor deposition involves three sequential unit operations: physical generation of individual molecules at the source, transport of these molecules as a vapor, and deposition of the molecules onto a substrate in a vacuum.

The next three sections describe different sources for physical vapor deposition: thermal evaporation (Section 10.3.1), sputtering (Section 10.3.2), and other techniques for generating a flux of molecules (Section 10.3.3). Sections 10.3.4 and 10.3.5 deal with deposition of the molecules onto the substrate and are concerned, respectively, with the effects of substrate condition and means for monitoring the thickness. We will pay particular attention to parameters that control nucleation and growth of the films on the substrate and parameters that control the rate of deposition.

We must also clearly define two other features important to the quality of our final thin-film coating: *uniformity* and *conformality*. Uniformity means that identical substrates located at various places in the coater should receive the same thickness of coating. Conformality means that for a given substrate, vertical faces (steps) and horizontal faces (terraces) should be coated with the same thickness. In most applications, it is essential that all facets of a substrate be covered equally, irrespective of orientation. Occasionally in the manufacture of microelectronic devices, it is desirable to coat only the horizontal surfaces that face the source.

In Section 10.1, Table 10.1, we noted how the level of impurities and contamination that can be tolerated in a thin film depends on the application. The rate of contamination depends on the balance between the flux of depositing atoms and the impingement due to residual gas. In optical, metallurgical, and simple electrical applications that can tolerate a moderately high level of impurities (10–1000 parts per million), HV conditions (10^{-5}–10^{-8} torr) suffice. In semiconductor films where impurities must be kept below 1 part per billion, UHV conditions (below 10^{-9} torr) are required. High-vacuum systems and pumps are a major capital cost item in PVD, and the selection of the appropriate equipment for the task becomes a foremost consideration.

10.3.1 Physical Vapor Deposition by Thermal Evaporation—Molecules Are Evaporated from a Thermally Heated Source

10.3.1.1 Evaporation Basics Are Independent of the Heating Method

In thermal evaporation, a stream of molecules is generated by heating a *source*. The temperature needs to be raised until the source's equilibrium vapor pressure reaches a value about 1–10 mtorr. Temperatures in excess of 1000 K are generally required to reach the needed vapor pressure. At these temperatures, if the source material is solid, the molecules sublimate. If it is liquid, the molecules evaporate. The energies of the evaporated molecules are characterized by the source temperature; they are relatively low, typically 0.1 eV. Molecules leave the source and travel in molecular flow (straight lines) to the *substrate* and other parts of the chamber, where they condense to form the thin film. Because evaporated atoms travel in a straight line, it is difficult to obtain conformal coverage of complicated shapes.

We can evaporate either individual pure elements, such as Al or Cu, or strongly bonded compounds. The crystal structure and morphology of the deposit may be different than that of the source. Thus, if we use diamonds as source material, we do not expect to get diamonds on the substrate. Compounds such as SiO or MgF_2 evaporate without dissociation; other compounds such as SiO_2 or ZnS dissociate and recombine on the substrate. In these latter cases, the stoichiometry of the deposited film generally differs from that of the source material, and evaporation in a reactive atmosphere may be necessary to regain the original stoichiometry.

Since we are at high vacuum, molecules leaving the source will travel to the substrate and not return to the source. In these conditions, we can write the evaporation flux, f_e, as in eq. 10.1.9 $f_e = P_o/\sqrt{2\pi mKT}$. The controlling variable in the evaporation flux is the vapor pressure P_o, which is determined by the temperature. We can write the Clausius–Clapeyron equation for vapor pressure

$$\ln P_o = -\Delta H_e/RT + \text{const.} \qquad (10.3.1)$$

where ΔH_e is the molar heat of evaporation, generally constant with temperature. Figure 10.6a shows the validity of the Clausius–Clapeyron equation, although Figure 10.6b is easier to read. Note that the vapor pressure changes very rapidly with temperature. We will find source temperature a critical parameter in determining the rate of deposition.

Thermal evaporation of alloys from a single source usually is impossible because the vapor pressures of the different elements are likely to differ vastly at the evaporation temperature. The next section discusses alternative means to evaporate alloys.

Ideally, the cosine law discussed in Section 10.2.2 gives directional evaporation from a flat surface. Consider the evaporation flux f_e from a flat source of area A_s shown in Figure 10.7. If a substrate is located a distance r and a polar angle θ from the normal to the source, we can determine the deposition flux f_{dep} as

$$f_{dep} = \frac{f_e A_s \cos\theta \cos\phi}{\pi r^2} \qquad (10.3.2)$$

where ϕ is the angle between the line from the source to substrate and the normal to the substrate. Distribution departs from ideal because sources may not be at a uniform temperature and may not be perfectly flat. Also the distribution changes with time as the source is consumed. In practice, therefore, we must calibrate the directional distribution for a given system experimentally.

We get the thin-film deposition rate, $T = dh/dt$, where h is film thickness, from eq. 10.3.2,

$$T = \frac{dh}{dt} = \frac{f_{dep}M}{\rho N_{Av}} \qquad (10.3.3)$$

where M is the molecular weight and ρ is the bulk density of the deposited material. Equation 10.3.3 presumes that all the material arriving at the substrate sticks to it. However, there is the caveat that the density of the deposited film may differ from the density of the bulk material, ρ. Rates of deposition can vary from less than 0.01 nm/s to more than 500 nm/s.

Deposition rates play a role in both nucleation and growth phenomena, and we will discuss their effect on the microstructure of vapor-deposited films in Section 10.6. For now, we simply note that deposition rate and substrate temperature are important determinants of the microstructure.

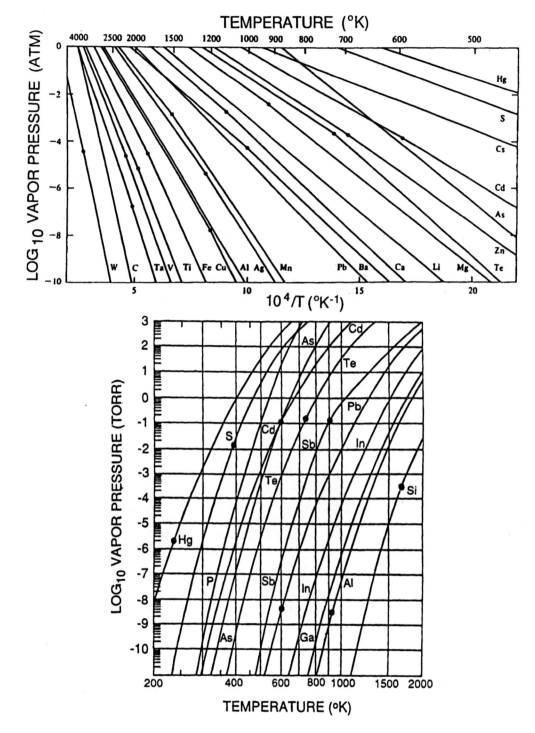

Figure 10.6
The vapor pressure of various elements as a function of temperature. Dots correspond to their melting points. (top) The linear relationship between log vapor pressure and inverse temperature is consistent with the Clausius–Clapeyron equation. (bottom) Vapor pressure of many elements employed in semiconductor devices. (M. Ohring, *The Materials Science of Thin Films*, Academic Press, Boston, 1992, p. 84.)

Figure 10.7
Thermal evaporation from a flat source. Assuming ideal cosine emission (Figure 10.5), the deposition flux at the substrate located as shown is $(f_e A_s / \pi r^2) \cos \theta \cos \phi$.

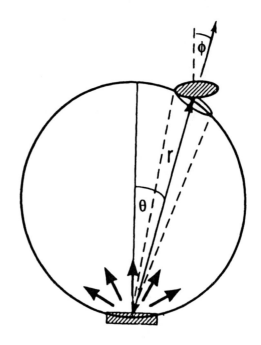

In summary, for thermal evaporation, the temperature of the source controls the deposition rate. Depositing molecules have a low energy of about 0.1 eV, and deposition rate and substrate temperature are the factors that determine nucleation and growth of the thin film.

10.3.1.2 Different Heat Sources Are Available for Thermal Evaporation

Different methods can be used to heat the source: resistance heating (the least expensive), electron-beam heating, and inductive heating. Molecular beam epitaxy uses a special form of resistance evaporation that we will also discuss here.

In *resistance heating,* as shown in Figure 10.8a, a refractory metal boat (or wire) made from tungsten, molybdenum, or tantalum holds the source material, which is also called the charge. The charge may be directly in contact with the metal or held in a ceramic crucible (alumina is popular). An electric current heats the boat, and some of this heat is conducted to the source. The efficiency of heat transfer depends on the ability of the source material to wet the boat. Wettability may be poor in some cases, especially for sublimation sources. A disadvantage of resistance heating is that the substrate is heated by infrared radiation from the source and may be contaminated by atoms of the refractory metal heater.

To deposit an alloy under steady-state conditions, multiple sources are held at different temperatures. When the melting points of the constituents are vastly different, flash evaporation may be used. In this process, a powder of the alloy is dropped onto a boat maintained at a very high temperature, where it evaporates almost instantaneously.

In *electron-beam heating,* as shown in Figure 10.8b, a high-power beam of electrons emitted by a heated filament is bent in a

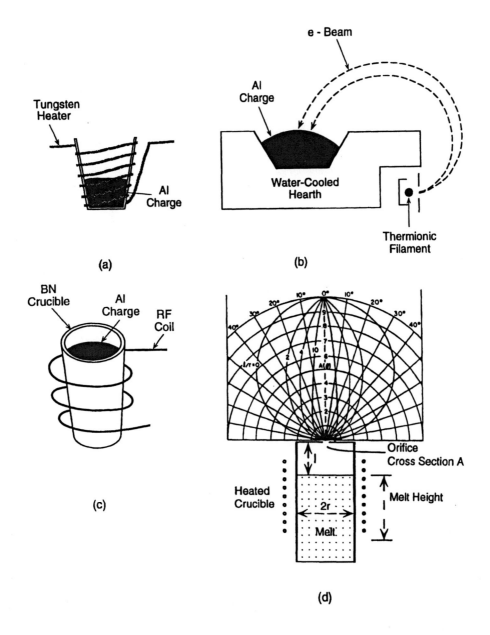

Figure 10.8
Thermal evaporation sources. (a) Electrical resistance heating. The tungsten wire is heated electrically; it melts and evaporates the aluminum source. (b) Electron beam heating. A focused beam of electrons melts and evaporates the aluminium source. The beam can be moved with a magnetic field. (c) Induction heating. A radio frequency electrical field induces currents that melt aluminum held in a boron nitride crucible. [(a)–(c) From S. M. Sze, *Semiconductor Devices: Physics and Technology,* Copyright © 1985, Bell Telephone Laboratories, p. 366. Reprinted by permission of John Wiley & Sons, Inc.] (d) Knudsen cell. Atoms from an electrically heated aluminum source are restricted to enter the ultra-high-vacuum MBE chamber (Figure 10.16) through a small orifice. The angular distribution of the flux depends on the source geometry and becomes more focused as the charge evaporates and l/r increases.

semicircle and focused to heat the source. The electron beam current may be 1 A at 5–15 kV, and moving it about with a magnetic field can heat a region of the source to an appropriate temperature for evaporation. Energy transfer to the source is direct in this process and more efficient than with resistance heating. Contamination issues are eliminated, since the outer cooled portions of the charge form a crucible for the heated portion. Disadvantages of the method are that it exposes the substrate to X-radiation and reflected electrons, and this ionizing radiation may affect the growing film in some applications. For semiconductors, a postdeposition anneal may be required to remove radiation damage.

Induction heating, as shown in Figure 10.8c, relies on a powerful radio-frequency (RF) power supply to induce currents in the source. This technique is limited to conducting sources. The charge is held in a graphite or boron nitride crucible. Larger charges are possible compared to resistance heating, and contamination from the source holder is reduced.

Molecular beam epitaxy (MBE) uses a specialized thermal evaporation source. The charge is generally resistance heated, but it is held within an enclosed chamber containing a small hole. The assembly shown in Figure 10.8d is called a Knudsen effusion cell. A large vapor pressure of evaporant is produced inside the cell, and some of the molecules are allowed to effuse into the vacuum chamber through the small hole. These molecules emerge into the ultra-high vacuum with the angular flux distribution shown in Figure 10.8d. Typically, the source is pointed directly towards the substrate (see Figure 10.16). We discuss MBE further in Section 10.5.

10.3.2 Physical Vapor Deposition by Sputtering— Ion Bombardment Ejects Molecules from a Source

In sputtering, depicted in Figure 10.9, we bombard the source, now called the *target,* with ions. Most commonly these are positively charged ions of an inert gas; generally argon is used. The ions arrive at the target with an energy of about 500 eV and set up a chain of collisions within it. Eventually, a target molecule retaining a substantial energy of 2–30 eV will be ejected into the vapor phase. The target, or source, molecule passes through the vacuum toward the substrate, where it sticks and the thin film grows. Sputtering is analogous to striking a rack of pool balls with an energetic cue ball: after several collisions, the target balls break in all directions; some even return toward the cue ball. We can generate sputtering ions either in a glow discharge plasma or in an ion gun.

10.3.2.1 Glow Discharge Sputtering, the Most Widespread Sputtering Method, Operates at High Vacuum

Glow discharge is the most common form of sputtering. An electrical potential of about 1000 V applied to argon gas at a pressure of 1 to 100 mtorr ionizes the argon and generates a dilute argon plasma. The discharge glows like a neon lamp except the color is blue for argon gas. Because the typical degree of ionization

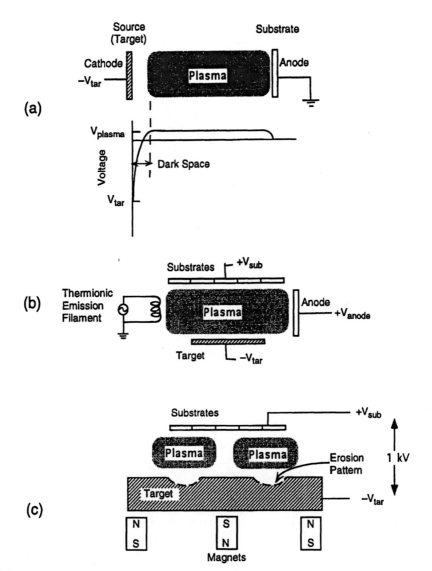

Figure 10.9
Glow discharge sputtering configurations. (a) Diode sputtering. A glow discharge is generated by applying about 500 V (V_{tar}) between the source (the target) and the substrate. The voltage drop occurs primarily in the "dark space" immediately adjacent to the cathode. The plasma voltage (V_{plasma}) is tens of volts positive. Positive ions (generally argon) entering the dark space accelerate toward the target, where source atoms are sputtered off. (b) Triode sputtering. Thermionic electrons accelerate toward the anode; the high-energy electrons ionize argon to generate a plasma; the argon ions accelerate toward the target, where source atoms are sputtered off. The substrate may be grounded or biased negatively. (c) Magnetron sputtering. Same configuration as (b) but the magnetic field stimulates electron collisions in the plasma. The magnetron consists of an outer annular magnet surrounding an inner magnet. Note the uneven erosion pattern.

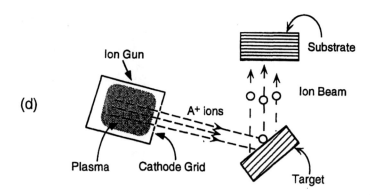

Figure 10.9 (*Continued*)
(d) Ion beam sputtering. Positively charged ions generated in the glow discharge within the "gun" emerge through holes in the cathode grid and enter the high-vacuum system to strike the target. Sputtered atoms travel in straight lines to deposit on the substrate.

is only 1 atom in 10^4, most of the argon gas atoms remain neutral. These neutral atoms have only room-temperature thermal energy; so they play a minor role in the process. On the other hand, the highly energetic argon ions are accelerated to strike the target and provide the sputter source.

Figure 10.9a shows a simple *diode sputtering* plasma and the voltage distribution between the electrode plates. Being a conductor, the plasma has negligible voltage drop across it. Electromagnetic theory dictates that the voltage of the plasma is the most positive in the chamber. This voltage—called the plasma potential—is usually a few tens of volts. A cathode held negative by several hundred volts is the target. The voltage drop from the plasma to the target occurs in a thin region called the dark space, whose dimensions are about one mean free path length (≈ 5 mm at 10 mtorr, eq. 10.1.6). Ions that drift and enter the dark space are accelerated toward the target, striking it with high energy at an angle close to normal.

Once the energetic ions strike the target, several things may happen. Ions may initiate the sputtering process and eject target molecules, or they may be reflected as ions or neutrals. Photons and electrons are also ejected, which exposes the substrates and growing film to substantial UV radiation as well as to electrons and ions. The radiation may cause structural modification to sensitive substrates such as polymers.

The ejected electrons are crucial to plasma maintenance. In a diode system these electrons are accelerated back across the dark space toward the anode and collide with sufficient numbers of neutral (argon) atoms to ionize them and maintain the plasma. The simple diode configuration requires a relatively high pressure, near 100 mtorr. To maintain glow discharge at lower pressures (1–10 mtorr), we use either *triode sputtering* (Figure 10.9b), a configuration in which a hot filament is used to provide a copious supply of electrons that are accelerated toward the anode and collide on their way with atoms to ionize the inert gas, or *magnetron sputtering* (Figure 10.9c), in which an applied magnetic field causes the electrons to follow a spiral path in the plasma to increase their chances of collision.

We have drawn the diode sputtering system in Figure 10.9a using a DC voltage. This arrangement works only for conducting targets. To sputter insulators, we must use high-frequency AC voltages, and the most common frequency for this purpose is the one the FCC allots for industrial use, a radio frequency (RF) at 13.56 MHz. This technique for sputtering insulators is called *RF sputtering*. RF sputtering has the side benefit that we can reduce the pressure to about 1 mtorr.

We can sputter elements, compounds, or alloys. With compounds, the sputtered target molecules may be dissociated and the deposited film stoichiometry can vary from that of the target. Sputtering in a background environment that contains a mixture of inert (argon) and reactive (oxygen or nitrogen) gases (called *reactive gas sputtering*) can reestablish stoichiometry. For example, titanium is reactively sputtered in nitrogen to produce a hard coating of titanium nitride, $Ti_3N_{4\pm x}$. We can vary the color of the coating from light gold to rose to brown by varying the partial pressure of the nitrogen because deviations from stoichiometry, $x \neq 0$, determine its color. Alloys can be sputtered easily either from an alloy target or from a target of one material with small chips of a second material sprinkled across it.

One advantage of sputter deposition is that we can clean the substrates by switching the potential and making them the target briefly (5 min). If the substrates are in the plasma too long, however, the substrate temperature will rise and damage may occur.

The operating inert gas pressure in glow discharge sputtering (10^{-2} to 10^{-3} torr) is high compared to that of thermal evaporation (10^{-6} to 10^{-9} torr), a feature that has both advantages and disadvantages. On one hand, higher pressure means that substantial numbers of inert gas atoms are incorporated into the growing film, which may be unacceptable in high-purity applications. (To minimize the incorporation of other undesirable residual gas atoms, the vacuum chamber must be pumped down to UHV before introducing the sputter gas.) On the other hand, the mean free path is smaller at higher pressures, and sputtered target molecules make several collisions with the sputtering gas prior to reaching the substrate. Therefore, the depositing flux of molecules is less directional than in thermal evaporation and coverage is more conformal. Sputtered molecules lose energy in the collisions and arrive at the substrate with substantially less energy than they had upon sputtering. Nevertheless, their energy is still an order of magnitude higher (approximately 1 eV) than that of atoms from a thermal source.

Low-energy sputtered molecules are generally desirable to minimize damage to the growing thin film and the substrate. However, in some instances—particularly those that require good adherence of the film to the substrate—high-energy impact of the target molecules is essential for their penetration into the surface layers of the substrate. To achieve high-energy impact, substrates are negatively biased to attract ions during deposition (called *bias sputtering*). In this case, the more weakly bound target molecules are resputtered off the substrate, leaving behind only the strongly adherent target molecules. With energetic bias, the final film is more adherent, but it is also highly stressed and may need a thermal anneal to reduce the residual stress (see Section 10.7.2).

With respect to coverage uniformity, in diode sputtering the substrates are frequently smaller than the sputter target and uniform coverage is easy to achieve; in magnetron sputtering the plasma density and sputtering rate are higher because the electrons are channeled into one region of the target, the target erodes unevenly, and uniformity is less easily achieved. To improve uniformity, we may need to move the substrates during deposition.

A sputtering system's deposition rate depends on the rate at which we remove material from the target (the sputter rate) and on the transport mechanisms between target and substrate. Sputter rate depends first on the flux of ions striking the target, which is proportional to the the current flowing in the discharge, and second on the sputter yield. We define *sputter yield* as the average number of sputtered molecules leaving the target per incoming ion. Sputter yield is a function of the sputter ion energy, the relative masses of the sputter ion and target molecule, and the binding energy of the target molecule. Table 10.4 gives some typical sputter yields; values range from 0.1 to 3. Sputter yields increase with the incoming ion energy, reaching a maximum somewhere between 500 and 1000 eV. At low voltages, the sputter yield is approximately proportional to the voltage; so the sputter rate is proportional to the product of the current and the voltage or the sputter power. As we have already noted, molecular transport from the target to the substrate depends on collision frequency. Although there is no simple dependence between deposition rate and pressure, the deposition rate decreases at higher pressures.

10.3.2.2 Ion Beam Sputtering Operates at Ultrahigh Vacuum

A second source of ions for sputtering is the ion gun shown in Figure 10.9d. The "gun" consists of a small cylindrical chamber in which a glow discharge plasma of the desired ion species,

TABLE 10.4 Sputtering Yields for Argon Ions[a]

Sputter target	Ion energy 500–600 eV	Ion energy 1000 eV	Ion energy (eV)
Al	1.05	1.0	
Au	2.4	3.6	7.9 (5000)
C	0.12		
Cu	2.35		
Si	0.5	2.85	
W	0.57	0.6	1.1 (5000)
Al_2O_3	0.18	0.04	
GaAs	0.9		
Pyrex 7740			0.15 (1100)
SiC (0001)	0.41		
SiC (poly)	1.8		
Cr	1.3		
Cr_2O_3	0.18		
Ti	0.58		1.7 (5000)
TiO_2	0.96		

M. Ohring, *The Materials Science of Thin Films,* Academic Press, Boston, 1992, p. 113.

[a]Units are atoms/sputter ion or molecules/sputter ion. Data should be viewed as having fairly large error bars.

typically argon, is maintained at a pressure of 1–10 mtorr. Some of the ions are extracted through small holes at the end of the cylinder and accelerated by a pair of grids. Electrons from a hot wire placed just outside the ion gun neutralize the resulting positive ion beam to prevent it from Coulombic divergence (due to electrostatic repulsion).

An advantage of this technique is that the target and substrate can be in a high vacuum, although the pressure is limited to about 10^{-6} torr because some neutral gas inevitably enters the system through the holes in the gun. At this pressure, the high-energy beam of ions travels in a straight line pointed toward the target to produce sputtering (called *ion beam sputtering*, Figure 10.9d). We can control the energy of the ions by the acceleration voltage, and we can adjust the angle of incidence of the ion beam on the target surface for optimum deposition rate at the substrate.

Several features of ion beam sputtering differ from glow discharge sputtering. Incorporation of sputtering gas into the growing film is dramatically reduced. The absence of collisions of atoms sputtered from the target means they have higher energy than in glow discharge sputtering, which leads to improved adhesion. Conformal coverage is difficult because the sputtered atoms move in straight lines from target to substrate. Because the ion beam current is considerably less than in a glow discharge sputtering system, the deposition rate is lower.

In summary, sputter deposition rates are low to moderate compared with thermal evaporation, reaching a deposition rate of 10 nm/s under optimum conditions. Sputtering is more energetic, and sputter-deposited atoms have energies around 1 eV. Sputtering provides many more process choices and process variables, such as target voltage and current, sputtering pressure, and substrate temperature. All these factors are important for nucleation and growth and must be optimized for the particular thin-film system being developed.

10.3.3 Other Physical Vapor Deposition Techniques Combine Features from Thermal and Sputtering Methods

A number of hybrid techniques use aspects of both thermal evaporation and sputtering. Reactive gases like nitrogen may be ionized in a plasma to generate free nitrogen that will combine with thermally generated gallium to deposit complex materials like gallium nitride. Ionized cluster beam deposition uses a special nozzle to create and ionize vapor-phase clusters of a few hundred to a thousand atoms that are then directed toward a substrate. Milton Ohring discusses these and other hybrid techniques in *The Materials Science of Thin Films*.

A comparatively novel technique that is still in the research and development phase but holds out exciting possibilities uses a UV laser beam to ablate material from the source. In this technique, a pulsed, high-energy excimer laser beam is focused onto the target, as shown in Figure 10.10, and molecules are emitted by a mechanism that includes thermal evaporation and

Figure 10.10
High-energy excimer laser
beam evaporation and
deposition.

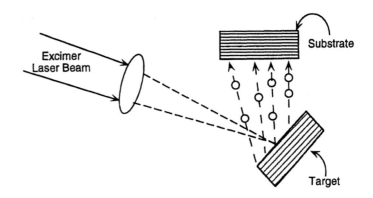

mechanical shock wave expulsion. The advantage of pulsed laser
ablation is that the material deposited on the substrate has exactly
the same composition as the source. It can be used for alloys,
refractory materials, and nonstoichiometric ceramics, as well as
for materials such as aluminum nitride and gallium nitride that
decompose below the melting temperature. It is a very high-
energy deposition process.

10.3.4 Substrates Must Be Carefully Prepared for Physical Vapor Deposition

Substrate preparation and handling are important steps in any
deposition process. Generally, substrates are cleaned chemically
prior to installation in the vacuum system using techniques
appropriate to the type of substrate and level of purity required.
Once inside the vacuum system, residual contaminants are removed by heating the substrate under vacuum. Heating may also
anneal out imperfections and relieve residual stresses in the
substrate that remain from mechanical polishing steps. In the case
of molecular beam epitaxy, substrate heating also is used to
roughen the single crystal substrate surfaces before deposition. As
discussed in Section 10.3.2.1, plasma cleaning of the substrate may
be carried out *in situ*. In the case of polymer substrates, plasma
treatment may dramatically alter the chemical nature of the surface,
sometimes improving the adhesion and interfacial bonding strength
of the thin film deposited on it (see also Section 7.4.3).

Control of substrate temperature during deposition is a
critical variable in physical vapor deposition. Depositing molecules bring a substantial energy flux to the surface of the growing
film, causing the substrate to heat up. High-energy sputtering
produces more heat than low-energy thermal evaporation. In
some applications, such as those involving polymers, the substrate needs to be maintained at ambient temperatures or lower
by water or liquid nitrogen cooling. In other applications, an
elevated substrate temperature of 300–500°C is desirable, and
additional heating is provided with a small electric heater contained in the substrate holder. Either way, the thermal conductivity of the substrate and thermal contact between the substrate and
the substrate holder are sensitive variables for substrate temperature control.

10.3.5 Deposition Rate and Thickness Can Be Monitored inside the Vacuum

Monitoring film deposition *in situ* is desirable for process control. The most widespread technique uses a quartz crystal thickness-rate monitor. A quartz crystal's mechanical resonant frequency is determined by its mass and thickness. A film deposited on one face of the crystal changes the resonant frequency due to its added mass. Because quartz is piezoelectric, we can determine the frequency electrically, and by measuring the change in resonant frequency, we can ascertain the deposition rate and thickness of a depositing film. We need to know the density and elastic modulus of the film for calibration purposes and usually take these calculations as the bulk values. Although the monitor is located close to the substrates, the deposition rate always differs between the two locations. As a result, a factor called the tooling ratio corrects for differences in location, density, and modulus. We use an *ex situ* thickness measurement to establish the tooling factor. A variety of methods measure film thickness *ex situ*, such as optical interference techniques, ellipsometry (for dielectrics), or etching away a section of the film and measuring the height of the step with a diamond stylus profilometer. Once calibrated, the quartz crystal monitor can be used in a feedback loop to control the power, ionizing gas pressure, and cutoff shutters in a physical vapor deposition system.

As shown in Table 10.1, special applications, such as optical interference filters or semiconductor devices, require films of a very precise thickness, sometimes down to a few atom spacings. To achieve this precision, it is necessary to terminate deposition abruptly. Ultra-high-vacuum physical vapor deposition methods suit this purpose well, since we can terminate deposition at any instant by moving the shutter shown in Figure 10.2 to cut off the beam .

10.4 Chemical Vapor Deposition Generally Employs Low-Vacuum Techniques

Now we focus our attention on thin-film processing by chemical vapor deposition (CVD) methods. In CVD a molecule in the vapor phase known as a precursor delivered to the substrate acts as the source of material. The precursor either undergoes chemical reaction with another vapor in the vicinity of the substrate or thermally decomposes on the substrate to form the thin film. Figure 10.11 sketches a typical although rather complex reactor system; it is designed to deliver a number of reactive gases to the reaction chamber concurrently or sequentially. We will discuss its essential features in the following sections. As with PVD, tradeoffs of purity, film quality, and expense enter the selection criteria for deciding the optimum process. However, unlike PVD—where generally we have only a single choice of source if we wish to achieve a given film—in CVD many different precursors can produce the same film.

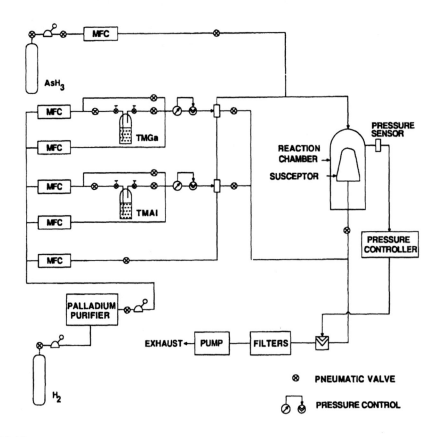

Figure 10.11
Schematic diagram of an organometallic vapor phase reactor illustrating both gas and liquid sources. Purified hydrogen gas bubbles through the liquid sources. Hydrogen also purges the sytem when the precursor is changed. Flow is regulated with electrically operated solenoid valves and mass flow controllers, MFCs. (S. K. Ghandi and I. B. Bhat, *Materials Research Society Bulletin*, **XIII** (11), 37, 1988.)

As with physical vapor deposition, chemical vapor deposition involves three sequential unit operations: generation of precursor molecules at the source; transport of precursor molecules to the substrate; and deposition of the thin-film molecules on the substrate in a vacuum.

10.4.1 Higher Operating Pressures Means CVD Gives Conformal Coatings

We usually conduct the chemical vapor deposition process in an enclosed Pyrex or quartz tube held at atmospheric pressure (760 torr, 10^5 Pa)—CVD, or at low pressure (0.1–1.0 torr, 1.3–13 Pa)–LPCVD. The fact that CVD operates at pressures much higher than PVD obviates the need for high-vacuum systems and presents other advantages. In PVD processing, conformal coverage of substrates is difficult. However, in CVD, higher pressures mean shorter mean free paths (0.05 mm at 1 torr), and the precursor molecules collide many times inside the reaction chamber before striking the surface. These collisions randomize the direction of arrival, and CVD growth gives much better conformality, depositing the thin-film material equally over all edges of steps and wells in the surface. Furthermore, this high conformality enables us to stack many substrates or wafers closely in a reactor and process them all at the same time to give uniform coatings with a high throughput and reduced cost.

10.4.2 Precursor Sources for Chemical Vapor Deposition Are Vapors, Liquids, or Solids

Next we discuss the generation and delivery of the precursor molecule(s) from the source to the substrate. Table 10.5 lists materials commonly deposited using CVD, and Table 10.6 lists typical molecular precursors. Precursors may be in vapor, liquid, or solid form.

Some precursors, such as silane (SiH_4) and arsine (AsH_3), are gases at room temperature. Obviously they can be stored and transferred to the reactor directly. For these precursors, therefore, the most critical components in the delivery system are the pressure regulators and mass flow controllers that must manage gas flow while maintaining high purity.

Other precursors, such as tetraethylorthosilicate (TEOS) or trimethylaluminum (TMAl), are liquids at room temperature, and they usually must be evaporated or delivered to the reactor with the aid of a carrier gas such as hydrogen, helium, nitrogen, or argon. The carrier gas is bubbled through the liquid precursor, and we can vary its delivery rate by adjusting the flow rate and temperature. This process also requires careful control over details such as the temperature of the liquid and maintaining a constant liquid level in the precursor vessel. Hydrogen is a popular carrier gas because by passing it through a palladium filter, we can purify it to a degree not affordable for other gases. In some depositions, the hydrogen carrier gas also acts as a scavenger by reducing impurities on the substrate.

Solid precursors are not as efficient as their liquid counterparts and create further difficulties of mass transport control.

10.4.3 Some Typical Chemical Vapor Deposition Reactions

After delivery to the reaction chamber, precursors react to form the desired product. Tremendous chemical diversity makes many CVD processes available for producing the same material. What procedure we actually choose depends on the temperature of the deposition, the purity required, and the cost. The following sections merely sample the diversity available.

10.4.3.1 Gas Sources Are Used to Deposit Metals, Polycrystalline Silicon, Silicon Dioxide, and Silicon Nitride

In some instances, LPCVD methods, such as the thermal reduction of tungsten hexafluoride, are preferred for depositing refractory metal contacts on silicon wafers.

$$WF_6 + 3H_2 \xrightarrow{\ 600^\circ C\ } W + 6HF \tag{10.4.1}$$

We can obtain polycrystalline silicon for electrical contacts by thermal decomposition of silane at low pressures. Hundreds of silicon wafers can be coated per run using the LPCVD configuration shown in Figure 10.12a operated at a pressure of 0.2–1.0 torr.

TABLE 10.5 Selection of Materials Grown by CVD and Applications

Material	Crystal structure	Film microstructure	Applications
Oxides			
SiO_2	Glass	Amorphous	Optical coatings, insulating layers
BPSilicate	Glass	Amorphous	Electronic packaging
TiO_2	Glass	Amorphous	Optical coatings
Elements (Metals)			
W	bcc	Polycrystalline	Diffusion barriers and interconnects
Al	fcc	Polycrystalline	Interconnects, reflective coatings, diffusion barriers
Cu	fcc	Polycrystalline	Interconnects
Au	fcc	Polycrystalline	Contacts
Carbides			
SiC	Cubic	Polycrystalline	Tribological coatings, high-band-gap semiconductors
WC		Polycrystalline	Tribological coatings
TiC		Polycrystalline	Tribological coatings
Nitrides			
Si_3N_4		Polycrystalline	Insulators, tribological coatings
AlN	Wurtzite	Polycrystalline	Tribological coatings
		Single crystalline	Piezoelectric sensor
BN	Hexagonal	Polycrystalline	High-temperature, chemically inert coating
TiN		Polycrystalline	Tribological coatings
Elements (nonmetals)			
C (diamond)	Cubic	Polycrystalline	Tribological coatings, thermal conductor
		Single crystalline	Semiconductors
Si	Cubic	Single and polycrystalline	Semiconductors
		Amorphous	Solar cells
Ge		Single crystalline	Semiconductors
Compound semiconductors			
III-V			
GaAs	Zinc blende	Single crystalline	Semiconductors
InP	Zinc blende	Single crystalline	Semiconductors
GaN	Wurtzite	Single crystalline	Semiconductors
II-VI			
ZnS	Zinc blende	Single crystalline	Semiconductors
ZnSe	Zinc blende	Single crystalline	Semiconductors
CdSe	Zinc blende	Single crystalline	Semiconductors
(Hg,Cd)T	Zinc blende	Single crystalline	Semiconductors

TABLE 10.6 Molecular Precursors for CVD Processing

Sources for Electropositive Elements			
Halides	*Alkyls*	*Hydrides*	*Alkoxides*
$AlCl_3$	$Al(CH_3)_3$	$[(CH_3)_2(CH_3CH_2)N]AlH_3$	$Al(O\text{-}i\text{-}C_3H_7)_3$
$SiCl_4$	CH_3SiCl_3	SiH_4	$Si(OC_2H_5)_4$
$GaCl_3$	$Ga(CH_2CH_3)_3$	$[(CH_3)_3N]GaH_3$	
BCl_3		B_2H_6	$B(OCH_3)_3$
WF_6	$Zn(CH_3)_2$	GeH_4	
$POCl_3$	SiH_2Cl_2	CH_4	
$TiCl_4$			

Sources for Electronegative Elements		
Hydrides	*Organic compounds*	*Elements*
NH_3	H_2O $(t\text{-}C_4H_9)PH_2$	O_2
PH_3	$(t\text{-}C_4H_9)AsH_2$	As_4
AsH_3	$(CH_2{=}CHCH_2)_2Se$	
H_2S, H_2Se	$(CH_3)TeTe(CH_3)$	

$$SiH_4 \xrightarrow{600^\circ C} Si + 2H_2 \qquad (10.4.2)$$

Adding oxygen and lowering the temperature provides for the LPCVD deposition of silicon dioxide dielectric layers.

$$SiH_4 + O_2 \xrightarrow{450^\circ C} SiO_2 + 2H_2 \qquad (10.4.3)$$

Dichlorosilane can be used as a source for LPCVD of silicon dioxide and silicon nitride dielectrics.

$$SiCl_2H_2 + 2N_2O \xrightarrow{900^\circ C} SiO_2 + 2N_2 + 2HCl \qquad (10.4.4)$$

$$3SiCl_2H_2 + 4NH_3 \xrightarrow{\sim 750^\circ C} Si_3N_4 + 6H_2 + 6HCl \qquad (10.4.5)$$

10.4.3.2 Liquid Sources Are Used to Deposit Silicon Dioxide and Metals

Silicon dioxide can also be formed by decomposing tetraethyl-orthosilicate (TEOS) vaporized from a liquid source.

$$Si(OC_2H_5)_4 \xrightarrow{500\text{--}800^\circ C} SiO_2 + H_2O, CO_2 \text{ byproducts} \qquad (10.4.6)$$

Thin films of aluminum can be prepared using triisobutylaluminum (TIBA). A typical LPCVD deposition conducted at 250°C at a total pressure of 2 torr results in a growth rate about 0.1 µm/min. Hydrogen is used as the carrier gas to transport the precursor into the reaction zone, and the balanced chemical reaction of the deposition is

$$2Al[CH_2CH(CH_3)_2]_3 \xrightarrow{250°C} 2Al(s) + 3H_2 + 6CH_2=C(CH_3)_2 \quad (10.4.7)$$

Figure 10.12
Chemical vapor deposition (CVD) reactor configurations. (a) Low-pressure CVD multiple wafer-in-a-tube reactor. Horizontal gas flow, hot-wall configuration. Used for conformal coating of multiple wafers. (b) CVD inclined-plane reactor. Horizontal gas flow, hot-wall configuration. Chemical reaction takes place above each wafer surface. (c) CVD pedestal reactor. Vertical gas flow, cold-wall configuration. RF induction heats the pedestal and the wafer. Thermal breakdown of precursor(s) leads to deposition on a single wafer surface. (d) CVD barrel reactor. Vertical gas flow, hot-wall configuration used for coating multiple wafers.

10.4.3.3 Multiple Sources Are Used to Deposit Complex Compounds

We can easily imagine the increased complexity of chemical vapor depositions involving two or more precursors. For example, growth of thin films of mercury cadmium telluride, which is used as an infrared detector, requires delivery of three precursors, usually Hg(vapor), vaporized $Cd(CH_3)_2$, and $(CH_3CH_2TeCH_2CH_3)$ in stoichiometric amounts.

Interest in the development of new molecular precursors for CVD processing that might offer advantages over current methods continues to expand. For example, tertiaryamine complexes of alane (AlH_3) have proved valuable for the growth of aluminum films and III—V semiconductor films that contain aluminum.

10.4.4 Reactor Design and Substrate Heating

Figure 10.12 displays several reactor designs used for different applications. Two important features enter into reactor design: the means of heating the substrate and the gas fluid dynamics. Re-

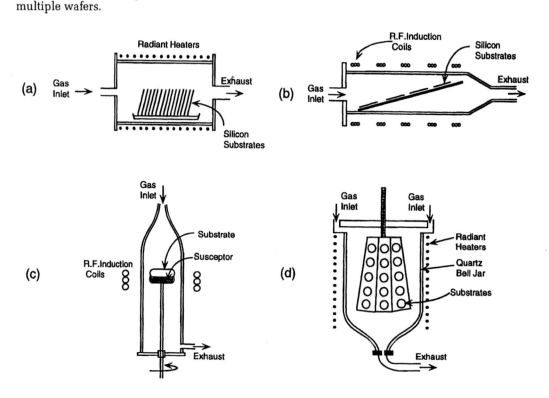

actors are optimized to (1) maximize the usefulness of the precursor, (2) minimize formation of stagnant recirculation flows, and (3) provide a constant flux of precursor to the entire surface of the substrate.

Depending on the application, we may heat the whole reactor (a hot-wall reactor) or just the substrate. Hot-wall reactors are used to grow amorphous or polycrystalline films of dielectrics such as SiO_2 or Si_3N_4. In a hot-wall reactor, many substrates can be stacked and coated simultaneously (Figures 10.12a and b). Cold-wall reactors are used to grow epitaxial layers of semiconductors such as gallium arsenide. In a cold-wall reactor, a single substrate is placed on top of a graphite susceptor that is inductively RF heated (Figure 10.12c).

To ensure uniform and conformal coverage in CVD, the gas flow must avoid regions of stagnant flow. Figure 10.13 shows the velocity profile of a single gas passing through a tubular reactor under ideal laminar flow conditions (see Section 3.8). At the gas–solid interface, the gas velocity is zero, giving rise to a region near the walls of a reactor (referred to as the boundary layer) where the concentration of precursor is depleted. We will discuss how this boundary layer affects the deposition process in Section 10.6.2.2.

Modeling the hydrodynamic behavior of real reactors confronts us with a complex challenge, but the results can be invaluable for optimizing reactor design and growth conditions. Numerical methods have been developed to solve the partial differential equations defining the conservation of energy, mass, and momentum for a given system (similar to the continuity equations in Appendix 3A). Figure 10.14 shows the flow pattern and thermal gradients for a typical inverted reactor used to grow single crystal films of GaAs and other semiconductors. Figure 10.14a illustrates a large recirculation cell at 500 torr. The recirculation can be substantially reduced by lowering the deposition pressure to 15 torr (Figure 10.14b). Precursor molecules trapped in recirculation cells spend a much longer time in the reactor than normal flow predicts, leading to unwanted side reactions that in turn generate impurities or particles with potentially detrimental effects on film quality. Configurations designed to improve the flow and overcome these stagnant regions include the barrel reactor shown in Figure 10.12d.

Figure 10.13
Gas velocity profile at different locations along a tubular reactor under ideal laminar flow conditions (see Figure 3.21). Parabolic profile is established by distance l.

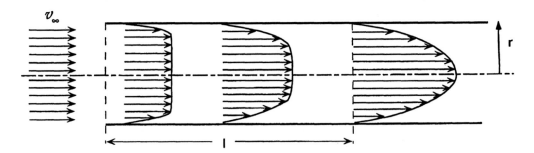

Figure 10.14
Hydrodynamic modeling predicts the mass flow pattern and thermal gradients for a typical inverted CVD reactor. (left) The model predicts a large recirculation cell when the pressure is 500 torr. (right) The model predicts almost pure laminar flow when the pressure is reduced to 15 torr.

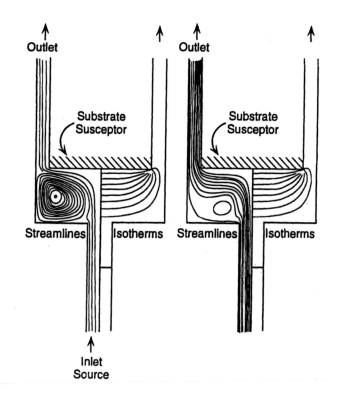

10.4.5 Plasmas and Lasers Stimulate Reaction at Lower Temperatures

One disadvantage of CVD arises from high substrate temperatures. Precursors stable enough to be stored for extended periods at room temperature require high temperatures for thermal decomposition. High substrate temperatures lead to solid-state diffusion and an undesirable smoothing of the composition profile at the interface, as illustrated in Figure 9.23. To overcome this limitation, new low-temperature CVD systems are being developed in which other energetic sources, such as plasmas (PECVD), lasers (LCVD), and electron beams, stimulate the chemical reaction at the substrate. Plasma excitation is particularly useful in depositing thin polymer films. Feeding monomer precursor into the reaction chamber in the presence of a plasma results in plasma polymerization and the deposition of a polymer film.

10.4.6 Techniques Are Needed to Monitor CVD Deposition Processes

One disadvantage of CVD is that we cannot monitor the process *in situ* in real time. Most CVD thin-film processing results are evaluated *ex situ* after the deposition is complete. In some situations, the result is known only after manufacture of the entire device, a fact that amplifies the cost of any errors that occur along the way. Attempts to change this situation have focused on the development of *in situ* methods to monitor the deposition process. Ideally, information gained by *in situ* real-time analytical probes would be fed back to control the deposition and correct any trends toward less than optimum processing conditions.

It is desirable to know the chemical status of the precursors in the different stages of reaction. Mass spectrometric methods, usually in the form of a small residual gas analyzer (RGA), are commonly used to probe the composition of the gas phase at the entry port. Monitoring the reaction at the substrate is more challenging. Spectroscopic methods identify molecular species by their characteristic absorption of electromagnetic radiation. High-intensity laser spectroscopic probes tuned to specific wavelengths hold promise for improving our capacity to identify and follow species of interest.

In situ probes of the growing surface topography and thickness are far less common. Light scattering has been used to probe surface roughness and to detect the presence of particles on surfaces. We can determine thickness by measuring the resistivity of metallic films by inductive methods without directly contacting the film. However, these results must be used with caution because the resistivity of the metal depends on both thickness and microstructure.

Currently, many of the methods for monitoring the growing film are expensive and not available for use with production deposition systems. Further research is needed to simplify these analytical methods.

10.4.7 For Certain CVD Systems, Safety Is an Important Issue

Significant expense involved in the design, construction, and maintenance of a CVD system arises from safety considerations. Some of the gases used as precursors for growing films (such as AsH_3 and PH_3) are very toxic, while others [such as SiH_4 and $Al(CH_3)_3$] are pyrophoric. A properly designed system not only must dispose of the normal effluent from the reactor in an approved fashion, but it must shut down automatically should an accident occur. Safety requirements usually involve the use of sensitive gas detectors that can actuate valves to stop the flow of gas and turn on the alarm system.

Reactor effluent contains a mixture of unreacted precursor (in some cases only a few percent of the precursor is deposited as the thin film) and reaction byproducts. CVD systems must be equipped to convert the residual gases into a form that can be disposed of properly. For example, highly toxic arsine can be passed through a furnace maintained at high temperature and converted into hydrogen and solid arsenic.

In spite of potential problems such as the difficulty of maintaining an ultrapure system, the high cost of certain precursors, doping level control, and safety issues, significant progress continues to be made in adapting the CVD process for manufacturing semiconductor devices. Figure 10.15 shows a complex gallium–aluminium–arsenide device array produced entirely by organo–metallic vapor-phase epitaxy (OMVPE).

Figure 10.15
High-performance thin film electronic devices are possible with organo-metallic vapor-phase (MOCVD) processing techniques. (top) This 3-inch gallium arsenide wafer contains over 5000 lasers with each cavity oriented for laser emission normal to the surface. The structure consists of two stacks of quarter-wave mirrors sandwiching a GaAs-based quantum-well region (see Section 10.8.2) and requires more than 100 distinct layers of AlAs/AlGaAs/GaAs. (bottom) Expanded view of surface in (a). (Photographs courtesy of Dr. Mary Hibbs-Brenner, Honeywell Technology Center.)

10.5 Epitaxial Growth on Single Crystal Substrates Requires Special Deposition Conditions

Successful epitaxial growth requires careful attention to all aspects of the vapor deposition processes we have described. There are several forms of epitaxy.[2] Processing strategy in semiconductor fabrication frequently involves *homoepitaxy*, the growth of a solid onto its own crystalline base, such as silicon on silicon, or

---

[2]*Epitaxy means different things to different people. Sometimes the term is used to describe any layer-by-layer growth resulting in an oriented thin film. Thus some workers discuss epitaxial growth onto glassy (noncrystalline) substrates. In this section we discuss epitaxial growth on single crystal substrates.*

gallium arsenide on gallium arsenide. This step serves many purposes. First, species adsorbed on a purchased substrate surface can be removed and replaced by pure material to provide a good, clean starting surface. Second, roughness on the atomic scale (illustrated in Figure 9.10) can be established to enhance subsequent growth. Third, and most important, the precise composition of the substrate immediately adjacent to the first device layer can be adjusted to suit the application. Depending on the device, small amounts of dopant may be added deliberately to convert the starting surface to n-type, p-type, or insulating material. Typically these so-called buffer layers are one micrometer thick. With good processing, the interface between the substrate and the buffer layer is featureless. Even high-resolution electron microscopy cannot detect structural discontinuity.

Growth onto a different crystalline base is known as *hetero-epitaxy*. The most important feature of this mode is that the contacting crystal planes of the substrate and film exhibit matching of symmetry. Any misfit between parallel lattice rows at the interface must be less than about 15%. We distinguish two different kinds of heteroepitaxy: one in which the film accurately maintains the crystal structure and crystallographic orientation of the substrate while only the chemistry changes—such as a layer of aluminum arsenide grown onto a substrate of gallium arsenide; and the other in which the film has a different crystal structure, crystallographic orientation, and chemistry yet the substrate controls the crystal orientation—such as a layer of superconducting ceramic yttrium–barium–copper oxide grown epitaxially on a magnesium oxide substrate. In the latter case, the {110} planes of the superconductor spatially match the {100} planes of the magnesium oxide with a 5% misfit. With good processing, the interface between the substrate and the epitaxial layer is clean and atomically sharp, as seen with high-resolution electron microscopy.

Epitaxial deposition is usually achieved through the vapor-phase—vapor phase epitaxy (VPE)—using the physical and chemical vapor deposition methods previously described. Epitaxial deposition also can be achieved through controlled solidification from the liquid phase—liquid-phase epitaxy (LPE). Because some modern semiconductor devices (see Table 10.1) require composition, thickness, and sharpness controlled down to an atomic level, more exacting vapor deposition techniques, such as molecular beam epitaxy (MBE), have evolved. MBE will be the focus of our attention in this section.

10.5.1 Molecular Beam Epitaxy Uses Evaporation under UHV Conditions

Figure 10.16 illustrates a typical molecular beam epitaxy system. To ensure epitaxial growth, the deposition rate is slow, and to obtain high-purity films, the residual gas pressure must be very low. Ultra-high vacuum systems operating below 10^{-8} Pa (10^{-10} torr) constructed out of stainless steel components with copper gaskets meet these conditions. Such a system contains three essential parts: molecular beam sources, the substrate support, and analytical tools for monitoring the process.

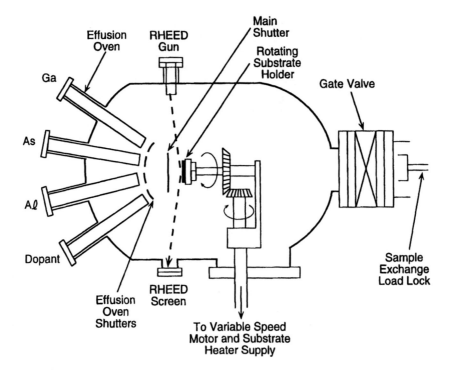

Figure 10.16
Schematic diagram of a
molecular beam epitaxy
(MBE) system used in the
fabrication of (GaAl)As
devices. Ultra-high-
vacuum system operates
below 10^{-8} Pa (10^{-10} torr).
The system contains four
Knudsen effusion cells
(Figure 10.8d) that provide
sources for gallium,
arsenic, aluminum, and a
dopant. Thickness of GaAs
and AlAs epitaxial layers
is controlled by shutters
and can be monitored by
the reflection high energy
electron diffraction
(RHEED) intensity (see
Figure 10.17). [A. Y. Cho
and K. Y. Cheng, *Applied
Physics Letters,* **38** (5), 360,
1981.]

10.5.2 Molecular Beam Sources

Molecular beams originate from carefully controlled evaporation
cells and rely on molecular flow in ultra-high vacuum. The
molecules travel in a straight line from source to substrate, and
by controlling the exposure, we can deposit an exact number of
molecules before switching to another source and repeating the
operation. Proper sequencing between sources can build up a
multilayer film of atomically precise thickness and composition.

The source is a Knudsen effusion cell, described in Section
10.3.1.2, Figure 10.8d. It consists of cylindrical crucible containing an evaporation source with a small orifice at one end. The
enclosure allows an equilibrium vapor pressure of the evaporant
to form, and a small portion of the evaporant effuses through the
orifice and streams toward the substrate in a straight line—molecular flow. Within the cell, the temperature of the source determines the vapor pressure. The cross-sectional area A of the orifice
determines the flux of material to the substrate. Ideally, the
angular distribution follows a cosine law, but in practice the
distribution is more peaked because $C > 1$ in eq. 10.2.4 and
because the flux profile varies with the geometry of the cell.
Specifically it is sensitive to the height of the melt and the radius
of the cell, and the ratio l/r in Figure 10.8d.

We can estimate the deposition flux f_{dep} of the molecular
beam near $\phi = 0$ by using eq. 10.1.10 (for the evaporation flux from
a source of area A), eq. 10.2.4 (for angular distribution of the flux),
and including the effect of the distance r from source to substrate
as in eq. 10.3.2. Then

$$f_{dep} = f\frac{(C+1)A}{2\pi r^2} = 2.63 \times 10^{20}\frac{P_o(\text{in Pa})(C+1)A}{\sqrt{MT}}\frac{(C+1)A}{2\pi r^2}\ \text{molecules/cm}^2\text{s} \qquad (10.5.1)$$

where P_o is the equilibrium vapor pressure within the cell at the source surface temperature T in K. M represents the molecular weight of the evaporating molecular species. For an aluminum molecular beam source, M = 26.98 g/mole, and $P_o \approx 0.1$ Pa at T = 1400 K; assuming a cell with orifice area 0.5 cm^2 placed about 10 cm from the substrate, the arrival rate is 2.15×10^{14} molecules/cm^2 s. If we assume that the aluminum atoms, which have a diameter of 2.864 Å, form a close-packed layer, then the surface density is 1.41×10^{15} atoms/cm^2 and the arrival rate corresponds to 0.15 atom layers per second or a deposition rate of 0.44 Å/s.

For precise control of the number of atomic layers deposited, molecular beams are switched on and off by means of mechanical shutters placed in front of each effusion cell. The shutters are operated electromagnetically in a switching time that is short (0.1–0.3 s) compared to the time (~10 s) needed to deposit a monolayer.

Under these circumstances, building up a film 1 μm thick requires about 6 h deposition time. During the course of such a long run, the flux profile from the Knudsen cell source changes due to exhaustion of the source material. Computer controls adjust the exposure during the lifetime of a newly charged source to compensate for gradual reduction in flux. The prime requirement for a source is a stable and reproducible flux, typically less than 1% drift over an 8-h day, with day-to-day variations of 2 to 5%. Maintaining such standards may require recharging the source before every run, a time-consuming and costly process.

More recently, replenishable external gas sources have been explored for some materials. Gaseous reactants like silane, SiH$_4$, or group V hydrides, such as arsine, are introduced into the system through a fine orifice to react at the substrate in same way as the CVD process discussed in Section 10.4.3. The term gas-source MBE (GSMBE) is used to designate such an approach.

10.5.3 Substrate Preparation and Temperature Control

As we might expect, depositing atomic layer by atomic layer on a crystalline substrate requires special attention to substrate surface preparation. A substrate must first be thoroughly cleaned both outside and inside the vacuum system using procedures already outlined in Section 10.3.4.

Because growth kinetics depend critically on the substrate surface topography, the surface must be roughened on an atomic scale to produce the desired distribution of ledges (steps) and kinks shown in Figure 9.10. As noted in Chapter 9, atomic roughness varies with crystallographic orientation and substrate temperature. Substrates are sold with the surface about 1° off the crystalline axis, and they may be roughened deliberately by raising the temperature to about $T_m/2$. Resistance heating elements are built into the substrate holder for this purpose and for cleaning and control of the substrate temperature during molecular beam deposition. Achieving uniform substrate temperatures at vacuum pressures of 10^{-8} Pa is not a trivial task. There is no convective heat transfer, and uniformity relies on good thermal contact. To obtain the desired thermal contact, liquid indium metal often is used to bond the substrate to the substrate holder.

10.5.4 Thickness Monitoring and Surface Analysis

Molecular beam fluxes are monitored by means of an ion gauge that can be placed in the sample position. Generally, fluxes are calibrated beforehand so that the whole deposition process (source temperatures, time of exposure, etc.) is preprogrammed and conducted under computer control.

In these ultra-high-vacuum systems, we can characterize the growth process *in situ,* using analytical tools such as Auger electron spectroscopy (AES) to determine the surface chemistry and low-energy electron diffraction (LEED) to determine crystal symmetry.

With certain materials, we can use reflection high-energy electron diffraction (RHEED) to monitor the deposition of individual atomic planes in real time. For layer-by-layer Frank–van der Merwe growth (see Section 10.6.3.1), the reflected electron beam intensity is at a maximum when the surface is flat (at the completion of a layer) and at a minimum (due to interference) when the surface is rough (at a half-monolayer coverage). Thus an oscillation in diffraction intensity corresponds to the growth of a single atomic plane. Figure 10.17 shows damped oscillations in the intensity of the RHEED pattern immediately following the start of growth of GaAs. In this example, each oscillation corresponds precisely to the growth of a single complete plane of Ga and As atoms equal to $a_0/2$ in the [001] growth direction, or 2.83 Å thick. By counting the oscillations, we can measure and control the

Figure 10.17
Intensity oscillations in the reflection high-energy electron diffraction (RHEED) pattern during AlAs and GaAs growth on (100)GaAs in an MBE system. The different stages of layer growth are depicted in the sketches at the top of the figure. Maximum diffracted intensity occurs when the layer is smooth, as in the first and last stage (see Figure 10.30). Each oscillation corresponds to the completion of a single layer of GaAs or AlAs.

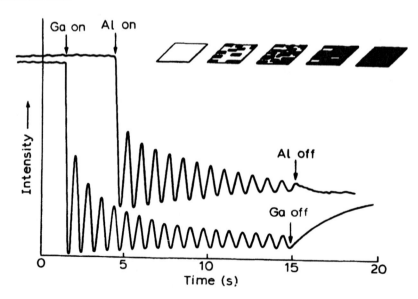

thickness of the layer to a prescribed number of atomic planes. Oscillations of the same type have been reported for other scattering techniques, including LEED, helium atom scattering, and Auger electron emission. These techniques can monitor thicknesses at the angstrom level.

10.5.5 Metallorganic Chemical Vapor Deposition Is an Alternative Technique for Processing Multilayer Epitaxial Thin Films

The CVD techniques described in Section 10.4 have been adapted for the growth of multiple epitaxial layers. In particular, gallium arsenide, aluminum–gallium arsenide, gallium phosphide, and indium phosphide layers have been grown on gallium arsenide substrates by CVD for various purposes, and Figure 10.15 presents one example. Gaseous arsine and phosgene and liquid or solid metallorganic sources are used. The acronym MOCVD (or in some circles OMCVD) identifies this technology. Arsine (AsH_3) is a highly toxic gas, and liquid sources of arsenic, such as tertiarybutylarsine, (t-C_4H_9)AsH_2, are available to avoid its use. These substances are more easily controlled, although they still retain high toxicity.

Multilayer processing with CVD involves switching from one vapor stream to another at preset intervals using mass flow controllers and complex systems like the one illustrated in Figure 10.11. Generally, an inert gas is used to flush the system between different vapor streams to avoid buildup of reaction products.

Because MOCVD uses relatively high pressures, we cannot monitor the growth process *in situ*. Examination of MOCVD epitaxial layers after deposition has shown them to be comparable in quality to films produced by MBE.

The relative merits of MBE and MOCVD have been hotly debated. MOCVD's high throughput makes it more attractive as a manufacturing process, but it lacks the precise control of composition and layer thickness that can be achieved with MBE.

10.6 Vapor-Phase Growth Mechanisms

So far, our discussion has been concerned with the apparatus and techniques used to form thin films. Next we focus our attention on what actually happens during vapor deposition from the point of view of the interaction between the substrate and the atoms or molecules that stream toward it. To simplify our discussion, we will refer to the incoming depositing species as atoms, even though they may be molecules. Incoming atoms are adsorbed onto the surface, whence they are called adatoms.

In the next three sections, we discuss first, in section 10.6.1, the thermodynamics of the critical nucleation stage; second, in Section 10.6.2, the kinetics that control the establishment of a continuous thin film and its growth rate for both physical and chemical vapor deposition processes; and finally, in Section 10.6.3, we bring the two together to describe alternative growth modes and their consequences with respect to the microstructure of thin film materials. The review by Venables, Spiller, and Handbücken contains an excellent presentation of this topic.

10.6.1 Capillarity Theory Describes Homogeneous and Heterogeneous Nucleation from the Vapor

10.6.1.1 Vapor Supersaturation and Substrate Condition Determine Nucleation on a Substrate of the Same Material

We begin by repeating the basic ideas of capillarity theory introduced in Section 3.4 for the nucleation of droplets from the vapor phase. These ideas provide a useful thermodynamic background for understanding nucleation of thin films, but we need to modify them to describe the formation of two-dimensional surface structures, such as disks, rather than three-dimensional spheres.

With respect to the nucleation of droplets, eq. 3.4.1 gives the free energy of forming a spherical drop of radius R in the vapor phase

$$\Delta G = -\frac{4}{3}\frac{\pi R^3}{V_L} RT \ln(P/P_o) + 4\pi R^2 \gamma_{lv} \qquad (10.6.1)$$

where V_L is the liquid molar volume. The first term in this equation is the product of the number of moles of material in the drop times the molar free energy of transferring material from the vapor into the drop. The second term gives the surface energy associated with creating the drop's surface area. Figure 3.9 shows how ΔG first increases with R, then goes through a maximum at R_c, where

$$\Delta G_{max} = \frac{4}{3}(\pi R_c^2 \gamma_{lv}) \qquad (10.6.2)$$

and subsequently decreases. Vapor-phase homogeneous nucleation requires the formation of a sufficient number of droplets with critical radius R_c, and this step in turn is very sensitive to the degree of supersaturation, P/P_o.

We can directly apply these ideas to *homogeneous nucleation* from the vapor of a solid onto a flat surface of the *same* solid. This is the case for homoepitaxy. Figure 10.18 shows the basic idea. We assume the new material takes the form of a disk of radius R and height a equal to one atomic layer area, where $a \approx D_{atom}$. The layer is added with perfect crystallographic continuity onto the planar substrate surface. No interfacial energy is required in homoepitaxy. (The same is not true for heteroepitaxy, which will be dealt with next.) In analogy with eq. 3.4.1, the free energy associated with forming the disk, ΔG_d, is

$$\Delta G_d = -\pi R^2 a \Delta G_s + 2\pi R a \gamma_{cv} \qquad (10.6.3)$$

where the first term is the product of the volume of the disk times the sublimation (or condensation) energy per unit volume, ΔG_s. The second term is the product of the extra surface area around the circumference of the disk times the surface energy of the

Figure 10.18
(a) Nucleation of a disk, radius R, one atomic layer high, during condensation from the vapor onto a flat surface of the same solid. (b) Cross-sectional view of the disk. (Reprinted with the permission of Macmillan College Publishing Company from *Electronic Thin Film Science: For Electrical Engineers and Materials Scientists*, by K. Tu, J. Mayer, and L. Feldman, p. 120. Copyright © 1992, Macmillan College Publishing Company.)

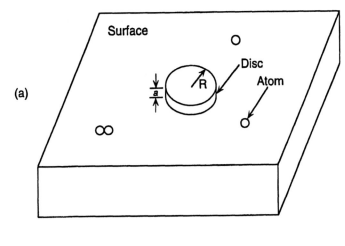

(a)

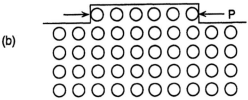

(b)

condensate per unit area, γ_{cv}. The maximum in ΔG_d, where $d\Delta G_d/dR = 0$, occurs when $R = R_c = \gamma_{cv}/\Delta G_s$, which gives

$$\Delta G_c = \pi R_c \, a \, \gamma_{cv} \qquad (10.6.4)$$

The vapor pressure above a disk is

$$\ln\left(\frac{P}{P_o}\right) = \frac{\gamma_{cv} v_m}{kTR} \qquad (10.6.5)$$

which differs from eq. 3.4.4 for spheres by a factor of two due to the difference in curvature H between the two geometries; for a sphere, $H = 2/R$, and for a disk, $H = 1/R$.

Substituting R from eq. 10.6.5 into eq. 10.6.4 (recognizing that for the critical condition $R = R_c$ and $P = P_c$) gives

$$\Delta G_c = \frac{\pi \gamma_{cv}^2 a v_m}{kT \ln(P_c/P_o)} \qquad (10.6.6)$$

We can view ΔG_c as the activation energy for homogeneous nucleation and R_c as the corresponding critical nucleus size. The disk nucleus is metastable when the vapor pressure is P_c; if P is lower than this value, the disk evaporates; if it is larger, the disk grows. The critical radius is

$$R_c = \frac{\gamma_{cv} v_m}{kT \ln(P_c/P_o)} \qquad (10.6.7)$$

We can obtain a feeling for the significance of this relationship by calculating the conditions required for homogeneous nucleation of silicon from the vapor onto a silicon substrate. Equation 10.6.6

provides an estimate of the critical energy needed to form the nucleus. The product $\gamma_{cv}a^2$ equals the surface energy per atom. For a planar silicon surface, this value is about 0.6 eV, but we expect it to be higher at the circumference of the disk because of the curvature and so assume a value of 1.0 eV. For a substrate temperature of 1223 K, $kT \approx 0.1$ eV. Assuming $v_m = a^3$, eq. 10.6.6 gives

$$\Delta G_c \approx \frac{10\pi}{\ln(P_c/P_o)} \text{ eV/atom} \tag{10.6.8}$$

When the supersaturation $P/P_o = 2.7$, $\Delta G_c = 31.4$ eV/atom, indicating that homogeneous nucleation is extremely unlikely to occur at such a low degree of supersaturation. The critical radius $R_c = 10a$ (estimated from substitutions in eq. 10.6.7) means that the number of atoms in the critical nucleus, $\pi R_c^2/a^2$, is 314—a number unlikely to congregate at random.

Just how much supersaturation does homogeneous nucleation require? Following the rationale of eq. 3.4.7, we can obtain the number of critical nuclei per unit area, N_c, from the product of the surface concentration of adatoms times the Boltzmann factor that describes their probability of accumulating to form a critical nucleus size. The surface concentration of adatoms is given by the product of the impinging flux of atoms f eq. 10.1.9, and their residence time τ_s. So that

$$N_c = f \tau_s \exp\left(-\frac{\Delta G_c}{kT}\right) \tag{10.6.9}$$

We can estimate the degree of supersaturation required for homogeneous nucleation to occur on a silicon surface from the following argument. In order for nuclei to grow, the distance between them must be less than the distance diffused by an adatom during the resident time period τ_s. This distance corresponds to about 10^8 nuclei/cm^2. When substituted into eq. 10.6.9, it requires a value of P_c/P_o from eq. 10.6.6 of about 10^4. In molecular beam epitaxy, a supersaturation of 10 is typical, meaning that no homogeneous nucleation is possible and no film growth should occur. Clearly, something is missing from our theory!

The huge discrepancy between theory and experience occurs because growth on a smooth surface rarely originates by homogeneous nucleation. Thin films nucleate *heterogeneously*. Real surfaces, as Figure 9.10 has emphasized, always contain a variety of surface sites where adatoms can be readily accommodated into the crystal structure, such as surface vacancies, kinks and steps, and chemical defects arising from impurities. Epitaxial growth occurs by surface diffusion of individual adatoms to these sites, at a rate determined by kinetics we will describe in Section 10.6.2.

Actually, the magnitude of the discrepancy is sensitive to substrate temperature. We analyzed the growth of silicon on silicon for a very high substrate temperature and a low level of supersaturation. The vapor pressure of silicon at the high temperature of 1223 K is 3×10^{-9} torr, a value comparable to the pressure in the high-vacuum system; so there is no supersaturation. But the degree of supersaturation over a cooled substrate can be enormous in the high-vacuum system. Vapor pressure is an exponential function of temperature; for silicon at 1000 K, the vapor pressure is 2.2×10^{-13} torr, and at 373 K, it is 5×10^{-51} torr—both values many orders of magnitude less than the pressure in the vacuum system. As supersaturation increases, the size of the critical nucleus decreases, and homogeneous nucleation becomes possible. For very high levels of supersaturation, the critical nucleus size reduces to the size of a single adatom and homogeneous nucleation is feasible. For dimensions this small, macroscopic thermodynamics is not applicable, and we should switch to atomistic theory to discuss nucleation. Actually, conclusions from atomistic theory are not qualitatively different from capillarity theory; so we will not pursue the atomistic theory any further here.

Even though homogeneous nucleation can occur on cooled amorphous substrates, it must compete with heterogeneous nucleation at chemically active sites. For real surfaces, nucleation takes place both homogeneously and heterogeneously.

10.6.1.2 Interfacial Energy, Vapor Supersaturation, and Substrate Condition Determine Nucleation on a Substrate of a Different Material

We now consider the more typical situation in which a thin-film material is being deposited from its vapor onto a crystalline substrate of *different* material, as is the case for heteroepitaxy. *Heterogeneous nucleation* occurs when adatoms accumulate at steps, defects, or impurities on the substrate surface. A change of phase takes place at the interface between the growing film and the substrate—an interphase interface is formed—and an interfacial energy term, $\pi R^2 \gamma_{sc}$, must be added to the right-hand side of the energy balance equation 10.6.3. In Section 3.4.3, we saw that nucleation of a new phase on a substrate reduced the free energy for critical nuclei formation from ΔG_c (the homogeneous value) to ΔG_{hetero} (the heterogeneous value) by an amount determined by the relative values of the surface energies of the substrate, γ_{sv}, and the condensate film, γ_{cv}, as well as the interface energy between the substrate and condensate, γ_{sc}. For liquid droplets, these energies are related to each other through the contact angle θ where

$$\gamma_{cv} \cos \theta = \gamma_{sv} - \gamma_{sc} \qquad (10.6.10)$$

$$\Delta G_{hetero} = \Delta G_c \frac{(2 + \cos \theta)(1 - \cos \theta)^2}{4} \qquad (10.6.11)$$

This result provides a guide for condensation from the vapor onto a solid substrate. The "contact" angle θ is related to the magnitude of the interface energy, γ_{sc}, between substrate and condensate, and γ_{sc} is related in turn to the effective interaction parameter W in eq. 3.1.8, between substrate and condensate atoms. We now consider two extreme conditions and then more typical conditions.

1. When the interaction is strong, W is small, $\gamma_{sc} \ll \gamma_{sv}$, $\theta \approx 0°-30°$ and $\Delta G_{hetero} \approx 0$. No barrier to direct nucleation on the substrate exists. Substrate and condensate adatoms are so strongly attracted that adatoms can accumulate any-

where on the substrate; no supersaturation is required; growth is controlled solely by kinetics.

2. When the interaction is weak, W is large, $\gamma_{sc} \gg \gamma_{sv}$, $\theta \approx 150°-180°$ and $\Delta G_{hetero} = \Delta G_c$. The substrate is totally inactive in the nucleation process. Substrate and condensate atoms do not interact; condensate atoms are preferentially attracted to each other. Under these conditions, nuclei must achieve the critical size for homogeneous nucleation, R_c in eq. 10.6.7, before growth can start; extremely high supersaturation is required.

3. When the interaction is moderate, $\gamma_{sc} \approx \gamma_{sv}$, $\theta \approx 90°$ and $\Delta G_{hetero} = \Delta G_c/2$. Substrate and condensate atoms interact, but with an energy about the same as self-attraction. Heterogeneous nucleation is preferred and is particularly favored at surface steps and kinks because there ΔG is smaller. This condition is the most prevalent and occurs for values of θ ranging from 50° to 130°.

In practice, control of nucleation sites is very important for the successful preparation of high-quality thin films by vapor deposition processes. It may be necessary at the outset to furnish a substrate with a high density of nucleation sites to establish uniform thin-film preparation. This roughness may be achieved outside the vacuum chamber through substrate treatments such as light mechanical abrasion or chemical etching to provide surface steps and inside the vacuum chamber through substrate treatments such as thermal desorption and/or sputter etching to clean and activate the surface (as mentioned in Section 10.3.4). As noted earlier, for some applications, single-crystal substrates are deliberately prepared off-axis to provide a high density of surface steps and kinks. Finally, the choice of substrate is important: a material that has a low contact angle with the condensate is most conducive to nucleation and growth.

10.6.2 Thin-Film Growth Kinetics Involves a Combination of Mass Transfer, Surface Adsorption, Desorption, and Surface Reaction Steps

In reality, nucleation and growth of a thin film is much more complicated than defined by simple capillarity theory. The surface is a very dynamic environment; adatoms driven by concentration gradients and thermal energy are coming and going and diffusing about the surface; adatoms are interacting with themselves to form dimers or larger aggregates or reacting with the surface to form new chemical compounds. As we shall show in Section 10.6.3, actual thin-film growth is the consequence of many competing events, but first we must consider in more detail the different aspects of surface kinetics as they pertain to growth of a film.

We can express the rate at which molecules in the vapor phase initially accumulate on the surface of a substrate by a simple mass-balance equation

$$\begin{bmatrix} \text{rate of mass} \\ \text{accumulation at} \\ \text{the surface} \end{bmatrix} = \begin{bmatrix} \text{rate of mass} \\ \text{in} \end{bmatrix} - \begin{bmatrix} \text{rate of mass} \\ \text{out} \end{bmatrix} \qquad (10.6.12)$$

This equation provides us with a way to organize and integrate our discussion of the kinetics of thin-film growth.

The general guidelines for using eq. 10.6.12 involve:

1. Postulating a sequence of steps that describes the deposition process;

2. Writing down a set of mass transfer and surface reaction rate equations for each step;

3. Combining these equations to obtain a general equation for the deposition process;

4. Solving the equation(s) to obtain an expression that relates deposition rate to measurable variables such as pressure, temperature, geometry, and flow.

Comparison with data then provides a way to determine which of the various possible mechanisms is operative during a deposition process and, more importantly, enables us to identify the rate-limiting step in the process.

We use CVD as the primary example in this section because it involves a more complex sequence of steps than PVD. Consider the basic steps in CVD: they involve decomposition of a gaseous precursor A into a solid deposit B and gaseous byproduct C. An example is CVD of silicon from silane

$$SiH_4 \rightarrow Si + 2H_2 \qquad (10.6.13)$$

$$(A) \rightarrow (B) + (C)$$

The gaseous precursor arrives at the boundary layer above the surface of the substrate and diffuses through it to the surface. Then the precursor adsorbs on the surface and either desorbs or reacts to form the new molecule that joins the growing film. Gaseous byproducts desorb and diffuse across the viscous boundary layer and are removed by the pumping system. Table 10.7 lists and Figure 10.19 shows the proposed six-step sequence whose individual steps we will analyze in turn. The subscripts refer to location: g is the flowing gas in the bulk; gs the gas in the boundary layer near the surface; and s the atoms on the surface. When we deal with more than one species, we also add a subscript to designate the component, such as the partial pressures, P_{gA} or P_{gC}, represented by P_{gSiH_4} or P_{gH_2} in this example.

Step 1. Mass transfer from the source to the substrate surface. In CVD, viscous flow leads to a boundary layer (thickness

TABLE 10.7 Reaction Steps Illustrated In Figure 10.19

Step No.	Symbols	Description	Rate
1	$A_g \rightarrow A_{gs}$	Mass transfer of A through the boundary layer	r_{mA}
2	$A_{gs} \rightarrow A_s$	Adsorption/desorption of A on surface	$r_{aA} - r_{dA}$
3	$A_s \rightarrow A^*_s$	Migration of A on surface	r_{sdiffA}
4	$A^*_s \rightarrow B_s + C_s$	Decomposition reaction	r_R
5	$C_s \rightarrow C_{gs}$	Adsorption/desorption of C from surface	$r_{dC} - r_{aC}$
6	$C_{gs} \rightarrow C_g$	Mass transfer of C through the boundary layer	r_{mC}

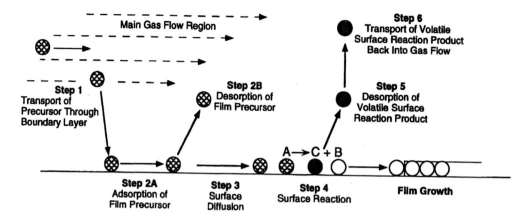

Figure 10.19
This schematic depicts the six different transport and reaction mechanisms underlying vapor deposition and thin-film growth.

δ) near the surface in which composition and temperature vary from P_g in the flowing stream to a value P_{gs} within the boundary layer. For chemical vapor deposition, diffusion controls the rate of mass transfer of the precursor molecule A, r_{mA}, across the boundary layer (Step 1). It can be written as

$$r_{mA} = k_{mA} (P_{gA} - P_{gsA}) \qquad (10.6.14)$$

where k_{mA} is the mass transfer coefficient and P_{gA} and P_{gsA} are the partial pressures of the *precursor* component A in the gas stream and near the surface, respectively. Using the relationship between mobility and diffusion coefficient in a fluid from eq. 3.8.35, we can write $k_m = D_g/kT\delta$, where D_g is the gas diffusion coefficient. We can estimate k_m if δ is known, and as a rough guide we approximate the distance δ as equal to the mean free path of the gas molecules, Λ in eq. 10.1.6. Generally, $k_m \sim P^{-1}$; so diffusion-controlled mass transfer is important as a rate-limiting step only for flow at high pressures such as those used for CVD. Thus eq. 10.6.14 forms the basis for modeling CVD processes outlined in Section 10.4.4.

In PVD, we can ignore boundary layers and estimate the mass transfer from the deposition flux. For thermal evaporation, discussed in Section 10.3.1.1, eq. 10.3.2

$$r_m = k_m P_g = \frac{P_g A_s \cos\theta}{\sqrt{2\pi m k T_g}} \frac{\cos\phi}{\pi r^2} \qquad (10.6.15)$$

where P_g and T_g are now the vapor pressure and temperature of the source, respectively. For sputtering, as discussed in Section 10.3.2.1, r_m depends on the specific sputter conditions.

Step 2A. Adsorption onto the Substrate Surface. When a gas atom or molecule impinges on a solid surface, a number of events can occur, depending on the energy and interaction between adsorbent and adsorbate. For low-energy deposition processes, such as chemical vapor deposition or thermal vapor deposition at room temperature, the molecule in the gas phase approaches the surface with an average thermal velocity of 500 m/s, eq. 10.1.4.

For high-energy deposition processes, such as sputter beam deposition, the molecular velocity is an order of magnitude higher. When an atom or molecule strikes the surface, it transmits some of its kinetic energy through an interaction that causes the surface atoms of the substrate to vibrate in a characteristic way. Eventually the gas atom or molecule may be ejected from the surface in a direction and with an energy that describes the nature of the interaction.

There are two extreme types of interaction. In one extreme, no attraction exists between the molecule and the surface. The time of interaction between the gas atom or molecule and the solid surface is limited to about one surface atomic vibration, 10^{-13} s, after which the atom or molecule rebounds off the surface with little energy exchange. A collimated beam of inert gas atoms or molecules such as He, Ne, and H_2 typically behaves this way. Atoms appear to be "reflected" from the surface and the scattering is said to be "specular." In the other extreme, the attraction is stronger. As a result, the period of interaction is much longer and may last for many atomic vibrations. An electrostatic attraction (Chapter 2) exists between the adatom and the substrate atoms, and the gas atoms are *adsorbed* onto the solid substrate. The adatom moves over the surface, jumping from one potential well to another, driven by its own residual kinetic energy parallel to the surface or in response to thermal activation from the surface. Should the molecule leave the surface, it does so in a direction that is independent of the direction of its arrival, with a cosine angular distribution, as discussed in Section 10.2.2. Adsorption is an important factor in the successful growth of thin films and one component of Step 2 in Figure 10.19.

We can describe these different energetic interactions with the substrate by an *accommodation coefficient* α. When no interaction takes place (specular reflection), $\alpha = 0$; when the interaction is complete and gas molecules adsorb on the surface, $\alpha = 1$. The magnitude of α is difficult to compute. It varies greatly depending on the type of interaction that takes place between the impinging atom and the surface. Some theoretical studies have examined the simple capture of an incident atom and its energy exchange in terms of van der Waals forces; others have considered head-on collisions of the adatom with a one-dimensional lattice of spring-connected masses. The net result is that α depends on the polarizability of the impinging adatom, the adatom—substrate interaction potential, and the substrate excitation spectrum due to phonons, electron excitation, etc. From this insight we might expect lower values of α for light adatoms of low polarizability interacting with a high-stiffness substrate. For example, alkali metals are difficult to deposit on a variety of substrates at room temperature because they have low atomic weight and polarizability. Saturation thicknesses for alkali metals are well below one monolayer.

For precursor adatoms (A) that do adsorb, we can obtain an expression for the rate of adsorption, r_{aA}, by modeling the surface as containing N_s surface sites per unit area, with N_A of them occupied and $N_s - N_A$ unoccupied. We assume that the rate of adsorption equals the product of the fraction of unoccupied sites, the accommodation coefficient, and the rate of mass transfer of A molecules, r_{mA}, eq. 10.6.14. Therefore, the rate of adsorption of the precursor molecule A is

$$r_{aA} = \frac{(N_s - N_A)}{N_s} \, \alpha_A \, r_{mA} \qquad (10.6.16)$$

Step 2B. Desorption from the Substrate Surface. Competing against Step 2A is the rate at which atoms leave or are *desorbed* from the surface. We focus first on one molecule adsorbed on the surface. Surface atoms vibrate with a characteristic frequency v_o, which is basically the same as the bulk vibration frequency ($v_o \approx 10^{13}$ Hz). If attractive forces exist between the molecule and the surface, then the frequency of desorption, v_{des}, is given by

$$v_{des} = v_o \exp\left(-\frac{\Delta G_{des}}{kT_s}\right) = k_{dA} \qquad (10.6.17)$$

where ΔG_{des} ($\approx \Delta G_{ads}$) is the change in free energy associated with adsorbing or desorbing one atom (see Section 9.3.1.2). T_s is the temperature of the *substrate surface*. We can think of the desorption frequency, v_{des}, as the frequency of the atom's or molecule's attempts to escape, v_o, multiplied by the probability of its success given by the Boltzmann exponential factor. The desorption frequency also defines the transfer rate per molecule, k_{dA}, by desorption.

The residence time before desorption, τ_s, of an atom on the surface is

$$\tau_s = \frac{1}{v_{des}} = \tau_o \exp\left(\frac{\Delta G_{des}}{kT_s}\right) \qquad (10.6.18)$$

where τ_o is the characteristic surface vibrational period of 10^{-13} s. The number of surface vibrations that it takes to shake the adsorbed atom loose is represented by τ_s/τ_o. When $\Delta G_{des}/kT_s$ equals 1.0, 3.0, or 10, then τ_s/τ_o is approximately 3, 15, and 22,000 surface vibrations, respectively. True adsorption occurs when τ_s becomes larger than several (~5) vibration periods. Then adsorbed molecules and surface atoms interact energetically and approach thermal equilibrium. As noted earlier, when a truly adsorbed molecule desorbs from the surface, it does so in a direction that is independent of the direction of its arrival; it demonstrates the cosine angular distribution.

The precursor desorption rate, r_{dA}, is given by the product of the desorption rate for a single molecule times the density of occupied sites, N_A,

$$r_{dA} = k_{dA} \, N_A = v_o \exp\left(-\frac{\Delta G_{des}}{kT_s}\right) N_A \qquad (10.6.19)$$

We have written desorption as a first-order process; it assumes the molecule desorbs from the surface without dissociation or association.

Step 3. Surface Migration during Residency on the Substrate Surface. We now consider the fate of a precursor molecule adsorbed onto a surface. During its period of residency, τ_s it does not remain at a fixed location, but moves dynamically from one site to a neighboring site by surface diffusion (see Section 9.5.6) with a frequency

$$\nu_{diff} = \nu_o \exp\left(-\frac{\Delta G_{surf\ diff}}{kT_s}\right) \qquad (10.6.20)$$

where $\Delta G_{surf\ diff}$ is the activation energy for surface diffusion. Using results developed in Sections 9.5.1 and 9.5.6, we can calculate the surface diffusivity as

$$D_{surf} = \lambda^2\ \nu_{diff} = \lambda^2\ \nu_o \exp\left(-\frac{\Delta G_{surf\ diff}}{kT_s}\right) \qquad (10.6.21)$$

where λ is the jump distance between neighboring surface sites, which will generally be equal to or related to the lattice parameter of the substrate.

As the molecule skitters over the surface, it encounters some sites where the binding energy is much stronger than it is at others; that is, the energy for desorption, ΔG_{des}, from those sites is higher. If the bonding is sufficiently strong, an adatom may form a nucleation site at that point. On crystalline substrates, these special sites include steps and kinks on the surface; on polymer substrates, they include particular portions of the chain where adatom–polymer chemical bonds are strong.

Step 3 involves an interplay between surface diffusion and desorption, therefore, that dictates the chances of a given adatom reaching a site where it is permanently bonded before desorbing from the surface and re-evaporating back into the gas phase. The critical feature in this process is the spacing between the sites of strong adsorption. To illustrate the interplay, we consider a surface containing steps with an average separation L_o, as depicted in Figure 10.20. For the adatom to be permanently adsorbed at the

Figure 10.20
Adatoms diffuse over the crystalline surface containing steps and kink sites. Adatoms become adsorbed at kink sites when their residence time on the surface is greater than the time required to diffuse the step separation distance L_o. (Reprinted with the permission of Macmillan College Publishing Company from *Electronic Thin Film Science: For Electrical Engineers and Materials Scientists*, by K. Tu, J. Mayer, and L. Feldman, pp. 130, 139. Copyright © 1992 Macmillan College Publishing Company.)

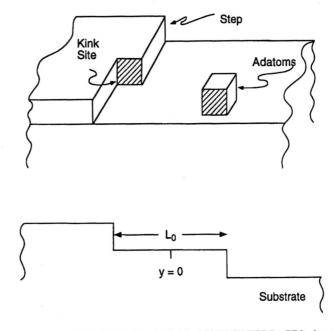

step and for net growth to occur, the residence time τ_s must be greater than the time required to diffuse to the step, τ_{diff}. We can describe this interplay between diffusion and desorption in terms of a coefficient S, defined by

$$S = \frac{\tau_s}{\tau_{diff}} \qquad (10.6.22)$$

When $S \geq 1$, the adatom has sufficient time to diffuse to a bonding site and to become incorporated in the growing film. For $S < 1$, the chances are greater that the adatom will desorb. From the Einstein random walk equation 3.8.25,

$$\tau_{diff} \approx L_o^2/D_{surf\ diff} \qquad (10.6.23)$$

Substituting eqs. 10.6.18, 10.6.21, and 10.6.23 into eq. 10.6.22 gives

$$S = \frac{\lambda^2}{L_o^2} \exp\left(\frac{\Delta G_{des} - \Delta G_{surf\ diff}}{kT_s}\right) \qquad (10.6.24)$$

For a noble metal surface with $\lambda^2 = 10^{-15}$ cm^2, $L_o = 10^{-5}$ cm, $\Delta G_{des} \approx$ 1 eV, $\Delta G_{surf\ diff} \approx 0.5$ eV, and $T_s = 298°$K, $S \approx 10^3$. When $T_s = 600°$K, $S \approx 10^{-1}$. Thus, at room temperature, sticking is complete while at elevated temperatures desorption dominates.

It should be noted that the accommodation coefficient α and the coefficient S are related but not identical. The accommodation coefficient describes the energetic interaction between the impinging ion and the substrate on arrival, while the coefficient describes the energetic interaction during residency.

The probability that an impinging atom will be permanently incorporated into the substrate is called the *sticking coefficient* σ. This pragmatic, measurable quantity is frequently used to characterize the efficiency of a vapor deposition process. We define it by the ratio of the first two terms in eq. 10.6.12

$$\sigma = \frac{\text{(rate of mass accumulation at the surface)}}{\text{(rate of mass in)}} \qquad (10.6.25)$$

The sticking coefficient incorporates features of both S and α. Clearly, σ is related to S and approaches 1 for higher values of S. The derivation of S in eq. 10.6.24 provides insight into ways to improve the sticking coefficient in a vapor deposition process. Choice and treatment of substrate and substrate temperature are the most obvious process variables that affect S. The derivation of α suggests that the energy of the incoming atom (or molecular) beam should also be a factor in determining σ. Atoms from low-energy CVD and thermal evaporation sources generally have a lower sticking coefficient than atoms from high-energy sputtering sources.

Step 4. Chemical Reaction on the Substrate. Permanent bonding of an adatom may result when a chemical reaction such as that described in eq. 10.6.13 takes place at a surface site. For chemical vapor deposition, this reaction occurs when the precursor molecule, A, decomposes into B + C, depicted as Step 4 in

Figure 10.19 and Table 10.7. It also occurs when the surface atoms react to form a new chemical compound and the adatom becomes chemisorbed on the surface. An example is the reaction of tungsten with a silicon substrate to form tungsten silicide.

When the precursor molecule A reaches a suitable reaction site, the rate of reaction to form B + C, r_R, is usually given in terms of an Arrhenius expression

$$r_R = k_r N_A = k_R \exp\left(\frac{U_R}{kT}\right) N_A \qquad (10.6.26)$$

where N_A is the density of adsorbed precursor molecules. Both the reaction rate constant k_R and the reaction energy U_R must be determined from chemical reaction data.

Steps 5 and 6. Desorption and Mass Transfer of the By-product Molecule. Completion of the CVD chemical reaction in eq. 10.6.13 also involves the desorption and mass transfer of the unwanted reaction product (C) away from the reaction site, Steps 5 and 6. These steps are represented by the rate constants r_{dC}–r_{aC} for the net rate of removal of C from the substrate and r_{mC} for mass transfer back into the vapor phase.

To analyze the kinetics of vapor-phase deposition, we need to combine the reaction rates of all the different steps to obtain a general equation for a given deposition process in terms of variables such as pressure, temperature, geometry, and flow rate. Actually, we often find that one step controls the overall film growth rate. Then steady-state approximation involves selection of the appropriate reaction rate from six possibilities:

$$
\begin{aligned}
r_{\text{film growth}} &= r_{mA} \\
&= r_{aA} - r_{dA} \\
&= r_{\text{surf diff A}} \\
&= r_R \\
&= r_{dC} - r_{aC} \\
&= r_{mC}
\end{aligned}
\qquad (10.6.27)
$$

as illustrated in the next two subsections.

10.6.2.1 Kinetics of the CVD Process Can Be Simplified by Focusing on Diffusion-Controlled and Reaction-Rate-Controlled Extremes

We can simplify our discussion of CVD kinetics by concentrating on the two major controlling steps: diffusion of the precursor through the viscous boundary layer and chemical reaction on the surface, Steps 1 and 4 in Table 10.7.

Equation 10.6.14 gives the rate of mass transfer through the boundary layer by

$$r_m = k_m (P_g - P_{gs})$$

and eq. 10.6.26 gives the rate of formation at the surface

$$r_R = k_r N_A = k_{rP} P_{gs}$$

where k_{rP} is in units appropriate for pressure.

At equilibrium, these two expressions are equal, and we can rearrange them to give the rate of formation in terms of rate constants and the pressure in the gas

$$r_R = \frac{(k_{rP}k_m)P_g}{(k_{rP} + k_m)} \qquad (10.6.28)$$

The reaction rate k_{rP} depends exponentially on temperature through a Boltzmann factor, while the value of k_m depends on a low power of temperature.

At low temperatures we are *reaction-rate limited*. Precursor gas diffuses to the surface but reacts slowly. Under these conditions,

$$r_R = k_{rP} P_{gs} \qquad (10.6.29)$$

and the film deposition rate is exponentially sensitive to temperature and linearly dependent on pressure.

At high temperatures and high pressures, we are *diffusion limited*. Precursor gas diffuses slowly through the boundary layer. Under these conditions,

$$r_R = k_m P_{gs} \qquad (10.6.30)$$

and the film deposition rate is less sensitive to temperature than with the reaction-rate-limited process and still increases linearly with the precursor gas pressure.

10.6.2.2 Kinetics of the PVD Process Can Be Simplified by Focusing on Adsorption and Cluster Growth

Discussion of PVD growth kinetics becomes simpler when we concentrate on the three basic controlling steps: arrival at the surface and adsorption of an adatom; incorporation of an adatom into the growing film; and desorption of the adatom, Steps 1, 2, and 3 in Table 10.7.

We outline the kinetics for island growth, also known as Volmer–Weber growth, one of three typical growth modes we will discuss in Section 10.6.3. Our discussion assumes the surface has N_1 adatoms per unit area and N_i clusters per unit area, with each cluster containing i atoms (referred to as an i-mer). Before growth starts, all the Ns are zero. To make the analysis tractable, we assume first that adatoms can diffuse on the surface, but clusters cannot; second, that adatoms can desorb, but clusters containing two or more atoms are stable and do not break up or desorb; third, that adatom residence on the surface is characterized by the time before desorption, τ_s from eq. 10.6.18; and fourth, we concentrate on the early stages of nucleation and growth and neglect incoming atoms that impinge on an existing cluster.

Using these simplifying assumptions, we can derive an expression for film growth as a function of the time by writing a simple mass balance equation for the change in the number of adatoms (i = 1) per unit time, dN_1/dt. The number of adatoms increases with the incoming deposition flux (r_m in eq. 10.6.15) and decreases for three reasons: desorption, meeting with another adatom to form a cluster of size i = 2, or diffusion to an existing cluster of size i to form a cluster of size i + 1. Taking each of these contributions in turn, for mass balance we have

$$\frac{dN_1}{dt} = r_m - \frac{N_1}{\tau_s} - k_1 N_1^2 - \sum_i k_i N_i N_1 \qquad (10.6.31)$$

To deal with the summation term, we turn our attention to clusters of size i. We gain an i-mer when an adatom joins a cluster of size (i − 1) and lose one when an adatom leaves to join an existing i-mer. Therefore, the mass balance equation for i-mers is

$$\frac{dN_i}{dt} = k_{i-1} N_1 N_{i-1} - k_i N_1 N_i \qquad (10.6.32)$$

Here the k_is represent rate constants for surface diffusion to clusters of size i. Various assumptions must be made to describe how these diffusion rate constants depend on the size of the clusters. If we assume that the ks are all equal and then substitute eq. 10.6.32 into eq. 10.6.31, we obtain a nonlinear second-order differential equation that can be solved numerically to give the density of clusters as a function of time. For the assumptions presented here, the total number of clusters per unit area is initially proportional to t^3, somewhat later is proportional to t, and yet later to $t^{1/3}$. Other assumptions give different time dependencies and distributions of sizes of clusters.

(Note that for *zero growth* conditions, we have $dN_i/dt = 0$ and the rates of adsorption and desorption are equal, $r_a = r_d$. Substituting eqs. 10.6.14 and 10.6.19 for r_a and r_d gives

$$k_{aj} P_{gj} (N_s - N_j) = k_{dj} N_j \qquad (10.6.34)$$

If we define Θ as the fraction of surface covered by gas molecules at equilibrium,

$$\Theta = \frac{N_j}{N_s} = \frac{KP}{1 + KP} \qquad (10.6.35)$$

where the reaction constant $K = k_a/k_d$. This result is the Langmuir adsorption isotherm for vapor on a solid surface, and its derivation is the one originally used by Langmuir. Other derivations produce the same relation, as already seen in eqs. 3.7.10 and 9.3.2. The Langmuir adsorbtion isotherm emphasizes the kinetic balance of adsorption/desorption processes at equilibrium.)

10.6.3 Growth Modes and Microstructures of Thin Films Result from One of Three Basic Mechanisms

Complete transformation of a flux of atoms or molecules from the gas phase into a thin film on a substrate involves a number of stages. In Section 10.6.1 we considered the nucleation events that provide the embryonic nuclei for thin-film growth, and in Section 10.6.2 we described the sequence of kinetic events that account for the accumulation of material on the substrate. Now we discuss how atoms organize on the substrate once they have arrived and been incorporated into the growing layer.

We find three broad categories of growth mode. Volmer–Weber or island growth occurs when the adatoms join three-dimensional *clusters* or *islands,* which eventually *coalesce* to form a continuous film. Frank–van der Merwe or layer growth occurs when the adatoms diffuse over the surface to locate at steps and kinks so that the film grows *one monolayer at a time.* Stranski–Krastanov or layer-plus-island growth occurs when the first or second monolayers grow as in Frank-van der Merwe growth, and then islands grow on top of these layers. Figure 10.21 schematically shows the three growth modes.

10.6.3.1 Volmer–Weber Growth Occurs by the Ripening and Coalescence of Clusters

Volmer–Weber, or *island, growth* is the most common growth mode. It prevails when the thin film and substrate are composed of dissimilar materials, have different crystal structure and chemistry, and no epitaxial relationship exists between them. Island growth occurs when the condensing film atoms interact more favorably with each other than with the substrate and the effective contact angle θ in eqs. 10.6.10 and 10.6.11 is about 90°.

Electron microscopy and scanning tunneling microscopy studies of thin-film growth have qualitatively identified similar stages of film evolution on a variety of substrates. These stages are:

1. Nucleation;

2. Growth of nuclei to form clusters;

3. Ripening or coalescence of clusters;

4. Contact of clusters or islands and filling of the channels between them.

Figure 10.22 illustrates these different stages of growth for a silver thin film on a sodium chloride substrate.

Figure 10.21
Schematic of the three growth modes for vapor-deposited thin films. (a) Volmer–Weber or island growth. Adatoms join three-dimensional clusters and eventually coalesce to form a continuous film. Note the microscopic voids and imperfections and the rough surface that results from this growth mode. (b) Frank–van der Merwe or layer growth. Growth proceeds one monolayer at a time. Dislocation lines can form at the interface to relieve strain. (c) Stranski–Krastanov or layer-plus-island growth. This is a mixture of the two previous growth modes.

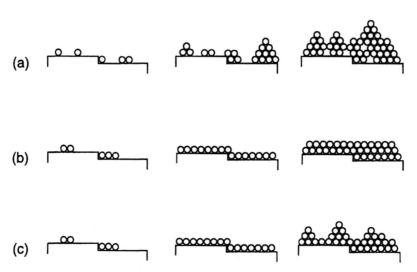

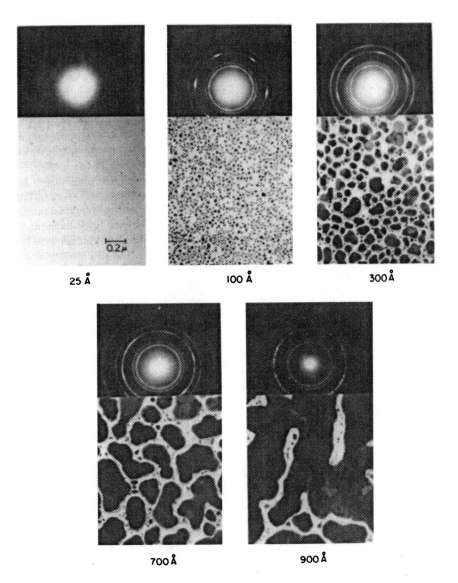

25 Å   100 Å   300 Å

700 Å   900 Å

Figure 10.22
Nucleation, growth, and coalescence of silver films grown by vapor deposition on sodium chloride (111) substrate. (R. W. Vook, *International Metals Reviews* **27**, 209, 1982. Photographs courtesy of Dr. R. W. Vook.)

Once nuclei have formed, they grow into islands or clusters by direct impingement of the incoming flux of atoms or by surface diffusion of adatoms. The differential equations given in Section 10.6.2.2 describe the initial stages of cluster growth.

Adatoms and small clusters are mobile over the surface, and when they approach each other and make contact, a kinetic exchange of material occurs between them. As a result, clusters constantly change their size and number. Two phenomena are responsible for this: *ripening*, in which large clusters grow at the expense of small ones, and *coalescence*, in which two clusters combine to form a single larger cluster. Reduction of the total surface-to-volume ratio of the deposited material drives both processes.

Ripening. The Kelvin equation, eq. 3.4.4, explains the ripening process. First, we assume all clusters to be hemispheres with $\theta = 90°$, so that small clusters have a higher surface curvature than large clusters. The Kelvin equation states that the vapor pressure above a small cluster must be higher than it is above a large cluster. Differences in vapor pressure set up a concentration gradient that causes preferential diffusion from small to large clusters. When an atom diffuses away from a small cluster to join a large cluster, as shown schematically in Figure 10.23, the curvature of the smaller cluster increases due to the loss of material, while the curvature of the large cluster decreases due to the gain in material. The equilibrium vapor pressure above the small cluster becomes even higher, and more atoms must be supplied from the small cluster to sustain the vapor pressure. This autocatalytic process causes the small cluster to shrink atom by atom, while the large one grows at the small one's expense.

We can envision two rate-limiting mechanisms for ripening: surface diffusion determines the first and the rate at which atoms detach themselves from the smaller islands determines the second. These mechanisms give different dependencies for island size versus time t. The size is proportional to $t^{1/4}$ for surface diffusion and to $t^{1/3}$ for detachment from the island. Figure 10.24 shows measurements of the radius R of tin islands on a silicon surface as a function of time. The results are consistent with the solid line that represents $t^{1/4}$ dependence, rather than the dotted line that represents $t^{1/3}$ dependence. In this instance, therefore, surface diffusion of adatoms between islands controls ripening. By measuring the rate at which islands ripen at different temperatures, we can determine the activation energy for the process. For Sn on Si, the activation energy for ripening is comparable to that for surface diffusion.

Coalescence. When the interaction energy between substrate and adatoms is small ($\theta > 90°$), clusters can detach themselves from any given location on the surface and diffuse as entities over the surface. They are very mobile, and when two clusters collide, they coalesce as shown schematically in Figure 10.25.

Reduction in surface energy drives coalescence. If we model the clusters as hemispheres (that is, $\theta = 90°$), then the cluster–vapor surface area $A_{cv} = 2\pi R^2$ and the cluster–substrate interface area $A_{sc} = \pi R^2$. The total surface plus interface energy U_s of the two isolated clusters of radius R_1 is given by

$$2U_{s1} = 2(2\pi R_1^2 \gamma_{cv}) + 2(\pi R_1^2 \gamma_{sc}) \tag{10.6.36}$$

and the surface plus interface energy of the combined cluster of radius R_T is given by

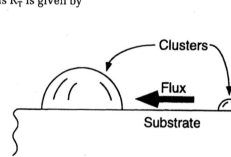

Figure 10.23
Ripening occurs when an atom diffuses away from a small cluster to join a larger cluster. Differences in vapor pressure set up a concentration gradient that drives diffusion from small to large clusters. (Reprinted with the permission of Macmillan College Publishing Company from *Electronic Thin Film Science: For Electrical Engineers and Materials Scientists,* by K. Tu, J. Mayer, and L. Feldman, p. 111. Copyright © 1992 Macmillan College Publishing Company.)

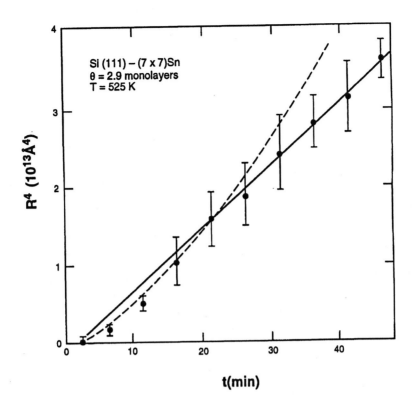

Figure 10.24
Ripening of tin clusters (islands) on a silicon (111) substrate as a function of time. The experiment, using 2.9 monolayers of tin, shows that the fourth power of the island radius is linear with time ($t^{1/4}$ dependence), consistent with surface-diffusion-controlled ripening. The broken curve corresponds to $t^{1/3}$ dependence for comparison. (Reprinted with the permission of Macmillan College Publishing Company from *Electronic Thin Film Science: For Electrical Engineers and Materials Scientists*, by K. Tu, J. Mayer, and L. Feldman, p. 109. Copyright © 1992 Macmillan College Publishing Company.)

$$U_{sT} = 2\pi R_T^2 \gamma_{cv} + \pi R_T^2 \gamma_{sc} \qquad (10.6.37)$$

Conservation of volume (mass) gives $2V_1 = V_T$, so that $(4/3)\pi R_1^3 = 2/3\pi R_T^3$ and $R_T = 2^{1/3} R_1$. Substituting this result in eq. 10.6.37 gives

$$U_{sT} = 2\pi \left(2^{1/3} R_1\right)^2 \gamma_{cv} + \pi \left(2^{1/3} R_1\right)^2 \gamma_{sc} \qquad (10.6.38)$$

Therefore, the ratio of the total surface energies before and after coalescence is

$$\frac{2U_{s1}}{U_{sT}} = \frac{2\pi R_1^2 (2\gamma_{cv} + \gamma_{sc})}{2^{2/3}\pi R_1^2 (2\gamma_{cv} + \gamma_{sc})} = \frac{2}{2^{2/3}} > 1 \qquad (10.6.39)$$

and coalescence is favored.

Most clusters are not hemispheres. They are spherical caps whose shape depends on the contact angle θ. Nevertheless, the arguments concerning ripening and coalescence apply to non-hemispherical shapes when we take into account corrections for the actual contact angle. For our purposes, the assumption of a hemispherical cap ($\theta = 90°$) for ripening and coalescence is not unreasonable, and the conclusions are valid.

Our discussion has implied that clusters behave more like liquid droplets than solid crystallites when they coalesce. Indeed, transmission electron microscopy studies show that crystalline clusters with faceted shapes behave like liquid droplets during coalescence. They may actually liquefy due to the release of

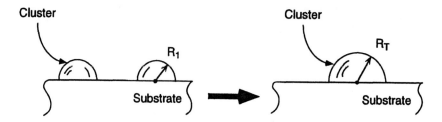

Figure 10.25
Coalescence occurs when two clusters collide and join. This schematic shows two clusters of radius R_1 joining to form a larger cluster of radius R_T having the same total volume as the two small clusters. (Reprinted with the permission of Macmillan College Publishing Company from *Electronic Thin Film Science: For Electrical Engineers and Materials Scientists*, by K. Tu, J. Mayer, and L. Feldman, p. 103. Copyright © 1992 Macmillan College Publishing Company.)

coalescence. They may actually liquefy due to the release of surface energy, but resolidify immediately afterwards to assume the crystalline orientation of the larger cluster. Other studies suggest that clusters remain crystalline during coalescence and that adatom diffusion or ripening fills up the spaces between the original clusters.

Filling in. Because all crystalline clusters grow in size through ripening or coalescence, eventually they come into contact to form an interconnected network with empty channels between them. Then a mechanism analogous to sintering takes place. At the re-entrant angles where the clusters touch, negative vapor pressure sets up a chemical potential gradient that causes material to fill the empty channels. The process of filling in occurs when adatoms separate from the clusters and diffuse to the most favorable surface locations.

10.6.3.2 The TMD Diagram Describes Thin-Film Microstructure Resulting from the Volmer–Weber Growth Mode

When all the clusters have connected and all the areas between them have been filled, growth continues due to the sustained atom flux. As illustrated in Figure 10.26, the resulting microstructure depends on competition between the growth and survival of existing clusters and the nucleation and growth of new crystalline clusters on the surface.

What alternatives are available for newly arriving adatoms? We have already discussed surface diffusion and desorption, each of which has a corresponding activation energy that appears in a Boltzmann factor. Adatoms can also incorporate into the microstructure by diffusing down the new grain boundaries or through the bulk to fill entrapped voids or porosity in the microstructure. Incorporation through grain boundary and bulk diffusion occurs most readily at higher temperatures. From Section 9.5.6, we expect the activation energies for surface diffusion, desorption, grain boundary diffusion, and bulk diffusion to follow the inequality $\Delta G_{\text{surf diff}} < \Delta G_{\text{des}} < \Delta G_{\text{gb diff}} < \Delta G_{\text{bulk diff}}$.

Empirically, all these energies scale with the melting point T_m of the thin-film material being deposited—higher energies generally being associated with materials that have high melting points. Therefore, it is not surprising to find that the ultimate structure and properties of thin films depend strongly on the ratio T_s/T_m, where T_s is the substrate temperature and T_m is the melting point of the thin-film material.

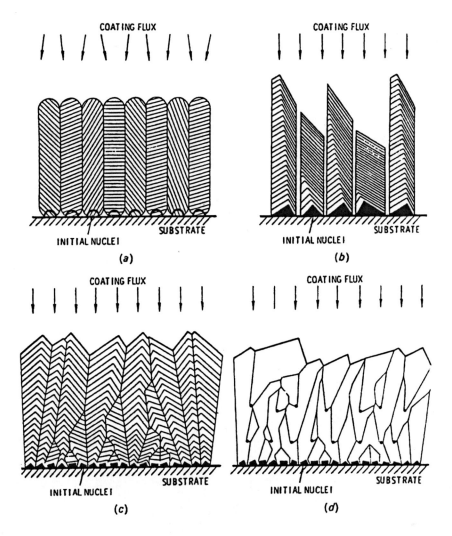

Figure 10.26
Computer modeling shows how the final microstructure of a vapor deposited film depends on competition between the growth and survival of existing clusters and the nucleation and growth of new crystalline clusters. (a) Growth structure at low temperatures when sticking coefficient is unity and surface diffusion coefficient is very low. (b) Growth structure at low temperatures when sticking coefficient is dependent on crystallographic orientation and surface diffusion coefficient is very low. (c) Growth structure at high temperatures when sticking coefficient is less than unity and surface diffusion coefficient is very high. (d) Growth structure at high temperatures when sticking coefficient is less than unity, surface diffusion coefficient is very high, and periodic renucleation occurs. (J. A. Thornton, *Annual Review of Materials Science*, 7, 239, 1977.)

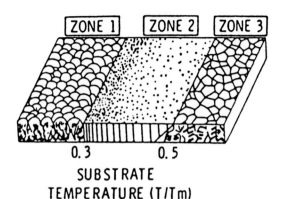

ZONE 1 ZONE 2 ZONE 3

0.3   0.5

SUBSTRATE
TEMPERATURE (T/Tm)

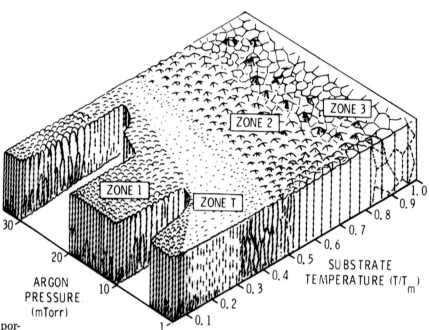

Figure 10.27
Growth structure of vapor-deposited thin films as a function of processing variables. (a) Zone model proposed by Movchan and Demchishin for the final structure of vapor-deposited coatings as a function of normalized substrate temperature (T_s/T_m, where T_m is the melting temperature). [B. A. Movchan and A. V. Demchishin, *The Physics of Metals and Metallography* **28**(4), 83 (1969). Translation of *Fizika Metallov I Metallovedenie* **28**, (4), 653, 1969.] (b) Zone model proposed by Thornton for the final structure of sputter-deposited coatings as a function of normalized substrate temperature and sputtering pressure. (J. A. Thornton, *J. of Vacuum Science and Technology* **11**, 666, 1974.)

Movchan and Demchysin initially looked at thick evaporated films and developed an empirical model for the final structure of the films as a function of T_s/T_m, as shown in Figure 10.27a. Thornton extended their model to cover sputtering, and the result is called the TMD *diagram*, which appears as Figure 10.27b. For evaporation the ratio T_s/T_m is the sole variable, while for sputtering this ratio and the sputtering pressure P are both variables.

At low temperatures, adatoms have little chance to diffuse or desorb, and they stick near where they deposit. The result is a fibrous film with a rough surface as depicted in Figures 10.26a and b. Such films exhibit poor mechanical and electrical properties. In Figure 10.27 this growth is represented in Zone 1, and it occurs for $T_s/T_m < 1/3$. At higher temperatures, surface diffusion takes place, and the adatoms can move to find more energetically favorable sites. The resulting film has a dense columnar structure, and a smoother surface as depicted in Figure 10.26c. Such films exhibit good mechanical and electrical properties. This growth is represented in Zone 2, and it occurs for $1/3 < T_s/T_m < 1/2$. At still higher temperatures, grain boundary diffusion, bulk diffusion, and recrystallization occur. The resulting film depicted in Figure 10.26d consists of equiaxed crystallites containing few defects such as dislocations. The film has a bright but faceted surface and is mechanically soft with good electrical properties. This growth is represented in Zone 3, and it occurs for $T_s/T_m > 1/2$.

In sputtering, a fourth transition zone, Zone T, occurs between Zones 1 and 2, as shown in Figure 10.27b. The extent of Zone T and the transitions between Zones 1, T, and 2 depend on the sputtering gas pressure. For many applications, Zone T or Zone 2 growth is the most desirable, since these zones produce films characterized by low density of voids, smooth surfaces, and good physical properties.

Figure 10.28 reproduces electron micrographs illustrating the different growth-zone modes. Computer modeling has confirmed the basic ideas that underlie the TMD diagram, and numerous experimental studies also have confirmed the diagram's utility. The approach has been applied to electroplated and to CVD films as well as the PVD films discussed here.

Some films grown on perfect crystal substrates initially produce a thin planar single crystal film whose orientation is epitaxial to the substrate or a thin planar polycrystalline film. These polycrystalline films are highly textured; that is, they have a common crystallographic axis driven by the orientation of the perfect crystal substrate. In later stages, growth switches to the Volmer–Weber mode. A columnar grain structure develops as the film thickens, but the substrate continues to control the orientation of the grains. This microstructure is important in magnetic thin films, where the easy direction of magnetization depends on film texture (see the example given in Section 10.8.1).

Shadowing also comes into play in determining the actual microstructure of Volmer–Weber thin films. If the atom flux arrives at an oblique angle, the taller clusters will shadow regions of the substrate, preventing atoms from reaching them. The result is a more pronounced columnar structure. Shadowing is strictly geometric and unaffected by substrate temperature or surface kinetics. A columnar structure is particularly sought in applications such as magnetic thin films in which the evaporated flux is deliberately introduced at an oblique angle to produce tilted columns.

Another strategy used to control the microstructure of thin films is to lower the interfacial energy between the condensing thin film and the substrate with intermediate layers. The role of these layers is similar to the one amphiphilic molecules play in stabilizing fluid systems, as discussed in Chapter 5. Indeed, the term *surfactant* is creeping into the solid-state literature to describe their use. For example, the surface energy of a ceramic is lower than that of a metal; so it is difficult to bond a metal thin film, such as copper, to a ceramic substrate, such as alumina. Attempts to deposit thin films of copper on alumina often result in balled-up droplets of copper but no continuous thin film. By comparison, a thin film of chromium easily wets and bonds to the alumina (because chromium atoms react chemically with the alumina to form a "spinel" crystalline phase). Since copper bonds

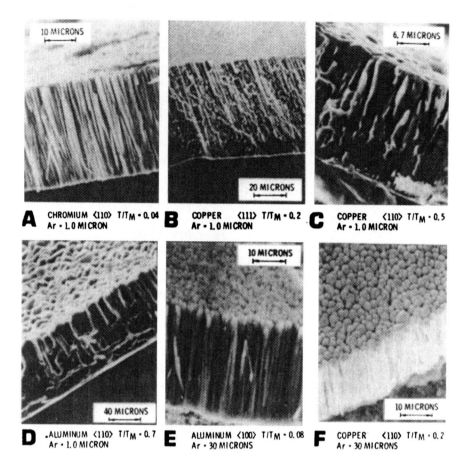

A CHROMIUM ⟨110⟩ T/T$_M$ · 0.04 Ar · 1.0 MICRON	**B** COPPER ⟨111⟩ T/T$_M$ · 0.2 Ar · 1.0 MICRON	**C** COPPER ⟨110⟩ T/T$_M$ · 0.5 Ar · 1.0 MICRON
D ·ALUMINUM ⟨110⟩ T/T$_M$ · 0.7 Ar · 1.0 MICRON	**E** ALUMINUM ⟨100⟩ T/T$_M$ · 0.08 Ar · 30 MICRONS	**F** COPPER ⟨110⟩ T/T$_M$ · 0.2 Ar · 30 MICRONS

Figure 10.28
Fracture cross sections of metal coatings sputter deposited at various normalized substrate temperatures and argon pressures. The scanning electron micrographs show both the surface texture and cross-section grain shape. (A), (E), and (F) have Zone 1 structure; (B) has Zone T structure; (C) has Zone 2 structure and (D) has Zone 3 structure. (J. A. Thornton, *J. of Vacuum Science and Technology* **11**, 666, 1974.)

to chromium, the introduction of a chromium intermediate layer enables us to produce a continuous copper film on the ceramic substrate. Intermediate buffer layers are commonly used to improve adhesion in the practical processing of thin films.

10.6.3.3 Frank–van der Merwe Growth Occurs by Monolayer Propagation

Figure 10.29 shows how the growth mode of a crystalline film on a single crystal substrate depends on the balance between the surface (or interface) energy and strain energy. We represent the surface energy axis by the ratio

$$\Omega = \frac{\gamma_{sv} - \gamma_{sc}}{\gamma_{sv}} \tag{10.6.40}$$

where γ_{sv} is the surface energy of the substrate and γ_{sc} is the substrate–condensate interfacial energy. The value of Ω is always positive and less than 1.0; it approaches a value of 1.0 as the

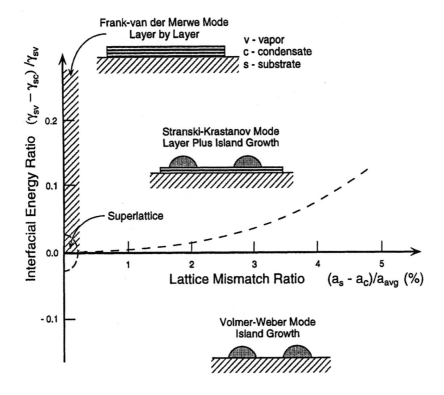

Figure 10.29
The initial growth mode of a crystalline film on a single crystal substrate depends on the balance between the interfacial interaction energy (represented by the ratio of surface and interfacial tensions) and the strain energy (represented by the lattice mismatch ratio). (*Electronic Thin Film Science: For Electrical Engineers and Materials Scientists,* by K. Tu, J. Mayer, and L. Feldman, p. 168. Copyright © 1992 Macmillan College Publishing Company.)

interfacial energy decreases to zero. We represent the strain energy axis by the lattice misfit strain between the crystalline substrate and the thin film, $\varepsilon_{\text{latt. misfit}}$, where

$$\varepsilon_{\text{latt. misfit}} = \frac{a_s - a_c}{a_{\text{avg}}} \tag{10.6.41}$$

In this equation, a_s and a_c are the lattice parameters of the substrate and the film materials, respectively, and $a_{\text{avg}} = (1/2)(a_s + a_c)$.

The *Frank–van der Merwe,* or *layer, growth* mode is prevalent when the thin film and substrate have similar crystal structure and chemistry. This mode is favored by epitaxial systems when the interaction between substrate and film atoms is favorable (that is, γ_{sc} is small and $\Omega \approx 1.0$) and the lattice misfit strain is small.

A slight misorientation of the substrate surface from a low-index plane creates a minimum-energy configuration in which the *vicinal* surface (that is, the actual higher-index surface) breaks up into low-index terraces separated by steps, as illustrated in Figures 9.10 and 10.30. Growth takes place atom layer by atom layer through surface diffusion of adatoms over the terraces. As the film grows, lattice mismatch leads to strain, and the total strain energy of the film increases as the film thickens. Above a specific thickness, the film becomes either unstable or metastable, and non-continuous film growth is preferred. If the film is unstable, transformation to the Volmer–Weber growth mode occurs spontaneously, and the combination is known as the *Stranski–Krastanov growth* mode. If the film is metastable, it may convert to isolated islands or clusters on heating.

Figure 10.30
Two alternative mechanisms for stable layer-by-layer growth. (a) Step propagation mechanism. Terraces remain smooth and flat throughout the deposition. The intensity of a diffracted high-energy electron beam (RHEED) remains constant. (b) Two-dimensional nucleation and growth mechanism. Terraces are rough when new disks form and smooth when disks grow and coalesce to form a complete layer. The intensity of (RHEED) beam fluctuates with this mechanism; intensity is at a minimum when rough and a maximum when smooth (see Figure 10.17).

Two alternative mechanisms for stable layer-by-layer growth relate to the adatom's surface diffusion length and its binding energy to the substrate and thus to the magnitude of S in eq. 10.6.22. Figure 10.30 illustrates these alternatives schematically. When S >> 1, adatoms can diffuse to the step, each layer grows laterally, and growth is controlled by the step propagation mechanism, Figure 10.30a. Terraces remain smooth and flat throughout the deposition. The intensity of a diffracted electron beam (reflected off the surface and used to monitor thin-film growth, as described in Section 10.5.4) remains constant with time with this mechanism.

When S < 1, nucleation and growth of new two-dimensional disks (islands one atom high) on the terraces controls growth, Figure 10.30b. The presence of many new disks roughens the terraces, but they become smoother as the disks grow and coalesce to form a complete layer. The intensity of a diffracted electron beam fluctuates with time with this mechanism. Diffracted intensity is at a minimum when the surface is roughest (because diffraction from adjacent layers is out of phase, and there is destructive interference) and at a maximum when the layer is complete. Each fluctuation corresponds to the completion of a single atomic layer.

In the epitaxial growth of compounds such as gallium arsenide, the two atoms can exhibit very different layer-by-layer growth characteristics. For example, homoepitaxial growth of GaAs onto a {100}GaAs substrate must occur by the sequential placement of gallium and arsenic layers (see Figure 9.5b).

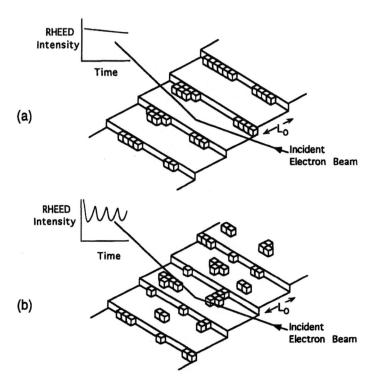

Arsenic's sticking coefficient (σ) on an arsenic-terminated surface is < 0.5, but it increases to 1 when a monolayer of gallium is present. Gallium's sticking coefficient on an arsenic-terminated surface also is 1. To grow gallium arsenide layers in an MBE chamber, we must maintain a constant vapor pressure of As_2 while gallium atoms are released from the evaporator cell at a controlled rate. As each gallium atomic layer is deposited, it is immediately covered by arsenic. The growth mode for this material occurs by the propagation of disks or steps two atoms high (Ga plus As). (Each oscillation in the diffraction intensity shown in Figure 10.17 corresponds to the completion of a layer pair.) Actually, this situation is fortunate for the processing of III–V arsenide and phosphide compounds because it means that we can accomplish epitaxial growth by controlling the supply of the group III element as long as we maintain an adequate arsenic or phosphorous background. Other multilayer epitaxial growth sequences are not as fortunate. For example, a silicon–germanium "superlattice" consisting of alternating layers of silicon and germanium must be fabricated sequentially: we must precisely regulate each element source to grow one atomic layer at a time.

10.6.3.4 Thin-Film Microstructure Resulting from Frank–van der Merwe Growth Mode Is Planar

Films grown by the Frank–van der Merwe mode generally produce planar single crystal films whose orientation is epitaxial with the substrate. When the substrate and film both have the same crystal structure (that is, when they are isomorphous) and the lattice mismatch is small, the film grows as a contiguous single crystal with the same orientation as the substrate. Imperfections at the interface become apparent only with high-resolution transmission electron microscopy, and they occur on a scale determined by the degree of misfit (see Section 10.7.1).

When the substrate and film are isomorphous yet the lattice mismatch is large, the film still grows on top of the substrate as a planar single crystal, but often it exhibits a different crystallographic orientation. The *Woods notation* describes the crystallographic relationship between the epitaxial overlayer and the substrate. In this notation, (a × b) Rθ means that registration of atoms in the ordered overlayer and in the substrate is achieved by a rotation (R) of θ degrees in the plane of the interface and at a location every a and b times the substrate lattice spacing.

In many applications, we seek heteroepitaxy between two materials of different crystal structures whose lattice registration for certain planes is close. For example, thin film layers of the yttrium–barium–copper oxide superconductor (tetragonal crystals with a modified perovskite crystal structure) are grown epitaxially on the {100} cube face of magnesium oxide (cubic sodium chloride crystal structure, Figure 9.4). These films are typically single-crystal layers, but they may be severely stressed or faulted.

10.6.3.5 Stranski–Krastanov Growth Combines Layer Plus Island Modes

We have cited a number of instances in which thin-film growth starts out as a planar layer and later switches to island growth. This effect is characteristic of the *Stranski–Krastanov growth* mode. Systems that favor this mode display strong interaction energy between atoms in the film and atoms in the substrate (that is, the effective interaction parameter, eq. 3.1.8, $W \approx 0$), but the interaction is not long-ranged. Stranski–Krastanov growth begins with the formation of a continuous film only a few atom layers thick through the layer growth mode. Subsequent growth on top of the first film occurs through the island growth mode because of reduced adatom–interface interaction. This growth mode is not so well understood as the other two mechanisms.

10.7 Imperfections and Failure Mechanisms in Thin Films

Structures produced by vapor deposition of thin films onto a substrate must have long useful lifetimes. Many factors adversely affect achieving this goal. The structures may be subject to high temperatures, high humidity, liquid immersion, abrasion, high electric fields, and currents. Such stressful environments can result in either initial failure, gradual decline in functionality, or delayed catastrophic failure as the thin film peels off the substrate. Initial failure requires immediate attention to the deposition process to remove its cause. Gradual decline and delayed failure are much more problematic to a manufacturer. Lengthy time delays between production and failure make it difficult to identify the cause of the failure, and customer goodwill is lost if problems show up in large amounts.

Here we discuss some of the factors that determine a thin film's longevity. For reasons of continuity with the previous section, we will begin by discussing imperfections in epitaxial films in Section 10.7.1. Common features of failures that occur in epitaxial layers include lattice imperfections such as dislocations, stacking faults, and twins. Section 10.7.2 discusses failures in Volmer–Weber films, which occur primarily due to stresses set up during film fabrication and due to diffusion along the grain boundaries between columns. We draw this distinction for convenience, and it should not be construed to mean that these modes of failure are exclusive to the respective types of solid–solid interface.

10.7.1 Failure of Epitaxial Interfaces Is Primarily Due to Mechanical Strain

10.7.1.1 Linear Imperfections—Misfit Dislocations—Appear at a Critical Strain

The Frank–van der Merwe growth mode illustrated in Figures 10.29 and 10.30 is limited by lattice misfit strain, $\varepsilon_{latt. misfit}$, eq. 10.6.41. If the misfit strain is greater than about 15%, Frank-van der Merwe growth becomes unstable and the growth mode switches to Stranski–Krastanov growth. If the misfit strain is less than about 15%, the epitaxial film strives to match the substrate at the interface through elastic distortion. As the film grows, however, the strain energy contained within the film volume increases until it reaches a level where accommodation of the lattice mismatch suddenly switches from elastic strain to misfit dislocations. Figure 10.31a illustrates the alternatives, and Figure 10.31b shows the misfit dislocations schematically. They constitute a square crossed grid of edge dislocations lying in the interface. Figure 10.31c reproduces a transmission electron micrograph picture of them. (Note: a crossed grid of edge dislocations accommodates a change in lattice parameter at an interface; a crossed grid of screw dislocations accommodates a pure twist rotation at an interface, Figure 9.16b.)

To estimate the critical film thickness h_c at which dislocations appear spontaneously at the interface, we compare the strain energy needed to stretch or compress the epitaxial layer with the dislocation energy. When the strain energy exceeds the dislocation energy, it is energetically more favorable for dislocations to reside at the interface. Lattice misfit strain energy (1/2 elastic modulus × strain^2 per unit volume) in a film of thickness h per unit area of the interface is

$$U_\varepsilon = \frac{E\varepsilon_{\text{latt. misfit}}^2 h}{2} \tag{10.7.1}$$

where E is Young's elastic modulus and $\varepsilon_{\text{latt. misfit}}$ is defined in eq. 10.6.41.

The energy per unit area of the interface due to the presence of misfit dislocations, U_d, is

$$U_d = \frac{2}{d} Gb^2 \tag{10.7.2}$$

where G is the shear modulus, b is the Burgers vector of the dislocation line, and the energy of a dislocation line is taken to be approximately Gb^2 per unit length (eqs. 9.4.9 and 9.4.10). To accommodate the change of lattice parameter, we assume the dislocations are regularly spaced a distance d apart in a square grid array. The value of d is given by

$$d = \frac{a_{\text{avg}}}{\varepsilon_{\text{latt. misfit}}} = \frac{b}{\varepsilon_{\text{latt. misfit}}} \tag{10.7.3}$$

where it is assumed that b, the Burgers vector of the dislocations, is approximately equal to the lattice spacing a_{avg}. Substituting for d, and assuming E to be about twice G, gives

$$U_d = Eb\varepsilon_{\text{latt. misfit}} \tag{10.7.4}$$

Equating and solving eqs. 10.7.1 and 10.7.4, we find that the dislocated interface is energetically preferred when the thickness exceeds a critical value approximately

$$h_c \approx \frac{2b}{\varepsilon_{\text{latt. misfit}}} \tag{10.7.5}$$

Figure 10.31
(a) Lattice misfit between substrate, a_s, and epitaxial film, a_c, leads to strain. For films thicker than a critical value h_c, strain is relieved by misfit dislocations. (J. M. Woodall, G. D. Pettit, T. N. Jackson, C. Lanza, K. L. Kavanagh, and J. W. Mayer, *Physical Review Letters* **51**, 1783, 1983.) (b) A square crossed grid of edge dislocations lying in the interface relieves strain in the epitaxial layer. (Reprinted with the permission of Macmillan College Publishing Company from *Electronic Thin Film Science: For Electrical Engineers and Materials Scientists,* by K. Tu, J. Mayer, and L. Feldman, p. 169. Copyright © 1992 Macmillan College Publishing Company.) (c) Transmission electron micrograph of an epitaxial interface between $Ga_{0.93}In_{0.07}As$ and GaAs showing crossed grid of edge dislocations. Dislocation spacing $\approx$ 140 nm. (J. M. Woodall, G. D. Pettit, T. N. Jackson, C. Lanza, K. L. Kavanagh, and J. W. Mayer, *Physical Review Letters* **51**, 1783, 1983.)

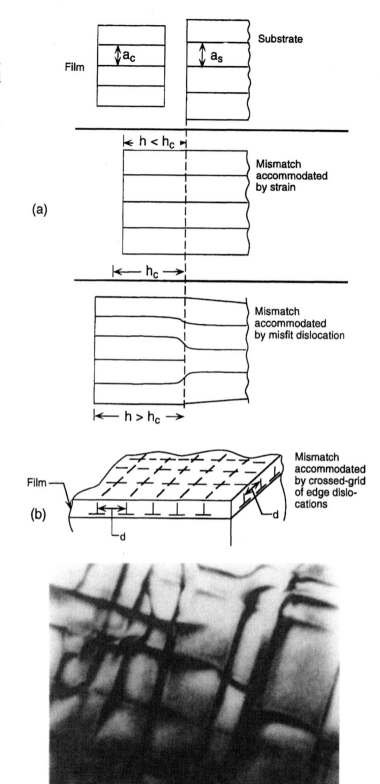

More precise calculations give

$$h_c = \frac{b}{8\pi(1 + \nu)\varepsilon_{\text{latt. misfit}}}\left(\ln\frac{h_c}{b} + 1\right) \qquad (10.7.6)$$

where ν is Poisson's ratio. Critical thickness values typically drop from about 100 to 5 nm as the lattice mismatch increases from 0.1 to 1%.

Ideally, these misfit dislocations are confined to the interface plane, but sometimes they thread out of the interface and into the film, hence the term "threading dislocations." Threading dislocations are undesirable for many applications of epitaxial structures. In modern semiconductor devices, such as high-speed transistors and electro-optic devices, dislocations act as centers for recombination of the electrons and holes that are the charge carriers responsible for signal generation; so special precautions are taken to avoid them. Three alternatives minimize the risk of threading dislocations: use of a substrate material with a close lattice match to the thin film material; introduction of intermediate layers to provide a more gradual change of lattice parameter; or keeping the thickness less than the critical value. The choice depends on the system.

10.7.1.2 Planar Imperfections—Stacking Faults and Twins—Appear in Epitaxial Films

Stacking faults and twins are a common result of epitaxial thin film processing. In view of the low surface energy of these planar defects (Section 9.4.3.2), it is not surprising that they appear along with dislocations in order to relieve strain at epitaxial interfaces. For example, in the growth of the III–V compounds InSb or GaAs on silicon, the lattice mismatch for InSb is 14%, while for GaAs it is only 4%. InSb/Si produces multiple microtwins at the interface whereas GaAs/Si produces both twins and dislocations. Twins are more prevalent with MBE than MOCVD processing, a fact that may be due to differences in energetics. Rough substrate surfaces also lead to twinning. Figure 10.32a shows severe twinning and lattice buckling in the CdTe layer of a CdTe/GaAs(100) interface for which the lattice mismatch $\approx 14.5\%$. Figure 10.32b reveals the detailed atomic structure of a microtwin and the dislocations introduced to relieve strain at the interface.

Even though lattice matching may be good across an interface, planar defects are generally undesirable because they reduce electronic response and are not acceptable for technological applications.

10.7.2 Residual Stresses Induce Failure of Interphase Interfaces

A principal failure mode in interphase interfaces is de-adhesion of the thin film from the substrate due to residual stresses. In some instances, these stresses can be so high that pieces of the thin film roll up or flake off the substrate surface, and the product must be discarded. In other instances, residual stress leads to a more gradual change in behavior. The film may crack or "craze" with

Figure 10.32
(top) Cross-sectional electron transmission micrograph of a MBE cadmium telluride film grown on a (100) gallium arsenide substrate. The large lattice parameter difference, ~14.5%, results in substantial twinning and elastic distortion of the upper CdTe layer. This cross section has been rotated 20° about the horizontal axis so that the interface projects as a narrrow strip containing Moire fringes (see A and B). Discontinuities in the fringes correspond to interface dislocations. Bar scale 100 Å. (J. E. Angelo, Ph.D. Thesis, University of Minnesota, 1993.) (bottom) High-resolution electron micrograph of the CdTe/GaAs (100) interface. This lattice image shows a microtwin (view bottom right to top left) and dislocations (view bottom left to top right) originating at the interface. (J. E. Angelo, Ph.D. Thesis, University of Minnesota, 1993.)

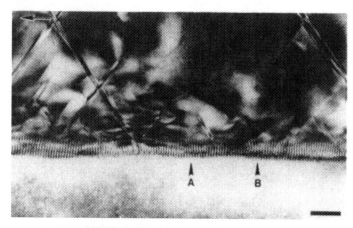

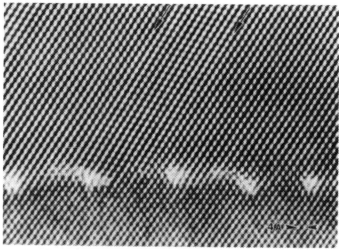

time, as shown in Figure 10.33, or the physical response may gradually change. Magnetic or electrical behavior is particularly susceptible to stress changes through coupled effects such as magnetostriction or electrostriction.

10.7.2.1 Intrinsic Residual Stresses Are Introduced during Deposition

Residual stresses in thin films arise from two mechanisms. A thin film may develop an intrinsic stress as it grows. We have just seen how intrinsic stress in epitaxial films arises from lattice mismatch. The origin of intrinsic stress in interphase interfaces is less clear. Evaporated films tend to have a residual tensile stress, whereas sputtered films may have tensile or compressive residual stresses. The magnitude of the stress depends on the growth mode and resultant microstructure discussed earlier in reference to the TMD diagram (Figure 10.27). Residual stress is typically smaller in Zone T. For sputter deposition, we can choose parameters such as power, bias voltage, sputter pressure, and substrate temperature to minimize stress. For thermal evaporation, we can use substrate temperature and ion-beam-assisted deposition to modify the stresses. If the substrate/thin film can withstand high

Figure 10.33
Failure due to residual compressive stresses in silver lithium thin films. (Photograph reproduced courtesy of Dr. Robert E. Cuthrell, Sandia National Laboratories.)

temperatures, we can also consider a postdeposition annealing step to remove stress.

10.7.2.2 Thermal Expansion Mismatch and Thermal Gradients Cause Thermal Residual Stresses

A film may also develop residual stress due to thermal expansion mismatch or temperature gradients.

Thin films and substrates contract by different amounts on cooling, and Figure 10.34 shows how a biaxial stress state (σ_{xx}, σ_{yy}) is set up in the thin film. Assuming completely isotropic material, simple elasticity theory gives

$$\sigma_{xx} = \sigma_{yy} = \frac{E_f \, \Delta\varepsilon}{(1 - \nu_f)} \tag{10.7.7}$$

where E_f and ν_f are the tensile elastic modulus and Poisson's ratio for the thin-film material, respectively, and $\Delta\varepsilon$ represents the strain produced by cooling a bimaterial interface with a thermal expansion mismatch, $\Delta\alpha$, through a temperature change, ΔT. $\Delta\alpha$ corresponds to the difference between the thermal expansion coefficients for film and substrate ($\Delta\alpha = \alpha_f - \alpha_s$), and ΔT corresponds to the difference between the substrate temperature during processing and the ambient value.

$$\Delta\varepsilon = \Delta\alpha \, \Delta T \tag{10.7.8}$$

Figure 10.34
Biaxial stress in a thin film deposited on a rigid substrate.

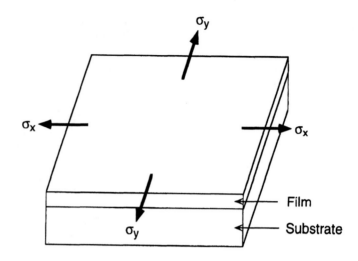

$$\sigma_{xx} = \sigma_{yy} = \frac{E_f \, \Delta\alpha\Delta T}{(1 - \nu_f)} \qquad (10.7.9)$$

In some instances, residual stresses originate due to temperature gradients. A thin-film system may be subject to localized heating from a radiant beam, as in laser-beam or electron-beam annealing of semiconductor films. Then residual thermal stress depends on the intensity of the incident beam and the rate at which power is absorbed. In this situation, the governing equation for heat transfer is the Fourier heat equation, given by

$$\frac{dT}{dt} = \frac{\kappa}{\rho C_p} \nabla^2 T + \frac{P_{abs}}{\rho C_p h} \qquad (10.7.10)$$

where κ stands for the thermal conductivity, P_{abs} is the power absorbed—in a single strike of an electron beam, for example—and ρ, C_p, and h refer to the density, specific heat, and thickness of the film. The term $\kappa/\rho C_p$ is often called the thermal diffusivity. If we know how P_{abs} depends on reflectance and absorption, we can obtain the temperature distribution of the sample as a function of time. Temperature gradients lead to thermally induced stresses. In monolithic films, the hot spot expands, the local material becomes compressed, and the far-field material develops tension. We can approximate the stress that develops in the plane of the film with a linear thermal expansion coefficient α by

$$\sigma = \alpha E_f \, \Delta T \qquad (10.7.11)$$

In a silicon film, pulsed beams can produce a local temperature difference of 100°C, and in a device, stresses of 100 MPa may easily develop. Although these stress levels are acceptable, materials with lower thermal diffusivities than silicon ($7.9 \times 10^{-5} \mathrm{\ m^2 \ s^{-1}}$ at room temperature) would produce much greater ΔTs and therefore larger stresses.

10.7.3 Mechanical Stresses due to Impact and Friction Cause Failure in Service

Interfaces are frequently subjected to mechanical stress during use. Inadvertent stresses may develop due to vibration during shipping; severe Hertzian impact stresses may develop in near contact or contact magnetic recording. Clearly, friction contacts or wear contacts can cause failure in many interfacial systems, and enormous efforts have been made to improve adhesion or to provide protective layers or boundary lubrication to increase component longevity.

As an example of mechanical stresses, we can outline the contact load problem. Consider a force F concentrated on a straight edge of a plate of thickness h which resembles the point force spreading out in two dimensions into AA′BB′ illustrated in Figure 10.35. Stress analysis gives a radial stress

$$\sigma_r = \text{const.} \, \frac{F \cos \theta}{hr} \qquad (10.7.12)$$

which we can easily solve for equilibrium, recognizing that the sum of the forces in the vertical direction is zero. From the discussion in the caption of Figure 10.35 and eq. 10.7.12, we can see that

$$F = h \int_0^\pi \sigma_r \cos \theta \, r \, d\theta = 2 \, \text{const.} \int_0^{\pi/2} F \cos^2 \theta \, d\theta \qquad (10.7.13)$$

so that

$$\text{const.} = \frac{1}{2\pi/4} = \frac{2}{\pi} \qquad (10.7.14)$$

Inserting this result in eq. 10.7.12 and letting r = z, the contact stress distribution just below the contact point is given by

$$\sigma_{zz} = \frac{2F \cos \theta}{\pi h z} \qquad (10.7.15)$$

Figure 10.35
Radial stress field due to point contact. The radial stress σ_r acting toward the load point over area dA (=hr dθ) contributes a vertical force component (= σ_r hr dθ cos θ) that when integrated for all θ counteracts the point force F.

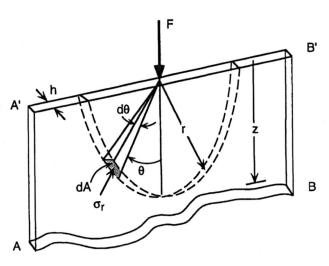

In analogous ways, we can find σ_{zz} stresses acting under contacts on a plane body of infinite thickness h and finite width AB, for point, spherical, or conical contacts. Table 10.8 gives the results of calculations made for these three types of contacts on a planar surface. For relatively high-strength materials—$\sigma_{yield\ strength} \sim 1500$ MPa—we obtain similar results for all three types of contact slightly below the contact. A 1 g load acting on a 1 μm contact point may generate stresses of 4000 MPa.

10.7.4 Thin-Film to Substrate Adhesion Is Important but Difficult to Measure

Of all the properties of a thin-film system, adhesion is one of the most important. However, the term adhesion is ill-defined and extremely difficult to quantify. We can describe fundamental adhesion in terms of the energy per unit area needed to separate a thin film from its substrate. Practical adhesion refers to a thin film's ability to resist delamination in a practical setting, under stresses appropriate to the application. Consider an evaporated metal on glass. Under an abrasive load, most metals adhere strongly to a glass substrate. However, if we apply a piece of adhesive tape to the metal as a test of its adhesion, the metal may pull off very easily. Both abrasive (tangential) and tape (normal) tests are legitimate measures of practical adhesion, but they give very different results.

We can consider the requirements for an ideal measure of adhesion. The test should be quick, sensitive to small changes in adhesion, able to measure a wide range of adhesion, reproducible, applicable to all film–substrate combinations, and if possible nondestructive. The ideal test should correlate well to other practical measures of adhesion or durability. Unfortunately no single test comes close to this ideal.

The following paragraphs describe some of the more pragmatic approaches to thin-film adhesion testing. For all tests, we must be concerned with the locus of failure. Is the failure adhesive (meaning that it occurs exactly at the interface), or cohesive (meaning it occurs in either the substrate or the film), or does it wander between the two locations? Surface analysis of the separated pieces is necessary to answer this question. Interpretation

TABLE 10.8 Stresses Due to Mechanical Contact

	Type of contact	σ_{zz}
Boussinesq (1885)	Point contact	$\dfrac{3F}{2\pi z^2}$
Hertz (1882)	Spherical contact	$\dfrac{3F}{2\pi(a^2 + z^2)}$
Hill (1950)	Conical contact	$\sigma_{ys}\left[2\ln\left(\dfrac{R_p}{z} + \dfrac{2}{3}\right)\right];$ $z \le R_p$
	Elasto-plastic	$\dfrac{2}{3}\left(\dfrac{R_p}{z}\right)^3; z \ge R_p$

becomes even more perplexing if the interface is diffused rather than sharp.

The first three tests are purely qualitative. The *tape test* is the most widely used test of practical adhesion. A piece of tape is attached to the film, burnished to make good contact, and then pulled off the substrate. A visual estimate is used to determine how much film has been removed. Tape tests frequently fail to remove the film, even in cases where adhesion differences are suspected.

The *abrasion test* uses an eraser, perhaps loaded with abrasives, which is rubbed across the film to determine how many strokes are required to remove a film. In abrasion tests it is not clear whether we are measuring the adhesion at the interface or the cohesiveness of the film.

The *boiling water test* requires a film to remain adherent after immersion in boiling water for some time. Again, it is unclear how this test relates to fundamental adhesion.

Among the more quantitative tests, Figure 10.36 illustrates a refinement on the tape test, the *peel test*. Here a tensile tester measures the force per unit width required to separate a film from its substrate. An adhesive backing tape may be used to pull off the film—in which case the elasto-plastic properties of the tape are confounded in the measurement—and measurements are limited to cases in which the adhesion is so weak that the film separates from the substrate before the tape separates from the film. Alternatively, the film may be made thick enough (tens or hundreds of micrometers) to allow it to be pulled off the substrate directly; however, in this case, the elasto-plastic properties of the film affect the reading. Both the thickness of the film and the rate of peeling affect the results.

The *scratch test* is a refinement on the abrasion test. Here a stylus of known shape is loaded normally and then pulled laterally across the film. The load is increased until the film separates from the substrate. This test works well for brittle coatings on hard substrates, but is less effective for ductile coatings applied to soft substrates such as polymers.

A recent improvement of the scratch test operates on a much finer scale. It uses a conical diamond *microindenter* to impress a microscratch on a thin film. The load on the microindenter (Figure 10.37a) is gradually increased as it is pulled across the film. Figure 10.37b shows how the groove widens until the film

Figure 10.36
Peel test to measure thin-film adhesion.

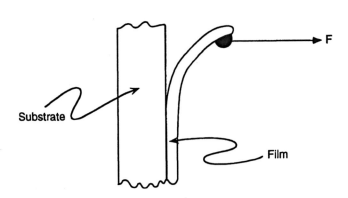

Figure 10.37
Nanoindenter tests characterize thin-film adhesion. (top) Load versus scratch length for an as-sputtered platinum thin film on nickel oxide. The film delaminates at the critical load L_{cr}. (bottom) Scratch trace corresponding to curve 3 above. The scratch deepens from left to right as the load increases until it delaminates at L_{cr}. Work of adhesion is estimated to be 53 mJ m^{-2}. [S. Venkataraman, D. L. Kohlstedt, and W. W. Gerberich, *Journal of Materials Research,* 7(5), 1126, 1992.]

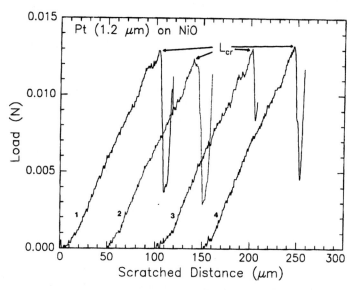

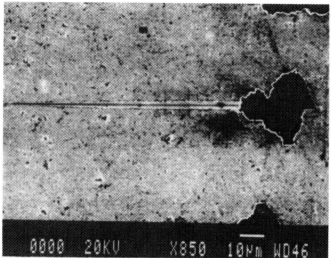

suddenly delaminates. There is a clear load drop when the lateral stress exceeds the shear strength of the interface and the film tears away. From the loads and areas involved it is possible to estimate the interface fracture toughness.

10.7.5 Interface Failure due to Surface Diffusion and Interdiffusion

Many applications of thin films depend on precisely patterned surfaces containing steps or grooves of the kind shown in Figure 10.38a. Associated with the large curvatures of steps and grooves are inhomogeneities in the local vapor pressure. At sufficiently high temperatures, surface diffusion causes atoms to move from regions of high curvature to regions of low curvature. As a consequence, the pattern profile changes with time and becomes smoother, as illustrated in Figure 10.38b. The change follows a

Figure 10.38
Scanning electron micrographs showing stages in the fabrication of an indium phosphide lenslet by mass transport. (a) Multilevel InP mesa structure formed by a sequence of precisely controlled chemical etches. (b) Smoothening of the surface by mass transport to produce a lenslet with a diameter of 45 µm and a front focal length of 20 µm useful for infrared integrated optical applications. (Z. L. Liau, V. Diadiuk, J. N. Walpole, and D. E. Mull, *Applied Physics Letters* **52**, 1859, 1988. Photographs courtesy of Dr. Z. L. Liau.)

(a)

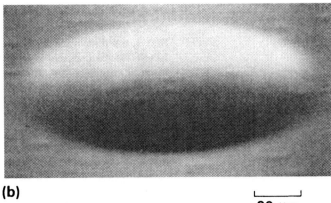

(b)

⊢————⊣
20 µm

$t^{1/4}$ relationship consistent with the surface diffusion mechanism outlined in our discussion of coalescence in Section 10.6.3.1. A major challenge in processing devices containing patterns is to find processing and application conditions that minimize pattern decay.

Smearing of sharp interfaces due to interdiffusion of the film and the substrate follows the kinetics outlined in Section 9.5.4.

10.7.6 Porosity of a Film

The columnar structure of films produced by the Volmer–Weber growth mode may include voids occupying up to several percent of the total volume. Their porosity provides an easy path for diffusion down the columnar boundaries to the thin film–substrate interface, with the result that these porous films may be attacked by water vapor or other vapors that are irreversibly pulled to the interface by capillary action. The effect is to alter the physical properties of these films, especially their index of refraction, and to degrade adhesion. This reliability issue must be addressed in practical applications.

10.8 Thin-Film Applications

10.8.1 Magneto-Optical Data Storage Disks

Digital storage of text and graphics requires large data storage capacity. While a simple page of text may contain 2 kilobytes of ASCII information, a graphic image on the same page with 200 dots per inch resolution requires 460 kilobytes of storage, and a high-quality digital photograph requires 1000 kilobytes. Trends towards increased resolution and TV-compatible moving images generates a compelling demand for ultra-high-capacity storage media. Current applications use magnetic media in the form of floppy disks, hard disks, or magnetic tape and optical media in the form of CD-ROM disks, and magneto-optical (MO) disks.

Table 10.9 compares the media in terms of storage capacity, data accessibilty, and cost. Some stored data needs to be *archival;* that is, the data, once written, must never be erased or changed. CD-ROMs are available to serve this need. Other stored data need to be changeable, and this requires *rewritable* media. Most rewriteable data should be stored on disks that can be removed from a system to allow expansion. This creates a demand for low-cost removable media. Magnetic hard drives and magneto-optical data storage systems are available with removable disks. In this section we review the fundamental aspects of magnetic data recording, the methods used to write and read, the materials involved, and thin-film processing techniques. The emphasis is on magneto-optical systems.

10.8.1.1 Magnetic Storage of Binary Information
The magnetic media in tapes (see Section 8.4.3) and hard disks are ferromagnetic materials that store the binary digits (bits) as magnetic domains with switchable directions of magnetization. In practice the media are designed to be anisotropic so that permanently magnetized domains are easier to create in one plane than others. The easy axis of magnetization can be arranged to lie

TABLE 10.9 Comparison of High Capacity Data Storage Media

	Removable hard disk	CD-ROM	Rare earth-transition metal magneto-optic drive[d]
Storage capacity (MByte)	270	650	230 1300
Access time[a] (ms)	13.5	280	28
Data rate[b] (MB/s)	1.7	1	1.4
Cost per unit hardware/ disk (Nov. 1994)	$430/75	$300/10–100[c]	$800/50 $2000/100

[a]Access time—time taken to locate the start of a data block.

[b]Data rate—readout rate of a data block.

[c]CD-ROM cost includes the content of the disk.

[d]Two sizes of magneto-optic drives are given.

either parallel to the plane of the tape or disk, giving rise to *longitudinal recording,* or perpendicular to the plane of the tape or disk, giving rise to *vertical recording.* These two alternatives are shown in Figure 10.39.

Within each domain, magnetic dipole moments of individual atoms couple to ensure a stable magnetization in a specific direction. As long as the temperature is below the Curie temperature T_C, the domains are stable and will not spontaneously disorient. If the temperature exceeds the Curie temperature, the permanent magnetization of the medium disappears. On recooling through the Curie temperature, the permanent magnetization returns with an orientation that can be dictated by a local magnetic field. In this manner a "bit" of information can be written into the magnetic storage medium, and depending on the orientation of the magnetic field (parallel or antiparallel) the "bit" can be a "1" or a "0," as illustrated in Figure 10.39.

The magnetic properties of ferromagnetic media can be described in terms of the magnetization (M–H) curve such as Figure 10.40. In response to an applied magnetic field H, the medium becomes magnetized to a magnetization M. Eventually all the domains align with the applied field, and the magnetization is complete at the *saturation magnetization* M_s. If the applied field is reduced to zero (H = 0), some thermal agitation and/or realignment reduces the magnetization slightly to its *remanent value* M_r. To reduce the magnetization to zero (M = 0), a field must be applied in the opposite direction. The magnitude of this demagnetizing field is the *coercivity,* H_c. A good magnetic recording medium will have a large remanent magnetization, making it easy to read, and a large coercivity, making it difficult to accidently change the bit of information on the disk.

Hard disks use longitudinal recording. Recording of the data is accomplished by applying a large magnetic field in the desired direction that briefly exceeds the coercivity. The domain magnetization switches to match the applied field. The field must be applied in a very small local region to ensure high bit density. This is accomplished by placing the writing head very close to the medium. To read the data, a read head, also in close proximity to the medium, senses the fringing magnetic fields where the direction of magnetization reverses between adjacent magnetic domains illustrated in Figure 10.39a.

With hard disks the writing and reading heads float on a cushion of air that forms as the disk rotates under them. The gap

Figure 10.39
(a) Longitudinal recording in a ferromagnetic thin film used for a magnetic hard disk. The magnetic domains are oriented in the plane of the disk and opposite directions at any location distinguish "1" from "0". (b) Vertical recording typical of magneto-optical disks. The magnetic domains are oriented perpendicular to the plane of the disk, and the direction of magnetization distinguishes "1" from "0". Higher bit density is achievable for vertical recording.

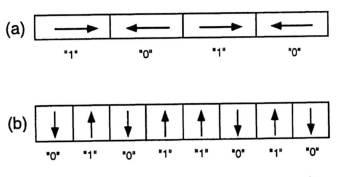

Figure 10.40
Magnetization M versus applied magnetic field H for a ferromagnetic material. M_s and M_r are the saturation and remanent magnetization, respectively. H_c is the coercivity.

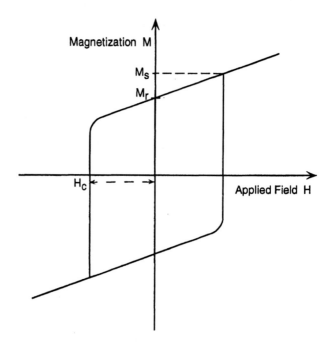

between the heads and the medium is 0.2 μm, less than the diameter of a human hair. This close proximity means that any dust particle present on the surface causes the head to crash. Since a head crash may damage the disk or the head or both, it must be avoided at all costs, and large-capacity hard disk drives are sealed in a dust-free environment for protection. This need for absolute cleanliness places some restriction on the transportability of hard disks.

Magneto-optical disks use vertical recording. The data are stored in magnetic domains, but the easy axis of magnetization is now arranged to be perpendicular to the plane of the disk. To write information onto a magneto-optic disk, a weak magnetic field in the desired vertical direction is applied over a large area. Then a laser is focused onto the region of a single bit, raising the temperature of the medium locally above the Curie temperature. When the laser is switched off, the heated spot cools down through the Curie temperature, and its magnetization aligns with the weak magnetic field. The direction of magnetization is frozen in, but it can be switched by reversing the field and re-pulsing the same spot with the laser. Reading the magnetic fields uses the *Kerr magneto-optical effect*. The Kerr effect rotates the plane of polarization of a light beam traveling in a magnetic field; the polarization rotates in a sense determined by the direction of the magnetic field. In the magneto-optic disk application a low-intensity polarized laser beam is focused onto the storage medium, and the rotation of the plane of polarization is sensed optically. In Figure 10.39b, for example, a "0" bit may rotate the plane of polarization clockwise, and a "1" counterclockwise.

With magneto-optic disks in either the writing or reading mode, the laser remains a large distance from the medium, eliminating the crash problem. Furthermore, dust may collect on the protective layer on top of the magneto-optic layer without any detrimental effects. As shown in Figure 10.41, the laser is unfocused at the top layer, and a dust particle has no effect on the writing or reading of digital data.

A CD-ROM uses nonmagnetic storage. The bits, 0 or 1, are represented as flat or pitted regions on the disk, respectively. The reflection of laser light from these two regions is different and is used to read the disk. The CD-ROM is, of course, a read-only storage medium. A CD-ROM has the advantage that many reproductions can be made inexpensively by pressing disks from a master.

In all types of storage disks the data are arranged on concentric circular tracks. Ingenious engineering is required to ensure that the magnetic heads or laser heads follow the data tracks accurately. This aspect of the technology will not be discussed here.

10.8.1.2 Magneto-Optical Quadrilayer Media

Current generation magneto-optic media are based on rare earth–transition metal amorphous alloys. Figure 10.42 shows a typical four-layer stack of thin films deposited onto a transparent polycarbonate substrate. Sequentially the layers on top of the substrate are optically transparent silicon nitride dielectric followed by the magneto-optic medium followed by another dielectric layer followed by a reflective metal coating. Finally the quadrilayer stack is covered with a protective overcoat. In operation, the write–read laser beam passes through the polycarbonate substrate (from below in the drawing) to interact with the magneto-optic medium. Note that the total thickness of the quadrilayer is only 225 nm, or 0.225 μm.

Technical criteria to be met by the assembly fall into three catagories: the stack must provide a good Kerr signal to allow the magnetization to be read easily, the stack must have appropriate thermal characteristics to allow repeated rewriting over the same bits without degradation, and the assembly must be stable over the device lifetime of several years.

Additional criteria must be met to make magneto-optic storage competitive with conventional hard disks. A high-volume-throughput manufacturing process with high yield is needed for inexpensive disks. Present manufacturing goals aim for gigabyte disks that retail for about $10.

Figure 10.41
Because dust particles or scratches are located on the surface where the laser is unfocused, such defects have little effect on the error rate during writing or reading.

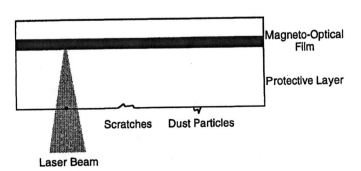

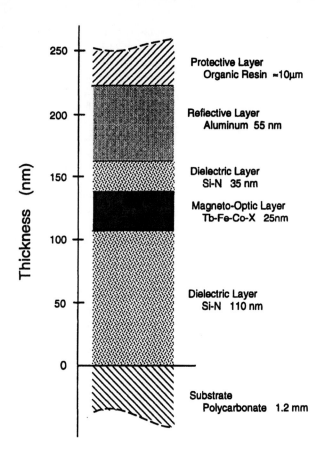

Figure 10.42
Rare earth–transition metal magneto-optic quadrilayer now at the heart of commercial magneto-optic media. The write/read laser is incident from the bottom in the drawing, passing through the polycarbonate substrate and the dielectric layer before interacting with the magneto-optic layer. (Adapted from Ikeda and co-workers, *J. Magnetic Society of Japan* **26,** 335, 1993.)

10.8.1.2A The Magneto-Optical Layer: Rare Earth–Transition Metal Alloys. The magneto-optic layer must be a material with a sufficiently large Kerr effect and a high coercivity at ambient temperature, and be capable of dense, low-noise storage. Dense storage requires vertical recording and therefore a direction of easy magnetization perpendicular to the film; low noise requires an amorphous or very small grain size deposit. Addition of rare earth elements such as terbium to transition metals such as cobalt–iron derives an alloy that satisfies all these requirements.

Adding a rare earth element to the transition metal alloy also makes the alloy ferrimagnetic. As a result, the coercivity of the alloy increases to infinity below the compensation temperature T_{comp}, which is lower than the Curie temperature T_C. Above T_{comp} and below T_C the coercivity drops off rapidly for a small increase in temperature. Hence rewriting these magneto-optic thin films can be accomplished by raising the temperature to a value between T_{comp} and T_C and reversing the magnetic field.

The Curie temperature of the alloy is adjusted by the cobalt-to-iron ratio. TbFe has $T_C = 135°C$, while TbCo has $T_C > 400°C$. A mixture of TbCo:TbFe::9:1 allows a Curie temperature of about 200°C. The compensation temperature is determined by the atomic percentage of the rare earth element. T_{comp} is typically 25°C for 22 atomic percent Tb in TbFeCo.

Thus far the magneto-optic and thermomagnetic properties of the medium have been emphasized, but other physical properties are important for the technological success of the system. In order to heat a small area of magneto-optic material rapidly with a diode laser, the layer must absorb the laser light. Radial and axial heat flow into the surrounding material and adjacent dielectric layers must be designed so that a bit can be repeatedly rewritten without degradation of the magneto-optic medium. Optical and thermal properties of the quadrilayer must be designed carefully to accomplish these goals.

The rare earth–transition metal magneto-optic thin film must be uniform and homogeneous through the thickness of the layer and across the whole disk. Such compositional requirements present a processing challenge. The magneto-optic layer is commonly deposited by DC sputtering from an alloy target. However, the angular emission of the three components of the alloy are different, and this can lead to a nonuniform radial distribution, as shown in Figure 10.43. The Co (not shown) is uniform, but the Tb and Fe concentration varies at the edges.

Uniformity can be achieved by careful choice of substrate temperature, background argon pressure, and composition of the alloy target. Targets may consist of a hot pressed mixture of intermetallic compound ($TbFe_2$), rare earth, and transition metal powders.

10.8.1.2B The Dielectric Layers. Rare earth transition alloys such as TbFeCo are very reactive with oxygen or oxides. So the magneto-optic layer must be surrounded by an impermeable layer containing no oxygen and with no pathways for oxygen diffusion. Fully dense silicon nitride, Si–N, serves this purpose.

Besides protecting the magneto-optic layer from oxides, the dielectric layers also establish proper optical and thermal prop-

Figure 10.43
Radial distribution of iron (upper) and terbium (lower) in Tb-Fe-Co films DC sputtered at different argon pressures (3 and 15 mtorr). The cobalt concentration is uniform at 12 atomic percent. [After Hatwar and co-workers, *IEEE Transactions (Magnetics)*, **24**, 2775, 1988.]

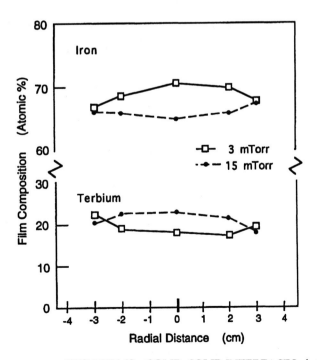

erties for the stack. The layers should be transparent to the laser radiation and absorb no energy, and have low thermal conductivity to limit heat flow away from the magneto-optic layer during writing. Si–N satisfies both these criteria.

The Si–N could be deposited with RF sputtering; however, the deposition rate is too slow, and the thermal load on the substrate (typically polycarbonate) is too great. DC reactive sputtering or variants of this technique are more common. Impermeable films with the proper optical and thermal characteristics are made by controlling the target voltage and current, the flow rate of reactive nitrogen, and the sputtering pressure.

10.8.1.2C The Reflecting Layer. The reflecting layer must have a high reflectivity with appropriate thermal conductivity. Aluminum is commonly chosen, since it is easy to handle. The aluminum is alloyed with other metals such as chromium to minimize corrosion and to allow a choice of conductivity.

10.8.1.3 Stack Design and Future Trends

The general procedure for designing and manufacturing a quadrilayer stack is first to determine the deposition conditions that produce a magneto-optical layer with optimum properties. Then the thicknesses of the dielectric layers and reflecting layer are chosen to provide a strong, low-noise optical signal. The layer thicknesses are fine-tuned together with the alloy composition of the reflector to give desired thermal conductivity behavior during the writing of information. Table 10.10 gives the material properties of the layers and substrate currently in use.

Considerable product development is underway to increase the throughput and reliability of the manufacturing process to reduce the cost of magneto-optic disks. Next-generation magneto-optic media are being designed to be compatible with new blue light diode lasers. The shorter wavelength allows an increase in bit density. New magneto-optic storage media such as Co–Pt multilayer superlattices that do not require dielectric or reflective layers are being explored. They offer the potential for superior performance at lower manufacturing cost.

TABLE 10.10 Specifications for a Typical Rare Earth–Transition Metal Quadrilayer Magneto-Optic System Shown in Figure 10.42

	Material	Thickness	Index of refraction	Thermal conductivity $(W/m^2 K)$
Substrate	Polycarbonate	1.2 mm	1.58	0.1
Dielectric layers	Si–N	110 nm 35 nm	$2 + i0.02$	1.5
Rare earth–transition metal layer	TbFeCo	25 nm	$n^+ = 3.20 - i3.50$ $n^- = 3.25 - i3.55$	7
Reflector	Al alloy	55 nm	$2.7 - i8.3$	25–100
Overcoat	Organic resin	10 μm	1.5	

10.8.2 Thin-Film Infrared Sensors

10.8.2.1 Infrared Sensors Require Response at Specific Wavelengths

To sense visible light and other electromagnetic radiation, a detector must be able to convert incident waves into a detectable electrical signal with low noise. Photoconductivity represents one approach to radiation detection. When the energy of the electromagnetic radiation incident on a photosensitive material is high enough to excite an electron across the energy gap from the valence band into the conduction band, the excited electrons contribute a photocurrent in response to the radiation. When the incoming radiation is in the infrared, the photon energy is so low that it is difficult to find a material with a narrow band gap that responds selectively to the incident radiation. For night vision applications, for example, the detector must respond to specific wavelengths within the atmospheric transmission window in the 8–14 μm wavelength range. In the past semiconductor alloy materials like mercury–cadmium telluride have been developed for this purpose with the wavelength specificity determined by the mercury-to-cadmium ratio. Another more recent approach makes use of so-called *multiple-quantum-well devices* (MQWs).

10.8.2.2 Multiple-Quantum Well Structures Can Be Tuned for Maximum Response at Specific Wavelengths

As illustrated in Figure 10.44a, a quantum well is formed when a two-dimensional semiconductor layer S is sandwiched between

Figure 10.44
(a) Quantum-well potential profiles within the superlattice. Electrons with both ground-state energy E_1 and excited-state energy E_2 are trapped within the superlattice. (b) When an electrical field is applied, the potential wells are skewed as shown. Electrons excited by incoming radiation to the energy level E_2 tunnel out of the well and contribute an external photocurrent.

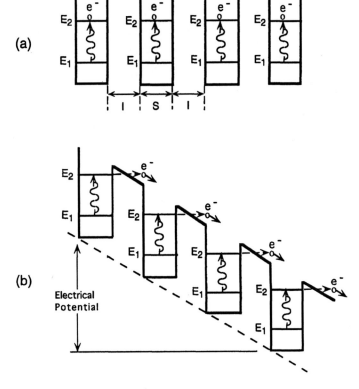

two insulating layers I. When the conducting layer is made very narrow, the electron energy states are quantized to very specific levels, say, E_1 and E_2, rather than the broad band associated with three-dimensional structures. Multiple-quantum-well structures are formed by bringing together many such conducting layers, each separated by thin insulating layers of identical thickness.

An applied electrical field changes the electron energy-level diagram for a multiple-quantum-well device to that shown in Figure 10.44b. Now electrons excited to the E_2 level in any quantum well are able to tunnel through the insulating barrier, as indicated by the horizontal arrows. They are freed to contribute an external electrical current. Obviously when E_2–E_1 corresponds to the energy ($h\nu$) of incoming radiation, then the device can be used as a photodetector. Without delving into the fundamental physics involved, it will suffice to say that the energy separation between E_2 and E_1 can be adjusted by the physical dimensions and chemical composition of the well structures. In this way a multiple-quantum-well device can be tailored to respond to a specific wavelength.

Figure 10.45 shows an example of such a device. It is based on a superlattice of GaAs/Al_xGa_{1-x}As consisting of 50 alternating layers of 40 Å GaAs doped with 2×10^{18} electron donors/cm^3 (the conducting layers) and 300 Å $Al_{0.31}Ga_{0.69}$As (the insulating layers), all sandwiched between electrical contacts of heavily doped GaAs. An applied electrical potential of 4 V sensitizes the material.

Analysis of the GaAs/Al_xGa_{1-x}As system predicts that E_2–E_1 varies linearly with mole fraction of AlAs. The wavelength λ of maximum response is given by

$$\lambda = \frac{(2.92 \times 10^4)}{x} \text{ Å} \qquad (10.8.1)$$

Figure 10.45
Schematic of a superlattice multilayered device. Incident radiation ($h\nu$) is refracted through the base to interact with the GaAs/Al_xGa_{1-x}As superlattice structure.

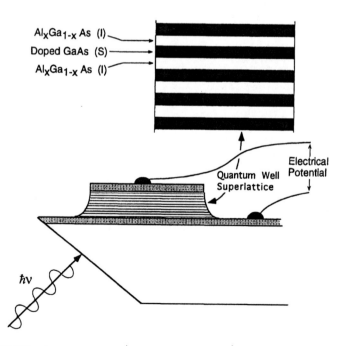

In the example x = 0.31, so eq. 10.8.1 predicts maximum response at 9.4 μm. The measured response curve in Figure 10.46 shows a peak at 8.2 μm. The slight discrepancy arises because eq. 10.8.1 as based on data from a single quantum well and effects of discontinuities in the conduction band at the interfaces are not fully accounted for. Response times of 30 ps have been estimated for these devices; however, due to instrument limitations, the response time of the device shown in Figure 10.46 has only been measured to be less than 10 μs.

10.8.2.3 The Precision Required for Multiple-Quantum-Well Structures Is Achievable with MBE Processing

The quality of the signal can be diminished in several ways. First, photons from energy level E_1 can also tunnel through the barriers. Electron motion from well to well in this manner is called *dark current*. As the applied potential is increased, the dark current also increases. A relatively large barrier thickness (300 Å) is designed to reduce tunneling between quantum wells. Also, the device is held at 77 K to minimize dark current.

Second, the detectors are sensitive only to light whose electrical field is polarized parallel to the growth direction or perpendicular to the superlattice. For that reason the gallium arsenide substrate is cut so that the incoming radiation v is refracted for optimum incidence on the superlattice.

Finally, reponsivity depends on layers having the same thickness, spacing, and composition. In view of the dimensions involved, this requires dimensional control at the atomic level. Impurities also have an adverse affect on device performance by attenuating the signal. Only by using the purest deposition materials and methods can impurities be reduced to an acceptable

Figure 10.46
Photoresponse curve for a GaAs/$Al_{0.31}Ga_{0.69}$As superlattice device. The maximum responsivity occurs for 8.2 μm wavelength radiation in the infrared portion of the spectrum. (From B. F. Levine, C. G. Bethea, G. Hasnain, J. Walker, and R. J. Malik, *Electronic Letters* **24**, 747, 1988.)

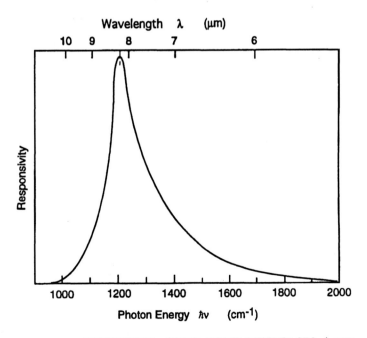

level. Such precision is achievable only with molecular beam epitaxy (MBE) thin-film deposition techniques. Because growth in certain crystallographic directions affects the quality of hetero-junctions in the superlattice, MBE growth on {100} planes is the preferred deposition orientation. Procedures used in the fabrication of multiple-quantum-well devices follow those outlined in Section 10.5.

Bibliography

R. F. Bunshah et al., *Deposition Technologies for Films and Coatings,* New Jersey: Noyes Publications, 1982

B. Chapman, *Glow Discharge Processes,* New York: John Wiley & Sons, Inc., 1980.

A. Y. Cho and J. R. Arthur, *Molecular Beam Epitaxy,* Progress in Solid-State Chemistry, Volume 10, pp. 157–91, 1975.

J. W. Mayer and S. S. Lau, *Electronic Materials Science,* New York: Macmillan, 1990.

T. McDaniel and R. Victora, *Magneto-Optical Data Recording—Materials, Subsystems, Techniques,* Noyes Publishing, 1995.

J. F. O'Hanlon, *A User's Guide to Vacuum Technology* (2nd. edition), New York: John Wiley & Sons, Inc., 1989.

M. Ohring, *The Materials Science of Thin Films,* New York: Academic Press, Inc., 1992.

E. H. C. Parker, *The Technology and Physics of Molecular Beam Epitaxy,* New York: Plenum Press, 1985.

S. M. Sze, *Semiconductor Devices: Physics and Technology,* New York: John Wiley & Sons, Inc., 1985.

K-N Tu, J. W. Mayer and L. C. Feldman, *Electronic Thin Film Science,* New York: Macmillan, 1992.

J. A. Venables, G. D. T. Spiller, and M. Hanbücken, *Nucleation and Growth of Thin Films,* Reports on Progress in Physics, Volume 47, pp. 399–459, 1984.

J. L. Vossen and W. Kern, *Thin Film Processes,* New York: Academic Press, Inc., 1978.

Exercises

10.1 A physical vapor deposition system is pumped down, baked out, backfilled with argon, and then pumped down to a pressure of 10^{-7} torr at room temperature. Assuming only argon atoms are present, calculate (a) the number density; (b) the mean free path; (c) the impingement flux; (d) the time to cover a surface with a monolayer of argon atoms (argon atomic weight 40; atomic diameter 0.376 nm; sticking coefficient 1.0).

10.2 In thermal equilibrium the fluxes of silicon atoms leaving and returning to a silicon wafer in a closed system are equal. What is the silicon flux at 1500 K? (Silicon atomic weight 28.1).

10.3 Given that the flux from a flat surface follows the cosine law (eq. 10.2.4), determine the polar angle of the cone into which 90% of the the flux of atoms is emitted. (The differential element of a solid angle $d\Omega = \sin\theta \, d\theta \, d\phi$, Figure 2.3.)

10.4 Consider designing a thermal evaporation deposition system for copper. We want to deposit 5000 Å of copper at a rate of 10 Å/s onto a small substrate located 50 cm directly above a flat circular source. The source temperature is uniform at 1600 K. We want the purity of the deposition to be 2 parts per thousand. Assume:

The film has the same density as the bulk, 8.96 g/cm^3;

Vapor pressure of copper at 1600 K = 2×10^{-2} torr;

Residual gas in the chamber is air, effective atomic weight 29;

Sticking coefficient onto substrate is 1.0 for copper, 0.1 for residual gas;

Emission of copper from a flat surface follows the cosine law.

Determine (a) the diameter of molten copper charge; (b) the depth of copper charge that is evaporated; (c) the required vacuum system background pressure; (d) the tooling factor for a thickness monitor placed at the same height as the substrate but at an angle 30° from the normal to the source.

10.5 A sputtering target is a 50:50 atomic % CoNi alloy. The sputter yield is 1.1 atoms/ion for cobalt and 1.4 atoms/ion for nickel, both measured at the sputtering voltage of 500 V. (a) Determine the initial stoichiometry of the deposited film. (b) After a long sputtering time, determine the stoichiometry of the surface of the target and of the deposited film. (c) Suppose instead that we try to evaporate the 50:50 alloy from a single source at 1520°C. The melting points are 1495°C for Co and 1450°C for Ni and at the chosen evaporation temperature the vapor pressures are 1×10^{-2} torr for Co and 7×10^{-2} torr for Ni. What will be the stoichiometry of the resulting film?

10.6 Sputtering can also be used to clean a substrate prior to deposition. Suppose we have a polycrystalline copper substrate. Sputtering yields for normally incident argon ions on copper at 100 V are 0.45 atoms per ion for the {100} planes and 0.20 atoms per ion for the {111} planes. What will happen to the surface after sputtering? What steps could you take to avoid these undesirable consequences?

10.7 Discuss the advantages and disadvantages of thermal evaporation versus sputter evaporation for processing (a) a reflective gold coating on Mylar; (b) an abrasion-resistant hard chromium–tungsten alloy coating on stainless steel.

10.8 Copper is being deposited at a rate of 10 Å/s from a source at 1600 K onto a substrate where the effective temperature of the copper adatoms is 900 K. Find the critical radius per atom diameter, R_c/D_{atom}, for homogeneous nucleation. Also find the number of atoms contained in the nucleus. Assume the surface energy per atom is 1 eV. Copper atomic weight 63.5.

10.9 Develop a relationship for the mean distance traveled by an adatom in terms of $\Delta G_{surf\ diff}$, ΔG_{des}, λ, and T.

10.10 At 895 K the residence time of Ga on a GaAs surface is 10 s, while at 915 K the residence time is 7.5 s. What is the residence time at 850 and 950 K?

10.11 Using answers from Exercises 10.9 and 10.10 and assuming a surface diffusion activation energy of 0.8 eV, how far does a Ga ion diffuse over the surface at the four temperatures, 850 K, 895 K, 915 K, and 950 K? Assume that the jump distance λ is 0.33 nm.

10.12 Explain the difference between reaction-rate-limited and diffusion-limited chemical vapor deposition. In particular, account for the differences in sensitivity to pressure and temperature for these two processing conditions.

10.13 A deposited film forms clusters whose average radius increases with time. Why do clusters form? What are the two mechanisms by which individual ions move from cluster to cluster? What steps would you take to avoid clustering in the early stages of thin-film growth?

10.14 Thin films of GaP (lattice parameter 5.45 Å), InAs (6.05 Å), and InSb (6.47 Å) are to be grown epitaxially on a GaAs substrate (lattice parameter 5.64 Å). Calculate the approximate value of the critical film thickness before dislocations appear at the film–substrate interface. Also calculate the approximate spacing between dislocations should the film exceed critical thickness. What steps would you take to alleviate the problems presented in attempting to grow a film of InSb on GaAs?

10.15 You grow a film, 1000 Å thick, of InGaAs on GaAs that has a lattice mismatch of 1.0 percent. A crossed grid of 100 equally spaced edge dislocations runs through the length or the width of the microbeam that is 10 μm long and 1000 nm wide. (a) What would the elastic strain energy stored in the film be if no dislocations were introduced? (b) How much of the residual strain energy is stored in the dislocation network ignoring any crossing points? (c) How many dislocations would be needed to accommodate the strain completely? (Assume shear modulus for InGaAs is 4.4×10^{10} Pa, Poisson's ratio is 0.3, and the Burgers vector for the edge dislocations is 0.4 nm; also assume elastic isotropy.)

10.16 A thin film of copper is deposited onto silicon at a temperature of 1000 K and allowed to cool to room temperature, 300 K. Use the following typical material properties:

	E (GPa)	α (K⁻¹)	ν
Silicon	160	2.6×10^{-6}	0.38
Copper	110	16.6×10^{-6}	0.34

(a) Estimate the thermal stress in the copper thin film. (b) Is the stress tensile or compressive? (c) Compare the stress to the ultimate strength of copper, 0.22 GPa.

11

GRAIN BOUNDARY SURFACES AND INTERFACES IN CRYSTALLINE SOLIDS

with W. Gerberich

LIST OF SYMBOLS

B	absolute mobility of an atom in a crystalline solid
c	molar concentration of solute in crystalline solid solution
d	grain diameter
D_{gb}	grain boundary diffusion coefficient
ΔG_p	change in free energy due to phase transformation
ΔH_{gb}	activation enthalpy for grain boundary diffusion
ΔP	capillary pressure difference due to a curved liquid surface
R	radius
T_m	melting temperature
v_m	atomic volume of atom m
V_S	solid molar volume
v_{gb}	velocity of grain boundary movement (eq. 11.2.8)
$w_{cohesion}$	work of grain boundary cohesion (eq. 11.2.1)
W_{AA}	molecular pair interaction energy
W	effective atomic interaction parameter (eq. 3.1.8)
z	coordination number
$\gamma_{\alpha\beta}$	interface energy between solid phases α and β
γ_{gb}	grain boundary surface energy
γ_{sl}	interface energy between crystalline solid and liquid phase
γ_{sv}	surface energy of a crystalline solid
Γ	excess concentration of impurity in the grain boundary (eq. 11.3.1)

φ	dihedral angle (eqs. 11.4.1 and 11.4.2)
λ	jump distance between grain boundary surface sites
ν	atomic jump frequency
ν_{diff}	frequency of jumps in diffusion
Σ	coincident lattice site parameter

CONCEPT MAP

General Properties of Grain Boundary Surfaces

- Grain boundaries are surfaces between two blocks of crystalline solid that differ only by their orientation in space. Grain boundary surfaces originate by grain impingement during vapor condensation or liquid solidification, or by solid-state recrystallization.

- Grain misorientations are described by the tilt and twist needed to bring adjacent grains into spatial registration.

- Grain boundary surfaces have excess surface free energy, γ_{gb}, due to unsatisfied bonds. For metals, γ_{gb} is about one-third the free surface energy; for ionic solids, γ_{gb} is about two-thirds the free surface energy.

- Grain boundary surface energy varies with misorientation. Certain symmetrical boundaries such as twin boundaries show grain boundary surface energy minima (Figure 11.5).

- For a given misorientation, grain boundary surface energy varies with the grain boundary plane. To reduce surface energy, a grain boundary surface contains low-energy steps.

- Grain boundaries reduce the work of cohesion of a solid, leading to preferential intergranular decohesion under mechanical stress (eq.11.2.1).

- Excess free energy of a grain boundary surface depends also on curvature and chemical segregation. By analogy with thermodynamic equations for fluids, the excess surface free energy can be written

$$dG = \gamma_{gb}dA + \Sigma\mu_i\, dn_i$$

$dG = \gamma_{gb} \, dA$ (Physical Effects)

- Curved grain boundary surfaces move toward their center of curvature.
- Curved grain boundary velocity is proportional to curvature and grain boundary diffusivity, and therefore increases exponentially with temperature (eq. 11.2.8).
- Grain boundary surfaces inside a solid are in equilibrium when they meet along triple lines at 120° and when triple lines meet at 109.5° at four grain corners.
- No three-dimensional shape exists that repeats to fill space and satisfies these requirements; so grain boundary surfaces are always curved and tend to move.
- Large grains grow, small grains shrink, and the average grain size, d, increases with $(time)^{1/2}$ (eq. 11.2.10).

$dG = \Sigma \mu_i \, dn_i$ (Chemical Effects)

- Defects and impurities segregate to grain boundary surfaces to reduce chemical energy, mechanical strain energy, and electrical energy (Figure 11.13).
- Grain boundary segregation can have disastrous effects on the fracture properties of bulk metals. Parts per million of P and S in steel segregate to grain boundaries and cause grain boundary embrittlement (Figure 11.14).
- Grain boundaries in ionic solids are charged. They attract oppositely charged imperfections and impurities to reduce the electrostatic potential.
- Grain boundary segregation can have advantageous effects in ceramic grain boundary layer devices, such as the ceramic varistor.

Grain Boundary Stabilization

- Segregation to grain boundaries curtails their movement. Impurities are added deliberately to restrict grain growth and control grain size during metallurgical and ceramic processing.

Grain Boundary Equilibrium Microstructures

- Grain boundaries come to equilibrium with vapor and liquid environments. The balance between surface and interface tensions can be used to define equilibrium configurations (Figure 11.15).

- The angle subtended by the solid–vapor interface along the line where the interface is intersected by a grain boundary is called the thermal groove angle φ. It depends on the relative values of grain boundary surface energy, γ_{gb}, and solid–vapor interface energy, γ_{sv}.

$$\cos \frac{\varphi}{2} = \frac{\gamma_{gb}}{2\gamma_{sv}} \text{ (eq. 11.4.1)}$$

- The angle subtended by the liquid–vapor interface along the line where the interface is intersected by a grain boundary is called the dihedral angle φ. It depends on the relative values of grain boundary surface energy, γ_{gb}, and solid–liquid interface energy, γ_{sl}.

$$\cos \frac{\varphi}{2} = \frac{\gamma_{gb}}{2\gamma_{sl}} \text{ (eq. 11.4.2)}$$

- When $\gamma_{gb} = 2\gamma_{sl}$, $\varphi = 0$ and the liquid phase completely wets the grain boundary surfaces (Figure 11.16).

- At high temperatures, complete wetting by a liquid glassy phase leads to grain boundary sliding and poor mechanical creep resistance, rendering materials unsuitable for high-temperature applications.

- At low temperatures, complete wetting leads to unique glass–ceramic grain boundary layer microstructures. Semiconducting ceramic grains, with an insulating glass layer isolating them, find application as gas sensors and varistor devices.

Solid-State Precipitation at Grain Boundaries

- During solid-state phase transformation, new phases prefer to nucleate and grow at grain boundaries. Precipitation in the grain boundary region is enhanced by segregation of solute, rapid grain boundary surface diffusion, reduction in strain energy, and grain boundary surface steps.

11

A typical metal or ceramic crystalline solid is not a single crystal but an aggregate of many crystals—it is *polycrystalline*. In a pure, single-phase material, each block (or grain) has the same crystal structure, the same composition, and therefore the same lattice parameter, as its neighbor. Grains differ only by their orientation in space. According to the definition contained in Chapter 9, the boundary that separates two identical grains of material in a pure solid creates a *grain boundary surface*. A grain boundary is unique because it represents a surface *inside* a crystalline solid.

This chapter reviews the origin and nature of grain boundaries and the characteristic behavior of grain boundary surfaces due to surface energy as well as chemical and stress effects. As we shall see, the grain boundary surface has a surface energy just like the free surface and experiences a "capillary" force due to curvature. Grain boundary surfaces are preferential sites for segregation of impurities or excess solute as well as for the nucleation and growth of second solid-state phases.

When a second phase forms at the grain boundary, it creates an interface between the host material and the new phase precipitating within it. Now the boundary separates two different materials; so by definition it has become an *interphase interface*. The distribution, size, and shape of constituent solid state phases—which we collectively refer to as a material's *microstructure*—determines the physical behavior of most polycrystalline metals and ceramics. Microstructure control is absolutely essential to enable the metallurgist or ceramist to attain desired technological goals. During the course of this chapter, we shall show how grain boundary surface energy and interphase interface energy influence the microstructure.

GRAIN BOUNDARIES IN CRYSTALLINE SOLIDS / 627

At the outset, it is important to stress that our diagrams and pictures treat grain boundaries as static entities. In reality, on an atomic scale, the grain boundary surface is a very dynamic region. Thermal agitation of an atom (ion) on one side of a grain boundary can cause it to become temporarily aligned with atoms on the other side. Atoms constantly move to and fro across the boundary and along the grain boundary surface. Solid-state diffusion determines how the solid and grain boundary surfaces respond kinetically to forces imposing change. At low temperatures, these changes take place very slowly (over months to years), but they can be speeded up tremendously (seconds to minutes) by raising the temperature above one-half the melting temperature, $T_m/2$. Thus microstructural changes driven by the reduction of grain boundary surface and interface energy are significant when we deal with the processing of solid-state materials at high temperatures.

11.1 Grain Boundaries Form during Condensation from the Vapor, during Solidification from the Melt, or by Solid-State Recrystallization

Grain boundaries occur naturally in solid crystalline materials. They can be introduced during formative processing steps, such as condensation from the vapor phase or solidification from the liquid phase, or during secondary processing steps, such as recrystallization. We consider each of these possibilities in turn.

In Chapter 10 we found that the formation of a solid by condensation from the vapor phase occurs through heterogeneous nucleation and growth on a substrate surface. Crystalline nuclei form preferentially at a step, impurity particle, or surface discontinuity and grow from the point of nucleation. Crystals originating from many different nuclei eventually impinge on each other and develop the columnar structure depicted in Figures 10.26, 10.27 and 10.28. The boundaries where impinging crystallites or grains meet are grain boundaries. As shown in Figure 10.26, the dimensions of the individual grains depend on the density of nucleation sites and the rate of growth, while the relative misorientation from one grain to another depends on the degree of epitaxy with the substrate template.

Crystalline solids solidified from the melt also contain many individual crystallites or grains that nucleate heterogeneously at sites on the surface of the casting mold and then grow together. In some instances, nucleating agents are deliberately added to a molten metal to stimulate nucleation and growth of many crystallites within the melt and thereby reduce the resultant grain size of the cast metal.

Another way to form a polycrystalline solid is through the *recrystallization* of a plastically deformed crystal. In Section 9.4.2, we discussed the role dislocation lines play in the permanent plastic deformation of crystalline solids. Appendix 11B.2 reviews the microscopic aspects of mechanical behavior in more detail. For the moment we should appreciate that a crystalline solid undergoing even a small amount of plastic strain becomes filled with entangled dislocation lines. These dislocations interact with each other through their respective stress fields. As plastic strain increases, the tangles become denser, the plastic strain becomes more difficult to sustain, the stress to continue deformation increases, and the solid undergoes strain hardening. As we noted in Section 9.4.2.2, each dislocation line is associated with a strain energy of about 4×10^{10} eV/m. After strain hardening, dislocation densities are typically 10^{12}–10^{16}/m^2, so that the stored dislocation energy density is of the order 10^{25} eV/m^3, or 10^6 J/m^3. Recrystallization releases this energy.

Actually, the tangled dislocations in a strain-hardened metal single crystal are not distributed totally at random throughout the crystal. They cluster to form the "cellular" structure illustrated in Figure 11.1a. When the temperature is raised above $T_m/2$, this heterogeneous dislocation distribution changes significantly. The dislocations rearrange by glide and climb (see Appendix 11B.3). Many of the dislocations are eliminated, others join the cellular network structure, and individual cell walls adopt the characteristics of the small-angle boundaries illustrated in Figure 9.16. The regions between these boundaries become free from distortion and relatively devoid of dislocations. Some cells are more perfect than others. Figure 11.1b illustrates how the most perfect become established as nuclei for the formation and growth of entirely new crystalline grains. Growth of these nuclei is driven by reduction of the total strain energy of the system—through the elimination of residual dislocations and debris generated by the plastic deformation and strain-hardening process. As a result of this sequence of events, a severely deformed *single crystal* becomes transformed into a *polycrystalline solid* consisting of many almost perfect crystals separated by clear grain boundaries: the single crystal has undergone *primary recrystallization*.

Figure 11.1
Electron transmission micrographs showing the transformation in dislocation line density in iron–3 wt. % silicon during recrystallization. (left) High density of dislocation tangles resulting from room temperature deformation (80% reduction by rolling). (right) Formation of clear cellular structure after five minutes at 600˚C. One of these cells can act as the nucleus for a new grain leading to recrystallization. (H. Hu, *Recovery and Recrystallization in Metals*, L. Himmel (ed.), New York: AIME and Interscience, Inc., 1963, pp. 321, 322.)

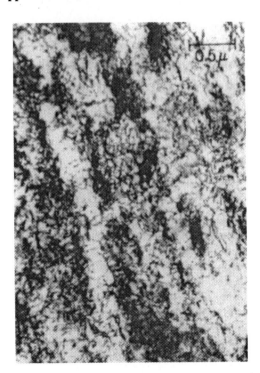 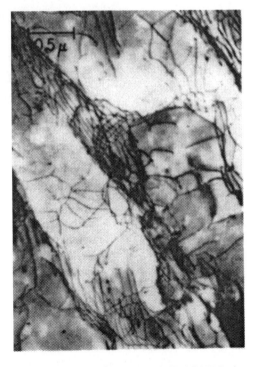

To take our analysis of recrystallization one stage further, the starting solid can be already polycrystalline (a metal prepared by solidification from the melt, for example). With plastic deformation, this polycrystalline solid also becomes filled with tangled dislocation arrays, and when heated to the primary recrystallization temperature (about 0.3 T_m), it recrystallizes and transforms into a polycrystalline solid with a new distribution of grains and grain boundaries. The final polycrystalline grain size depends on the number of nuclei produced and their rate of growth, factors that are related to the starting state of the material, the amount of plastic strain, and the temperature and time required for recrystallization.

Practical metallurgical processing often employs complex deformation processes followed by heat treatments to regulate the grain size of the finished product. Processes such as forging subject a cast ingot to greater than 25% plastic deformation before it is heated to a high temperature to promote recrystallization. In hot forging, the two processes of deformation and recrystallization take place concurrently. Forging may be followed by additional mechanical processing, such as rolling through a rolling mill or extrusion through an extrusion die, interspersed with heat treatments to promote recrystallization before the metal reaches its final shape. After such a succession of mechanical/thermal combinations, the metal consists of many small grains separated by grain boundaries. Mechanical and thermal treatments break up the original cast ingot structure and produce prescribed distributions of grain size, grain shape, and preferred grain orientation (texture) for a great variety of applications. For high mechanical strength it is generally desirable to retain a very small, micrometer-sized, grain dimension. For magnetic alloys it is sometimes desirable to produce a larger grain size with a strongly oriented grain texture.

11.2 In Pure Polycrystalline Solids, the Surface Energy of Grain Boundaries Gives Rise to Curvature Effects and Grain Boundary Motion

11.2.1 Grain Boundary Misorientation Is Represented by Tilt and Twist Components or by the Inverse of the Coherent Lattice Site Density Σ

We can describe a grain boundary in terms of the misorientation of the two grains on either side of its surface. First we must fix the crystallographic plane of the boundary in the two grains by the normal vectors $\hat{n}_1$ and $\hat{n}_2$, as shown in Figure 11.2. When the second grain can be aligned with the first grain by simple rotation about an axis lying in the plane of the grain boundary surface, then the boundary is an asymmetrical tilt boundary of the kind shown in Figure 11.2a. When the second grain can be aligned with the first grain by simple rotation about an axis normal to the plane of the grain boundary surface, then the boundary is an asymmet-

Figure 11.2
Creation of a grain boundary (a) by an asymmetrical tilt about an axis in the plane of the boundary, (b) by an asymmetrical twist about an axis, $\hat{n}_2$, normal to the boundary.
[K. L. Merkle and D. Wolf, *Materials Research Society Bulletin*, **XV**(9), 42 (1990) and *Journal of Materials Research* 5(8), 1708 (1990.]

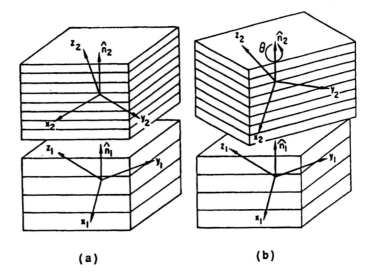

(a)　　　　　　　　　(b)

rical twist boundary of the kind shown in Figure 11.2b. For a general grain boundary, a combination of tilts and twists brings the two grains into alignment. Full mathematical treatment of grain boundary misorientation shows that we must specify five macroscopic and three microscopic degrees of freedom to define a boundary fully.

Another way to represent the misorientation across a grain boundary involves extending the three-dimensional lattice from one crystalline grain across the boundary. We can observe the density of atoms in the adjacent grain that coincide with and lie on the extended lattice points. For a random misorientation between two grains, instances of coincidence will occur periodically to form a repetitive array called the *coincident site lattice* (CSL). The symbol Σ represents the *inverse* of the coincident lattice site density (the inverse density of CSL). For example, $\Sigma = 15$ means that one atom site in fifteen is common to the extended crystal lattice of the other grain. In the extreme case of a small-angle tilt boundary, almost total coincidence of lattice sites exists on either side of the boundary, the two grains are completely commensurate with each other, and $\Sigma = 1$. As the tilt angle increases, Σ assumes values greater than one. For two randomly misoriented grains, the values for Σ can exceed $\Sigma = 100$. The symmetry inherent in twin boundaries suggests that Σ will have lower values for them. Table 11.1 shows how Σ varies for different angles of misorientation in simple twist boundaries in cubic crystal systems. It indicates the plane of the boundary and the twist angle (θ) about an axis normal to the plane. For a 60° twist across {111}, we see that $\Sigma = 3$ due to the symmetry. For a 36.9° twist across {100}, $\Sigma = 5$. Figure 11.3 reproduces a high-resolution electron transmission micrograph of a $\Sigma = 27$ twin boundary in germanium, showing the location of atoms on either side of the boundary.

For a 36.9° *tilt* boundary across the {100}, once again $\Sigma = 5$. Figure 11.4 presents a possible cross section of this boundary for an ionic solid with the sodium chloride crystal structure. In this

TABLE 11.1 θ–Σ Combinations for Selected Angles of (111), (100), and (110) Twist Boundaries; θ in Degrees

(111)		(100)		(110)	
θ	Σ	θ	Σ	θ	Σ
13.17	57	12.68	41	13.44	51
16.43	147	18.92	37	20.05	33
21.79	21	22.62	13	26.53	19
27.80	13	28.07	17	31.59	27
38.21	7	36.87	5	38.94	9
44.82	129	43.60	29	45.98	59
46.83	19	53.13	5	50.48	11
50.57	37	59.49	65	58.99	33
52.66	61	69.38	13	70.53	3
60.00	3	79.61	61	80.63	43

diagram the structure has been relaxed to minimize interionic interaction potential.

Generally, the change in crystallographic order from one grain to the next is not accomplished in the width of a single interatomic spacing. Individual atoms in the grain boundary region adjust or relax their positions to maximize their coordination with neighboring atoms. The grain boundary surface may be two to five atoms thick, depending on the bond character and crystal structure of the material and on the misorientation across the surface. This situation should be compared with the free surface, where adjustments of the kind illustrated in Figures 9.8 and 9.9 are accomplished in the first atomic layer. For certain

Figure 11.3
A high-resolution electron transmission micrograph of a Σ = 27 twin boundary in germanium showing the location of atoms on either side of the boundary. (Photograph courtesy of Dr. Stuart McKernan, University of Minnesota.)

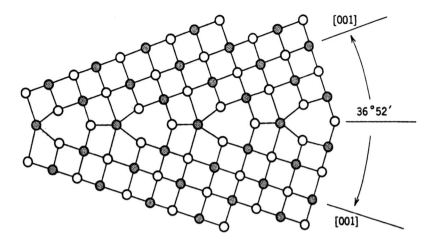

Figure 11.4
Representation of a symmetrical 36.9° tilt boundary in the sodium chloride crystal structure for which $\Sigma = 5$. Tilt occurs about a $\langle 100 \rangle$ axis in the $\{100\}$ plane. The proposed configuration has been relaxed to minimize the interionic interactions. (W. D. Kingery, H. K. Bowen, and D. R. Uhlmann, *Introduction to Ceramics,* 2nd ed., Copyright © 1976, John Wiley & Sons, p. 191. Reprinted by permission of John Wiley & Sons, Inc.)

symmetrical grain boundaries, such as small-angle tilt boundaries (Section 9.4.3) and certain twin boundaries, individual atoms share membership in both periodic arrays. Then the boundary has no thickness at all.

11.2.2 Grain Boundary Surface Energy Is Determined by the Density of Unsatisfied Bonds and Related to Misorientation

Just as a solid has surface energy because some of the intermolecular bonds are not satisfied at the free surface (as discussed in Section 9.2.2), so a grain boundary has a surface energy due to loss of coordination [the number of atoms (ions) in contact] and incomplete bonding across the boundary. We want to estimate the magnitude of the grain boundary energy per unit area, γ_{gb}.

One indirect approach for computing grain boundary energy assumes that grain boundaries are simply extensions of small-angle boundaries to higher and higher angles of misorientation. We computed the energy of a small-angle boundary in Section 9.4.3.1, eq. 9.4.12, by summing the strain energy of the dislocations that compose the boundary. There we argued that the asymptotic value of γ_{sgb} (approached for angles greater than 15°) corresponds to the energy of a grain boundary surface of large misorientation, γ_{gb}. Values for γ_{gb} obtained by this extrapolation agree surprisingly well with experimental values.

Many theoretical studies and experimental measurements of grain boundary energy have been conducted. Modern computer simulations estimate grain boundary surface energy directly in a two-step approach. First, all the pair-potential interactions between atoms (ions) at a boundary are summed and relaxed to obtain the minimum-energy configuration. Then the density of unsatisfied bonds is calculated. Figure 11.5 shows results for copper as a function of misorientation, specified in terms of tilt and twist. As might be expected, for certain misorientations, such as the twin configurations of Figure 9.18b and other low-Σ coincident site lattice (CSL) configurations, the high degree of conti-

Figure 11.5
Grain boundary energy
computations as a function
of misorientation. Compu-
tations are presented for
symmetrical boundaries in
copper, the misorientation
is defined in terms of tilt
and twist across the {110}
grain boundary plane. The
computations assume a
Lennard–Jones (LJ) inter-
action potential between
atoms (see Section 2.6.1).
[K. L. Merkle and D. Wolf,
MRS Bulletin, **XV**(9), 42
(1990) and *Journal of Materials Research* 5(8), 1708
(1990).]

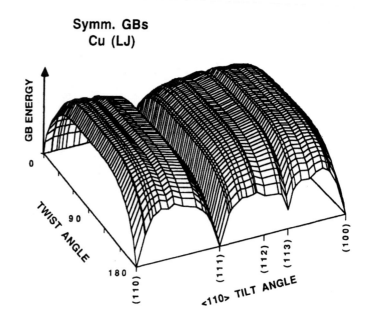

nuity across the boundary results in an especially low grain
boundary surface energy. These orientations correspond to the
cusps in the curves shown in Figure 11.5. Results for Cu and Mo
similarly show strong energy minima at the $\Sigma = 3$ boundaries.

Figure 11.6 summarizes a feature that emerges from all
computations of grain boundary energy and free surface energy
in solids: surface energy rises linearly with the density of unsatisfied bonds per unit area. This conclusion is valid to a first
approximation. Exceptions include boundary types with no
unsatisfied bonds—like twins and stacking faults—that should
exhibit no surface energy if unsatisfied bonds are the only source
of surface energy. But twins and stacking faults do have ener-

Figure 11.6
Correlation between
computed grain boundary
energy and free surface
energy and the density of
unsatisfied nearest-
neighbor bonds per unit
area, C/a^2. Computations
for gold using the embed-
ded atom model (EAM).
[K. L. Merkle and D. Wolf,
MRS Bulletin, **XV**(9), 42
(1990).]

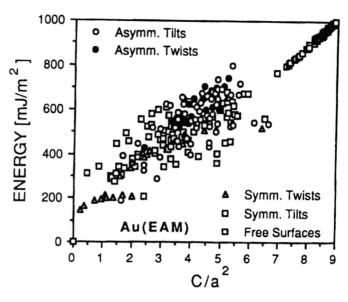

gies ranging from 5 to 100 mJ/m^2 (see Table 11.2a). Recent calculations have become more sophisticated and incorporate other contributions, such as electron interactions, into surface energy calculations.

For our purpose, it is sufficient to know that grain boundaries exhibit a surface energy about one-third of the free surface energy found in metals and as high as two-thirds of that found in ionic solids (see Table 11.2). We can explain the higher relative value for ionic solids by the fact that the boundary region incorporates electrostatic charge discontinuity as well as structural discontinuity, particularly when the boundary is asymmetric, as drawn in Figure 11.7. In regions where ions of the same sign are in close proximity, they must clearly relax from their periodic position to reduce the electrostatic interaction potential. Earlier we found that the *free surface* energy of metals is approximately twice that of ionic solids; so the absolute values of the grain boundary energies for both classes of materials should be close. Grain boundary energy values range from 150 to 1050 mJ/m^2. Note that the high end of the range corresponds to a surface energy 15 times the surface tension of water (see Table 9.2). It should come as no surprise that grain boundaries in pure crystalline solids rearrange rapidly at a high enough temperature (within minutes at temperatures > 0.8T$_m$) as they reduce the total grain boundary surface energy. Section 11.2.5 develops this topic.

TABLE 11.2a Selected Values of Intrinsic Stacking-Fault Energy, γ_{sf}, Twin-Boundary Energy γ_{twin}, Grain-Boundary Energy, $\gamma_{gb,,}$ and Crystal-Vapor Surface Energy, $\gamma_{sv,,}$ for Various Materials - mJ/m^2

Metal	γ_{sf}	γ_{twin}	γ_{gb}	γ_{sv}
Ag	16	8	790	1140
Al	166	75	325	980
Au	32	15	364	1485
Cu	45	24	625	1725
Fe			780	1950
Ni	125	43	866	2280
Pd	180			
Pt	322	161	1000	3000
Rh	~750			
Th	115			
Ir	300			
W				2800
Cd	175			
Mg	125			
Zn	140		340	

J. P. Hirth and J. Lothe, *Theory of Dislocations*, 2nd ed., Copyright © 1982, John Wiley & Sons, p. 839. Reprinted by permission of John Wiley & Sons, Inc.

TABLE 11.2b Free-Surface Energies, High-Angle Grain Boundary Energies, and Thermodynamic Cohesive Energy - mJ/m^2

		$d(hkl)/a$	γ_{sv}	γ_{gb}	$2\gamma_{sv}-\gamma_{gb}$
(111)	Cu(EAM)	0.433	1012	290	1734
(100)	Cu(EAM)	0.5	1125	700	1550
(110)	Mo(F–S)	0.707	1829	945	2713
(100)	Mo(F–S)	0.5	2100	2170	2030
(111)	Si(S–W)	0.433	1361	650	2072
(110)	Si(S–W)	0.354	1667	1400	1934

Free-surface energies, γ_{sv}, and 'typical' high-angle grain boundary energies, γ_{gb}, (derived from the asymptotic value of the boundary energy curve in Figure 9.17). Also shown is the thermodynamic cohesive energy, $2\gamma_{sv}-\gamma_{gb}$, for brittle bicrystal fracture. d(hkl)/a is the interplanar spacing of (hkl) planes in units of the lattice parameter a.

11.2.3 Grain Boundary Surface Energy Also Varies with the Boundary Plane

For a given misorientation, the grain boundary surface energy also depends on the *orientation of the grain boundary plane*. In Section 9.2.1.2, we emphasized that free surface energy is anisotropic; grain boundary surface energy is anisotropic, too. If we imagine one grain to be a sphere immersed totally within the other

Figure 11.7
Representation of an asymmetrical boundary in the sodium chloride crystal structure. Charge discontinuity as well as structural discontinuity results in high grain boundary surface energy.

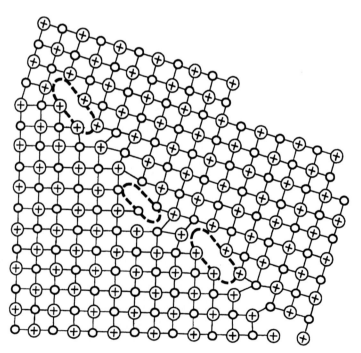

grain, then once more we can represent the grain boundary surface energy variation with orientation by a Wulff plot. Grain boundary surface energy is not independent of the grain boundary plane, and the Wulff figure for the equilibrium shape of the immersed grain is not a sphere. The equilibrium shape is a polyhedron.

The total free energy of the boundary can be reduced if the higher-energy planar surfaces become stepped to be composed of a mixture of lower-energy planar surfaces. Grain boundary steps have an origin similar to the steps in the free surface illustrated in Figure 9.10.

11.2.4 Grain Boundary Surface Energy Lowers the Cohesive Strength of a Solid—Grain Boundary Decohesion

In Section 9.2.1 we saw that we can connect the cohesion of a single crystal solid to the surface energy by a simple relationship

$$W_{cohesion} = 2\, \gamma_{sv}$$

In the presence of a grain boundary, the work of cohesion of the grain boundary surface is reduced to

$$W_{cohesion} = 2\, \gamma_{sv} - \gamma_{gb}$$

or more precisely to

$$W_{cohesion} = \gamma_{svA} + \gamma_{svB} - \gamma_{gbAB} \qquad (11.2.1)$$

where γ_{svA} is the surface energy of the exposed crystallographic surface of grain A after decohesion, γ_{svB} is the surface energy of the exposed crystallographic surface of grain B, and γ_{gbAB} is the surface energy of the grain boundary surface between them. This more precise version of the equation is necessary because we know from Section 9.2.1.2 that the free surface energy of a solid depends on its crystal planar orientation, and the previous sections showed that grain boundary surface energy depends on the misorientation across the grain boundary and the plane of the grain boundary. Table 11.2b includes some typical values for $W_{cohesion}$.

From this general relationship we conclude that:

1. Under stress a solid containing a grain boundary will preferentially separate at the boundary. This is known as *grain boundary decohesion*.

2. The higher the grain boundary surface energy relative to the free surface energy of the solid, the lower the relative cohesive strength and the more likely it is that boundary decohesion will occur. Thus we expect ionic solids [$\gamma_{gb} \approx (2/3)\gamma_{sv}$] to show boundary decohesion more readily than metals [$\gamma_{gb} \approx (1/3)\gamma_{sv}$].

3. Boundaries with the highest surface energy will separate first. Thus we expect high-angle boundaries to fail first because their higher grain boundary surface energy substantially lowers $W_{cohesion}$.

4. For a given grain boundary surface, we expect the regions of greatest misfit to separate first and the coincident sites to give way last.

Intergranular decohesion of polycrystalline sodium chloride demonstrates these features. Figure 11.8 shows separated grain boundaries within a polycrystalline block of pure sodium chloride that has been stressed in tension. This transparent polycrystalline specimen clearly reveals the different stages of boundary decohesion.

11.2.5 Minimizing Grain Boundary Surface Energy Determines the Equilibrium Configuration of Grain Boundary Junctions

Pure crystalline solids consist of many grains, and their grain boundary surfaces spread throughout the material to form a three-dimensional mesh like the foam in a bubblebath or the froth on a head of beer. Now we examine what happens along the *triple lines* where three grains and thus three grain boundaries meet, and at

Figure 11.8
Stressed polycrystalline sodium chloride illustrates grain boundary decohesion. (top) This rod of polycrystalline NaCl has been stressed horizontally. The sample is transparent and we can see grain boundary separation within the solid. (bottom) Higher magnification (100×) shows that decohesion initiates along the triple lines where the bonding is weakest; then grain boundary surfaces partially separate as at X and Y before complete decohesion.

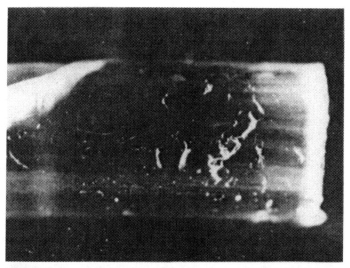

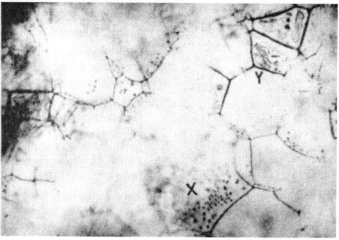

the *four-grain corners* where four grains and six grain boundaries meet as illustrated in Figure 11.9.

When three grain boundary surfaces intersect each other at an angle of 120° along triple lines, as in Figure 11.9a, the free energy of the polycrystalline system is minimized. We obtain this result by making a number of assumptions about boundary energy and "tensions," and mobility of grain boundary surfaces. First, we assume that the grain boundary surface energy per unit area is equivalent to a grain boundary surface "tension" per unit length, just like the fluid in Figure 3.1. In Section 9.2.1.3, however, we noted that surface energy and tension become equivalent in solids only at high temperatures. Because we are interested in determining the equilibrium of polycrystalline solids heated close to their melting point, our assumption is acceptable in this instance. Second, we assume that we can represent the surface tension (energy) of a particular boundary by a vector whose direction lies tangential to the grain boundary surface it represents and whose magnitude equals the grain boundary suface energy. Third, we assume that the total energy of the system is minimized when the tensions are balanced at every pont along the triple line. Fourth, we assume that boundaries are so mobile that they can rearrange their configuration to assume the equilibrium configuration. Finally, for this computation we assume that each grain boundary has the same surface energy, corresponding to the asymptotic value of Figure 9.17. Thus we can assume that the three grain boundary surface tensions, γ_{gbAB}, γ_{gbBC}, and γ_{gbAC}, for the boundaries between the three grains, A, B, and C, are equal. Balancing them in Figure 11.9a gives

$$\frac{\gamma_{gbBC}}{\sin \theta_1} = \frac{\gamma_{gbAC}}{\sin \theta_2} = \frac{\gamma_{gbAB}}{\sin \theta_3} \qquad (11.2.2)$$

from which it can be seen that at equilibrium $\theta = 120°$.

At a four-grain corner, six equal grain boundary surface tensions come into equilibrium, and the free energy of a polycrystalline system is minimized when all six tensions are balanced. Equilibrium occurs when the triple lines meet at 109.5° (the Miraldi angle). As Figure 11.9b shows these angles form where the medians intersect at the centroid of a regular tetrahedron.

11.2.6 Grain Boundary Curvature Causes Grain Boundary Movement and Increases the Average Grain Size

The free energy difference that exists between atoms placed on either side of a *curved* grain boundary surface causes it to move. Following an analysis similar to that used in Section 3.3, eq. 3.3.8, to calculate the pressure difference across a curved fluid surface, we can show that the effective pressure acting on a grain boundary is

$$\Delta P = \gamma_{gb}\left(\frac{1}{R_1} + \frac{1}{R_2}\right) \qquad (11.2.3)$$

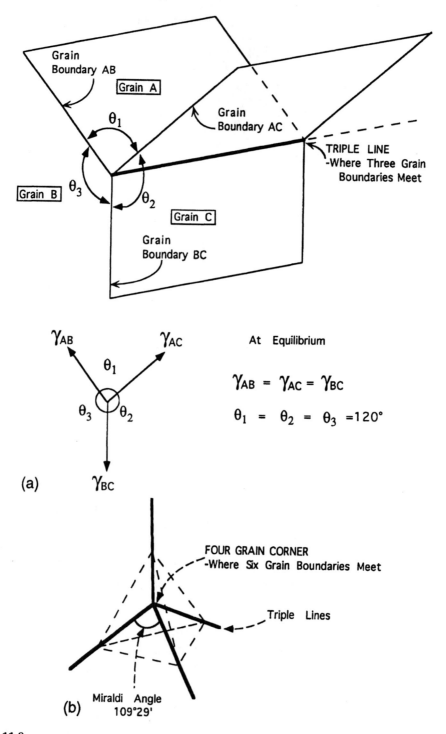

Figure 11.9
Three grain boundary surfaces meet inside a polycrystalline solid along triple lines as shown in
(a), and six grain boundary surfaces (associated with four grains) meet at a point as shown in
(b). Balancing grain boundary surface energies to minimize the total energy requires angles of
120° between surfaces along triple lines and 109.5° between triple lines at a four-grain corner.

where R_1 and R_2 are the respective principal radii of curvature of the grain boundary surface. The change in free energy, ΔG, that occurs when an atom moves from one side of the boundary to the other is given by

$$\Delta G = V\, \Delta P = \gamma_{gb}\, v_m \left(\frac{1}{R_1} + \frac{1}{R_2} \right) \qquad (11.2.4)$$

where v_m is the atomic volume.

Figure 11.10 helps us appreciate this difference in energy from an "atomic" perspective by demonstrating the change in coordination of atoms on either side of the boundary. Atoms on the convex side at sites such as B have a slightly higher coordination and thus a slightly stronger attractive interaction potential than those on the concave side at A. Therefore, these atoms exhibit a statistically significant preference to stay at B during the dynamic fluctuations back and forth across the grain boundary surface. Once an atom has successfully transferred from the concave side to the convex side, it becomes integrated as a member of the adjacent grain. The net result of attracting many atoms to preferred locations such as B is to cause the boundary to move to the left in the diagram. In other words, *a curved grain boundary surface moves toward its center of curvature.*

Figure 11.10
Difference in coordination of atoms on either side of a curved grain boundary. Atoms on the convex side at B have a slightly higher coordination and thus a slightly lower free energy than those on the concave side at A, making the B site more stable. The net flux of atoms from A to B causes the boundary surface to move toward the center of curvature.

11.2.6.1 Grain Boundary Velocity Is Proportional to Curvature

Grain boundary movement is governed by the net movement of all the individual atoms oscillating to and fro in the boundary region. The velocity of this grain boundary movement, v_{gb}, is equal to the velocity of an individual atom across the boundary. The velocity of a particular atom is given by the product of the force (F) acting on it and its mobility (B), eq. 9.5.14.

$$v_{gb} = FB \qquad (11.2.5)$$

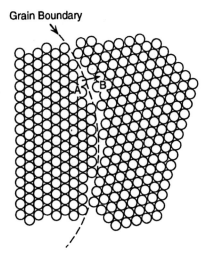

Grain Boundary

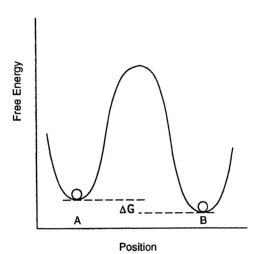

Because the force F times the distance moved λ equals the change in energy ΔG going from one side of the boundary to the other, we have $F = \Delta G/\lambda$; and because the movement is diffusion controlled we have $B = D/kT$ from the Nernst–Einstein eq. 9.5.18. The velocity of the individual atoms therefore is given by

$$v_{gb} = FB = \frac{\Delta G}{\lambda} \frac{D}{kT} \qquad (11.2.6)$$

The diffusivity D is determined by the frequency of jumps across the boundary. It is assumed to equal the grain boundary diffusion coefficient D_{gb}. From eq. 9.5.12 we can write

$$D_{gb} = v\, \lambda^2\, A''' \exp\left(-\frac{\Delta H_{gb}}{kT}\right) \qquad (11.2.7)$$

where v is the atomic vibration frequency at the boundary surface (approximately equal to the lattice vibration frequency, $v_o \approx 10^{13}\, s^{-1}$), λ is the jump distance or the interatomic spacing at the boundary (approximately equal to the lattice parameter), A''' contains the entropy term and is approximately 1. ΔH_{gb} is the enthalpy for grain boundary diffusion and is about half the enthalpy for bulk diffusion, eq. 9.5.25.

Substituting eqs. 11.2.7 and 11.2.4 in eq. 11.2.6 gives

$$v_{gb} = A'''\, v_o\, \lambda\, \frac{\gamma_{gb} V_m}{kT}\left(\frac{1}{R_1} + \frac{1}{R_2}\right)\exp\left(-\frac{\Delta H_{gb}}{kT}\right) \qquad (11.2.8)$$

Important features of eq. 11.2.8 are that the velocity of a boundary toward its center of curvature is proportional to the boundary curvature and depends exponentially on temperature.

11.2.6.2 To Minimize Surface Energy, Grain Size Increases Gradually with Time

We are now in the position to understand the factors that dictate grain size stability in a pure polycrystalline solid. Equation 11.2.2 shows that a polycrystalline aggregate is in equilibrium only if the grain boundary surfaces meet at the proper angles—120° along triple lines and 109.5° at the four-grain corners. Equation 11.2.8 shows that a boundary remains stationary only if the surface is flat. In two dimension a hexagonal "microstructure" consisting of "grains" of identical size and shape that meet at 120° at the triple points and that have straight edges satisfies these criteria. This idealized, metastable two-dimensional "microstructure" appears in Figure 11.11a.

Any deviation from the ideal arrangement of Figure 11.11a introduces curvature into the boundaries and instability into the system. Geometry requires that if a regular polygon has seven sides or more, the angle contained at the corners is >120°. If the boundaries of a seven-sided "grain" are to meet at 120°, then the edges of the polygon must be bowed and become convex. If a "grain" has five sides or fewer, then the edges of the polygon must become concave. So the edges of grains with more than six sides move outward, while the edges of grains with less than six sides

Figure 11.11
Two-dimensional view of grain boundary configurations. (a) Idealized two-dimensional stable "microstructure." The equiaxed hexagonal grain boundary configuration fulfills the energy balance requirement that boundary faces meet at 120°. (b) If a "grain" has five sides or less, the grain boundary surfaces become concave to fulfill the energy balance requirement that they meet at 120°. If a "grain" has seven sides or more, the grain boundary surfaces become convex to fulfill the energy balance requirement that they meet at 120°. Grains with more than six sides grow; grains with fewer than six sides shrink. The average grain size of this "microstructure" increases. (After J. E. Burke.)

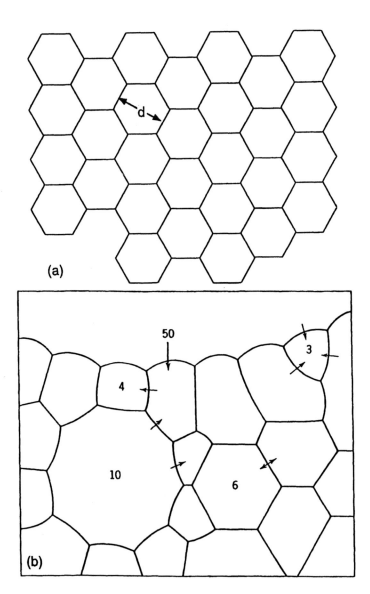

move inwards. Furthermore, the rate of movement depends on the curvature. The von Neumann "six minus n" rule summarizes these results. It states that if all grain boundaries have the same energy, the rate of change in area of each grain is proportional to −(6−n), where n represents the number of corners. As Figure 11.11b shows, this relationship causes large grains to grow and small grains to shrink. Eventually the smaller grains disappear, they are consumed, and so the average grain dimension increases.

In the real world of three dimensions, the situation is more complicated. Unlike the two-dimensional hexagon, no three-dimensional polyhedron exists that can be repeated to fill space that has flat surfaces and satisfies the angular requirements of 120° along the edges and 109.5° at the corners. The closest solution comes with truncated octahedra, each with eight hexagonal faces and six square faces (shown in Figure 11.12), arranged on a body-centered cubic lattice. Grains with this shape still must have

Figure 11.12
The truncated octahedron is a solid figure that can be joined on a body-centered cubic lattice to fill space. Surfaces between joined octahedra come close to fulfilling the grain boundary energy balance requirements that surfaces meet at 120° along the edges and at 109.5° at the corners.

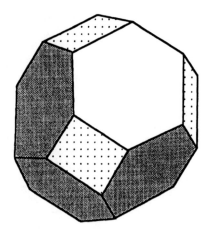

surfaces that are slightly distorted and curved to meet at the equilibrium angle. Since curved grain boundaries move, three-dimensional microstructures of pure materials are always unstable. Again, the sense of motion is such that big grains grow and small grains are consumed. So the average grain size gradually increases, and the phenomenon known as *grain growth* occurs. Transformation of a polycrystalline material from one average grain size to another larger grain size through grain growth is called *secondary recrystallization* to distinguish it from primary recrystallization, the mechanism associated with the initiation of new grains described in Section 11.1.

Grain growth does not occur at a constant rate. In a polycrystalline material, the average grain size, represented by the grain diameter d in Figure 11.11a, shows a nonlinear dependence on time. We can derive this dependence using eq. 11.2.8 with two assumptions. First we assume that the rate of grain diameter growth, $d(d)/dt$, is proportional to v_{gb} and second, that the grain diameter is related to the radius of curvature at the grain surface. Substituting these assumptions into eq. 11.2.8 gives

$$\frac{d(d)}{dt} \sim \frac{C}{d} \qquad (11.2.9)$$

where C contains all the parameters of eq. 11.2.8 other than d. Integrating eq. 11.2.9 gives

$$\frac{d^2 - d_o^2}{2} = Ct \qquad (11.2.10)$$

for the grain size d as a function of time t. Here d_o is the starting grain size. When d becomes much larger than d_o this equation reduces to a simple parabolic relationship.

In principle, grain growth ceases only when all the grain boundaries have been driven out of the system and the material reverts to a perfect single crystal. This phenomenon is demonstrated by model single-phase systems containing high-energy boundaries, such as pure polycrystalline sodium chloride and potassium chloride. Also it is used to prepare pure metal single crystals by the "strain-anneal" method, in which a polycrystalline metal is slightly strained (5–10% strain) to introduce sufficient dislocations to stimulate grain growth and then annealed in a temperature gradient so that one grain takes over and gradually consumes all the other grains.

Because most physical properties benefit from a small uniform grain size, generally the challenge to the materials engineer is to stop grain boundary motion and grain growth rather than stimulate it. The presence of very large grains in the midst of a fine-grained polycrystalline microstructure is the root cause of many disastrous structural failures. Precise control over grain growth to develop prescribed stable microstructures lies at the heart of metallurgical and ceramic materials processing. In Section 11.3.2, we discuss one way to impede the motion of grain boundaries and control grain size by introducing impurities or precipitate phases to change the chemistry of the grain boundary regions.

11.2.7 Grain Boundary Surfaces Act as the Source and Sink for Vacancies; As a Result, Grain Boundary Surfaces in Ionic Solids Can Become Charged

In Section 9.4.1.1, we noted that the concentration of point imperfections in a crystalline solid comes into thermodynamic equilibrium when vacancies generated at a free surface diffuse into the solid by exchanging places sequentially with nearby ions. Point imperfections in a polycrystalline solid achieve thermodynamic equilibrium more rapidly than a single crystal of the same size because the vacancies can travel over the grain boundaries before diffusing into the crystalline grains. Enhanced diffusivity over grain boundary surfaces means that grain boundaries act as sources (and sinks) for vacancies and interstitials in the bulk of the grains.

In Section 9.4.1.2 we showed that because vacancies or interstitials of one ion have a slightly smaller formative energy than the other, the free surface of an ionic solid becomes charged and a space charge of oppositely charged imperfections forms just beneath the free surface. Likewise, grain boundaries in ionic solids become charged, and to redress the charge imbalance, they attract a space charge of vacancies of the opposite sign or impurity ions with a valence different from that of the host. Equilibrium distribution of these vacancies or impurities in the vicinity of grain boundaries also follows the concentration profile described by the Gouy–Chapman equations 4.3.6 and 4.3.12 and shown in Figure 9.13.

11.3 Solute Atoms (or Impurities) Segregate Preferentially in the Grain Boundary Region

11.3.1 Solute Atoms Segregate at Grain Boundaries to Reduce the Chemical, Mechanical, or Electrical Energy of the System

In Sections 3.5.2 and 9.3.2, we saw how solute atoms (or impurities) concentrate at the free surfaces of liquids and solids to reduce the surface energy. Similarly the grain boundary surface is a preferred location for solute (or impurity). Grain boundary energy is reduced by *segregation* of impurities to the boundary region. In fact, segregation is more likely to occur at a grain boundary surface than at a free surface because no competition from vapor adsorption to reduce the surface energy occurs inside the solid. Furthermore, two additional driving forces, mechanical strain energy and electrical energy, enhance segregation to grain boundaries at low temperatures. Together with the reduction in chemical energy all three factors contribute to reduce the free energy of the system when an impurity moves to the vicinity of the grain boundary.

The chemical term arises from the fact that impurity elements change the character, strength, and density of the dissatisfied bonds across the boundary and thereby reduce the grain boundary energy by a small amount. We can describe the excess concentration of impurity in the grain boundary, Γ, due to this chemical factor by substituting γ_{gb} in the Gibbs adsorption isotherm, eqs. 3.5.13 and 9.3.6

$$\Gamma = -\frac{1}{RT}\frac{d\gamma_{gb}}{d(\ln c)} \qquad (11.3.1)$$

where c is the molar concentration of the impurity (or solute).

The mechanical term arises from the elastic distortion produced within a crystal structure by an impurity atom that has a slightly larger or smaller radius than the host atom that it replaces. If we locate the impurity at the grain boundary where the structural regularity is broken and the density is less than in the crystalline grain, we can reduce strain energy due to this distortion.

The electrical term was described in Section 11.2.7 for ionic solids. Impurity ions are drawn to the boundary by the electrical potential due to the surface charge. A similar effect exists for metals when an "embedded" impurity ion has a different electronic structure than the host and therefore experiences an electrical attraction to the intergranular surface.

These three terms are not independent. Depending on chemistry, size, valence, and electronegativity, a particular impurity may experience one, two, or all three forces driving it to segregate.

Segregation proceeds at a rate determined by how quickly the solute element diffuses through the bulk to become trapped at the grain boundary. At high temperatures, diffusion occurs more rapidly and speeds up the segregation process. However, it is the *total free energy* of the system that determines equilibrium, and at high temperatures, entropy favors having unsegregated impurites dispersed at random throughout the grain. At low temperatures, the total free energy favors segregation, but diffusivity is much reduced. Optimum conditions for segregation exist at intermediate temperatures, where the entropy term favors segregation and the temperature is high enough for it to proceed at a reasonable rate.

Generally, segregation at intermediate temperatures results in an excess concentration of the second element at the boundary. Figure 11.13 reproduces computer simulations that predict a tenfold increase in nickel concentration in the vicinty of a simple,

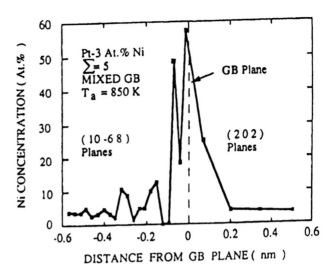

Figure 11.13
Segregation to grain boundaries in fcc metals. Calculated nickel concentration profile in the vicinity of a $\Sigma = 5$ grain boundary in a platinum–3 at. % nickel alloy. Computation predicts a tenfold increase in concentration at 850 K, in agreement with experimental measurements. [S. M. Foiles and D. N. Seidman, *MRS Bulletin*, **XV**(9), 51 (1990).]

$\Sigma = 5$, grain boundary in platinum–3 at. % nickel. Direct experimental measurements are in substantial agreement with the predictions. Later Figure 11.24a shows experimental evidence for the segregation of bismuth to zinc oxide grain boundaries.

11.3.2 Impurity Segregation Is Used to Control Grain Growth

Now we are able to understand how a small amount of impurity or solute can be used to control grain growth in metals and ceramics. We want to stop grain boundaries from moving, and one mechanism to reduce grain boundary velocity results from the segregation of impurities. When an impure boundary moves, it must either drag the "cloud" of impurities along or escape from the segregated impurities altogether. Each option involves additional energy and creates an opposing force that retards the forward movement of the boundary. Another mechanism to reduce grain boundary velocity involves actual precipitation of a second phase within the material, as discussed in Section 11.5. For a boundary to move under these conditions, it must either cut through the precipitate particles or wrap itself around them to pass by. In this way pores also impede grain boundary motion.

A classic example of the use of a small amount of impurity to enhance the processing of a material through grain growth control is the manufacture of the ceramic Lucalox™. Lucalox™ is a fully dense (completely devoid of any porosity), optically transparent aluminum oxide ceramic used as the lens for high-intensity street lamps. In the early stages of the alumina sintering process, pores form as the loose powder particles bond together. At a later stage the pores become preferentially located at grain boundaries in the evolving microstructure. If the grain boundaries move too quickly, these pores are left behind trapped within the grains: they can escape only by slow bulk-diffusion mechanisms. Restraining grain boundary motion during sintering gives the pores a chance to move out of the material more rapidly by diffusion over the grain boundary surfaces to escape at the free surface. To make Lucalox™, a very small percentage (less than 1%) of magnesium oxide powder is added to the starting alumina powder to enhance its sinterability and eliminate porosity. Magnesium oxide is a *grain growth inhibitor* when added to alumina. Differences in ionic valence (Al^{3+} versus Mg^{2+}) and size lead to very limited solubility of magnesium oxide in alumina (less than 0.01% at the sintering temperature 1500°C). Magnesium ions segregate to the grain boundaries at the sintering temperature where they restrain grain growth and provide the opportunity for vacancies and porosity to escape by grain boundary diffusion during the sintering process. With the right combinations of magnesium oxide content, temperature, and time the final stages of sintering produce a fully dense ceramic material free from pores.

11.3.3 Impurity Segregation Modifies the Properties of Metals and Ceramics

Grain boundary segregation exerts a profound effect on physical properties of crystalline solids. As a result, it is a very important phenomenon in the application of many metallurgical and ceramic materials.

In low-alloy steels, for example, segregation often leads to grain boundary decohesion, otherwise known as *grain boundary embrittlement*. Undesirable foreign atoms inevitably find their way to the grain boundaries. Examples are legion in which extremely small concentrations (parts per million range) of foreign atoms—particularly metalloids such as P, S, Sb, Sn, As, or Te—provide near-monolayer coverages of the grain boundary surfaces. Contamination of the bulk as low as 10 parts per million results in segregation as high as 1% in the vicinity of the boundary, a 1000-fold increase in local concentration. While segregation lowers the grain boundary surface energy (γ_{gb}), it lowers the free surface energy (γ_{sv}) even more—by a factor of two, according to some estimates. So segregation of certain impurities almost halves the grain boundary cohesion strength in eq. 11.2.1.

Elements such as phosphorus or sulfur reduce the mechanical tensile strength of steel. Even more insidious is the change in fracture mode. In pure or refined steels, fracture is gradual and energy absorbing; in impure steels it is sudden and catastrophic. Grain boundary embrittlement reduces the fracture toughness (see Appendix 11B.4) of the steel by orders of magnitude. One measure of structural material performance relates to the temperature at which a *ductile-to-brittle transition* (T_{DBT}) takes place. Above the T_{DBT} temperature, the material is tough, and the fracture process absorbs a great deal of mechanical energy. Below the T_{DBT}, the material is brittle, and relatively little energy is absorbed in fracture. The *lower* the T_{DBT}, the *better* the low-temperature fracture resistance of a material and the higher its fracture toughness. Figure 11.14 shows impact energy transition curves for a steel heated (aged) different lengths of time at 500°C. Segregation of P and Sn during aging increases T_{DBT} by 300°C. This extreme sensitivity emphasizes the need for purity control in high-performance steels. Often we can trace brittle rupture of large pressure vessels and large structures to trace impurity segregation

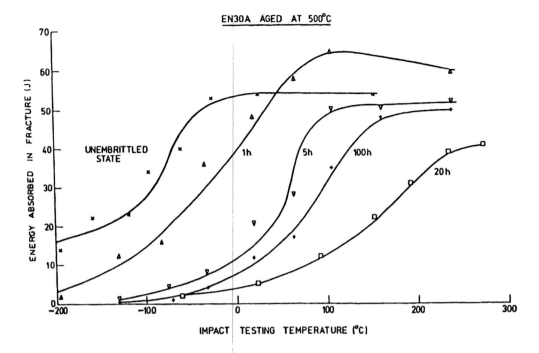

Figure 11.14
Grain boundary embrittle-
ment due to segregation of
phosphorous and tin at the
grain boundaries of a steel.
The ductile-to-brittle
transition temperature has
increased by about 300°C
following a thermal
treatment that enhances
segregation.

and intergranular embrittlement. The challenge is to find a way
to localize the undesirable elements in the microstructure away
from the grain boundaries, yet at the same time foster the
segregation of the desirable elements needed for grain growth
control.

Grain boundary segregation does not always have disastrous consequences on the physical properties of a crystalline
solid. In fact, grain boundary segregation is absolutely essential
in the performance of certain devices. For example, Section
11.6 discusses the characteristics of the ceramic varistor, a
grain boundary layer electronic device.

11.4 Grain Boundary Surfaces in Equilibrium with Other Phases Lead to Different Microstructural Configurations

When a polycrystalline solid system is held at a high temperature
for a period of time, the grain boundaries come to equilibrium
with their environment. This environment may be composed of
ambient gases, a molten liquid phase, or another solid phase
within the polycrystalline solid. Section 11.2.5 set out the procedures we can use to derive the equilibrium configurations for
two-phase structures. Briefly stated, we minimize surface energies associated with grain boundaries and other phase boundaries
when we balance surface or interfacial tensions at the line of
intersection of the phases with the boundary surface. We illustrate
the procedure by considering three different types of grain bound-

ary interfaces: those in equilibrium with vapor, liquid, and solid phases.

11.4.1 Grain Boundary/Vapor-Phase Interaction Leads to Thermal Grooving

First, we consider the consequences of heating a solid containing a grain boundary in a nonreactive vapor or vacuum. Exposure to high temperature for a prolonged period results in a groove where the boundary meets the surface. We can understand thermal grooving by balancing the surface tensions drawn in Figure 11.15a to give the relationship

$$2\,\gamma_{sv}\,\cos\frac{\varphi}{2} = \gamma_{gb}$$

or

$$\cos\frac{\varphi}{2} = \frac{\gamma_{gb}}{2\gamma_{sv}} \tag{11.4.1}$$

Figure 11.15
Grain boundary surfaces in equilibrium with their environment. (a) A grain boundary comes to equilibrium with a vapor phase at the free surface when the surface tensions balance the grain boundary tension. A thermal groove forms at high temperatures. (b) A grain boundary comes to equilibrium with a liquid phase at the free surface when the interface tensions balance the grain boundary tension. (c) Three grain boundary surfaces come to equilibrium with a liquid phase in the interior of the polycrystalline solid when the interface tensions balance the grain boundary tensions.

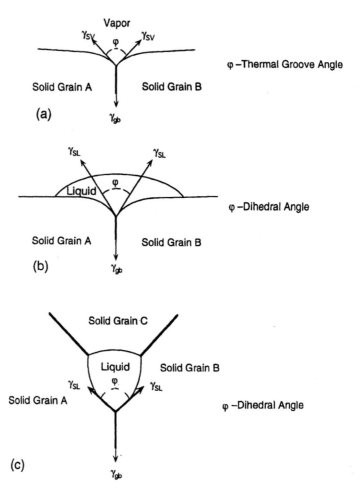

ϕ is referred to as the *dihedral angle* (or as the *thermal grooving angle* when it occurs at the free surface). This relationship defines the shape and depth of the thermal groove formed during high-temperature exposure.

Thermal grooving may be harmful because the groove represents a surface notch that can decrease the mechanical strength of the material at low temperatures (see Appendix 11B.4). To reduce the sharpness and stress concentration associated with the groove, we must increase ϕ. For a given solid this means changing the vapor in the environment, coating or treating the surface of the solid to increase its surface energy γ_{sv}, or adding impurities to the solid to reduce γ_{gb}.

Sometimes thermal grooving is useful. For example, it provides a means to mark grain boundaries to measure grain size in high-temperature metals and ceramics that are otherwise hard to etch chemically. Grooves also reflect the nature of the boundary. The depth of the grooves depends on γ_{gb}, and they are shallower for twin boundaries than for regular grain boundaries.

11.4.2 Grain Boundary/Liquid-Phase Interaction Leads to Two-Phase Microstructures of Varying Complexity

Figure 11.15b shows how a groove develops at the surface of a solid material containing a grain boundary along the line where the boundary comes into contact with a liquid at high temperature. By balancing the surface tensions for a liquid phase in contact with a grain boundary at a free surface, we derive a similar relationship for this type of groove as in eq. 11.4.1

$$2\,\gamma_{sl}\cos\frac{\phi}{2} = \gamma_{gb}$$

or

$$\cos\frac{\phi}{2} = \frac{\gamma_{gb}}{2\,\gamma_{sl}} \tag{11.4.2}$$

The only difference is that γ_{sl} replaces γ_{sv}. Again, ϕ is known as the dihedral angle.

Of much greater technological significance is the situation in which the polycrystalline solid contains a liquid phase *within* the microstructure. Figure 11.15c shows the construction for balancing tensions at internal interfaces. The mathematical expression for the dihedral angle ϕ within the microstructure is the same as that given in eq. 11.4.2. The value of ϕ and the consequent distribution of the liquid phase in the polycrystalline material depend critically on the relative values of γ_{sl} and γ_{gb}. Table 11.3 lists the value of ϕ corresponding to specific values of γ_{sl}/γ_{gb}. The table also lists the resulting microstructures, some of which are illustrated in Figure 11.16.

Figure 11.16a shows the situation in which the solid–liquid interface energy is much larger than the grain boundary surface energy, $\gamma_{sl}/\gamma_{gb} > 1.5$, and the dihedral angle is 135° or greater [condition (a) in Table 11.3]. To minimize the total energy, the

TABLE 11.3 Liquid-Phase Distribution in Polycrystalline Materials

Relative values of γ_{sl} and γ_{gb}	Value of φ	Equilibrium configuration
(a) $\gamma_{sl} > (3/2)\,\gamma_{gb}$ $\gamma_{sl}/\gamma_{gb} > 1.5$	$\varphi \geq 140°$	Liquid phase confined to four-grain corners with convex interface
(b) $\gamma_{sl} = \gamma_{gb}$ $\gamma_{sl}/\gamma_{gb} = 1.0$	$\varphi = 120°$	Liquid phase confined to four-grain corners with flat interface
(c) $\gamma_{sl} = (1/\sqrt{2})\,\gamma_{gb}$ $\gamma_{sl}/\gamma_{gb} = 0.707$	$\varphi = 90°$	Liquid phase spreads out from four-grain corners with concave interface; penetrates triple lines
(d) $\gamma_{sl} = (1/\sqrt{3})\,\gamma_{gb}$ $\gamma_{sl}/\gamma_{gb} = 0.577$	$\varphi = 60°$	Liquid phase confined to grain boundary triple lines with flat interface
(e) $\gamma_{sl} = 0.541\,\gamma_{gb}$ $\gamma_{sl}/\gamma_{gb} = 0.541$	$\varphi = 45°$	Liquid phase spreads out from triple lines; penetrates grain boundary surfaces
(f) $\gamma_{sl} < (1/2)\,\gamma_{gb}$ $\gamma_{sl}/\gamma_{gb} \leq 0.5$	$\varphi = 0°$	Liquid phase completely wets grain boundary surfaces

liquid phase stabilizes in one of two ways: either it balls up on the grain boundary surface to form an almost spherical droplet, or it locates at the four-grain corners, where the closest match to $\varphi \approx 135°$ occurs. Actually the ratio $\gamma_{sl}/\gamma_{gb} > 1.5$ more accurately describes the situation for a vapor trapped inside the solid. As noted in Section 11.2.2, the typical value of γ_{gb} is one-third to two-thirds the surface energy γ_{sv}; so $\gamma_{sv}/\gamma_{gb} > 1.5$. Therefore, the configuration shown in Figure 11.16a is more characteristic of pores than of liquids within the solid. Internal porosity during the midstages of ceramic powder sintering is located primarily on the grain boundary surfaces, but as sintering proceeds and porosity is reduced in the final stages, the pores prefer to locate along the triple lines and in the four-grain corners. Finally the last few pores remain as spheres in the four-grain corners. The microstructure shown in Figure 11.16a typifies the final stages of ceramic densification.

When the two interfacial energies are equal, $\gamma_{sl}/\gamma_{gb} = 1.0$, and the dihedral angle is 120° [condition (b) in Table 11.3]. As in Figure 11.16a, the preferred location for the liquid phase in the microstructure providing the closest match to $\varphi \approx 120°$ occurs at the four-grain corners. The solid–liquid interfaces are now flat rather than convex.

Conditions (c) through (f) in Table 11.3 describe four situations in which the solid–liquid interface energy is less than the grain boundary energy, $\gamma_{sl}/\gamma_{gb} < 1.0$. For condition (c), $\gamma_{sl}/\gamma_{gb} = 0.707$, and the dihedral angle is 90°. The liquid phase prefers to locate at the four-grain corners, but the interface has become concave and starts to penetrate into the adjacent triple lines, as

Figure 11.16
Different equilibrium
microstructures exist
inside a two-phase poly-
crystalline material as the
ratio of interfacial energy
to grain boundary energy,
and as a consequence the
dihedral angle φ, varies.

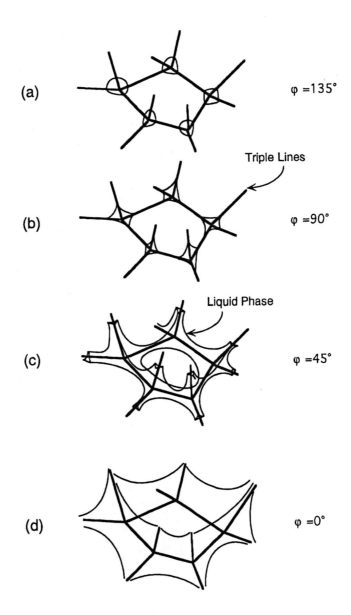

(a)   φ =135°

Triple Lines

(b)   φ =90°

Liquid Phase

(c)   φ =45°

(d)   φ =0°

illustrated in Figure 11.16b. For condition (d), $\gamma_{sl}/\gamma_{gb} = 0.577$ and
the dihedral angle is 60°. The liquid phase now spreads along all
the triple lines with a flat interface to create a triangular prism
shape. For condition (e), $\gamma_{sl}/\gamma_{gb} = 0.541$ and the dihedral angle is
45°. The liquid phase lies along the triple lines with a concave
interface that starts to penetrate into the adjoining grain boundary
surface, as illustrated in Figure 11.16c. For condition (f), the
solid–liquid interface energy is equal to or less than one-half the
grain boundary energy, $\gamma_{sl}/\gamma_{gb} \leq 0.5$, and the dihedral angle is
0°. Thermodynamic equilibrium is achieved when the liquid
phase *completely wets* the intergranular surfaces, as illustrated
in Figure 11.16d.

    Note how slight variations in the values of γ_{sl} for a fixed γ_{gb}
radically change the microstructure in conditions (c) through (f)
in Table 11.3. Minor impurities added to the liquid phase can

exert a profound effect on γ_{sl} and on the microstructure. In fact, active elements like titanium often are deliberately added as impurities to ceramic powders to manipulate the liquid–solid interface equilibrium during processing.

11.4.2.1 When Liquid Phases Wet Grain Boundaries, the Grains Become Disjoined, Leading to Grain Boundary Layer Microstructures

Interaction between a polycrystalline solid and a liquid that wets its grain boundaries can be so strong that the solid actually imbibes the liquid and swells in size. The process is necessarily slow because the displaced solid must diffuse out of the path of the advancing liquid. As liquid is imbibed, it initially penetrates down the triple lines and then over the intergranular surfaces. Ultimately, the individual grains become isolated by the liquid phase. They are said to be *disjoined* by the interfacial forces. Fully dense polycrystalline alumina placed in contact with molten glassy calcium silicate at 1000°C swells by about 10% due to imbibed liquid. The glassy phase freezes when cooled to room temperature, and the original single-phase polycrystalline alumina becomes transformed into a two-phase alumina–glass composite. A similar situation prevails when magnesium oxide is cofired with monticellite, a calcium magnesium silicate, the magnesia grains are wetted by the silicate and become completely disjoined, as shown in Figure 11.17, which should be compared with Figure 11.16d.

Figure 11.17
Dense, polycrystalline magnesium oxide (dark phase) is completely wetted and disjoined by liquid monticellite ($CaMgSiO_4$) at 1700°C. Note how the intergranular separation varies with grain boundary misorientation; some boundaries are still in contact. (Scanning electron micrograph courtesy of Sundar Ramamurthy and Prof. C. Barry Carter, University of Minnesota.)

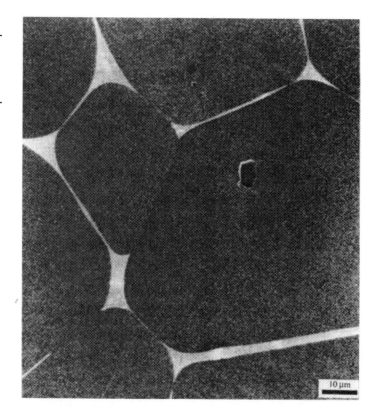

The presence of very small amounts of liquid phase is an important aspect of ceramics processing and properties. For example, consider the sintering of silicon nitride at high temperatures (>1500°C). Pure silicon nitride is difficult to sinter because its diffusion coefficients are low. (Covalent materials have a high activation energy for the formation and movement of vacancies.) But pure silicon nitride powder is impossible to obtain anyway because an atomically thin surface layer of silicon oxide covers the silicon nitride powder particles. At high temperatures all the silicon oxide coalesces to form a minor amount of silica (silicon oxide) that liquefies to form a glassy liquid phase. In some instances, a small amount of magnesium oxide is deliberately added to the starting powder to react with the silicon oxide to form liquid magnesium silicate. Surface tension of liquid silica or the liquid silicate is very low, and interfacial tension between liquid silicate and solid silicon nitride is lower than the grain boundary energy; so $\gamma_{sl}/\gamma_{gb} \leq 0.5$. The liquid silicate wets and covers the sintered silicon nitride grains with a very thin layer as the system comes to equilibrium. The distribution of the liquid phase at high temperatures is like that shown in Figure 11.16c or d, and on cooling to room temperature, the liquid silica phase solidifies as an amorphous phase at the grain boundaries. Sintered silicon nitride is generally a two-phase ceramic, consisting of crystalline silicon nitride grains separated by a thin layer of glassy silicate.

Unfortunately, this silicate layer weakens the high temperature mechanical strength of silicon nitride. The liquid silicate layer causes the grains to slide apart at high temperatures when the glass liquefies. *Grain boundary sliding* makes a significant contribution to the high-temperature creep of these materials. Ingenious processing strategies to eliminate the liquid phase in silicon nitride (or to crystallize or to stabilize it) are being pursued in the development of this ceramic for applications in high-temperature gas turbines.

Other applications use the liquid phase to advantage. Microstuctures in which a grain is isolated from its neighbors by a very thin grain boundary insulating layer are used in *grain boundary layer devices*. For example, in processing ceramics like strontium titanate for gas and humidity sensors, grains are isolated from each other by boron oxide or bismuth oxide glassy phases. The liquid oxides completely wet the ceramic grains and on freezing at room temperature, they generate microstructures with a very thin continuous grain boundary layer.

11.4.2.2 To Reduce Interfacial Energies Further, Grain Boundary/Liquid-Phase Interfaces Can Become Stepped

So far, in our discussion of the interaction between grain boundaries and a liquid-phase environment, we have neglected the effects of orientation on grain boundary energy. In all our computations, we assume that the grain boundary surface energy γ_{gb} has a single average value. This assumption needs to be qualified for two reasons.

First, grain boundary energy varies with the misorientation between two grains on either side of the boundary. As shown in Figure 11.5, grain boundary surface energy in the vicinity of small-angle misorientations and twin cusps is sensitive to the relative tilt and twist components of the misorientation. As a consequence, not all grain boundary surfaces are in equilibrium when they meet at 120° along triple lines (eq. 11.2.2). In particular, when a twin boundary intersects a grain boundary, the low energy of the twin boundary prevails and the orientation of the twin plane dictates the angle of interception.

Second, as noted in Section 11.2.3, for a given misorientation, the grain boundary energy depends on the orientation of the grain boundary plane. For some planar orientations the intergranular surface becomes stepped. Likewise, in the presence of a liquid phase, the equilibrium configuration of the liquid phase in contact with grain boundaries as defined by eq. 11.4.2 changes with planar orientation. Studies of alumina bicrystals containing liquid calcium silicate glass show how the interface changes with grain boundary planar orientation. In Figure 11.18, the boundary plane between the upper and lower grains is stepped; the almost horizontal planar segments are well separated by the glass phase g, while the inclined planar segments have only a very thin liquid interface.

11.4.3 Grain Boundaries between Mixed Solid Phases Lead to Unusual Microstructures

Next we consider the interfacial interactions between two solid phases. We can generate a polycrystalline blend of two different solids in two ways. One procedure blends two immiscible solid powders together and then sinters them to form a two-phase polycrystalline solid. In the other procedure, a solid solution in thermodynamic equilibrium at high temperature is cooled into a two-phase region, precipitation is initiated within the solid, and a two-phase solid–solid microstructure evolves. Diphasic solids

Figure 11.18
Transmission electron micrograph of an aluminum oxide bicrystal partially wetted by liquid calcium silicate. The boundary is divided into regions that are fully wetted by the glass, g, and regions that appear to contain no liquid phase at all, depending on the orientation of the boundary. Glass pockets, g, are bounded by stepped low index planes. [D-Y. Kim, S. M. Wiederhorn, B. J. Hockey, C. A. Handwerker, and J. E. Blendell, *Journal of the American Ceramic Society,* **77**, 444 (1994). Photomicrograph courtesy of Dr. Sheldon Wiederhorn, N.I.S.T.]

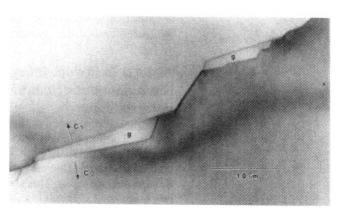

formed by the first method best illustrate our present topic; solid-state precipitation reactions are discussed in Section 11.5.

Two ceramic solids that show little or no solid solubility at any temperature are alumina and zirconia. They simply do not mix. If we synthesize and blend powders of the two materials using the colloidal processing techniques described in Section 4.7.3 and then sinter them together to form a polycrystalline material, three types of boundaries form between the respective grains: alumina–alumina grain boundary surfaces (gbaa), zirconia–zirconia grain boundary surfaces (gbzz), or alumina–zirconia grain boundary interfaces (gbaz). Balancing the grain boundary surface energies, γ_{gbaa} and γ_{gbzz}, and the grain boundary interfacial energy, γ_{gbaz}, along a triple line yields angles of equilibrium other than the 120° found for a pure single-phase solid. In the microstructure of sintered alumina–zirconia reproduced in Figure 11.19, it can be seen that the interfaces curve and meet at angles greater and less than 120°. The microstucture does not exhibit the classical hexagonal shapes shown in Figure 11.7a. Computer simulation shows the variety of microstructures that can result from this conformation for different interfacial energy regimes. Four-grain junctions along certain junction lines are among the posssiblities. Microstructure evolution of diphasic structures—particularly the stability of interpenetrating phases—is an important topic in materials research.

Figure 11.19
Scanning electron micrograph of the microstructure of an alumina–15 vol % zirconia ceramic composite. (D. J. Green in *Industrial Materials Science and Engineering*, L. E. Muir (ed.), Marcel Dekker, Inc., New York, 1984, p. 84.)

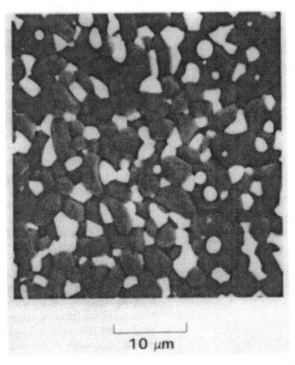

10 μm

11.5 Grain Boundaries Form Preferential Sites for Solid State Precipitation Reactions

11.5.1 Most Technologically Significant Crystalline Solids Undergo Phase Transformations and Precipitation at Low Temperatures

Until now, we have focused our discussion on the behavior of single-phase polycrystalline solids. This emphasis is satisfactory because most pure solids, such as polycrystalline gold (face-centered cubic crystal structure) or polycrystalline sodium chloride (rock salt structure), maintain the same crystal structure at all temperatures from absolute zero to the melting point. However, a few *pure* solids such as iron or silica switch from one crystalline form to another at specific transformation temperatures—they are *polymorphic* solids. For example, pure iron switches from its body-centered cubic crystal structure at room temperature to face-centered cubic at 910°C and back to body-centered cubic at 1400°C before melting at 1539°C. Pure crystalline silica switches from quartz to tridymite to cristobalite forms as its temperature is raised. Only the crystalline form changes in these *allotropic phase transformations;* chemical composition remains the same.

Other kinds of phase transformation occur in *impure* solids. Indeed most technologically significant crystalline solids are neither pure nor single phase. Materials such as steels, aluminum alloys, and refractory ceramics consist of two or more phases in thermodynamic equilibrium over a range of temperatures. These two phases differ in both chemistry and crystalline form. At high temperatures, alloyed materials generally exist as a single-phase solid solution, but as the temperature is lowered, the solid solution undergoes phase separation into two phases of different chemical composition and crystal structure. *Solid-state precipitation* occurs.

Because all three crystallographic parameters (crystal structure, lattice parameter, and orientation) usually change across the interface between the new precipitate and the host, an *interphase boundary* forms between the precipitate and the matrix phases in the solid. The tremendous body of literature on solid–solid phase transformations, interphase boundaries, and interface structures in metals and ceramics lies beyond the scope of this text. Here we give only an overview to highlight the role grain boundaries play in solid-state precipitation.

11.5.2 Regular Solution Theory Describes Solid Solution Stability

Thermodynamic description of solid solutions and their stability as a function of composition and temperature is based on the regular solution theory outlined in Appendix 6A. The tendency for a solid solution to phase separate depends on temperature as well as on the sign and magnitude of the effective atomic interaction parameter W defined in eqs. 3.1.8 and equation 6A.5.

$$W = z \left(W_{AB} - \frac{(W_{AA} + W_{BB})}{2} \right) \qquad (11.5.1)$$

When $W = 0$, the affinity of two different atoms (A and B) for each other equals their self-affinity, and we have an ideal solid solution. When $W < 0$, or when the interaction between A and B is stronger than thermal fluctuations, $W < 2kT$, the solid solution is stable. For this condition, at a high temperature, say T_1, the total free energy (ΔG) curve as a function of composition is smooth, as illustrated in Figure 11.20a. Phase separation does not occur at temperature T_1. On the other hand, when $W > 2kT$, the solute and solvent atoms prefer to be self-associated. At a low temperature, say T_2, the total free energy curve as a function of composition contains a double minimum, as shown in Figure 11.20b. For compositions between α and β (defined by the common tangent to the two minima), the mixture of the two phases α plus β has a lower total free energy than the solid solution. Phase separation does occur at temperature T_2.

Compositions between A and B form a stable solid solution over the whole composition range at sufficiently high temperatures, such as T_1. Compositions between α and β undergo a solid-state precipitation reaction when the temperature is lowered to T_2. Examining phase stability at different temperatures enables us to construct the phase diagram. When the two phases α and β differ in chemical composition but not in crystal structure, a phase diagram with a simple miscibility gap results, like the one shown in Figure 11.20c (see also Figure 6.10). When the two phases α and β differ in crystal structure as well as composition, more complex phase diagrams result. We shall not discuss them here.

Some elements or compounds differ so much in chemistry and structure that we find little or no solid solution of one in the other. Such is the case for the alumina and zirconia discussed in Section 11.4.3.

11.5.3 Heterogeneous Nucleation Theory Describes Initiation of Solid-State Precipitation

Nucleation and growth of a new solid phase (say β) from the parent solid solution (say α) follows a similar theoretical development to those used in Section 3.4 to describe nucleation of liquid droplets from a supersaturated vapor and in Section 10.6.1 for condensation of vapor onto solid substrates. In solid-state transformations, the free energy change, ΔG_p, on forming a spherical precipitate nucleus of radius R becomes

$$\Delta G_p = -\frac{4}{3}\pi R^3 (\Delta G_v + \Delta G_\varepsilon) + 4\pi R^2 \gamma_{\alpha\beta} \qquad (11.5.2)$$

We can compare this equation with eqs. 3.4.1 and 10.6.3. Here ΔG_v is the free energy change per unit volume involved in the phase transition and $\gamma_{\alpha\beta}$ is the interphase interfacial energy. The additional term, ΔG_ε, is the strain energy per unit volume of the host material due to the structural misfit between α and β; ΔG_ε can be represented by $E'\varepsilon_t^2$, where E' is the elastic modulus of the

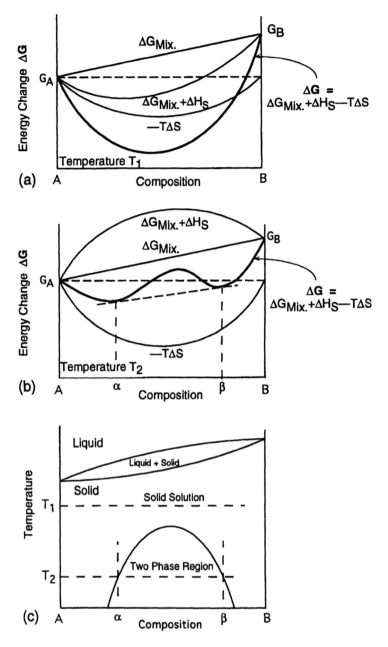

Figure 11.20
Free energy–composition curves and simple phase diagram for real solid solution of A and B.
(a) Free energy curve for high-temperature (T_1) solid solution for all compositions. (b) Free
energy curve for low-temperature (T_2). Phase separation occurs between compositions α and β.
(c) Phase diagram showing miscibility gap.

matrix modified to account for the crystallographic and orientational relationship between the matrix α and the precipitate β and ε_t is the transformation misfit strain. As we might expect, the value of ΔG_ε dictates the morphology of the precipitate. A strong dependence of ΔG_ε on interface boundary orientation leads to nonspherical precipitate particle shapes. Acicular, needlelike, ellipsoidal, or platelike precipitates with stepped planar interfaces between the precipitate and the matrix are favored energetically.

Proceeding as in Sections 3.4.1 and 3.4.2, the critical size of the nucleus becomes

$$R_c = \frac{2\gamma_{\alpha\beta}V_S}{(\Delta G_v + \Delta G_\varepsilon)} \tag{11.5.3}$$

and the height of the activation barrier for solid state nucleation becomes

$$\Delta G_{max} = -\frac{16\pi\gamma_{\alpha\beta}^3 V_S^2}{3(\Delta G_v + \Delta G_\varepsilon)} \tag{11.5.4}$$

where V_S is now the solid molar volume.

11.5.4 Precipitate Particles Form from the Matrix Phase in Different Ways

11.5.4.1 Shear of the Parent Phase Controls Displacive Transformations

The kinetics of nucleus growth is limited by the fact that atoms or ions must move through the parent solid and gather together in large enough quantities for the critical size nucleus to form and grow. Transport to and from the new phases usually occurs by the diffusion mechanisms outlined in Section 9.5, but in instances in which the chemical composition does not change, the transformation can be sudden and occur at the speed of sound by shear displacement within the crystal structure. This event is known as a *displacive transformation*. As the name implies, atoms are sheared from their original lattice sites to form the new crystal structure. Iron–carbon solid solutions quenched from the high-temperature face-centered cubic phase (austenite) undergo a displacive transformation to form the low-temperature distorted body-centered cubic phase (martensite). In fact, all displacive transformations are commonly known as *martensitic transformations*.

11.5.4.2 Atomic Diffusion Controls Reconstructive Transformations; Precipitate–Matrix Interfaces May Be Coherent or Incoherent

Most transformations involve a complete change in chemical composition and crystal structure and occur by diffusion-controlled nucleation and growth of the new solid phase within the old. This event is referred to as *reconstructive transformation*. The new phase may or may not retain a structural identity with the matrix. When an epitaxial relationship exists between the precipitate and the host (see Section 10.5), the interphase boundary is said to be *coherent* (or commensurate). For coherent precipitates the transformation strain ε_t is small. When the two phases have totally different crystal structures and lattice accommodation is spatially impossible, the interphase interface is said to be *incoherent* (or incommensurate) and the transformation strain is large. The interface energy $\gamma_{\alpha\beta}$ for a coherent precipitate particle is an order of magnitude less than that for an incoherent precipitate particle.

Once nucleated, two circumstances favor the continued growth of a coherent phase: first, the interfacial energy, the epitaxial strain (see Section 10.7.1) and other energetic factors such as electron wave scattering at the interface should be at a minimum; and second, the phase change should be accomplished with an economy of atom movement. Examples of coherent reconstructive transformation include the precipitation of $CuAl_2$ intermetallic particles out of aluminum–magnesium–copper solid solutions. The earliest manifestations of precipitation in this alloy are thin needles and sheets of $CuAl_2$ intermetallic oriented epitaxially (commensurate) with the {100} planes of the host aluminum. In Figure 11.21, transmission electron microscopy clearly reveals the anisotropy and elastic strain associated with these precipitate particles.

Again, as we might expect the interfaces between the matrix and the new precipitate particles are not atomically flat but stepped. Growth takes place by the propagation of step fronts.

Figure 11.21
Precipitation of $CuAl_2$ platelets out of aluminum–4 wt% copper solid solution aged at 200°C. Electron transmission micrograph down the $\langle 100 \rangle$ zone axis shows how the precipitate favors an interface with the {100} planes of the aluminum matrix. Edge-on views reveal elastic distortion in the aluminum matrix immediately adjacent to the platelets; face-on views reveal interfacial misfit dislocations introduced to relax the epitaxial strain (see Figure 10.31). (Photomicrograph (50,000×) courtesy of Prof. James M. Howe, University of Virginia.)

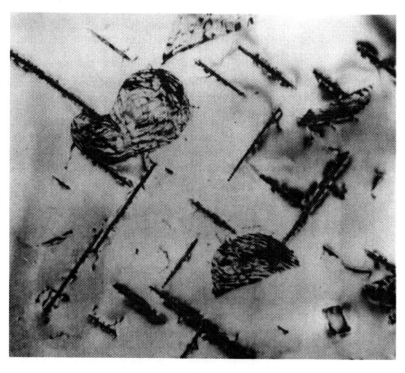

11.5.4.3 Grain Boundaries Are the Preferred Sites for Initial Precipitation of New Phases

Nucleation and growth of an incoherent phase is particularly sensitive to the magnitude of strain energy and interfacial energy. Reduction of the energy associated with both these terms explains why precipitation generally occurs preferentially in the vicinity of grain boundaries. Grain boundaries are attractive locations for the initiation of a new phase for a number of reasons. First and foremost the boundary is a region where structural order is relaxed, and the elastic strain due to differences in crystal structure between precipitate and matrix is reduced. Second, solute atoms segregate to the grain boundary region; so their concentration there is higher than normal, favoring coalescence. Third, grain boundary diffusion is more than an order of magnitude faster than bulk diffusion, which facilitates rapid transport of solute atoms to the growing precipitate. Fourth, stepped facets on the grain boundary surface provide a ready template for precipitate growth by the propagation of steps. Actually the grain boundary region is so attractive for nucleation and growth that it depletes solute in the surrounding material making precipitation less likely in the immediate vicinity of the boundary. These features are illustrated in Figure 11.22.

Because reconstructive transformation is limited by solid-state diffusion, the kinetics of precipitate growth is sensitive to temperature. At low temperatures, change takes place very slowly, but it can be accelerated by exposing the alloy to the highest temperatures possible without redissolving the precipitate. Varying the length of time a material is held at different temperatures makes it possible to manipulate the chemistry of the grain boundary surface and the distribution of new phases precipitating within the solid. As we saw earlier, thermal treatments also control the size of the grains. So, for specific applications, alloy heat treatments must be prescribed carefully to control the distributions, size, and shape of precipitated phases as well as grain size, shape, and orientation.

Figure 11.22
Precipitation from an aluminum–zinc–magnesium alloy solid solution. Initial heating for 2 hours at 180°C produced large precipitate particles within the grains and along the grain boundary (running from top to bottom in the photograph) and resulted in a broad precipitate free "depletion" zone. Subsequent heat treatment for 2 weeks at 20°C produced the additional fine-scale precipitate particles within the zone, and a much narrower region close to the boundary completely free from precipitate. [R. B. Nicholson in *Phase Transformations*, H. I. Aaronson and M. Cohen (eds.), Metals Park, Ohio: American Society For Metals, 1970, p. 308.]

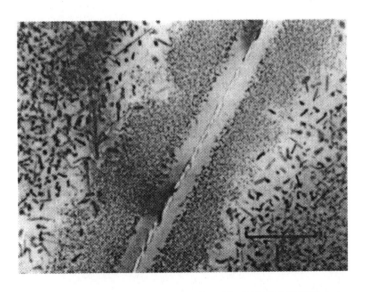

11.6 Grain Boundary Devices Called Ceramic Varistors Protect Electrical Circuits from Voltage Surges

There are a number of ceramic materials whose physical properties depend critically on the performance of the grain boundary region. These are known collectively as *grain boundary devices*. The ceramic varistor is one example.

A varistor (variable resistor) is a voltage surge protection device. It has the property of behaving as a high-resistance insulator up to a certain critical voltage level where it suddenly converts to a low-resistance conductor. Figure 11.23 reproduces a set of characteristic electrical field–current curves for a zinc oxide varistor at different operating temperatures near room temperature. When the applied electrical field exceeds about 10^3 volts per centimeter, a slight increase in field increases the current by 10 orders of magnitude. Under normal conditions, placing such a device across the input to an electrical circuit has no effect on circuit operation, but in the event of a voltage surge, the varistor short circuits the load and quenches the surge. Such protective devices find application at all levels of electrical circuitry from high-voltage power transmission lines to household appliances to memory chips for computers. The physical size of the varistor determines the voltage and current it can handle. As the scale of the application is reduced, its size decreases from meters to centimeters to micrometers. First discovered in 1978, the zinc oxide varistor has found many applications. At present, about one billion zinc oxide varistor pieces of various sizes are manufactured per year.

Figure 11.24b illustrates the principles of operation of the zinc oxide varistor. Pure zinc oxide is a semiconductor and its electrical behavior is ohmic (the electrical current increases linearly with applied voltage). A relatively low electrical resistivity

Figure 11.23
Current–voltage characteristics of a metal oxide varistor for a range of temperatures at and above room temperature. [After L. M. Levinson and H. R. Philipp, *American Ceramic Society Bulletin,* **65,** 639 (1986) and T. K. Gupta, *Journal American Ceramic Society,* **73,** 1817 (1990).]

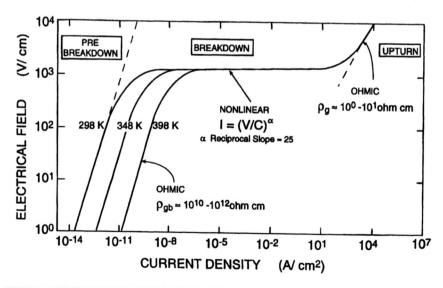

Figure 11.24
The chemical composition and electrical structure of a grain boundary in a bismuth-doped zinc oxide varistor. (a) The bismuth distribution increases adjacent to the boundary. [W. D. Kingery in *Grain Boundary Phenomena in Electronic Ceramics*, Vol. I of *Advances in Ceramics*, L. M. Levinson (ed.), Columbus, Ohio: American Ceramic Society, 1981, p. 17.] (b) The proposed electron energy band diagram in the vicinity of a grain boundary in bismuth-doped zinc oxide. [W. C. Richmond and M. A. Seitz in *Grain Boundary Phenomena in Electronic Ceramics*, Vol. I of *Advances in Ceramics*, L. M. Levinson (ed.), Columbus, Ohio: American Ceramic Society, 1981, p. 433.]

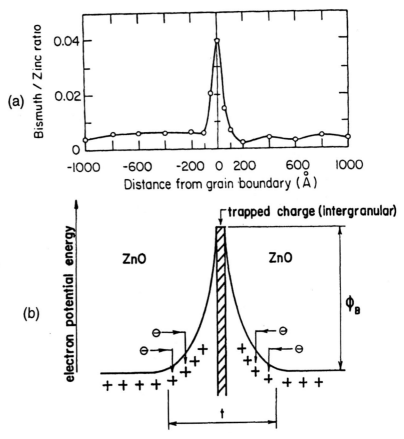

(in the range 1 to 10 ohm cm at room temperature) characterizes the behavior of the individual zinc oxide grains. Grain boundaries in zinc oxide are charged, and when impurities are introduced, they segregate to the grain boundaries. In particular, aliovalent impurity ions like bismuth with a valence different from the host increase the effective charge at the boundary. Because impurity ions cannot move at normal temperatures, the charges associated with them are fixed and act as traps that repel the mobile charge carriers. As a result the intergranular region behaves like an insulator. The band structure describing the energy distribution of conducting electrons becomes distorted in the vicinity of the boundary, as indicated in Figure 11.24b. When the applied electrical potential is high enough, electrons have sufficient energy that they can tunnel through the energy barrier, and the grain boundary insulation breaks down.

The current–voltage curve displays three segments. In the prebreakdown segment, the potential drop is confined to the grain boundary, and the resistivity of the grain boundary region, indicated by ρ_{gb} in Figure 11.23, controls the slope of the current–voltage curve. In the nonlinear second segment, the grain boundary insulation breaks down, and the electrical current increases suddenly, with very little increase in electrical field. In the subsequent upturn segment, the resistivity of the grains, ρ_g in Figure 11.23, controls the current through the device. We can

Figure 11.25
Scanning electron micrographs of bismuth-doped zinc oxide ceramics from which the zinc oxide has been leached to leave only the grain boundary triple line depositions of excess bismuth oxide. (top), K. Mukae and I. Nagasawa in *Grain Boundary Phenomena in Electronic Ceramics*, Vol. I of *Advances in Ceramics*, L. M. Levinson (ed.), Columbus, Ohio: American Ceramic Society, 1981, p. 337.; (bottom), M. Matsuoka in *Grain Boundary Phenomena in Electronic Ceramics*, Vol. I of *Advances in Ceramics*, L. M. Levinson (ed.), Columbus, Ohio: American Ceramic Society, 1981, p. 302.]

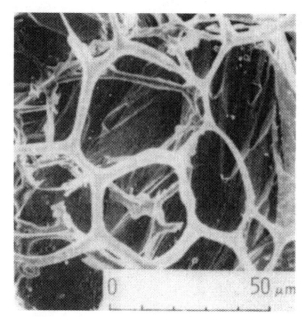

measure the sharpness of the transition from insulator to conductor by the reciprocal of the slope in the breakdown region, α, where α needs to be large, on the order of 25 for an effective device. Because the behavior of the varistor depends on the existence and insulating properties of grain boundaries, it is a grain boundary device.

To process ceramic varistor materials, zinc oxide powder is deliberately doped with small amounts (about 3 wt. %) of bismuth oxide. Bismuth oxide is added during formulation of the ceramic powders prior to the sintering stage (see Figure 4.29). As demonstrated by the bismuth–zinc ratio shown in Figure 11.24a, bismuth ions segregate to the grain boundaries. Because the valence of bismuth (Bi^{3+}) differs from that of zinc (Zn^{2+}), the grain boundary region becomes charged. The Bi^{3+} sites trap the electrons, and as a result, the electron potential energy bands in the vicinity of the boundary bend and curve upward, as illustrated in Figure 11.24b. To an electron, this configuration is similar to that of two back-to-back Zener diodes. Much research has been directed at finding the optimum chemical composition and distribution of the impurity in the zinc oxide.

Transmission electron micrographs of bismuth-doped zinc oxide varistors show no second-phase precipitate in the grain boundary interface. The liquid bismuth oxide phase does not completely wet the zinc oxide, and no glassy layer forms at the grain boundary layer. However, if we leach the zinc oxide out of the structure, a three-dimensional bismuth oxide–rich phase remains, as shown in Figure 11.25. We can compare the microstructure in this example with that shown in Figure 11.16c. This observation is consistent with the model in Figure 11.24b, that suggests the segregation of impurity to the grain boundary region is crucial to the operation of the varistor rather than formation of a continuous glassy phase.

Bibliography

H. I. Aaronson and M. Cohen, Eds., *Phase Transformations*, Metals Park, Ohio: American Society for Metals, 1970.

W. Bollmann, *Crystal Defects and Crystalline Interfaces*, New York: Springer Verlag, 1970.

J. W. Cahn, E. A. Holm, and D. J. Srolovitz, *Modeling Microstructural Evolution in Two-Dimensional Two-Phase Microstructures*, *Proceedings of the International Conference on Grain Growth in Polycrystalline Materials*, 1991.

J. W. Christian, *The Theory of Transformations in Metals and Alloys*, Oxford: Pergamon Press, 1965.

W. D. Kingery, H. K. Bowen, and D. R. Uhlmann, *Introduction to Ceramics*, New York: John Wiley, Inc., 1976.

P. G. Shewmon, *Diffusion in Solids*, New York: McGraw-Hill Book Company, 1963.

C. S. Smith, *Metallurgical Reviews*, Volume 9, p. 1, 1964.

Exercises

11.1 For polycrystalline copper, the surface energy γ_{sv} is 1750 mJ/m^2; the grain boundary energy γ_{gb} is 600 mJ/m^2; and the interfacial energy between copper and liquid lead is 350 mJ/m^2. Show that, so far as interfacial energies are concerned, there is little difference between transgranular and intergranular fracture of copper in air, whereas in the presence of liquid lead intergranular fracture is favored by a factor of 7:1 (this phenomenon is known as liquid metal embrittlement).

11.2 A sample of polycrystalline copper contains a small spherical grain 1 μm in diameter. Estimate the grain boundary velocity (cm per minute gives an appreciation for the effect) as this grain shrinks in size during an anneal at 1150 and 600K. Repeat the estimate for a 10 μm grain size. What steps would you recommend to reduce the grain growth rate in copper? (Copper has fcc crystal structure, a ≈ 0.36 nm. Assume γ_{gb} = 600 mJ/m^2, and ΔH_{gb}= 1 eV.)

11.3 The rate of growth of a grain with 8 sides is equal to _ ; faster than _ ; or slower than _ the rate of growth of a grain with 12 sides. Why?

11.4 Draw the metastable shape of a four-sided rectangular grain in a pure polycrystalline solid and illustrate the sequence as this grain shrinks in size and disappears.

11.5 (a) Distinguish between the origins of surface charge in (i) solid–fluid colloidal systems; and (ii) solid–solid grain boundary systems. (b) What is the difference in time scales for counterions to assume their equilibrium distribution in the vicinity of the charged surfaces in (i) and (ii)? (c) How does the electrical potential vary with distance away from a charged grain boundary surface?

11.6 Describe the three factors involved in the segregation of impurities to a ceramic (ionic solid) grain boundary.

11.7 The surface tension of liquid copper is 1300 mJ/m^2. The interfacial tension between liquid copper and solid alumina is 2080 mJ/m^2. The contact angle of liquid copper on alumina is 138°. Calculate the dihedral angle for liquid copper in equilibrium with a grain boundary at the surface of polycrystalline alumina. Repeat the calculation for liquid calcium silicate where the interfacial tension between liquid silicate and solid alumina is 350 mJ/m^2. What are the microstructural implications of these two results? (Assume grain boundary surface energy to be about two thirds surface energy.)

11.8 When a drop of liquid calcium silicate is heated and remains molten on the top of a thin slice of polycrystalline alumina for a few minutes, the alumina gradually becomes curved and warped. (a) Why does this happen? (b) In what direction do you expect the warping to occur? (c) What relationship would you anticipate between the radius of curvature and the time that the liquid droplet remains on the surface?

11.9 Calculate the critical radius of a new solid phase, β, precipitating out of a super heated matrix phase, α, at 1200 K. Assume that the new phase is spherical in shape and that strain effects may be negelected. At 1200 K the molar volume free energy = 6.52 × 10^6 J per m^3 and the interfacial energy $\gamma_{\alpha\beta}$ = 230 mJ/m^2.

11.10 If the new solid phase β in Exercise 11.8 strains the matrix α, the size of the critical nucleus will

increase _ ; decrease _ ; stay the same _ .
Why?

11.11 If the new solid phase β is commensurate with the parent phase α rather than incommensurate, would you expect the size of the critical nucleus to
increase _ ; decrease _ ; stay the same _ .
Why?

APPENDICES FOR CHAPTER 11

SYMBOLS

b	Burgers vector, slip vector, of dislocation lines
c	depth of surface flaw
E	static tensile elastic modulus; Young's modulus
G	elastic shear modulus under static conditions
G', G''	elastic shear modulii under dynamic conditions
v	dislocation velocity
γ_f	fracture surface energy
γ_{yx}	shear strain
δ	mechanical loss angle; tan δ mechanical loss tangent
ε_{xx}	tensile strain
$\dot{\varepsilon}_p$	plastic strain rate
Λ	logarithmic decrement due to mechanical dissipation
v	Poisson's ratio
ρ_m	density of mobile dislocation lines
σ_{xx}	tensile stress; σ_{flow}-plastic flow stress; σ_f-ultimate fracture stress
η	viscosity
τ_r	relaxation time
τ_{yx}	shear stress
ω	angular velocity; $\omega = 2\pi f$. f-frequency of oscillating mechanical strain

APPENDIX 11A MECHANICAL BEHAVIOR OF MATERIALS UNDER STRESS—MACROSCOPIC VIEW

11A.1 Elastic Deformation

As depicted in Figure 11A.1a, when a tensile force F is applied to the ends (area A) of a block of material of length L it elongates by amount ΔL. For a purely elastic response the material obeys Hooke's law and the uniaxial tensile stress, $\sigma_{xx} = F/A$, and the tensile strain, $\varepsilon_{xx} = \Delta L/L$, are proportional, at least for small deformations. We can express this linear relationship by

$$\frac{F}{A} = E\left(\frac{\Delta L}{L}\right) \quad \text{or} \quad \sigma_{xx} = E\varepsilon_{xx} \tag{11A.1}$$

where the constant of proportionality, E, is the static *Young's modulus*. E is a measure of the stiffness of the material: when E is high, the material undergoes very little strain for a large applied stress; when E is low, the material is easily distorted by an applied stress. For a given material the magnitude of Young's modulus is related to the nature of the interatomic bonding. It is highest for covalent solids, followed by ionic solids, metals and polymers. Although for polymer materials the value of E is very sensitive to microstructure and temperature, E is generally a characteristic property of a material.

A second property that also describes the response of a sample to a stress is *Poisson's ratio*. Possion's ratio relates the contraction of the sample in the dimensions perpendicular to the direction of elongation to the strain. If w and h represent the width and breadth of the cross-sectional area A in Figure 11A.1a, then for a homogeneous material the contraction in both dimensions is equal and is given by

$$-\frac{\Delta w}{w} = -\frac{\Delta h}{h} = v\frac{\Delta L}{L} \tag{11A.2}$$

where the constant of proportionality, v, is Poisson's ratio. Poisson's ratio is also a characteristic property of a material and can be expressed as

$$v = \frac{1}{2}\left(1 - \frac{1}{V}\frac{dV}{d\varepsilon}\right) \tag{11A.3}$$

Figure 11A.1
Schematic illustrating the effect of applying a load F to an elastic solid (a) in tension, (b) in shear.

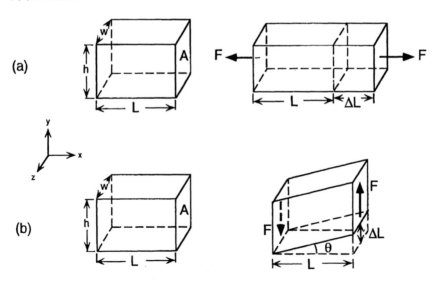

where V is the volume of the sample. If the volume does not change under stress, $dV/d\varepsilon = 0$ and $v = \frac{1}{2}$, the fraction contraction in the directions perpendicular to the elongation is half the fraction increase in the length. For most materials $v \approx \frac{1}{3}$, which means they dilate slightly under tensile stress.

Equation 11A.1 gives $F/A = E\,\Delta L/L$ when the deformation is simple tension or compression. We can develop a similar expression when the deformation is simple shear, as illustrated in Figure 11A.1b. When we apply a shear stress, $\tau_{xy} = F/A$, we generate a shear strain, $\gamma_{xy} = \Delta L/L$, where L and ΔL now are measured at right angles to each other.[1] For an elastic distortion, which is instantaneous and independent of time, the stress and strain are linearly related by

$$\tau_{xy} = G\gamma_{xy} = G\frac{\Delta L}{L} = G \tan \theta \cong G\,\theta \qquad (11A.4)$$

where G is called the static *shear modulus* or the modulus of rigidity. The last approximation in eq. 11A.4 results from $\tan \theta \cong \theta$ for small values of θ.

We can relate Young's modulus (eq. 11A.1) and the shear modulus (eq. 11A.4) by

$$G = \frac{E}{2(1 + v)} \qquad (11A.5)$$

where v is Poisson's ratio, eq. 11A.3. For constant volume deformations, $v = \frac{1}{2}$ and $G = E/3$; when $v = \frac{1}{3}$, $G = 3E/8$.

Since the modulus is the ratio of the stress to the strain, it makes no difference whether we apply a constant force and measure the strain or induce a constant strain and measure the resulting force. If the ratio of the strain to stress were measured, the reciprocal of the modulus would result; this is known as the *compliance*. As long as moduli and compliance for a particular deformation are constant *and independent of time,* they are simply reciprocals of each other.

11A.2 Fracture and Plastic Deformation

The stress-strain relationship for purely elastic materials is linear and reversible, as shown in Figure 11A.2, curve (a). When the force is removed, the material shrinks back to its original shape. The energy used to produce the deformation is completely recovered; no energy is consumed in this reversible deformation process. However, as we increase the applied stress to higher and higher levels as in curves (b) and (c), we reach a limit at which the material behaves in one of two ways: either it breaks or it yields.

---

[1] *In tensor notation, τ_{xy} signifies shear stress due to a force in the y-direction acting over an area normal to the x-direction; γ_{xy} signifies a shear strain due to a displacement in the y-direction relative to a dimension in the x-direction.*

Figure 11A.2
Macroscopic stress-strain relationships for different kinds of solids. (a) Reversible elastic deformation. (b) Brittle behavior. The material deforms elastically up to the stress (the fracture strength or tensile strength) where it fractures catastrophically. (c) Superplastic behavior. The material starts to deform plastically at the yield strength, σ_y, and undergoes more than 100% strain with no increase in flow stress. (d) Plastic behavior. The material starts to deform plastically at the yield strength, but the stress to sustain flow must increase with strain as the material 'strain-hardens'. Eventually the stress reaches the tensile strength and the material fractures. The area under the stress-strain curve represents the mechnical energy consumed in fracture and is related to the toughness of the material. (e) Elastomeric behavior. The material deforms elastically by more than 100% strain with a very low

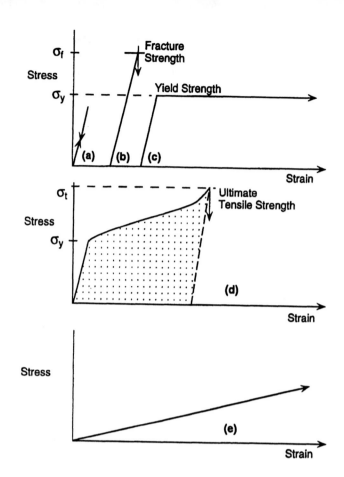

Completely brittle materials are stiff (high modulus) and deform elastically right up to the breaking stress, at which point they fail suddenly and catastrophically. This behavior is illustrated in Figure 11A.2, curve (b). The breaking stress defines the *fracture strength* or *tensile strength*, σ_f, of the material. Materials like glass, cast iron, and ceramics behave this way.

Superplastic or *ductile* materials remain elastic up to the yield stress, σ_y, and then elongate continuously at an almost constant stress. This behavior is illustrated in Figure 11A.2, curve (c). The stress at the onset of the plastic region defines the *yield strength* of the material. Weak metals, such as aluminium and gold, behave this way.

Most solids demonstrate a mechanical behavior involving different degrees of brittleness and plasticity. Typically, solids show some plasticity, but the elongation produces hardening and so the stress must be increased to continue deformation. Eventually, the stress reaches a level at which the material breaks. This behavior is illustrated in Figure 11A.2, curve (d) and Figure 11B.1. Strong metals, such as steels and alloys (see Appendix 11B), as well as many polymers behave this way.

It can be shown that the energy stored in a mechanical system per unit volume is equal to one half the product of stress and strain ($\frac{1}{2}\sigma\varepsilon$) per unit volume. Thus, the integrated area under a stress-strain curve corresponds to twice the energy consumed in the process of deformation. Completly brittle materials (curve (b) consume a relatively small amount of energy before fracture. It is all elastic energy and is used to generate the two new fractured surfaces ($2\gamma_{sv}$). Completely plastic materials, curve (c), consume plastic energy by an amount that increases with the elongation. It is dissipated as heat. Usually, the area under the stress-strain curve (the shaded area in curve (d)) is a measure of the energy that it takes to fracture the material; it relates to the *toughness* of the material. Metal superalloys, some modern ceramics, fibrous materials like wood and fiber composites, and special polymer composites are examples of tough materials.

11A.3 Elastomer Deformation

Rubbery materials exhibit another kind of mechanical behavior. The distinguishing feature of elastomers (see Section 6.2.3) is that they may be stretched elastically to several times their original length (several hundred percent strain) before they finally harden and fracture. (Figure 11A.2, curve (e) illustrates this behavior).

11A.4 Time Dependent Behavior— Viscoelastic Deformation

So far, we have limited our discussion to the mechanical behavior of solid systems under static conditions in which we apply a stress (or strain) and wait momentarily for the system to come to equilibrium before measuring the resulting strain (or stress). Now we generalize our discussion to systems in which the conditions are dynamic and the relations between stress and strain are time dependent.

For fluids and solids in general there are two extreme responses to the imposition of a mechanical force. At one extreme we have Newtonian fluids and at the other extreme Hookean solids.

Imposing a shear stress on a Newtonian fluid results in fluid flow. From Section 3.8.1.1, eq. 3.8.1,

$$F/A = \eta \; dv/dy \quad \text{or} \quad \tau_{xy} = \eta \; d\gamma_{xy}/dt \qquad (11A.6)$$

so that in Newtonian fluids the shear stress, τ_{xy}, and shear strain rate, $d\gamma_{xy}/dt$, are linearly related. We also found in Section 3.8.1.1 that energy is dissipated during fluid flow, and the rate of energy dissipation is proportional to the viscosity.

Imposing a shear stress on a Hookean solid results in elastic strain. From eq. 11A.4,

$$\tau_{xy} = G \; \gamma_{xy} \qquad (11A.7)$$

so that in Hookean solids the shear stress, τ_{xy}, and shear strain, γ_{xy}, are linearly related. Because elastic strain is reversible, all the strain energy stored on loading is recovered when the stress is released and no dissipation of energy occurs as the result of loading and unloading.

Figure 11A.3
The response of a material to a small fixed shear strain varies according to type. (a) A small fixed shear strain, γ, is suddenly imposed on a system. (b) For a Newtonian viscous fluid the shear stress, τ_{xy}, exhibits a pulse that relaxes to zero immediately when the strain is complete. (c) For a Hookean solid the shear stress exhibits a jump in value to τ_{xy} that remains fixed when the strain is complete. (d) For a viscoelastic solid the shear stress exhibits a jump to τ_{xy} and relaxes asymptotically to an equilibrium value, τ_e. For a viscoelastic liquid, τ_{xy} relaxes asymptotically to zero. (After C.W. Macosko, *Rheology-Principles, Measurements, and Applications,* VCH Publishers, New York, 1994, p. 110)

Another way we can distinguish these two very different responses to the application of force is to compare what happens when we subject them to a sharp and sudden distortion, as illustrated in Figure 11A.3a. The strain associated with the distortion may be a shear strain, γ_{xy}, or an elongation or compaction, ε_{xx} or a combination of both. On one hand, the ideal Newtonian fluid experiences a momentary stress of infinitesimal duration while the distortion takes place, but the stress relaxes to zero as soon as the strain becomes constant, as illustrated in Figure 11A.3.b. This instantaneous reponse occurs because the liquid molecules move instantly in response to the applied stress, and the relaxation time for an ideal Newtonian fluid is *zero*. On the other hand, an ideal Hookean solid experiences a sudden increase in stress while the distortion takes place, but the stress remains constant with time as soon as the strain becomes constant, as illustrated in Figure 11A.3.c. The magnitude of the stress on the Hookean solid is dictated by the value of the shear modulus, G, for shear strain and by the value of the Young's modulus, E, for elongation or compaction. Once the distortion is complete, the solid molecules do not move from the positions prescribed by the elastic distortion, and the relaxation time for an ideal Hookean solid is *infinite*.

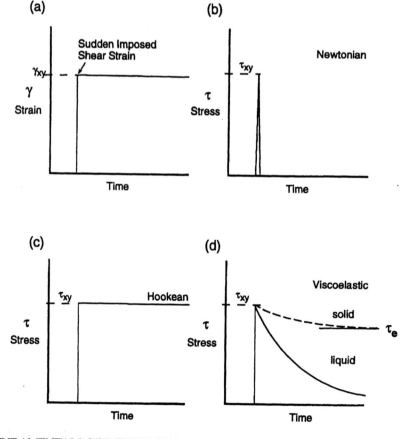

In reality no liquid exhibits pure Newtonian behavior and no solid exhibits pure Hookean elastic behavior; all materials exhibit varying degrees of both. Polymers simultaneously display characteristics of both a viscous fluid and an elastic solid; they are *viscoelastic* materials. When a polymer system is subjected to a sudden distortion, at first the stress increases to a maximum value dictated by the elastic moduli, but then it relaxes to an equilibrium value. The mechanical response is time dependent. The stress drops gradually with time, t, following an inverse exponential relationship, as illustrated in Figure 11A.3.d. The stress and (by implication) the moduli are time dependent, that is

$$\sigma(t) \propto \exp\left(-\frac{t}{\tau_r}\right) \qquad (11A.8)$$

We can associate this time dependent behavior with a finite *relaxation time*, τ_r.

By considering the molecular origins of relaxation times in more detail, we gain further insight into viscoelastic behavior. Stress exerted on a Newtonian fluid results in flow in which molecules move past one another, and the characteristic relaxation or response time for molecular movement is rapid compared to the shear rate. In a polymer system, the relaxation times are finite because it takes time for the molecules in the system to rearrange as they respond to the step change in strain. The reptation relaxation time τ given in eq. 6.6.15 provides one specific example of a finite relaxation time. Other relaxation times may be associated with other polymer molecular movements, such as chain coiling and uncoiling or molecular chain alignment in the stress direction.

With elastic (Hookean) systems we associate a storage of mechanical energy that is completely recoverable on unloading, so that no energy is lost during a stress (or strain) cycle. With viscous (Newtonian) systems, on the other hand, we associate a complete dissipation of mechanical energy during a strain cycle; it is converted to heat. Viscoelastic systems combine both types of behavior, so that during a strain cycle, energy adsorption depends on the time taken for the cycle. We can illustrate this fact by bouncing a ball off a perfectly elastic floor. A perfectly elastic (Hookean solid) ball bounces back to its initial height, whereas a liquid ball splats out on the floor. A viscoelastic ball bounces back to a lower height that corresponds with the stored elastic energy. The loss of height corresponds to the energy lost as heat when the ball collides with the floor. Because viscoelastic properties are time dependent, we can expect the ball's behavior to depend on the length of time it is in contact with the floor. The faster it is thrown at the floor, the more elastic is the bounce—a well-known behavior exhibited by 'Silly Putty'™.

The behavior of a viscoelastic system can be represented by a number of 'models' consisting of variuous combinations of springs (to represent the elastic component) and dashpots (to represent the viscous component). The classic Maxwell model consists of a spring and dashpot in series as illustrated in Figure 11A.4. In response to an imposed load, the spring stretches with a spring constant equivalent to the shear modulus G and the

Figure 11A.4
The Maxwell spring and dashppot model of a visco-elastic system.

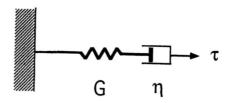

G   η

dashpot displaces against a friction equivalent to the viscosity η. The characteristic relaxation time for the spring-dashpot system is proportional to η/G.

We can study the time dependence of viscoelastic systems more systematically by cycling the stress or strain in a periodic manner and following the frequency response of the system. Special instruments (rheometers) designed for this purpose are used to explore the *rheological behavior* of polymers. If the shear strain γ_{xy} is varied sinusoidally

$$\gamma_{xy} = \gamma_{xyo} \sin \omega t \qquad (11A.9)$$

where $\omega = 2\pi f$, and f is the frequency, then the shear stress τ_{xy} also varies sinusoidally, but with a phase lag, or loss, δ, so that

$$\tau_{xy} = \tau_{xyo} \sin (\omega t + \delta) \qquad (11A.10)$$

The modulus, G, is frequency dependent, therefore, and can be written in terms of real and imaginary components

$$G(\omega) = G'(\omega) + iG''(\omega) \qquad (11A.11)$$

where $G'(\omega)$ is associated with elastic storage and $G''(\omega)$ with dissipative loss. For an ideal Hookean elastic solid $G'(\omega) = G$ and $G''(\omega) = 0$, and for a Newtonian fluid $G'(\omega) = 0$ and $G''(\omega) = \eta\omega$. It can also be shown that $G''/G' = \tan \delta$ where $\tan \delta$ is known as the *loss tangent*.

By measuring the frequency dependence of the storage and loss moduli for a viscoelastic polymer, it is possible to identify characteristic frequencies (f_c) where these moduli undergo a transition in value, as illustrated in Figure 11A.5. The storage modulus, $G'(\omega)$, goes through a stepwise increase in value as the frequency increases; and the dissipative modulus, $G''(\omega)$, goes through a peak in value corresponding to the energy dissipation associated with the specific relaxation event. As shown in Figure 11A.5a, $\tan \delta$ also peaks as a function of frequency but at a lower value than $G''(\omega)$. Analysing material behavior in this way is known as dynamic mechanical spectroscopy.

Solids that are purely elastic show no relaxation peaks and such materials absorb little mechanical energy. Solids that are viscoelastic show relaxation peaks and have a loss angle, δ, that is typically a few degrees and can approach 30° at frequencies corresponding to the relaxation peaks. Viscoelastic materials such as polymers absorb mechanical energy, they show very high mechanical damping at the relaxation frequency. This feature finds practical application in frequency selective sound absorbers for example.

Figure 11A.5
Variation of the elastic
modulus, G′, the loss modulus, G″, and the loss tangent, tanδ, as a function of
frequency in a viscoelastic
material. The figure also
includes the equivalent
compliances J. (N. G.
McCrum, C. P. Buckley,
and C. B. Bucknall, *Principles of Polymer Engineer-
ing*, Oxford University
Press, Oxford, 1988,
pp. 128, 130. By permis-
sion of Oxford University
Press.)

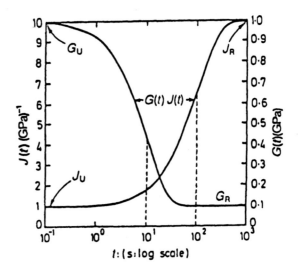

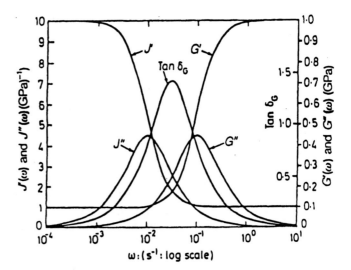

The inverse of the characteristic relaxation frequencies give characteristic relaxation times, $1/f_c = \tau_c$. Each corresponds to the relaxation time for a specific molecular transport phenomenon contributing to the viscoelasticity of the system. In this way dynamic mechanical spectroscopy provides a tool that can be used to probe relaxation and transport phenomena in polymeric systems.

Further extensions of dynamic mechanical spectroscopy keep the frequency constant while the temperature is varied. Relaxation peaks then appear at characteristic temperatures. By following both the frequency and temperature response it is possible to derive thermal activation energies for the specific molecular transport phenomena.

Experimentally it is often easier to determine energy dissi- pation by measuring the decline in the magnitude of oscillation of a sample set into vibration, The logarithm of the ratio of successive amplitudes is known as the *logarithmic decrement*, Λ,

GRAIN BOUNDARIES IN CRYSTALLINE SOLIDS / 677

and $\Lambda = \pi \tan \delta$. Examples of mechanical spectra are included in Figure 6.29, they show the variation of G′ with temperature at a fixed frequency (1 Hertz) for different polymer blends.

APPENDIX 11B
MECHANICAL BEHAVIOR OF MATERIALS UNDER STRESS—MICROSCOPIC VIEW

As noted in Appendix 11A crystalline solids react to mechanical stress in different ways depending on the interatomic bond character, the crystal structure, temperature and method of loading. The typical response may be represented by the three behaviors depicted in Figure 11.B1 elastic deformation, plastic deformation, and brittle fracture. In this appendix we briefly review plastic deformation and brittle fracture of crystalline solids in terms of atomic behavior.

Figure 11B.1
Crystalline solids respond to the application of stress in one of three ways.
(a) Purely elastic behavior
(b) Plastic deformation above the flow stress
(c) Brittle fracture after little or no plastic deformation

11B.1 Elastic Deformation

Elastic deformation, curve (a) in Figure 11.B1, is the recoverable strain experienced by all crystalline solids when a stress is applied. Elastic strain completely reverses when the stress is removed and the solid reverts to its' original shape. A linear relationship exists between the tensile stress (σ_{xx}) and the elongation (ε_{xx}) and between the shear stress (τ_{xy}) and the shear strain

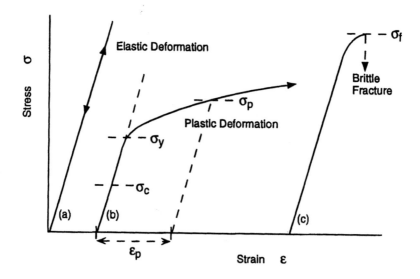

(γ_{xy}) it produces. The constants of proportionality are Young's modulus of elasticity (E) and the shear modulus of elasticity (G) respectively, where

$$E = \frac{\sigma_{xx}}{\varepsilon_{xx}} \quad \text{and} \quad G = \frac{\tau_{xy}}{\gamma_{xy}} \qquad (11B.1)$$

Elastic deformation measures the stretching of interatomic bonds in response to the applied stress, As we might anticipate from Chapter 2 the modulus of elasticity is related to the bond character. Covalent solids have the larger values then metallic, then ionic solids, and finally loosely bonded van der Waals solids and polymers have the lower values.

11B.2 Plastic Deformation at Low Temperatures

Plastic deformation, curve (b) in Figure 11B.1, is the permanent strain of a material stressed above the yield strength, σ_y, also referred to as the flow stress, σ_{flow}. This plastic strain is not recovered when the stress is removed. In Figure 11B.1, curve (b), for example, the material deformed to the stress level σ_p has a permanent strain of ε_p when it is unloaded.

Plastic deformation of crystalline solids is due to the motion and multiplication of dislocation loops of the kind described in Chapter 9.4.2. At low temperatures dislocations simply move over the slip (or glide) planes to produce permanent offset between the upper and lower portions of the crystals as illustrated in Figure 9.14.

The process actually starts when dislocations just begin to move as the stress rises above the critical stress σ_c in Figure 11B.1 curve (b). The plastic *strain rate* of the crystal, $\dot{\varepsilon}_p$, at any instant is given by the product

$$\dot{\varepsilon}_p = \rho_m b \, v \qquad (11B.2)$$

where ρ_m is the density of mobile dislocations, b their Burgers vector, and v their velocity. (The density of dislocations is measured by summing the total length of dislocation lines within unit volume or by counting the number of dislocations that thread a unit cross sectional area; both methods give the same result in number per area, m^{-2}.) Initially this product is not high enough for the plastic strain rate of the crystal to keep up with the imposed strain rate of the system and so the stress continues to increase. As the stress goes higher the dislocations move faster and faster. In fact their velocity v rises rapidly with stress according to $v \propto \sigma^m$, where the stress exponent m is approximately 10. The density of mobile dislocations, ρ_m, must also regenerate to replace those that have moved out of the crystal. Various multiplication mechanisms (for example the Frank-Read mechanism) have been proposed that lead to an increase in the value of ρ_m. Ultimately the product $\rho_m bv$ reaches a value equal to the imposed deformation rate and there is measurable permanent strain. This occurs at the *flow stress*, σ_y or σ_{flow}.

The strongest materials are those that require the highest stress to make dislocations move and multiply to produce measurable strain. Various strategies for strengthening materials, therefore, are designed to place obstacles in the path of dislocations to restrain their motion. One process is self-imposed. So many dislocations are generated by the multiplication mechanism that they get in each others' way. This is the basic mechanism of *strain-hardening* (or work-hardening) and is the reason why the plastic stress strain curve in Figure 11B.1 curve (b) has a positive slope. Another effective barrier to dislocation movement is the crystallographic misorientation across a grain boundary. The stress needed to sustain flow in the presence of grain boundaries increases as the grain size (d) decreases, following an inverse one-half power relationship,

$$\sigma_{flow} \propto d^{-1/2} \tag{11B.3}$$

Other barriers to dislocation motion are precipitate particles like those illustrated in Figures 11.21 and 11.22 that cause precipitation-hardening.

11B.3 Plastic Deformation at High Temperatures

At high temperatures vacancies play an increasing role in mechanical plasticity. Vacancies contribute to plastic deformation through their interaction with dislocations, by the mechanism known as *dislocation climb*. Dislocation climb may be understood by referring to the edge dislocation in Figure 9.15a. Adding vacancies removes atoms from the extra half plane and so the dislocation core that was located along A→D 'climbs' in a vertical direction. As a result parts of the edge dislocation become located in adjacent slip planes. By climbing far enough edge dislocations manoeuver over and around obstacles and are then free to move over new slip planes to continue plastic deformation. For this reason plastic deformation becomes easier in spite of impediments at high temperatures. The process of dislocation 'glide and climb' also makes it possible for dislocations to become disentangled and to reorganize into ordered small angle boundaries like Figure 9.16. Climb is a major feature in the formation of cell structures that act as recrystallization nuclei in Figure 11.1b.

Under stress at high temperatures, a crystalline material flows continuously until it ruptures; it is said to *creep*. Figure 11B.2 reproduces a typical *strain versus time* creep curve. Note that it contains three stages; the first transient stage where strain increases rapidly, the second linear stage where the creep rate is constant, and the third stage leading up to creep rupture. The constant slope in the second stage defines the creep strain rate, $\dot{\epsilon}_{creep}$. For excellent creep resistance the creep strain rate should be low with the curve in Figure 11B.2 almost hoprizontal.

Dislocation 'glide and climb' is one of the mechanisms contributing to the creep of a material. To improve creep resistance it is necessary to restrict dislocation motion and to present

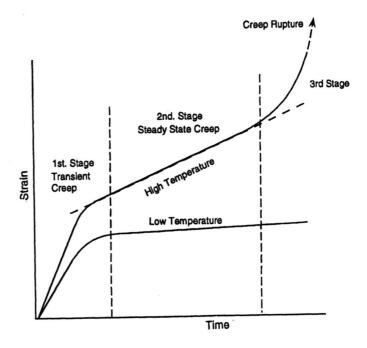

Figure 11B.2
At high temperatures crystalline solids undergo creep, that is, they continue to deform with time under a constant stress Plastic strain persists at a strain rate that is constant in the steady state, second, stage.

permanent obstacles to dislocation climb. A fine distribution of precipitate particles is most effective in this respect.

For many creep mechanisms the creep strain rate is directly proportional to the applied shear stress—Newtonian creep. In Section 9.5.7.2 we derive the expression, eq. 9.5.39, for the creep strain rate due to the simple flow of vacancies from surfaces in tension to surfaces in compression. This mechanism predicts Newtonian behavior. It also predicts that the creep rate is inversely dependent on the sample size, or in the case of polycrystalline material on the grain size. Grain boundaries act as a source and a sink for vacancies, and an easy diffusion path, so a small grain size increases the creep deformation rate. One strategy for improving high-temperature creep resistance is to increase the grain size. Unfortunately this reduces low temperature strength. A better strategy utilizes precipitate particles in the grain boundaries to reduce viscous flow in the boundary region.

11B.4 Brittle Fracture

Brittle fracture, curve (c) in Figure 11.B1, is the decohesion of a crystalline material at the fracture stress, σ_f, either by rupture of intermolecular forces across a crystal plane (transgranular fracture) or by separation of the grain boundaries (intergranular fracture). In most brittle materials, fracture is associated with the extension of flaws either in the surface or in interfaces within the bulk. The conditions for flaw extension are most simply described by the Griffith theory, based on the diagram in Figure 11B.3. The Griffith criterion states that a flaw (or crack) of length c will grow when the decrease in elastic strain energy exceeds the energy for creating additional surface. Assuming the stress falls to zero in

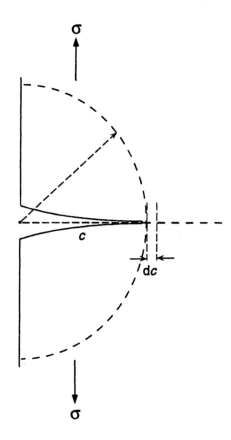

Figure 11B.3
Griffith flaw model used to estimate conditions for crack extension in a brittle solid.

the semicircular volume containing the crack (outlined by the dashed curve in Figure 11B.3) the decrease in strain energy ($\frac{1}{2}\sigma\varepsilon$ × volume = $\frac{1}{2}E\sigma^2$ × volume) is $\dfrac{1}{2}\dfrac{\sigma^2}{E}\dfrac{\pi c^2}{2}$ per unit distance normal to the page; the increase in surface energy is $2c\gamma_{sv}$ per unit distance normal to the page. For the crack to extend by a small amount dc, the Griffth criterion states

$$\frac{d}{dc}(\pi c^2\sigma^2/4E + 2c\gamma_{sv}) \geq 0 \qquad (11B.4)$$

which leads to

$$\sigma_f = \sqrt{\frac{2E\gamma_{sv}}{\pi c}} \qquad (11B.5)$$

where σ_f is the ultimate applied stress for fracture of the material containing a surface flaw (crack) of length, c. E and γ_{sv} are the tensile modulus of elasticity and surface energy of the solid respectively. With a typical surface tension of 2000 mJ/m^2 and a tensile modulus of 100 GPa, a stress of 2.5 GPa would produce catastrophic failure of a microcrack 20 nm deep.

A more general derivation for the fracture strength due to flaws leads to

$$\sigma_f = Y \sqrt{\frac{2E\gamma_f}{c}} \qquad\qquad (11B.7)$$

where Y is a geometrical parameter that varies with the shape and location of the flaw and the method of applying the load and γ_f is the fracture surface energy a term that includes all the energy dissipating processes that accompany crack propagation.

$$\gamma_f = \gamma_{sv} + \gamma_p + \gamma_t + \gamma_d \cdots \qquad\qquad (11B.7)$$

Processes such as plastic deformation, γ_p, twinning or a phase transformation, γ_t, or crack deflection, γ_d, greatly increases the tolerance of a material to surface flaws.

The strategy for increasing the fracture strength of brittle materials is to drastically reduce the dimension of the flaws, c, or to increase the *fracture toughness parameter*, $\sqrt{2E\gamma_f}$. The size of the microstructure often determines the crack size. This is obviously true if the crack originates due to intergranular decohesion, then the flaw size equals the grain size and a small grain size improves the fracture strength. Fracture toughness increases if weak interfaces or other energy-absorbing obstacles deflect the growing crack in its path. Composite structures are deliberately designed for this purpose, as described in Section 7.4.

INDEX

The letter f following a page number indicates a figure: the letter t following a page number indicates a table.

Concentration of point imperfections, 492
Condensation polymerization, 281
Consistency index (fluid flow), 95, 418
Contact
    angle, liquid-solid
        dynamic, 373, 454
        static, 59, 368
    line, 59, 410, 455
    zone, 455
Coordination number, 32, 54, 466, 468
Copolymers
    block, 302, 331, 335, 337, 342
    grafted, 331, 335
Core repulsion
    colloid particles, 121
    intermolecular, 15, 30, 32f
Cosine angle law of thermal emission, 543, 544f, 579
Cosurfactant, 241
Couette flow, 408
Coulomb's electrostatic interaction equation, 17–18, 468
Counter-ions in electrolytes, 131, 223
Coupling agents for interfacial bonding, 307, 378–80, 379t
Covalent
    bond, 14, 471
    crystals, 471
Crack growth, 375, 681–83
Creaming of emulsions, 96, 247
Creep of solids, 519, 680
    Newtonian, 680
Critical aggregation concentration (CAC), 313, 314t
    polymer-surfactant micelles, 313
Critical coagulation concentration (CCC), 147
Critical film thickness for epitaxial growth, 599
Critical flocculation concentration (CFC), 147
Critical flocculation temperature (CFT), 311, 313t
Critical length (fiber composites), 382, 390
Critical micelle concentration (CMC), 212
Critical point drying, 71, 181
Critical radius for nucleation
    disk on surface, 573
    droplet from vapor, 65f, 66
    solid precipitate, 661
Crystal imperfections, 491–505
    dislocations, 496–501
    interstitials, 491–96
    small-angle grain boundaries, 501–04
    stacking faults, 504, 601, 602f
    twins, 505, 601, 602f
    vacancies, 491–96
Crystals, molecular

liquid, 235
polymers, 283
    lamellae, 285
    spherulites, 285
surfactant hydrated, 218
Crystals, solid
    crystal structures, 464–74
        cesium chloride, 468
        diamond, 471f
        face-centered cubic, 32, 467f
        hexagonal close-packed, 32, 467f
        sodium chloride, 468, 470f
        zinc blende, 471f
    inert gas atoms, 32, 466
Curie temperature, magnetic, 611
Curvature effects
    amphiphile-water interface, 230
    surfactant number, 230
    emulsion oil-water interface, 241
    liquid-vapor interface
        pressure change, 60–62
Cylindrical micelles (amphiphile aggregates), 230

D

Dangling bonds at surfaces, 476, 633
Darcy's law for flow through a porous medium, 362
Darken's interdiffusion equation, 515
Debye equations
    dipole-induced dipole interaction, 28
    screening length, 134, 191
Debye length, 134, 191
Decohesion, at grain boundaries, 637
Defects. *See* Coating processes, defects; Crystal imperfections; Thin film, imperfections
Deformation
    elastic, 669
    plastic, 671
    viscoelastic, 673–78
De Gennes reptation model (polymer transport), 320
Degree of polymerization, 277
Deposition rate of thin films, 546, 554, 569
Derjaguin approximation, 36–38
Desorption from solid surfaces, 580–83
    energy, 580
    kinetics, 580
Detergency, 248–57
    detergents, 250–54
        additives, 252–54
    mechanisms, 255–57
    rollup mechanism, 255, 256f
    soils, 250
    stains, 250
Dewetting, 428

plasticizers, 259
tackifiers, 259
viscoelastic behavior, 262
Primary energy minimum (colloid particle interaction), 121, 122f, 146
Primary recrystallization in crystalline solids, 629
Principal radii of curvature, 62
Printing processes, 435
Proteins
denaturing, 301
structure, 300, 302t
Pseudoplastic fluids, 95
Pultrusion (polymer processing), 357f, 391
PVD. *See* Physical vapor deposition

Q

Quadrupole, electrical, 14
quadrupole interaction, 22
Quantum well semiconductor structures, 617

R

Radius
atomic (ionic), 15, 466
of curvature, 60–63
of gyration, 290
Random coil (polymer molecule), 290
Random walk, equation, 99, 582
Raoult's law (vapor pressure of solutions), 141
Rare gas atom crystals. *See* Inert gas atom solids
Rate constant
amphiphiles
association, 227
dissociation, 226
colloid coagulation
rapid, 149–50, 196
slow, 152, 198
thin film growth
chemical vapor deposition, 583
physical vapor deposition, 584
Rate of shear
by creep, solids, 521, 680
by viscous flow, liquids, 85
Read-Shockley grain boundary energy equation, 503
Real solution (dilute) theory, 79–81, 344–46
Real-time imaging in coating process control, 429–33
Recirculation cell in fluids
in coating flows, 416, 456
in vapor deposition, 563
Reconstruction of solid surfaces, 483
Reconstructive phase transformation, 661
Recrystallization of crystalline solids, 628

primary, 629
secondary, 643
Reflection high-energy electron diffraction (RHEED), 570, 596
Regular solution theory, 295, 344–46
Relaxation processes
fluid viscosity, 96
viscoelastic solids, 674
Relaxation time
amphiphiles
bilayer transformation, 232
micelle association, 227
micelle dissociation, 226
polymer segment reptation, 322
viscoelastic materials, 324, 675
Reptation model (polymer transport), 320
Repulsive force. *See* Force of interaction
Residence time
adatom on surface, 580–82
amphiphiles in micelle, 226
Residual stress in thin films, 601–06
contact stress, 605
deposition, 602
thermal expansion mismatch, 603
thermal gradients, 604
Resin, liquid polymer, 356
Retardation effect, intermolecular, 33
Reynolds number, 88, 113
Rheology, 676
Rheopectic fluids, 96
Ribbing in coatings, 427
Ripening of clusters in thin film formation, 587
Rollup mechanism of detergency, 255
Root mean square velocity of atoms in gases, 536
Rovings, fiber 355
Rubber
elastic behavior, 288
natural, 288
synthetic, 329–33

S

Scanning electron microscopy (SEM), 182, 267, 377, 378, 594, 603, 609, 654, 657, 666
Scanning tunneling microscopy (STM), 486
Scattering (light, neutron) by polymers, 282, 293t
Schottky crystal imperfection, 492f, 494
Schrödinger's equation, 14
Schulze-Hardy rule for colloid coagulation, 147
Screening length. *See* Debye length

Printed in the United States
86990LV00004B/11-14/A